Geophysical Monograph Series

Including

IUGG Volumes

Maurice Ewing Volumes

Mineral Physics Volumes

GEOPHYSICAL MONOGRAPH SERIES

Geophysical Monograph Volumes

1 **Antarctica in the International Geophysical Year** *A. P. Crary, L. M. Gould, E. O. Hulburt, Hugh Odishaw, and Waldo E. Smith (Eds.)*
2 **Geophysics and the IGY** *Hugh Odishaw and Stanley Ruttenberg (Eds.)*
3 **Atmospheric Chemistry of Chlorine and Sulfur Compounds** *James P. Lodge, Jr. (Ed.)*
4 **Contemporary Geodesy** *Charles A. Whitten and Kenneth H. Drummond (Eds.)*
5 **Physics of Precipitation** *Helmut Weickmann (Ed.)*
6 **The Crust of the Pacific Basin** *Gordon A. Macdonald and Hisashi Kuno (Eds.)*
7 **Antarctic Research: The Matthew Fontaine Maury Memorial Symposium** *H. Wexler, M. J. Rubin, and J. E. Caskey, Jr. (Eds.)*
8 **Terrestrial Heat Flow** *William H. K. Lee (Ed.)*
9 **Gravity Anomalies: Unsurveyed Areas** *Hyman Orlin (Ed.)*
10 **The Earth Beneath the Continents: A Volume of Geophysical Studies in Honor of Merle A. Tuve** *John S. Steinhart and T. Jefferson Smith (Eds.)*
11 **Isotope Techniques in the Hydrologic Cycle** *Glenn E. Stout (Ed.)*
12 **The Crust and Upper Mantle of the Pacific Area** *Leon Knopoff, Charles L. Drake, and Pembroke J. Hart (Eds.)*
13 **The Earth's Crust and Upper Mantle** *Pembroke J. Hart (Ed.)*
14 **The Structure and Physical Properties of the Earth's Crust** *John G. Heacock (Ed.)*
15 **The Use of Artificial Satellites for Geodesy** *Soren W. Henricksen, Armando Mancini, and Bernard H. Chovitz (Eds.)*
16 **Flow and Fracture of Rocks** *H. C. Heard, I. Y. Borg, N. L. Carter, and C. B. Raleigh (Eds.)*
17 **Man-Made Lakes: Their Problems and Environmental Effects** *William C. Ackermann, Gilbert F. White, and E. B. Worthington (Eds.)*
18 **The Upper Atmosphere in Motion: A Selection of Papers With Annotation** *C. O. Hines and Colleagues*
19 **The Geophysics of the Pacific Ocean Basin and Its Margin: A Volume in Honor of George P. Woollard** *George H. Sutton, Murli H. Manghnani, and Ralph Moberly (Eds.)*
20 **The Earth's Crust: Its Nature and Physical Properties** *John C. Heacock (Ed.)*
21 **Quantitative Modeling of Magnetospheric Processes** *W. P. Olson (Ed.)*
22 **Derivation, Meaning, and Use of Geomagnetic Indices** *P. N. Mayaud*
23 **The Tectonic and Geologic Evolution of Southeast Asian Seas and Islands** *Dennis E. Hayes (Ed.)*
24 **Mechanical Behavior of Crustal Rocks: The Handin Volume** *N. L. Carter, M. Friedman, J. M. Logan, and D. W. Stearns (Eds.)*
25 **Physics of Auroral Arc Formation** *S.-I. Akasofu and J. R. Kan (Eds.)*
26 **Heterogeneous Atmospheric Chemistry** *David R. Schryer (Ed.)*
27 **The Tectonic and Geologic Evolution of Southeast Asian Seas and Islands: Part 2** *Dennis E. Hayes (Ed.)*
28 **Magnetospheric Currents** *Thomas A. Potemra (Ed.)*
29 **Climate Processes and Climate Sensitivity (Maurice Ewing Volume 5)** *James E. Hansen and Taro Takahashi (Eds.)*
30 **Magnetic Reconnection in Space and Laboratory Plasmas** *Edward W. Hones, Jr. (Ed.)*
31 **Point Defects in Minerals (Mineral Physics Volume 1)** *Robert N. Schock (Ed.)*
32 **The Carbon Cycle and Atmospheric CO_2: Natural Variations Archean to Present** *E. T. Sundquist and W. S. Broecker (Eds.)*
33 **Greenland Ice Core: Geophysics, Geochemistry, and the Environment** *C. C. Langway, Jr., H. Oeschger, and W. Dansgaard (Eds.)*
34 **Collisionless Shocks in the Heliosphere: A Tutorial Review** *Robert G. Stone and Bruce T. Tsurutani (Eds.)*
35 **Collisionless Shocks in the Heliosphere: Reviews of Current Research** *Bruce T. Tsurutani and Robert G. Stone (Eds.)*
36 **Mineral and Rock Deformation: Laboratory Studies —The Paterson Volume** *B. E. Hobbs and H. C. Heard (Eds.)*
37 **Earthquake Source Mechanics (Maurice Ewing Volume 6)** *Shamita Das, John Boatwright, and Christopher H. Scholz (Eds.)*
38 **Ion Acceleration in the Magnetosphere and Ionosphere** *Tom Chang (Ed.)*
39 **High Pressure Research in Mineral Physics (Mineral Physics Volume 2)** *Murli H. Manghnani and Yasuhiko Syono (Eds.)*
40 **Gondwana Six: Structure, Tectonics, and Geophysics** *Gary D. McKenzie (Ed.)*
41 **Gondwana Six: Stratigraphy, Sedimentology, and Paleontology** *Garry D. McKenzie (Ed.)*
42 **Flow and Transport Through Unsaturated Fractured Rock** *Daniel D. Evans and Thomas J. Nicholson (Eds.)*
43 **Seamounts, Islands, and Atolls** *Barbara H. Keating, Patricia Fryer, Rodey Batiza, and George W. Boehlert (Eds.)*

44 Modeling Magnetospheric Plasma *T. E. Moore and J. H. Waite, Jr. (Eds.)*

45 Perovskite: A Structure of Great Interest to Geophysics and Materials Science *Alexandra Navrotsky and Donald J. Weidner (Eds.)*

46 Structure and Dynamics of Earth's Deep Interior (IUGG Volume 1) *D. E. Smylie and Raymond Hide (Eds.)*

47 Hydrological Regimes and Their Subsurface Thermal Effects (IUGG Volume 2) *Alan E. Beck, Grant Garven, and Lajos Stegena (Eds.)*

48 Origin and Evolution of Sedimentary Basins and Their Energy and Mineral Resources (IUGG Volume 3) *Raymond A. Price (Ed.)*

49 Slow Deformation and Transmission of Stress in the Earth (IUGG Volume 4) *Steven C. Cohen and Petr Vaníček (Eds.)*

50 Deep Structure and Past Kinematics of Accreted Terranes (IUGG Volume 5) *John W. Hillhouse (Ed.)*

51 Properties and Processes of Earth's Lower Crust (IUGG Volume 6) *Robert F. Mereu, Stephan Mueller, and David M. Fountain (Eds.)*

52 Understanding Climate Change (IUGG Volume 7) *Andre L. Berger, Robert E. Dickinson, and J. Kidson (Eds.)*

53 Plasma Waves and Instabilities at Comets and in Magnetospheres *Bruce T. Tsurutani and Hiroshi Oya (Eds.)*

54 Solar System Plasma Physics *J. H. Waite, Jr., J. L. Burch, and R. L. Moore (Eds.)*

55 Aspects of Climate Variability in the Pacific and Western Americas *David H. Peterson (Ed.)*

56 The Brittle-Ductile Transition in Rocks *A. G. Duba, W. B. Durham, J. W. Handin, and H. F. Wang (Eds.)*

57 Evolution of Mid Ocean Ridges (IUGG Volume 8) *John M. Sinton (Ed.)*

58 Physics of Magnetic Flux Ropes *C. T. Russell, E. R. Priest, and L. C. Lee (Eds.)*

59 Variations in Earth Rotation (IUGG Volume 9) *Dennis D. McCarthy and Williams E. Carter (Eds.)*

60 Quo Vadimus Geophysics for the Next Generation (IUGG Volume 10) *George D. Garland and John R. Apel (Eds.)*

61 Cometary Plasma Processes *Alan D. Johnstone (Ed.)*

62 Modeling Magnetospheric Plasma Processes *Gordon R. Wilson (Ed.)*

63 Marine Particles: Analysis and Characterization *David C. Hurd and Derek W. Spencer (Eds.)*

64 Magnetospheric Substorms *Joseph R. Kan, Thomas A. Potemra, Susumu Kokubun, and Takesi Iijima (Eds.)*

65 Explosion Source Phenomenology *Steven R. Taylor, Howard J. Patton, and Paul G. Richards (Eds.)*

66 Venus and Mars: Atmospheres, Ionospheres, and Solar Wind Interactions *Janet G. Luhmann, Mariella Tatrallyay, and Robert O. Pepin (Eds.)*

67 High-Pressure Research: Application to Earth and Planetary Sciences (Mineral Physics Volume 3) *Yasuhiko Syono and Murli H. Manghnani (Eds.)*

68 Microwave Remote Sensing of Sea Ice *Frank Carsey, Roger Barry, Josefino Comiso, D. Andrew Rothrock, Robert Shuchman, W. Terry Tucker, Wilford Weeks, and Dale Winebrenner*

69 Sea Level Changes: Determination and Effects (IUGG Volume 11) *P. L. Woodworth, D. T. Pugh, J. G. DeRonde, R. G. Warrick, and J. Hannah*

70 Synthesis of Results from Scientific Drilling in the Indian Ocean *Robert A. Duncan, David K. Rea, Robert B. Kidd, Ulrich von Rad, and Jeffrey K. Weissel (Eds.)*

71 Mantle Flow and Melt Generation at Mid-Ocean Ridges *Jason Phipps Morgan, Donna K. Blackman, and John M. Sinton (Eds.)*

72 Dynamics of Earth's Deep Interior and Earth Rotation (IUGG Volume 12) *Jean-Louis Le Mouël, D.E. Smylie, and Thomas Herring (Eds.)*

73 Environmental Effects on Spacecraft Positioning and Trajectories (IUGG Volume 13) *A. Vallance Jones (Ed.)*

74 Evolution of the Earth and Planets (IUGG Volume 14) *E. Takahashi, Raymond Jeanloz, and David Rubie (Eds.)*

75 Interactions Between Global Climate Subsystems: The Legacy of Hann (IUGG Volume 15) *G. A. McBean and M. Hantel (Eds.)*

76 Relating Geophysical Structures and Processes: The Jeffreys Volume (IUGG Volume 16) *K. Aki and R. Dmowska (Eds.)*

77 The Mesozoic Pacific: Geology, Tectonics, and Volcanism—A Volume in Memory of Sy Schlanger *Malcolm S. Pringle, William W. Sager, William V. Sliter, and Seth Stein (Eds.)*

78 Climate Change in Continental Isotopic Records *P. K. Swart, K. C. Lohmann, J. McKenzie, and S. Savin (Eds.)*

79 The Tornado: Its Structure, Dynamics, Prediction, and Hazards *C. Church, D. Burgess, C. Doswell, R. Davies-Jones (Eds.)*

80 Auroral Plasma Dynamics *R. L. Lysak (Ed.)*

81 Solar Wind Sources of Magnetospheric Ultra-Low Frequency Waves *M. J. Engebretson, K. Takahashi, and M. Scholer (Eds.)*

82 **Gravimetry and Space Techniques Applied to Geodynamics and Ocean Dynamics (IUGG Volume 17)** *Bob E. Schutz, Allen Anderson, Claude Froidevaux, and Michael Parke (Eds.)*

83 **Nonlinear Dynamics and Predictability of Geophysical Phenomena (IUGG Volume 18)** *William I. Newman, Andrei Gabrielov, and Donald L. Turcotte (Eds.)*

84 **Solar System Plasmas in Space and Time** *J. Burch, J. H. Waite, Jr. (Eds.)*

85 **The Polar Oceans and Their Role in Shaping the Global Environment** *O. M. Johannessen, R. D. Muench, and J. E. Overland (Eds.)*

86 **Space Plasmas: Coupling Between Small and Medium Scale Processes** *Maha Ashour-Abdalla, Tom Chang, and Paul Dusenbery (Eds.)*

87 **The Upper Mesosphere and Lower Thermosphere: A Review of Experiment and Theory** *R. M. Johnson and T. L. Killeen (Eds.)*

88 **Active Margins and Marginal Basins of the Western Pacific** *Brian Taylor and James Natland (Eds.)*

89 **Natural and Anthropogenic Influences in Fluvial Geomorphology** *John E. Costa, Andrew J. Miller, Kenneth W. Potter, and Peter R. Wilcock (Eds.)*

90 **Physics of the Magnetopause** *Paul Song, B.U.Ö. Sonnerup, and M.F. Thomsen (Eds.)*

91 **Seafloor Hydrothermal Systems: Physical, Chemical, Biological, and Geological Interactions** *Susan E. Humphris, Robert A. Zierenberg, Lauren S. Mullineaux, and Richard E. Thomson (Eds.)*

92 **Mauna Loa Revealed: Structure, Composition, History, and Hazards** *J. M. Rhodes and John P. Lockwood (Eds.)*

93 **Cross-Scale Coupling in Space Plasmas** *James L. Horwitz, Nagendra Singh, and James L. Burch (Eds.)*

Maurice Ewing Volumes

1 **Island Arcs, Deep Sea Trenches, and Back-Arc Basins** *Manik Talwani and Walter C. Pitman III (Eds.)*

2 **Deep Drilling Results in the Atlantic Ocean: Ocean Crust** *Manik Talwani, Christopher G. Harrison, and Dennis E. Hayes (Eds.)*

3 **Deep Drilling Results in the Atlantic Ocean: Continental Margins and Paleoenvironment** *Manik Talwani, William Hay, and William B. F. Ryan (Eds.)*

4 **Earthquake Prediction—An International Review** *David W. Simpson and Paul G. Richards (Eds.)*

5 **Climate Processes and Climate Sensitivity** *James E. Hansen and Taro Takahashi (Eds.)*

6 **Earthquake Source Mechanics** *Shamita Das, John Boatwright, and Christopher H. Scholz (Eds.)*

IUGG Volumes

1 **Structure and Dynamics of Earth's Deep Interior** *D. E. Smylie and Raymond Hide (Eds.)*

2 **Hydrological Regimes and Their Subsurface Thermal Effects** *Alan E. Beck, Grant Garven, and Lajos Stegena (Eds.)*

3 **Origin and Evolution of Sedimentary Basins and Their Energy and Mineral Resources** *Raymond A. Price (Ed.)*

4 **Slow Deformation and Transmission of Stress in the Earth** *Steven C. Cohen and Petr Vaníček (Eds.)*

5 **Deep Structure and Past Kinematics of Accreted Terrances** *John W. Hillhouse (Ed.)*

6 **Properties and Processes of Earth's Lower Crust** *Robert F. Mereu, Stephan Mueller, and David M. Fountain (Eds.)*

7 **Understanding Climate Change** *Andre L. Berger, Robert E. Dickinson, and J. Kidson (Eds.)*

8 **Evolution of Mid Ocean Ridges** *John M. Sinton (Ed.)*

9 **Variations in Earth Rotation** *Dennis D. McCarthy and William E. Carter (Eds.)*

10 **Quo Vadimus Geophysics for the Next Generation** *George D. Garland and John R. Apel (Eds.)*

11 **Sea Level Changes: Determinations and Effects** *Philip L. Woodworth, David T. Pugh, John G. DeRonde, Richard G. Warrick, and John Hannah (Eds.)*

12 **Dynamics of Earth's Deep Interior and Earth Rotation** *Jean-Louis Le Mouël, D.E. Smylie, and Thomas Herring (Eds.)*

13 **Environmental Effects on Spacecraft Positioning and Trajectories** *A. Vallance Jones (Ed.)*

14 **Evolution of the Earth and Planets** *E. Takahashi, Raymond Jeanloz, and David Rubie (Eds.)*

15 **Interactions Between Global Climate Subsystems: The Legacy of Hann** *G. A. McBean and M. Hantel (Eds.)*

16 **Relating Geophysical Structures and Processes: The Jeffreys Volume** *K. Aki and R. Dmowska (Eds.)*

17 **Gravimetry and Space Techniques Applied to Geodynamics and Ocean Dynamics** *Bob E. Schutz, Allen Anderson, Claude Froidevaux, and Michael Parke (Eds.)*

18 **Nonlinear Dynamics and Predictability of Geophysical Phenomena** *William I. Newman, Andrei Gabrielov, and Donald L. Turcotte (Eds.)*

Mineral Physics Volumes

1 **Point Defects in Minerals** *Robert N. Schock (Ed.)*

2 **High Pressure Research in Mineral Physics** *Murli H. Manghnani and Yasuhiko Syona (Eds.)*

3 **High Pressure Research: Application to Earth and Planetary Sciences** *Yasuhiko Syono and Murli H. Manghnani (Eds.)*

Geophysical Monograph 94

Double-Diffusive Convection

Alan Brandt
H. J. S. Fernando
Editors

American Geophysical Union

Published under the aegis of the AGU Books Board.

Cover photograph: J. Stewart Turner, Australian National University

Library of Congress Cataloging-in-Publication Data

Double diffusive convection / Alan Brandt, H.J.S. Joe Fernando, editors
p. cm. -- (Geophysical monograph ; 94)
Includes bibliographical references.
ISBN 0-87590-076-3
1. Ocean mixing. 2. Turbulence. 3. Salinity. I. Brandt, A. (Alan) II. Fernando, H. J. S. Joe. III. Series.
GC299.D68 1995
551.47--dc20 95-26425
CIP

ISBN 0-87590-076-3
ISSN 0065-8448

Printed in the United States of America.

CONTENTS

CONTENTS

PREFACE

The existence of double-diffusive convection and the associated, visually dramatic and dynamically significant salt fingers (as a molecular instability mechanism that can naturally arise in the ocean) was first recognized in the late 1950s. Since then, research in this area has increased almost exponentially, and new applications of the basic phenomenology continue to arise. At this time the importance of double-diffusive convection (DDC) has been recognized in fields as diverse as geophysics, astrophysics, metallurgy and chemistry as well as in the parent field—ocean physics. In each of these fields the small-scale, DDC phenomenology has been shown (or at least postulated) to be a critical driver for large, even global scale processes. Examples include DDC as a mechanism for maintaining the ocean thermocline and thus the global circulation pattern and DDC as a factor in convection of the Earth's mantle and at the core-mantle boundary.

This volume presents a cross-section of contemporary research on DDC in these diverse fields as well as recent theoretical, numerical, and fundamental laboratory studies of DDC. Recent fundamental efforts include, for example, studies of the effects of shear on the kinematic tilting and dynamical evolution of salt fingers, the formation of traveling interfacial waves at finger boundaries and nonlinear DDC instability mechanisms. This is the first published volume in which the broad scope of current DDC research is presented.

Although historical references indicate that salt fingering was observed in the mid-nineteenth century, it was not until the mid-1950s that Stommel and Arons recognized that fluid transport could result from differences in molecular diffusivities in a multi-component system (e.g., salt and temperature in the ocean), confined in a vertical tube—a salt fountain. A few years later Melvin Stern recognized that such DDC fingers can be produced naturally in the ocean as a result of the difference between the thermal and saline diffusivities and the effective range of heat conduction. This marked the beginning of the explosive growth in DDC research. Laboratory studies illustrating and enhancing our understanding of these phenomena were conducted in the late 1960s by Stern, Stommel and Turner. These studies led to major ocean experiments in the mid-1970s to confirm the existence of DDC in the Mediterranean outflow and in the Caribbean Sea. Direct evidence for DDC salt fingering was found, and analysis and interpretation of data of these experiments continues to the present. A general overview of salt fingering in the ocean is given by Schmitt (The Oceans Salt Fingers, Scientific American, 272(5), 70-75, 1995). Moreover, these oceanographic results stimulated researchers in other fields, who have increasingly found evidence for the existence and potentially important role of double and multiply diffusive convection.

As DDC applications in such areas as fluid dynamics, astrophysics, geology, oceanography, limnology, and energy technology were realized, major review articles covering different facets were authored by Turner (Double-diffusive phenomena, Annu. Rev. Fluid Mech., 6, 37-56, 1974) and Huppert and Turner (Double diffusive convection, J. Fluid Mech., 106, 299-329, 1981). The latter article motivated C. F. Chen and D. H. Johnson to convene an Engineering Foundation Conference on Double-Diffusive Convection in Santa Barbara, California, during May 6-15, 1983 (Double diffusive convection: A report on an engineering foundation conference, J. Fluid Mech., 138, 405-416, 1984). This first general conference was followed by a major review by Turner (Multicomponent convection, Annu. Rev. Fluid Mech., 17, 11-44, 1985) and a specialized conference on oceanic DDC at Woods Hole Oceanographic Institution during September 14-18, 1989 (Schmitt, R.W., Double-diffusion in oceanography: Proceedings of a meeting, Woods Hole Oceanographic Institution, Rep. WHOI-91-20, 1991). Another major review in an oceanic context has appeared (Schmitt, R.W., Double diffusion oceanography, Annu. Rev. Fluid Mech., 26, 255-285, 1994). The second general conference was held in Scottsdale, Arizona, during November 3-6, 1993, as an AGU Chapman Conference. The aim was to foster cross-communication among a variety of research groups and to review the significant advances over the past decade. About 60 scientific papers covering sub-areas of laboratory experiments, oceanography, geophysics-astrophysics, metallurgy-chemistry, and theoretical-numerical aspects were presented. A review of the Chapman Conference has appeared (Fernando and Brandt 1994a), and a brief summary is presented in Eos (Fernando and Brandt 1994b).

This volume is an outgrowth of the second general DDC conference and covers major sub-areas of contemporary DDC research. Papers submitted after the conference were vetted by an independent peer review process. Appropriately, this volume begins with an insightful introduction by Melvin Stern, quite arguably the father of this field of endeavor. Following this introduction, Ray Schmitt provides a historical perspective on close encounters with discovery of DDC. Stewart Turner then gives an overview of the laboratory experiments and models that have contributed immeasurably to our understanding of DDC. The remainder and greater part of the volume contains papers on recent results, including the broad range of DDC phenomenology and applications: laboratory, theoretical and numerical studies, metallurgy, geophysics, and oceanography. These papers collectively illustrate the potential importance of DDC in an interdisciplinary context and confirm that fundamental questions still remain, particularly regarding the relative importance of DDC as compared to other instability mechanisms, the behavior of DDC in the presence of ambient convective and turbulent flows, and the role of nonlinear mechanisms in DDC.

The editors wish to acknowledge the Office of Naval Research for financial support and the American Geophysical Union and Arizona State University for the organization and support of the Chapman Conference.

Alan Brandt
Applied Physics Laboratory
The Johns Hopkins University
Laurel, Maryland

H. J. S. Fernando
Environmental Fluid Dynamics Program
Arizona State University
Tempe, Arizona

Introduction: The Varieties of Turbulent Experiences

Melvin E. Stern

Department of Oceanography, Florida State University, Tallahassee

"We may not be able to define turbulence, but we know it when we see it." This pronouncement seems to be biased by the experience of engineers and other folk who go down to the sea in ships to watch the wind and the waves and the weather. This kind of turbulence has large Reynolds number as a distinctive feature; and one may want to recognize another variety of stochastic motion characterized instead by its very large Peclet number.

One revelation of micro-structure measurements (e.g., mechanical dissipation rates) is the relatively small vorticity and high intermittency of the non-isentropic motions responsible for mixing in the main ocean thermocline, where the average value of the nominal vertical eddy diffusivity for heat is only one hundred (or less) times the molecular viscosity. The smallness of the vertical heat flux merely reflects the long (decade-century) relaxation time scale of the global temperature and salinity (T-S) distributions; it is in this context that the possibility of the direct parametric importance of the molecular coefficients for the global T-S distributions must be viewed.

In speaking of small scale turbulence one usually has in mind "rapidly" evolving eddies such as breaking Kelvin-Helmholtz waves, but this may be unhelpfully restrictive. In fully developed thermal (Rayleigh-Benard) turbulence, for example, the key element is the thin thermal boundary layer with small Reynolds number, which supplies the energy for the disordered large eddies in the overlying convective region. The counterpart in double diffusion is an internal boundary layer (e.g., containing salt fingers), which provides the energy for larger scale motions on either side of that layer. The latter motions can disrupt the fingers and also enhance their flux by thinning the layer in which they exist. Moreover, double-diffusive interfaces appear to be produced at water mass boundaries by large scale lateral intrusions whose vertical shear systematically strains synoptic scale horizontal T-S gradients on isopycnal surfaces, eventually resulting in vertical T-S gradients much larger than the mean values. Thus the smallest scale in double diffusion is coupled with the larger energy bearing scales in the ocean circulation. The task of measuring and understanding non-isentropic motions in the ocean is as formidable as it is important, e.g., in numerical models of the general circulation.

Double-Diffusive Convection
Geophysical Monograph 94

A simpler but still relevant problem appears in early experiments which approximate an ideal initial state of uniform vertical T-S gradients in a completely unbounded fluid; in these experiments, quasi-vertical and almost uniformly spaced salt fingers form. Their width (L) and the flux ratio in the realized quasi-equilibrium can be explained by the well-known instability theory in which the exponentially growing vertical velocity (W) happens to be an exact solution of the nonlinear Boussinesq equations. But the convective heat flux (F) in this "underdeveloped" experiment is not explained, since the realized fingers bifurcate or fracture (due to some kind of Floquet instability) in a vertical distance O(10) L; the Reynolds number (based on W and L) is apparently O(1). Because of the long-range disorder in (T,S) fluctuations, and because all nonlinear Eulerian terms in the heat/salt equations are important, I will call this state high Peclet number turbulence. In addition to describing the (T,S,W) statistics, one wants to know the magnitude of F. On dimensional grounds it might be given by an eddy diffusivity [O(10)L]W, and this equals O(10)O(1) times the molecular viscosity.

The laboratory experiments that Stewart Turner and I did (1969, Deep-Sea Res., 16) show that when the initial vertical T-S gradients are changed in a suitable way, the foregoing "underdeveloped" equilibrium changes to the "sheets and layer" regime, in which the asymptotic "4/3 flux law" is realized. When double diffusion was first suggested as a mechanism for thermocline mixing, there was much doubt that delicate fingers and interfaces could survive in the turbulence of the great outdoors.

Even after salt fingers were photographed, and appropriately thin "sheets" were observed in the thermocline, skepticism persisted because application of the laboratory "4/3 law" gave heat fluxes which were much too large! Then the task became one of reducing the effect by considering ambient shear, or other effects which would produce sheets thicker than the asymptotic value, and of finding the heat flux law appropriate to this underdeveloped case. For this oceanic purpose as well for purely fluid dynamical interest, it may be worthwhile to direct more laboratory and theoretical attention to the aforementioned vertically unbounded regime of stochastic salt fingers at high Peclet number, low Reynolds number, and without mean shear.

Why Didn't Rayleigh Discover Salt Fingers?

Raymond W. Schmitt

Woods Hole Oceanographic Institution, Woods Hole, Massachusetts

Over a century before Melvin Stern discovered salt fingers, W. Stanley Jevons performed the first salt finger experiment in an attempt to model cirrus clouds. Remarkably, he seemed to realize that the more rapid diffusion of heat relative to solute played a role in the experiments. However, he incorrectly assumed that the "interfiltration of minute, thread-like streams" that he observed was a general result of superposing heavy over light fluid. Interestingly, Lord Rayleigh became aware of these experiments more than two decades later. Recently uncovered evidence is here presented that he performed his own salt finger experiments in 1880. These led Rayleigh to the discovery of the expression for the buoyancy frequency of internal waves and the convective process now known as the Rayleigh-Taylor instability. However, his neglect of diffusion meant that he missed an opportunity to discover double-diffusive convection. Similar near misses were achieved by Ekman and of course, Stommel, Arons and Blanchard. Consideration of this tortuous intellectual history reveals that salt fingers were one of the first convective phenomena to be studied, but among the last to be understood.

INTRODUCTION

Melvin Stern is credited with the discovery of double-diffusive convection [*Stern*, 1960]. His derivation is one of the few instances in oceanography where a new phenomenon was discovered theoretically, without the motivation of preceding experiments or observations. However, it is now recognized that there were, in fact, several prior salt finger experiments, some more than 100 years earlier. The earlier investigators include Jevons, Rayleigh, Ekman, and Stommel. All failed to fully appreciate the physics of the process or its oceanographic implications.

It is appropriate that we here consider these historical missteps on the road to double diffusion, for several reasons:

1. This volume represents a partial survey of recent progress in the field and reflects the wide range of disciplines where double diffusion now has been found to be important,
2. A knowledge of the history of a concept can be a useful guide for our own deliberations, and
3. The story of how other distinguished scientists might have made the discovery, but did not, increases our respect for Stern's insight. As Stern has just been recognized by the American Meteorological Society with the first Stommel Award for his achievements, it is timely that we now consider these stories of "what might have been".

Moreover, developing evidence indicates that internal waves provide only weak mixing in the ocean interior [*Toole et al*, 1994]. This means that the modest fluxes ascribed to salt fingers must now be considered one of the most important processes in determining the temperature-salinity structure of the main pycnocline [*Schmitt*, 1994].

In this report, several near discoveries of salt fingers are discussed. The first is due to *Jevons* [1857], who performed and described a heat/sugar finger experiment and came close to understanding it. This work led to theory and experiments by Rayleigh in 1880, who also failed to grasp the double diffusive nature of the phenomenon, though he did make other important discoveries in the dynamics of buoyant fluids. A fuller report on these two contributions is given by *Schmitt* [1995]. Additional material presented here discusses a footnote of *Ekman* [1906], the work of *Stommel, Arons and Blanchard* [1956] and Stern's discovery itself.

W. STANLEY JEVONS

Veronis [1981], in discussing the work of Stommel on the salt fountain, noted that *Jevons* [1857] had performed the first salt finger experiment. Here I provide an abbreviated treatment of the description given in *Schmitt* [1995].

Double-Diffusive Convection
Geophysical Monograph 94

William Stanley Jevons was born in 1835 in Liverpool, England, into a middle-class merchant family. He was well educated and displayed an early interest in the sciences, particularly chemistry, biology and mathematics. A family financial crises forced him to leave University College, London, to take a position as Assayer at the new Royal Mint in Sydney, Australia, in 1854, when he was 19. While in Australia he developed an interest in meteorology, making routine weather observations for several years before persuading the government to maintain the record. He also performed some laboratory experiments to model clouds; these are discussed below. After nearly five years in Sydney he returned to England to complete his education. He took up economics and achieved prominence with what was one of the first discussions of an energy crisis [*Jevons*, 1865]. He went on to hold Professorships at Owens College, Manchester, and University College, London, and was a noted author on economics, logic and the philosophy of science. Biographical sketches of Jevons are available in *Black* [1972], *Keynes* [1951] and *Schabas* [1990]. *Inoue and White* [1993] provide an extensive list of his publications.

In the experiments of interest *Jevons* [1857], motivated by observations of clouds, introduced warm, sugar water over cold, fresh water. He reported an "interfiltration of minute, thread-like streams". Figure 1 is from his paper. Displaying remarkable insight, he seems to appreciate the importance of thermal conduction to the instability when he states:

"The parts of these strata, however, which are immediately in contact, soon communicate their heat and tend to assume a mean temperature; and it is evident that whenever this is the case, the portions of liquid containing sugar must always be slightly denser than those that are pure, and must consequently sink below and displace the latter."

This description shows an implicit understanding that heat diffuses faster than solute. One can imagine that this insight simply reflected his experience as a working chemist; temperature contrasts in fluids equilibrate rather quickly compared to the slow diffusion of salts. The statement itself is a reasonably accurate description of the salt finger instability, which can release the potential energy contained in an unstable solute distribution by virtue of the two order-of-magnitude difference between the thermal conduction and the diffusivities of dissolved species. Unfortunately, he next incorrectly dismisses the temperature difference as being only a means of laying a denser strata over a lighter one:

"[It is evident that the difference of temperature of the strata in this experiment is not a material point, being simply a means employed to enable us to lay one stratum upon another of a slightly greater density when of the same temperature, so that we may afterwards observe the mixing process and change of place in the most gradual manner possible.]"

In fact, the large scale stability of the water column is

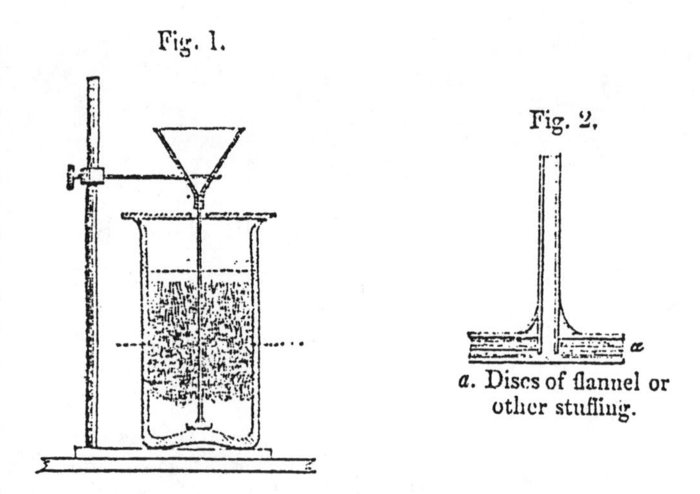

Fig. 1. A sketch by Jevons of the experiment described in his 1857 article in the *Philosophical Magazine.*

maintained by the temperature difference, and actually increases as the fingers grow and redistribute the solute. Apparently, Jevons believed that the thermal conduction was only a vertical process, and failed to appreciate that lateral conduction of heat between the "interfiltrating streams" was what drove the observed motion.

Jevons also performed a control experiment in which he introduced sugar to the cold water. This was completely stable, with a flat interface separating the fluids. He believed that the experiments were an accurate analog of the atmospheric processes controlling cirrus clouds, because of the wispy appearance of the "minute, threadlike streams". He thought that the slow, small-scale exchange in parallel bands that he observed was the natural form of all convection. He focussed on the simpler problem of a heavy fluid overlying a light fluid, and failed to develop the initial insight into the role of thermal conduction. This appears to have been sufficient to later distract Rayleigh from the opportunity to discover double-diffusive convection.

In a longer but less available paper in the *Sydney Magazine of Art and Science*, *Jevons* [1858] performed some further experiments in which he injected an intermediate density fluid between two layers. One of the runs appears to have been the first demonstration of a double-diffusively driven lateral intrusion, as it closely resembles the modern intrusion experiments of *Turner* [1978]. *Stern* [1967] first showed the potential for such motions in the ocean as a larger scale consequence of salt finger fluxes.

These two papers were Jevons's first serious scientific contributions. They are remarkable for the maturity of his insight; he was only 22 when they were published! Though he struggled in his early years after returning to England, he was eventually successful in the field of economics, and became a Fellow of the Royal Society in 1872. In 1880, weary of teaching, he retired to devote more time to his writing and research. Unfortunately, he drowned 2 years later, in a somewhat mysterious accident, while on holiday with his family, just short of his 47th birthday. The timing of his death is sadly ironic, for if he had lived another year or so he would have had the satisfaction of seeing his earliest scientific work recognized by England's most prominent physicist of the time, Lord Rayleigh.

RAYLEIGH'S 1883 PAPER

Jevons's work on clouds seems to have attracted little attention until 1883, when Rayleigh published a theoretical analysis of the two experiments. Lord Rayleigh (John William Strutt), of course, was one of the most accomplished scientists who ever lived. He was seven years younger than Jevons, Cambridge educated, and succeeded his father as Baron Rayleigh at the age of 31. He used his wealth to great scientific advantage, devoting his life to the pursuit of a wide variety of topics in optics, acoustics, electricity, fluid dynamics and chemistry. He followed Maxwell as Cavendish Professor of Experimental Physics at Cambridge University, received the Nobel Prize in 1904 for the discovery of argon, was secretary (editor) of the Royal Society for many years and later its president.

In 1883 he published an "Investigation of the character of the equilibrium of an incompressible heavy fluid of variable density" in the *Proceedings of the London Mathematical Society*. Rayleigh's paper, which he says was motivated by Jevons experiments, is a short mathematical treatment of the stability of a stratified fluid with density either increasing or decreasing with depth. The three cases considered included two superposed layers, exponential stratification, and two layers with a transition zone. He neglects diffusion, assuming that the density of every particle remains unchanged. Viscosity is also ignored and he makes no attempt to relate the theory to observations.

Rayleigh [1883] finds that in the stably stratified case, there is a "limit on the side of rapidity of vibration but none on the side of slowness". The square of this limit is defined by:

$$n^2 = -g\beta, \qquad \text{where} \qquad \beta = \sigma^{-1} d\sigma/dz,$$

and σ is the density.

It is surprising to see the expression for the buoyancy frequency of internal waves derived over 40 years prior to *Brunt* [1927] and *Vaisaila* [1925]; even more so because the notation is so standard! As *Gill* [1982] has pointed out, Rayleigh certainly deserves credit for priority in this discovery.

In the two layer case of heavy fluid over light, Rayleigh finds that the growth rate "is greater the smaller the wavelength". This result was not rediscovered until 1950 by Sir Geoffrey Taylor who also treated the case of arbitrary accelerations [*Taylor*, 1950]. Now known as the Rayleigh-Taylor instability, it is important in plasma dynamics, super-novae explosions, and heavy-nuclei collisions [*Petrasso*, 1994]. While the two results to come out of this paper are notable achievements, Rayleigh, by ignoring Jevons's hint about the role of thermal conduction, missed an ideal opportunity to discover double-diffusive convection as well. The preference for small scales in the two layer unstable case is an obvious point of similarity to Jevons's "minute, threadlike streams". However, Rayleigh neglects to discuss this, or any other physical application of the theory. Indeed, the only reference to Jevons is contained in the footnote to the title of his paper. The footnote states:

"These calculations were written out in 1880, in order to illustrate the theory of cirrus clouds propounded by the late Prof. Jevons [*Phil. Mag*. xiv. p. 22, 1857]. Pressure of other work has prevented me hitherto from pursuing the subject."

If the Jevons results were common knowledge it might be reasonable to assume that the reader would draw the obvious connections between theory and experiment. However, this seems unlikely, as the experiments had been published more than 25 years earlier, and Jevons was known mostly as an economist. While it may be that such a spare style was typical of papers of Rayleigh and/or the *London Mathematical Society*, one comes away with a sense of incompleteness with the lack of discussion of motivation or application of the theoretical results.

THE RAYLEIGH-SIDGWICK EXPERIMENTS OF 1880

Curious about Rayleigh's motivation for the analysis, and the possibility of any communication between he and Jevons, I recently examined his original laboratory notebooks. These are available in the Rayleigh Collection at the Phillips Laboratory Research Library, Hanscom Air Force Base, Bedford MA. The notebooks are described by *Howard* [1964a]. There are twelve notebooks in Lord Rayleigh's hand and one by his sister-in-law, Eleanor M. Sidgwick. The Collection also includes the original hand-written manuscripts of many of Rayleigh's papers, though not for the 1883 paper of interest here. However, no mention of Jevons or any notes relating to the paper were found in Rayleigh's own notebooks. Fortunately, the Sidgwick notebook provides one page of highly relevant material.

Eleanor Balfour Sidgwick was the sister of Evelyn Balfour, Lord Rayleigh's wife, and sister of Prime Minister Arthur Balfour. Her husband Henry Sidgwick was lecturer and professor of moral philosophy at Cambridge and wrote many works on morals, ethics, and political economy. Eleanor displayed interest in, and talent for, physics and mathematics, and enjoyed discussing science with Rayleigh. While Rayleigh was at the Cavendish from 1879 to 1884 she served as his laboratory assistant. She helped with the redetermination of the value of the ohm. She was coauthor on five of his dozen papers on electrical standards and was acknowledged as coworker in three others [*Howard*, 1964b]. Her notebook covers the period from April 1880 to March 1881, early in Rayleigh's tenure at the Cavendish.

Figure 2 is a copy of the notebook entry, which is the first page of the notebook. My transcription follows:

"Cavendish Laboratory April 1880

We repeated several times the experiment of W. W. S. Jevons [see *Phil. Mag.* for July 1857] on the formation of cirrous clouds. A glass funnel connected with a straight glass tube by a short piece of india rubber tube was used and the current, which had to be very slow, regulated by nipping the india rubber tube. The experiment seemed to answer best when the lower end of the tube rested on a small disc of thick flannel laid on the bottom of the beaker. The apparatus in other respects was similar to Jevons. The effects obtained resembled those described by him only generally the filaments seemed not fine enough to correspond with his description and drawing. In all cases moreover the extremities of the filaments were expanded in a mushroom like form. We also tried the experiment with water coloured with aniline & uncoloured water - still adding the sugar solution to that which was ultimately to fall. The appearance produced was much the same, but visible only in the uncoloured liquid."

This notebook entry provides proof that Rayleigh had taken the time to repeat Jevons's experiments. The text implies that they duplicated his visualization technique (silver chloride precipitate), and also used a dye as a simpler means of observing the flow. The apparently wider fingers they report could have been due to a weaker vertical temperature gradient, though the dependence is only on the fourth root of the temperature gradient [*Stern*, 1960]. However, it is difficult to judge the significance of their statement, since neither Jevons or Rayleigh/Sidgwick quantified the scale of the fingers. Their "mushroom like form" is corroborated by recent laboratory experiments [*Taylor and Bucens*, 1989] and numerical models

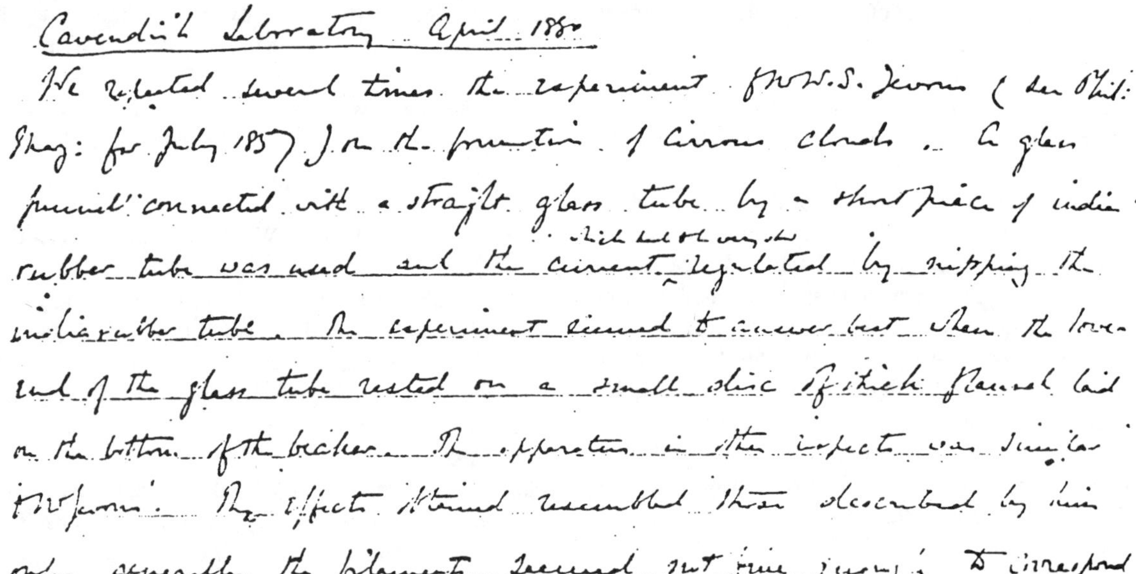
Cavendish Laboratory April 1880
We repeated several times the experiment of W.W.S. Jevons (see Phil: Mag: for July 1857) on the formation of Cirrous clouds. A glass funnel connected with a straight glass tube by a short piece of india rubber tube was used and the current which had to be very slow regulated by nipping the india rubber tube. The experiment seemed to answer best when the lower end of the glass tube rested on a small disc of thick flannel laid on the bottom of the beaker. The apparatus in other respects was similar to W Jevons. The effects obtained resembled those described by him only generally the filaments seemed not fine enough to correspond with his description & drawing. In all cases moreover the extremities of the filaments were expanded in a mushroom like form. We also tried the experiment with ~~liquid~~ water coloured with aniline & uncoloured water — still adding the sugar solution to that which was ultimately to fall. The appearance produced was much the same, but visible only in the uncoloured ~~fluid~~ liquid

Fig. 2. The first entry in the Elanor Sidgwick Notebook in the Rayleigh Collection, Phillips Research Library, Hanscom Air Force Base, Bedford, MA.

[*Shen*, 1989] on salt finger convection. Given this new evidence that Rayleigh performed his own salt finger experiments, it is puzzling that he makes no mention of them in his 1883 paper. Presumably he decided that Jevons's description was sufficiently complete and accurate.

While the performance of a second set of salt finger experiments by Rayleigh at the Cavendish in April, 1880, is of historical interest, it provides no insight into the question of how Rayleigh came to be aware of Jevons work. There exits no correspondence between them in the archives of either Jevons [*McNiven*, 1983] or Rayleigh (John Armstrong, personal communication, 1994). Three scenarios can be suggested:

1. Jevons mentioned the experiments to Rayleigh in conversation at the Royal Society. Jevons became a Fellow in 1872, Rayleigh in 1873. Jevons would certainly have identified Rayleigh as having a potential interest.

2. Others alerted him to the Jevons paper. One possibility is his brother-in-law, Henry Sidgwick, who was in the same field as Jevons, and knew him and his work [*Schabas*, 1990]. However, as "the last of the utilitarians" and in the same school of thought as J. S. Mill, he might not have been sympathetic to Jevons' new quantitative economics. Sir John Herschel, who was aware of Jevons's papers, died in 1871, so was not a likely conduit to Rayleigh.

3. Rayleigh came across the paper while reading back issues of the *Philosophical Magazine*. This was his favorite journal [*Howard*, 1964a], he read widely, and maintained a collection of the serial in his home (and would not lend it out even to Lord Kelvin [*Strutt*, 1968]!)

By whatever means Rayleigh came to know about Jevons's experiments, the timing of his paper, three years after performing the work and one year after Jevons' death, is very curious. Having duplicated the experiments and developed the theory in 1880, he may have deferred publication in order to discuss the results with Jevons. But prevented from contact with Jevons by his untimely death, Rayleigh must have decided to simply publish what he had. One can only speculate that with greater opportunity for exchange between these two distinguished scientists, much greater insight could have been achieved. However, Jevons had a reputation as a rather reclusive character [*Keynes*, 1951], and may have been difficult to approach. Moreover, Rayleigh was a keen experimentalist and should have realized the role of thermal conduction in the process himself, especially given Jevons marvelous hint, and having witnessed the fingers in his own laboratory. Had Jevons survived long enough to discuss the experiments with Rayleigh they could well have reached a more complete understanding. How much more would we know about double diffusion if it had been discovered over a century ago instead of just three decades?

However, given the lack of knowledge of fluid dynamics in the last century, it might not have made any difference if Rayleigh had come to understand the physics of double diffusion. After all, his derivation of the buoyancy frequency of internal waves was overlooked for years, and the Rayleigh-Taylor instability was not rediscovered until 1950. Progress in fluid dynamics in this century has been heavily dependent on experimental, observational and computational tools that did not exist then. Rayleigh makes reference to the 1883 paper (but not Jevons's work) in his much later treatment of Be'nard convection, specifically recognizing it as the inviscid case of the more general problem [*Rayleigh*, 1916]. Indeed, one cannot fault Rayleigh's logical development of the theory, even if the motivating experiments were misinterpreted. The 1883 paper treated inviscid, nondiffusive, stable and unstable stratifications. The 1916 paper added viscosity and thermal diffusion (though the experiments of Be'nard were actually affected by surface tension). The addition of a second buoyancy-affecting solute with a different diffusivity would have been a logical next step. Perhaps if Rayleigh had focussed more on fluid dynamics he might have had time to realize that step, and also appreciate the true nature of the original experiments which introduced him (and the world) to buoyancy effects in fluids.

V. W. EKMAN

Since the development of the foregoing Jevons/Rayleigh story, Henry Charnock has informed me of a similar near discovery of salt fingers by *Ekman* [1906] in his famous "dead water" paper [*Charnock*, 1983]. In the course of his careful laboratory experiments on the generation of internal waves by moving ship models, Ekman had attempted to use a two layer system of milk over seawater as a means of visualizing the interface displacements. This was unsuccessful, as he describes in a footnote:

"Milk, or milk and water, does not float very long above salt water. The salt diffuses into the milk; and milk being heavier than freshwater, salt milk is heavier than salt water and consequently sinks. New quantities of milk come into contact with the water, become salt, and sink in their turn; and after a little while, the milk is seen to fall through the salt water as a shower of small vortex-rings. The same thing takes place in the case of all precipitates in fresh water, but at different velocities."

Clearly, Ekman's shower of vortex rings are actually milk fingers. He, unlike Jevons and Rayleigh, understood the physics of the process he observed. However, since he had used a salt/precipitate system, rather than heat/salt, the connection with the ocean was not apparent. Without the motivation of an oceanic application he did not pursue the phenomenon, though doubtless he could have made significant progress, given his insight and analytic skills.

STOMMEL, ARONS AND BLANCHARD

The *Stommel, Arons and Blanchard* [1956] paper on the "perpetual salt fountain" is occasionally cited as the origin of double diffusion, but in fact it was not. *Arons* [1981] recounts how it arose as he and Stommel were discussing the problem of measuring pressure accurately at the bottom of the sea:

"In desperation we were considering the brute force technique of making a 3-mile-long manometer, i.e., literally extending a tube from the ocean surface to the bottom and drawing abyssal water up into the tube. Since the salinity of the abyssal water is invariably lower than the average salinity of the water above it, the water in the tube, on coming to thermal equilibrium with its surroundings, would stand above the level of the surrounding ocean surface, and we would watch the level in the tube go up and down with variations of pressure at the bottom. We had this picture sketched on the chalkboard and were entirely focused on the manometric aspect when my own mind took a divergent turn. In some astonishment, I added a faucet to the upper level of our manometer and said, 'Hank, if we open the faucet, it will run forever.'

After we satisfied ourselves concerning the nature and temporal limitations of the physical phenomenon, Stommel ran down to Duncan Blanchard's laboratory and recruited this ready and skillful gadgeteer to our party. Blanchard quickly set up a large beaker with a layer of hot salty water floating on cold fresh, and we blissfully watched the little fountain that spurted for a long time out of the glass tube in which the cold fresh water had been drawn upward to start the sequence.

This was the genesis of our oft-cited short letter to *Deep-Sea Research* describing the 'perpetual' salt fountain. We recognized that the key lay in blocking salt transfer while allowing thermal equilibrium; we recognized that if surface water were initially drawn downward in the tube, there would be a steady downward flow; but we did not perceive a deeper significance. ..."

"Not long afterward, Melvin Stern, in his quite independent investigation of the stability problem, became aware of the dynamic significance of the huge difference between the molecular diffusion coefficients of heat and salt and thus discovered double-diffusive convection and 'salt fingers'. I believe that Stommel was a bit chagrined about having missed this himself, but he was strongly supportive of Stern's priority for the discovery and unstinting in his praise and enthusiasm."

We can only speculate on what might have resulted if Stommel and coworkers had put dye into one of their layers, and seen something more than just the fountain. They had insight to most of the physics, and lacked only an appreciation of the large difference between the diffusivities of heat and salt. Knowledge of the Jevons paper, with its strong hints in this direction, also might have been sufficient to tip them off. In any case, a colleague was soon to make the discovery.

Thus, we now know that at least four prior experimentalists had the opportunity to discover salt fingers, but did not. Stern accomplished this through direct theoretical considerations of the effects of the very different diffusivities of heat and salt.

STERN'S DISCOVERY

We can learn a bit about Stern's accomplishment from the description provided by *Faller* [1992] of the day of the discovery. Though he sets the day in the spring of 1960, Stern's paper was received at *Tellus* on January 4 of that year, so it was more likely the fall of 1959. Quoting from the text:

"I was across the hall from Stommel and Bolin, waiting for the 10 o'clock coffee that we all shared in Andy Bunker's lab, when Melvin Stern breezed into Stommel's office in an excited state that was unusual for the pipe-smoking contemplative Mel. Animated voices drifted across the hall, Willem Malkus dropped in, the decibel level rose, and I went across the hall to investigate the excitement of Stern, Stommel, Bolin and Malkus. As Stern talked, I quickly grasped that he had quite independently developed a new theory, that of double-diffusive convection. He had then realized the relation of this presumed natural phenomenon to the 'Perpetual Salt Fountain' of Stommel, Arons and Blanchard, and had come to get Stommel's opinion.

You may well ask how one would come to conceive of the double-diffusive phenomenon in the first place, but this is not hard to understand in the context of the related studies then taking place within our group. Veronis and Malkus, in particular, were deeply concerned with theories of the onset of convection in which thermal diffusivity and kinematic viscosity, differing by an order of magnitude, play leading roles. It is not surprising then that some inquisitive scientist concerned with oceanography should ask about the additional effect of the relatively low diffusivity of salt when both thermal and salinity gradients are present. But it is one thing to inquire and quite another to carry through all of the detailed mathematical manipulations, to discover the various types of double-diffusive phenomena that may occur, as did Melvin Stern.

Based on Stern's concepts and equations, Malkus went to the chalkboard and in about two minutes produced an estimate of the scale of the proposed phenomenon: less than 1 centimeter. Stommel suggested that we do an experiment, whereupon I produced a 1,000-milliliter graduated cylinder and some potassium permanganate [as both a salt and a colorful tracer]. We prepared a warm solution of the salt, of lesser specific gravity than the cool tap water that now half filled the graduate, and we carefully poured the solution on top. Voila! In a few seconds, descending fingers of salty solution appeared as Stern had predicted, about 5 millimeters in diameter. These estimates and experiments, of course, would have been done by Stern himself in due time, but in Stommel's office the pace was markedly accelerated."

DISCUSSION

The historic record shows that at least four distinguished scientists had performed salt finger experiments prior to *Stern* [1960]. *Jevons* [1857] performed careful experiments and seemed to possess intuition about the difference between heat and salt diffusivities. *Rayleigh* [1883] ignored Jevons's hints and focussed on the simpler [but also important] buoyancy physics of internal waves and nondiffusive convection. *Ekman* [1906], also working on internal waves, observed and understood the phenomenon of precipitate-salt fingers, but did not consider the much larger difference between the diffusivities of heat and salt. *Stommel, Arons and Blanchard* [1956] appreciated the potential energy available in the salt distribution and the need to separate the salt and heat transfers, but, for the lack of flow visualization in their experiment, did not realize it would occur naturally.

Thus, it was theory that ultimately triumphed in the discovery of double-diffusive convection. *Stern* [1960] derived the conditions for the salt finger instability and also, in an often overlooked footnote, identified the potential for an oscillatory instability when a stable salt gradient is heated from below.

What are the lessons to be learned from this record? A cynic would say that theoreticians and experimentalists have a hard time understanding one another. But communication must occur before understanding. In Woods Hole of the 1950's there was no problem with communication, and the understanding came rather quickly. Ekman's isolated observations are hidden in a large monograph, and it is not surprising that they remained obscure. Surely, the most significant failure in communication involved Jevons and Rayleigh; no record of correspondence between them has been found. Jevons had good intuition about some aspects of the experiments, and Rayleigh certainly could have developed the theory. Perhaps they did talk at a Royal Society meeting but did not quite sort it out. Modern scientists could be faulted for failure to read the earlier work of Jevons and Rayleigh. However, in a sense, discoveries must await their time; the improved observational capabilities of the present day, and the pressing need to understand ocean mixing, are enabling and motivating factors in developing our knowledge of double-diffusive convection. Ultimately, the passage of time, and a measure of serendipity, are both necessary to synthesize the unique perspectives of individual scientists into a coherent picture of the natural world.

Acknowledgements. Colleen Hurter of the WHOI/MBL Library located key references. Herbert Huppert provided leads on Jevons, Rayleigh, and the Royal Society. Stewart Turner located a copy of the *Sydney Magazine* article. Michael White and Margaret Schabas provided valuable information on Jevons. John Howard, former Curator of the Rayleigh Collection, provided insight to Rayleigh and guidance through the Rayleigh Notebooks. John Armstrong, present Curator of the Rayleigh Collection at the Phillips Laboratory Research Library, kindly checked for correspondence with Jevons, and granted access to the Rayleigh Notebooks. Henry Charnock alerted me to the Ekman footnote. Arnold Arons and Alan Faller shared their memories of Woods Hole in the 1950's. Alan Brandt and Joe Fernando organized the Chapman Conference on Double-Diffusive Convection, where part of the Jevons-Rayleigh story was first presented, and encouraged me to prepare this manuscript. AGU and the Office of Naval Research (Grant N00014-89-J-1073) supported my participation in the Conference.

REFERENCES

Black, R. D. C. (ed.), *Papers and Correspondence of William Stanley Jevons*, Macmillan Press, London, 7 vols., 1972-1981.

Brunt, D., The period of simple vertical oscillations in the atmosphere. *Q. J. R. Meteorol. Soc.*, **53**, 30-32, 1927.

Charnock, H., Advances in physical oceanography. *Recent Advances in Meteorlogy and Physical Oceanography,* Royal Meterological Society, 67-81, 1983.

Ekman, V. W., On Dead Water. *Scientific Results of the Norwegian North Polar Expedition, 1893-1896,* **5**, (15), 1-152, 1906.

Faller, A., J., A tribute to Henry Stommel. *Oceanus*, **35**, (special issue), 30-33, 1992.

Gill, A. E., *Atmosphere-Ocean Dynamics*, Academic Press, 662 pp, 1982.

Howard, J. N., The Rayleigh Notebooks. *Applied Optics,* **3** (10), 1129-1133, 1964a.

Howard, J. N., Eleanor Mildred Sidgwick and the Rayleighs. *Applied Optics*, **3** (10), 1120-1122, 1964b.

Inoue, T. and M. V. White, Bibliography of published works by W. S. Jevons. *Journal of the History of Economic Thought*, **15**, 122-147, 1993

Jevons, W. S., On the cirrous form of cloud. *London, Edinburgh and Dublin Philosophical Magazine and Journal of Science*, 4th Series, **14**, 22-35, 1857.

Jevons, W. S., On clouds; their various forms, and producing causes. *Sydney Magazine of Science and Art,* **1**, (8), 163-176, 1858.

Jevons, W. S., *The Coal Question*, Macmillan, Cambridge, England, 1865.

Keynes, J. M., William Stanley Jevons. in: *Essays in Biography,* (2nd edition), Rupert Hart-Davis, London. pp 255-309, 1951.

McNiven, P., Hand-list of the Jevons archives in the John Rylands University Library of Manchester. *Bulletin of the John Rylands University Library of Manchester*, **66**, (1), 213-248, 1983.

Petrasso, R. D., Rayleigh's challenge endures. *Nature*, **367**, 217-218, 1994.

Rayleigh, Lord, Investigation of the character of the equilibrium of an incompressible heavy fluid of variable density. *Proceedings of the London Mathematical Society*, **14**, 170-177, 1883.

Rayleigh, Lord, On convection currents in a horizontal layer of fluid, when the higher temperature is on the under side. *Philosophical Magazine*, **32**, 529-546, 1916.

Schabas, M., *A World Ruled by Number: William Stanley Jevons and the Rise of Mathematical Economics*, Princeton Univ. Press, Princeton, 192pp, 1990.

Schmitt, R. W., Double diffusion in oceanography. *Annual Review of Fluid Mechanics*, **26**, 255- 285, 1994.

Schmitt, R. W., The salt finger experiments of Jevons (1857) and Rayleigh (1880). *Journal of Physical Oceanography*, **25**, (1), 1995.

Shen, C., The evolution of the double-diffusive instability: salt fingers. *Physics of Fluids*, **A1**(5), 829-844, 1989.

Stern, M. E., The 'salt fountain' and thermohaline convection. *Tellus*, **12**, 172-175, 1960.

Stern, M. E., Lateral mixing of water masses. *Deep-Sea Res.* **14**, 747-753, 1967.

Stommel, H., Aarons, A. B. and D. Blanchard, An oceanographical curiosity: the perpetual salt fountain. *Deep-Sea Res.* **3**, 152-153, 1956.

Strutt, R. J., . *Life of John William Strutt, Third Baron Rayleigh.* University of Wisconsin Press, Madison, 439 pp., 1968.

Taylor, G. I., The instability of liquid surfaces when accelerated in a direction perpendicular to their planes. I. *Proc. Royal Society*, **A201**, 192-196, 1950.

Taylor, J. and P. Bucens, Laboratory experiments on the structure of salt fingers. *Deep-Sea Res.*, **36**(11), 1675-1704, 1989.

Toole, J. M., K. L. Polzin and R.W. Schmitt, Estimates of Diapycnal Mixing in the Abyssal Ocean. *Science*, **264**, 1120-1123, 1994.

Turner, J. S., Double-diffusive intrusions into a density gradient. *J. Geophys. Res.*, **83**, 2887-2901, 1978.

Vaisaila, V., Uber die Wirkung der Windschwankungen auf die Pilotbeobachtungen. *Soc. Sci. Fenn. Commentat. Phys.-Math.*, **2** (19), 19-37, 1925.

Veronis, G., A theoretical model of Henry Stommel. in: *The Evolution of Physical Oceanography*, B. A. Warren and C. Wunsch, editors, MIT Press, Cambridge, MA, xx, 1981.

Raymond W. Schmitt, Department of Physical Oceanography, Woods Hole Oceanographic Institution, Woods Hole, MA 02543.

Laboratory Models of Double-Diffusive Processes

J.S. Turner

Research School of Earth Sciences, Australian National University, Canberra, ACT, Australia

This review traces the development of ideas about double diffusion from their origins in the field of oceanography to the more recent applications in other contexts. The influence of laboratory experiments has been great, and this has been illustrated in a very visual way using many photographs (and also, in the oral presentation, time-lapse movies) of the experiments. These experiments have been carried out using both temperature and salinity differences, which are the properties of direct relevance to the ocean, and the analogue system consisting of two solutes having much closer molecular diffusivities. Of particular interest to the author and his colleagues has been the extension to geological problems, which requires the addition of crystallization to the double-diffusive processes treated previously. Compositional convection, accompanying the release of light residual fluid when denser crystals form at a boundary, then becomes significant. Experiments modelling flows occurring in very different natural contexts, which are nevertheless dynamically closely related, demonstrate the value of a broad interdisciplinary perspective in this field.

INTRODUCTION

In preparing the opening address and related publication for a specialist conference of this kind, a keynote lecturer always faces a dilemma. All those coming to a meeting on Double-Diffusive Convection will take the basic principles for granted, and they will have read several previous reviews, so one cannot start right at the beginning. On the other hand, an introductory speaker should not attempt to cover comprehensively the most recent material. Though the subject has expanded rapidly in the thirty-five years or so of its existence, and many new ideas and disciplines are now included by this title, these developments will be reviewed in detail by others in subsequent papers in this volume. At an early stage I decided that my role should be to give a general historical review of the subject, and from a very personal perspective.

Two themes are emphasized here. The first aim is to document the influence that simple laboratory experiments have had in increasing our understanding of the basic physical processes underlying double-diffusive phenomena. This is done in a very visual way, using photographs of laboratory experiments, taken mainly by me and various colleagues in Woods Hole, Cambridge and Canberra (many of which have previously been published), and (in the oral presentation) time-lapse movies of some of these. The second related aim is to demonstrate the importance for the development of this subject of interactions between researchers with very different backgrounds. The initial motivation for the study of double-diffusive convection came from the field of oceanography, where it was conjectured that the differential diffusion of heat and salt could affect the distribution and vertical transports of these two properties in the ocean. Detailed discussions and close collaborations between sea-going oceanographers, theoreticians and laboratory experimenters played a vital role in the rapid spread and acceptance of ideas in this field.

The subject has broadened in recent years to include applications to fields in which there had been no previous indication that double-diffusive effects could be significant. The author's research interests have shifted towards geological applications, and the way in which information and ideas can be transferred between different fields will also be considered. Laboratory experiments have proved to be important in this field too, with the addition of crystallization to the double-diffusive processes treated previously.

FUNDAMENTAL IDEAS ABOUT DOUBLE DIFFUSION

Convective motions occurring in a fluid when only one property affects the density, for example temperature

Double-Diffusive Convection
Geophysical Monograph 94

differences driving convection in a fluid layer heated from below, or a compositional difference producing a plume from an isolated source, are now very well understood. When gradients of more than one diffusing property are introduced simultaneously, however, a whole new range of phenomena can arise, and our intuition based on observations of simple thermal or compositional convection may be of little help in understanding what we observe.

The basic requirements for double-diffusive convection to occur are now well understood, and they have been reviewed many times [*Turner*, 1973, 1974, 1985, *Huppert and Turner*, 1981a]; most recently this has been done in the oceanographic context by *Schmitt* [1994]. The necessary minimum conditions for a double-diffusive system to be unstable are that the two substances stratifying the fluid should have different molecular diffusivities and opposing effects on the vertical density gradient. (See Fig. 1.) The definition can be broadened and extended to multicomponent systems [*Turner*, 1985] but the essential point is that in every case differential or coupled diffusion can produce convective motions that are associated with a decrease in the gravitational potential energy of the system. A striking feature is that instabilities can arise even when the horizontally averaged net density distribution is decreasing upwards.

The original application was to the ocean, where of course heat and salt are the two stratifying properties. Following an early exploratory study [*Stern and Turner*, 1969] much of the laboratory work has been carried out using two solutes with much closer molecular diffusivities. Besides being of direct interest in various applications, this system has the advantage that we do not have to be concerned about heat losses through the side walls of the experimental tank. The striking differences introduced by a modest variation of diffusivity are exemplified by the flow of fluid into homogeneous surroundings of nearly the same density. Dyed salt solution flowing slowly into a solution of the same solute (say NaCl) forms a laminar plume which gently rises or falls without mixing. A plume of sugar solution of the same density, released at the same rate into homogeneous sodium chloride which has a molecular diffusivity about three times as great, results in a much more vigorous flow (see Fig. 2). Because of the different diffusion rates across the plume boundary, more salt is added to the plume than sugar is removed, so the plume fluid becomes heavier and its immediate surroundings lighter, generating vigorous turbulent convection which further enhances the rate of diffusion. The final result of this process is to produce a strong vertical density gradient in an originally homogeneous region.

Thus we have already identified what are perhaps the three most surprising and counter-intuitive features of double-diffusive convection experiments:

i) many phenomena associated with fluid motions are observed in systems which are hydrostatically stable;

ii) coupled *molecular* diffusion can produce large vertical velocities and fluxes;

iii) the potential energy always decreases and therefore the vertical density gradient increases as a result of double-diffusive convection .

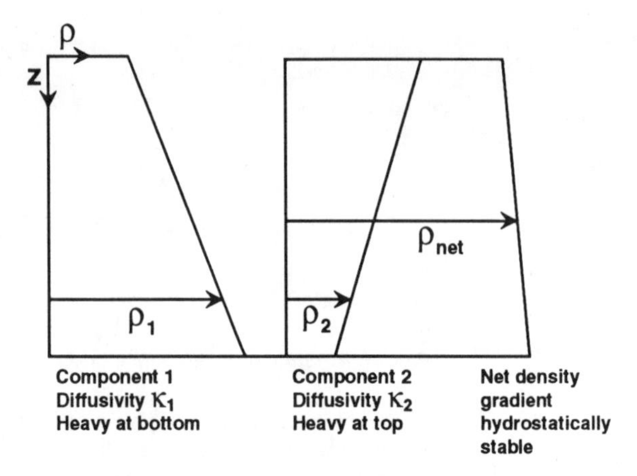

Fig. 1. The relation between the density gradients and the diffusivities of two solutes required for double-diffusive convection to occur.

ONE-DIMENSIONAL EXPERIMENTS

The form of the motion set up by differential diffusion of two properties depends on the relation between the two diffusivities and the density gradients i.e. on which component is 'heavy at the top' and thus provides the energy to drive convection.

The "Diffusive" Regime

A simple example of a system in which this energy comes from the component having the *larger* diffusivity is a stable salinity gradient, heated uniformly from below. The system does not immediately overturn from top to bottom. A thermal boundary layer becomes unstable through an oscillatory instability and breaks down to form a thin convecting layer, which grows by incorporating fluid from the gradient region above it. When the thermal boundary layer ahead of this convecting layer in turn becomes unstable, a second layer forms above the first, in which convection is sustained by the more rapid diffusion of heat relative to salt across this "diffusive" interface. *Turner* [1968] showed how the scale of the layers can be determined; it varies directly with the heating rate and inversely with the initial density gradient. An example of a series of convecting layers and heat-salt diffusive interfaces is shown in Fig. 3. *Huppert and Linden* [1979] followed up this study with a combined laboratory and numerical investigation which treated the sequential formation of new layers at the top of the region and the merging of the lowest

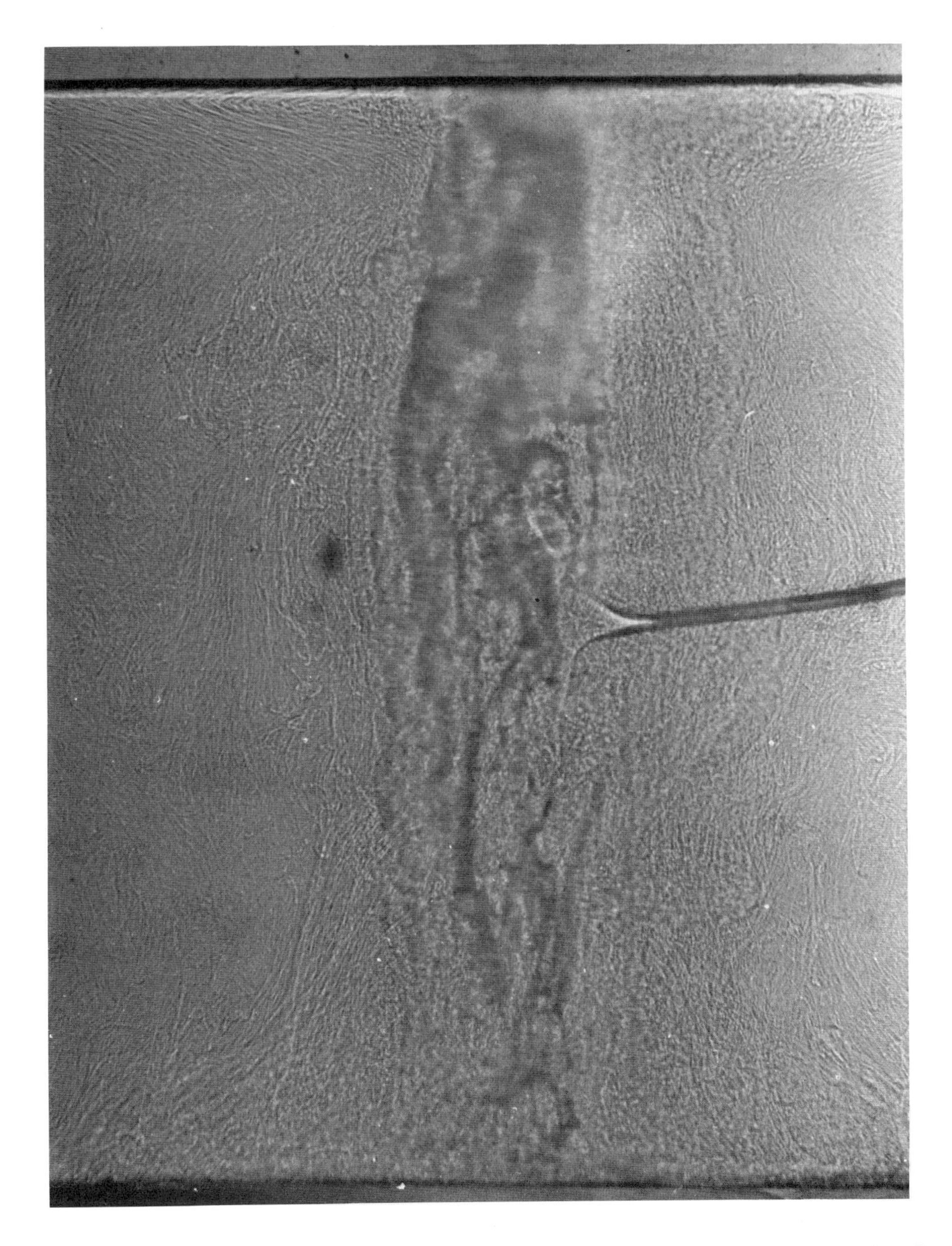

Fig. 2. A plume of sugar solution flowing slowly into homogeneous salt solution of the same density. Double diffusion causes separation and strong upward and downward convection.

layers in pairs. A different criterion for the formation of persistent layers has since been suggested by *Fernando* [1987]. Layers corresponding to this mechanism of formation have been observed in the ocean, and *Linden* [1976] showed that analogous layers and interfaces can also be produced when a destabilizing salt layer is placed above a sugar gradient.

The more direct method of studying the properties of diffusive interfaces is to set up two layers either by using a dam break technique or by carefully pouring a layer of salt solution on top of a denser layer of sugar solution. Both techniques were illustrated by time-lapse movies, as were the various methods of visualization which have been developed over the years: the use of dye markers,

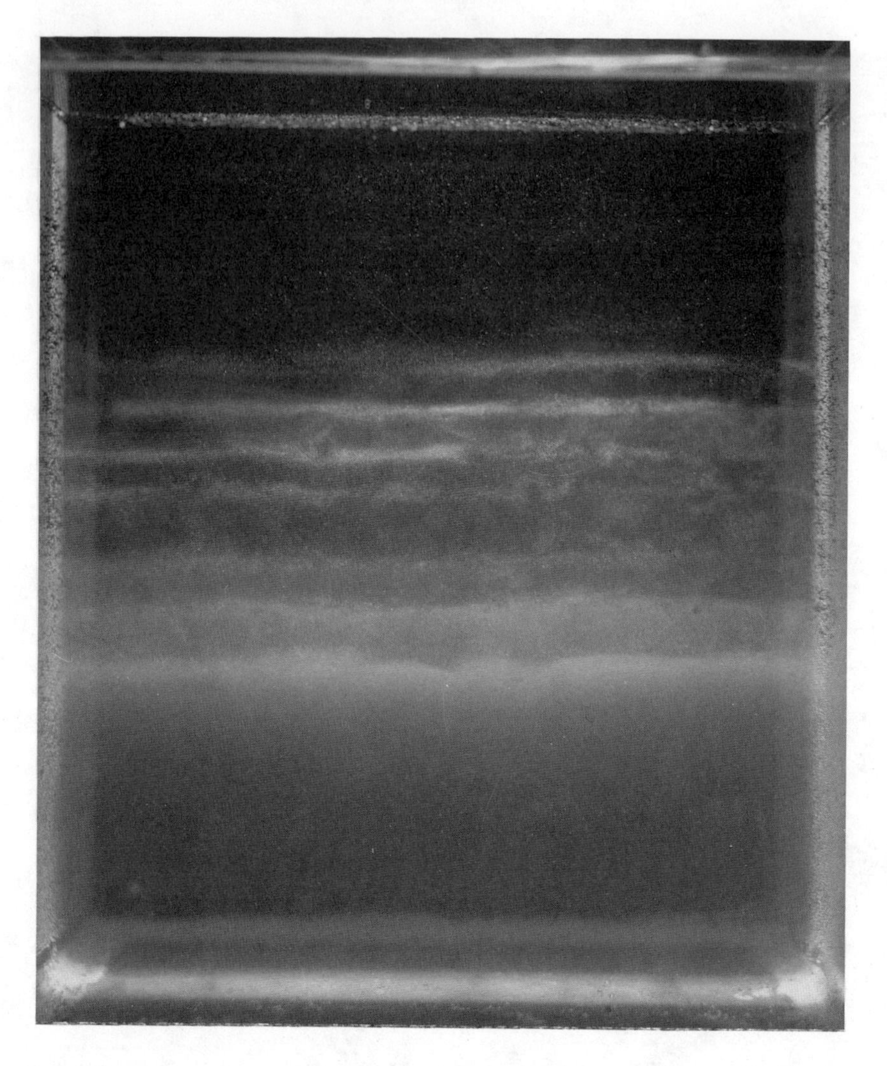

Fig. 3. A series of 'diffusive' interfaces set up by heating a stable salinity gradient from below. The layers are marked by fluorescent dye, and lit from the top.

shadowgraph, colour schlieren, and optical rotation (to monitor sugar concentration). A shadowgraph of a salt/sugar diffusive interface is shown in Fig. 4.

The "Finger" Regime

In the opposite situation, where the driving potential energy comes from the more slowly diffusing component (for example when a small amount of hot salty water is placed carefully on top of a stable temperature gradient), a very different phenomenon is observed. Long narrow convection cells or "salt fingers" are formed (Fig. 5) due to the more rapid sideways diffusion of heat relative to salt, as first predicted by *Stern* [1960]. Similar phenomena are again readily observed using two solutes; Fig. 6 shows an interface containing small-scale fingers between convecting layers of sugar above salt solution. When they are considered on the scale of the layers, there is in fact not as big a difference between the "diffusive" and "finger" cases as there appears to be at first sight. Convecting layers can also be formed from a gradient in the finger configuration, and layers corresponding to this distribution (hot salty on top of colder fresher water) have been observed in the ocean (for example, under the outflow of Mediterranean water into the Atlantic). Convection in the layers on each side of the interface are driven by the net unstable buoyancy flux in both cases, with only the mechanism of transport through the interface being different.

When a two-layer double-diffusive system is set up, and allowed to "run down" until the destabilizing component is uniformly distributed in the vertical, the density difference between the two layers is increased, as shown in Fig. 7. This is because, for energetic reasons, the flux of the

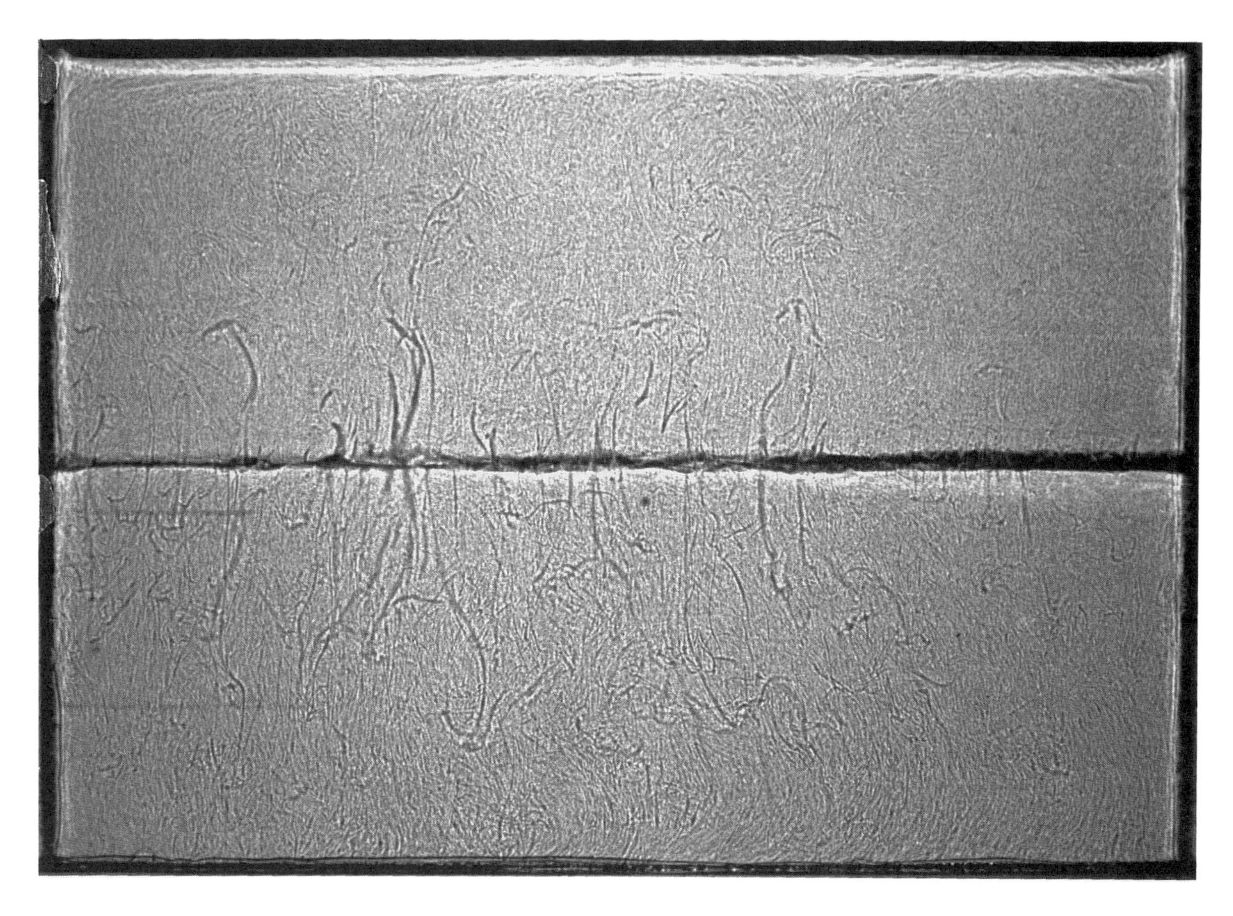

Fig. 4. Shadowgraph of a diffusive interface set up by pouring a layer of NaCl solution (the more rapidly diffusing solute) on top of a layer of denser sucrose solution.

driving component must always be greater (in density terms) than the flux of the lifted, stabilizing property, so that the potential energy decreases. This is quite different from the effect of a single diffusive or turbulent mixing process, which always leads to a decrease in the interfacial density difference.

Relation to Observations in the Ocean

Since this review is particularly concerned with the influence of double-diffusive laboratory experiments on other fields, it is relevant to note here that quantitative laboratory measurements of the fluxes across both types of interface were made before there were any field observations with which to compare them. *Turner* [1965] measured the relative rates of transport of heat and salt through a diffusive interface, and showed that both the heat flux and the ratio $\mathcal{R}$ of the salt flux to the heat flux depend systematically on the density ratio $R_\rho = \beta\Delta S/\alpha\Delta T$. A particularly striking result is shown in Fig. 8; for $R_\rho > 2$, $\mathcal{R}$ is nearly constant, with a value of 0.15. (This is approximately the square root of the molecular diffusivities, and is consistent with a mechanistic theoretical model of the interfacial transport mechanism which cannot be described in detail here.) Observations of interfaces which could be identified with this process began to be reported shortly afterwards, in Antarctic Lakes [*Hoare*, 1966], at the bottom of the Red Sea [*Munns et al.*, 1967] and underneath an ice island in the Arctic [*Neal et al.*, 1969].

In the case of a finger interface, *Turner* [1967] measured the relative fluxes of salt and heat in the laboratory, and that paper also referred to recent observations of steps underneath the Mediterranean outflow which were published shortly afterwards [*Tait and Howe*, 1968]. It was not until 1973 that salt fingers were directly detected in the ocean [*Williams*, 1975], using a sophisticated variant of the shadowgraph technique. Fingers were observed in regions of strong gradients of temperature and salinity (i.e. in the interfaces, as predicted by the laboratory work). For a discussion of the more recent developments in the observation of salt fingers, see the review referred to above

Fig. 5. Salt fingers formed by pouring hot, dilute salt solution, dyed with fluorescein, on top of a stable temperature gradient.

[*Schmitt*, 1994].

TWO-DIMENSIONAL EFFECTS

Conditions are especially favourable for double-diffusive convection to occur when a layer with compensating temperature and salinity differences intrudes at its own density level into an environment with different properties. This became clear first in the ocean, where the strongest layering was observed near intrusions and fronts. It took some time for the laboratory experimenters to break away from their preoccupation with one-dimensional processes and take horizontal non-uniformities properly into account.

Intrusions from Discrete Sources

The basic intrusion process with which other phenomena can be compared is the two-dimensional flow of a uniform fluid at its own density level into a linear density gradient set up using the same property. Fig. 9 shows the behaviour of a dyed source of salt solution released into a salinity gradient. The intruding fluid, as we might intuitively expect, just displaces its surroundings upwards and downwards but remains confined to a horizontal layer by the density gradient. The "upstream wake" effect, revealed by the displacement of dye streaks ahead of the intruding fluid, is an interesting phenomenon, which has been studied for example by *Manins* [1976], who also gives references to previous related work.

Turner [1978] has compared this behaviour with that of a source of sugar solution, having exactly the same density and flowrate as that of the salt source in Fig. 9, flowing into the same salinity gradient at its neutral buoyancy level. As shown in Fig. 10, there is now strong vertical double-diffusive convection near the source, produced by the same mechanism which led to the effects shown in Fig. 2 in homogeneous ambient fluid. The vertical spread is now limited by the stratification, and intruding "noses" begin to spread out at levels above and below the source. The total volume of fluid affected by mixing is many times that of the input. Each individual nose as it spreads contains an excess of sugar relative to its environment, so that conditions are favourable for the formation of a diffusive interface above and fingers below, as can be seen in Fig.

Fig. 6. Shadowgraph of a thickened sugar/salt finger interface formed by placing sucrose solution on top of denser NaCl solution, and leaving for three hours.

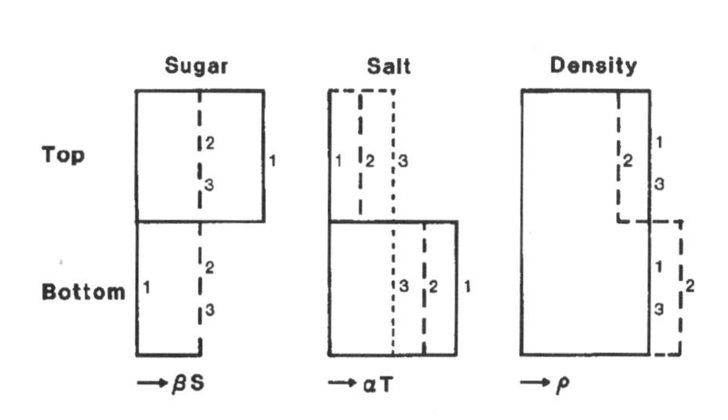

Fig. 7. The distribution of sugar, salt and density at various stages of a run-down finger experiment, with the flux ratio assumed to be 1/2 for the purposes of illustration. 1 is the initial state, 2 the distribution after the double-diffusive rundown and 3 the final state achieved by molecular diffusion after a much longer time.

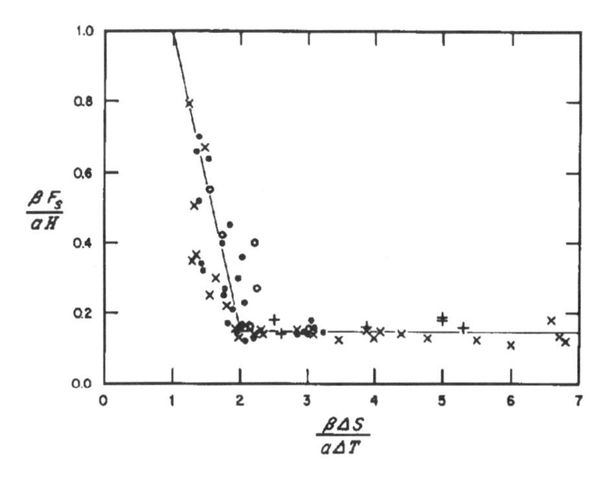

Fig. 8. The measured ratio of salt to heat fluxes (in density units), across an interface between a hot salty layer below a cooler fresher, less dense layer, plotted as a function of the density ratio. (From Turner (1965).)

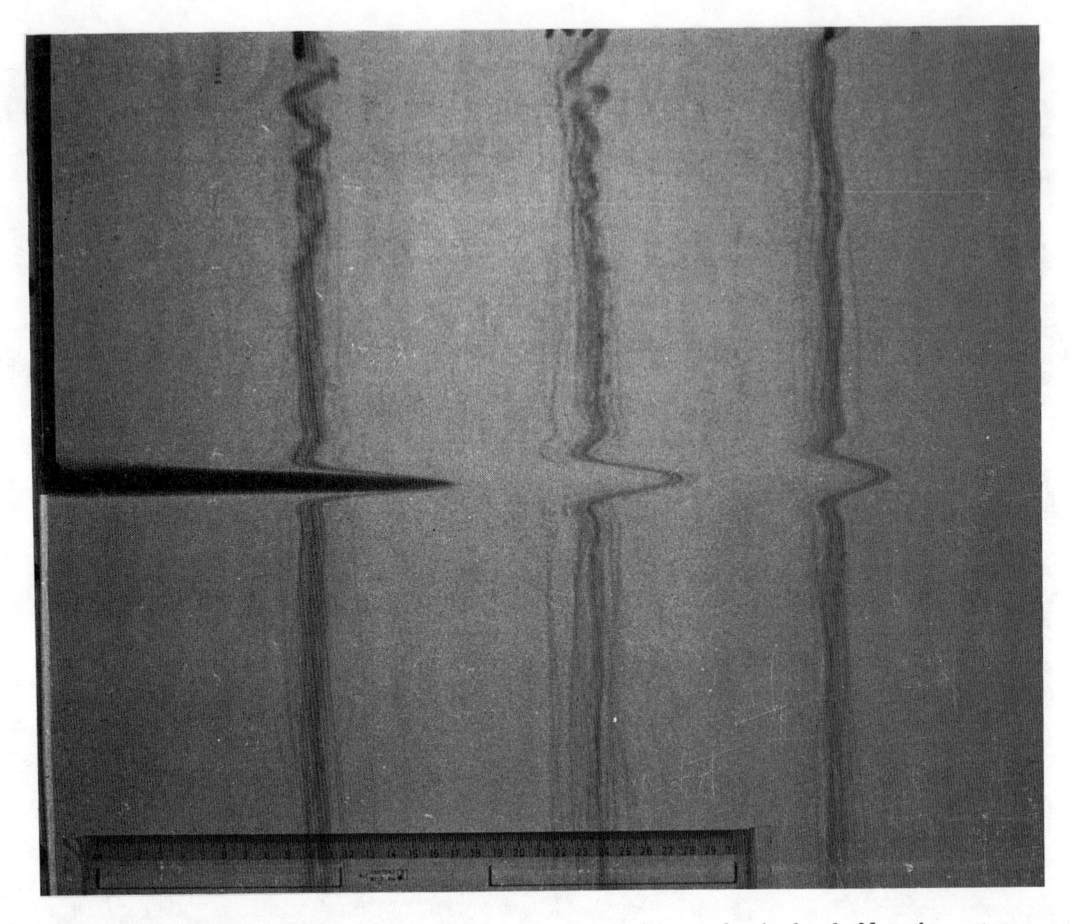

Fig. 9. The intrusion of dyed salt solution into a salinity gradient at its own density level. Note the upstream wake, shown by the distortion of initially vertical dye streaks.

10. Note too the slight upward tilt of each layer as it extends, which implies that the net density flux due to fingers is larger than that across the diffusive interfaces, so a layer becomes lighter and rises across isopycnals. This interpretation is supported by the corresponding experiments with a source of salt released into a sugar gradient, where there are diffusive interfaces below and fingers above each intrusion and a systematic downward tilt.

The examples given above have all used the convenient salt/sugar analogue system, and intrusions driven by temperature and salinity differences have been observed in both the laboratory and the ocean. A particularly well-documented oceanic example has been given by *Gregg* [1980], who described an intrusion with a systematic tilt in the sense which supports salt-fingering as the most probable driving mechanism. A remarkable general implication of these and other oceanic measurements is that not only can molecular diffusion affect the convective motions on the layer scale i.e. tens of metres in the vertical, but it can also, by driving intrusions across fronts, influence the large-scale mixing between water masses.

Horizontal Property Gradients and Fronts

When the side-wall boundary conditions do not match the conditions in the interior of a fluid containing several diffusing properties, *Turner and Chen* [1974] showed that instabilities can develop rapidly, and lead to layers propagating away from the slope. Fig. 11 is a photograph of an experiment in which smooth opposing gradients of salt and sugar solutions were set up, with a maximum salt concentration at the top and a maximum of sugar at the bottom. With vertical side walls, the surfaces of constant concentration were normal to the boundaries, and the no-flux condition for the two properties was automatically satisfied. When the sloping boundary was inserted, as shown, the concentration and density surfaces were distorted so that they could remain normal to the slope. This upset the hydrostatic equilibrium, and flows parallel to the slope developed, with a thin downflow just above the boundary and above that a thicker upslope counterflow. These flows did not continue indefinitely, but they rose until they reached the level where they were neutrally buoyant and then turned outward to form a series of layers

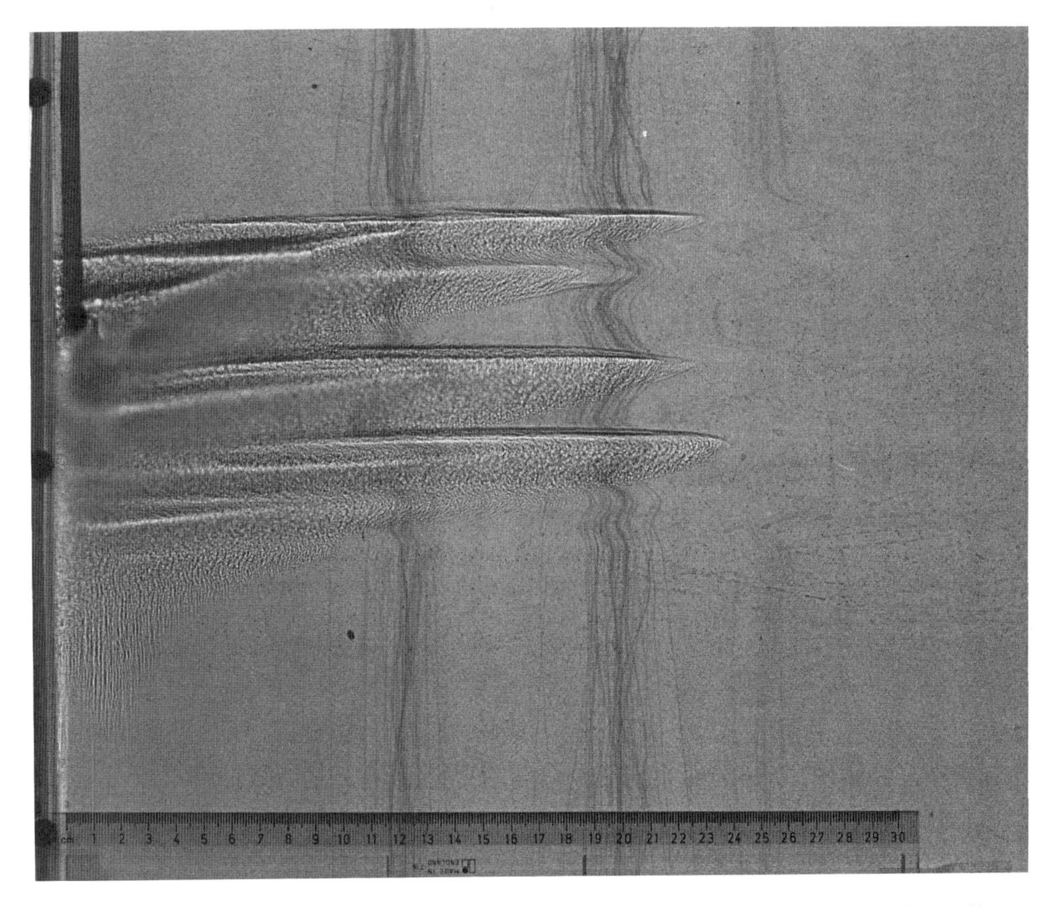

Fig. 10. The flow produced by releasing sugar solution at its own density level into the same density gradient as used for the experiment shown in Fig. 9. There is strong vertical double-diffusive convection near the source, followed by the intrusion of several layers having a slight upward tilt.

propagating into the interior of the fluid.

The equivalent of the above in the heat-salt system is achieved by heating a salinity gradient through a side wall; this produces a series of equally spaced, extending layers. Another example in which both the temperature and salinity are unmatched at the side wall is a block of ice melting into a salinity gradient. Motivated by the problem of "harvesting" fresh water from icebergs, *Huppert and Turner* [1978, 1980] carried out an extensive series of experiments to determine the distribution of the fresh water. They showed that there is a thin boundary layer of fresh meltwater which flows upwards against the block, and a thicker cooled layer outside this which flows downwards. These flows do not continue indefinitely, but as illustrated in Fig. 12, the melt water spreads out into a series of slightly inclined convecting layers along the whole depth of a vertical wall of melting ice, rather than convecting right to the surface and collecting in a pool as it would if the ambient fluid were unstratified. In the latter case too there would be a great deal of mixing with the ambient salt water, so the water reaching the surface would not be fresh. Thus whatever the stratification of the surroundings the collection of fresh melt water from icebergs is likely to be much more difficult than was implied by those who suggested that it might be technically feasible to tow icebergs from the Antarctic to coastal cities in arid regions.

Experiments using discrete sources or solid boundaries do not properly model the conditions at an oceanic front, where there are horizontal property gradients but no physical barriers. *Ruddick and Turner* [1979] investigated the motions across a front formed by withdrawing a dam separating regions having the same vertical density gradient but produced using different solutes, i.e. on one side there is salt solution and on the other sugar solution. When the dam is removed there are small wavelike adjustments of the residual density anomalies, and then a series of regular interleaving layers develops more slowly, as shown in Fig. 13. The vertical spacing and speed of advance of the layers are proportional to the horizontal property differences, consistent with driving by the local density anomalies produced by double-diffusive

Fig. 11. The growth of layers at a sloping boundary in a double-diffusive sytem with opposing linear gradients (sugar stabilizing, NaCl destabilizing).

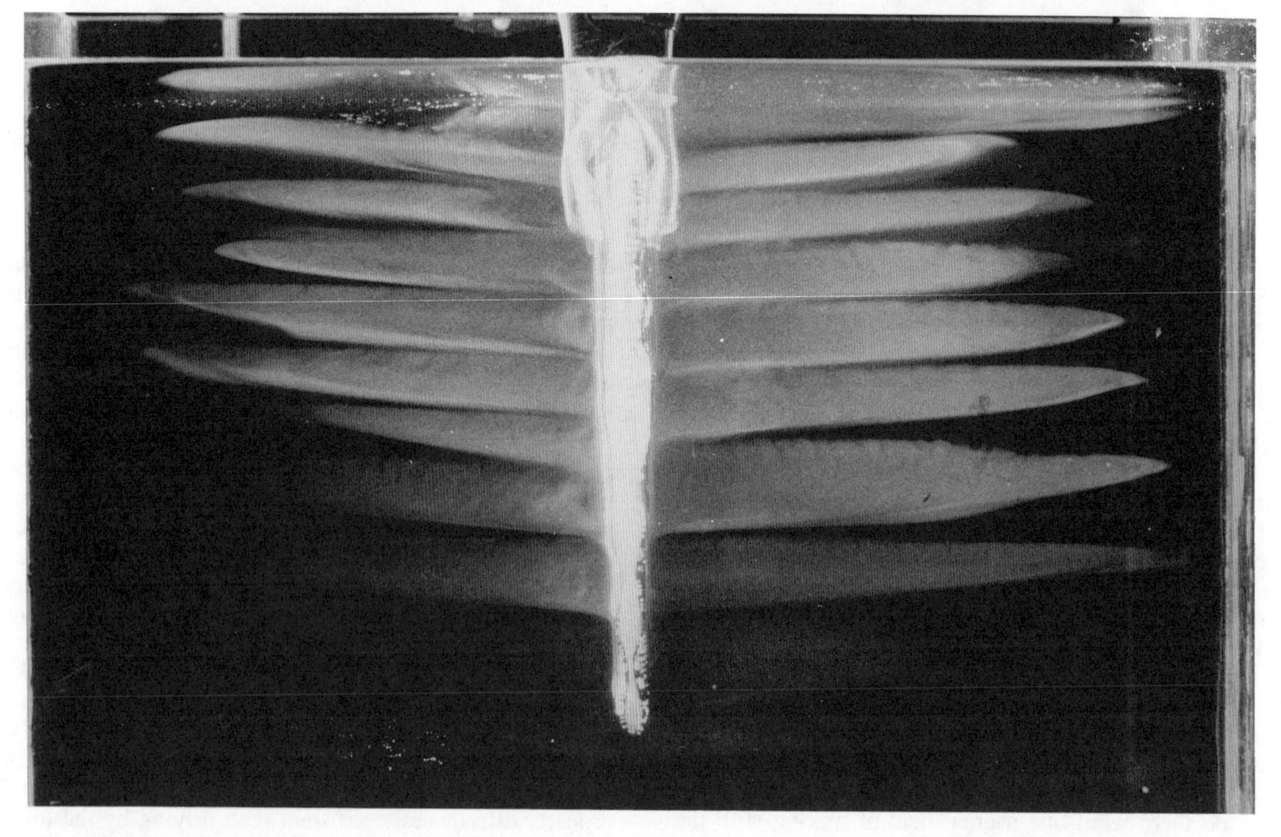

Fig. 12. A series of layers formed by the melting of an iceblock, containing fluorescein dye, into a salinity gradient.

Fig. 13. A system of interleaving layers produced by removing a barrier separating sugar solution (left) and salt solution (right) which have identical linear vertical density distributions.

convection.

APPLICATIONS TO GEOLOGY

All the experiments described so far were motivated by oceanographic applications, and we now consider the way the ideas have been applied in other contexts, particularly geology. The review by *Huppert* [1986] treats the geological developments in more detail than there is space for here.

Turner and Gustafson [1978] reviewed the various phenomena that can result from the efflux of hot, salty water from vents in the sea floor, spurred by suggestions that this could be the mechanism of formation of ancient massive sulphide ore deposits. This extension from the previous oceanographic studies of coupled salinity and temperature anomalies was straightforward, but it is worth noting that it was made before "black smokers" had been observed on the sea floor. With that discovery, there has been an acceleration of interest in these processes as active examples of contemporary ore deposition, including further laboratory experiments aimed at understanding the deposition mechanism [*McDougall*, 1984; *Campbell et al.*, 1984].

The paper by *Turner and Gustafson* [1978] also pointed out, however, that double-diffusive effects can be important not only in aqueous solutions, but also in liquid rocks. Magmas and lavas have a wide range of chemical constituents with different molecular diffusivities, and there are also large temperature variations. Many igneous rocks have strong layering on various scales, as seen for example in Fig. 14. This raised the question: could such layers have formed first in the liquid state, by a mechanism related to double-diffusive layering in the ocean, and then have been preserved during cooling and solidification? The process is very different from the long-held idea that crystal settling dominates layer formation in igneous rocks.

This novel approach to the problem has led to a whole new range of experiments, as described below and in the reviews by *Huppert and Turner* [1981] and *Huppert and Sparks* [1984]. It can now be confidentially asserted that double-diffusive convection probably plays a role in all major igneous processes. Clearly crystallization needs to be added to the double-diffusive processes previously investigated, but it has been found that aqueous solutions are again convenient and appropriate analogues to model dynamical phenomena in solidifying magmas.

The Effects of Crystallization in Various Geometries

Most of the convective phenomena studied, including those responsible for setting up compositional gradients from which layering can develop, depend on density changes produced by crystallization. In magmas (and many common solutions which can be used as laboratory analogues) crystallization causes much larger changes in melt (or solution) density than the associated temperature changes. When a dense mineral crystallizes out, the residual fluid next to the boundary can be lighter than the bulk magma even though it is colder. Similarly, for certain aqueous solutions, for example sodium carbonate solution more concentrated than the eutectic composition, the residual fluid is cold but less dense, and upward "compositional convection" can occur. The effects of this have now been studied in many different geometries.

Top cooling. Chen and Turner [1980] used Na_2CO_3 solution, chosen because of the strong dependence of saturation concentration on temperature, to examine the effect of cooling and crystallizing a density gradient from the top. A series of convecting fluid layers was first produced, separated by sharp interfaces - the mechanism of formation is physically equivalent to the formation of layers by heating a salinity gradient from below, shown in Fig. 3. With the top boundary held below the eutectic

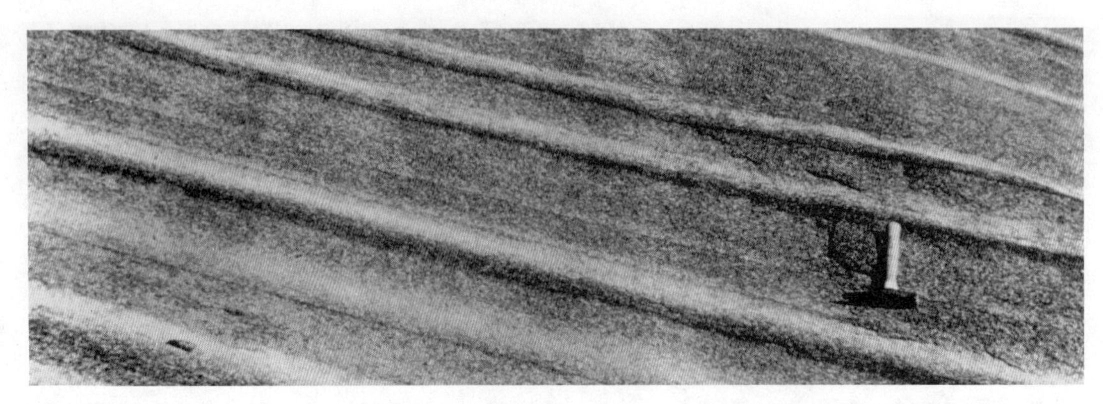

Fig. 14. Layers in the Upper Zone of the Skaergaard igneous intrusion, made visible by contrasting light and dark colours of different minerals (from McBirney and Noyes 1979).

temperature, dendritic crystals grew down nearly vertically in a static fluid layer generated by lighter residual fluid, while the eutectic layer advanced more slowly and filled the spaces between the dendrites (Fig. 15). The growth of both crystals and eutectic was clearly influenced by the position of the interfaces in the fluid, and the advancing front of the crystals remained nearly horizontal. These experiments have been followed up by *Turner et al*. [1986] with an application to komatiite (very high temperature) lavas in mind.

Side cooling. When crystallization occurs at a cooled vertical boundary in a tank of finite size containing homogeneous Na_2CO_3 the upflowing residual fluid produces stratification in the interior. The temperature and compositional variations are compensating and in the "diffusive" sense, and this produces layering in the fluid (as shown in Fig. 16, where the cooling is at a central pipe with coolant circulating through it). When there is a stable gradient in the tank initially, a series of convecting layers first forms, extending away from the boundary, and the growth of crystals on the wall is affected by the position of these layers. The dynamical processes of stratification and layer formation are exactly the same as those studied by *Huppert and Turner* [1978] with a melting iceberg in mind. Cold, light residual fluid is produced in each case, the only difference being that the boundary is advancing slowly in one case and receding in the other. This analogy shows the value of cross-fertilization, but it also points to a problem in transferring ideas from one field to another: what geologist, without personal guidance, would be motivated to read a paper about melting icebergs?

If a second solute is added to the initial fluid being cooled at a vertical boundary, the process of "differentiation" in a magma chamber, i.e. the production of vertical gradients of mineral composition, can be modelled explicitly. *Turner* [1980] used a homogeneous mixed solution of Na_2CO_3 and K_2CO_3 with cooling at a vertical metal pipe. Again pure hydrated sodium carbonate crystallized out over the range of temperatures and concentrations used. Thus, as shown in Fig. 17, the upward flowing boundary layer was relatively enriched in potassium as well as being depleted in sodium, and a corresponding density stratification was also produced. *Turner and Gustafson* [1981] extended these experiments to explore cases where more than one component crystallizes out on the boundary, leading to vertical zonation in the solid as well as in the fluid remaining in the chamber.

Cooling at inclined roof. Related experiments have also been carried out in a geometry which more closely models a magma chamber. Fig. 18 shows a tank with an inclined roof which is cooled below the eutectic temperature, all other boundaries being insulated. Compositional convection produced by the crystallization leads to an upflow along the inclined roof which ponds under the high point of the roof. With further crystallization a stable "differentiated" region is built up, stratified in temperature and composition. Some crystals fall to the floor, and as they grow they cause strong compositional convection and turbulent mixing in the lower part of the chamber, but this region remains quite distinct from the stratified region above, with no penetration through the sharp double-diffusive interface separating them.

The implication for magma chambers is clear: compositional convection is a powerful mechanism giving rise to zonation in an initially well-mixed magma chamber. Eruptions from vents tapping the upper and lower layers could produce silicic and basaltic lavas simultaneously from the same chamber. The systematic variation of mineral composition in successive ash-flow deposits, such as the Bishop Tuff discussed by *Hildreth* [1981], can also be readily explained by the zoning at the top of the chamber produced by this convective mechanism.

Bottom cooling. Cooling of compositionally mixed or layered solutions through a horizontal bottom boundary can produce compact layers of dense crystals, the composition of which changes abruptly in response to the evolution of the fluid in contact with them. Fig. 19, reproduced from

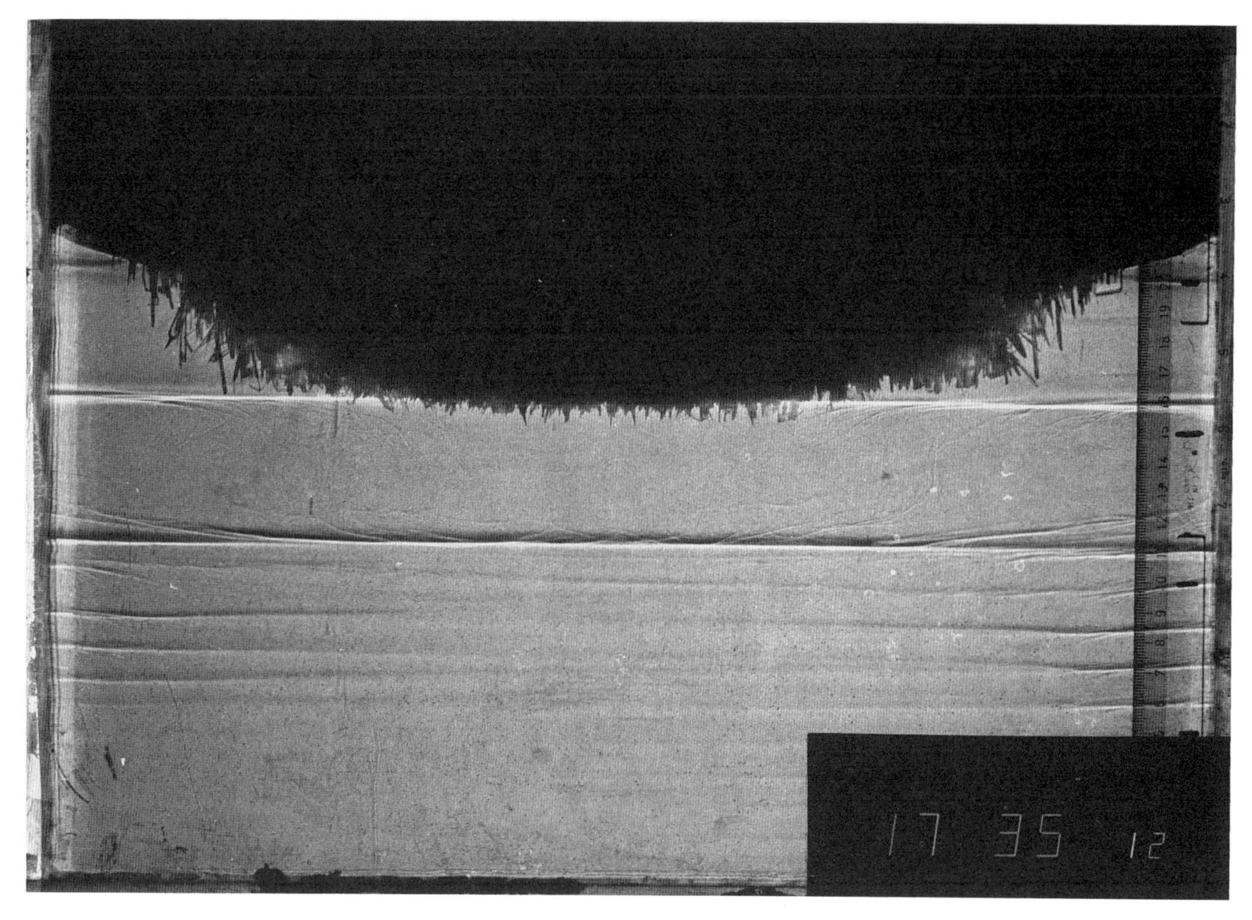

Fig. 15. Shadowgraph picture showing the effect of strong cooling at the top of a tank containing a stable gradient of Na_2CO_3 solution. Convecting double-diffusive layers formed first, and these influenced the subsequent growth of crystals.

Kerr and Turner [1982], shows three crystal layers formed by cooling the bottom of a tank filled initially with two layers containing different mixtures of sodium and copper sulphates. First the copper salt crystallizes, and the composition of the lower fluid layer evolves until sodium sulphate becomes the preferred crystallizing phase. At the same time, the density of the lower layer is decreasing, until eventually the diffusive interface above it breaks down, producing mixing with the upper layer. This suddenly brings the top of the crystals into contact with a copper-rich solution, and another layer of copper sulphate crystals builds up on top of the previous layers. Note that though the fluid layers undoubtedly controlled the growth and sequence of the crystal layers in this experiment, the thicknesses in the two states were not identical, and only indirectly related.

Replenished Magma Chambers

When a magma chamber is replenished from below with magma which is compositionally denser (and usually hotter) than the resident magma, a wide range of dynamical and double-diffusive phenomena can be identified. The behaviour depends on the rate of inflow and the relation between the viscosities of the two fluids. The initial mixing processes and subsequent evolution can again be simply modelled in the laboratory using aqueous solutions.

*Turbulent inflow.*For a rapid turbulent inflow of denser fluid, directed upwards, and comparable viscosities of the inflowing and resident fluids, the behaviour is shown in Fig. 20. When the input Froude number is large (or the upward momentum is large in relation to the density difference), this flow forms a fountain which rises and mixes vigorously with a much larger volume of the surrounding fluid before falling back. A hybrid layer of this mixture builds up at the bottom, producing a stratified region even in a tank or magma chamber which was originally homogeneous. If the input fluid is hot as well as compositionally dense, a series of double-diffusive layers can form, each of which is convecting and well-mixed. All of these features have analogues in the prototype magma chambers, and the experiments have been applied by

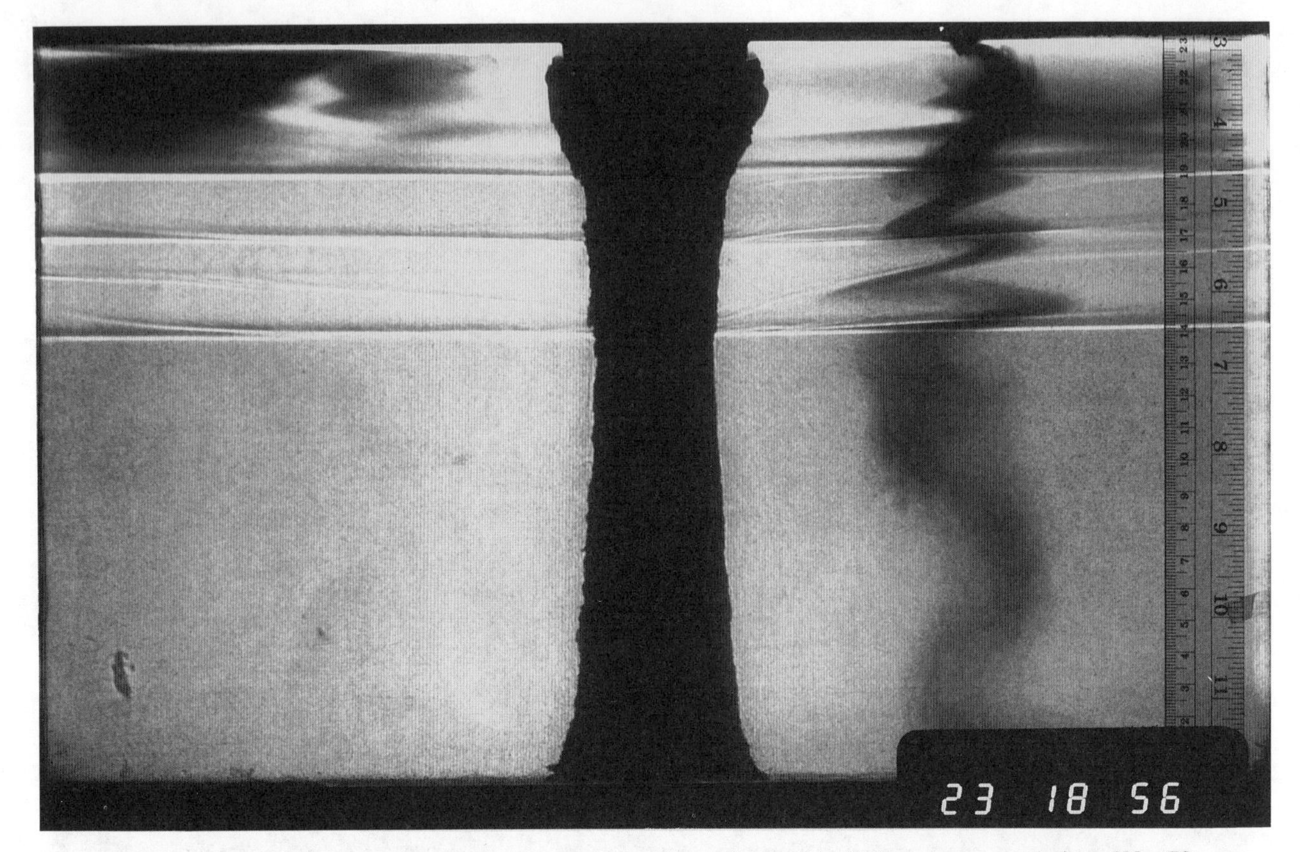

Fig. 16. Shadowgraph picture of a crystal column produced by cooling an initially homogeneous solution of Na_2CO_3 at a central pipe. Upflow in the boundary layer due to crystallization has produced stratification and layering.

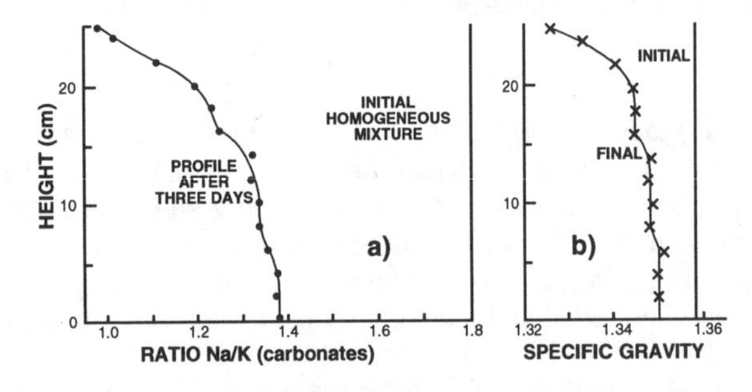

Fig. 17. 'Differentiation' or the compositional gradients produced by cooling and crystallizing at a vertical rod. Sodium carbonate alone crystallized from an originally homogeneous mixture of Na and K carbonates, leading to upflow and stratification. a) shows the ratio of the compositions and b) the corresponding density distribution as functions of height.

Turner and Campbell [1986] to understand the formation of layers of chromium and platinum ores in large igneous intrusions. Incidentally, the more detailed study of fountains by *Baines et al.* [1990] is an example of the further spread of ideas across very different fields. Professor Baines' interest was in the heating of large buildings by blowing hot air in through ducts located near the roof, which is a dynamically identical, but inverted, process.

It is worth noting in passing that, when the kinematic viscosity of the fluid in a tank (or magma chamber) is much greater than that of the input fluid, mixing between an inflowing "fountain" and its surroundings can be inhibited, even though the inflow remains turbulent. *Campbell and Turner* [1985] have shown theoretically and experimentally that mixing is entirely suppressed when a Reynolds number formed from the velocity and diameter of the inflow and the viscosity of the *surrounding* fluid falls below a critical value. Physically this implies that inertial forces in the turbulent flow must be able to generate large enough normal stresses in the surroundings to overcome the viscous stresses, and distort the boundary in the manner necessary for mixing to be initiated.

Slow inflow, with crystallization. We now turn to the opposite extreme of a slow inflow rate, following *Huppert et al.* [1982]. When a hot, nearly saturated solution (of KNO_3 for example) is emplaced very slowly below a cold less dense solution of Na NO_3, much of the input fluid is immediately "quenched" and crystallizes during

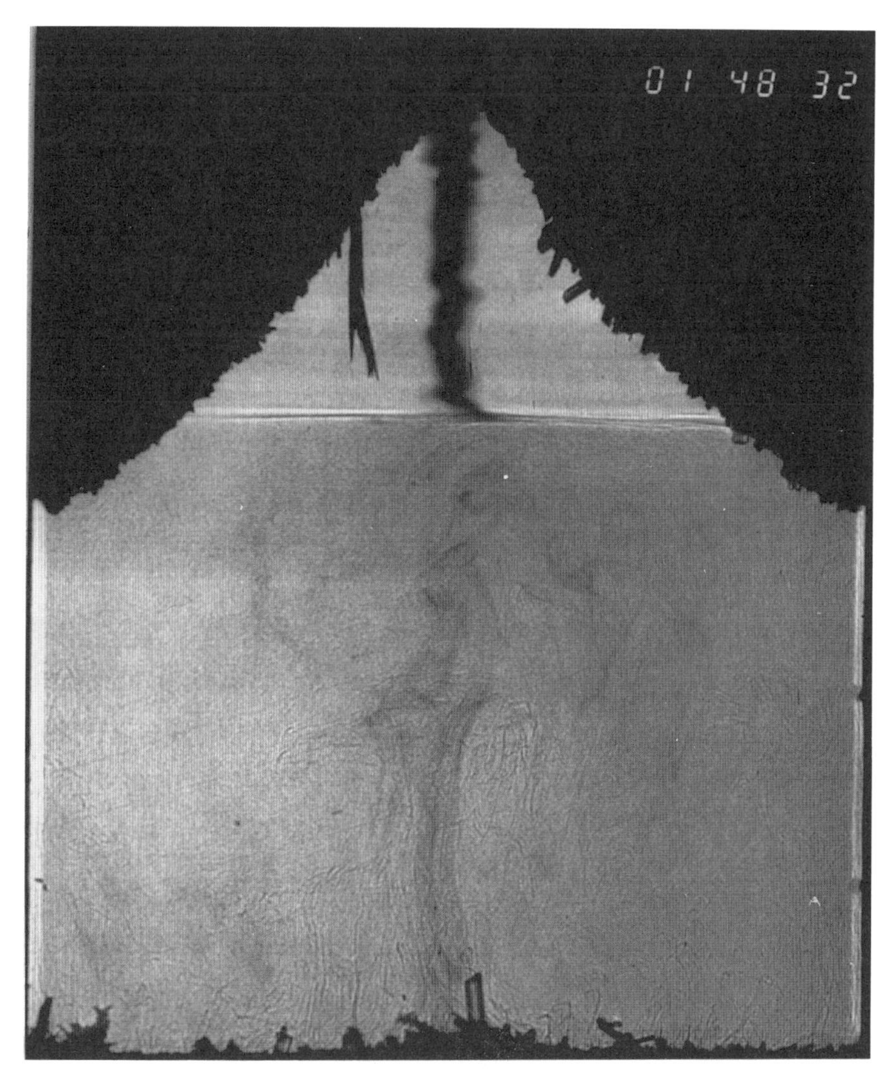

Fig. 18. A laboratory model of a differentiated magma chamber, showing a nearly static, stratified region under a cooled, sloping roof, on which a layer of crystals is growing. There is a strongly stirred lower layer in which crystallization on the floor is driving compositional convection.

replenishment. Not all the input fluid crystallizes, however, and some lighter residual fluid is released to mix convectively with the homogeneous layer above. In time, some hot dense input fluid spreads along the bottom to form a layer as before. When this crystallizes further, some of the input channels between the crystals become blocked, and columns of crystals or "chimneys" build up above the input vents. Hot dense fluid is forced upwards through a cluster of thin-walled conduits, and it ponds under gravity to form a nearly horizontal top to the column, with further crystal growth occurring as saturated fluid flows over the edges and down the sides.

Hot saturated lower layer. Intermediate rates of input lead to perhaps the most interesting and practically important phenomena. Following *Huppert and Turner* [1981b], consider what happens when a hot saturated layer of KNO_3 is inserted under cold, less dense $NaNO_3$ solution, slowly enough for there to be little mixing, but fast enough for negligible crystallization to occur during replenishment. A double-diffusive interface forms, through which heat is transferred much faster than the solutes. The lower layer cools (much more rapidly than it could just by conduction through the side walls) and crystallizes as it does so, and the density of the residual fluid in contact with the crystals decreases because of the strong dependence of the density of the saturated solution on temperature. The lower layer evolves towards a state where its density approaches that of the upper layer, but the interface remains sharp during this time, and there is little transfer of composition between the layers. As the densities become equal, the interface

Fig. 19. Successive layers of Cu and Na sulphates formed by cooling a two-layer system from below, as it evolved in composition and eventually overturned. The scale bar is 5 cm long.

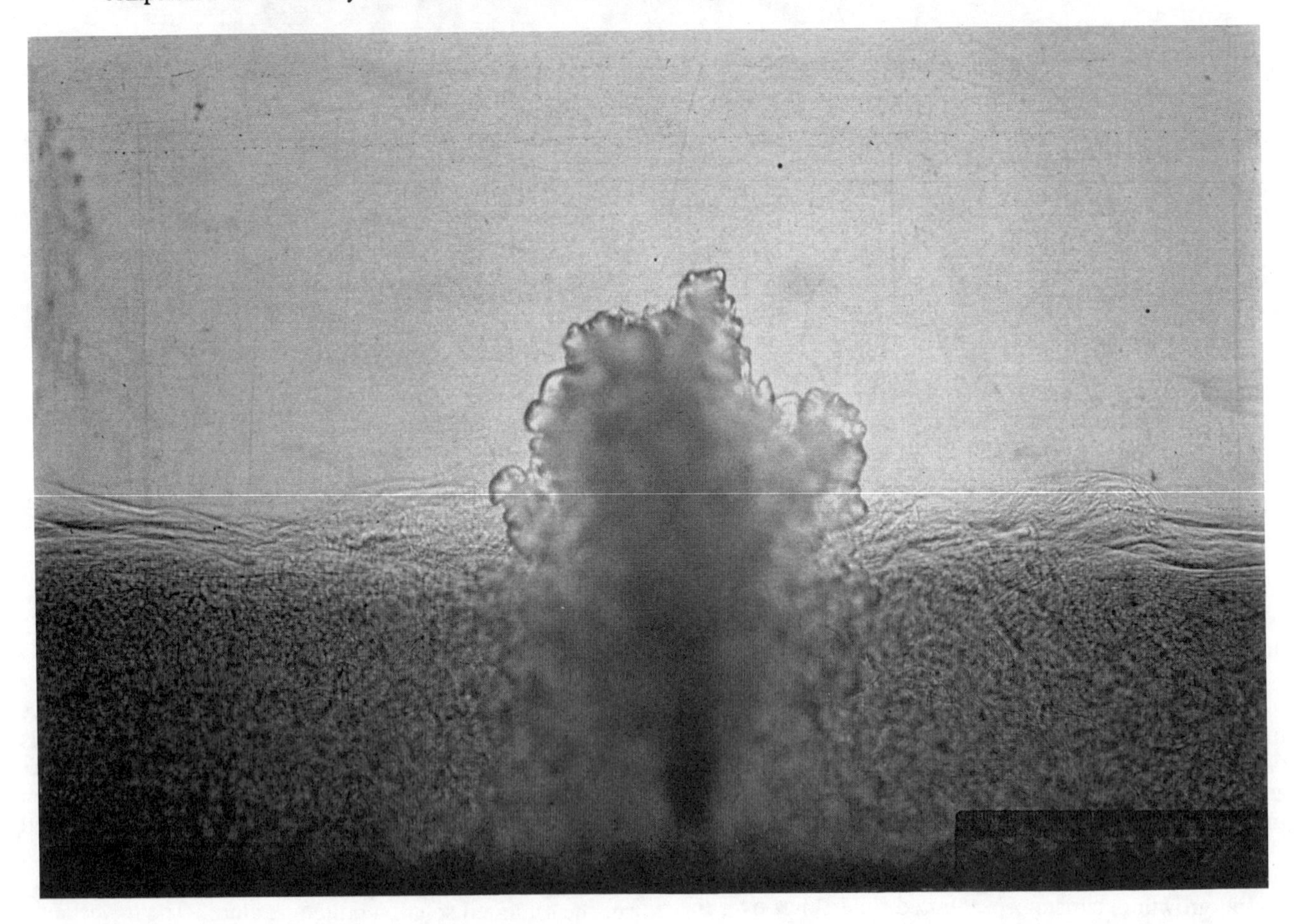

Fig. 20. A turbulent 'fountain' formed by the input of dense salt solution at the bottom of a tank of fresh water, showing the development of a stratified layer .

Fig. 21. A hot dyed layer of KNO_3 solution has been inserted below a stable gradient of K_2CO_3 solution, to simulate the influx of new magma into the base of a magma chamber. Heating from below produces a series of convecting layers and diffusive interfaces in the gradient region, while cystallization reduces the density of the input layer. Overturning is confined to the lowest layers.

suddenly breaks down and the two layers mix intimately together, leaving a layer of KNO_3 crystals at the base. Note that there would have been no crystallization at all if the two layers had been immediately mixed together, rather than being separated by the double-diffusive interface as they evolved.

The above model explains how a magma chamber can act as a buffer between new input of basaltic magma, containing, say, 18% MgO, and the magma (of comparable viscosity) erupted from a chamber below a mid-ocean ridge, which is observed to contain 10% MgO. The behaviour of a layer of dense, less viscous basaltic magma at the bottom of a chamber containing very viscous rhyolitic magma has also been modelled by *Huppert et al.* [1983] using hot KNO_3 solution emplaced below cold glycerine. In this case, rather than suddenly overturning, the lower layer releases less dense fluid continuously into the upper layer as crystals are deposited against the cold interface.

Overturning into a density gradient. Several variations on these experiments have also been carried out with magma mixing in mind, and two of these were illustrated using time-lapse movies. When a layer of hot, nearly saturated KNO_3 is put in below a fluid having a vertical compositional (and hence density) gradient, for example a gradient of K_2CO_3, the lower layer at first cools and evolves as before. The upward heat flux also produces a series of convecting layers and diffusive interfaces (c.f. Fig. 3) in the initially smooth gradient region. Crystals grow on the bottom, and the density of the input layer decreases until it reaches the density of the next layer above it. When overturning occurs, the rise of the lower-layer fluid is constrained by the gradient to the lower part of the tank, as shown in Fig. 21.

Overturning with gas release. The final example is a laboratory model of the exsolution of gas following the quenching of water-rich magma in the lower layer, as it overturns and mixes with colder magma in the chamber above. The physical process used by *Turner et al.* [1983] to generate the gas release is quite different, but the dynamical effects are closely analogous. The lower layer again consists of hot saturated KNO_3 solution, but with an added component such as HNO_3, which can react with a dilute extra solute in the upper layer e.g. a small amount of K_2CO_3 added to the cold $NaNO_3$. As the system evolves because of the heat transfer through the interface and the associated crystallization at the bottom there is little transfer of solutes across the multiply-diffusive interface, and no chemical reaction. When the densities of the layers become equal and rapid overturning and mixing occurs, the reaction between the acid and carbonate leads to the copious release of CO_2 which increases the vigour of the

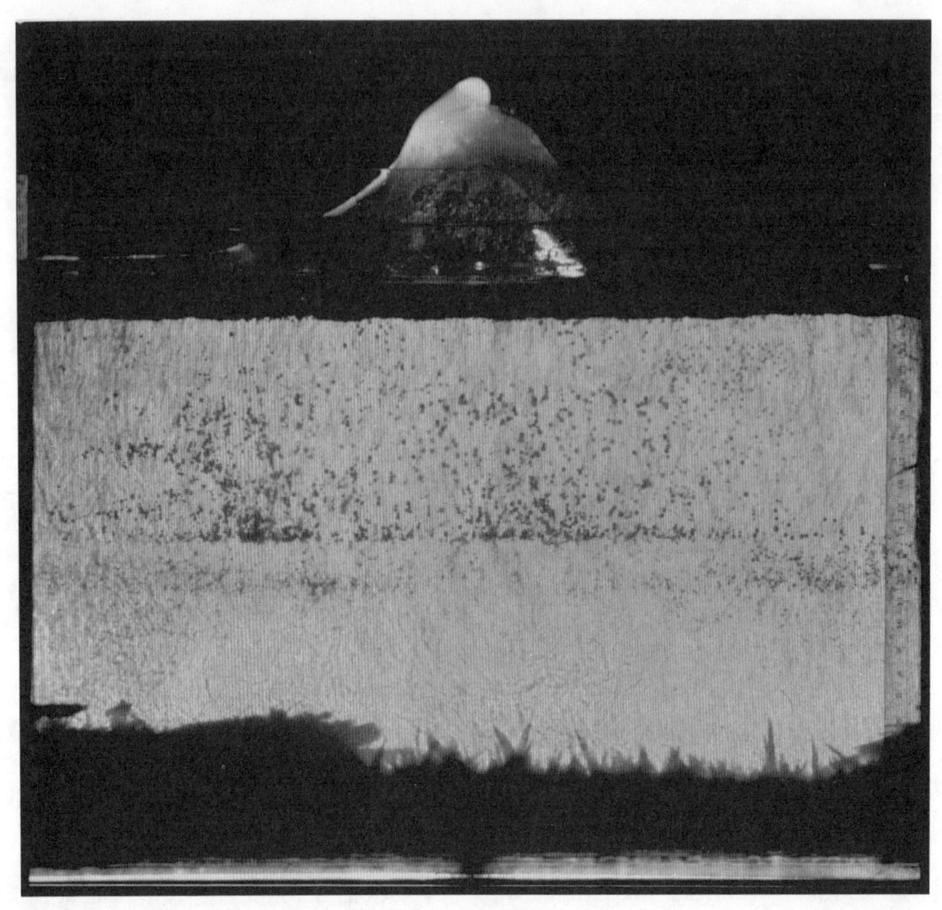

Fig. 22. Model of an overturning magma chamber, in which 'quenching' of gas-rich magma in the lower layer, by mixing with colder magma in the upper layer, leads to the release of gas. The analogue system uses the reaction between acid in the lower layer and a carbonate in the upper layer to release CO_2, following the sudden overturn and mixing. This increases the pressure and drives the fluid-gas mixture out through a vent in the top.

convection. When the top of the chamber is closed except for a small vent, the increased pressure generated by the gas release forces the fluid-gas mixture out the top, as shown in Fig. 22. This looks like an erupting volcano, but the superficial resemblance is less important than the fact that the dynamical processes have been appropriately modelled.

SUMMARY AND CONCLUSIONS

This paper has traced the evolution of ideas about double-diffusive convection, from the earliest thoughts about the possible oceanographic consequences to some of the more recent extensions to other fields. Double diffusion is a fundamental physical process, and once the principles are thoroughly understood, they can readily be transferred to many different contexts. The present review has shown how laboratory experimenters have played a significant role in the development of the subject, by identifying new phenomena and illuminating the underlying physics. Simple experiments are particularly effective when they are studied in close contact with both observers of the natural phenomena, and theoreticians who can develop more detailed interpretations of the relationship between the laboratory and prototype processes.

Double-diffusive, or more generally, multicomponent convection undoubtedly is of importance in many different fields, some of which are covered by other articles in this volume, but further potential applications have probably not yet been identified. Many of the phenomena observed can properly be described as being 'solutions in search of a specific problem'. In spite of the spur to interaction given by this conference and its predecessors, the transfer of information between very different research areas will always be difficult and to some extent a matter of chance. I do not believe it can be programmed from the top down. It must grow from the bottom up, through direct contacts between individuals who recognise their mutual interests and the need for complementary knowledge. Only when it is clear that 'there is something in it for everyone' can specialists be persuaded to step across the boundaries of their own well-trodden field to collaborate with others having apparently unrelated interests and skills.

Acknowledgments. I thank Karen Buckley for her expert production of the camera-ready copy, and Ross Wylde-Browne for his assistance both with the original photography and the preparation of the illustrations for publication.

REFERENCES

Baines, W.D., J.S. Turner, and I.H. Campbell, Turbulent fountains in an open chamber. *J. Fluid Mech. 212*, 557-592, 1990.

Campbell, I.H., and J.S. Turner, Turbulent mixing between fluids with different viscosities. *Nature*, *313*, 39-42, 1985.

Campbell, I.H., T.J. McDougall, and J.S. Turner, A note on fluid dynamic processes which can influence the deposition of massive sulphides. *Econ. Geol.*, *79*, 1905-1913, 1984.

Chen, C.F., and J.S. Turner, Crystallization in a double-diffusive system. *J. Geophys. Res.*, *85*, 2573-2593, 1980.

Fernando, H.J.S., Formation of a layered structure when a stable salinity gradient is heated from below. *J. Fluid Mech.*, *182*, 525-541, 1987.

Gregg, M.C., The three-dimensional mapping of a small thermohaline intrusion. *J. Phys. Oceanogr.*, *10*, 1468-1492, 1980.

Hildreth, W., Gradients in silicic magma chambers: implications for lithospheric magmatism. *J. Geophys. Res.*, *86*, 10153-10192, 1981.

Hoare, R.E., Problems in heat transfer in Lake Vanda, a density-stratifiedAntarctic lake. *Nature*, *210*, 787-789, 1966.

Huppert, H.E., The intrusion of fluid mechanics into geology. *J. Fluid Mech.*, *173*, 557-594, 1986.

Huppert, H.E., and P.F. Linden, On heating a stable salinity gradient from below. *J. Fluid Mech.*, *95*, 431-464, 1979.

Huppert, H.E., and R.S.J. Sparks, Double-diffusive convection due to crystallization in magmas. *Ann. Rev. Earth Planet. Sci.*, *12*, 11-37, 1984.

Huppert, H.E., and J.S. Turner, Melting icebergs. *Nature*, *271*, 46-48, 1978.

Huppert, H.E., and J.S. Turner, Ice blocks melting into a salinity gradient. *J. Fluid Mech.*, *100*, 367-384, 1980.

Huppert, H.E., and J.S. Turner, Double-diffusive convection. *J. Fluid Mech.*, *106*, 299-329, 1981a.

Huppert, H.E., and J.S. Turner, A laboratory model of a replenished magma chamber. *Earth Planet. Sci. Lett.*, *54*, 144-172, 1981b.

Huppert, H.E., J.S. Turner, and R.S.J. Sparks, Replenished magma chambers: effects of compositional zonation and input rates. *Earth Planet. Sci. Lett.*, *57*, 345-357, 1982.

Huppert, H.E., R.S.J. Sparks, and J.S. Turner, Laboratory investigations of viscous effects in replenished magma chambers. *Earth Planet. Sci. Lett.*, *65*, 377-381, 1983.

Kerr, R.C., and J.S. Turner, Layered convection and crystal layers in multicomponent systems. *Nature*, *298*, 731-733, 1982.

Linden, P.F., The formation and destruction of fine-structure by double-diffusive processes. *Deep-Sea Res.*, *23*, 895-908, 1976.

Manins, P.C., Intrusion into a stratified fluid. *J. Fluid Mech.*, *74*, 547-560, 1976.

McBirney, A.R., and R.M. Noyes, Crystallization and layering of the Skaergaard intrusion. *J. Petrol.*, *20*, 487-554, 1979.

McDougall, T.J., Fluid dynamic implications for massive sulphide deposits of hot saline fluid flowing into a submarine depression from below. *Deep-Sea Res.*, *31*, 145-170, 1984.

Munns, R.G., R.J. Stanley, and C.D. Densmore, Hydrographic observations of the Red Sea brines. *Nature* , *214*, 1215-1217, 1967.

Neal, V.T., S. Neshyba, and W. Denner, Thermal stratification in the Arctic Ocean. *Science*, *166*, 373-374, 1969.

Ruddick, B.R., and J.S. Turner, The vertical length scale of double-diffusive intrusions. *Deep-Sea Res.*, *26*, 903-913, 1979.

Schmitt, R.W., Double diffusion in oceanography. *Ann. Rev. Fluid Mech.*, *26*, 255-285, 1994.

Stern, M.E., The 'salt fountain' and thermohaline convection. *Tellus*, *12*, 172-175, 1960.

Stern, M.E., and J.S. Turner, Salt fingers and convecting layers. *Deep-Sea Res.*, *16*, 497-511, 1969.

Tait, R.I., and M.R. Howe, Some observations of thermohaline stratification in the deep ocean. *Deep-Sea Res.*, *15*, 275-280, 1968.

Turner, J.S., The coupled turbulent transports of salt and heat across a sharp density interface. *Int. J. Heat Mass Transfer*, *38*, 375-400, 1965.

Turner, J.S., Salt fingers across a density interface. *Deep-Sea Res.*, *14*, 599-611, 1967.

Turner, J.S., The behaviour of a stable salinity gradient heated from below. *J. Fluid Mech.*, *33*, 183-200, 1968.

Turner, J.S., *Buoyancy Effects in Fluids*. Cambridge University Press, 367pp., 1973.

Turner, J.S., Double-diffusive phenomena. *Ann. Rev. Fluid Mech.*, *6*, 37-56, 1974.

Turner, J.S., Double-diffusive intrusions into a density gradient. *J. Geophys. Res.*, *83*, 2887-2901, 1978.

Turner, J.S., A fluid-dynamical model of differentiation and layering in magma chambers. *Nature*, *285*, 213-215, 1980.

Turner, J.S., Multicomponent convection. *Ann. Rev. Fluid Mech.*, *17*, 11-44, 1985.

Turner, J.S., and I.H. Campbell, Convection and mixing in magma chambers. *Earth-Science Reviews*, *23*, 255-352, 1986.

Turner, J.S. and C.F. Chen, Two-dimensional effects in double-diffusive convection. *J. Fluid Mech.*, *63*, 577-592, 1974.

Turner, J.S. and L.B. Gustafson, The flow of hot saline solutions from vents in the sea floor: some implications for exhalative sulfide and other ore deposits. *Econ. Geol.*, *73*, 1082-1100, 1978.

Turner, J.S. and L.B. Gustafson, Fluid motions and compositional gradients produced by crystallization or melting at vertical boundaries. *J. Volcanol. and Geothermal Res.*, *11*, 93-125, 1981.

Turner, J.S., H.E. Huppert, and R.S.J. Sparks, Experimental investigations of volatile exsolution in evolving magma chambers. *J. Volcanol. and Geothermal Res.*, *16*, 263-277, 1983.

Turner, J.S., H.E. Huppert, and R.S.J. Sparks, Komatiites II: experimental and theoretical investigations of post-emplacement cooling and crystallization. *J. Petrol.*, *27*, 397-437, 1986.

Williams, A.J., Images of ocean microstructure. *Deep-Sea Res.*, *22*, 811-829, 1975.

J.S. Turner, Research School of Earth Sciences, The Australian National University, Canberra, Australian Capital Territory 0200, Australia.

Thermal Diffusion Phenomena in Thick Fluid Layers

J. Tanny

Center for Technological Education Holon, POB 305, Holon 58102, Israel

Z. Harel and A. Tsinober

Faculty of Engineering, Tel-Aviv University, Tel-Aviv, Israel

Various salt solutions are characterized by a strong tendency of the salt to diffuse from lower to higher temperature zones. This phenomenon, commonly known as thermal diffusion (Soret effect), was studied extensively either for the determination of its own characteristics in various solutions, or in the context of double diffusive convection, where its inclusion introduces peculiar effects. All the experiments reported so far considered a very thin (about few mms) horizontal layer of initially uniform solution exposed to a carefully controlled vertical temperature difference. The existence of a vertical temperature gradient at the impervious horizontal boundaries, gave rise to the establishment of a solute gradient whose magnitude and sign depended on the thermal diffusion properties of the solution. In many situations, however, much thicker fluid layers are encountered and therefore, in this work, thermal diffusion processes were studied experimentally by heating a large tank of salt solution from below while its upper free surface was exposed to room conditions. It is shown that depending on the initial solute distribution, a solution of salt with a large enough negative Soret effect, heated from below, can develop into (at least) two different states. If the process starts with a two-layer stratified solution, a double diffusive system with uniform gradients in equilibrium is established, but if initially the solution is uniform, the resulting system consists of a very thick mixed layer, topped by a relatively thin zone with large gradients of temperature and solute.

INTRODUCTION

In a wide variety of natural and industrial double diffusive systems, besides the regular diffusion of each component down along its gradient, cross diffusion between the two (or more) agents plays an important role. The most familiar example of this phenomenon is the Soret effect (also known as thermal diffusion) where molecular solute diffusion is caused by a temperature gradient. Diffusional flux of heat due to a salinity gradient (Dufour effect) is very small and usually negligible in the analysis of heat-salt systems in liquids [De Groot & Mazur 1962].

Double-Diffusive Convection
Geophysical Monograph 94

The magnitude and direction of these extra fluxes depend on the coupled characteristics of the diffusing components and on their concentrations (e.g. salinity and temperature) in the solution. A large number of experimental studies has been devoted to double diffusive stability phenomena, associated with thermal diffusion effects, using various solutions each characterized by different cross diffusion properties [e.g. Caldwell 1973, 1974 and 1976, Hurle & Jakeman 1971, Platten & Legros 1984]. Most of the results were obtained for fluids with negative Soret effect, where a stabilizing solute gradient is established when a layer of uniform solution is heated from below. This gradient produces a (regular) solute flux which balances the Soret flux so that equilibrium is achieved throughout the layer. The solute gradient is initially formed at the boundaries owing to the zero solute flux there and then extends into the fluid by diffusion. Therefore, the experimental cell used by all investigators was rather thin (about 0.5 cm thick), so as to attain equilibrium throughout the cell within a reasonable time. The thin cell also allows the assumption of constant fluid properties, when the experimental results are compared with the theory.

The motivation for the present research is twofold. Many natural and industrial systems consist of fluid layers, much thicker than those examined so far, and are exposed to much less controlled boundary conditions (e.g. solar ponds, magma chambers). For example, in such a thick fluid layer, convection driven by the bottom heating is much more vigorous than that in a thin layer, and it is not at all obvious that cross diffusion processes will dominate and a stable concentration gradient will be induced. It is therefore questioned under which conditions a stable steady state double diffusive system, in equilibrium, can be developed within a thick fluid layer. There are three necessary conditions for the establishment of a stabilizing concentration gradient in equilibrium within a fluid layer heated from below. These are: (i) the Soret effect should be large enough and negative, (ii) the applied temperature gradient should be large enough ,and (iii) a sufficient amount of salt should exist in the system. In a very thick fluid layer, however, other factors, e.g. the initial solute distribution, can significantly influence the characteristics of the eventually established system.

The second motive is that the very thin cell utilized so far [e.g. Hurle & Jakeman 1971, Caldwell 1974] does not allow for a detailed study of the transient build up process. In the present experiments, therefore, a much thicker fluid layer is studied in which the transient process can be monitored.

The objective of the present study is to consider these two issues within a single experimental framework. A thick fluid layer, about 45 cm in height, consisting of a KNO_3 - water solution at uniform temperature and two different initial solute distributions, is heated from below to a prescribed temperature above the ambient, while the upper free surface is exposed to the room temperature which was kept nearly constant by the air conditioning system. The transient build up process was studied by means of vertical temperature and density profile measurements, from which concentration distributions were extracted.

In §2 we describe the experimental apparatus and procedures,in §3 the experimental results are presented and discussed, and the paper is summarized in §4.

2. APPARATUS AND MEASUREMENTS

The experimental facility consisted of a square glass tank $50 \times 50 \times 80$ cm high with a bottom made of a 3 mm brass plate. Beneath this plate was a series of 10 flat electrical heating elements (Chromalox SE-17), 48 cm long, 2.5 cm wide and 0.8 cm thick. The sidewalls of the tank were made of 12 mm thick glass plates, and were insulated by special transparent thermal insulation units, each consisting of five thin (1.5 mm thick) perspex plates, separated by 4.5 mm air gaps. This special insulation prevented horizontal temperature gradients within the test fluid, due to sidewall cooling, and allowed flow visualization throughout the experiment. The heating elements were insulated from below by a 5 cm glass wool insulation.

Temperatures were measured by thermocouples type T (copper - constantan) with an accuracy of $\pm 0.3^{\circ}C$. The output of the thermocouples was measured, linearized and recorded by a data acquisition system, consisting of a data logger (ADP 65) and a personal computer.

The bottom temperature was measured by 9 thermocouples, distributed over the brass plate. Individual thermocouple measurements indicated that the bottom temperature was uniform within $\pm 1.0^{\circ}C$, which is about 2.5% of a typical difference between the bottom and the ambient temperature. In a typical experiment, the average measured bottom temperature was compared with a programmed value; on the basis of this comparison the heating elements were turned on or off by the data-logger. This feedback and control system allowed the bottom temperature to rise to a prescribed value and afterwards maintained the bottom at this temperature throughout the rest of the experiment.

The local fluid density was measured by utilizing Archimedes' principle. The buoyant force exerted on a glass rod (150 mm length, 4 mm diameter) was measured by a force transducer ("LVDT"), that is accurate to 0.001g. The glass rod was hung on the force transducer by a very thin (about 0.2 mm diameter) nylon thread. The thread was attached to the two edges of the rod to keep it in a horizontal position. The rod was submerged horizontally within the test fluid and traversed vertically at increments of 2 cm from the surface down to the bottom. No significant disturbance to the fluid was noticed during the vertical motion of the rod. Its measured local apparent mass $m(z)$ was used to calculate the local density $\rho(z)$ by: $\rho(z) = \left(m_g - m(z)\right)/V_0$, where m_g is the mass of the glass rod, and V_0 is its volume, calculated from m_g and the glass density $\rho_g = 2.203 g/cm^3$. The maximum error in the measurement of $\rho(z)$ was estimated as $0.003 g/cm^3$. The local temperature was measured simultaneously with the density measurement at each level by a single thermocouple, mounted on a small lead cone and coated by a special heat conducting adhesive to prevent erosion by the salt solution. On the basis of the density and temperature distributions, the concentration profile was calculated, using the physical properties of KNO_3 given in Washburn [1929] and Gmelin [1955].

The temperature and density probes were mounted on an arm, and were traversed vertically through the fluid, completing one vertical scan within a period of about 40 minutes. The traversing mechanism was operated by a step motor and an indexer, controlled by the data logger and the personal computer.

3. EXPERIMENTAL RESULTS AND DISCUSSION

Three types of experiments were performed to study the conditions under which a steady state double diffusive system in equilibrium can be established in a large tank. The influence of the Soret effect on the establishment of a system in equilibrium, was determined by two experiments: one with KNO_3- water solution for which the Soret effect is large and negative (see the appendix), and the other with aqueous solution of NaCl (common salt) with a small positive Soret effect. Both experiments were started with a two-layer stratified system. The influence of the initial solute distribution on the final state was studied by an additional experiment with KNO_3 solution, with an initial state of uniform concentration. In all experiments, a sufficient amount of salt was dissolved within the system so as to satisfy the third condition (§1) for build up of a steady state system of gradients in equilibrium.

This chapter is divided into three parts. The first is devoted to the influence of the initial solute distribution on the build up of gradients in a thick layer of a solution with a negative and large Soret effect, in the second we demonstrate what happens when the Soret effect is positive, and in the third we introduce the chemical potential as a means for the analysis of the transient build up process.

3.1 The Influence of the Initial Solute Distribution

Two experiments were carried out with KNO_3 - water solution to study the influence of the initial solute profile on the build up of gradients within a very thick layer heated from below. In the first experiment (No. 1) a two - layer system was considered with a bottom layer of concentrated solution and a top layer of fresh water, while in the second one (No. 2), a thick layer of concentrated uniform solution was examined. In Figures 1 and 2 the evolution of the (a) temperature, (b) density and (c) concentration fields during each experiment are shown. It is noticed that the overall salt content in the second experiment was larger than that in the first one.

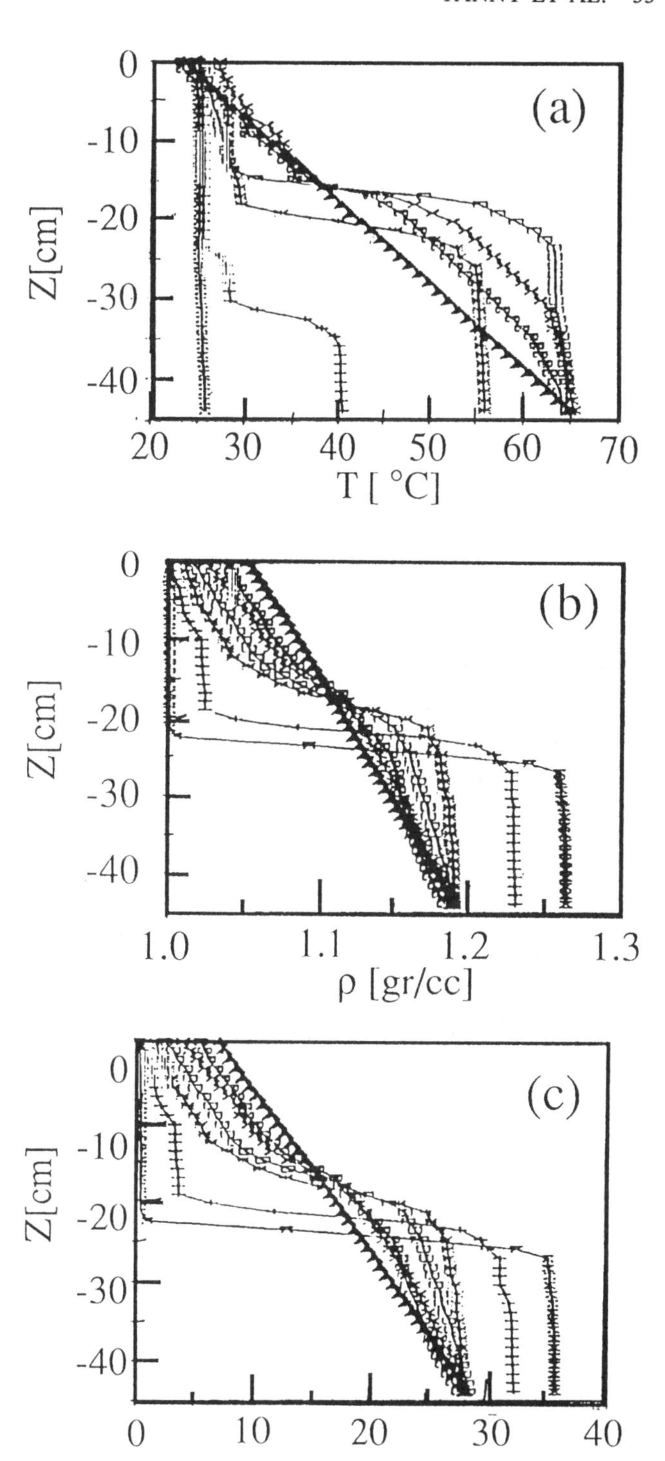

Fig. 1. Profiles evolution in experiment no. 1 with a two layer system of saturated KNO_3 solution-fresh water. a) Temperature profiles, b)Density profiles, c) Concentration profiles. t (in hrs) = ⌧0, +24, *48, □72, ×96, ⊠120, Δ312.

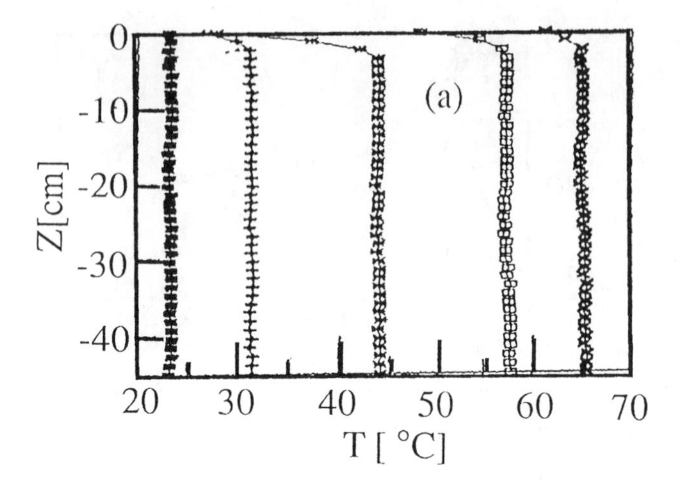

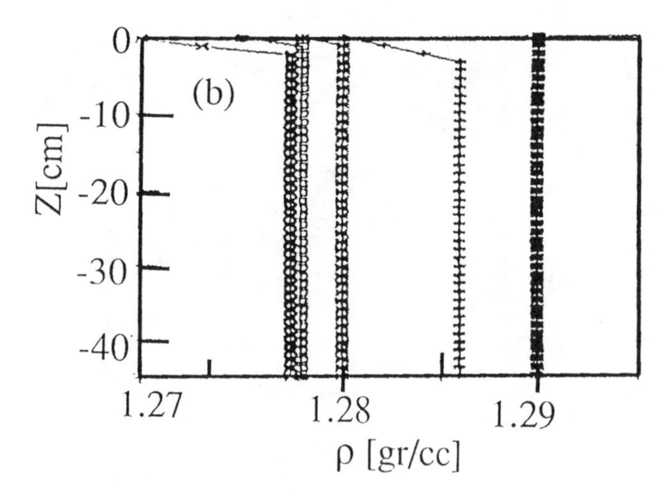

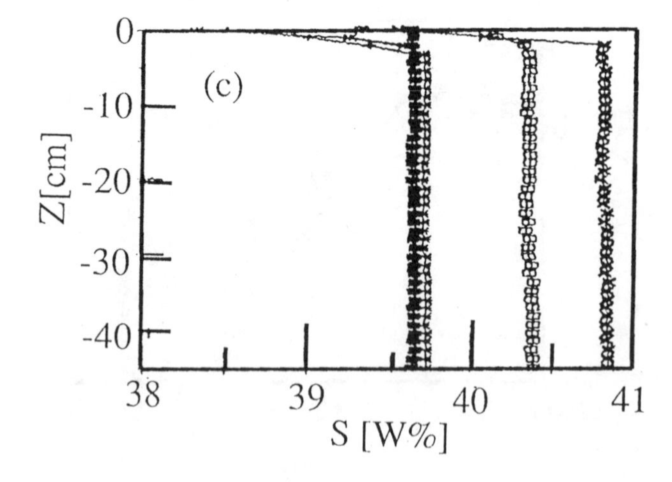

Fig. 2. Profiles evolution in experiment No. 2 with a uniform concentrated layer of KNO_3 heated from below. a) Temperature, b) Density, c) Concentration. t (in hrs) = ■0, +6, *12, □18, ×24.

A first observation of these two figures shows that in experiment No. 1 a stable, steady state double diffusive system in equilibrium was established; this system consists of a destabilizing temperature gradient and a stabilizing concentration profile. On the other hand, in experiment No. 2 such a system was not developed and instead, a thick uniform mixed layer, bounded by a thin zone of large negative temperature and concentration gradients at the top was established. This result implies that the initial density step between the two layers in experiment No. 1 plays an important role in the build up process. To further clarify this issue, the transient build up process in experiment No. 1, as shown in Figure 1, is now discussed in more detail.

Up to t = 72 hrs (where t is the time elapsed from the initiation of heating), the bottom temperature (Figure 1a) increases continuously to its prescribed value of 65°C; for t > 24 hrs, the temperature of the lower layer is almost uniform and equal to the instantaneous bottom temperature, and a relatively sharp temperature step exists between the upper and lower layer.

The density profiles (Figure 1b) show that the lower layer density is almost uniform and decreases with time up to t = 48 hrs. The initial density interface migrates upwards, and remains relatively sharp up to 24 hrs. The corresponding concentration profiles (Figure 1c) indicate that upwards transport of salt is associated with the density decrease in the lower layer.

In the temperature profile (Figure 1a) at t = 24h two "steps" are observed indicating a layered structure. These two steps are also apparent in the corresponding concentration profile (Figure 1c). These steps are probably caused by the transport of salt to the upper layer. Because of this transport a weak concentration gradient is formed below the initial density interface. Due to the bottom heating this gradient breaks down to the observed two-step structure within the lower layer.

At t = 48 hrs, a density (and concentration) gradient develops within the upper layer and this is associated with the evolution of a destabilizing temperature profile within the upper layer, as seen on Figure 1a at t =72 hrs. Along with the development of the density gradient in the upper layer, weak density and concentration gradients (Figures 1b and 1c) are established below the interface at t = 48 hrs. At t = 72 hrs, however, the large negative Soret effect of the KNO_3 solution dominates the system, by establishing a stable, almost linear density (and concentration) distribution within the lower layer.

It is suggested that due to the large temperature step across the interface, the negative Soret effect acts as a 'salt pump', transporting salt from the upper to the lower (more hot) layer, in a direction opposite to that of the regular diffusion. This transport reduces the upwards salt flux, in such a way that at the vicinity of the bottom of the tank the salt flux becomes close to zero (at the bottom the flux is always zero). Consequently, the concentration at the bottom of the fluid remains almost unchanged throughout the rest of the experiment (Figures 1b and 1c) and this facilitates the further build-up of the stable density profile within the fluid.

The density profiles for t >72 hrs (Figure 1b) indicate that once the bottom density becomes fixed, density and temperature gradients will be gradually established throughout the fluid. The density profile developed at t = 72 hrs stabilizes the lower layer, thus allowing the evolution of a destabilizing temperature gradient as observed at t = 96 hrs (Figure 1a). At the upper layer, the density close to the free surface increases with time, but this increase eventually stops due to the reduced salt flux upwards.

Very small variations are observed in the profiles from t = 120 up to 312 hrs, where a steady state system in equilibrium is achieved, consisting of almost constant gradients of temperature and concentration with opposing contributions to the density distribution. It is noteworthy that experimentally the system was observed to be highly stable. Besides some evaporation at the free surface, this double diffusive system remained unchanged for a period of at least one week as was tested in this experiment.

The stability of this double diffusive system can be considered by applying the results of the linear stability analysis [e.g. McDougall 1983] which takes into account the cross diffusion effects. For the diffusive stratification, neglecting the Dufour effect, the system is stable provided:

$$\left(1-\frac{\beta}{\alpha}S_T\right)Ra_T < \frac{\tau}{\Pr+1}(\Pr+\tau)Ra_S + \frac{27\pi^4}{4}\frac{1+\tau}{\Pr}(\Pr+\tau) \qquad (1)$$

where:

$Ra_T = g\alpha\Delta TH^3/\upsilon K_T$ and $Ra_S = g\beta\Delta SH^3/\upsilon K_S$. Here g is the acceleration due to gravity, α and β are the coefficients of thermal and solutal expansion respectively, ΔT and ΔS are the temperature and salinity differences across the layer of thickness H, υ is the kinematic viscosity and K_T and K_S are the coefficients of heat and salt diffusion respectively. The diffusivity ratio is $\tau = K_S/K_T$, the Prandtl number is $\Pr = \upsilon/K_T$ and S_T is the Soret coefficient.

For the temperature and concentration profiles measured at t = 312 hrs (Figures 1a and 1c) the left hand side of equation (1) is equal to 1.88×10^{11} while its right hand side is 6.52×10^{11}. This result supports the experimental observation regarding the very high stability of the system.

The above estimate is based on the analytical solution with 'free' boundary conditions. Knobloch & Moore [1988] have made computations for realistic boundary conditions but for much lower Ra_T than ours. It is noteworthy, however, that their results regarding the critical Ra_T corresponding to our boundary conditions differ less than 10% from those corresponding to the 'free' boundary conditions.

At this stage it is instructive to compare the process observed here to the build-up process occurring in the thin cell experiments [e.g. Caldwell 1974, Hurle & Jakeman 1971]. In these experiments, a vertical temperature gradient was quickly established throughout the cell, and its existence at the impervious horizontal boundaries combined with the negative Soret effect, gave rise to a local establishment of a concentration profile at the boundaries, which afterwards extended into the whole fluid layer. The essential difference between this process and the one occurring in the present two-layer experiment is that here, the temperature gradient, which is responsible for the cross-diffusion flux, is located far from the boundary, in our case at a distance of about 25-30 cm above the bottom. Close to the lower boundary itself no significant temperature gradient is observed up to about t =120 hrs (Figure 1a) at which moment a considerable density gradient already exists (Figure 1b). This observation implies that for the build-up of the concentration profile, *a sufficiently large temperature gradient must exist in the bulk of the fluid, but not necessarily at the boundaries.*

The characteristic time scale of salt diffusion in a layer of thickness H is $t_d = H^2/K_S$. For the two-layer stratified system considered in the present experiments, the characteristic time for the build up of a concentration gradient by diffusion only is about $t_d = 10,000$ hrs, much larger than the actual time during which the system attained equilibrium $(\approx 300\ \text{hrs})$. This implies that the build up process is governed by an additional faster mechanism, presumably thermal convection which transports the salt upwards, at least during the initial stages of the experiment. A typical convective timescale can be defined as:

$$t_c = \left(\frac{H}{g\alpha\Delta T}\right)^{1/2}$$

where ΔT is a typical temperature difference across the layer. In the present experiments we obtain t_c of about a few seconds, much shorter than the actual time of build up. Thus, one arrives at the conclusion that in a thick layer, the build up process is governed by a combined mechanism consisting of both convection and diffusion. The profiles in Figure 1 suggest that during the initial stages of the process (up to about 72 h) thermal convection is dominant. However, at later stages diffusion due to the Soret effect becomes more important and thus dominates over the thermal convection.

In experiment No. 2 (Figure 2) the initial concentration difference between the top and bottom layers was set to zero, i.e. a thick layer of uniform concentrated solution was heated from below. The overall salt content in this experiment was larger than that in experiment No. 1, so that for the same final temperature gradient, a similar concentration gradient is allowed to develop, but with a larger average concentration. As observed in Figure 2, however, the final state now is very different from that of experiment No. 1.

Instead of uniform gradients, most of the fluid consists of a homogeneous mixed layer in which convection is driven by the bottom heating. At the top, however, where a considerable temperature gradient inevitably exists, a concentration gradient is induced to balance the cross diffusion salt flux due to Soret effect. Thus, at the top free surface, a thin layer with strong gradients is formed, and coexists with the mixed layer below. Measurements within this thin layer were rather difficult. This problem was most pronounced over the few top mms where the glass rod, by which density was measured, was either not fully submerged within the fluid or subjected to significant surface tension effects. Therefore at present we indicate the existence of such a non-convective stratified layer, presumably at equilibrium, but we cannot make any quantitative conclusions regarding its detailed properties. In particular, the real concentration gradient at this top layer is much larger in order to compensate for the slight increase in the concentration of the thick mixed layer (Figure 2c).

It is clear from the results of experiments 1 and 2 that some restraint to thermal convection should exist within the system to allow for a sufficiently large temperature gradient to drive the cross diffusion salt flux. In a very thin cell, the fluid itself may restrain convection, but in a very thick layer where the thermal Rayleigh number is much larger, the restraint should be provided by some other factor. In experiment No. 1 such a restraint was introduced in the form of an initial density interface through which the transports of heat and solute were mainly by conduction. In experiment No. 2, on the other hand, such a restraint did not exist and therefore thermal convection dominated and prevented the build up of uniform gradients throughout the fluid layer. Further work is needed to study quantitatively how strong the restraint should be in order to obtain a final equilibrium state like that in Figure 1.

3.2 A Comparison with an Experiment with NACℓ

To demonstrate the significance of the large negative Soret effect of the KNO_3-water solution to the build-up process, an experiment similar to experiment No. 1 was performed with common salt NaCl (experiment No.3), whose Soret coefficient in water at temperatures larger than 12°C is positive and small [Caldwell 1973, 1974]. A two layer system consisting of a fresh water layer and a concentrated layer of NaCl solution at 26 wt%, at room temperature was heated from below up to 65°C. The process of the gradient development for this experiment is shown on Figure 3.

The temperature (Figure 3a) density (3b) and concentration (3c) profiles exhibit the regular entrainment process typical of a stratified two-layer system with one layer heated from below or mixed by other means [see Turner 1986, 1991 and references therein]. The density is uniform throughout the lower layer, while salt diffusion into the upper layer produces a moderate density gradient within that layer. This diffusion process facilitates the generation of a destabilizing temperature gradient within the upper layer, as seen in Figure 3a. Due to the mixing, however, the density interface migrates upwards, until eventually, at t = 96 hrs the whole fluid is well mixed and almost uniform.

It is noteworthy that at the final stage (t = 96 hrs) a thin layer with a negative temperature gradient exists at the top of the thick mixed layer, as was also observed in experiment No. 2. In contrast with the latter, however, the very small (and positive) Soret effect of the NaCl solution does not induce a concentration gradient within this layer and the salinity (Figure 3c) is uniform throughout.

3.3 The Chemical Potential

One of the most appropriate thermodynamic properties, by which the transient build-up process can be analyzed is the chemical potential μ. The definition of the chemical potential and some related physical properties are given in the Appendix. Here we will just mention that the gradient of the chemical potential of a solution is the driving force for solute diffusion due to the concentration and temperature gradients and therefore represents both regular

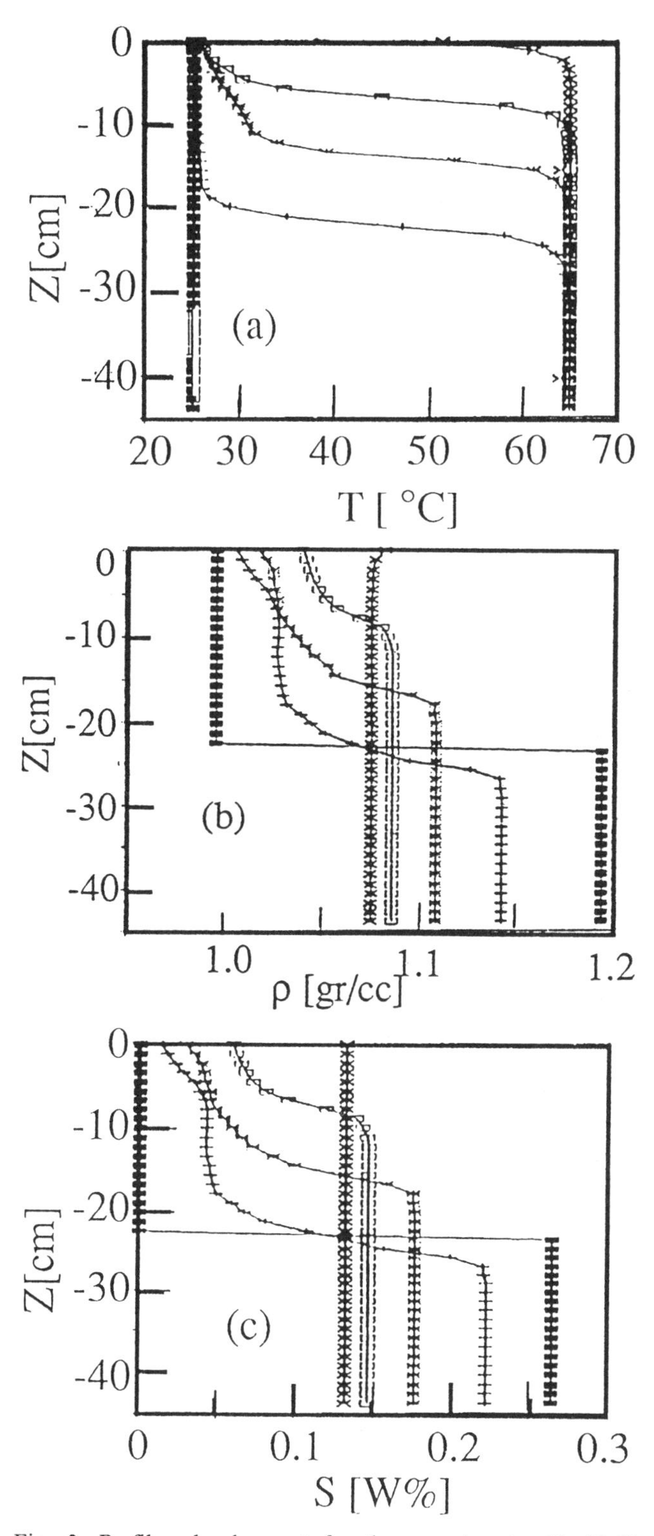

Fig. 3. Profiles development for the experiment with NaCl solution - fresh water, two layer system.
a) Temperature b) Density c) Concentration t (in hrs) = ■0, +24, ∗48, □72, ×96.

and cross diffusion (Soret effect) fluxes. At equilibrium $\nabla\mu = 0$, and the net salt flux is zero.

In Figure 4 the chemical potential μ as a function of height for various stages of experiment No. 1 is shown. These profiles show that initially, a strong gradient of the chemical potential exists at the density interface. This gradient, and hence the driving force for solute flux, decreases with time and at t = 48 hrs, the chemical potential at the bottom becomes fixed. Furthermore, the profile of μ at the final state (t = 312 hrs) shows that at t = 48 hrs, close to the bottom, μ equals its equilibrium value and $\nabla\mu$ is approximately zero. This suggests that the fluid close to the bottom of the tank attains local equilibrium already at about t = 48 hrs, and the subsequent gradient build-up process is associated with the extension of this equilibrium state into the rest of the fluid. Eventually, at t = 312 hrs, $\nabla\mu = 0$ throughout, and the driving force for the diffusion of salt is zero. At this stage, the regular diffusion of salt down its gradient is exactly balanced by the cross diffusion of salt due to the Soret effect so that the net salt flux vanishes throughout the system.

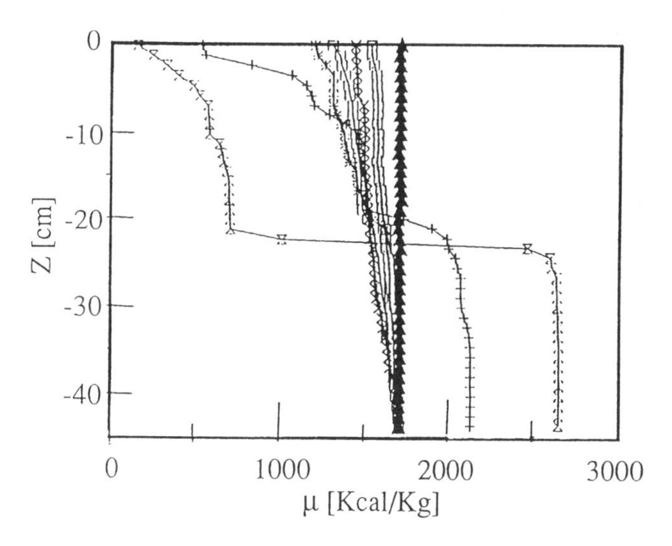

Fig. 4. The development of the chemical potential profiles for experiment No. 1. t (in hrs) = ⧗0, +24, ∗48, □72, ×96, ⊠120, Δ312.

CONCLUSIONS

In the present experiments thermal diffusion phenomena were investigated in thick fluid layers. The following conclusions can be drawn on the basis of the results.

- In a solution with a strong enough negative Soret effect, heated from below in a large tank, the initial solute distribution has a crucial influence on the final state. It is shown that if initially a large enough density step exists within the layer, build up of a stable uniform density gradient is possible, but if the initial solution is uniform (or if the initial density step is not large enough to restrain the thermal convection at the bottom layer), a different state is achieved, consisting of a thick uniform mixed layer bounded by a thin gradient zone at the top. Thus, in a thick layer, the establishment of uniform gradients is far from being obvious.

- The (local) temperature gradient associated with the density step mentioned above, can drive the build up of gradients from the bulk of the fluid and not necessarily from the vicinity of the impervious boundaries, as is the case in a thin fluid layer.
- In contrast with the thin layer experiments, where molecular cross-diffusion was the only mechanism of gradients build up [Hurle & Jakeman 1971], in a thick layer thermal convection plays a crucial role. Analysis of characteristic time scales shows that without convection, the build up of uniform gradients would have lasted about 10,000 hrs in the thick layer considered here.

APPENDIX: THERMODYNAMIC PROPERTIES OF POTASSIUM NITRATE SOLUTION (KNO_3).

One of the reasons for the choice of KNO_3 as the solute in the present study is its strongly temperature dependent solubility as shown in Figure A1. This phenomenon is associated with a strong negative Soret effect as shown in figures A3 and A4 presented below.

The gradient of the chemical potential is the driving force for solute diffusion due to the concentration and temperature gradients [De Groot & Mazur, 1962] and therefore represents both regular and cross diffusion (Soret effect). The chemical potential is defined as follows:

$$\mu = \mu_0(T) + RT\ell n(a) \quad (2)$$

where $\mu_0(T)$ is the standard chemical potential of the solid salt, R is the universal gas constant, T is the temperature, a is the activity of the solution which depends on T and S where S is the solute concentration. The activity of the solution can be expressed as the product of the activity coefficient γ and the concentration, i.e. $a = \gamma C$ where C is the concentration in moles of solute per kg solution. It is noteworthy that the activity coefficient γ is a function of T and S as shown on Figure A2 for KNO_3

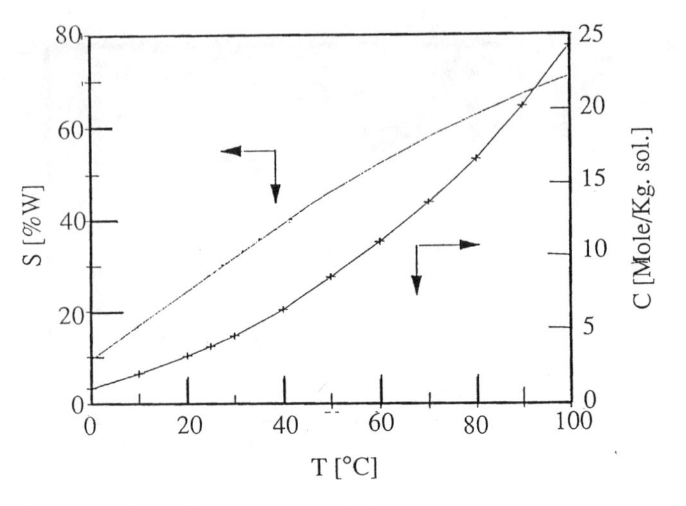

Fig. A1. Solubility, S and molality, C of saturated KNO_3 solution as a function of temperature. Solubility, Molality.

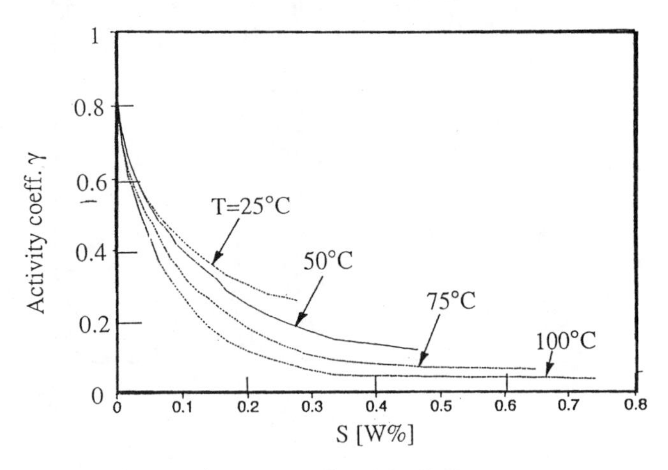

Fig. A2. The activity coefficient γ of KNO_3 solution as a function of concentration, S for various temperatures.

solution. Thus for given temperature and concentration profiles, one can obtain the distribution of the chemical potential υ.

At equilibrium (quasi steady state) conditions the vertical diffusive flux of solute is zero and this is achieved when the spatial gradient of the chemical potential vanishes i.e $\nabla\mu = 0$. This condition may be expressed as:

$$\nabla\mu = \frac{\partial\mu}{\partial S}\nabla S + \frac{\partial\mu}{\partial T}\nabla T = 0 \quad (3)$$

Since in the present work ∇S and ∇T have the same sign,this condition can be satisfied only if $\partial\mu/\partial T$ is opposite in sign to $\partial\mu/\partial S$. Figure A3 shows the chemical potential as a function of concentration S at two typical

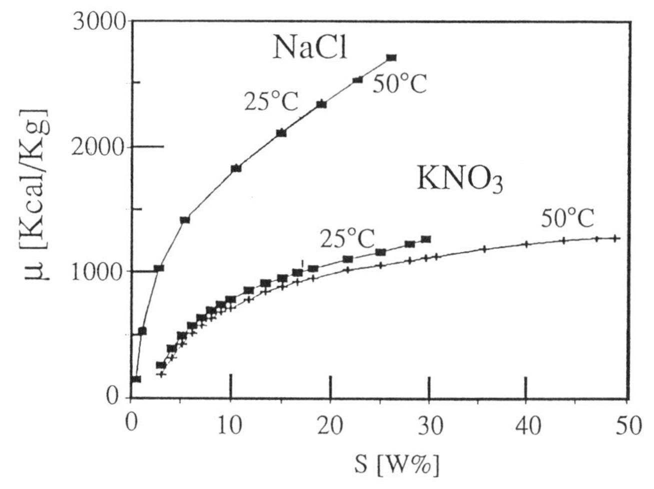

Fig. A3. The chemical potential μ as a function of the concentration S at two typical temperatures.

temperatures for KNO_3 and NaCl solutions. Figure A4 shows the chemical potential as a function of the temperature at two typical concentrations for KNO_3 and NaCl solution. It is seen that for KNO_3 solution $\partial\mu/\partial T$ is negative and $\partial\mu/\partial S$ is positive, while for NaCl, $\partial\mu/\partial T$ is practically zero and $\partial\mu/\partial S$ is positive. Therefore, for a given temperature profile, an equilibrium with $\nabla\mu = 0$ (zero salt flux) can be reached only for KNO_3 solution and not for NaCl. This phenomenon is associated with the opposite signs of the Soret effect of KNO_3 and NaCl solutions.

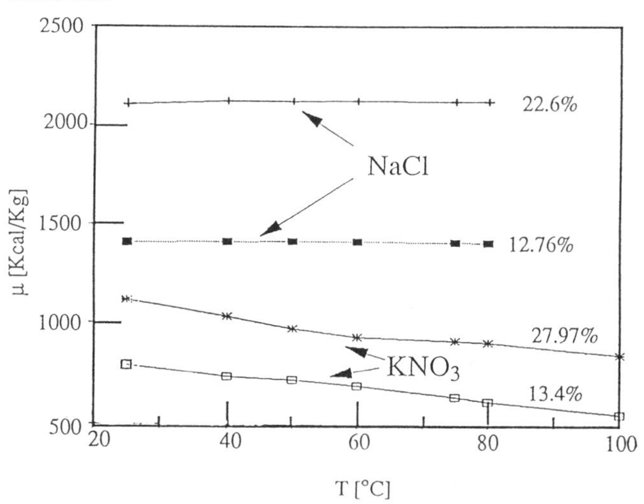

Fig. A4. The chemical potential μ as a function of the temperature T at two typical concentrations.

Acknowledgement. This work was partially supported by the Gordon Center for Energy Studies, Faculty of Engineering and Faculty of Exact Sciences, Tel Aviv University.

REFERENCES

Caldwell, D.R. 1973 Measurement of negative thermal diffusion coefficients by observing the onset of thermohaline convection. *J. Phys. Chem.* 77, 2004-2008.

Caldwell, D.R. 1974 Experimental study on the onset of thermohaline convection. *J. Fluid Mech.* 64, 347-367.

Caldwell, D.R. 1976 Thermosolutal convection in a solution with large negative Soret coefficient. *J. Fluid Mech.* 74, 129-142.

De Groot, S.R. and P. Mazur, 1962, *Non-Equilibrium Thermodynamics*. North Holland.

Gmelin 1955 *Handbuch Der Anorganische Chemie*, Verlag Chemie, Berlin, Vol. 22, p. 297.

Hurle, D.T.J. and E. Jakeman, 1971 Soret-driven thermo solutal convection. *J. Fluid Mech.* 47, 667-687.

Knobloch, E. and D.R. Moore, 1988 Linear stability of experimental Soret convection. *Physical Review A* 37, 860-870.

McDougall, T.J. 1983 Double diffusive convection caused by coupled molecular diffusion. *J. Fluid Mech.* 126, 379-397.

Platten, J.K. and J.C. Legros, 1984 *Convection in Liquids*. Springer.

Turner, J.S. 1986 Turbulent entrainment: the development of the entrainment assumption, and its application to geophysical flows. *J. Fluid Mech.* 173, 431-471.

Turner, J.S. 1991 Convection and mixing in the oceans and the earth. *Phys. Fluids A*, 3(2), 1218-1232.

Washburn, E.W. 1929 *International Critical Tables of Numerical Data*, Physics, Chemistry & Technology, McGraw-Hill, New-York.

Z. Harel, Faculty of Engineering, Tel-Aviv University, Tel-Aviv 69978, Israel.

J. Tanny, Center for Technological Education Holon, POB 305, Holon 58102, Israel.

A. Tsinober, Faculty of Engineering, Tel-Aviv University, Tel-Aviv 69978, Israel.

The Onset of Double-Diffusive Instability in a Layer with Temperature-Dependent Fluid Properties

J. TANNY and V. A. GOTLIB

Center for Technological Education Holon, Holon, Israel

A. TSINOBER

Faculty of Engineering, Tel-Aviv University, Tel-Aviv, Israel

The linear double-diffusive stability of a two-dimensional infinite horizontal layer stratified vertically by temperature and solute concentration is analyzed numerically by the Galerkin method for the case of temperature dependent kinematic viscosity and molecular salt diffusivity (aqueous solution of NaCl). The horizontal boundaries are shear-free and perfectly conducting. The results for the direct mode ('finger regime') show that, in contrast to the constant properties case, the critical wave-number increases with the solute Rayleigh number and the critical thermal Rayleigh number is reduced from its corresponding constant properties value. Two different branches exist for solute Rayleigh number larger than some fixed value in the case of the oscillatory mode ('diffusive regime'). The less stable branch is characterized by a high wave-number.

INTRODUCTION

Double-diffusive convection takes place in a wide variety of technological application (e.g. solar ponds, crystallization and solidification processes, nuclear engineering) and in other scientific branches (e.g. geology, oceanography, astrophysics). One of the fundamental problems of double-diffusive convection is the stability of a statically stable horizontal fluid layer, stratified by two buoyancy components with different molecular diffusivities (e.g. heat and salt) which make opposite contributions to the overall vertical density distribution. In such systems, motion can arise even when the basic state density distribution is gravitationally stable [Turner, 1979]. The early theories pertinent to this problem (e.g. [Baines and Gill, 1969]) analyzed the stability of the above described layer under conditions of constant fluid properties.

Double-Diffusive Convection
Geophysical Monograph 94

In any realistic double-diffusive system the temperature or concentration gradient can cause considerable spatial variations of the physical properties of the fluid which, in turn, vary the gradient itself. For example, in a normal solar pond temperature gradient, the kinematic viscosity and salt diffusivity are varied by a factor of 3 from top to bottom. The present investigation was motivated by the conjecture that the assumption of constant fluid properties can be one of the causes for the discrepancies between the theoretically predicted values of the stability parameters and those observed in real systems [Zangrando and Bertram, 1985]. The aim of the present work is to analyze the stability of a steady state double-diffusive layer, *simultaneously* exposed to the effects of temperature dependent viscosity and salt diffusivity and a non-linear basic state salinity distribution.

FLUID PROPERTIES AND BASIC STATE

The ranges of variation of the temperature ($T = 20 - 90^\circ C$) and salt concentration ($S = 0 - 0.2$ in fraction) utilized here are those appropriate for NaCl solar pond conditions. We have taken into account only the effect of the temperature on the viscosity (ν) and salt diffusivity (k_S) because this effect is the dominant one over the concentration effect.

The fluid properties (measured in cm^2 s^{-1}) were approximated as:

$$\nu = \exp(-4.099 - 0.0269T + 0.991 \times 10^{-4}T^2), \quad (1)$$

$$k_S = 1.35 \cdot (1 + 0.033 \cdot (T - 20)) \cdot 10^{-5}, \quad (2)$$

$$k_T = \text{const} = 0.0014, \quad (3)$$

where k_T is the coefficient of heat diffusivity.

At the basic state we consider a horizontally infinite stationary fluid layer, stratified by vertical distributions of non-dimensional temperature $\theta_b(z)$ and solute concentration $\sigma_b(z)$, where z is the vertical coordinate axis directed downwards and the subscript "b" stands for the basic state.

The dimensionless steady state conductive solution with zero velocity in our case reduced to:

$$\theta_b = z,$$

$$\sigma_b = 0.8355 \ln(2.31z + 1).$$

The dimensionless variables are defined as $\theta = (T - T_0)/(T_1 - T_0)$ and $\sigma = (S - S_0)/(S_1 - S_0)$ where the subscripts "0" and "1" correspond to the top and bottom of the layer respectively. The coordinate z is normalized by the layer depth h.

THE SMALL PERTURBATION EQUATIONS

The linearized dimensionless small perturbation equations (under the Boussinesq approximation, as usual) are:

$$\left(\frac{1}{Pr}\frac{\partial}{\partial t} - 2\nu_b'\frac{\partial}{\partial z} - \nu_b\nabla^2\right)\nabla^2\psi +$$

$$+\nu_b''\left(\frac{\partial^2}{\partial x^2} - \frac{\partial^2}{\partial z^2}\right)\psi = Ra_S\frac{\partial\sigma}{\partial x} - Ra_T\frac{\partial\theta}{\partial x}; \quad (4)$$

$$\left(\frac{\partial}{\partial t} - \nabla^2\right)\theta = -\frac{\partial\psi}{\partial x}; \quad (5)$$

$$\left(\frac{\partial}{\partial t} - \tau(1 + \gamma\theta_b)\nabla^2\right)\sigma =$$

$$= \tau\gamma\left(\sigma_b'\frac{\partial\theta}{\partial z} + \sigma_b''\theta + \frac{\partial\sigma}{\partial z}\right) - \sigma_b'\frac{\partial\psi}{\partial x}. \quad (6)$$

In this set of equations ψ, θ and σ are the two-dimensional time dependent perturbations of the stream function, temperature and concentration respectively, $Ra_S = (g\beta\Delta Sh^3)/(\nu_r k_T)$ and $Ra_T = (g\alpha\Delta Th^3)/(\nu_r k_T)$ are the solute and thermal Rayleigh numbers respectively, $Pr = \nu_r/k_T$ – the Prandtl number, t is the time, x is the horizontal coordinate and $\tau = k_{S,r}/k_T$ – the diffusivity ratio. The normalized basic state viscosity $\nu_b = \nu/\nu_r$ is expressed as a function of z, the *prime* symbol denotes the derivative operator d/dz related to functions of single argument and the subscript "r" denotes a reference constant value. To obtain equations (4)–(6), distance was normalized by the layer depth h, time by h^2/k_T, stream function by k_T, and temperature and concentration differences by $\Delta T = T_1 - T_0$ and $\Delta S = S_1 - S_0$. It is also noticed that hereafter τ, Pr, Ra_S and Ra_T are based on ν_r and $k_{S,r}$; α and β are coefficients of heat and solute expansion respectively, $\gamma = (dk_S/dT)\Delta T/k_{S,r}$. The linear stability problem is studied under ideal free-free boundary conditions of zero shear stress and zero perturbations at the bottom and top surfaces of the fluid layer:

$$\psi''(x,0) = \psi''(x,1) = 0; \quad (7)$$

$$\psi(x,0) = \psi(x,1) = \theta(x,0) = \theta(x,1) =$$

$$= \sigma(x,0) = \sigma(x,1) = 0. \quad (8)$$

THE NUMERICAL PROCEDURE

To solve our problem we follow the Galerkin method. At the first stage separation of variables is employed in the form:

$$\psi(t,x,z) = f_1(z)\exp pt \sin \pi ax,$$

$$\theta(t,x,z) = f_2(z)\exp pt \cos \pi ax,$$

$$\sigma(t,x,z) = f_3(z)\exp pt \cos \pi ax,$$

where f_1, f_2 and f_3 are some smooth functions, a is a constant (wave-number) and p is in general complex, $p = p_r + ip_i$ (p_i represents the frequency of marginal oscillations).

After substituting these expressions into equations (4)–(8), the latter are transformed into a set of ordinary differential equations with the appropriate boundary conditions.

At the second stage, the functions f_i ($i = 1,2,3$) are expanded into a truncated series on a complete linearly-independent system of functions $\{\sin \pi nz\}$, $(n = 1, M)$, which satisfy the necessary boundary conditions and are orthogonal on the non-dimensional interval $z \in [0,\ 1]$.

Our problem is reduced to the set of homogeneous algebraic equations and the stability criterion satisfies the requirement $D(Ra_S, Ra_T, a, p) = 0$ where D is the determinant of the homogeneous system. The stability limits are found numerically by solving the last equation $D = 0$.

RESULTS AND DISCUSSION

The results presented below were obtained with the reference values of the variable fluid properties at $T = 20°C$, which yields $Pr = 7.20$, $\tau = 0.0096$ and $\gamma = 2.31$.

Direct Instability Mode

The calculated critical thermal Rayleigh number Ra_T^c was divided by Ra_T^0 – its value corresponding to the constant properties case. The resulting ratio $\eta = Ra_T^c / Ra_T^0$ is shown in figure 1. The values of η indicate that for the variable properties case Ra_T^c is smaller than its value corresponding to constant properties, and the ratio of the two decreases as Ra_S increases. This suggests that the asymptotic behaviour is $Ra_T^c \propto Ra_S$ like in the case with constant properties and gradients. The same asymptotic behaviour was obtained by Zangrando and Bertram [1985] for a cubic concentration profile under constant properties.

The wave-number at the onset of direct instability is plotted in figure 2. It is shown that the value of the wave-number a increases with Ra_S in contrast to the constant properties case where $a = \text{const} = 2^{-1/2}$. For $Ra_S > 10^3$ the wave-number is varied as $a \propto Ra_S^{0.18}$ (all exponents displayed hereinafter in *decimal form* have been estimated from our numerical results using the least squares method). This relation is very close to $a \propto Ra_S^{1/6}$ obtained by Zangrando and Bertram [1985]. The proximity of these results suggests that the dominant reason for the variation of a is the nonlinearity of the salinity profile. The variation of a with Ra_S suggests that the dimensional wavelength depends not only on h (as in the case $a = \text{const}$), but also on the concentration difference ΔS and the fluid properties.

Oscillatory Instability Mode

Essentially, there exists a bifurcation point at $Ra_S \simeq 4000$ from which two separate branches emerge. Figure 3 shows the calculated values of η for both branches. For the oscillatory mode also $\eta = Ra_T^c / Ra_T^0$ but here Ra_T^0 is related to oscillatory mode of the constant properties case (see [Turner, 1979]). It is observed that on both branches the values of Ra_T^c are smaller than the ones for constant properties and gradients. However,

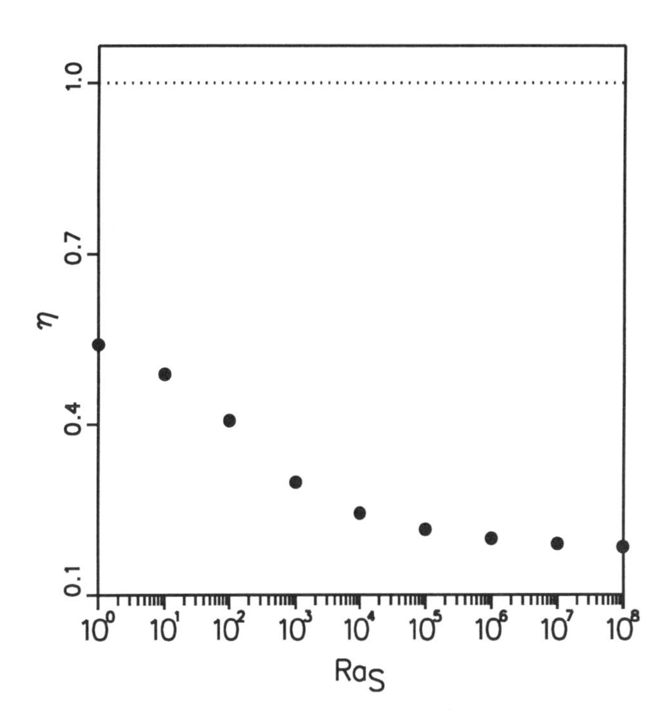

Fig. 1. The ratio $\eta = Ra_T^c / Ra_T^0$ at the onset of direct instability as a function of Ra_S.

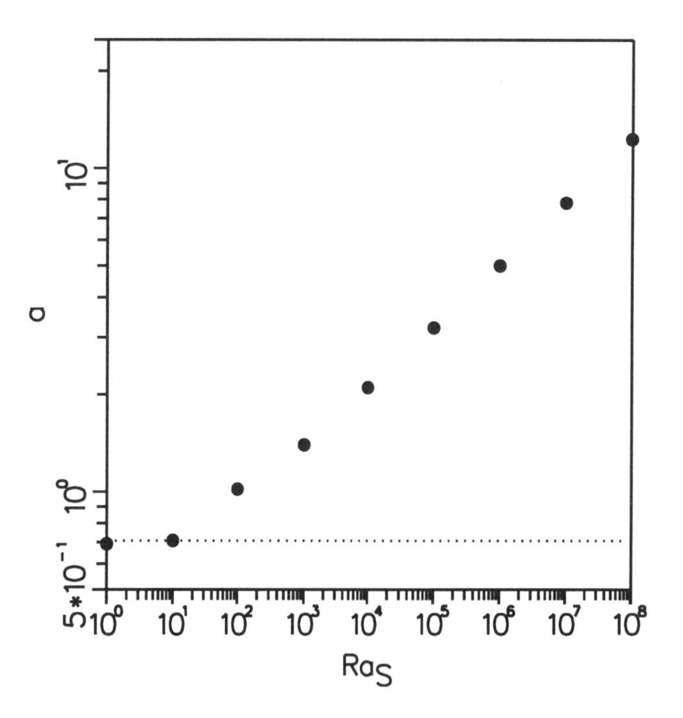

Fig. 2. The wave-number a at the onset of direct instability as a function of Ra_S. The dotted line indicates the case of constant properties and gradients.

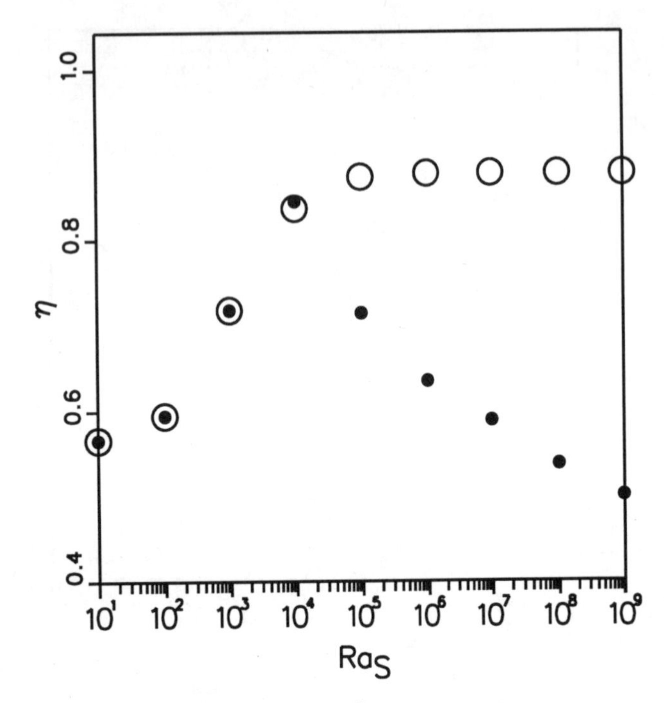

Fig. 3. The ratio $\eta = Ra_T^c/Ra_T^0$ at the onset of oscillatory instability as a function of Ra_S. The two different symbols indicate the different branches.

one of these branches has lower η values and therefore this branch is the critical one.

The graph of the wave-number a at the onset of oscillatory instability is shown in figure 4. Up to the bifurcation point $a \simeq 0.69$, slightly smaller than its value in the case of constant properties, but afterwards it bifurcates into an increasing and a decreasing branch. Basically the most unstable branch (the lower one in figure 3) corresponds here to higher values of a, which varies like $a \propto Ra_S^{0.25}$.

However it seems that at the narrow neighbourhood of $Ra_S = 10^4$ the branch with lower a is the critical one since it has a slightly smaller η. The horizontal wavelength $\lambda = 2\pi h/a$ on the critical branch can be expressed in dimensional variables as

$$\lambda \propto \left(\frac{\nu_r k_T}{g\beta} \cdot \frac{h}{\Delta S}\right)^{0.25} = \xi. \quad (9)$$

This relation indicates that λ does not depend on the layer depth h only, but also depends on the concentration difference (ΔS) and the fluid properties. This result is different from the constant properties and gradients case where $\lambda \propto h$ ([Baines and Gill, 1969]). Furthermore, since $Ra_S = (h/\xi)^4$, for $Ra_S > 1$, $\xi < h$ so that for a fixed layer depth the horizontal wavelength of the unstable motion scales down with increasing Ra_S.

Zangrando and Bertram [1985] have shown that large horizontal wave-numbers in a solar pond with a variable salinity gradient are associated with vertical localization of the unstable zone. This vertical localization gives rise to the formation of isolated thin horizontal layers in the interior of the pond.

It is noteworthy that a scale similar to ξ was found in four other double-diffusive problems investigated by Stern [1960], Walton [1982], Tanny and Tsinober [1988] and Kerpel et al. [1992]. This diversity suggests the universality of this scale in double-diffusive phenomena.

The analysis by Zangrando and Bertram [1985] was carried out under different conditions from the present ones: the fluid properties were constant while the quasi-steady concentration profile was a cubic with z. Nevertheless Zangrando and Bertram [1985] obtained exactly the same relations for the variation of a with Ra_S in the oscillatory mode. Thus it is likely that for these two cases (ours and Zangrando and Bertram's [1985]) the characteristic wavelength of oscillatory instability is insensitive to the detailed shape of the concentration profile and to the variability of the fluid properties.

One may further conjecture that in a double-diffusive horizontal layer with any kind of a smooth non-linear

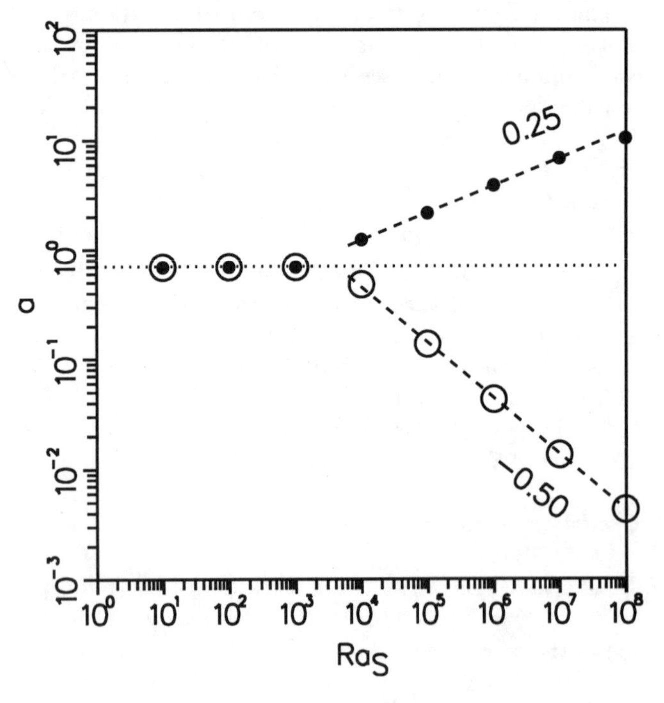

Fig. 4. The wave-number a at the onset of oscillatory instability as a function of Ra_S. The symbols of the different branches correspond to those in figure 3. The dotted line indicates the case of constant properties and gradients.

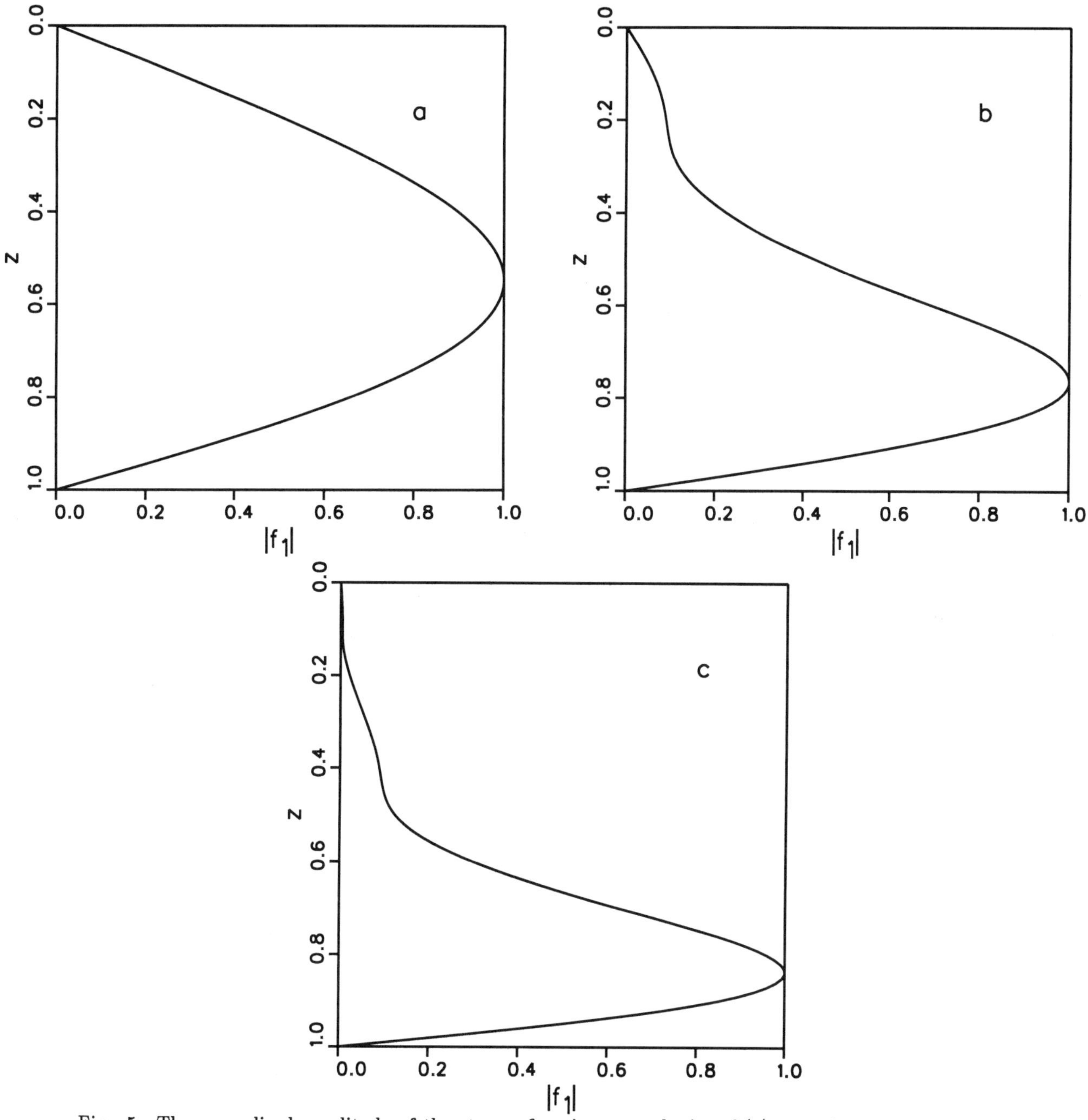

Fig. 5. The normalized amplitude of the stream function perturbation $f_1(z)$, as a function of the layer depth z. (a) $Ra_S = 10^3, a = 0.69$. (b) $Ra_S = 10^5, a = 2.09$. (c) $Ra_S = 10^6, a = 3.88$.

concentration profile, the wave-number of the oscillatory mode would be similar to that shown in figure 4.

The ratio η as shown in figures 1 and 3 indicates, that for both direct and oscillatory modes the fluid layer with variable properties and a variable salinity gradient would be more unstable than that with constant properties and gradients.

Physically it is expected that when the fluid is heated and salted from below, the unstable motion would be most intense in the lower part of the layer, where both the viscosity and the salinity gradient are smaller. This phenomenon is illustrated in figure 5, which shows the vertical distribution of the normalized amplitude of the stream function perturbation within the layer. This parameter is directly related to the vertical component of the velocity perturbation at the instability onset. The corresponding amplitudes of the temperature and concentration perturbations have basically the same shape as that of the stream function.

These observations suggest that up to the bifurcation point, the variable properties and the non-linear concentration profile have almost no effect on the structure of the flow, although they considerably reduce the critical Ra_T as was shown in figure 3. However, for $Ra_S > 4000$ (beyond the bifurcation point), the structure of the flow is essentially changed, and the perturbations become localized in the vertical direction. It should be emphasized that the vertical localization observed in figure 5 is associated with the increase of a or the decrease of the critical horizontal wavelength, a phenomenon which was also observed by Zangrando and Bertram [1985].

Another important stability parameter is the switch from the direct to the oscillatory mode on the plane Ra_T–Ra_S. For constant properties and for the values of Pr and τ used here, this point is at $Ra_S^w \simeq 0.07$ (here the superscript 'w' denotes the switching point), while for variable properties our calculations indicate $Ra_S^w \simeq 11$. This suggests, that within the range $0.07 < Ra_S < 11$, the direct mode can become critical in a stratification, which – on the basis of the constant properties theory – is favorable to oscillatory instability. Such small values of Ra_S are not of practical interest, but some further calculations revealed that the values of Ra_S, over which such a phenomenon exists, depend strongly on τ. For example, for $\tau \simeq 0.4$ (and the same $Pr = 7.2$), the value of Ra_S^w for constant properties is nearly 180, while in the variable properties case $Ra_S^w > 10^4$, a practically achievable value. The latter result was obtained by increasing $k_{S,r}$ and leaving all other properties unchanged. Although we are presently not aware of a realistic double-diffusive system with such a set of properties, it is important to recognize that with variable properties and a non-linear salinity gradient, switching from oscillatory to direct instability can occur at $Ra_S \gg 1$.

The frequency of unstable oscillations is shown in figure 6. For the most unstable branch and high a the frequency increases as $Ra_S^{0.53}$ which is similar to the variation found in [Zangrando and Bertram, 1985]. It is also noteworthy that on this branch the calculated frequency is very close to the one predicted by the theory for constant fluid properties and gradients where $p_i \propto Ra_S^{1/2}$ ([Veronis, 1968]). Thus variable properties and gradients have a very small effect on the frequency of the critical oscillations.

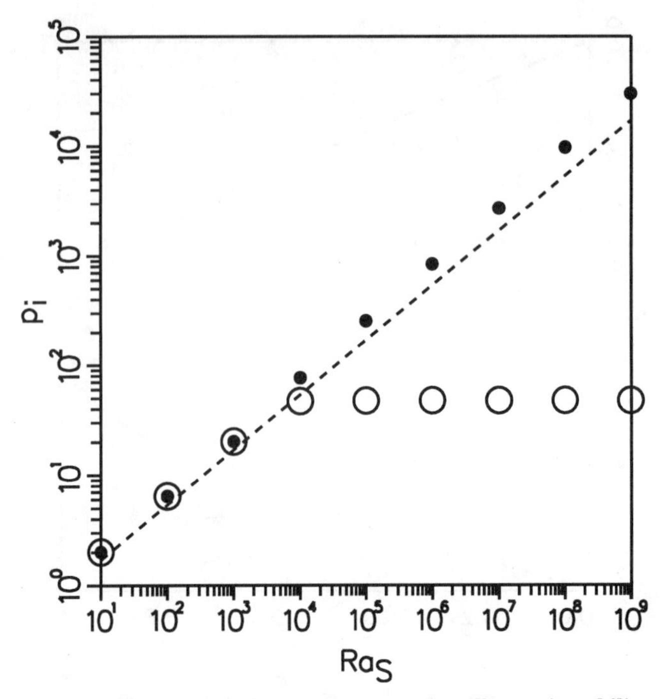

Fig. 6. The frequency p_i at the onset of oscillatory instability as a function of Ra_S. The symbols of the different branches correspond to those in figure 3. The dotted line indicates the case of constant properties and gradients.

CONCLUSIONS

In this work we have analysed the linear stability of a horizontal double-diffusive layer in which the kinematic viscosity and the coefficient of salt diffusivity are temperature dependent, and consequently the steady basic state salinity distribution is non-linear. The main conclusions of the stability analysis are:

- Variable fluid properties and a non-linear salinity gradient reduce the critical Ra_T with respect to its constant properties and gradients value, and vary the wave-number at the onset of instability.

- The stability parameters of the direct mode are characterized by a single critical curve while for the oscillatory mode bifurcation to two separate branches appears (in the investigated case it occurs at $Ra_S \simeq 4000$).

- For both direct and oscillatory modes, the horizontal non-dimensional wave-number a at the instability onset increases with Ra_S in contrast to the case of constant properties and gradients where $a =$ const. Hence, the horizontal wavelength depends not only on the layer depth but also on the concentration difference across the layer and on the fluid properties. For the oscillatory mode the increase of a is associated with vertical localization of the flow.

- The variations of the critical wave-numbers and frequencies with Ra_S on the two branches of the oscillatory mode are similar to those found by Zangrando and Bertram [1985] for a different basic state salinity profile and for constant properties. This implies that the variable properties affect these stability characteristics mainly through the nonlinearity of the basic state profile.

Acknowledgements. This investigation was supported by the Israeli Ministry of Science and Technology under grant No. 3394-1-91 and by the Israeli Ministry of Absorption.

REFERENCES

Baines, P. G., and A. E. Gill, On the Thermohaline Convection with Linear Gradients, *J. Fluid Mech.*, *37*, 289–306, 1969.

Kerpel, J., J. Tanny, and A. Tsinober, On the Secondary Layers in a Stable Solute Gradient Heated From Below, *Fluid Dyn. Res.*, *10*, 141–147, 1992.

Stern, M. E., The 'Salt Fountain' and Thermohaline Convection, *Tellus*, *12*, 172–175, 1960.

Tanny, J., and A. Tsinober, The Dynamics and Structure of Double Diffusive Layers in Sidewall Heating Experiments, *J. Fluid Mech.*, *196*, 135–156, 1988.

Turner, J. S., *Buoyancy Effects in Fluids*, 368 pp., Cambridge University Press, Cambridge, 1979.

Veronis, G., Effect of a Stabilizing Gradient of Solute on Thermal Convection, *J. Fluid Mech.*, *34*, 315–336, 1968.

Walton, I. C., Double Diffusive Convection with Large Variable Gradients, *J. Fluid Mech.*, *125*, 123–135, 1982.

Zangrando, F., and L.F. Bertram, The Effect of Variable Stratification on Linear Doubly Diffusive Stability, *J. Fluid Mech.*, *151*, 55–79, 1985.

V. A. Gotlib, Department of Mechanics and Control, Center for Technological Education Holon, P.O.B. 305, Holon 58102, Israel

J. Tanny, Department of Mechanics and Control, Center for Technological Education Holon, P.O.B. 305, Holon 58102, Israel

A. Tsinober, Faculty of Engineering, Tel-Aviv University, Tel-Aviv 69978, Israel

'Wave-Convection Coupling' in Multicomponent Convection: an Experimental Study

G.O. Hughes[1] and R.I. Nokes

Department of Engineering Science, University of Auckland, Auckland, New Zealand

Large amplitude wave-like motions have been observed in a double-diffusive laboratory experiment. A simple two-layer system was stratified in the 'diffusive' sense using solutions of salt and sugar for the upper and lower layers respectively. 'Waves', corresponding to a local thickening of the otherwise sharp diffusive interface between the two layers, were observed to propagate back and forth along the length of the tank. The horizontal variation in interfacial thickness gives rise to a horizontal differential in flux. This is seen to drive large-scale cellular convective motions in the layers and maintain the quasi-periodic wave motion. Results are presented showing the dependence of time and length scales for the wave motion upon parameters of the diffusive system. Measurements made of the sugar and salt fluxes through the diffusive interface during this and previous studies have shown a systematic deviation from the theoretical predictions of Linden & Shirtcliffe [1978]. It is postulated that the incorporation of the wave motion into such a model would help to reconcile this discrepancy between theory and observations.

1. INTRODUCTION

During a simple double-diffusive laboratory experiment initiated by placing a layer of salt solution on top of a layer of sugar solution, Turner [1974] noted the existence of a 'systematic interfacial wave motion, coupled to the large scale convection in the layers.' It is this large amplitude wave-like motion of the interface which is referred to here as 'Wave-Convection Coupling.'

Although Turner [1974] and Linden & Shirtcliffe [1978] both commented that the presence of such a 'wave' on the interface may increase the fluxes of salt and sugar across the diffusive interface, there has been little further investigation of this phenomenon until recently. In the current study, a series of experiments was conducted using sugar and salt solutions to construct a diffusive system. The methods and apparatus used are described in §3. The coupling of the convection in the layers with the interface led to a quasi-periodic wave motion and measurements were taken of the associated time and length scales. These results are presented in §4, along with a scaling analysis showing the dependence of the time and length scales upon the important experimental parameters. A more extensive series of experiments investigating the same problem has also been undertaken (A.P. Stamp, G.O. Hughes, R.W. Griffiths, R.I. Nokes, in preparation, 1994).

Measurements were also made during this study of the sugar and salt fluxes through the diffusive interface, but these did not agree with the predictions made by the theoretical model of Linden & Shirtcliffe [1978]. This theoretical model is reviewed briefly in §2, followed by a discussion in §5 of how the wave-convection coupling might account for the differences between predictions and experimental measurements.

2. A THEORETICAL MODEL FOR THE DIFFUSIVE INTERFACE

A model for single component convection at high Rayleigh number was proposed by Howard [1964] and extended to the double-diffusive case by Linden &

[1]Now at DAMTP, University of Cambridge, Silver Street, Cambridge CB3 9EW, England.

Double-Diffusive Convection
Geophysical Monograph 94

Shirtcliffe [1978].

Howard [1964] considered a fluid layer heated from below through a metal plate. Molecular conduction was assumed to heat a thin boundary layer immediately above the plate. This buoyant boundary layer was postulated to grow by diffusion until it becomes sufficiently unstable to erupt, releasing all the buoyant fluid from the vicinity of the plate and allowing the cycle to start again. Replacing the plate by an interface which permits heat or solute diffusion forms the basis of the Linden & Shirtcliffe [1978] model for the diffusive system.

The cyclic process is illustrated in Figure 1. As in Figure 1(a), a continuous density profile was supposed to exist, with a high gradient region through the interface. Distributions of the solute components having the higher and lower diffusivities, which give rise to the density profile illustrated, are also shown, these components being denoted by T and S respectively. Diffusion of T and S is assumed to occur independently so that as time passes the T profile develops a buoyant region either side of the interface. At the same time the S component also diffuses, albeit at a slower rate, tending to stabilise an inner region of the unstable boundary layer as shown in Figure 1(b).

Linden & Shirtcliffe [1978] assumed at a time t_* the boundary layer becomes sufficiently unstable, as measured by a critical Rayleigh number criterion, so that all the buoyant material is swept away to drive convection in the layers, while leaving a stable interfacial 'core' region ($-z_1 \le z \le z_1$) untouched. The large gradients of T and S in this region were assumed linear and constant in time, with the flux release of T and S from the boundary layers (denoted by F_T and F_S respectively) arising from molecular diffusion through the core. The core thickness z_1 was assumed to be much greater than that of the unstable boundary layer, leading to several useful analytical results. The two of most importance here are:

(i) the ratio of mass fluxes R_F, given in equation (1), and

(ii) the non-dimensional flux F_T^* defined by Turner [1965], as in equation (2),

$$R_F = \beta F_S / \alpha F_T = \tau^{1/2}, \tag{1}$$

$$F_T^* \equiv \frac{\alpha F_T}{c(\alpha \Delta T)^{4/3}} = \frac{1}{\pi^{1/3}} \frac{\left(1 - \tau^{1/2} R_\rho\right)^{4/3}}{\left(1 - \tau^{1/2}\right)^{1/3}}. \tag{2}$$

Here α and β are coefficients of expansion for the T and S components, c is a constant, τ is the ratio of diffusivities for the components T and S,

$$\tau = \kappa_S / \kappa_T, \tag{3}$$

and R_ρ is the density anomaly ratio,

$$R_\rho = \beta \Delta S / \alpha \Delta T, \tag{4}$$

where $\alpha\Delta T$ and $\beta\Delta S$ are the non-dimensional density differences due to salt and sugar across the interface, respectively.

For the sugar/salt system (τ = 0.33) the predicted mass flux ratio R_F = 0.58 agrees well with the value of R_F = 0.6 measured by Shirtcliffe [1973]. However experimental measurements of the non-dimensional fluxes F_T^* exceed the predicted values by up to a factor of three at low R_ρ and again to a lesser degree at higher R_ρ values. In the latter range, as $R_\rho \to \tau^{-1/2}$, the predicted flux reaches zero because the diffusive flux through the core is not sufficient to generate unstable boundary layers. As a result the model predicts that the diffusive core thickness will increase

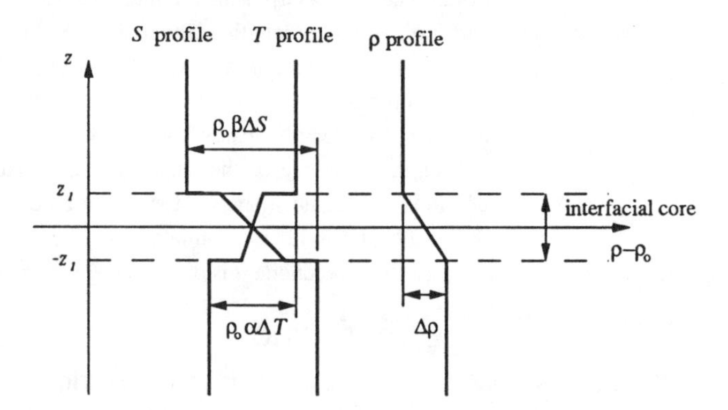

Fig. 1(a). Schematic diagram of the assumed distribution for the T and S components through a diffusive interface at the beginning of the cycle. The horizontal axis has the units of density such that the density profile is represented by $\rho - \rho_0 = \rho_0(\alpha\Delta T + \beta\Delta S)$.

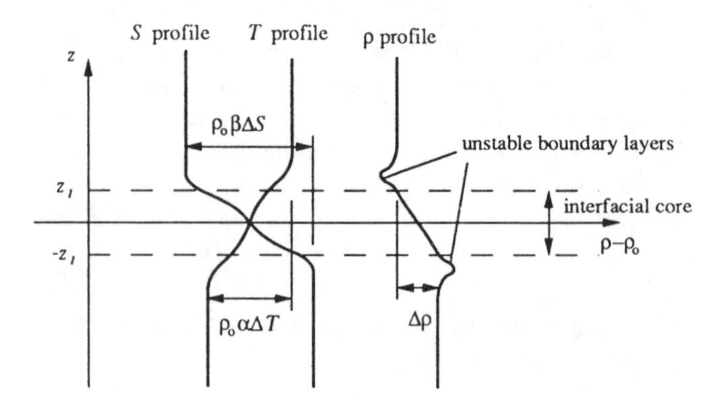

Fig. 1(b). As for (a), some time later into the cycle, during which diffusion of T and S has taken place. The profiles through the gravitationally stable interfacial core between ($-z_1 \le z \le z_1$) are assumed to remain almost constant in time.

steadily in time, contradicting the assumption of a constant gradient core region. Meanwhile at low R_ρ values ($R_\rho \to 1$), Linden & Shirtcliffe [1978] also remark that their assumption, that the unstable boundary layer is much thinner than the core, does not appear to hold.

We return to discuss this model in §5 in light of the observed wave-convection coupling in a diffusive system.

3. EXPERIMENTS

The experiments to be described were all conducted in a tank of horizontal cross-section 25 × 10 cm. A diffusive system was constructed by filling the tank with a layer of sugar solution 12 cm deep and carefully floating a layer of salt solution of the same depth on top.

During preliminary experiments conducted in tanks 150 × 10 × 25 cm deep and 25.4 × 25.4 × 12 cm deep, a very complex three-dimensional system of interfacial waves had been observed. The smaller tank used was selected in an attempt to simplify this complex behaviour by exciting only two-dimensional motions, corresponding to a single interfacial wave.

A dimensional analysis highlighted numerous independent parameters of the system. If the properties of the sugar and salt solutions are assumed to be approximately constant and the tank geometry fixed, then the behaviour should be governed by just two quantities:

(i) a density anomaly ratio R_ρ, already defined in equation (4), and

(ii) the non-dimensional density difference $\alpha\Delta T$, for the unstably distributed T component which drives the convection.

The Rayleigh number Ra defined below, and commonly used to characterise convection, is equivalent to $\alpha\Delta T$ in this instance since the experimental geometry and fluid properties were assumed constant;

$$Ra = g\alpha\Delta T d^3 / \nu\kappa_T, \tag{5}$$

where d is the layer depth, g gravity and ν the fluid viscosity.

During an experiment, a density probe, a conductivity probe and a thermistor were placed in one of the layers and the output voltages continually logged on a computer via an A/D board. The density probe consisted of an extraction tube in the layer, through which fluid was passed into a continuous flow densimeter (Anton Parr DMA 602) before being returned to the same layer, via another tube. As the Rayleigh number during experiments was of $O(10^{10})$ or greater, each layer was assumed to be well mixed by the turbulent convection. Therefore point measurements were used to characterise layer properties. After being corrected for temperature variations, the conductivity and the measured density were substituted in polynomials, similar to those of Ruddick & Shirtcliffe [1979], in order to calculate the concentrations of sugar and salt in the measurement layer.

The equations for conservation of mass and solute [e.g., *McDougall*, 1981] were then applied to give a complete record of the density, sugar and salt concentrations in each layer. Thus the independent parameters, R_ρ and $\alpha\Delta T$, could be calculated along with the fluxes of salt and sugar, F_T and F_S respectively, across the interface.

The flow was observed using the Schlieren method. A parallel beam of light passed through the flow is distorted by gradients in refractive index produced by perturbations in the flow density field. In this diffusive system the buoyant plumes released from the interface and the very high density gradient interfacial region were highlighted against a black background. The wave motion was clearly evident using this visualisation method, permitting accurate measurements of its period to be taken. However the interfacial thickness and fluid velocities could not be accurately measured by this method, although qualitative observations were possible.

4. EXPERIMENTAL RESULTS

The initial density anomaly ratio R_ρ at which an experiment was started was varied between 1.05 and 1.5 by adjusting the densities of the salt and sugar solutions used to set up a diffusive system. Experiments were of the 'rundown' type, being started with fixed quantities of salt and sugar which are redistributed through the system during an experiment. Because T is the more diffusive component, $\alpha\Delta T$ decreases more rapidly than $\beta\Delta S$, and as a result R_ρ values increase from their initial value. This allowed experimental behaviour to be investigated for R_ρ ranging between 1.05 and 1.8. Accordingly a six-fold variation in $\alpha\Delta T$ was also obtained and this information is summarised on the $\alpha\Delta T$–R_ρ regime plot shown in Figure 2 for the experiments (see also Table 1). Experimental measurements of these parameters were generally stopped after 20 hours or more had elapsed after beginning an experiment, when any observable wave motion or activity had decayed.

The diffusive interface between the layers of salt and sugar was for the most part very thin. Accordingly the gradients in solute concentration and of density are very large within the interface, giving rise to large diffusive fluxes through the core which drive the turbulent convection in each layer. However this high Rayleigh number convection was seen to become organised and coupled with the motions or 'waves' on the diffusive interface. The waves correspond to a local thickening of the otherwise thin

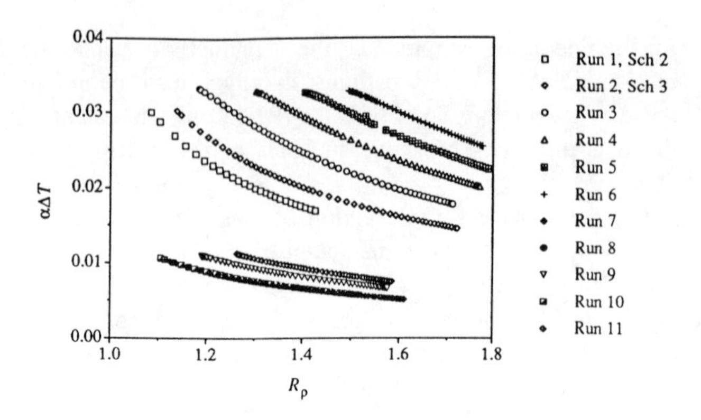

Fig. 2. $\alpha\Delta T$ versus R_ρ regime plot for the experiments conducted, the nominal initial and final conditions ($\alpha\Delta T$, R_ρ) being listed in Table 1 for later reference.

TABLE 1. Summary of Experiments

Experiment	Initial R_ρ	Initial $\alpha\Delta T$	Final R_ρ	Final $\alpha\Delta T$
Run 1	1.05	0.033	1.426	0.0169
Run 2	1.1	0.033	1.722	0.0145
Run 3	1.2	0.033	1.712	0.0178
Run 4	1.3	0.033	1.772	0.0202
Run 5	1.4	0.033	1.790	0.0225
Run 6	1.5	0.033	1.777	0.0294
Run 7	1.05	0.01	1.611	0.0051
Run 8[a]	1.1	0.01		
Run 9	1.2	0.01	1.576	0.0066
Run 10	1.1	0.01	1.517	0.0075
Run 11	1.3	0.01	1.585	0.0075
Sch 2[a]	1.05	0.032		
Sch 3[a]	1.1	0.032		

[a]Refers to experiments where the complete time histories of $\alpha\Delta T$ and the layer densities was not available. The partial record of $\alpha\Delta T$–R_ρ plotted in Figure 2 for Run 8 overlies that of Run 10, which was started with the same initial conditions. For the purposes of plotting measured wave lengthscales and periods in Figures 6(a), 6(b), 7(a) and 7(b), the $\alpha\Delta T$–R_ρ data from Run 10 was therefore used. A similar assumption was used in the case of Sch 2 and Sch 3, which were started with the same initial conditions as Run 1 and Run 2 respectively.

diffusive interface and were observed to propagate back and forth along the length of the tank in a quasi-periodic manner.

The manner in which the coupling between convection in the layers and the interfacial waves operates is considered here in terms of how the waves modify the interfacial structure. At points where the interface is thicker, diffusive fluxes will be small as the local solute gradients are reduced in comparison to the thin sections of the diffusive interface.

It is apparent for the wave system shown in Figure 3 that the horizontal flux differential will modify the residual large-scale circulation. Observations show a particularly strong release of buoyant plumes near the front of the wave as unstable fluid from either side of the thin interface is transported towards the front by this residual circulation. Almost immediately after the situation depicted in Figure 3, these strong plumes divide the large single cells into two, resulting in a reversal in circulation above and below the wave. A second and much weaker cell is maintained in each layer either side of the thin interface. These two weaker cells remain until the state shown in Figure 4, which evolves from Figure 3 as a combination of gravitational forces and the stress exerted on the thicker interface by the dominant cell causes the wave to propagate to the left. The situation shown in Figure 5 is reached as these forces continue to drive the thicker interface towards the opposite wall, causing the weaker cell to disappear. The interface at the right of Figure 4 seems to be thinned in response to the stress exerted on it by the dominant circulation cell.

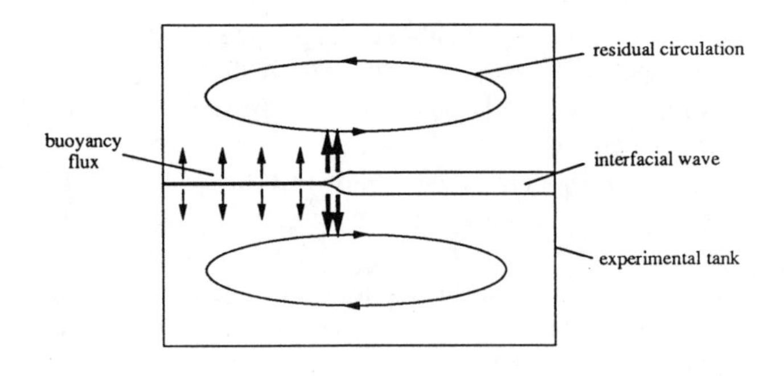

Fig. 3. Schematic side view of experimental tank. The interfacial wave structure is shown, sandwiched between an upper layer of salt solution and a lower layer of sugar solution.

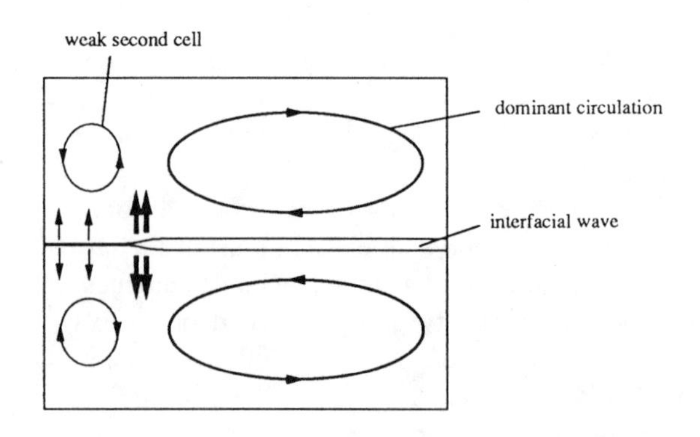

Fig. 4. As for Figure 3, except at a time approximately quarter of a cycle later. The strong buoyancy flux near the front of the wave has tended to form a weak second cell ahead of the wave while reinforcing the dominant circulation.

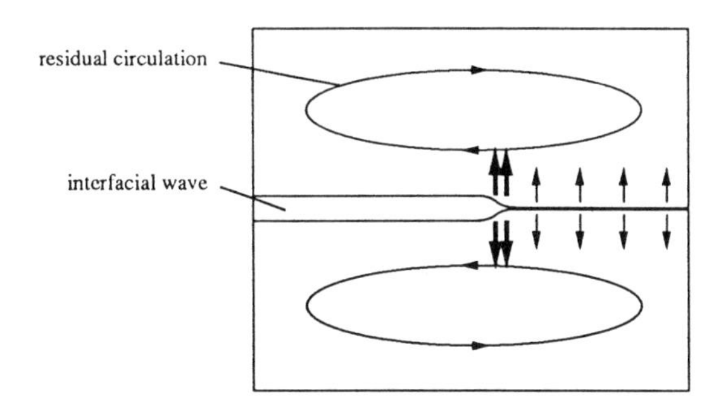

Fig. 5. A mirror image of Figure 3, halfway through the cycle when the wave has reached the opposite wall of the tank.

The process described above is now reversed, forcing the wave to propagate back to the situation depicted in Figure 3. Hence the cycle can begin again.

Measurements were made of the length of the thickened interface and also of the period of this cyclic motion $\mathcal{T}$ to determine how these quantities evolved over time and their dependence on the experimental parameters. This data is plotted in Figures 6(a) and 6(b) as a function of the parameter R_ρ. It is evident that the strongest wave-convection coupling, as measured by shorter cycle periods $\mathcal{T}$, occurs at low R_ρ and high $\alpha\Delta T$; the flux-driven convection being stronger by equation (2) when the driving density difference ($\alpha\Delta T$) is greatest and when the stability of the interface is least ($R_\rho \rightarrow 1$). Stronger coupling also coincides with the regime in which shorter wave lengthscales are observed because the larger convective stresses on the thickened interface are able to force a greater compression of the wave against the wall. As the system runs down, less convective energy is available to drive the wave and so the period becomes longer and the lengthscale is increased.

To examine these claims in a more formal fashion we form non-dimensional quantities for the period and the lengthscale, namely $\mathcal{T}u/2L$ and l/L, where u and L are suitable velocity and length scales. In keeping with the postulate that the wave is coupled to the convection, u is chosen to scale like the convective velocity scale u' of Hunt [1984],

$$u \sim u' = \left(\frac{g\alpha F_T d}{\rho_0} \right)^{1/3}, \tag{6}$$

and L was taken to be the length of the large cells shown in Figure 3 and Figure 5, over which the wave motion took place. In nearly all of the experiments a single cell was observed either side of the interface, hence the cell length L was just the length of the tank. However in the case of Run 7, it was taken to be half of this value since two cells formed along the tank length. The reason for the different behaviour in this experiment is not clear.

Plots of the rescaled wavelength and period are shown in Figures 7(a) and 7(b). The wavelength l appears to be well described by a relation of the form,

$$\frac{l}{L} = \mathcal{F}\left(R_\rho\right), \tag{7}$$

with no dependence on $\alpha\Delta T$. If a similar relation is assumed for the period $\mathcal{T}$,

$$\frac{\mathcal{T}u'}{2L} = \mathcal{G}\left(R_\rho\right), \tag{8}$$

then from equations (2) and (6) we see the following dependence on $\alpha\Delta T$ may be deduced;

$$\mathcal{T} \sim (\alpha\Delta T)^{-4/9}. \tag{9}$$

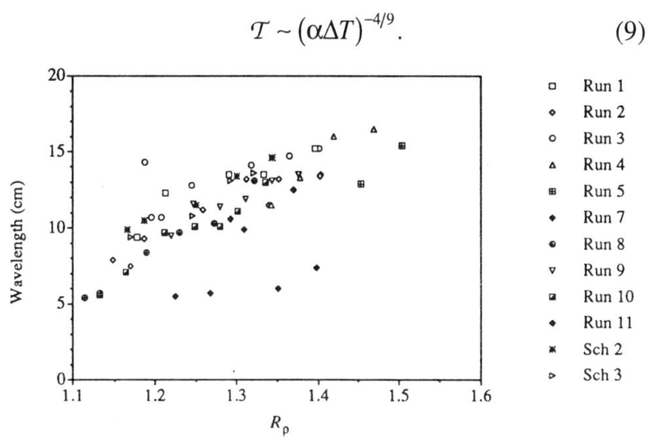

Fig. 6(a). Plot of the measured length of thickened section l versus R_ρ. The measurement was made when the wave appeared to be at maximum compression against the tank wall.

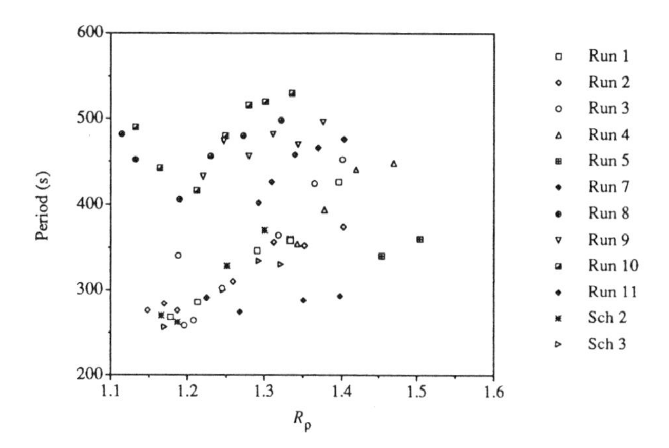

Fig. 6(b). Plot of the measured period $\mathcal{T}$ of the cyclic wave motion versus R_ρ.

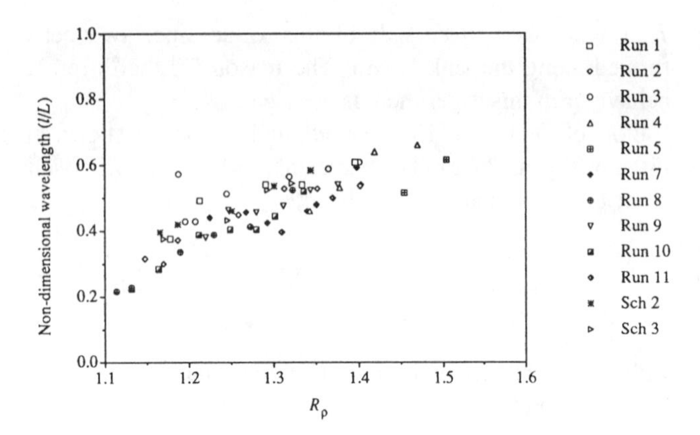

Fig. 7(a). Plot of the scaled length l/L from Figure 6(a) versus R_ρ.

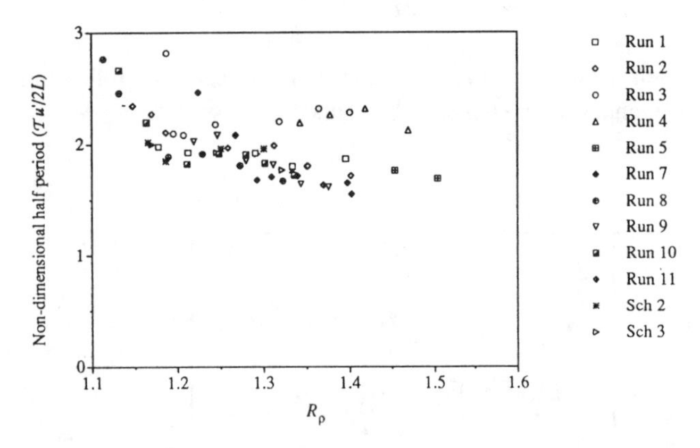

Fig. 7(b). Plot of the rescaled period $\mathcal{T}u'/2L$ from Figure 6(b) versus R_ρ.

Although Runs 3 and 4 show some systematic deviation from the rest of the data in Figure 7(b) for $1.3 \lesssim R_\rho \lesssim 1.5$, the assumption in equation (8) collapses the majority of points onto a single curve. This scaling is equivalent to assuming that the period $\mathcal{T}$ only depends upon $\alpha\Delta T$ insofar as the driving flux F_T changes with $\alpha\Delta T$. Under the coupling model proposed, where it is the flux that drives the convection, this would seem reasonable.

The solute fluxes were also measured during these experiments. The non-dimensional flux F_T^* is plotted as a function of R_ρ in Figure 8. Our experimental measurements are compared with those predicted theoretically by equation (2) and a numerical model. However the previous measurements of F_T^* found by Shirtcliffe [1973] are not plotted. This is because we were unable to reconcile them with the non-dimensional values of F_T^* obtained from the current experiments, despite the measured time variations of $\alpha\Delta T$, and therefore the dimensional fluxes F_T, being in good agreement. The numerical simulation was based closely on the ideas and criteria for eruption of the unstable boundary layers proposed by Linden & Shirtcliffe [1978] and is discussed in greater detail in §5. Apart from the experimental points connected by dotted lines, which are due to transient effects when an experiment was started, it can be seen that the theoretical/numerical results compare reasonably well with experiment in the range $1.3 \lesssim R_\rho \lesssim 1.5$. For $R_\rho \lesssim 1.3$, experimentally determined fluxes are up to a factor of three higher than predicted, and for $R_\rho \gtrsim 1.6$, where the theoretical model becomes invalid, the experimental fluxes are again larger.

5. WAVE-CONVECTION COUPLING AND A MODEL FOR THE DIFFUSIVE INTERFACE

Observations, together with the results presented in §4, show that wave-convection coupling is most active at low R_ρ. This seems to correspond with the region in which the observed flux discrepancy between theory and experiment is greatest and suggests that the coupling might account for this. We discuss qualitatively how this might occur.

It was found that at low R_ρ the traditional picture of the diffusive interface requires modification. A numerical model which simulated the diffusive interface using a time-stepping finite difference scheme to solve the one-dimensional diffusion equation for the profiles of T and S was implemented. No attempt was made to incorporate two-dimensional effects, such as a wave motion, since a direct comparison with the Linden & Shirtcliffe [1978] theory was desired. The model was initialised with step discontinuities of ΔT and ΔS in the T and S distributions at the interface

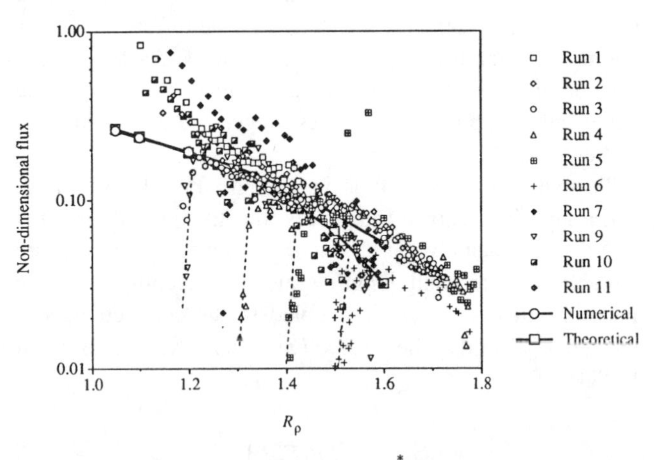

Fig. 8. Plot of non-dimensional flux F_T^* versus R_ρ. Experimental measurements of F_T are non-dimensionalised using equation (2) and the dotted lines show how the measured flux F_T^* approaches the true curve as transient effects decay after an experiment is set up. The solid line through the squares represents the theoretical curve obtained from equation (2), while the solid line through the circles are the predictions made by the numerical model described in the text.

with the diffusion equation being resolved at every time step. The same Rayleigh number criterion defined by Linden & Shirtcliffe [1978] for instability of the boundary layers was used to determine the time of eruption of the boundary layer, at which stage all buoyantly unstable material was removed. The T and S profiles remaining in the 'diffusive core' after eruption were then used as initial conditions for the next cycle. Several cycles were simulated numerically in this manner until a steady cyclic behaviour was established.

Two observations are particularly worthy of note at low R_ρ. The boundary layers, defined by Linden & Shirtcliffe [1978] as that region which is gravitationally unstable with respect to the layer above/below it, were considerably thicker than the diffusive core sandwiched between them. Secondly, the diffusive core could not be considered to be of constant thickness with fixed linear solutal gradients through it. These two points are not unrelated however.

If the boundary layers were much thinner than the core, then the accumulation and release of unstable material in the boundary layer over a cycle occurs almost independently of the solutal profiles in the core. This is not true however when the boundary layers are thicker than the diffusive core. The solutal profiles are then significantly modified over the course of a cycle, and this is apparent in Figure 9 where the salinity gradient at the centre of the core from numerical simulations is plotted as a function of time. The discontinuity in the T profile at the edge of the core immediately after eruption (t = 0) causes a very large increase in the gradient, which then gradually decreases as diffused unstable material accumulates in the boundary layers during the cycle. The core gradient can also be used as an indicator of the core thickness since the two quantities should vary almost inversely. Hence it can be inferred that neither the core thickness nor the diffusive flux through the core remains constant during the cycle.

Despite this behaviour violating some assumptions made by Linden & Shirtcliffe [1978] the differences in fluxes predicted by their theoretical model and our numerical simulation are not great, as can be seen in Figure 8. However we believe that these modifications to the traditional picture play a key role in increasing the flux in the presence of an interfacial wave. Firstly, a wave produces thick and thin sections of the interface, the contrast between which is most evident at low R_ρ. Hence the solutal gradients in the thin section will already be higher than might be predicted from theory. Secondly, the organised convection cells resulting from the wave-convection coupling will tend to have a 'sweeping' action, removing unstable material from the boundary layers either side of the thin interfacial section. The combination of these two effects will therefore

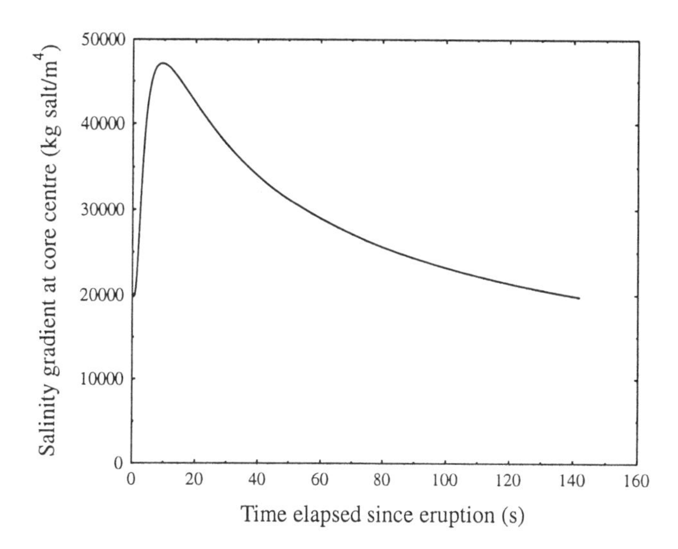

Fig. 9. Example plot of how the salinity gradient at the core centre varies over a cycle between eruptions. The data shown was obtained from a numerical simulation of the diffusive interface based closely on the Linden & Shirtcliffe [1978] model for $R_\rho = 1.2$ and $\alpha\Delta T = 0.03$.

serve to maintain very high solutal gradients in the thin section of the interface, because unlike the cycle in Figure 9, unstable material will be prevented from accumulating in the boundary layer and thus lowering the gradient. It is proposed that this mechanism could therefore explain the significant enhancement of fluxes observed at low R_ρ.

6. CONCLUSIONS

A series of experiments in a two-layer sugar/salt diffusive system have been described, where the convection has been observed to force interfacial waves. The strength of the coupling between the convection and the wave is dependent on the fluxes across the interface, being strongest at low values of R_ρ and at high values of $\alpha\Delta T$.

Although the double-diffusively driven convection was turbulent, an organised circulation cell was observed to form in each layer. The action of such cells caused regions of the diffusive interface to thicken, taking on the appearance of an interfacial wave, while other parts of the interface became thinner. However as the fluxes through the thinner section of the diffusive interface were greatest, a buoyancy flux was generated in this region to oppose and subsequently reverse the organised circulation which originally produced the thinner interface. This feedback mechanism could then act to thin regions of the interface which were previously thickened, giving rise to a quasi-periodic wave motion. Such behaviour was observed in a diffusive system naturally without any need for artificial forcing. The period and lengthscales associated with this wave motion have been measured and their dependence upon the parameters R_ρ and $\alpha\Delta T$ found experimentally.

Qualitative evidence has been presented to show how wave-convection coupling is a possible mechanism to explain the discrepancy between experimental measurements and the theoretical predictions of fluxes made by Linden & Shirtcliffe [1978]. Although traditional theory assumes the interface to consist of a steady diffusive core bounded by relatively thin boundary layers, this does not appear to be the case in the parameter regime of current interest. The existence of an interfacial wave creates very thin regions of the interface with high local solutal gradients and fluxes which are maintained by convective 'sweeping' of the boundary layers either side, consistent with observations.

Acknowledgements. One of the authors (GH) acknowledges the financial support of IPENZ and McConnell Dowell Corporation Ltd while carrying out this study.

REFERENCES

Howard, L.N., Convection at high Rayleigh number, in *Proc. 11th Int. Cong. Appl. Mech.*, Springer Verlag, Berlin, pp. 1109-1115, 1964.

Hunt, J.C.R., Turbulence structure in thermal convection and shear-free boundary layers, *J. Fluid Mech., 138,* 161-184, 1984.

Linden, P.F., & Shirtcliffe, T.G.L., The diffusive interface in double-diffusive convection, *J. Fluid Mech., 87,* 417-432, 1978.

McDougall, T.J., Double-diffusive convection with a non-linear equation of state, Part 1: The accurate conservation of properties in a two-layer system, *Progr. Oceanogr., 10,* 71-89, 1981.

Ruddick, B.R., & Shirtcliffe, T.G.L., Data for double-diffusers: Physical properties of aqueous salt-sugar solutions, *Deep-Sea Res., 26A,* 775-787, 1979.

Shirtcliffe, T.G.L., Transport and profile measurements of the diffusive interface in double-diffusive convection, *J. Fluid Mech., 57,* 27-43, 1973.

Turner, J.S., The coupled turbulent transports of salt and heat across a sharp density interface, *Int. J. Heat and Mass Trans., 8,* 759-767, 1965.

Turner, J.S., Double-diffusive Phenomena, *Ann. Rev. Fluid Mech., 6,* 37-56, 1974.

G.O. Hughes & R.I. Nokes, Department of Engineering Science, University of Auckland, Private Bag 92019, Auckland, New Zealand.

A Quasi-Lagrangian Analysis of Particle Motion in Different Kinds of Background Stratification

A.Y. Benilov[1] and D.W. Collins[2]

Davidson Laboratory, Stevens Institute of Technology, Hoboken, New Jersey

The dynamic properties of vertical motion of a fluid particle in a fluid stratified by heat and salt are studied. The stratification is favorable for development of double diffusive convective motion. The densities of the fluid particle and the background fluid are governed by the same equation of state, which for the present problem is linear. The particle can exchange heat and salt with the background fluid, which is why buoyancy forces arise, leading to vertical motion. With the addition of viscosity as a resistive force, the motion in an arbitrarily stratifed fluid is described by a nonlinear ordinary differential equation of 3rd order. This equation has two asymptotic linear forms. The first describes the motion of the particle with small values of displacement compared with the characteristic vertical scale of background temperature and salinity changes. This asymptotic solution corresponds to background stratification with constant Brunt-Vaisala frequency. The second asymptotic solution corresponds to the case where the particle moves with such value of displacement that it escapes from local non-uniformity and arrives in an area of uniform heat and salt distribution. This case corresponds to zero Brunt-Vaisala frequency. In the first case the characteristic equation has two complex and one real root or three real roots. Our model predicts the stability boundaries for the salt finger and diffusive oscillatory regimes. In the second case all roots are real and negative, and motion far from the interfacial layer will decay. When displacement of the fluid particle is of the order of interface thickness, we get diffusive oscillatory convection tending to stable nonlinear oscillations or limit cycles. By increasing the first integral of our system we can transform these limit cycles to exponentially decaying solutions. Numerical experiments will be discussed.

INTRODUCTION

Historically, the analysis of the stability of thermohaline systems has proceeded from earlier work on convection due to heat alone [Rayleigh 1916, Pellew and Southwell 1940]. In this, it was generally assumed that the fluid was of finite depth and was constrained between two horizontal boundaries. By applying the governing Navier-Stokes equations in the Boussinesq approximation to a small perturbation, one was led to relations which depended on a Rayleigh number, or the ratio of buoyancy forces in a fluid to the retarding effects of viscosity and diffusion. When the Rayleigh number was exceeded, convection would follow.

Double-Diffusive Convection
Geophysical Monograph 94

Extended into the domain of two stratifying components by Stern [1960], Veronis [1965], Baines and Gill [1969] and others, the form of the results changed little, except that now a new parameter, the saline Rayleigh number, had to be reckoned with. Over the past twenty years, much fruitful work has been done exploring the boundaries of stability in terms of Rayleigh numbers. Descriptions of the onset of both stationary and oscillatory instability have been found [Turner 1973], and using truncated models this has been carried into the nonlinear domain [Veronis 1968, Huppert and Moore 1976, Moore et al 1983], with aperiodic or chaotic behavior at large values of the driving force. A variety of geometries have been considered, including sidewall heating, sloping boundaries, and lateral intrusions [Manins 1976, Turner 1978, Paliwal and Chen 1980]

It is somewhat surprising, however, that the literature scarcely contains mention of the behavior in time of an individual fluid particle. Surely this is a valid approach, since the bulk properties of fluids can be viewed as the sum of the individual fluid particles which compose it. If an analysis of motion in a stratified fluid were made based on an individual fluid particle, what form would it take? More importantly, what would be the advantages, in our understanding, of doing it?

To our knowledge, such an enterprise has never been undertaken. But we can expect that fluid particle motion will not depend on boundaries when particle displacements are much less than the distance to the boundaries. This situation can be observed for many real oceanographic cases. Stability solutions will depend only on local properties

In what follows we will describe the effects of forces on a fluid particle in different kinds of background stratification, in a system where two components contribute to density. We will consider vertical motions of the particle in an arbitrary stratified fluid. Linear stratification as a particular case of our analysis plays a fundamental role for the classification of particle behavior. We compare our results with those from a Eulerian approach, in particular with some of the results of Baines and Gill (1969). Nonlinear effects are discussed, and computer simulations are presented which describe a case modeling a situation when a fluid partical moves through an interface separating two quasi-homogeneous layers. Previous studies of nonlinear stratification have been limited to an examination of specific forms of polynomial density distribution (Walton, 1982; Zagrando and Bertram, 1985)

STATEMENT OF PROBLEM

We take as density components, for convenience and to better fix our ideas, heat and salt, denoted by T and S, in a fluid that is unbounded both vertically and horizontally. The density of the fluid particle and the density of the background fluid share the same equation of state. We assume a body force g on the particle, but no external forces apart from those created by the density distribution. Our analysis is for vertical motion only.

We consider a fluid in which the temperature T and salinity S are functions of a vertical coordinate z taken as positive downwards, so that $T = T(z)$, $S = S(z)$. We set the origin at $z = 0$ without loss of generality. The values of T and S at the origin will be termed the reference levels for those quantities, and are identified by a zero subscript. Thus $T_0 = T(0)$, $S_0 = S(0)$. Deviations from the reference levels are indicated by a tilde, $\widetilde{T}$ $\widetilde{S}$, so that the background distribution of temperature and salinity can be written as

$$\left.\begin{aligned} T &= T_0 + \widetilde{T}(z) \\ S &= S_0 + \widetilde{S}(z) \end{aligned}\right\}. \tag{1}$$

Temperature and salinity profiles which are conducive to double diffusion can be portrayed as pair-wise increasing or decreasing curves, as shown in Figure 1. We will investigate the forces acting on a fluid particle embedded in this distribution. As illustrated in Figure 2, the particle will be characterized by a temperature T_1 and salinity S_1 which are functions of time t, and are defined in the following form:

$$T_1(t) = T_0 + T_e(t), \quad S_1(t) = S_0 + S_e(t), \tag{2}$$

where T_e , S_e represent the local deviations of these properties in the neighborhood of the particle.

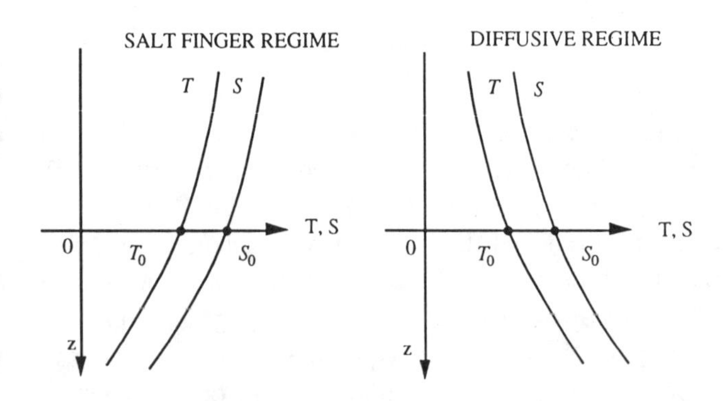

Figure 1. Temperature and salinity profiles.

To establish the basic equations governing particle motion, we write the equation of state for the background fluid density and for the fluid particle as, respectively,

$$\rho(z) = \rho_0 + \widetilde{\rho}(z) \text{ and } \rho_1(t) = \rho_0 + \rho_e(t), \tag{3}$$

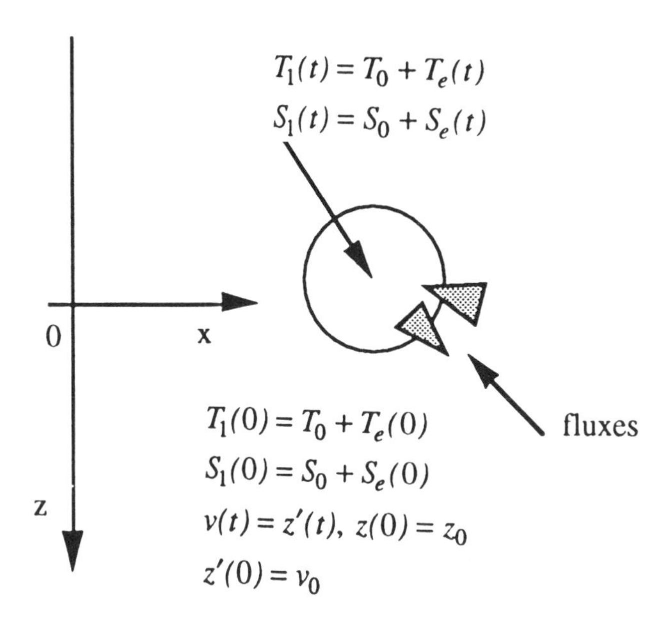

Figure 2. A fluid particle embedded in the distribution given in Figure 1.

or

$$\rho(z) = \rho_0 - \alpha_T \tilde{T}(z) + \beta_S \tilde{S}(z)$$

$$\text{and} \tag{4}$$

$$\rho_1(t) = \rho_0 - \alpha_T T_e(t) + \beta_S S_e(t),$$

where α_T and β_S are the coefficients of thermal expansion and haline contraction. When the Brunt-Vaisala frequency is given by

$$N^2(z) = -\alpha_T d_z \tilde{T}(z) + \beta_S d_z \tilde{S}(z) > 0, \tag{5}$$

heat and salt transfer between the particle and the background fluid can be represented in the form

$$d_t T_e = a(\tilde{T} - T_e), \tag{6}$$

$$d_t S_e = b(\tilde{S} - S_e), \tag{7}$$

in which we introduce new quantities α, a heat transfer coefficient and b, a salt transfer coefficient, having dimensions of frequency and assumed constant. The left hand side of equations (6) and (7) represents the changes with time in the local deviations of properties in the neighborhood of the particle, while the right hand side is the flux of those properties arising from the difference between the background and fluid particle.

Assume the particle has a density $\rho_1 \neq \rho$. The local force balance will depend on local position, which is a function of time. Let us therefore introduce a new variable Z(t) which describes an immediate position of the particle. To describe viscosity effects, we write the viscous force $\mathbf{f}_v$ as $\mathbf{f}_v = -\alpha_v\ \mathbf{v}$ where $\mathbf{v}$ is fluid particle velocity, α_v is a viscous drag coefficient, and the sign is understood to show that the viscous force acts opposite to velocity. In the present case $\mathbf{v} = v_z = d_t Z$. Taking these observations into account and changing everywhere in the background states the variable z to Z(t), the local balance of forces results in

$$Z'' + a_v Z' = \rho_1^{-1}(\rho_1 - \rho) g \approx \rho_0^{-1}(\rho_1 - \rho) g, \tag{8}$$

where we require that

$$(\tilde{\rho}/\rho) << 1, \quad (\rho_e/\rho) << 1, \tag{9}$$

and the primes indicate differentiation with respect to time. We now have three equations. One describes particle motion and two describe heat and salt transfer with the background fluid. For the second order dynamic problem the initial conditions are $Z(0) = Z_0$, $Z'(0) = v_0$, while for the transport problem they are $T_e(0) = T_{e0}$, $S_e(0) = S_{e0}$. But using the equation of state for ρ_e in terms of T_e, S_e we can rewrite (8) as

$$Z'' + a_v Z' = \rho_o^{-1} g\left(-\alpha_T (T_e - \tilde{T}) + \beta_S (S_e - \tilde{S})\right). \tag{10}$$

We now have a fourth order system governing the motion of a fluid particle, consisting of equations (6), (7) and (10). If we express the temperature and salinity differences through the time derivatives of T_e and S_e, equation (10) becomes

$$Z'' + a_v Z' + \rho_o^{-1} g\left(-\alpha_T a^{-1} d_t T_e + \beta_S b^{-1} d_t S_e\right) = 0, \tag{11}$$

which then can be integrated to yield

$$Z' + a_v Z + \rho_o^{-1} g\left(-\alpha_T a^{-1} T_e + \beta_S b^{-1} S_e\right) = E = \text{constant} \tag{12}$$

From this we see the constant of integration is entirely defined by the initial conditions. Once found, this constant determines motion for all time forward. We will see shortly to what extent the value of E affects the behavior of the system.

First, we seek to eliminate dependence on the local values T_e and S_e, using (10) and (12) with the requirements that $a \neq b$. After manipulation, we obtain a single equation describing particle motion:

$$Z''' + (a_v + a + b)Z'' + \left(N^2(Z) + a_v a + a_v b + ab\right)Z' + a_v abZ + \\ + \rho_o^{-1} g\left(-b\alpha_T \tilde{T}(Z) + a\beta_S \tilde{S}(Z)\right) = Eab. \qquad (13)$$

Thus, we have a third order nonlinear equation, where our only assumption has been that N^2 (Z)>0. We can solve this equation with the initial conditions $Z(0) = Z_0$, $Z'(0) = v_0$, $Z''(0) = A_0$, where any two conditions can be chosen freely, while the remaining one is determined by equation (10) and (12).

LINEAR STRATIFICATION: STABILITY OF SOLUTIONS

To analyze equation (13), we begin with small values of particle displacement. If we Taylor expand N^2, $\tilde{T}$ and $\tilde{S}$ around the reference levels the gradients can be assumed constant. By writing $\tilde{T}(z)$, $= z\, d_z\tilde{T}$, $\tilde{S}(z) = z\, d_z\tilde{S}$, we introduce a new constant parameter

$$\tilde{N}_*^3 = \rho_o^{-1} g\left(-b\alpha_T d_z\tilde{T} + a\beta_S d_z\tilde{S}\right), \qquad (14)$$

which allows us to reformulate (13) more conveniently:

$$Z''' + (a_v + a + b)Z'' + \left(\mathbf{N}^2 + a_v a + a_v b + ab\right)Z' \\ + \left(\tilde{\mathbf{N}}_*^3 + a_v ab\right)Z = Eab. \qquad (15)$$

We nondimensionalize as $t = \tau N^{-1}$, $a_v = \alpha_v N$, $a = \alpha N$, $b = \beta N$, $z = \eta L$, since we want to refer all our parameters to the Brunt-Vaisala frequency, the outer time dimension of our system. Furthermore, what we will call temperature and salinity density fractions in n_T and n_S defined implicitly by

$$\rho_o^{-1} g\alpha_T d_z\tilde{T} = n_T N^2, \rho_o^{-1} g\beta_S d_z\tilde{S} = n_S N^2 \qquad (16)$$

Then we observe that $E = LNC$, $\tilde{N}_*^3 = N_*^3 N^3$, $N_*^3 = \alpha + (\alpha-\beta)n_T$, where C is a new constant, and in particular that

$$-n_T + n_S = 1, \qquad (17)$$

which is just the normalization of

$$N^2 = \rho_o^{-1} g\left(-\alpha_T d_z\tilde{T} + \beta_S d_z\tilde{S}\right). \qquad (17')$$

Equation (15) becomes

$$\eta''' + (\alpha_v + \alpha + \beta)\eta'' + (1 + \alpha_v\alpha + \alpha_v\beta + \alpha\beta)\eta' + \\ \left(N_*^3 + \alpha_v\alpha\beta\right)\eta = C\alpha\beta, \qquad (18)$$

which is our basic equation for motion in a linearly stratified fluid.

We search for solutions of the form $\eta = \Phi\, exp(\lambda t)$, where Φ is a constant determined from initial conditions and λ is an exponent to be found by solving the characteristic equation

$$\lambda^3 + (\alpha_v + \alpha + \beta)\lambda^2 + (1 + \alpha_v\alpha + \alpha_v\beta + \alpha\beta)\lambda + \\ N_*^3 + \alpha_v\alpha\beta = 0. \qquad (19)$$

This cubic equation can have either three real roots, or one real root and two complex conjugates. If any of the roots has a positive real part, the system described by (18) is asymptotically unstable, and any disturbance to it will grow in time . Conversely, if all roots have only negative real parts, the system is asymptotically stable.

Signs of the roots can be determined any one of several classical methods. We choose the Routh-Hurwitz method for its ease analytically, and we form the determinants Δ_i in the standard way (for example, see [Korn and Korn, 1961]):

$$\Delta_1 = 1 > 0,$$

$$\Delta_2 = (\alpha_v + \alpha + \beta)(1 + \alpha_v\alpha + \alpha_v\beta + \alpha\beta) - \left(N_*^3 + \alpha_v\alpha\beta\right)$$

which is > 0 if

$$N_*^3 < (\alpha_v + \alpha + \beta)(1 + \alpha_v\alpha + \alpha_v\beta + \alpha\beta) - \alpha_v\alpha\beta,$$

$$\Delta_3 = (N_*^3 + \alpha_v\alpha\beta)[(\alpha_v + \alpha + \beta)(1 + \alpha_v\alpha + \alpha_v\beta + \alpha\beta) - (N_*^3 + \alpha_v\alpha\beta)]$$

which is > 0 if $N_*^3 + \alpha_v\alpha\beta > 0.$ (20)

The condition for all roots of the characteristics equation to have only negative real parts is that determinants Δ_i corresponding to roots λi be positive. Since the number of roots with positive real parts is equal to the number of sign changes in the Δ_i's, we suspect that in certain regions of the n_T, n_S plane we will find unstable solutions. What regions are they?

From the third determinant, we find that $N_*^3 = N_{*_1}^3 = -\alpha_v\alpha\beta$ is the first transition point between stable and unstable solutions, with $N_*^3 < -\alpha_v\alpha\beta$ representing instability. From the definition of N_*^3 we see that $d_z\tilde{S} < 0$ and $d_z\tilde{T} < 0$, which can happen only if warm salty water is on top, i.e., if we are in a salt finger regime. At this transition point we also notice that equation (18) can be integrated to yield an arbitrary forcing constant on the right hand side, which implies the instability is direct, not oscillatory.

From the second determinant another transition point is found, this one in a diffusive region. For N_*^3 values greater than $N_{*_2}^3 = (\alpha_v + \alpha + \beta)(1 + \alpha_v\alpha + \alpha_v\beta + \alpha\beta) - \alpha_v\alpha\beta$ we have instability of the diffusive type. Expressing $N_*^3 = -\beta n_T + \alpha n_S$, we can construct lines of constant N_*^3 for the onset of both salt finger and diffusive instability, with intercepts on the n_T, n_S axis as shown in Figure 3. These lines intersect lines of constant Brunt-Vaisala frequency at points A and B, from which the coordinates of marginal stability can immediately be found. Thus, by knowing the background stratification and the values of parameters α_v, α, β, we have a way of deciding the stability of a system for any given n_T and n_S.

The considerations above suggest a more direct way of calculating the points of transition from salt finger regime to stable damping to diffusive oscillatory regime. By identifying the real root with salt fingers and the complex conjugate roots with the diffusive regime,

$$\lambda_1 = \lambda_{SF}, \lambda_2 = \lambda_D + i\omega, \lambda_3 = \lambda_D - i\omega, \qquad (21)$$

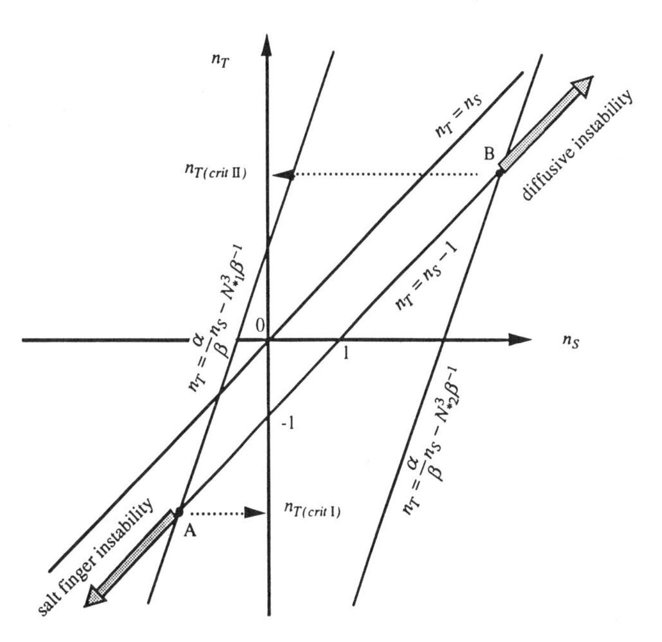

Figure 3. Stability diagram, non-dimensionalized.

we can solve in a simple manner for the transition points. We will assume that $\alpha > \beta$ throughout to maintain a common ground for analysis. To illustrate our method, since $\lambda_1 > 0$ represents stable damping and $\lambda_1 > 0$ instability, the value of n_T when $\lambda_1 = 0$ is a transition point, or point of marginal stability. By forming the product of the roots $(\lambda - \lambda_1)(\lambda - \lambda_2)(\lambda - \lambda_3)$, setting it to zero, and comparing with the characteristic equation (19), we find that quantities λ_{SF}, λ_D, and ω have following properties:

$$\lambda_{SF} + 2\lambda_D = -(\alpha_v + \alpha + \beta) = -A \leq 0, \qquad (22)$$

$$\lambda_D^2 + 2\lambda_D\lambda_{SF} + \omega^2 = 1 + \alpha_v\alpha + \alpha_v\beta + \alpha\beta = B \geq 1, \qquad (23)$$

$$\lambda_{SF}(\lambda_D^2 + \omega^2) = -(\alpha + \alpha_v\alpha\beta + (\alpha - \beta)n_T). \qquad (24)$$

Using equations (22)-(24) it is easy to find the stability boundaries. The salt finger boundary is when $\lambda_1 = \lambda_{SF} = 0$, and then $\lambda_D = -A/2$, and $\omega = (B - A^2/4)^{1/2}$ if $B > A^2/4$.

In the case of all real roots when $B < A^2/4$, there are $\lambda_1 = 0$, $\lambda_2 = -(A - (A^2 - 4B)^{1/2})/2$, $\lambda_3 = -(A + (A^2 - 4B)^{1/2})/2$. The critical temperature component of background density is given by

$$n_T = n_{T(critI)} = -\alpha(1+\alpha_v\beta)/(\alpha-\beta) \le -1, \qquad (25)$$

and also when λ_1 equals the maximum of the three real roots and is zero. For values below this critical value, the salt finger root exhibits linear instability. Similarly, for the onset of diffusive-type instability, we set $\lambda_D = 0$, to obtain a critical value of n_T at which neither growth nor damping occur:

$$n_{T(critII)} = \left[(\alpha_v+\alpha+\beta)(1+\alpha_v\alpha+\alpha_v\beta+\alpha\beta) - \alpha(1+\alpha_v\beta)\right]/(\alpha-\beta) > 0. \qquad (26)$$

Here we note that the salt finger root is negative, as expected

$$\lambda_{SF} = -(\alpha_v+\alpha+\beta) = -A \qquad (27)$$

and the frequency of the stable oscillations is

$$\omega = (1+\alpha_v\alpha+\alpha_v\beta+\alpha\beta)^{1/2} = B^{1/2} \ge 1 \qquad (28)$$

which reduces to the Brunt-Vaisala frequency when all parameters vanish.

The range n_T of three real roots existence also can be found from (22)-(24) if we set ω=0. The first conclusion is that all of the three roots are real when.

$$A^2 - 3B = D > 0. \qquad (29)$$

From the definitions of A and B it follows that an existence of this situation depends on numerical values of the parameters α_v, α, β. Let us take as $\alpha_1 = \max(\alpha_v,\alpha,\beta)$ and two others normalized by α_1 as x_1 and x_2 (for example when $\alpha_1 = \alpha_v > \alpha > \beta$, then $x_1 = \alpha/\alpha_v$, and $x_2 = \beta/\alpha_v$). Then the case D=0 corresponds to a surface in the space of parameters α_1, x_1, x_2 or α_v, α, β given by

$$\bar{\alpha}_1(x_1,x_2) = \left(3/\left(1-x_1-x_2-x_1x_2+x_1^2+x_2^2\right)\right)^{1/2}, \quad 0 \le x_1, x_2 \le 1, \qquad (30)$$

and shown in Figure 4. The space above the surface (30) (or $\alpha_1 > \bar{\alpha}_1$) is a range of parameters α_v, α, β for which the roots λ_1, λ_2, λ_3 are real and satisfy inequalities:

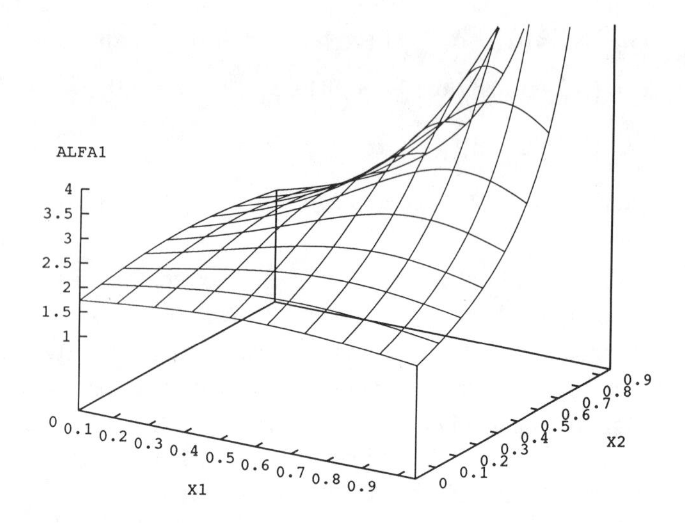

Figure 4. The boundary of real roots in the space of parameters α_v, α,β.

$$\lambda^*_{SF1} \ge \lambda_1 \ge \lambda^*_{D2}, \lambda^*_{D2} \ge \lambda_2 \ge \lambda^*_{D1}, \lambda^*_{D1} \ge \lambda_3 \ge \lambda^*_{SF2}, \qquad (31)$$

where

$$\left.\begin{aligned} \lambda^*_{SF1} &= \left(-A+2D^{1/2}\right)/3, \lambda^*_{SF2} = \left(-A-2D^{1/2}\right)/3 \\ \lambda^*_{D1} &= \left(-A-D^{1/2}\right)/3, \lambda^*_{D2} = \left(-A+D^{1/2}\right)/3, \end{aligned}\right\}, \qquad (32)$$

and the temperature contribution n_T in the background density is in the range $n^*_{T1} \le n_T \le n^*_{T2}$, where n^*_{T1}, n^*_{T2} are given by

$$\left.\begin{aligned} n^*_{T1} &= \left[A^3 - 3AD - 2D^{3/2} - 27\alpha(\alpha_v\beta+1)\right]/27(\alpha-\beta) \\ n^*_{T2} &= \left[A^3 - 3AD + 2D^{3/2} - 27\alpha(\alpha_v\beta+1)\right]/27(\alpha-\beta) \end{aligned}\right\}. \qquad (33)$$

This case corresponds to the behavior of the roots given in Figure 5a. The case D=0 or $\alpha_1 = \bar{\alpha}_1$ corresponds to Figure 5b when the difference $\left(n^*_{T2} - n^*_{T1}\right) \to 0$,

$\lambda^*_{SF1} \to \lambda^*_{D2} \to \lambda^*_{D1} \to \lambda^*_{SF2} \to \lambda^* = -A/3$, and $\omega = 0$. The case $D < 0$ or $\alpha_1 < \bar{\alpha}_1$ corresponds to the situation when only one of three roots is real, and then $n_T = n^{**}_T$ is given by

$$n^{**}_T = \left[\left(A\left(9B-2A^2\right)/27\right) - \alpha(\alpha_v\beta+1)\right]/(\alpha-\beta) \qquad (34)$$

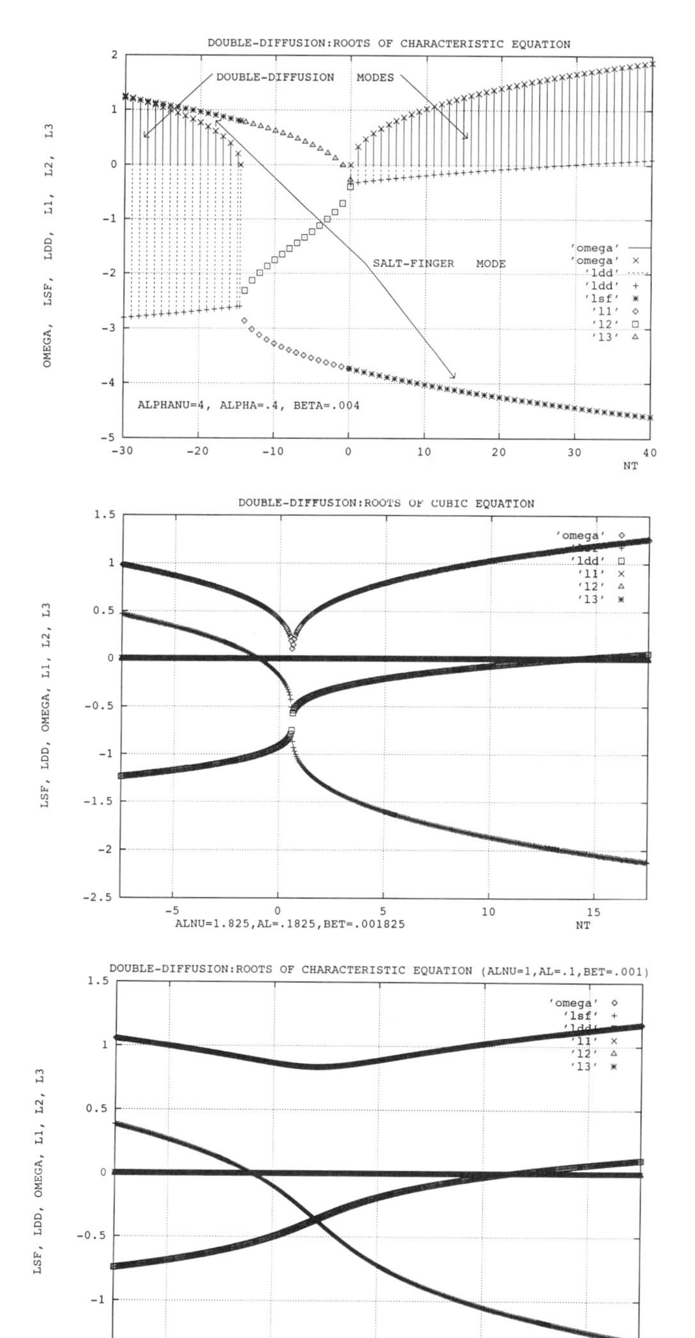

Figure 5. Roots of the characteristic equation, linear stratification: (a) $\alpha_v = 4$, $\alpha = 0.4$, $\beta = 0.004$; (b) $\alpha_v = 1.825$, $\alpha = 0.1825$, $\beta = 0.001825$; (c) $\alpha_v = 1$, $\alpha = 0.1$, $\beta = 0.001$.

where the frequency ω has a minimal value $\omega=\omega_{min}=((3B-A^2)/3)^{1/2}$ and $\lambda_{SF}=\lambda_D=\lambda^*=-A/3$. An example of this case is presented in Figure 5c.

To show the effect of parameter changes, let us assign a proportion to them that approximates what is found in the ocean:α_v: α: β will take the ratio 1000:100:1. As we successively increase our parameters through Figure 5c to 5a, we lose the central region of stable solutions and discover a region of real roots (Figure 5a). In the Appendix, where we demonstrate the equivalence of our approach ad the Navier-Stokes approach, it is shown that this is equivalent to increasing the wavenumber when the vertical component of wavenumber is negligible. At a value of n_T around -15 in Figure 5a, the diffusive root bifurcates into two negative real roots, which regain nonzero oscillatory components near $n_T = 0$ before one of them grows unstable at an n_T of 28. The salt finger root is positive until the real root region, then reappears at a value of -3.7 and continues to slope negatively. In Figure 5b the parameter values are low enough to collapse the real root region, setting into relief the salt finger root, positive to the left, and the diffusive root, positive to the right. The fluid particle can oscillate with a frequency either greater or less than the Brunt-Vaisala frequency, and its motion will either decay or grow unstable diffusely depending on values of n_T. Figure 5c shows what happens when the parameters decrease still further: as we anticipate from equations (25) and (26), the transition points move closer together, and the minimum frequency of oscillation translates slightly in the positive direction. If we were to decrease our parameters even beyond the values displayed, we would recover, as they approached zero, stable stratification at the Brunt-Vaisala frequency.

NONLINEAR STRATIFICATION

We wish now to model the situation where a fluid particle may move through an interface separating two quasi-homogeneous layers. The simplest way to describe this is to use a buoyancy frequency through the interface that reflects interface depth h and the difference between temperature and salinity in the upper and lower layers, as shown in Figure 6. This buoyancy frequency will be constant, and is given by:

$$N_n^2 = g\left(-\alpha_T \Delta T + \beta_S \Delta S\right) / \rho_0 h \qquad (35)$$

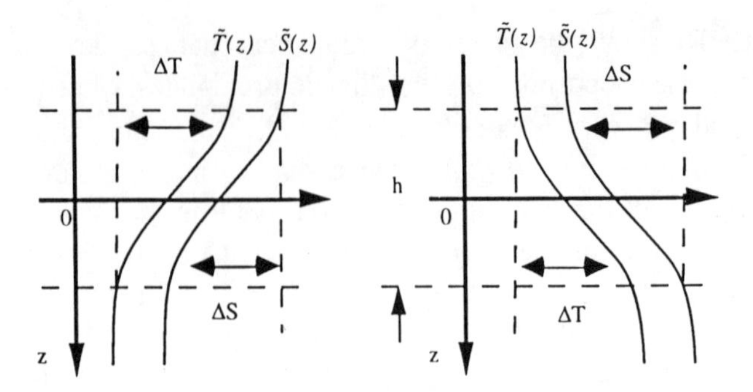

Figure 6. Nonlinear Stratification

where ΔT and ΔS are the differences of temperature and salinity between the upper and lower liquids.

The distribution of temperature and salinity in the interface can be used to regulate interface thickness. We then represent the background buoyancy by

$$\rho_0^{-1} g\left(-\alpha_T \widetilde{T}(z) + \beta_S \widetilde{S}(z)\right) = N_h^2 h \tanh(z/h) \quad (36)$$

and the distribution of buoyancy frequency by

$$N^2(z) = N_h^2 / \cosh^2(z/h). \quad (37)$$

By substituting (37) into (13), using (14) and nondimensionalizing, with the length scale L now associated with h, the general equation of motion becomes

$$\eta''' + (\alpha_v + \alpha + \beta)\eta'' + \left(\left(1/\cosh^2(\eta)\right) + \alpha_v\alpha + \alpha_v\beta + \alpha\beta\right)\eta' + \alpha_v\alpha\beta\eta + N_*^3 \tanh(\eta) = C\alpha\beta. \quad (38)$$

In the case where $\eta \ll 1$, in the vicinity of the origin, this reduce to equation (18), which describes linear stratification. Here the particle does not feel the nonlinear effects of background stratification at all.

In the opposite case, where $\eta \gg 1$, we have $\tanh(\eta) \rightarrow \pm 1$ and buoyancy frequency $1/\cosh(\eta) \rightarrow 0$. Here equation (38) is transformed into the asymptotic form, which is linear and describes the final stages of particle trajectory into the uniform layers

$$\eta''' + (\alpha_v + \alpha + \beta)\eta'' + (\alpha_v\alpha + \alpha_v\beta + \alpha\beta)\eta' + \alpha_v\alpha\beta\eta \pm N_*^3 = C\alpha\beta. \quad (39)$$

Notice that N_*^3 = constant. When $\eta \sim exp(\lambda t)$ we substitute into (39) and get a new form of the characteristic equation:

$$\lambda^3 + (\alpha_v + \alpha + \beta)\lambda^2 + (\alpha_v\alpha + \alpha_v\beta + \alpha\beta)\lambda + \alpha_v\alpha\beta = 0 \quad (40)$$

or

$$(\lambda + \alpha_v)(\lambda + \alpha)(\lambda + \beta) = 0, \quad (41)$$

which gives the roots to (39) as $\lambda_1 = -\alpha_v$, $\lambda_2 = -\alpha$, $\lambda_3 = \beta$. Therefore all roots are real and negative, and motion far from the interface region of high gradients will decay.

In the final case, we consider intermediate values of particle displacement, on the order of interface thickness h. Using computer simulations, we get unstable double diffusive convection tending to stable nonlinear oscillations. If we increase the first integral, the nonlinear stable oscillations (limit cycles) will transform to exponentially decaying solutions.

In Figure 7 we show a series of phase portraits of particle motion in different regimes, where the initial driving force is made to vary. In Figure 7a we have an initially high velocity motion of the salt finger regime (n_T = -30) which undergoes nonlinear effects before transforming to linear damping. In Figure 7b we have stable solutions with essentially linear behavior. At n_T = 10, with weak initial excitation (Figure 7c), we have oscillatory damping. If this same point is sought in Figure 5a, it is found just before the real part of the diffusive root changes sign. If we increase the excitation as in Figure 7d, this transforms to full damping after one cycle, showing the strong effects of the asymptotic solution since we are outside the interface layer. In Figure 7e with small initial excitation, motion tends to a limit cycle whose center is determined by initial conditions. When the excitation is increased as in Figure 7f, the limit cycle shrinks, and its center moves farther from the origin.

In Figure 8 we examine the behavior of particle motion in the diffusive regime for different values of the first integral. In Figure 8a, the particle first experiences linear instability before asymptotically reaching a nonlinear limit cycle. As the first integral is increased in Figures 8b through 8e, the nonlinear limit cycle decreases in amplitude, while the centers of the cycles are gradually displaced to the right.

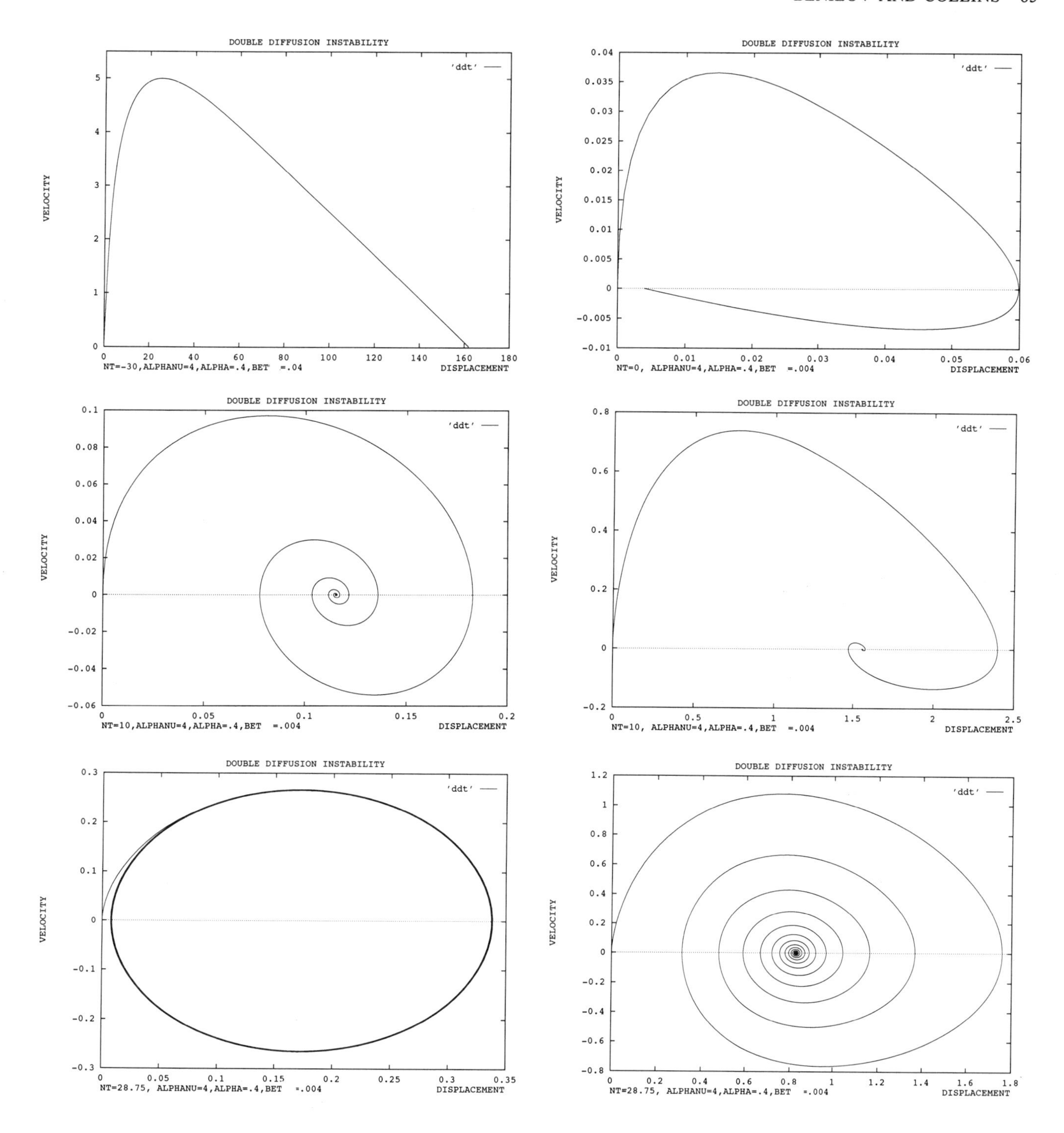

Figure 7. Phase portraits: (a) n_T = -30 (salt finger regime); (b) n_T = 0; (c)n_T = 10, weak initial excitation; (d) n_T = 10, strong initial excitation; (e)n_T = 28.75, weak initial excitation; (f) n_T = 28.75, strong initial excitation.

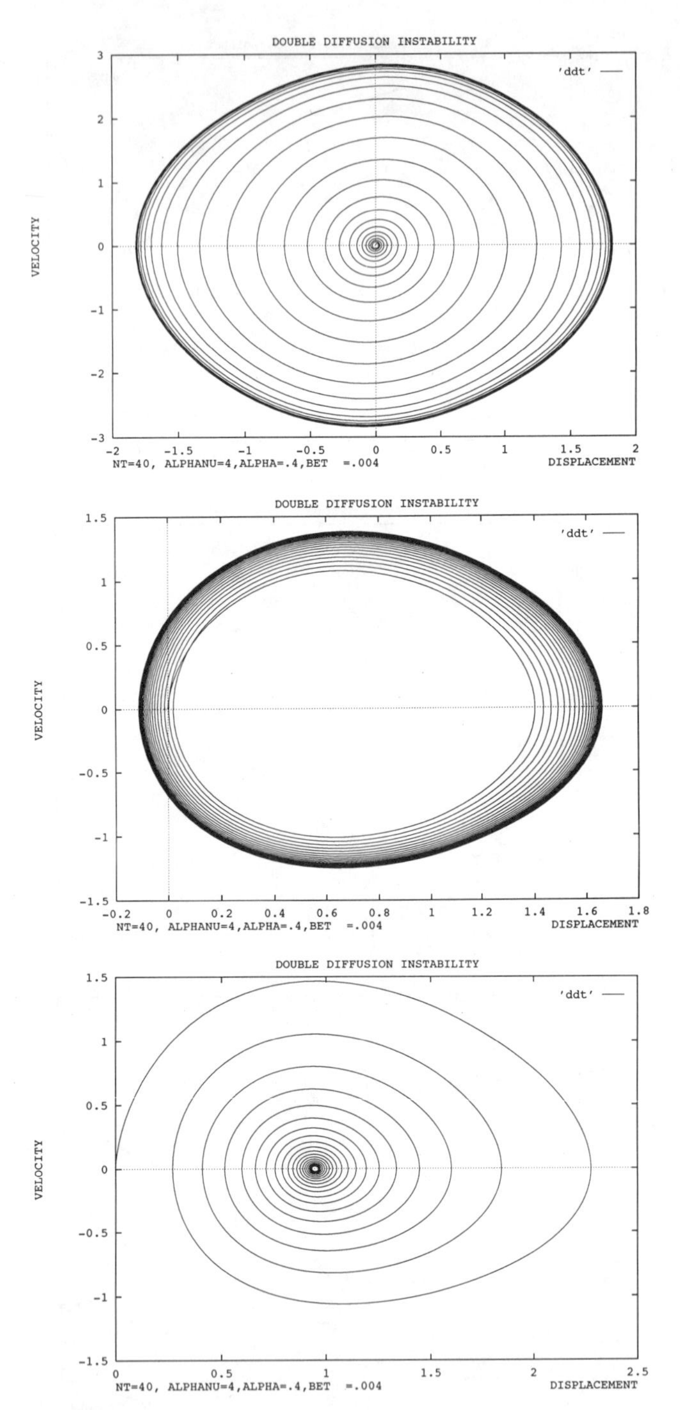

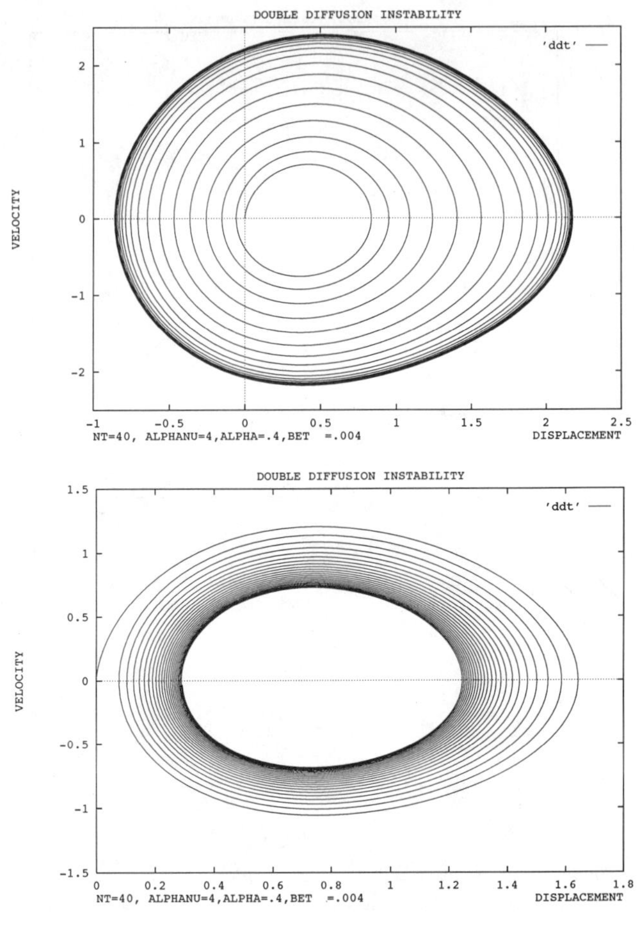

Figure 8. Diffusive regime: Transformations of the nonlinear limit cycle by increasing the first integral through C = 1, 3750, 5625, 6250, 7500.

CONCLUSIONS

We have described the motion of a fluid particle from a simple model of force balances which produce convective movements. If sharp interfaces of fluid properties separate uniform layers above and below, the interfaces yield unstable solutions while the uniform layers give us strong damping. The penetration of the fluid particle into the uniform layer depends on strength of the initial driving force, and once there either comes to complete rest or oscillates in small limit cycles. In either case the initial instability degenerates. We have presented the behavior of the roots of the characteristic equation of particle motion in elementary pictures representing all possible situations. Our enterprise is justified by the development of stability diagram consisting of one line, not a plane as in previous work.The linear problem corresponds completely to the linear Navier-Stokes approach, as shown in the Appendix.

APPENDIX

Using our nomenclature, we write the Navier-Stokes equations for the conservation of momentum in a linear approach as

$$\partial_t \mathbf{u} = -\rho_0^{-1}\nabla p + \nu\Delta\mathbf{u} + \rho_o^{-1}\mathbf{g}(-\alpha_T T + \beta_S S), \qquad (A1)$$

where p is the pressure of fluid, **u** is the velocity vector, and ν is the coefficient of kinematics viscosity. The equation of continuity is

$$\nabla\mathbf{u}=0, \qquad (A2)$$

and the linear expressions for heat and salt diffusion are

$$\partial_t T + (\mathbf{u}\nabla)\widetilde{T} = \kappa_T \Delta T, \qquad (A3)$$

$$\partial_t S + (\mathbf{u}\nabla)\widetilde{S} = \kappa_S \Delta S, \qquad (A4)$$

with κ_T, κ_S representing the coefficients of thermal and haline diffusivity. If we let w be the vertical component of velocity and $\Delta_L = (\partial/\partial x)^2 + (\partial/\partial y)^2$ be the horizontal Laplacian, equations (A1) to (A4) can be combined to

$$\begin{aligned}&\partial^3_{ttt}\Delta w - (\nu+\kappa_T+\kappa_S)\partial^2_{tt}\Delta^2 w + \\ &\partial_t\left[(\nu\kappa_T+\nu\kappa_S+\kappa_T\kappa_S)\Delta^3 + N^2\Delta_L\right]w - \qquad (A5)\\ &-\left[\rho_0^{-1}\mathbf{g}(-\kappa_S\alpha_T d_z\widetilde{T} + \kappa_T\beta_S d_z\widetilde{S})\Delta_L\Delta + \nu\kappa_T\kappa_S\Delta^4\right]w = 0.\end{aligned}$$

Setting $w \sim exp(\lambda t + i\mathbf{k}\mathbf{x})$ and $k_L \sim k$, where $k = (k_x^2+k_y^2+k_z^2)^{1/2}$ is the wavenumber and $k_L = (k_x^2+k_y^2)^{1/2}$, this has the same characteristics equation as (18).

To associate our solutions with the wavenumber of small perturbations in the linear dynamic problem described by (A1), we can make the following correspondences:

$$\left.\begin{aligned}\nu k^2 &\rightarrow \alpha_\nu \rightarrow \alpha_\nu\\ \kappa_T k^2 &\rightarrow \alpha \rightarrow \alpha\\ \kappa_S k^2 &\rightarrow b \rightarrow \beta\end{aligned}\right\} \qquad (A6)$$

which shows that wavenumber is proportional to our basic parameters.

To continue the analogy, the density ratio takes the following form in our perspective:

salt finger regime $(n_T, n_S < 0)$: $R_{SF} = n_T/(n_T+1)$, (A7)

diffusive regime $(n_T, n_S > 0)$: $R_D = (n_T+1)/n_T$. (A8)

The thermal and saline Rayleigh numbers are

$$Ra_T = n_T/(\alpha_\nu\alpha), Ra_S = n_S/(\alpha_\nu\alpha) \qquad (A9)$$

and the Prandtl number σ and the Lewis number τ become

$$\sigma = \alpha_\nu/\alpha, \tau = \beta/\alpha. \qquad (A10)$$

If we substitute (A9) and (A10) into our equations (22) and (23) for stability boundaries , we obtain the equations used by [Baines and Gill, 1969] to construct their well-known stability diagram:

$$\tau Ra_T - Ra_S = \tau, \qquad (A11)$$

$$\sigma(\sigma+1)Ra_T - \sigma(\sigma+\tau)Ra_S = (\sigma+1)(\tau+1)(\sigma+\tau). \qquad (A12)$$

Thus the present approach is equivalent to the classical methods.

REFERENCES

Baines PG and Gill AE. 1969. On Thermohaline Convection with Linear Gradients, *J. Fluid Mech. 37,* 289- 306.

Huppert HE and Moore DR. 1976. Nonlinear Double-Diffusive Convection. *J. Fluid Mech. 78*, 821-854.

Korn GA and Korn TM. 1961. *Mathematical Handbook for Scientists and Engineers*. McGraw-Hill, NY.

anins PC. 1976. Intrusion into a Stratified Fluid. *J. Fluid Mech 74*, 547 - 560.

Moore DR, Toomre J, Knobloch E and Weiss NO. 1983. Period Doubling and Chaos in Partial Differential Equations for Thermosolutal Convection. *Nature 303*, 663 -667.

Paliwal RC and Chen CF. 1980. Double-Diffusive Instability in an Inclined. Fluid Layer.,2. Theoretical Investigation. *J. Fluid Mech. 98*, 769-785.

Pellew A and Southwell FRS. 1940. On Maintained Convective Motion in a fluid. Heated from Below. *Proc. Royal Soc.London Ser. A 176*, 312-343.

Rayleigh Lord, 1916. On Convection Currents in a Horizontal Layer of Fluid. When the Higher Temperature is on the Underside. *Phil. Mag. 32*, 529-546.

Stern ME. 1960. The "Salt-Fountain" and Thermohaline Convection. *Tellus 12*, 172-175.

Turner JS. 1973. *Buoyancy Effects in Fluids*. Cambridge Univ. Press, Cambridge.

Turner JS. 1978. Double Diffusive Intrusions into a Density Gradient. *J. Geophys. Res. 83*, 2887-2901.

Veronis G. 1965. On Finite Amplitude Instability in Thermohaline Convection. *J. Mar. Res. 23*, 1-17.

Veronis G. 1968. Effect of a Stabilizing Gradient of Solute on Thermal Convection. *J. Fluid Mech. 34*, 315-336.

Walton IC. 1982. Double-Diffusive Convection With Large Variable Gradients. *J.Fluid Mech. 125*, 123-135.

Zangrado F. and Bertram Lee A. 1985. The Effect of Variable Stratification on Linear Doubly Diffusive Stability. *J. Fluid Mech., 151*, 55-79.

---

[1]Benilov, A, Davidson Laboratory, Stevens Institute of Technology, Castle Point on the Hudson, Hoboken, New Jersey 07030

[2]Collins, D. W., Davidson Laboratoty, Stevens Institute of Technology Castle Point on the Hudson, Hoboken, New Jersey 07030

Numerical Experiment on Double Diffusive Currents

Jiro Yoshida[1], Hideki Nagashima[2] and Moh Nagasaka[1]

1) Department of Marine Science and Technology, Tokyo University of Fisheries, Tokyo, Japan
2) Laboratory of Earth Science, The Institute of Physical and Chemical Research(RIKEN), Saitama, Japan
Present address: Department of Marine Science and Technology, Tokyo University of Fisheries, Tokyo, Japan

We investigated the behavior of double diffusive density current numerically. We released cold/fresh water on to the ambient warm/salty water, by parameterizing the effective heat and salt transport in double diffusive convection by an experimentally determined flux law due to Linden[1974]. It is found that the behavior of the current is well described by the combination of three parameters. One is the turbulent Prandtl number ε ; when ε is large, the density current is suppressed. A second parameter is the turbulent Rayleigh number Ra ; when Ra is large, the induced convection becomes vigorous and the resulting density current becomes strong. The last parameter is the Turner number $R\rho$; when $R\rho$ is sufficiently small, the activity of double diffusive convection becomes large and is strongest when $R\rho$ = 1.01, then becomes weaker. The results of the numerical experiments on double diffusively induced density current not only explain well Maxworthy's laboratory experiments but also cover a wide range of the parameters.

1. INTRODUCTION

Gravity current, sometimes called density current, is essentially a horizontal convection induced by a horizontal density gradient. When the density of the current and its environment is determined by properties such as temperature or salinity alone, its behavior has been investigated extensively in the wide variety of environments, such as in the ocean, atmosphere, volcano and so on [for details, see *Simpson,* 1987].

If the density field is determined by two properties having different diffusivities, double diffusive convection should affect the behavior of the gravity current. *Thangam and Chen* [1981] first investigated such effects. They released a light hot salty water on to the relatively heavier salt stratified cold body of water. The resulting current has a salt finger interface below, and the advancing speed of the current seems to be retarded by the salt finger convection. However, stable stratification complicated the description of the current, and no systematic conclusion was obtained. *Maxworthy* [1983] simplified the environments by using salt and sugar as density contribution components to the environment. He released salt (sugar) water on to the homogeneous sugar (salt) water, which corresponds to the oceanic situation that cold/fresh (warm/salty) water on to warm/salty (cold/fresh) water. He found that the velocity of the current is considerably reduced by double diffusive convection. He considered a force balance between accelerating buoyancy force and retarding double diffusively induced force, and obtained current length and time relationships of the current. Interesting phenomena observed in his experiment was that, when the density differences of two solutions were extremely low, a secondary current was generated near the bottom of tank by the descending plume from the original surface current. This secondary current was also observed when the double diffusive current was produced by the classical dam break method [*Yoshida et al.*, 1987].

When doing laboratory experiments such as those mentioned above, there are some restrictions in the experiments. One is the control of heat when we use the

Double-Diffusive Convection
Geophysical Monograph 94

heat and salt for density contribution components. Heat is easily lost through the side wall of tank. This can be avoided by using salt and sugar as was done by *Maxworthy* [1984] and *Yoshida et al.* [1987]. However, the difference in the density flux ratio between heat/salt system and salt/sugar system might affect the behavior of gravity current [*McDougall*, 1986]; here the density flux ratio is defined as $\alpha F_T/\beta F_S$, for salt finger convection; αF_T is the vertical density anomaly flux due to heat and βF_S is that due to salt. See, for example, *Turner* [1973]. A second restriction in the experiment is the wide range of parameters. We must be very careful to measure the parameters to reproduce the same experiment. A powerful tool to overcome such difficulties is the use of numerical experiments. If we use a suitable formulation, it is possible to examine a whole range of parameter values [*Nagashima et al.,* 1990]. In the present paper, we restricted our attention to the case when the cold/fresh water was released on to the warm/salty water as a typical example of behaviors of ice melting water and spreading of rain pools. We investigated numerically the behavior of the resulting current in various parameter ranges. In section 2, the basic equations are presented. These equations are then non-dimensionalized to obtain some important governing parameters. In section 3, we present the results of the numerical experiments and in section 4, we compare these results with those of *Maxworthy's* laboratory experiments [1983].

2. FORMULATION

2.1. *Governing Equations*

The governing equations used here are presented as follows:

$$\frac{\partial\xi}{\partial t}+J(\phi,\xi)=g(-\alpha T_x+\beta S_x)+\nu\nabla^2\xi \tag{1}$$

$$\frac{\partial T}{\partial t}+u\frac{\partial T}{\partial x}+w\frac{\partial T}{\partial z}=K_T\nabla^2 T \tag{2}$$

$$\frac{\partial S}{\partial t}+u\frac{\partial S}{\partial x}+w\frac{\partial S}{\partial z}=K_S\nabla^2 S \tag{3}$$

$$\xi=\frac{\partial u}{\partial z}-\frac{\partial w}{\partial z}=\nabla^2\phi . \tag{4}$$

Here, t is time, u and w are velocities in the x, z directions, respectively, g is the gravitational acceleration and ϕ is stream function; T represents the property for the faster diffusing component and S is that for the slower diffusing component (In the oceanic case, T is heat and S is salinity; hereafter, for convenience we consider T as heat, and S as salt). Moreover, ν , K_T and K_S are viscosity, and the diffusivities of T and S , respectively. J is the Jacobian, ξ is vorticity, and α is the expansion coefficient for the faster diffusing property and β is that for the slower diffusing property.

The equations (2) and (3) represent the conservation of heat and salt. In the previous modeling of the double diffusive interleaving problem, the efficient vertical mixing due to double diffusive convection is sometimes parameterized by using the density flux ratio. This parameterization was first introduced by *Stern* [1967] in his analytical modeling of double diffusive interleaving, in which salt finger convection dominates the interleaving dynamics. This is traditionally called Stern's parameterization. In the present case, however, diffusive type convection should be anticipated, because cold/fresh water lies on the warm/salty water. The density flux ratio γ in this case is defined as

$$\gamma=\frac{\beta F_S}{\alpha F_T}=\frac{\beta K_S\frac{\partial S}{\partial z}}{\alpha K_T\frac{\partial T}{\partial z}} . \tag{5}$$

This relation is used to rewrite the equations (2) (Heat equation) and (3) (Salinity equation) as follows:

Heat Equation

$$\frac{\partial T}{\partial t}+u\frac{\partial T}{\partial x}+w\frac{\partial T}{\partial z}=K_{TH}\frac{\partial^2 T}{\partial x^2}+K_{TV}\frac{\partial^2 T}{\partial z^2} . \tag{6}$$

Salinity Equation

$$\frac{\partial S}{\partial t}+u\frac{\partial S}{\partial x}+w\frac{\partial S}{\partial z}=K_{SH}\frac{\partial^2 S}{\partial x^2}+\frac{\partial}{\partial z}\left(\frac{\alpha}{\beta}\gamma K_{TV}\frac{\partial T}{\partial z}\right) . \tag{7}$$

In these conservation equations, horizontal and vertical diffusivities are described separately. This generalized description of diffusion was introduced by considering the possibility of inhomogeneity in turbulent diffusion. However, in the present experiment, for simplicity, the ratio of vertical to horizontal diffusion is taken to be unity.

The density flux ratio γ is usually a function of the stability ratio of the double diffusive system in terms of the Turner number $R\rho$. From the experimental results of *Linden* [1974], γ is given as

$$\gamma=\frac{(R\rho-1)^{3/2}+0.5R\rho}{10(R\rho-1)^{3/2}+0.5} \quad where,\ R\rho=\frac{\beta\Delta S}{\alpha\Delta T} . \tag{8}$$

Fig. 1. Schematic view of the model. Numerals in the figure represents the non-dimensional length of the domain.

Here, ΔS and ΔT are the vertical salinity and temperature difference.

2.2. *Non-Dimensionalization*

We non-dimensionalize the above equations by introducing the following scales;length scale H (Depth of the calculating domain); temperature scale ΔT (initial temperature difference); salinity scale ΔS (initial salinity difference) and reduced gravity $g^*(=-g\alpha\Delta T(1-R\rho))$.

Based on these scales, the dimensional variables are transformed by the following relations;

$$x=Hx',\ z=Hz',\ u=\sqrt{g^*H}\cdot u',\ w=\sqrt{g^*H}\cdot w',\ \phi=\sqrt{g^*H}\cdot H\phi$$
$$\xi=\sqrt{g^*H}/H\cdot\xi',\ T=\Delta T\cdot T',\ S=\Delta S\cdot S',\ t=H/\sqrt{g^*H}\cdot t'$$

where, $(')$,represents non-dimensional variables. After these procedures, we have the following non-dimensional equations;

$$\frac{\partial\xi}{\partial t}+u\frac{\partial\xi}{\partial x}+w\frac{\partial\xi}{\partial z}=-\frac{1}{1-R\rho}\left(-\frac{\partial T}{\partial x}+R\rho\frac{\partial S}{\partial x}\right)+\varepsilon Ra^{-\frac{1}{2}}\nabla^2\xi \quad (9)$$

$$\frac{\partial T}{\partial t}+u\frac{\partial T}{\partial x}+w\frac{\partial T}{\partial z}=Ra^{-\frac{1}{2}}\left(\frac{\partial^2 T}{\partial x^2}+\delta\frac{\partial^2 T}{\partial z^2}\right) \quad (10)$$

$$\frac{\partial S}{\partial t}+u\frac{\partial S}{\partial x}+w\frac{\partial S}{\partial z}=Ra^{-\frac{1}{2}}\left(\tau^{-1}\frac{\partial^2 S}{\partial x^2}+\frac{\delta}{R\rho}\frac{\partial}{\partial z}\left(\gamma\frac{\partial T}{\partial z}\right)\right) \quad (11)$$

$$\xi=\frac{\partial u}{\partial z}-\frac{\partial w}{\partial z}=\nabla^2\phi. \quad (12)$$

In these non-dimensional equations, we have the following five dominant parameters which govern the motion of the current:

$R\rho=\dfrac{\beta\Delta S}{\alpha\Delta T}$ Turner Number,

$Ra=\dfrac{g^*H^3}{K_{TH}^2}$ Turbulent Rayleigh Number,

$\varepsilon=\dfrac{\nu}{K_{TH}}$ Turbulent Prandtl Number,

$\tau=\dfrac{K_{SH}}{K_{TH}}$,

$\delta=\dfrac{K_{TV}}{K_{TH}}$.

In the present numerical experiments, τ and δ are fixed to be unity. Ra is taken to be 10^4,10^6,10^7; ε is 1, 10, 100; $R\rho$ is 1.01 and 2. With these parameters changed in various combinations, equations (9) to (12) are solved numerically by using the finite difference method. The model domain is shown in Fig. 1, where the horizontal grid number is 201 and the vertical one is 25. Cold fresh water, which occupies half of the depth and one-tenth of the horizontal length, are set on the hot salty water. All fluid is initially at rest. All boundary conditions are rigid lid and no-slip.

3. RESULTS

Time sequences of numerical calculations when $R\rho$ =2, Ra =10^7 and ε =10 are shown in Figure 2a through 2e. Here, density, salinity, temperature fields and stream function are shown together. It is noted that time shown in the figures has non-dimension. The real time can be calculated by multiplying the dimensional scale $H/\sqrt{g^*H}$ (See section 2.2). Even in the early stages of development (Figure 2a), the surface density current is well recognized in density, temperature and salinity fields. It is noted that the thickness of current deduced from the temperature field is larger than that from the salinity field. This could be explained by the fact that the vertical heat transport is larger than the vertical salt transport in diffusive type convection. Clockwise circulation clearly dominates the density current, which indicates downward flow near the head of the current and upward flow behind the head. Anti-clockwise circulation is seen to exist in the lower layer near the source region. We can see a slight tendency that the interface becomes unstable near the source region in both the temperature and salinity fields. As time goes on (Figure 2b), this tendency becomes more distinct, especially in temperature field, to form a plume-like structure. This plume does not descend directly from the upper layer, because lower anti-clockwise circulation does not reach beyond the upper layer. This plume is generated by the process that relatively warm/fresh water in the lower layer near the

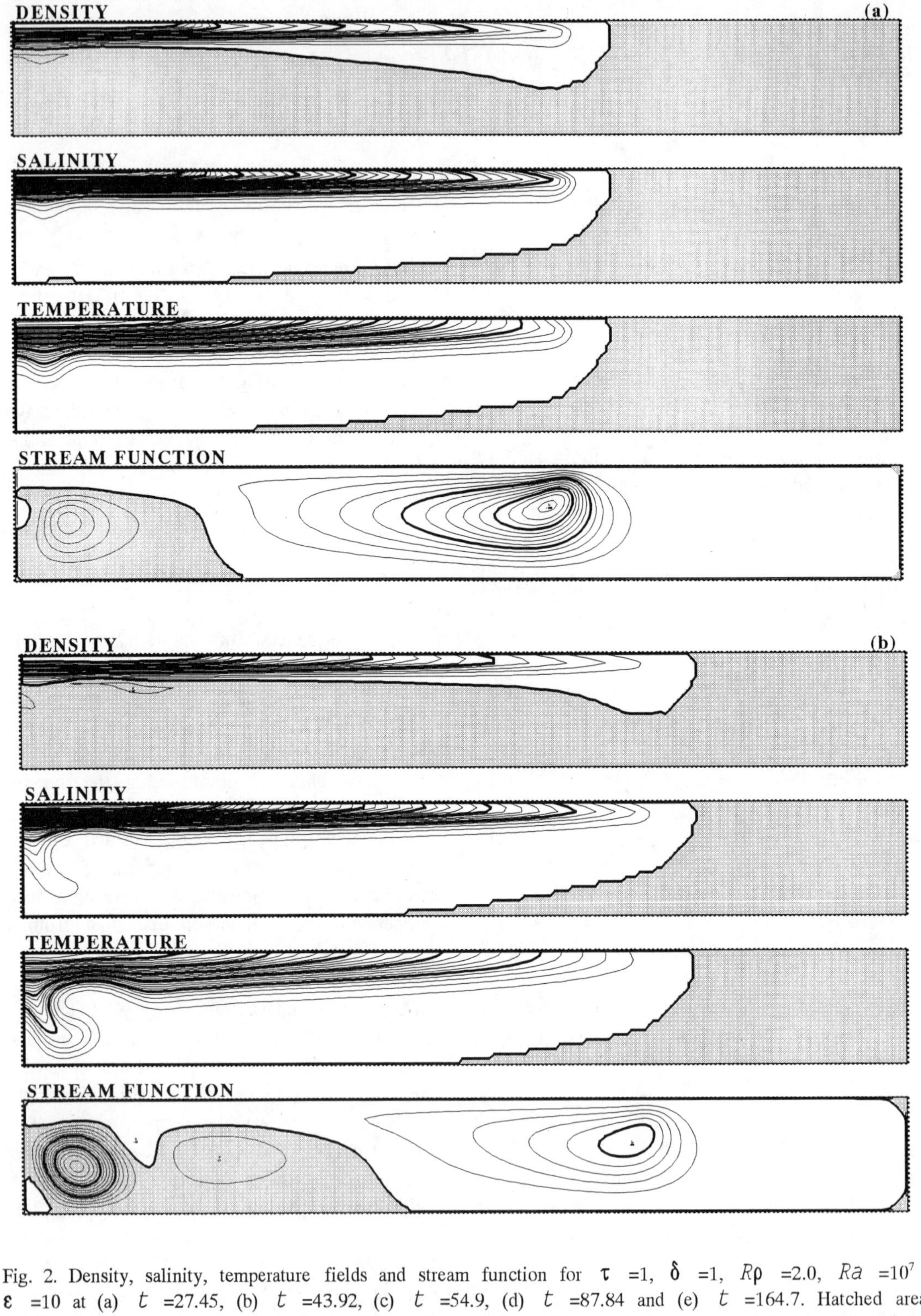

DENSITY (a)

SALINITY

TEMPERATURE

STREAM FUNCTION

DENSITY (b)

SALINITY

TEMPERATURE

STREAM FUNCTION

Fig. 2. Density, salinity, temperature fields and stream function for τ =1, δ =1, $R\rho$ =2.0, Ra $=10^7$ and ε =10 at (a) t =27.45, (b) t =43.92, (c) t =54.9, (d) t =87.84 and (e) t =164.7. Hatched area in density, salinity and temperature field shows the regions of the initial warm/salty water, and that in stream function shows the positive value region which corresponds to anti-clockwise circulation region. Case when the double diffusive convection is turned off (Ra $=10^7$, ε =10 and t =87.84) is shown in Fig. 2f for comparison.

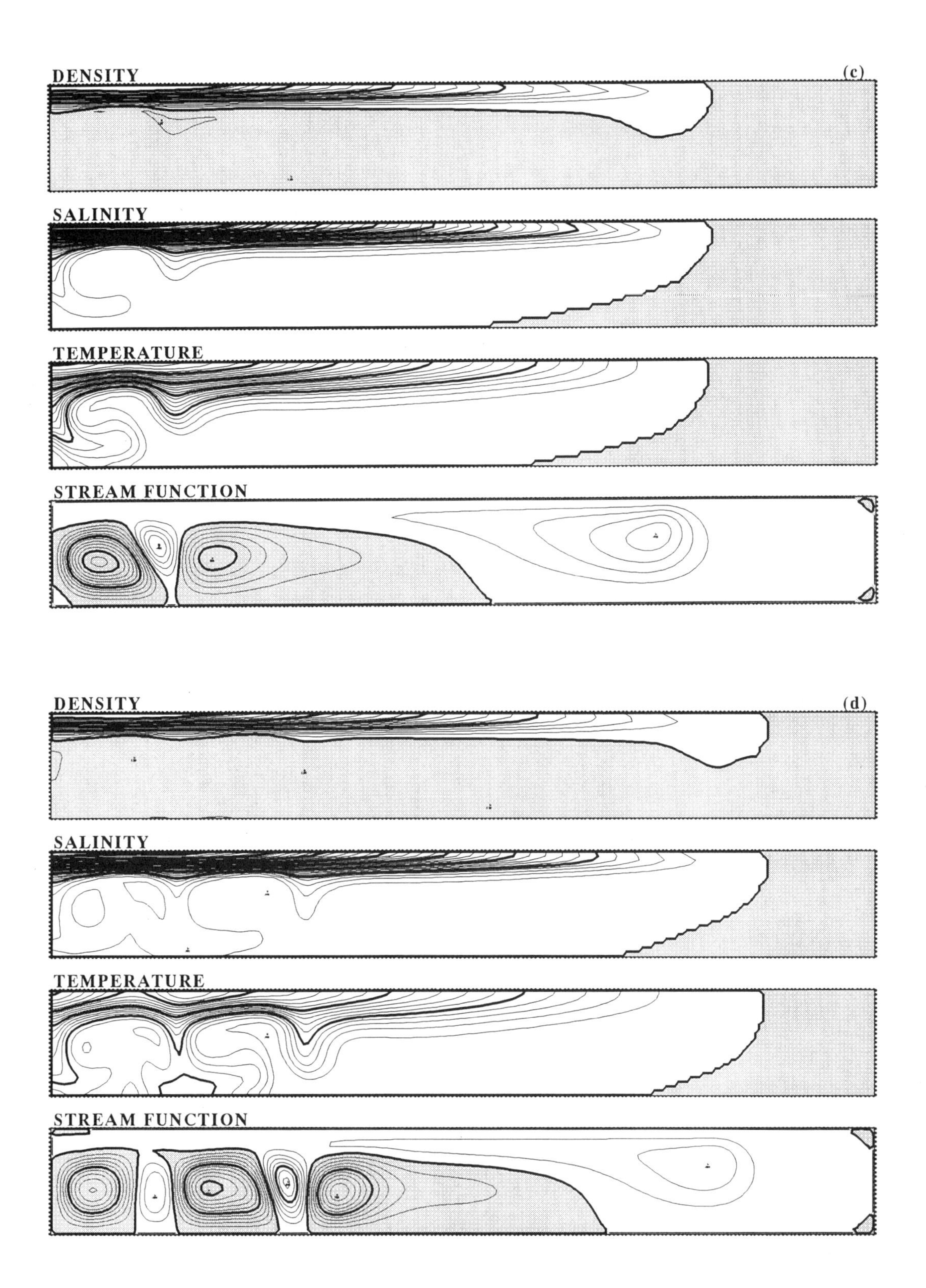

Figure 2 (continued)

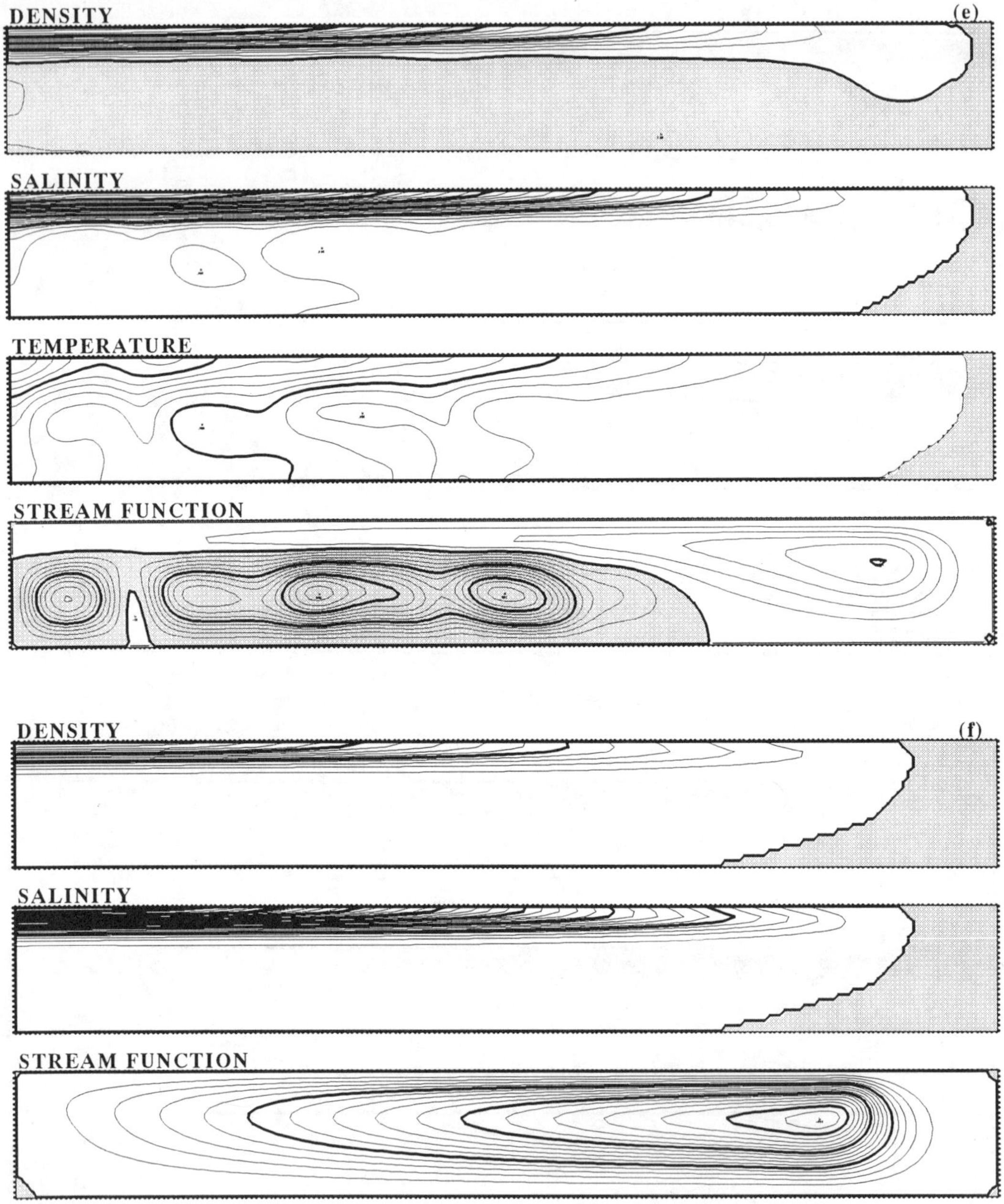

DENSITY (e)

SALINITY

TEMPERATURE

STREAM FUNCTION

DENSITY (f)

SALINITY

STREAM FUNCTION

Figure 2 (continued)

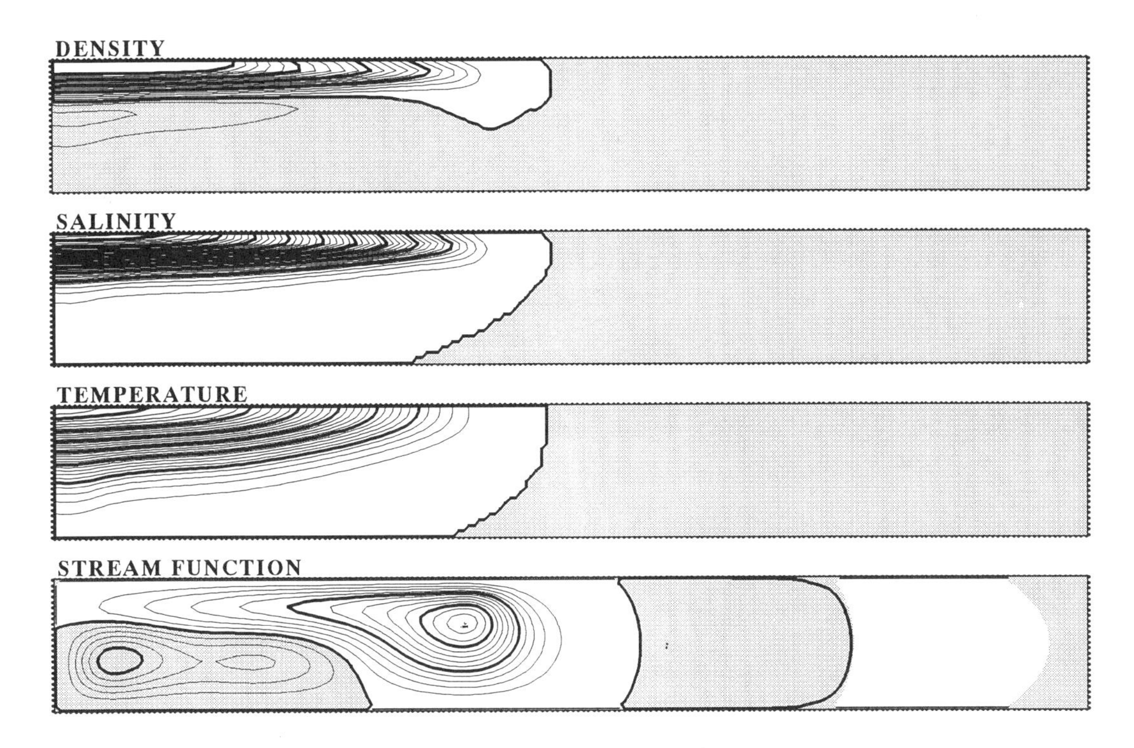

Fig. 3. Same as in Fig. 2, but Ra $=10^7$ and t $=27.78$.

interface loses its heat through contacting upper cold/fresh water, and eventually descends downward. This feature is characteristic of diffusive type convection, and is not seen in the fingering type convection. Another remarkable feature seen here is the splitting of the anti-clockwise circulation in the lower layer. This feature is clearly seen in Figure 2c. There appears another clockwise circulation between anti-clockwise circulations. As this circulation structure in the lower layer develops, wavy motion appears at the interface in the salinity field. Contour lines of salinity extend downwards in the downwelling region and vice versa in upwelling region. This wavy motion is slightly observed in the density field. Such wavy structure is sometimes seen in laboratory experiments [Yoshida et al. 1987], and could be closely related to these circulation structures. As time goes on (Figure 2d), clockwise and anti-clockwise circulations exist alternatively in the lower layer, in which case downward plumes in the temperature field and wavy structure in the salinity field become more distinct. These circulations are seen to be merged in Figure 2e, and accordingly the wavy motion in the salinity field disappeared above this merging region.

For comparison, we calculated the case when the double diffusive convection is turned off, and shown in Figure 2f. It is clear that the ordinal gravity current appears and the behavior is quite different from the double diffusive case (see Figure 2d).

In Figure 3, we show the case when Ra is taken to be 10^6 and other parameters are the same as in Figure 2. The effect of vertical diffusion of temperature is stronger in this case, temperature contour lines extend downward more (see Figure 2a for comparison), and, as a result, surface density current is weaker in this case.

In Figure 4 we show the effect of $R\rho$ variation on the behavior of the current. In this figure, $R\rho$ is changed to be 1.01 and other parameters are the same as in Figure 2. Temperature fields are almost the same as those in Figure 2d, but, the salinity field becomes almost identical to the temperature field. This is due to the fact that if $R\rho$ equals unity, the density flux ratio γ equals unity (see Equation (8)), and vertical salt transport becomes comparable to heat transport (Equations. (10) and (11) become identical when equals unity). It is curious to note here that density field and stream function are almost identical in both cases. This feature was seen in all other cases when $R\rho$ alone is changed. We will discuss this point later.

In Figure 5, ε is taken to be 100, but the other parameters are same as in Figure 2. Apparently, the surface density current is weaker in this case. Temperature and

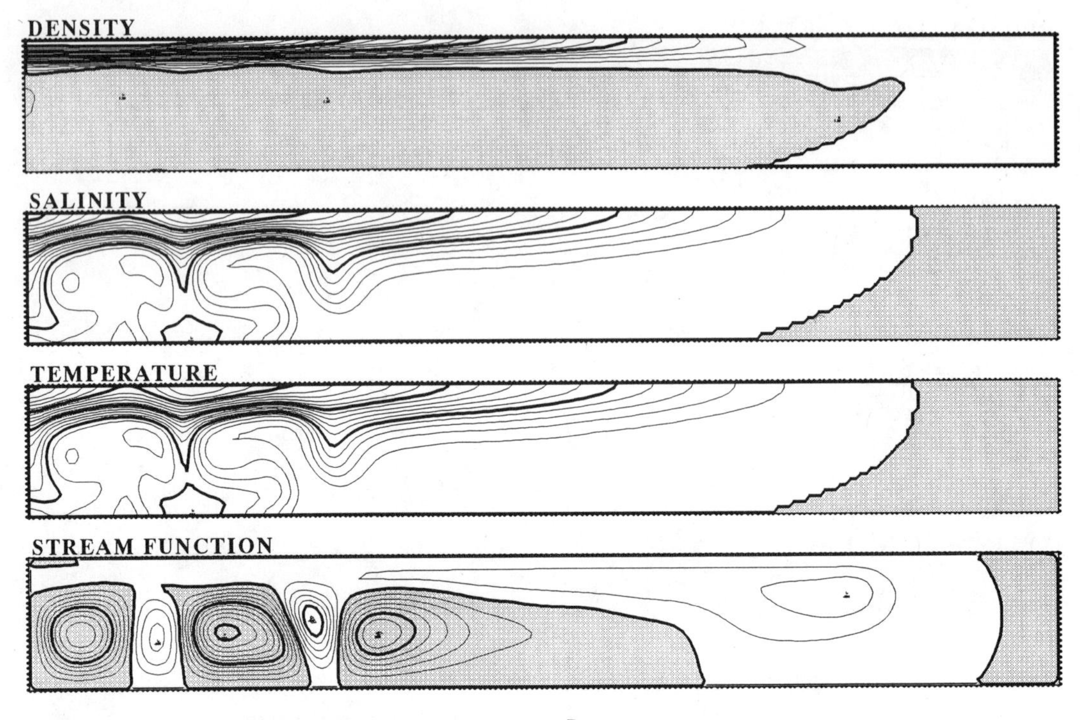

Fig. 4. Same as in Fig. 2, but $R\rho$ =1.01 and t =87.84.

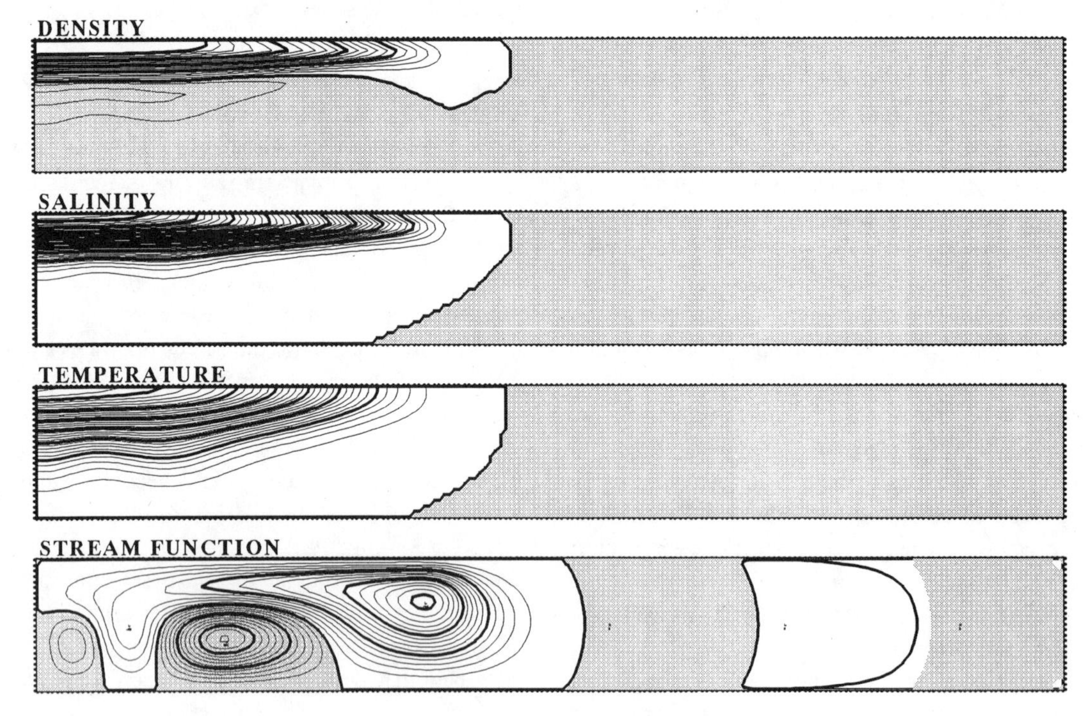

Fig. 5. Same as in Fig. 2, but ε =100 and t =87.84.

salinity fields resemble those in Figure 3 when Ra is relatively low, but wavy motion appears in both fields as in Fig. 2. This could be explained by examining Equations. (9) to (11), where ε only appears in the second term in the right hand side of Equation (9) as a product with $Ra^{-1/2}$. This means that vorticity is diffused more when Ra becomes small and ε becomes large.

TABLE 1. $L-t$ Relation by *Maxworthy* [1983]

Force Balance	$L-t$ Relation
1. slumping phase	$L \propto t$
2. buoyancy --- inertia	$L \propto t^{2/3}$
3. buoyancy --- D.D. (U dominant)	$L \propto t^{1/4}$
4. buoyancy --- viscous	$L \propto t^{1/5}$
5. buoyancy --- D.D. (U_∞ dominant)	L :const

4. DISCUSSION

As was pointed out in the introduction, the dynamics of double diffusive gravity current was investigated experimentally by *Maxworthy* [1983]. His results were interpreted by considering the force balances among the buoyancy force, the viscous force and the double diffusively induced retarding force. He obtained various relationships between current length *vs* time ($L-t$ relation) at various phases of force balance, and are summarized in Table 1. Here, U dominant and U_∞ dominant mean that the double diffusively induced retarding force is determined by the gravity current velocity itself (U) or the external flow field (U_∞). It is noted that the buoyancy force is also created by the vertical transport of heat and salt by double diffusive convection so that at any phase of the force balance, double diffusive convection has an important effect on the behavior of the current. $L-t$ relations in our experiment are shown in Figure 6 for the case $R\rho$ =2. It is seen that for relatively high values of ε and Ra (Figures 6a and 6b), the current was accelerated rapidly at first and later the powers of the $L-t$ relation tend to 1/5. These features were also observed in *Maxworthy's* experiment [1983]. However, for the case of low ε values (Figure 6c), the power is 2/3, which indicates that the buoyancy and inertia force balance is achieved in this case. The case when Ra is 10^4 is clearly different from other cases. The power of the $L-t$ relation is at first small and then increases to 1/4, which means that buoyancy *vs* double diffusive retarding force balance (U dominant) is achieved in this case. This phenomena was not observed in Maxworthy's experiment. This might be due to the fact that enhancement of the vertical diffusion of heat interrupts the horizontal spreading of current at first, but then current is accelerated later by the increase of the current velocity itself.

A curious feature noticed in our numerical experiments is that no substantial differences in the behavior of the current were found between the case when $R\rho$ =2 and $R\rho$ =1.01. This tendency is somewhat curious because the activity of double diffusion depends on the magnitude of $R\rho$; when $R\rho$ is near to unity, the double diffusively induced density anomaly becomes compatible or sometimes larger than the initial stable density anomaly. However, in our experiments, the behavior of the density currents are almost similar in the two cases except for the salinity field. This could be understood by considering Equations. (9) to (11). When Ra and ε are constant, and the non-linear terms are neglected, Equation (9) is approximated by

$$\frac{\partial \xi}{\partial t} \sim -\frac{1}{1-R\rho}\left(-\frac{\partial T}{\partial x}+R\rho\frac{\partial S}{\partial z}\right). \tag{13}$$

If this equation is differentiated with respect to t we have

$$\frac{\partial^2 \xi}{\partial t^2} \sim -\frac{1}{1-R\rho}\left(-\frac{\partial}{\partial x}\left(\frac{\partial T}{\partial x}\right)+R\rho\frac{\partial}{\partial x}\left(\frac{\partial S}{\partial z}\right)\right). \tag{14}$$

Here, $\frac{\partial T}{\partial t}$ and $\frac{\partial S}{\partial t}$ can be approximated from Equations. (10) and (11), and are

$$\frac{\partial T}{\partial t} \sim \frac{\partial^2 T}{\partial z^2}, \tag{15}$$

$$\frac{\partial S}{\partial t} \sim \frac{1}{R\rho}\frac{\partial}{\partial z}\left(\gamma\frac{\partial T}{\partial z}\right). \tag{16}$$

If equations (15) and (16) are used to rewrite Equation (14), we have finally

$$\frac{\partial^2 \xi}{\partial t^2} \sim \frac{1-\gamma}{1-R\rho}\frac{\partial}{\partial x}\left(\frac{\partial^2 T}{\partial z^2}\right) \tag{17}$$

This last equation means that the variation of vorticity (stream function, and then flow field) is determined by the horizontal variation of temperature diffusion and the coefficient $C=(1-\gamma)/(1-R\rho)$. As we used the flux law (Equation (8)) by *Linden* [1974], γ is a function of $R\rho$, so that, when $R\rho$ becomes nearer to unity, γ becomes also near to unity. From Equation (10), the temperature field does not depend on $R\rho$ variations so that the stream function could be determined solely by the behavior of C . In Figure 7, the variation of C is plotted against $R\rho$. C decreases at first, and takes a minimum value of −2.8 at $R\rho$ = 1.01; it then increases and tends to zero as $R\rho$ increases, as is predicted by the form of C in Equation (17). It is interesting to note that the value of C is accidentally almost the same when

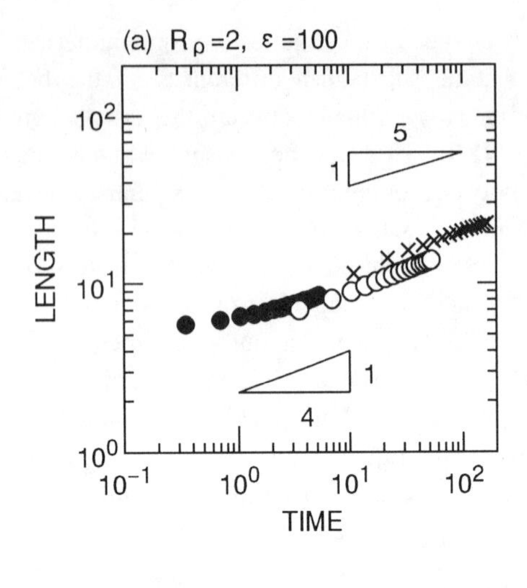

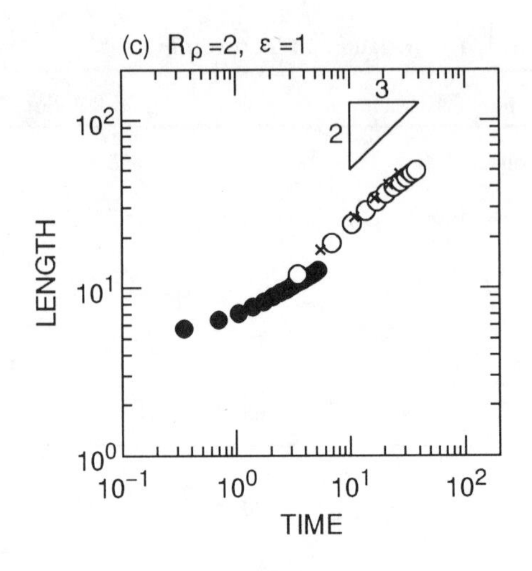

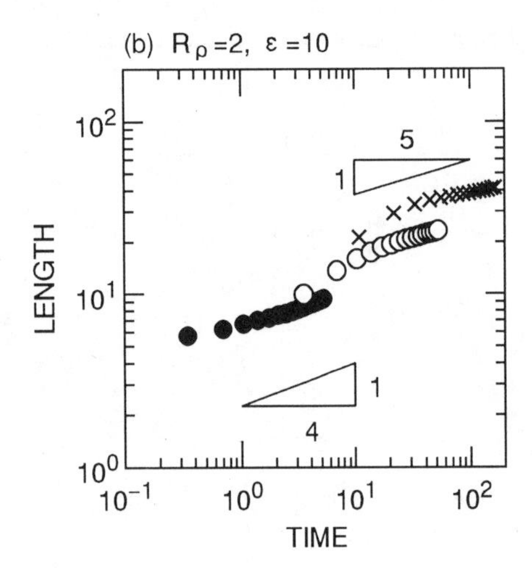

● Ra=10^4
○ Ra=10^6
× Ra=10^7

Fig. 6. *L–t* relations are shown in *log–log* frame for (a) $R\rho$ =2.0, ε =100, (b) $R\rho$ =2.0, ε =10 and (c) $R\rho$ =2.0, ε =1. Triangles in these figures indicate the slope of power of t in *L–t* relations which were obtained in Maxworthy's experiment[1983].

$R\rho$ = 1.01 and 2 (~ 0.8). Therefore, the stream function (the velocity field) and the density fields for both values of $R\rho$ resemble each other. As we mentioned before, the equation of heat balance does not depend on $R\rho$ so that the temperature fields of the two cases are also the same, although there is a considerable difference in the salinity field.

5. SUMMARY

Numerical experiments on a double diffusive density current are carried out when cold/fresh water is released on to the ambient warm/salty water.

It is found that the phenomena are governed by the combination of three parameters. One is the turbulent Prandtl number (ε); when ε is large, the density current is suppressed. A second parameter is the turbulent Rayleigh number (Ra) defined by $Ra=g^*H^3/K_{TH}^2$; when Ra is large, the induced convection becomes vigorous and the resulting density current becomes strong. The last parameter is the Turner number ($R\rho$) defined by $R\rho=\beta\Delta S/\alpha\Delta T$; when $R\rho$ is sufficiently small, the activity of double diffusive convection becomes large and is strongest when $R\rho$ = 1.01, then becomes weaker. The results of numerical experiments on double diffusively induced density current not only explain well *Maxworthy's* laboratory experiments [1983] but also cover a wide range of the parameters.

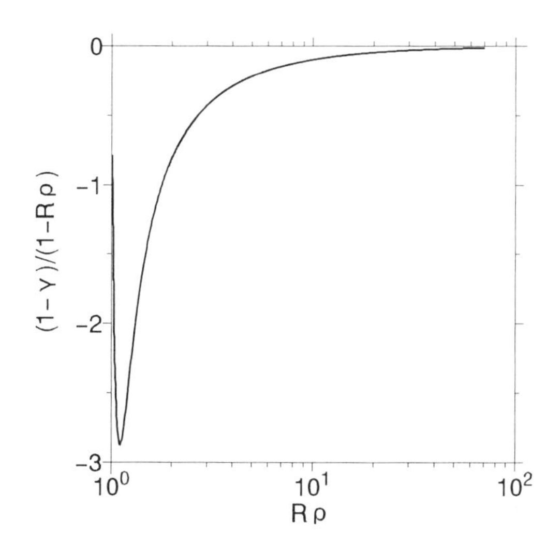

Fig. 7. The variation of $C=(1-\gamma)/(1-R\rho)$ is plotted against $R\rho$. C takes a minimum value of −2.8 at $R\rho$ =1.01.

In the real ocean, as in a fjord, if we take representative values of the scales as $H\sim100m$, $K_{TH}\sim10^{-2}m^2/\mathrm{sec}$, $g^*\sim10^{-2}m/\mathrm{sec}^2$, $\Delta T\sim10^\circ C$ and $\Delta S\sim4PSU$, then Ra becomes of order 10^8 and $R\rho\sim2$. It can be safely concluded that double diffusive convection occurs and plays an important role in the mixing process of such a field.

Acknowledgments. One of authors (J.Y.) was supported by the Grant in Aid for Scientific Research from the Ministry of Education and Culture in Japan. We express our sincere thanks to anonymous reviewers for their invaluable comments.

REFERENCES

Linden, P.F., A note on the transport across a diffusive interface, *Deep Sea Res., 21,* 283–287, 1974.

Maxworthy, T., The dynamics of double diffusive gravity currents, *J. Fluid Mech., 128,* 259–282, 1983.

McDougall, T.J., Oceanic intrusions: Some limitations of the Ruddick and Turner(1979) mechanism, *Deep Sea Res., 33,* 1653–1664, 1986.

Nagashima, H., J. Yoshida and M. Nagasaka, The behavior of double diffusive gravity current: Laboratory and numerical experiments, In *Double Diffusion in Oceanography: Proc. of a Meeting September 26–29, 1989,* compiled by R. W. Schmitt, pp. 193–200, Woods Hole Oceanog. Inst. Tech. Rept, WHOI–91–20, 1990.

Simpson, J.E., *Gravity Currents:In the Environmental and the Laboratory,* 244 pp., John Wiley and Sons.

Thangam, T., and C.F. Chen, Salt–finger convection in the surface discharge of heated saline jets, *Geophys. Astrophys. Fluid Dyn., 18,* 111–146, 1981.

Yoshida, J., H. Nagashima and Wen–Ju Ma, A double diffusive lock–exchange flow with small density difference, *Fluid Dynamics Res., 2,* 205–215, 1987.

J. Yoshida, H. Nagashima and M. Nagasaka, Department of Marine Science and Technology, Tokyo University of Fisheries, 4–5–7, Konan, Minato–ku, Tokyo, Japan 108

Double Diffusively Induced Intrusions into a Density Gradient

Moh Nagasaka[1], Hideki Nagashima[2] and Jiro Yoshida[1]

1) Department of Marine Science and Technology, Tokyo University of Fisheries, Tokyo, Japan
2) Laboratory of Earth Science, The Institute of Physical and Chemical Research(RIKEN), Saitama, Japan
Present address: Department of Marine Science and Technology, Tokyo University of Fisheries, Tokyo, Japan

Double diffusive intrusion into a density gradient is investigated experimentally. We focus our attention on the dynamic structure of intrusion, although experimental conditions were essentially same as *Turner* [1978]. We released a sugar (salt) water body at its neutral buoyancy level into salt (sugar) stratified water. As an initial density anomaly of intruding water becomes large, a released water splits to form multiple intrusions. The total layer thickness(H) of multiple intrusion is proportional to the predicted vertical scale " H " [*Ruddick & Turner*, 1979].The averaged velocity of intrusion is proportional to NH (N ;buoyancy frequency), but the proportional constant is smaller when split occurs. This result suggests that there is some dissipation associated with the occurrence of split. The "collective instability" mechanism could be responsible for this phenomena "split", because the layer splitting occurred when the Stern Number is much greater than order 1.

1. INTRODUCTION

Double diffusive intrusions into a density gradient were first investigated experimentally by *Turner* [1978]. He released a body of water at its neutral buoyancy level into another body of water stratified with different property. He explained qualitatively the behavior of resulting intrusion in terms of convection induced by differences in molecular diffusivities between two fluids, that is, the double diffusive convection. *Ruddick and Turner* [1979] later produced a lateral interleaving by removing a barrier placed between two fluid layers stratified with different properties, but density distribution is same(salt and sugar was used in their experiment).

In these experiments, salty intrusions sink and sugar ones rise as they advance. In real ocean, this phenomenon corresponds to that warm/salty intrusions rise and cold/fresh ones sink. This somewhat curious behavior is explained by the differences between the magnitude of vertical density flux through the salt finger interface and that through the diffusive interface. In addition, *Ruddick and Turner* [1979] showed that the thickness of intrusive layer is proportional to the local horizontal density anomaly of slowly diffusing component and inversely proportional to vertical density gradient.

However, the dynamics governing the propagation of such intrusions was not discussed. Unpublished results of *Ruddick and Turner* [1979] suggest that the intrusion velocity(U) should be proportional to Nh (N :buoyancy frequency, h :layer thickness) and is about $0.005Nh$ [*Ruddick and Hebert*, 1988].

Yoshida et al. [1987] carried out a double diffusive lock exchange flow experiment and showed that even when two water masses initially separated by the lock gate have the same density, a significant horizontal flow can be induced by double-diffusive effects, provided that these water masses have contrasts in temperature and salinity. They also considered a simple force balance on the double diffusive current, and briefly discussed the dynamics of current.

Bormans [1992] recently investigated the *Ruddick and Turner* [1979] type experiment in more detail. She confirmed the $U \propto Nh$ relation which was suggested in the *Ruddick and Turner* [1979] experiment.

In the present paper, we carried out *Turner* [1978] type

Double-Diffusive Convection
Geophysical Monograph 94

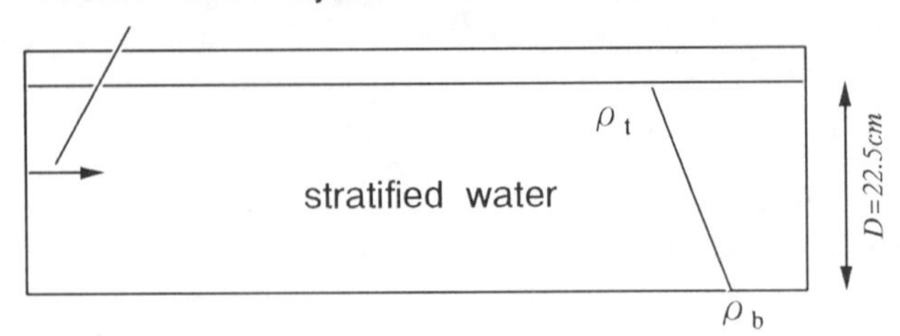

Figure 1. Schematic view of experimental apparatus. We released a sugar (salt) water into a salt (sugar) stratified water in Case 1 (Case 2).

experiments extensively under various experimental conditions and tried to find the mechanism which interpret the behavior of intrusion.

2.EXPERIMENT

Sketch of experimental apparatus is shown in Figure 1. The tank made of perspex is 200cm long, 30cm deep and 20cm thick. We released a sugar(slowly diffusing component) water at its neutral buoyancy level into a stratified salt (faster diffusing component) water from a two-dimensional line source at the end of tank. The sugar water is supplied continuously by the rate of 30ml/min. Hereafter, we call this experiment Case 1.

We also released a salt water into a stratified sugar water. We call this experiment Case 2, where another tank made of perspex of 100cm long, 30cm deep and 20cm thick was used.

A typical example of resulting intrusion of Case 1 is shown in Plate 1. It is found that a vigorous convection is first formed near the source region, and then the source water begins to intrude into an ambient fluid. In some cases, a new intrusive layer is formed just below the first intrusion layer(we call this phenomena "split"), and in turn other new intrusive layers are successively formed below. Each intrusion has 'diffusive interface' above and 'finger interface' below and tilts upward as it advances. In Case 2, the behavior of resulting intrusion has an opposite sense. Namely, each intrusion has 'finger interface' above and 'diffusive interface' below, new layers are formed above and tilts downward and so on.

The densities of released sugar water (ρ_T , Case 1) and salt water (ρ_S , Case 2) are defined as

$$\rho_S=\rho_0(1+\beta\Delta S_0) \tag{1}$$

and

$$\rho_T=\rho_0(1+\alpha\Delta T_0)\ . \tag{2}$$

Here, ρ_0 is the reference density, $\beta\Delta S_0$ and $\alpha\Delta T_0$ are initial density anomalies due to sugar and salt, respectively. β and α are contraction coefficients of sugar and salt,respectively.

The ambient stratification is characterized by the buoyancy frequency N , and is

$$N^2=-\frac{g}{\rho_0}\frac{\partial\rho}{\partial z}=\frac{g}{\rho_0}\frac{(\rho_b-\rho_t)}{D}\ . \tag{3}$$

Here, g is the gravitational acceleration, ρ_b the density near the bottom, ρ_t the density near the surface and D the total depth.

Shadowgraph images of the tank were taken successively at intervals of 5 minutes(Case 1) or 2 minutes (Case 2). The length of intrusions (L) were measured against time. Each experiment was terminated when the furthest intrusion reaches at 60cm from the source. At that time, the total thickness (H) of intrusion was measured at the source position. The thicknesses of each intrusion (h) were measured at the position 30cm source-ward from the head of intrusion. There are generally thin gaps between layers at the points of measurement so that $\Sigma h_i<H$. Definitions of measured quantities are shown in Figure 2.

In our experiment $\beta\Delta S_0$, $\alpha\Delta T_0$ and N^2 are changed in various combination as listed in Table 1a and 1b.

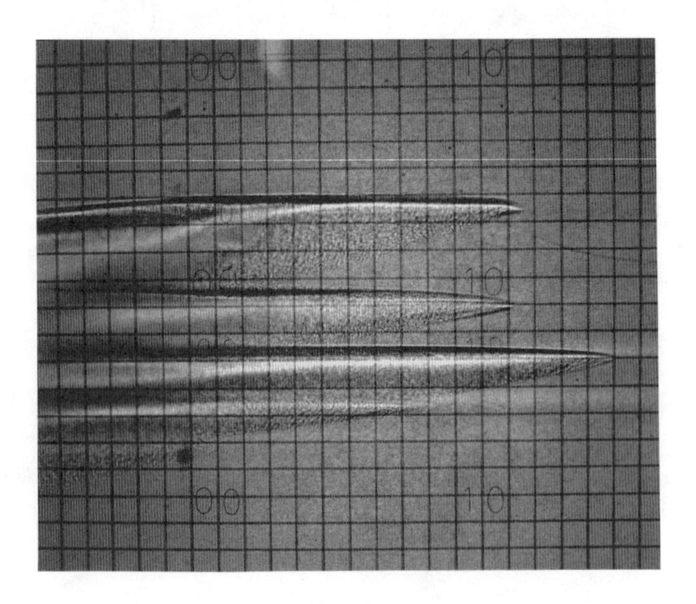

Plate 1. A typical example of intrusion for Case 1. This photograph is taken at 20 minutes after the injection.

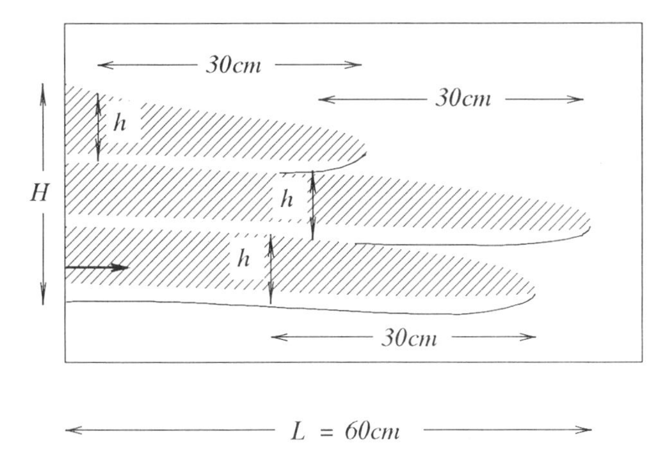

Figure 2. Definition of measured quantities.

3. RESULTS

Vertical length scales of double diffusive intrusions were discussed by *Ruddick and Turner* [1979]. They considered the difference in potential energies between the initial state and the final state of intrusion. At the final state, a salt finger convection has run down, and the initial density anomaly of slowly diffusing component $\beta\Delta S_0$ vanishes. They obtained the vertical length scale d as

$$d=\frac{g\beta\Delta S_0(1-\gamma)}{N^2}, \tag{4}$$

where γ is the density flux ratio defined as

$$\gamma=\frac{\alpha F_T}{\beta F_S}. \tag{5}$$

Here, αF_T and βF_S are vertical density fluxes of faster and slower diffusing component. Equation (4) means the vertical length scale in which fluid particle, having implicit potential energy, can move vertically in the stratification of $\partial\rho/\partial z$.

In case 1, we used the length scale " d " as the representative scale for the thickness of intrusion. In Case 2, we calculated d by replacing $\beta\Delta S_0$ by $\alpha\Delta T_0$ in equation (4). The value of density flux ratio 0.88 [*Griffiths and Ruddick*, 1980] is used in equation(4). This predicted vertical scales (d) and experimental results are summarized in Table 1a and 1b.

As we see in Plate 1, intrusion sometimes splits to form multiple intrusions. Then, we define the average thickness of intrusion h_{av} in each experiment as

$$h_{av}=\frac{\Sigma h_i}{K} \qquad (i=1,2,\cdots,K). \tag{6}$$

Here, K is the number of split. In Figure 3, h_{av} is plotted against d together with the results of *Ruddick and Turner* [1979] and Bormans [1992]. For relatively small values of d , our results are well coincided with them. But when d becomes large, our results show different tendency from their results.

The velocity of each intrusion (V) is defined as L/T , where L is the length of intrusion at the end of each run and T is the elapsed time. We plotted V against Nh in Figure 4. V seems to be proportional to Nh , however the correlation is very low.

Next we focus our attention on the number of split layers. In Figure 5, K is plotted against the initial density anomaly $\beta\Delta S_0$ ($\alpha\Delta T_0$). It can been seen that K increases with the increasing $\beta\Delta S_0$ ($\alpha\Delta T_0$).

4. DISCUSSION

As seen in previous section, the occurrence of the split should have an important effect on the behaviors of intrusions. Here we discuss several characteristics of our experimental results with taking the phenomena "split" into consideration.

4.1 *Thickness of Intrusion*

In Figure 6, h_{av} is re-plotted against d with taking the number of split into consideration. It is seen that h_{av} are almost proportional to d and the proportional constant decreases with increasing the number of split K . Thus, the phenomena "split" is probably essential to decide the thickness of intrusion.

In *Ruddick and Turner* [1979], the "split" was not observed. In our case, however, the "split" is sometimes observed and the thicknesses of split layers are generally smaller than d . Another observed vertical scale is the total thickness H which represents the sum of each layer thickness. In Figure 7, H is plotted against d together with *Ruddick and Turner* [1979] and *Bormans* [1992]. It can been seen that H is almost proportional to d when split does not occur or the number of split is small. There is a little difference in magnitude of H between *Ruddick and Turner* [1979] and ours. This might be caused by the difference of the supply of the released water; in our case, the flow rate is constant. It is noted that when the "split" occurs, there is a change of tendency from the case of a single intrusion.

TABLE 1a. Summary of experimental results for Case 1. H :Total thickness of intrusion. $h_{av}(=(1/K)\Sigma h_i)$:Averaged thickness of each split intrusion(h_i :thickness of i th intrusion). $d(=g\beta\Delta S_0(1-\gamma)/N^2)$:Predicted thickness scale of intrusion by *Ruddick and Turner* [1979]. K :Number of split of intrusions. $V_{av}(=\Sigma V_i h_i/\Sigma h_i)$:Averaged velocity of split intrusion(V_i :thickness of i th intrusion).

N 1/sec	$\beta\Delta S_0$	H cm	h_{av} cm	d cm	K	V_{av} cm/min
0.48	0.0222	22.0	4.2	11.2	4	0.54
0.79	0.0865	16.5	3.3	16.2	5	0.88
0.82	0.0612	11.9	3.0	10.6	5	0.86
0.93	0.0345	12.5	2.6	4.7	4	1.10
1.01	0.0089	5.8	2.3	1.0	1	0.91
1.27	0.0751	13.0	2.8	5.4	5	0.80
1.28	0.1139	13.4	2.7	8.2	5	1.09
1.30	0.0489	11.8	2.8	3.4	4	0.73
1.42	0.0209	5.7	3.4	1.2	2	0.65
1.56	0.1139	13.2	2.7	5.5	5	1.01
1.60	0.0868	11.1	2.4	4.0	4	0.88
1.64	0.0612	10.0	2.6	2.7	4	0.70
1.69	0.0332	6.0	3.2	1.4	2	0.73
1.98	0.0791	9.5	2.4	2.4	4	0.69

TABLE 1b Summary of experimental results for Case 2 . Quantities are same as in Table 1a, except that $\beta\Delta S_0$ is replaced with $\alpha\Delta T_0$.

N 1/sec	$\alpha\Delta T_0$	H cm	h_{av} cm	d cm	K	V_{av} cm/min
1.16	0.0486	12.1	3.4	4.3	4	0.98
1.18	0.0135	7.5	4.5	1.1	1	1.66
1.19	0.0330	11.6	3.2	2.7	3	0.75
1.54	0.0682	12.0	3.1	3.4	4	1.04
1.58	0.0330	8.0	4.0	1.6	2	1.00
1.64	0.0135	4.9	3.0	0.6	1	1.92
1.69	0.0682	12.0	3.0	2.8	4	1.04
1.82	0.0135	3.8	2.9	0.5	1	2.92
1.83	0.0330	7.3	3.2	1.2	2	0.97
1.83	0.0544	9.1	4.0	1.9	2	0.92
1.97	0.0544	7.0	2.3	1.6	3	0.95
2.07	0.0330	6.1	2.5	0.9	2	1.30
2.08	0.0135	3.1	2.3	0.4	1	3.54
2.12	0.0686	8.3	3.7	1.8	2	1.10

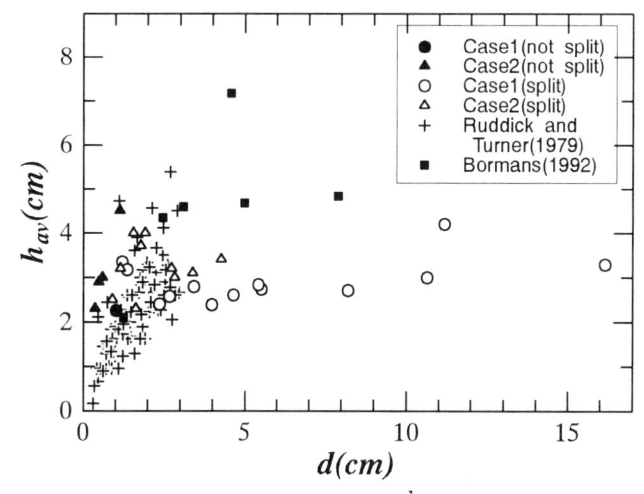

Figure 3. Averaged layer thickness h_{av} of each intrusion against d .

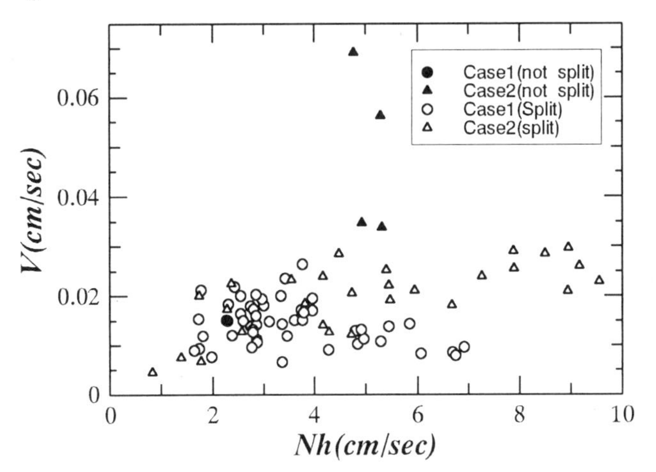

Figure 4. Velocity of each intrusion V against Nh .

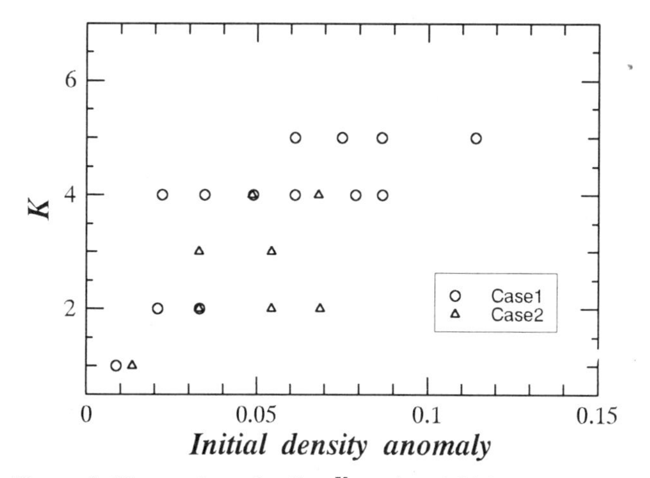

Figure 5. The number of split K against initial density anomaly.

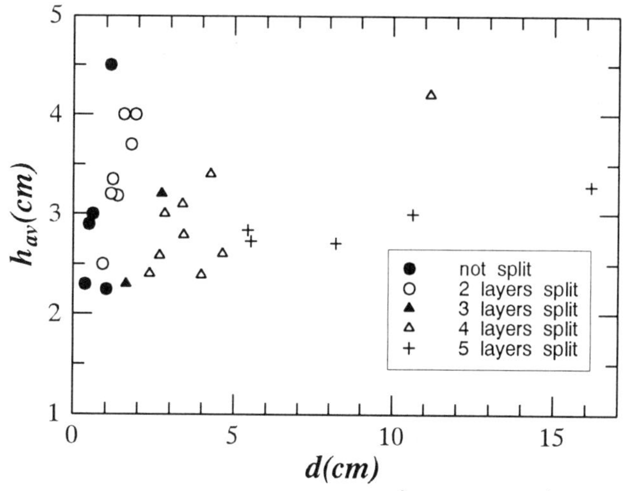

Figure 6. Averaged layer thickness h_{av} against d as in Figure 3, however, the symbols are changed according as the number of split.

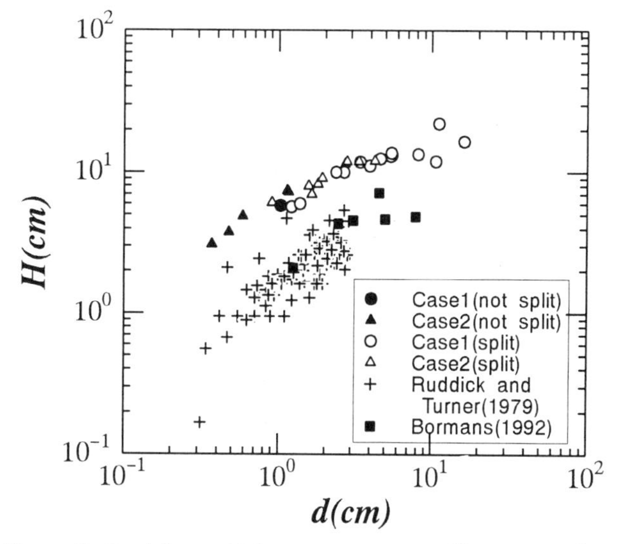

Figure 7. Total layer thickness of intrusion H against d .

4.2 *Velocity of Intrusion*

Ruddick and Turner [1979] and*Bormans* [1992] have suggested that the velocity of intrusive layer is proportional to Nh . Our results in Figure 4 suggest the same tendency with *Ruddick and Turner* [1979]. However, there is a substantial difference between the case whether split occur or not. In the previous section, we also see that velocities of each intrusions are weakly proportional to Nh except the case when intrusion splits. In the case of split, each split layer advances with a little different propagation speed. Thus, we calculate the averaged velocity of each split intrusion by the definition:

$$V_{av}=\frac{\Sigma V_i h_i}{h_i} \qquad (i=1,2,\cdots,K). \tag{7}$$

V_{av} is plotted against NH in Figure 8 together with results of *Bormans* [1992]. We can see that V_{av} is proportional to NH in both cases, but the proportional constant is quite different each other in the cases whether the "split" occurs or not. When split does not occur, the tendency is similar to those by *Bormans* [1992]. When the "split" occurs, however, V_{av} becomes small.

Yoshida et al. [1987] conducted double diffusive lock exchange flow experiments. In their experiment, resulting surface or bottom current also spread at a constant velocity. They explained this results in terms of the force balance between double diffusively induced buoyancy force and double diffusive retarding force. This force balance was first introduced by *Maxworthy* [1983]. As was done by *Yoshida et al.* [1987], we only consider the case that salt finger convection dominate the vertical transport. From *Yoshida et al.* [1987] double-diffusively induced buoyancy force F_g should be written as

$$F_g=2\rho_0(\beta\Delta S_0)(1-\gamma)gd^2(1-(1+\frac{t}{\tau})^{-3}), \tag{8}$$

where, t is time and $\tau(=\frac{1}{3}\rho_0 d/C(\beta\Delta S_0)^{1/3})$ is a characteristic time scale. And double diffusive retarding force F_{DD} ($=\rho u v^* L$:where, v^* is double diffusively induced vertical velocity, L is a length of current) is

$$F_{DD}=C(\beta\Delta S_0)^{\frac{1}{3}}(1+\frac{t}{\tau})^{-1}L^2 t^{-1}, \tag{9}$$

where C is a constant to be determined experimentally. If F_g is balanced with F_{DD} , it becomes that

$$L^2=\frac{2\tau}{3}(\beta\Delta S_0(1-\gamma))gd\cdot f(t). \tag{10}$$

Here,

$$f(t)=(1+\frac{t}{\tau})(1-(1+\frac{t}{\tau})^{-3}). \tag{11}$$

If $t\ll\tau$, we get $f(t)\doteq 3t^2/\tau$ and if $t\gg\tau$, we get $f(t)\doteq t^2/\tau$. then we obtain the following relation for the current length L

$$L\propto((\beta\Delta S_0)(1-\gamma)gd)^{\frac{1}{2}}\cdot t. \tag{12}$$

This is easily differentiated by t , and get the expression for the current velocity V

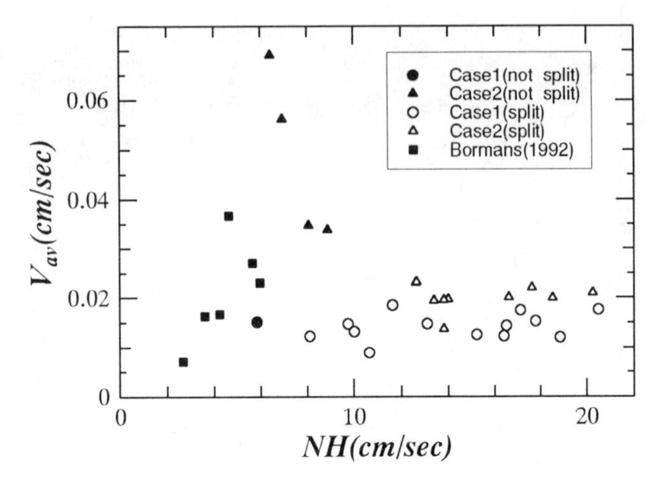

Figure 8. Averaged layer velocity V_{av} against NH .

$$V\propto((\beta\Delta S_0)(1-\gamma)gd)^{\frac{1}{2}}. \tag{13}$$

If we use equation (4), we get a simple expression for V as

$$V\propto Nd. \tag{14}$$

As shown in Figure 8, V_{av} is proportional to NH . When the "split" does not occur, V_{av} is proportional to d , because H is proportional to d (Figure 7). In this case, the retarding force F_{DD} must be balanced with the buoyancy force F_g . In case of "split", however, the intrusion velocity V_{av} is clearly smaller than the expected velocity above. This fact suggests that the energy probably dissipated when the split occurs.

4.3 *Split*

As shown in Figure 5, we found that when the initial density anomaly becomes large, the number of split increases. Such splitting seems to be concerned with the activity of salt finger convection from its appearance. Thus, the active salt finger is closely related to the magnitude of initial density anomaly $\beta\Delta S_0$, and cause the split of intrusion. The strong fingering plume becomes probably unstable, and mix the surrounding water and produce the water body whose density is a little different from that of firstly developed intrusion. Accordingly, mechanism of the split seems to be concerned with stability of finger plumes.

Stability of salt finger convection was first discussed by *Stern* [1969] in terms of a balance between the net mass transport and viscous dissipation. This mechanism is named "collective instability". If salt finger convection was once developed, energy is transferred from groups of fingers to large scale internal wave field. In some conditions, this transfer of energy exceed the viscous dissipation, and salt finger convection is destructed to form new convection layer. He proposed a non-dimensional number (St :Stern Number) for a plume stability as follows:

$$St=\frac{g\beta F_S-g\alpha F_T}{\nu N^2}=\frac{g\beta F_S(1-\gamma)}{\nu N^2}, \qquad (15)$$

where ν is the molecular viscosity. The critical value of stability is

$$St\sim O(1) \qquad (16)$$

In our experiment, using a value 0.88 for γ and substituting it in *Kelly*'s formula [1988], we can estimate density flux βF_S . Hence, the Stern number can be calculated by using above values together with the value of N listed in Table 1a and 1b. As shown in Figure 9, it is found that the number of split increases when Stern number becomes large. Therefore, it is considered that if initial density anomaly is large, salt finger convection below (or above) the intrusive layer becomes so vigorous and transfer of energy to internal wave field might exceed the viscous dissipation. Then salt finger convection is to break down to form new intruding layer below (or above).

5. SUMMARY

In the present paper, laboratory experiments on double diffusive intrusions into a density gradient are conducted. We focus our attention on the dynamic structure of intrusion, although experimental conditions are essentially same as *Turner* [1978] experiment. In general, when an initial density anomaly of intruding layer is small, the thickness of intrusion (h) is proportional to Ruddick and Turner[1979] scale " d " and the velocity of intrusion is proportional to Nd . When an initial density anomaly of intruding water becomes large, a released water splits to form multiple intrusions. In this case, the total layer thickness (H) of multiple intrusions is proportional to d . Moreover, the averaged velocity of intrusion is proportional to NH : the proportional constant is smaller when split occurs. Such experimental results suggest that when the multiple intrusion occurs, there should be some energy dissipation. The "collective instability" mechanism could be responsible for this phenomena "split", because the layer splitting occurred when Stern Number is much grater than order 1.

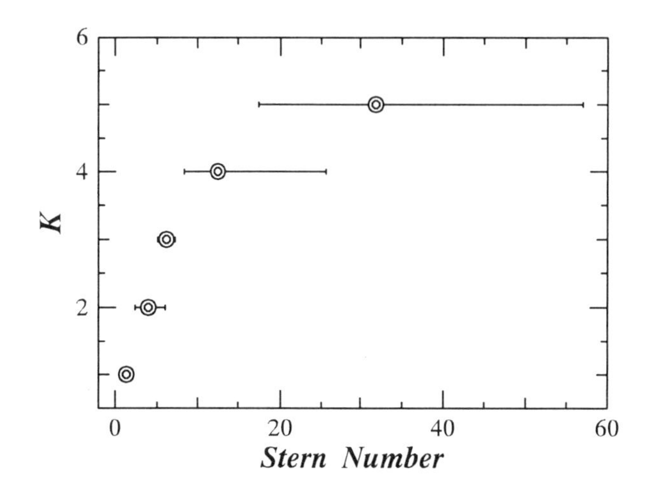

Figure 9. The number of split K against Stern Number. In the figure, bars show the limit of distribution of data.

Acknowledgments One of authors (J.Y.) was supported by the Grant in Aid for Scientific Research from the Ministry of Education and Culture in Japan. They are grateful to Mr.A.Miura, a student at Tokyo University of Fisheries, for his assistance in a part of the experiment. We express our sincere thanks to anonymous reviewers for their invaluable comments.

REFERENCES

Bormans, M., Effect of R_ρ on double diffusive interleaving, Deep Sea Res., 39(5), 871–884, 1992.

Griffiths, R.W., and B.R. Ruddick, Accurate fluxes across a salt–sugar finger interface deduced from direct density measurements, J. Fluid Mech., 99, 85–96, 1980.

Kelly, D., Oceanic Thermohaline staircases, Ph.D Thesis, Dalhousie University, Halifax, Nova Scotia, 1986.

Maxworthy, T., The dynamics of double diffusive gravity currents, J. Fluid Mech., 128,259–282,1983.

Ruddick, B.R., and J.S. Turner, The vertical length scale of double–diffusive intrusions, Deep Sea Res., 26A, 903–913, 1979.

Ruddick, B.R., and D. Hebert, The Mixing of Meddy "Sharon", In Small scale turbulence and mixing in the ocean, edited by C.J. Nihoul and B.M. Jamart, pp. 249–261 Elsevier, 1988.

Stern, M.E., Collective instability of salt fingers, J. Fluid Mech., 35, 209–218, 1969.

Turner, J.S., Double–diffusive intrusions into a density gradient. J. Geophy. Res., 83, 2887–2901, 1978.

Yoshida, J., H. Nagashima, and Wen–Ju Ma, A double diffusive lock–exchange flow with small density difference, Fluid Dyn. Res., 2, 205–215. 1987.

J. Yoshida, H. Nagashima and M. Nagasaka, Department of Marine Science and Technology, Tokyo University of Fisheries, 4–5–7, Konan, Minato–ku, Tokyo, Japan 108

Double Diffusive Flow Patterns in the Unicellar Flow Regime: Attractor Structure and Flow Development

E. Jurjen Kranenborg and Henk A. Dijkstra[1]

Institute for Marine and Atmospheric Research,
Department of Physics and Astronomy,
University of Utrecht, The Netherlands

A numerical study is performed of the two-dimensional double diffusive layer formation process in a laterally heated cavity containing a liquid which is stably stratified through a constant vertical salinity gradient. This gradient is imposed by fixing the salt field at the horizontal walls, such that the configuration allows for steady state solutions. The structure of these stationary solutions and their linear stability is determined, as the Rayleigh number is varied, using continuation methods for a fixed value of the buoyancy ratio. This bifurcation structure is complicated at relatively small Rayleigh number where multiple stable solutions exist. At much larger Rayleigh number, the structure is relatively simple in that only a few solutions were determined of which only one is stable. Transient simulations of the evolution of the flow show that on a relatively fast time scale, the flow approaches one of the unstable equilibria and remains near it for a long time. The instability of this state occurs at a larger time scale and eventually the stable steady state is reached. Hence, a connection between (unstable) steady flow patterns and quasi-stationary flows is demonstrated. The physics of the slow time scale instability is still not precisely clear, but is likely to be related to the intercellular diffusion of salt.

1. INTRODUCTION

When a lateral temperature gradient is applied to a motionless liquid layer which is stably stratified through a vertical salinity gradient, double diffusive layer formation may occur. Parcels of liquid near the heated wall initially start to move upwards. These parcels retain almost all of their salt due to its very small diffusivity but lose their heat relatively fast. Eventually the parcel rises to a level where its density is equal to that of the surrounding fluid. Because of continuity these parcels are then forced to move laterally and thus a layered convection pattern emerges [*Huppert and Turner*, 1980b]. Since these layers significantly change the transport of heat and salt, double diffusive convection is a potentially important transport mechanism e.g. for heat and salt in the ocean [*Schmitt*, 1994].

Much information on the layer formation process was obtained from laboratory experiments. These were performed either in narrow slots or in wide tanks. They differ also in the way the heating is imposed on the lateral walls, for example, slowly [*Chen et al.*, 1971; *Thorpe et al.*, 1969; *Wirtz et al.*, 1972] or through a prescribed time dependence [*Tanny and Tsinober*, 1988]. One of the main results of the latter studies was that a layered convection pattern with a vertical lengthscale

$$\eta = \frac{\alpha \Delta T}{\beta(\partial S_0/\partial z)} \qquad (1)$$

only appears when a Rayleigh number Ra_η based on this length scale, exceeds a critical value given approximately by $Ra_{\eta,cr} = 1.5*10^4$. Here $\partial S_0/\partial z$ is the initially constant vertical salt gradient and the length scale η is

[1]E. Jurjen Kranenborg & Henk A. Dijkstra, Institute for Marine and Atmospheric Research, Department of Physics and Astronomy, University of Utrecht, Princetonplein 5, 3584 CC Utrecht, The Netherlands

Double-Diffusive Convection
Geophysical Monograph 94

directly related to the movement of a heated liquid parcel to its neutrally buoyant level. This indicates that a layered structure arises due to a diffusive instability of a buoyancy driven background flow. Also in experiments in wide tanks, eventually layers with the characteristic scale η develop [*Huppert and Turner*, 1980a; *Jeevaraj and Imberger*, 1991; *Tanny and Tsinober*, 1988]. The evolution to this pattern depends strongly on the heating applied to one of the sidewalls.

The layer formation problem has been studied theoretically in idealized models. Thangam et al. [*Thangam et al.*, 1981] determine the linear stability boundary of a steady parallel flow in a narrow slot and show that there is a transition from shear induced instabilities (at small solutal Rayleigh number) to double diffusive instabilities (at large solutal Rayleigh number). A rigorous series of studies on the instability of the boundary layer valid for gradual wall heating has been performed by Kerr [*Kerr*, 1989; *Kerr*, 1990]. He demonstrated that for this case, using a weakly nonlinear analysis, the instability is oscillatory and that small amplitude flows exist below the instability boundary (although these were shown to be unstable).

Although much is known on the diffusive instability, there have been only a few studies dealing with finite amplitude flows in laterally heated stably stratified liquids at large Rayleigh number. Lee and Hyun [*Lee and Hyun*, 1991a; *Lee and Hyun*, 1991b] studied the evolution of the layered structure in a laterally heated narrow slot for different buoyancy ratios by transient integration of the governing equations. They found flow patterns with a characteristic length scale η for Ra_η values larger than $Ra_{\eta,cr}$, but also found other solutions with a much larger length scale. Recently, multiple stable cellular flow patterns at values of Ra_η much smaller than $Ra_{\eta,cr}$ were found [*Tsitverblit and Kit*, 1993; *Kranenborg and Dijkstra*, 1994]. Although the larger scale convection patterns are induced by the restraining effects of the top and bottom walls, there are also smaller scale (unstable) flows where double diffusion might be important.

It appears that one of the stable flow patterns found at small Ra_η [*Kranenborg and Dijkstra*, 1994] is very similar to one of the large length scale solutions at very large Ra_η found by Lee and Hyun [*Lee and Hyun*, 1991a]. The question arises whether the other solutions, as found for small Ra_η, connect up with the solutions found by Lee and Hyun [*Lee and Hyun*, 1991a] with characteristic length scale η. In this paper, we investigate this for the particular case of fixed buoyancy ratio. The salt boundary conditions at the horizontal walls are chosen to allow for steady states to exist (as in [*Lee and Hyun*, 1991a]). We use techniques from numerical bifurcation theory to follow the branches of steady solutions, as computed in [*Kranenborg and Dijkstra*, 1994], to very large Ra_η. Moreover, time integration is used to study the evolution of the flow at particular locations in parameter space. We demonstrate that indeed a connection as suggested above exists. The solutions with characteristic length scale η correspond to continuations of unstable steady states. During the evolution of the flow, they are approached on a relatively short time scale. On a much longer time scale the instability occurs and (in our particular example), a stable steady state is reached.

2. FORMULATION AND NUMERICAL METHODS

A two-dimensional rectangular container (length L and height H) is filled with a Newtonian liquid with a constant thermal diffusivity κ_T and viscosity μ. A stable vertical salinity gradient is maintained in the liquid by imposing a constant salinity difference ΔS between the horizontal walls of the container; the vertical heat flux at these walls vanishes. A constant horizontal temperature difference ΔT is applied between the vertical walls, which are impervious to salt. The density ρ depends linearly on temperature and salinity and given by $\rho = \rho_0(1 - \alpha(T^* - T_0) + \beta(S^* - S_0))$. The relative importance of saline versus thermal buoyant forcing is given by the buoyancy ratio $R_\rho = \beta\Delta S/\alpha\Delta T$. The governing equations are non-dimensionalized using scales H, H^2/κ_T and κ_T/H for length, time and velocity, respectively. A dimensionless temperature and salinity are defined by $T = (T^* - T_0)/\Delta T$ and $S = (S^* - S_0)/\Delta S$. In terms of the streamfunction ψ and vorticity ω, where

$$u = \frac{\partial\psi}{\partial z},\ w = -\frac{\partial\psi}{\partial x},\ \omega = -\nabla^2\psi \tag{2}$$

the full equations , with the usual Boussinesq approximation, are given by:

$$Pr^{-1}(\frac{\partial\omega}{\partial t} - J(\psi,\omega)) = \nabla^2\omega + Ra_T(\frac{\partial T}{\partial x} - R_\rho\frac{\partial S}{\partial x}) \tag{3}$$

$$\frac{\partial T}{\partial t} - J(\psi,T) = \nabla^2 T \tag{4}$$

$$\frac{\partial S}{\partial t} - J(\psi,S) = Le^{-1}\nabla^2 S \tag{5}$$

where the Jacobian J is defined as

$$J(a,b) = \frac{\partial a}{\partial x}\frac{\partial b}{\partial z} - \frac{\partial a}{\partial z}\frac{\partial b}{\partial x} \tag{6}$$

At all boundaries no-slip conditions for velocity are prescribed and for the temperature and salinity the following boundary conditions hold:

$$x = 0: \ T = -\tfrac{1}{2}, \ \tfrac{\partial S}{\partial x} = 0,$$
$$x = A: \ T = \tfrac{1}{2}, \ \tfrac{\partial S}{\partial x} = 0, \qquad (7)$$
$$z = 0: \ S = 1, \ \tfrac{\partial T}{\partial z} = 0,$$
$$z = 1: \ S = 0, \ \tfrac{\partial T}{\partial z} = 0. \qquad (8)$$

The dimensionless parameters which appear in the equations above are defined as

$$Ra_T = \frac{g\alpha\Delta T H^3}{\nu\kappa_T}, \ R_\rho = \frac{\beta\Delta S}{\alpha\Delta T},$$
$$Pr = \frac{\mu}{\kappa_T\rho_0}, \ Le = \frac{\kappa_T}{\kappa_S}, \ A = \frac{L}{H} \qquad (9)$$

The solutal Rayleigh number is defined as

$$Ra_S = Ra_T R_\rho = \frac{g\beta\Delta S H^3}{\nu\kappa_T} \qquad (10)$$

A relation between the Rayleigh numbers Ra_η, used in [*Chen et al.*, 1971], and Ra_T is given by

$$Ra_\eta = Ra_T/(R_\rho)^3 \qquad (11)$$

Using $\partial S_0/\partial z = \Delta S/H$, a straightforward relation exists between the lengthscales H, η and the buoyancy ratio, i.e.

$$\frac{H}{\eta} = R_\rho \qquad (12)$$

In a liquid layer of height H, solutions with characteristic length scale η therefore correspond to R_ρ cells.

The equations and boundary conditions were discretized using a finite volume finite difference method as in [*Dijkstra*, 1992]. We use three type of numerical codes to study steady and transient solutions of the system of equations above. Steady states and their linear stability are calculated as a function of parameters using the continuation code as presented in [*Dijkstra et al.*, 1994]. A non- equidistant grid was used near the vertical boundaries in order to get an accurate representation of boundary layers. The mapping used to obtain an non-equidistant grid in x from an equidistant one in $\tilde{x}$ is given by:

$$x = \frac{1}{2} + tanh(q\ (\tilde{x} - \frac{1}{2}))/(2\ tanh(\ \frac{q}{2})) \qquad (13)$$

where q is a stretching factor. No stretching was applied in z because in addition to the boundary layers at the horizontal walls, also large internal vertical gradients, for instance in salinity, may occur.

Two time-dependent numerical solvers were used. First, an explicit code using a fast Poisson solver was used, having the advantage that it can be run at high resolution. However, its disadvantage is the restriction of the time step because of numerical instability. To use larger time steps, an implicit Crank-Nicolson method was used. Here the time step is only limited by accuracy and much larger time-steps than allowed by the explicit code can be taken. Hence, the explicit code was used in the initial stage of the development of the flow and the implicit code in the approach to (quasi) steady state. The codes were verified using standard problems and the choice of resolution and time step was based on extensive testing of the accuracy of the solutions.

3. RESULTS

The Prandtl and Lewis numbers are fixed at values corresponding to the heat/salt system: $Pr = 6.7, Le = 101$. The aspect–ratio A is fixed at $A = 1/2$ similar to that chosen by Lee and Hyun [*Lee and Hyun*, 1991a].

3.1. *Branches of Steady Solutions*

In these computations, we fixed the buoyancy ratio R_ρ and used Ra_η as the bifurcation parameter. Note that in this case, both Ra_T and Ra_S vary as Ra_η is changed. In Figure 1 the bifurcation diagram for $R_\rho = 3$ is shown for relatively small Ra_η, i.e. much smaller than the critical value as determined in [*Chen et al.*, 1971]. This figure was computed using a 25 x 41 grid. On the vertical axis, a value of the streamfunction at a particular gridpoint is plotted. Drawn (dotted) lines indicate stable (unstable) branches and singularities are indicated by heavy dots. When Ra_η is increased from zero, there exists a unique steady state consisting of two cells. This structure is thermally dominated and expected in a container with $A = 1/2$. This 2-cell pattern becomes unstable through a subcritical pitchfork bifurcation at the point labelled P_1. Two asymmetric solutions (but related through point-symmetry) branch off. Both remain unstable up to the limit point L_1, stabilize with increasing Ra_η and remain stable up to the bifurcation point P_2. Along the branch $L_1 - P_2$, the flow pattern changes from a 2-cell to a 1-cell solution. For values of Ra_η, larger than P_2, the 1-cell pattern is the only stable pattern. There are other branches on which the flow patterns have a smaller length scale, but these solutions are all unstable.

In this regime there are multiple stable steady states over an interval $L_1 - P_2$ in Ra_η. These patterns are

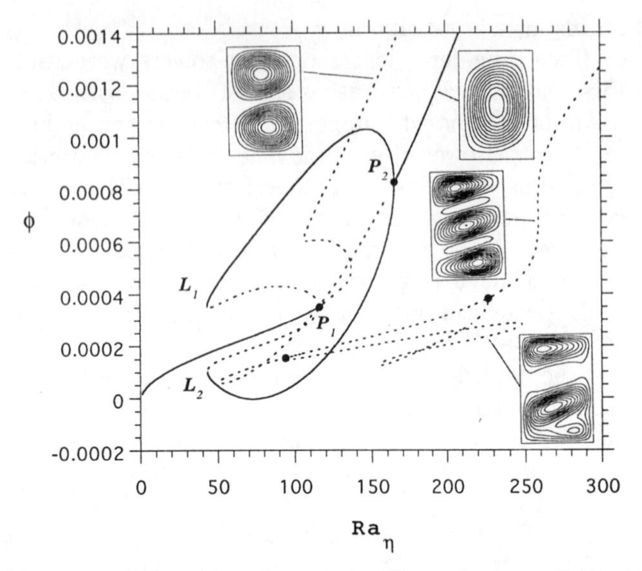

Figure 1. Bifurcation diagram for $R_\rho = 3$ at small Ra_η. Drawn branches are stable, dotted branches are unstable and large dots denote singularities. On the vertical axis ϕ corresponds with the value of ψ in a particular gridpoint. Note that the unstable branch starting at P2 is not connected to P1; the crossing is a visual effect caused by the particular choise of monitor function.

caused by the restraining influences of the container, just as in ordinary thermal convection. The unstable branches containing smaller scale patterns may be viewed as double diffusive modifications of the larger scale flow patterns. The three branches in figure 1 are continued to larger Ra_η and the result in shown in figure 2. A relatively simple bifurcation structure appears consisting of a stable branch of 1-cell solutions and an unstable 2-cell and 3-cell branch. Hence, there is no difference in the number and stability of steady states as $Ra_{\eta,cr}$ is crossed. One would expect long-term time-dependent calculations to approach the only stable solution. This is in agreement with the results of Lee and Hyun [*Lee and Hyun*, 1991a] for large Ra_η. For example, they find at $R_\rho = 3$ and $Ra_T = 8*10^7$ such a steady solution. The bifurcation diagrams, however, do not indicate how the steady state is approached. This is studied in the next section, using direct time integration of the governing equations.

3.2. *Transient Flows*

As an example of approach to steady state, $Ra_\eta = 2.37 * 10^4$, was chosen slightly larger than $Ra_{\eta,cr}$. According to experiments [*Chen et al.*, 1971] flow patterns with a lengthscale η may be expected at this location of parameter space. In this case, the saline and thermal Rayleigh numbers have values $Ra_S = 1.92 * 10^6$ and $Ra_T = 6.4 * 10^5$. The initial conditions correspond to a motionless linearly salt-stratified solution, i.e.

$$\psi = \omega = T = 0 \tag{14}$$

$$S = 1 - z \tag{15}$$

At $t = 0$ a constant horizontal temperature difference is imposed.

In Figure 3a, the development of the flow as a function of time is presented by plotting the maximum of the streamfunction. Also the dimensionless heat flux at $x = 0$, the Nusselt number

$$Nu = A \int_0^1 \frac{\partial T(0,z)}{\partial x} \, dz, \tag{16}$$

and the dimensionless salt flux at $z = 0$, the Sherwood number

$$Sh = -A^{-1} \int_0^A \frac{\partial S(x,0)}{\partial z} \, dx \tag{17}$$

are shown.

Three stages can be distinguished in the evolution of the flow. First a four cell solution is reached (figure 3b, panel a). This pattern is apparently unstable because the flow quickly changes, whereby the upper and lower cell weaken (figure 3b, panel b), to a two-cell pattern (figure 3b, panel c). The flow remains in this state for a long time (figure 3b, panel d). The flow eventually changes, whereby the lower cell is weakened (figure 3b, panel e), and approaches the stable steady 1-cell solution

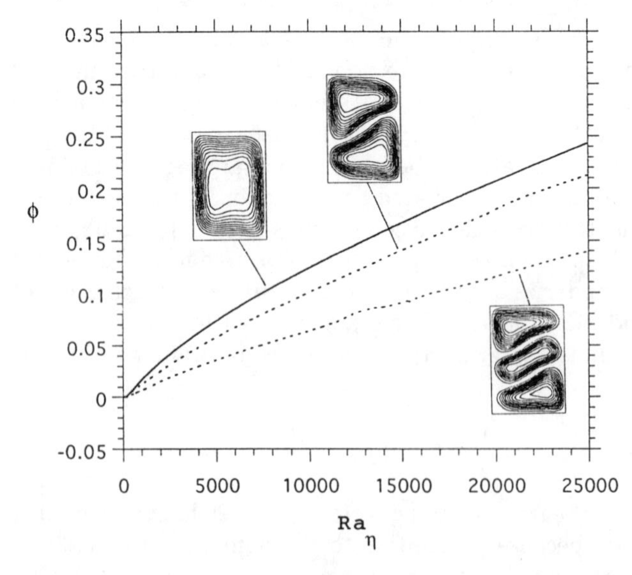

Figure 2. Bifurcation diagram for $R_\rho = 3$ and large Ra_η.

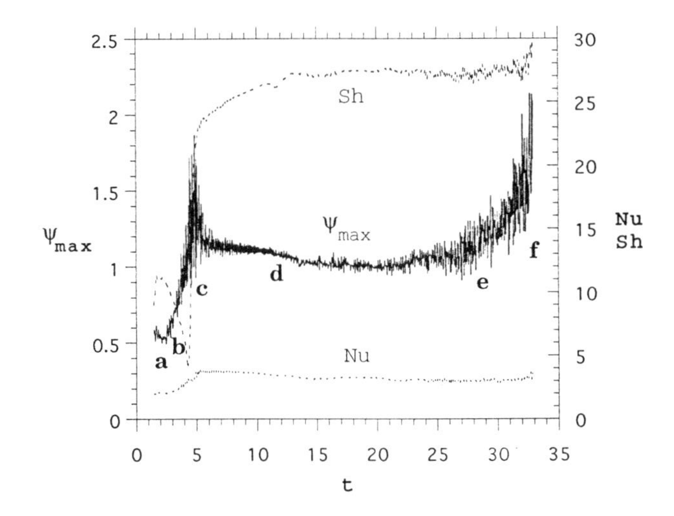

Figure 3a. Plot of the maximum of ψ, Nusselt number and Sherwood number as a function of time, $R_\rho = 3$.

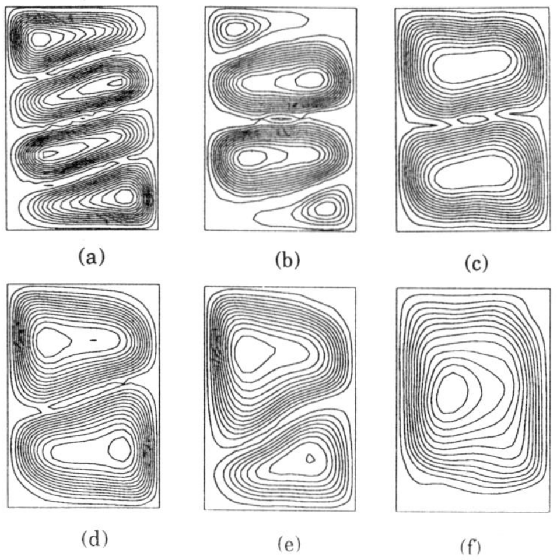

(a) (b) (c)

(d) (e) (f)

Figure 3b. Plots of the stream function at selected points in Figure 3a.

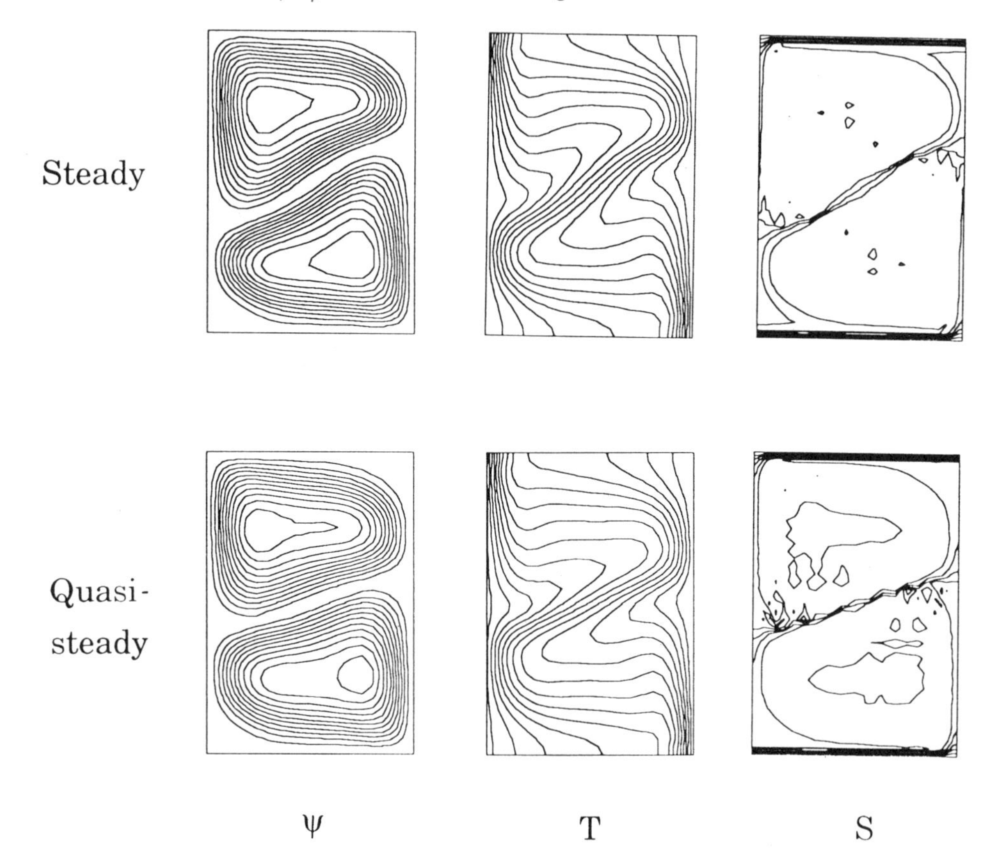

Figure 4. Comparison of steady-state and quasi-steady solutions (ψ, T, S) at $t = 11.5$.

already found with the bifurcation analysis (figure 3b, panel f).

A comparison of ψ, T and S (figure 4) shows a good agreement between the quasi steady-state (at point d in figure 3a) and the steady solution on the unstable 2-cell solution branch. Although this state is unstable, it is reached along its stable manifold. Hence, the instability of this pattern must be associated with a slow time

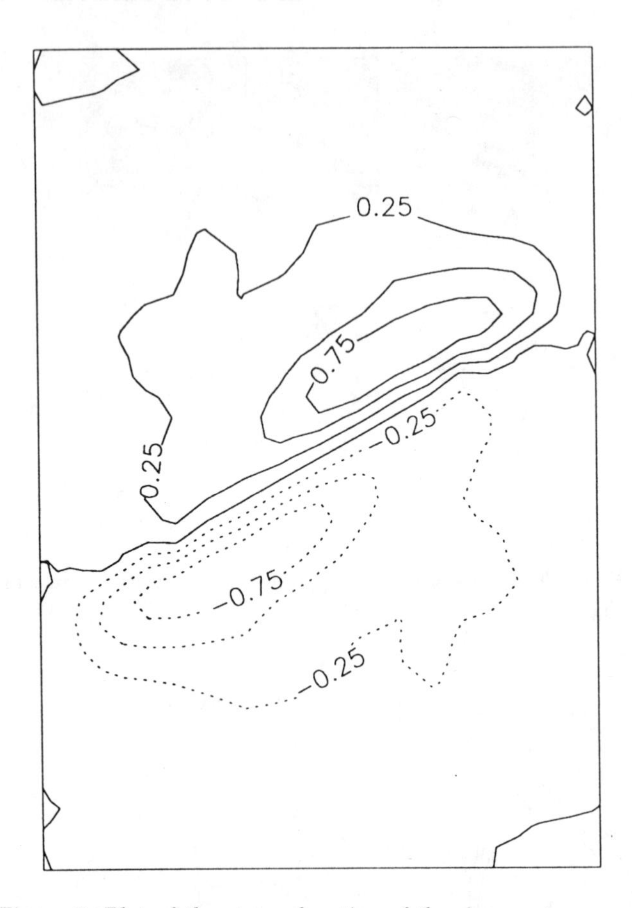

Figure 5. Plot of the streamfunction of the eigenvector corresponding to the most unstable mode of the 2-cell branch (eigenvalue $\lambda = 2.0 * 10^{-2}$).

scale. The eigenvector corresponding to the most unstable mode of the 2-cell branch (at the particular value of Ra_η as used in the time integration), is plotted in figure 5. This mode corresponds to the unstable manifold of the equilibrium solution. Clearly, the eigenmode superposed on the steady state weakens the lower cell. On a longer time scale, the flow follows this unstable manifold and the pattern change along this manifold is towards the 1-cell solution.

As an example that these type of quasi steady solutions are also present at higher buoyancy ratio, we present (figure 6) the transient flow for $R_\rho = 6$ and the same value of Ra_η as above. A flow pattern with vertical lengthscale η appears after some time (figure 6b, panel a), which goes unstable and changes to a four cell solution (figure 6b, panel b). It is likely, that the 6-cell solution also corresponds to an unstable equilibrium solution.

4. DISCUSSION

The structure of equilibrium solutions corresponding to finite amplitude steady flow patterns were studied for a laterally heated stably stratified liquid. It was demonstrated that, for a buoyancy ratio equal to 3, multiple steady states exist not only for small Ra_η, but also at very large Ra_η. Multiple stable steady states exists at small buoyancy forcing (figure 1) and these stable patterns occur due to the presence of the walls. However, there are many more steady states in this regime, which can be thought of as double diffusive modifications of the stable patterns. The flow patterns clearly show the signature of double diffusion, for example, the characteristic tilt and some of them correspond to flow patterns

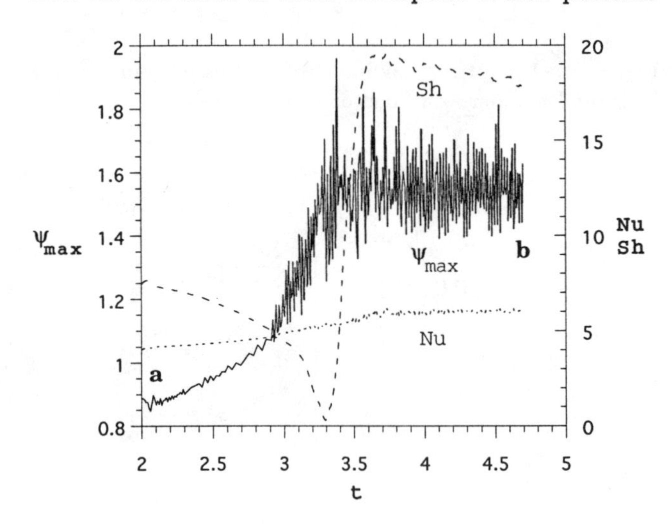

Figure 6a. Plot of the maximum of ψ, Nusselt number and Sherwood number as a function of time, $R_\rho = 6$.

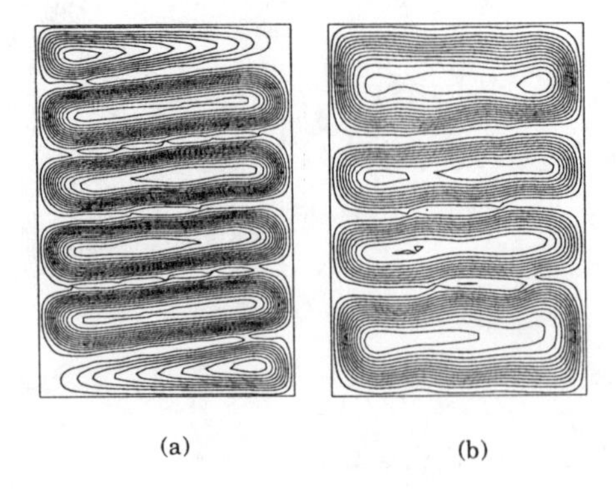

Figure 6b. Plots of the stream function at selected points in Figure 6a.

with a characteristic scale η. All these solutions appear to be unstable for the case considered.

At small Ra_η our results compare fairly well with those of Tsitverblit and Kit [*Tsitverblit and Kit*, 1993] in that a large number of steady states exist (most of which were shown to be unstable [*Kranenborg and Dijkstra*, 1994]). Since they kept Ra_S fixed instead of R_ρ, at larger Ra_η the buoyancy ratio becomes very small leaving a unique thermally dominated flow; other steady branches are absent. Hence, for large Ra_η they did not find the transitions in the flow described in Section 3.2.

A few of the branches in figure 1 extend into the large Ra_η regime (figure 2). In fact, there might be more of these branches, since a complete bifurcation diagram can never be claimed to be computed. For instance, there may always be isolated branches somewhere in parameter space. In the particular case considered, a time evolution at values larger than $Ra_{\eta,cr}$ shows that one of the unstable steady states is approached on a relatively fast time scale. The correspondence of the quasi-steady flow and the unstable equilibrium was clearly demonstrated (figure 4). Hence, although the flow is unstable, it is reached because the physical process causing the instability occurs on a much larger time scale. The correspondence of the pattern of the most unstable mode and the development of the transient flow is a strong support for this dynamical picture.

Although not enough analysis is done at this moment to clearly identify the physics of the instability, one of the possibilities is the slow diffusion of salt. This would explain why these type of layered solutions are found in just double diffusive systems.

The combination of bifurcation techniques and time-integration is shown to be a fruitful approach to study the layer formation problem. In most experiments, the salt flux at the horizontal walls is zero; hence only trivial steady states exist. However, it is still very likely that points in phase space remain, where the flow associated with the vector field is slow, mimicing the (unstable) equilibrium states as discussed in this paper. Although also the slowly varying background stratification can become important to generate quasi stationary flows, they may also be caused through the same dynamics as sketched above. The existence of these 'slow' regions in phase space is currently under investigation.

Acknowledgments. All computations were performed on the CRAY C98 at the Academic Computing Centre (SARA), Amsterdam, the Netherlands within the project SC212. Use of these computing facilities was sponsored by the Stichting Nationale Supercomputer faciliteiten (National Computing Facilities Foundation, NCF) with financial support from the Nederlandse Organisatie voor Wetenschappelijk Onderzoek (Netherlands Organization for Scientific Research, NWO). The authors thank Jeroen Molemaker for the use of the explicit CFD code and for valuable comments on the results.

REFERENCES

Chen, C. F., Briggs, D. G., Wirtz, R. A., Stability of thermal convection in a salinity gradient due to lateral heating, *Int. J. Heat Mass Transfer*, 14, 57–65, 1971.

Dijkstra, H.A., On the structure of cellular solutions in Rayleigh-Bénard-Marangoni flows in small-aspect-ratio containers, *J. Fluid Mech.*, 243, 73–102, 1992.

Dijkstra, H.A., Molemaker, M.J., Van der Ploeg, A., Botta, E.F.F., An efficient code to compute non-parallel flows and their linear stability, *submitted to Comp. Fluids*, 1994.

Huppert, H. E., Turner, J. S., Ice blocks melting into a salinity gradient, *J. Fluid Mech.*, 100, 367–384, 1980a.

Huppert, H.E., Turner, J.S., Double-diffusive convection, *J. Fluid Mech.* , 106, 299-329, 1980b.

Jeevaraj, C., Imberger, J., Experimental study of double-diffusive instability in sidewall heating, *J. Fluid Mech.*, 222, 565–586, 1991.

Kerr, O. S., Heating a salinity gradient from a vertical sidewall: linear theory, *J. Fluid Mech.*, 207, 323–352, 1989.

Kerr, O. S., Heating a salinity gradient from a vertical sidewall: nonlinear theory, *J. Fluid Mech.*, 217, 529–546, 1990.

Kranenborg, E. J., Dijkstra, H. A., The structure of (linearly) stable double diffusive flow patterns in a laterally heated stratified liquid, *Phys. Fluids A*, 7, 680–682, 1995.

Lee, J. W. & Hyun, J. M., Double diffusive convection in a cavity under a vertical solutal gradient and a horizontal temperature gradient, *Int. J. Heat Mass Transfer*, 34, 2423–2427, 1991a.

Lee, J. W. & Hyun, J. M., Time–dependent double diffusion in a stably stratified fluid under lateral heating, *Int. J. Heat Mass Transfer*, 34, 2409–2421, 1991b.

Schmitt, R.W., Double diffusion in oceanography *Ann. Rev. Fluid Mech.*, 26, 255-285, 1994.

Tanny, J., Tsinober, A. B., The dynamics and structure of double-diffusive layers in sidewall-heating experiments, *J. Fluid Mech.*, 196,135–156, 1988.

Thangam, S., Zebib, A. & Chen, C. F., Transition from shear to sideways diffusive instability in a vertical slot, *J. Fluid Mech.*, 112, 151–160, 1981.

Thorpe, S.A., Hutt, P.K., Soulsby, R., The effect of horizontal gradients on thermohaline convection, *J. Fluid Mech.*, 38, 375-400, 1969.

Tsitverblit, N., Kit, E., The multiplicity of steady flows in confined double-diffusive convection with lateral heating, *Physics Fluids A.*, 5, 1062–1064, 1993.

Wirtz, R. A., Briggs, D. G., Chen, C. F., Physical and numerical experiments on layered convection in a density-stratified fluid, *Geophysical Fluid Dynamics*, 3, 265–288, 1972.

Layer Formation in a Salt Stratified Liquid Cooled From Above

M. Jeroen Molemaker and Henk A. Dijkstra

Institute for Marine and Atmospheric Research, Department of Physics and Astronomy, University of Utrecht, The Netherlands

By direct numerical simulation, the double diffusive layer formation process in a stably salt stratified layer cooled from above is studied in a two-dimensional geometry. The validity of several scalings, previously proposed and based on experiments, is considered. In the example studied, the buoyancy jump which builds up ahead of the first layer, appears to be much larger than previously estimated. The thickness of the first layer is determined by a balance of kinetic energy and potential energy, as previously suggested, but the length scale associated with the latter is not the thickness of the first layer, but the thickness of the interface separating the layers. The formation of the second layer does not solely arise from an instability of the thermal boundary layer ahead of the first layer. Erosion of the stable stratification due to mechanical mixing appears to be a necessary preconditioning phase.

1. INTRODUCTION

Layer formation is a characteristic feature of double diffusive convection which may occur in stably stratified liquids in which two substances, e.g. heat and salt, diffuse at a different rate. Therefore, when vertical and/or horizontal gradients of both components are present in a stably stratified liquid, double diffusive induced layer formation may change the transport of each component significantly. This makes double diffusive convection a potentially important transport mechanism e.g. for heat and salt in the ocean [*Schmitt*, 1994] and for different components in a magma chamber [*Huppert and Turner*, 1981]

Layer formation typically occurs when a vertical temperature gradient is imposed on an initially vertical salinity gradient. In case the fastest diffusing substance is (de)stabilizing, layers may form through (diffusive) finger instabilities [*Huppert and Linden*, 1979; *Turner and Stommel*, 1964]. In this paper, we consider layer formation in the 'diffusive' favourable case. A typical experiment consists of heating a layer, which has constant temperature but otherwise is stably stratified through a constant salt gradient, from below. In the late sixties these experiments were first performed by Turner [*Turner*, 1968]. A constant heat flux $\mathcal{H}$ $[Wm^{-2}]$ was applied to the bottom of the layer. First a well-mixed layer develops near the bottom and the thickness of this layer grows with time. After a while, a second layer develops which is separated from the first by a relatively sharp interface. Subsequently, this process repeats itself and finally a series of well-mixed layers develops separated by sharp (diffusive) interfaces.

Let ρ_0 $[kg\ m^{-3}]$, C_p $[J\ kg^{-1}\ K^{-1}]$ and α $[K^{-1}]$ indicate the reference density, the specific heat and the coefficient of thermal expansion of the liquid, respectively. The buoyancy flux q_0 $[m^2 s^{-3}]$ which is applied to the bottom of the liquid is given by

$$q_0 = \frac{g\alpha\mathcal{H}}{\rho_0 C_p} \qquad (1)$$

Double-Diffusive Convection
Geophysical Monograph 94

where g is the gravitational acceleration. The other quantities important to describe the physics of the layer formation are the coefficient of salinity contraction β, the initial constant salinity gradient dS_0/dz and the thermal and solutal diffusion coefficients κ and D. In experiments [*Turner*, 1968; *Huppert and Linden*, 1979; *Fernando*, 1987] it is found that the thickness h of the first layer increases initially as

$$h = C{q_0}^{1/2}N^{-1}t^{1/2} \tag{2}$$

where C is a constant having values between 1 and 2 and N the buoyancy frequency associated with the initial stable stratification. For example, when the buoyancy jump across the interface is assumed to be zero, the constant C in (2) is equal to $\sqrt{2}$ [*Turner*, 1968]. After some time t_c, the thickness of the first layer is nearly constant with time. This critical layer thickness h_c is proportional to a power n of q_0 and a power m of N. Turner could correlate his experiments well with $n = 3/4$, $m = -2$ while Fernando [*Fernando*, 1987] was able to do this with $n = 1/2$, $m = -3/2$.

Both Fernando and Turner propose simple models, giving a scaling of the thickness of the first layer and a physical picture of the formation of the second layer. The initial growth can be explained using both simple models. An essential difference between both models is the explanation of the critical layer thickness h_c. Turner assumes the buoyancy jump at the top of the layer to be zero, i.e. the interface is always marginally stable. Turner then explains the critical layer thickness h_c through an instability of a thermal boundary layer, which is present just ahead of the interface. As soon as this instability occurs, the first layer stops growing. The latter ideas have also been used in the work of Huppert and Linden [*Huppert and Linden*, 1979], in which total staircase growth was studied. In contrast with Turner, Fernando does assume a buoyancy difference over the layer and the final thickness of the first layer is determined by a balance between the kinetic energy flux of the eddies and potential energy production associated with buoyancy differences over the first layer.

The approach followed in this work is numerical; a high resolution two-dimensional code is used to simulate the time-evolution of the layer formation process. In the simulation a liquid is cooled from above which is equivalent to the case of heating from below. In this paper preliminary results are presented by showing one typical example and analyzing the evolution of several averaged quantities. In this way, the physics of the layer formation process can be studied in quite detail. First, we focus on how the thickness of the first layer is determined. Subsequently, the initial formation of the second layer is addressed.

2. FORMULATION

Consider a two - dimensional incompressible liquid, with constant kinematic viscosity ν in a rectangular box of aspect ratio **A** (ratio of length L to height H, figure 1). At the top of the liquid a constant heat flux is prescribed which cools the layer from above. This set-up is equivalent to that in experiments where a liquid is heated from below. The governing equations are non-dimensionalized using scales H, H^2/κ, κ/H, T_∞, S_∞ for length, time, velocity, temperature and salinity where H is the heigth of the container and κ is the thermal diffusivity of the liquid. If the horizontal and vertical velocities are u and w, respectively, the dimensionless equations, with the usual Boussinesq approximation, become in a streamfunction - vorticity $(\psi - \omega)$ formulation

$$\mathbf{Pr}^{-1}\left(\frac{\partial \omega}{\partial t} + u\frac{\partial \omega}{\partial x} + w\frac{\partial \omega}{\partial z}\right) = \nabla^2\omega + \mathbf{Ra}\left(\frac{\partial T}{\partial x} - \lambda\frac{\partial S}{\partial x}\right) \tag{3a}$$

$$\omega = -\nabla^2\psi \tag{3b}$$

$$\frac{\partial T}{\partial t} + u\frac{\partial T}{\partial x} + w\frac{\partial T}{\partial z} = \nabla^2 T \tag{3c}$$

$$\frac{\partial S}{\partial t} + u\frac{\partial S}{\partial x} + w\frac{\partial S}{\partial z} = \tau\nabla^2 S \tag{3d}$$

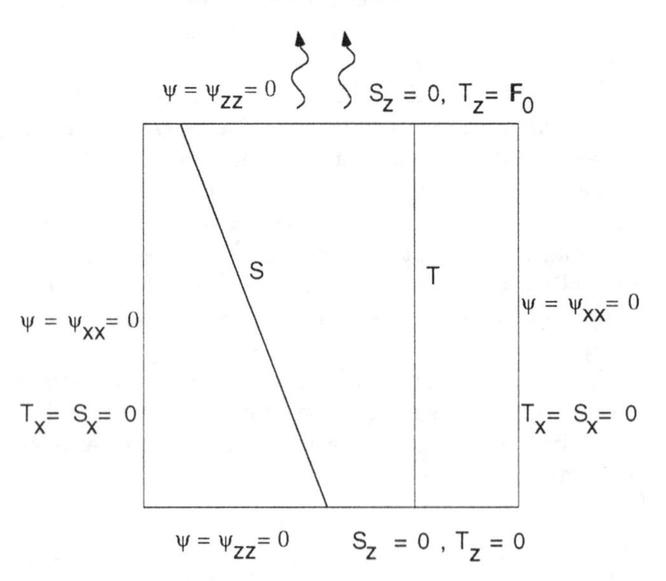

Fig. 1. Geometrical set-up of the problem. A two-dimensional container with a stable salt stratified liquid is cooled at the top.

where $u = \partial\psi/\partial z$, $w = -\partial\psi/\partial x$) and $\omega = \partial w/\partial x - \partial u/\partial z$). Furthermore, T and S are temperature and salinity, **Pr** is the Prandtl number, τ the inverse Lewis number, **Ra** the Rayleigh number and λ the buoyancy ratio based on the reference values of salinity and temperature. These parameters are defined as

$$\mathbf{Ra} = \frac{gH^3\alpha T_\infty}{\nu\kappa}, \quad \lambda = \frac{\beta S_\infty}{\alpha T_\infty}, \quad \mathbf{Pr} = \frac{\nu}{\kappa}, \quad \tau = \frac{D}{\kappa} \qquad (4)$$

All boundaries are assumed to be stress-free. At the top wall the salt flux is zero and the heat flux prescribed and equal to $\mathcal{H}$. All other walls satisfy no-flux conditions for heat and salt. The dimensionless boundary conditions then become

$$x = 0, \mathbf{A}: \qquad \psi = \omega = \frac{\partial S}{\partial x} = \frac{\partial T}{\partial x} = 0 \qquad (5a)$$

$$z = 0: \qquad \psi = \omega = \frac{\partial S}{\partial z} = \frac{\partial T}{\partial z} = 0 \qquad (5b)$$

$$z = 1: \qquad \psi = \omega = \frac{\partial S}{\partial z} = 0, \quad \frac{\partial T}{\partial z} = \mathbf{F}_0 \qquad (5c)$$

where $\mathbf{F}_0$ is the dimensionless heat flux, given by

$$\mathbf{F}_0 = \frac{q_0 H}{g\alpha T_\infty \kappa} \qquad (5d)$$

A classsical explicit Euler method is used as time discretization and central differences are used as space discretization on an equidistant grid for $i = 0, \ldots, N; j = 0, \ldots, M$. This scheme is second order accurate in (Δx, Δz) and first order accurate in Δt. To compute the solution at a new time level, first Euler time stepping is performed for ω, T and S and then the Poisson equation (2b) is solved for the streamfunction. This explicit formulation enables one to use a very high resolution, which ensures that although only molecular diffusion is used all relevant scales can be resolved. A drawback of this method is that, due to numerical stability limitations, the size of the time step is quite severely constrained [*Peyret and Taylor*, 1983]. The parameters **Pr** and τ are fixed for the water-heat- salt system ($\mathbf{Pr} = 7$ and $\tau = 10^{-2}$) and the aspect ratio of the container is set to $\mathbf{A} = 1$. The code was verified through comparison with an implicit code (as used in [*Dijkstra*, 1988]) for several test problems.

3. RESULTS

The simulation discussed in this paper is a numerical version of the experiments in [*Fernando*, 1987] and [*Huppert and Linden*, 1979]. The case considered is $\mathbf{Ra} = 10^9$, $\lambda = 1$ and a non-dimensional heat flux $\mathbf{F}_0 = -4.8$. The numerical resolution used is $N = 200$, $M = 300$. The time step ($\Delta t = 10^{-7}$) and resolution were chosen after testing with different grid sizes and time steps. The initial conditions are a motionless solution with a stable salt stratification together with a homogeneous temperature profile, similar to laboratorium experiments. The choice of the parameters and initial conditions corresponds to the following dimensional values: total depth of the container $H = 10^{-1}$ m, a prescribed buoyancy flux $q_0 = 10^{-6}$ $m^2 s^{-3}$ and an initial buoyancy frequency $N = 1.25$ s^{-1}. To derive the dimensional values we used constant values for kinematic viscosity and diffusivity of heat $\nu = 10^{-6}$ $m^2 s^{-1}$, $\kappa = 1.44\ 10^{-7}$ $m^2 s^{-1}$. The simulation was calculated until a second layer was well established and a third layer started to develop ($t = 3\ 10^3$ s). This simulation took about 20 CPU hour on a Cray C98.

3.1. Description

Due to cooling at the top, the thermal boundary layer becomes unstable and rapidly convection develops, forming a well mixed layer. The evolution of the, horizontally averaged salinity, temperature and buoyancy profiles is shown in the figures $2a - 2c$ for several times. Here the buoyancy b is defined as $b = (\rho_0 - \rho)/\rho_0 = g(\alpha T - \beta S)$. As the thickness of the mixed layer increases through turbulent entrainment at the bottom, a density jump develops leading to an interface between the mixed layer and the underlying liquid. This can be observed in figure 2c or in figure 2d where the same buoyancy profiles are plotted, but for many more times, showing an increase and a steepening of the density jump between the mixed layers. When the density jump is sufficiently large the eddies do not have sufficient energy to penetrate the interface and the turbulent entrainment ceases. The thickness of the first layer thereafter remains nearly constant (at approximately $t = 2500$ s in the figures 2); it can only change through diffusion, which is slow. For times t larger than $t = 500$ s, a certain amount mixing is present below the first layer. During the turbulent growth phase, this region of mixing is incorporated by the first layer. Only when the depth of the first layer does not increase significantly anymore a second mixed layer is established. This second layer is smaller than the first and also the intensity of mixing in this layer is much smaller. In figure 3 the salinity distribution is shown at $t = 2750$ s when the first layer is nearly stationary and a second mixed layer is present. The interface separating the two layers can be clearly identified in the figure.

The parameters of the simulation are chosen in such

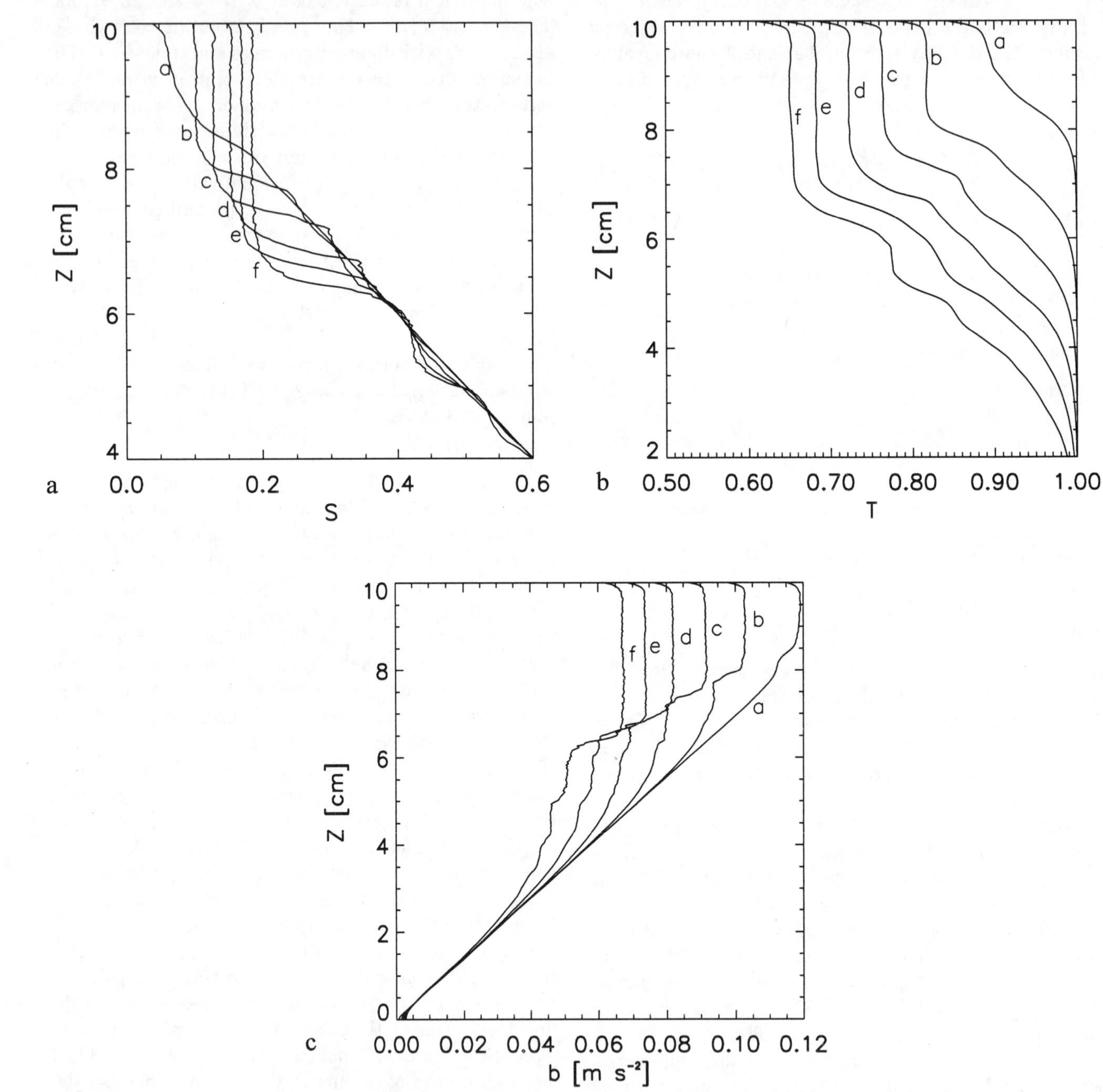

Figs. 2. Transient development of horizontally averaged profiles. Profiles are shown at a: $t = 70\ s$, b: $t = 550\ s$, c: $t = 1040\ s$, d: $t = 1528\ s$, e: $t = 2380\ s$ and f: $t = 2940\ s$

a) Salinity

b) Temperature

c) Buoyancy $(\rho_0 - \rho)/\rho_0$

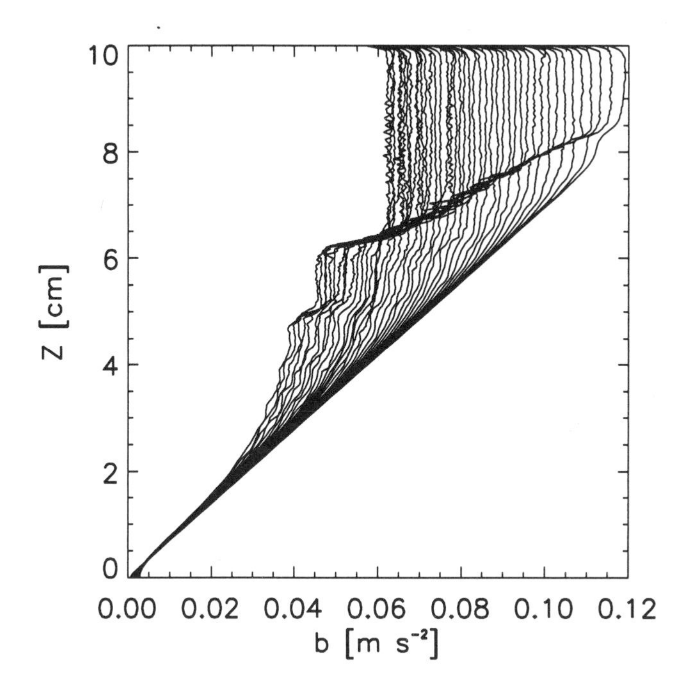

Fig. 2*d*. Buoyancy, for many times between $t = 70\ s$ and $t = 2920\ s$. The development and sharpening of the interface is clearly shown as well as the development of a second interface between the second and third layer.

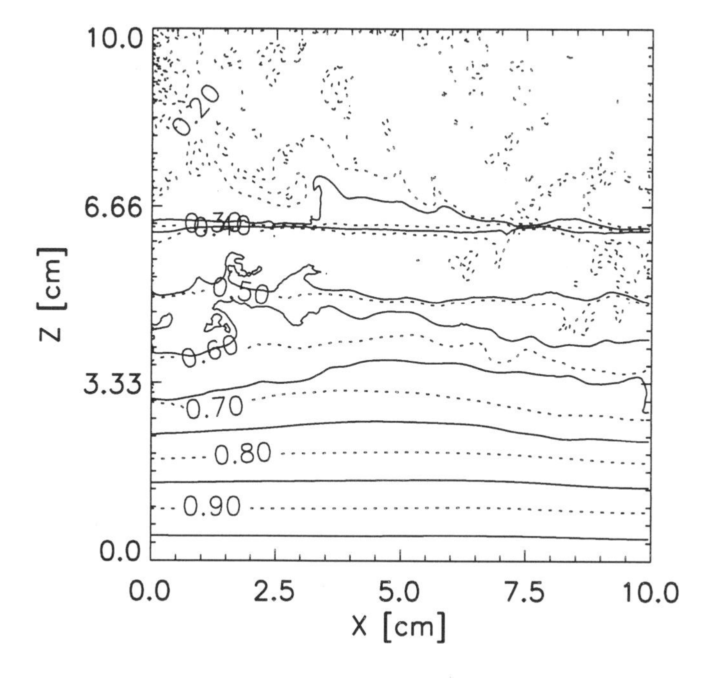

Fig. 3. Salinity distribution at $t = 2750\ s$.

a way to represent one of the laboratory experiments of Fernando [*Fernando*,1987]. If our results are compared to these experiments, the qualitative agreement is good. Without any imposed constraints the same layered structure as in the experiments is found and the first layer is much larger than the next layers. More quantitatively, in [*Fernando*,1987] it was found that, for $q_0 = 10^{-6}\ m^2\ s^{-3}$ and $N = 1.25\ s^{-1}$, the depth of the first layer is $h \approx 3.5\ cm$ (figure 6 in his paper). This compares very well with the result of the numerical simulation where a final depth $h \approx 3.7\ cm$ of the first layer is found (cf. figure 2*d*).

3.2. Analysis

In this section we analyse the growth rate of the first layer, the scale of vertical velocities in this layer and whether a simple balance between kinetic energy and potential energy determines the final thickness of the first layer.

To estimate the thickness of the mixed layer we have to choose a criterium to determine the location of the interface. In figure 4*a* the intensity of vertical mixing $<w^2>$ (horizontally averaged) is plotted. From this figure it is clear that $<w^2>$ has a maximum in the first mixed layer together with a smaller local maximum in the second layer. The layers are separated by a minimum of $<w^2>$ (in figure 4*a* located at $0.064\ m$) coinciding with steep gradients in S and T. The location of this minimum in $<w^2>$ is used as a measure of the depth of the first layer. Results of layer thickness versus time are plotted in figure 4*b* on a logarithmic scale. Our best fit to the initial growth is a power law dependence with exponent 0.36. The exponent proposed by Fernando to fit the experimental data is slightly larger. However when the experimental results are considered in more detail, quite a range of power dependencies would fit the data of figure 4 in [*Fernando*, 1987]. A power 0.5 as proposed overestimates the overall growth rate for the range of q_0 and N shown. Both experiments and our numerical results show that the growth rate decreases with increasing time.

A scaling for the r.m.s. vertical velocity in terms of the imposed buoyancy flux q_0 at the boundary has been proposed by Hunt [*Hunt*, 1984]

$$< {w_*}^2 >^{1/2} = C_1 (q_0 h)^{1/3} \tag{6}$$

where we interpret h as the thickness of the mixed layer and C_1 denotes an $\mathcal{O}(1)$ constant. If we take the maximum of $<w^2>$ (see figure 4*a*) as a typical scale for $<w_*^2>$

Fig. 4*a*. A snapshot of the horizontally averaged intensity of vertical mixing $<w^2>$ at $t = 2170\ s$. The minimum located at $z = 0.064\ m$ coincides with the interface separating the first and the second layer.

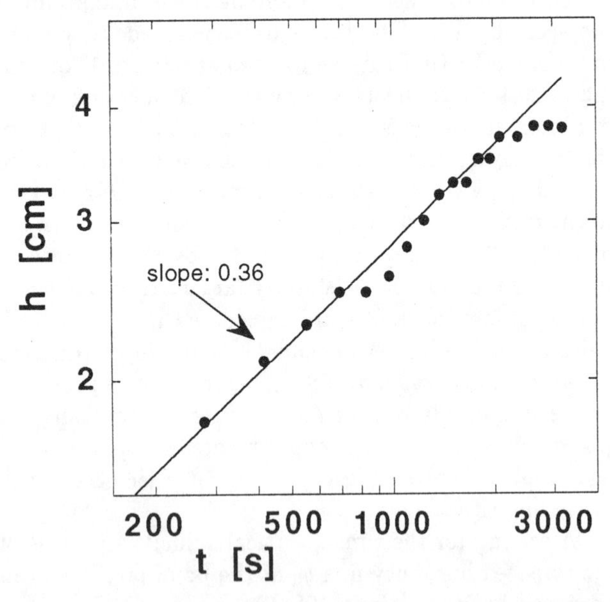

Fig. 4*b*. The evolution of the thickness of the first layer h based on the minimum in $<w^2>$ as shown in Fig. 4*a*.

we can verify this scaling in our simulation by plotting this maximum as a function of time (figure 5). Using the figure we find a value of $<w_*^2>^{1/2} \approx 2.5\ 10^{-3}\ m\,s^{-1}$ and we can compare this value against the quantity $(q_0 h)^{1/3}$. With $q_0 = 10^{-6}\ m^2\ s^{-3}$ and $h \approx 3\ 10^{-2}\ m$ we find $(q_0 h)^{1/3} \approx 3\ 10^{-3}\ m\,s^{-1}$ implying that $C_1 = \mathcal{O}(1)$. This suggests that this scaling is valid, at least for this example. Alternatively, one can view this as a confirmation of sufficient accuracy of the numerical results. The fluctuations in the graph of $<w^2>^{1/2}$ increase during the development of the mixed layer. This is due to the fact that the convective eddies increase in size during the growth of the mixed layer. At later stages they have a size comparable with the width of the container. Therefore the fluctuations in these eddies are not averaged out over the width of the container as is the case in the beginning of the simulation.

Fernando [*Fernando*, 1987] suggests that the limiting height of the mixed layer can be found from [*Long*, 1978]

$$< w_* >^2 = C_2 \Delta b\, h_c \tag{7}$$

which represents a balance between kinetic energy and potential energy associated with opposing buoyancy forces. Δb is the buoyancy jump across the interface, which is assumed to be also the characteristic buoyancy variation of the eddies in the mixed layer. From figure 4*b* we find that the first layer reaches its final depth at $t \approx 2500\ s$. It is at this moment that the balance (7) should be valid. From figure 5 we find $w_*^2 \approx 10^{-5}\ m^2 s^{-2}$ and from figure 2*c*, $\Delta b \approx 1.5\ 10^{-2}\ m\ s^{-2}$. With $h_c = 3.7\ 10^{-2}\ m$ we get a value $C_2 = \mathcal{O}(2\ 10^{-2})$ giving no support for the balance (7).

From the data we infer that buoyancy variations within the mixed layer are very small. The major part of the density variations takes place within the interface,

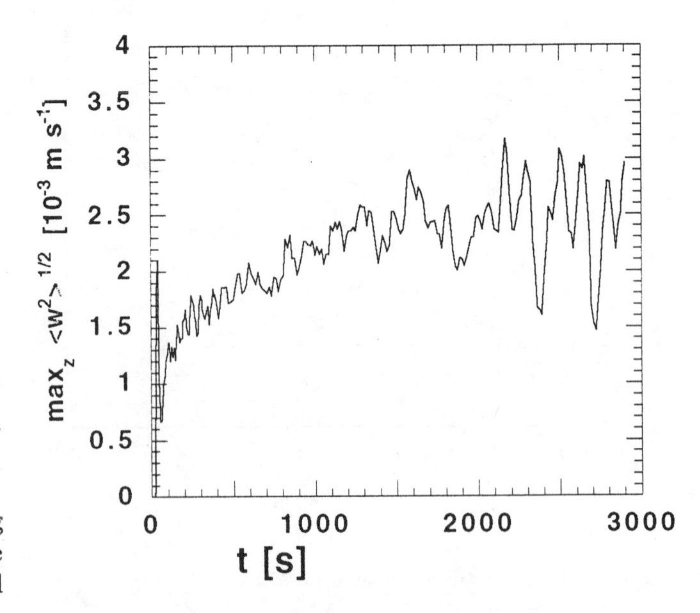

Fig. 5. The evolution of the maximum of $<w^2>$ in time.

separating the mixed layer from the underlying fluid. A different scaling is therefore proposed, also based on a balance between kinetic and potential energy, but now the interface thickness δ is used a characteristic value over which the buoyancy variations occur, hence

$$w_*{}^2 = C_3 \Delta b\, \delta_c \tag{8}$$

C_3 is a dimensionless constant which should be $\mathcal{O}(1)$. Again from figure 2c, we obtain $\delta_c \approx 3\ 10^{-3}\ m$, and using the same values for w_*^2 and Δb, with (8) it follows that the constant $C_3 \approx 0.2$. This suggests that (8) is the proper scaling in this example.

If the formation of a second mixed layer underneath the first layer is carefully studied, it follows (figures $2a - 2c$) that for $t > 500\ s$ a region of mixing is present underneath the first mixed layer. Initially this second layer is entrained by the first layer and this continues until the growth rate of the first layer becomes negligibly small due to the build up of the density jump. During all times there is some mixing underneath the first layer (see figures 2), either caused by eddies which do break through the interface or by eddies which are viscously driven by the deformation of the interface. Although the amount of 'pre-mixing' is small it does have two effects. By transporting salt upward it enhances the buoyancy jump at the interface, slowing down the growth of the first layer. Moreover it erodes the salinity gradient underneath the first layer. Due to the much smaller diffusion coefficient the salt gradient recovers only very slowly back to its diffusion solution, whereas the (destabilising) temperature gradient remains mostly intact. In this way the 'pre-mixing' allows for sustained mixing in the second layer by reducing the background buoyancy frequency N locally.

4. DISCUSSION

In a high resolution numerical study, we presented an example of layer formation in a stratified layer cooled from above. The results are qualitatively in correspondence with experiments and confirm the physical picture given by Fernando [*Fernando*, 1987] that '...the mixed layer grows by the action of large eddies until they are not energetic enough ...'. Although the intensity of the eddies increases with time as $h^{2/3}$, implied by the Hunt scaling (5) as confirmed in figure 5, the buoyancy jump Δb increases at a faster rate.

Eventually, the buoyancy jump will become large enough, such that the most energetic eddies will not be able to entrain fluid below the interface. The growth of the mixed layer will then become negligibly small. The final thickness of the first layer is determined by a simple energy balance (8), which is different than the one proposed by Fernando. The reason for this is that the buoyancy jump across the interface is not a characteristic buoyancy variation within the mixed layer, as assumed by Fernando. Our numerical results clearly illustrate that the buoyancy differences within the mixed layer are substantially smaller than those across the interface. Moreover, the results indicate that in [*Fernando*, 1987] (in the case of heating from below) the buoyancy jump Δb is underestimated. In estimating Δb, Fernando (following [*Turner*, 1968]) neglects heat and salt fluxes through the interface at the top of the mixed layer. Although this is probably correct for the salt flux, it is certainly not so for the heat flux. A heat flux from the mixed layer to the overlying fluid does increase the buoyancy difference at the top of the mixed layer.

It seems that before the actual formation of the second layer, a preconditioning process is necessary. At the stage at which the most energetic eddies are able to penetrate the interface, they induce a net upward salt transport below the interface. The stability of the layer below the interface is decreased. Hence, a smaller temperature difference at the front of the interface is sufficient to induce an instability, leading to the formation of a second layer. Upward transport of salt also increases the density at the interface giving rise to a larger buoyancy jump. Although it might be unimportant for the final thickness of the first layer, it could influence the growth rate of the first layer. The consequences of these results are still under investigation and require simulations at different parameter values.

Acknowledgments. All computations were performed on the CRAY C98 at the Academic Computing Centre (SARA), Amsterdam, the Netherlands within the project SC212. Use of these computing facilities was sponsored by the Stichting Nationale Supercomputer faciliteiten (National Computing Facilities Foundation, NCF) with financial support from the Nederlandse Organisatie voor Wetenschappelijk Onderzoek (Netherlands Organization for Scientific Research, NWO).

REFERENCES

Dijkstra, H.A., Transient Marangoni convection in a square container. *Physico Chemical Hydrodynamics* **10**, 493-515, 1988.

Fernando, H.J.S., The formation of a layered structure when a stable salinity gradient is heated from below. *J. Fluid Mech.* **182**, 525-541, 1987.

Hunt, J.C.R., Turbulence structure in thermal convection and shear-free boundary layers. *J. Fluid Mech.* **138**, 161-184, 1984.

Huppert, H.E., and P.F. Linden, On heating a stable salinity gradient from below. *J. Fluid Mech.* **95**, 431-464, 1979.

Huppert, H.E., and J.S. Turner, Double-diffusive convection. *J. Fluid Mech.* **106**, 299-329, 1981.

Long, R.R., A theory of mixing in a linearly stratified fluid. *J. Fluid Mech.* **84**, 113-124, 1978.

Peyret, R., and T.D. Taylor, Computational Methods for fluid flow. Springer-Verlag, New York, 1983.

Schmitt, R.W., Double diffusion in oceanography. *Ann. Rev. Fluid Mech.* **26**, 255-285, 1994.

Turner, J.S., The behavior of a stable salinity gradient heated from below. *J. Fluid Mech.* **33**, 183-200, 1968.

Turner, J.S., and H. Stommel, A new case of convection in the presence of vertical salinity and temperature gradients. *Proc. Natl. Acad. Sci.* **52**, 49-53, 1964.

M.J. Molemaker and H.A. Dijkstra, Institute for Marine and Atmospheric Research, Princetonplein 5, 3584 CC Utrecht, the Netherlands.

Double Diffusive Instabilities at a Vertical Boundary

Oliver S. Kerr

Department of Mathematics, City University, London, U.K.

Double diffusive instabilities in a otherwise stably stratified fluid can be driven by the presence of horizontal temperature and salinity gradients. In many situations there exist such gradients which are effectively confined within a finite distance of an isolated boundary. These gradients could arise because there is a flux of heat or salt through the bounding wall, or, in the case of no-flux boundary conditions, just because the wall is not vertical. In situations where the wall is far from other boundaries these horizontal temperature and salinity gradients typically evolve with time. In this paper we will concentrate on the archetypical example of the heating of a salinity gradient from a single vertical wall. We will review the theoretical analysis that has been conducted for this problem, and highlight those areas that are not yet well understood. It is these areas which would provide us with a better insight into the complex mechanism involved, and which provide us with challenges for the future theoretical analysis.

1. INTRODUCTION

In this paper we look at the situation when there is an isolated non-horizontal boundary to a body of stratified fluid next to which there are localised horizontal temperature and salinity gradients. This situation can give rise to double-diffusive instabilities that are confined to a region near the boundary and that are driven by these gradients. In particular we will review what is known theoretically about this problem, and focus on those important areas that are not well understood and remain as theoretical challenges. We concentrate mainly on the case where the boundary is vertical, although much of what is said holds for sloping boundaries as well.

It has been known for some time that horizontal temperature and salinity gradients can drive double-diffusive instabilities in fluid [see, for example, *Stern*, 1967]. These instabilities have been investigated experimentally by various people in a variety of configurations. The stability of a finite front between two different bodies of stratified fluid has been investigated by *Ruddick and Turner* [1979], *Holyer et al.* [1987] and *Ruddick* [1992]. The stability of a stratified fluid between two parallel non-horizontal walls has been looked at experimentally by *Thorpe et al.* [1969], *Chen and Sandford* [1977] and *Paliwal and Chen* [1980a]. The experiments that most concern us here are those which have instabilities near one boundary only, i.e. instabilities that occur at a wall of an effectively semi-infinite body of fluid. This situation has been looked at by *Thorpe et al.* [1969], *Chen et al.* [1971], *Chen and Skok* [1974], *Linden and Weber* [1977], *Huppert and Turner* [1980], *Huppert and Josberger* [1980], *Narusawa and Suzukawa* [1981], *Huppert et al.* [1984], *Chereskin and Linden* [1986], *Tanny and Tsinober* [1988, 1989] and *Schladow et al.* [1992]. These experiments have looked at the effects of horizontal compositional gradients near the boundary in fluids with either vertical salinity gradients or mixed vertical salinity and temperature gradients. The sidewalls in these experiments may be either vertical or sloping. Mostly these experiments looked at heating (or cooling) from the boundary, however *Linden and Weber's* experiments looked at the effect of sloping insulating impermeable boundaries, while those of *Chereskin and Linden* [1986] looked at the additional effect of rotation on the instabilities at a heated wall.

The theoretical analysis of double diffusive instabili-

Double-Diffusive Convection
Geophysical Monograph 94

ties driven by horizontal gradients falls into four classes: instabilities driven by infinite uniform horizontal gradients [*Stern*, 1967; *Toole and Georgi*, 1981; *McDougall*, 1985; *Holyer*, 1983; *Walsh and Ruddick*, 1994], instabilities in a slot [*Thorpe et al.*, 1969; *Hart*, 1971, 1973; *Chen and Sandford*, 1977; *Thangam et al.*, 1982], instabilities in a finite front in an infinite body of stratified fluid [*Niino*, 1986; *Yoshida et al.*, 1989] and lastly instabilities at isolated boundaries [*Kerr*, 1989, 1990, 1991]. It is this last class of instabilities that concerns us here. In many practical applications it is these instabilities that would occur. For example when an iceberg melts into a stratified part of an ocean, when the wall of a salt-gradient solar pond is of a different temperature from the bulk of the water in the pond (or even if it is just sloping), where the walls of a magma chamber are cooler than the intruding hot magma, or when any large container enclosing a fluid with a vertical compositional gradient is subject to external temperature variations that are fast compared to the time taken for the heat to diffuse into the core of the fluid.

In this paper we will concentrate on the archetypical single boundary problem, the heating of a salinity gradient from a vertical sidewall. This is the configuration that has been most studied both theoretically and experimentally. Unlike either the case of a slot or infinite uniform gradients in an unbounded fluid, in this situation there is no steady state solution to the governing equations for the motion of the fluid and the distribution of the heat and salt even when instabilities are not present. In this case, and in nearly all double-diffusive single wall problems, there is an evolving region whose thickness is proportional to the square root of the time since the onset of the heating at the boundary. It is this time-evolution of the basic background state that causes difficulty in some of the theoretical analysis. When looking at the theoretical work that has previously been done we will highlight some of these difficulties. In this review we will look at the problem in two parts, firstly in §2 we look at the onset of instability, and then in §3 we will concentrate on the finite amplitude behaviour of the instabilities.

2. ONSET OF INSTABILITY

A theoretical study of double-diffusive instabilities due to the presence of lateral temperature and salinity gradients was first given by *Stern* [1967] in the context of examining the possible driving of observed layered intrusions in frontal regions of the oceans. In his study he looked at uniform infinite horizontal and vertical temperature and salinity gradients, and assumed that the vertical fluxes of heat and salt are driven by the presence of salt fingers. This analysis was extended by *Toole and Georgi* [1981] who included viscous effects, and by *Holyer* [1983], *McDougall* [1985] and *Walsh and Ruddick* [1994] who used different models for the heat and salt fluxes. In all these cases it was assumed that all the gradients present were uniform and infinite in extent. It was found that in each case the fluid was always unstable provided that horizontal gradients were present, and the background vertical gradients supported the mechanisms that were required to drive the fluxes. The stability of fluids with horizontal gradients present with boundaries was first looked at theoretically by *Thorpe et al.* [1969], who analysed the instabilities that are driven by temperature differences applied across vertical parallel walls enclosing a salinity gradient. This analysis was extended by *Hart* [1971] who included the effects of horizontal diffusion and the wall boundary layers, and *Thanga et al.* [1981] who investigated the stability of the background state in a vertical slot numerically. The stability of a stratified fluid in an inclined slot was examined by *Paliwal and Chen* [1980b].

The linear stability of a stratified fluid with lateral temperature and salinity gradients near a single boundary suffers from an intrinsic difficulty that is not present in the stability analysis of stratified fluids enclosed between parallel boundaries with lateral gradients. This crucial difference is that the background state is not time-independent. This can lead to the mathematical difficulty that the growth rate of any predicted instabilities is comparable to the rate of the evolution of the background state and so the mathematical separation of the two processes of instability and background evolution is not possible. However, this is not always the case. For many of the experiments that have been carried out [for example *Chen et al.*, 1974, and some of those of *Tanny and Tsinober*, 1988] the instabilities do not appear instantly, but some time after the horizontal gradients have been established. In such circumstances it can be the case that once the instabilities have first appeared their growth rate is on a much faster time-scale than the time-scale for the evolution of the horizontal gradients. This differing time-scale facilitates a conventional stability analysis.

Of fundamental importance in the context of instabilities due to the lateral heating of a salinity gradient are two length scales. The first of these is the depth of penetration of the heat into the fluid due to thermal diffusion:

$$L = (\kappa_T t)^{\frac{1}{2}}, \tag{1}$$

where κ_T is the thermal diffusivity, and t the time since the onset of heating. The second important scale is the Chen length-scale [*Chenet al.*, 1971]

$$H = \frac{\alpha \Delta T}{|\beta \overline{S}_z|}. \tag{2}$$

This height represents the distance that an element of fluid would have to rise when its temperature was increased by an amount ΔT when the vertical salinity gradient is $\overline{S}_z$ in order to be in ambient fluid of its new density. Here α is the coefficient of thermal expansion and β is the coefficient of density increase with respect to the addition of salt. There is a further length-scale associated with the adjustment of the horizontal salinity gradients on the thermal length-scale to the no-flux boundary condition for the salt which is often appropriate at the wall. This adjustment layer is of constant thickness, and for $\tau \ll 1$ separates from the thermal layer thickness, L, after only a few seconds in typical experiments. This thin boundary layer plays no part in the leading order stability analysis. The effect of this boundary layer was also examined by *Hart* [1971] for the case of a vertical slot.

In many of the experiments that have been conducted with relatively gentle heating at the wall the initial appearance of the convection cells is of almost horizontal flat thin cells whose height is very close to the Chen scale, and whose horizontal extent corresponds to the horizontal diffusion scale, L. In these circumstances, where $L^2 \gg H^2$, *Kerr* [1989] has shown that the growth rate of the instabilities scales with the diffusion time-scale over the Chen scale, while the background horizontal temperature and salinity profiles evolve over the longer time-scale of the diffusion time over the horizontal length-scale, L, and hence that it is possible to make a quasi-static assumption based on this difference. In this quasi-static limit the horizontal temperature and salinity gradients balance at leading order, resulting in a negligible horizontal density gradient. The stability of a salinity gradient heated from a single vertical wall in this regime is governed by the single nondimensional parameter

$$Q = \frac{(1-\tau)^6 g (\alpha \Delta T)^6}{\nu \kappa_S \kappa_T t \left(-\beta \overline{S}_z\right)^5} = \frac{(1-\tau)^6 g \alpha \Delta T H^5}{\nu \kappa_S L^2}, \tag{3}$$

where κ_S is the salt diffusivity, ν is the kinematic viscosity, t the time since the onset of heating and $\tau = \kappa_S/\kappa_T$ is the salt heat diffusivity ratio. This nondimensional number is related to a Rayleigh number but has the important difference in that it incorporates both the horizontal scale, L, and the Chen scale, H. The additional factor of $(1-\tau)^6$ which arises from the analysis ensures that this stability parameter remains of similar magnitude at marginal stability for all values of the Prandtl number, $\sigma = \nu/\kappa_T$, and in particular the salt/heat diffusivity ratio, τ. Below is a sketch of a mechanistic argument showing why this is the important parameter in this situation. It is based on the argument of *Turner* [1973, pp. 208–209] for the rôle of the Rayleigh number in Rayleigh-Bénard convection. It has been adapted to take into account the constraint that the instabilities are limited in size in the vertical direction by the Chen scale, H, but extend horizontally over the thermal layer thickness, L. It also assumes a two dimensional flow that was considered in the analysis and often observed in experiments. A similar argument based on three dimensional flows yields the same final result.

A parcel of fluid with typical dimension δ has a density anomaly $\rho'(t)$. Initially this is of order $\rho'(0) \sim \rho_0 \alpha \Delta T$. This density anomaly is made up of both temperature and salt variations. The diffusivity of salt is much less than that of heat, and so the loss of density from the anomaly will be controlled approximately by the diffusion of salt out of the parcel. This flux will be proportional to the diffusivity, the typical gradients of the density anomaly, and the boundary size, so $\delta^2 \times d\rho'/dt \sim -\kappa_S \times \rho'/\delta \times \delta$, giving $\rho'(t) \sim \rho'(0) \exp\left(-\kappa_S t/\delta^2\right)$. The component of the excess gravitational force on this parcel in the direction of motion is of order $g\rho'(t)\delta^2 \times H/L$. The last factor, H/L, is due to the magnitude of the slope of the path along which the parcel is constrained to move. The viscous drag on the parcel is proportional to the dynamic viscosity, a typical velocity gradient and the length of its boundary. Hence the drag goes as $\mu \times u/\delta \times \delta = \mu u = \mu\, dx/dt$ where u and x are the horizontal velocity and displacement of the parcel. This drag balances the buoyancy force and so $dx/dt \sim g\rho'(0)\exp(-\kappa_S t/\delta^2)\delta^3/L\mu$. Hence the horizontal distance moved by the parcel will scale like $x \sim g\rho'(0)\delta^5/(\mu\kappa_S L)$. This distance is maximised when δ is as large as possible, i.e. when $\delta \approx H$. For convection to take place this distance must be greater than L and so we require that

$$Q' = \frac{g \alpha \Delta T H^5}{\nu \kappa_S L^2} \tag{4}$$

is sufficiently large. This is just the non-dimensional number Q given by (3) but without the factor of $(1-\tau)^6$. The exact value that this parameter will take at the onset of instability is governed by the shape of the temperature profile, which in turn is related to history of the wall heating. This value is obtained by finding the eigenvalues of a second order differential equation [*Kerr*,

1989]. For the case of the archetypical error function temperature and salinity profile the critical value of Q is 147 700 with a corresponding vertical wavenumber $m =$ 6·244, giving a vertical periodicity approximately equal to the Chen scale.

When this theory is used to predict the onset of instability in the experiments of *Tanny and Tsinober* [1988], allowance must be made for the temporal evolution of the wall temperature. In their experiments they followed the instantaneous values of the thermal and solutal Rayleigh numbers, $Ra_T = g\alpha\Delta T L^3/(\nu\kappa_T)$ and $Ra_S = g\alpha\overline{S}_z L^4/(\nu\kappa_T)$. Each experiment then followed a trajectory in the Ra_T–Ra_S plane and the point at which instabilities appeared was noted. When the varying nature of the wall heating is taken into account it can be shown [*Kerr*, 1989] that the experiments should go unstable as the trajectory of each experiments cross a narrow band from below in the Ra_S–Ra_T plane used by *Tanny and Tsinober* [1988] to display their results. The points in this plane where the experiments went unstable can be seen in Figure 1 along with the band where the experiments should become unstable. It can be seen that for the larger values of Ra_S there is good agreement between the expermental observations and the theory (even allowing for the automatic enhancing effects of using a log–log plot).

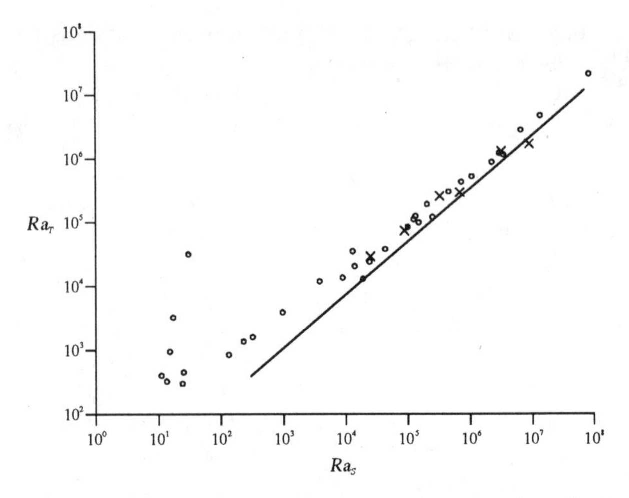

Fig. 1. Comparison of the linear theory of *Kerr* [1989] with the experimental results of *Tanny and Tsinober* [1988] for heating a salinity gradient from a sidewall. The points at which the experiments became unstable marked o for a wall temperature evolution of the form $\Delta T(1 - \exp(-st))$ and $\times$ for wall temperature proportional to time. [From *Kerr*, 1989.]

Care must also be taken when applying any stability criterion globally. Because of the nonlinear nature of the true equation of state, and the dependency of the other physical quantities on temperature and salinity [see *Ruddick and Shirtcliffe*, 1979] the local value of Q may vary greatly along a heated wall in a typical experiment. The local use of this stability criterion derived from linear theory can give predictions, both qualitatively and quantitatively, of how the instabilities in a typical experiment may be seen to appear at different portions of the heated wall at different times. This gives good agreement with the experimental observations of *Chen et al.* [1971] where the convection cells appeared to migrate up the heated wall. In their experiments the critical value of Q was attained towards the base of the wall first, and subsequently higher up. In some of their experiments the critical value of Q was only attained towards the base of the wall, and so instabilities only appeared in this region and not at all in the higher regions of their apparatus which remained stable. Further details can be found in *Kerr* [1989].

Although this linear stability analysis describes the onset of the instabilities in the strongly stratified regime, it is clear from the experimental results of *Tanny and Tsinober* [1988] shown in Figure 1 that for several of the experiments with lower values of Ra_S this linear theory is totally inappropriate. In their presentation of the experimental results they superimposed the stability curve derived by *Thangam et al.* [1981] for the case of instabilities in a laterally heated slot. This marginal stability curve has several sections, only one of which corresponds approximately to the stability boundary shown here. The other sections correspond to regions where the instabilities in the slot are affected or dominated by the shear flow along the walls. This situation is more important when the vertical salinity gradients are weaker, or equivalently the heating rate is faster. In these cases *Thangam et al.* [1981] found in a vertical slot that the fluid may become unstable to steady or oscillatory instabilities. How this structure for instabilities in a slot really transfers to the single sidewall case is not clear. The analysis is certainly more complicated as the quasi-static assumption is no longer valid. In addition the relevant temperature, salinity and velocity profiles for the background state will no longer be adequately described by the large-time asymptotic description, with its separated length and time-scales. A better understanding of this regime may enable a better comparison between many of the experiments that have been conducted for faster wall heating rates. Currently it seems difficult to compare these different sets of results in any meaningful way [see, for example, *Schladow et al.*, 1992].

The quasi-static approach can also be used for other

related situations. For example the heating or injection of salt at a vertical or sloping boundary into a fluid with a combined vertical temperature and salinity gradient, or for investigating the stability of the evolving horizontal gradients found when a sloping impermeable and insulating boundary is present in a fluid [*Kerr*, 1991]. In the latter case this approach shows that instabilities should always arise, but that they may take a long time before they first appear. *Linden and Weber* [1977] only observed instabilities when their parameter $\lambda = \alpha\overline{T}_z/\beta\overline{S}_z$ was greater than 0·7. As λ decreases the predicted time taken for the onset of instabilities becomes rapidly longer and exceeds the time that they were able to continue their experiments. If, for example, $\lambda = 0{\cdot}1$ in their experimental system the predicted time before the onset of instabilities is almost a decade! It would be hard to imagine being able to secure a research grant to investigate this small λ limit experimentally.

Another area in which this approach could be used is in the investigation of the stability of finite fronts. This has been looked at for the case where the vertical heat and salinity fluxes are dominated by salt fingering by *Niino* [1986] and *Yoshida et al.* [1989], but not for molecular diffusivities. One problem is that for the situation where a barrier is removed suddenly from between two bodies of fluid the instantaneous value of Q would be infinite, and more importantly the quasi-static assumption would be invalid. However for the case where two differing bodies of fluid were converging, as they do in the oceans, the horizontal extent of the region with lateral gradients may actually decrease with time, leading to an *increase* in Q which may lead to the subsequent onset of instabilities in a regime where the quasi-static approximation is valid.

3. NONLINEAR DYNAMICS OF CONVECTION CELLS

Although the linear stability analysis highlighted in the previous section provides a good prediction for the onset of instability in cases where the quasi-static assumption is valid, the form of convection cells that the linear stability analysis predicts consists, by its very nature, of a series of counter-rotating eddies. No physical experiments reported in the literature have ever yielded such behaviour. In all cases co-rotating cells are observed. The nature of the bifurcation from the evolving background state has been analysed using a weakly nonlinear analysis [*Kerr*, 1990] which is compatible with the the quasi-static assumption. This analysis showed that for heating a salinity gradient from a single sidewall, just as in the case of heating a salinity gradient between two parallel vertical walls [*Hart*, 1973] the bifurcation is subcritical. This is a result which holds for all values of σ and τ. Although this analysis cannot show the form taken by the fully nonlinear instabilities, the initial trends along the unstable subcritical branches are in general agreement with the form of the observed instabilities; the convection cells with the fluid rising near the wall are enhanced at the expense of the cells with the opposite sense of rotation, and the velocity of the cells along the wall is diminished. However this unstable subcritical branch has not been followed to large amplitudes by, say, numerical means, and so it is not known whether this branch becomes stable at larger amplitudes and these trends are subsequently reflected in the form of the actual disturbances observed, or whether these trends are only coincidental.

Analytic results have also been obtained for nonlinear sidewall convection cells uses energy stability analysis. With this technique it is possible to show that disturbances with arbitrary amplitude and form, but with a vertical periodicity of the Chen scale or less will decay if the instantaneous value of Q falls below a critical value:

$$Q < 64\pi^5(1-\tau)^4 \approx 19\,585(1-\tau)^4. \qquad (5)$$

This result does not use a quasi-static assumption. The importance of this result lies in the horizontal length scale in the denominator of (3). For any situation when the wall is heated so that the wall temperature increase is bounded, the instantaneous value of Q will eventually decay as t^{-1}. The subcritical nature of the instabilities means that the convection will persist when the wall heating is below the critical value required by linear stability, but this result shows the time when this convection *must* be decaying.

The other technique which has been used to examine the nonlinear behaviour of instabilities due to horizontal gradients with sidewalls has been numerical simulations of the full system of governing equations. These have been used to investigate the nonlinear dynamics of instabilities in vertical or inclined slots [*Wirtz et al.*, 1972; *Chen and Sandford*, 1977; *Thangam et al.*, 1982] or in rectangular cavities with applied temperature differences across the vertical walls [*Lee and Hyun*, 1991; *Schladow et al.*, 1992; *Tsitverblit and Kit*, 1993]. The simulations that are of particular interest are those that involve a transient heating of a wide slot. In these cases the initial instabilities that appear are effectively located at a single boundary and hence are of relevance here. Such calculations have been conducted by *Lee and*

Hyun [1991] and *Schladow et al.* [1992]. Unfortunately because of the difficulty and cost or performing such calculations, especially for realistic values of σ and τ, it has not yet been possible to conduct wide ranging systematic investigations of parameter space, and as such the use of numerical methods as a tool applied to this problem is still in its infancy. As yet no broad conclusions of the physics of the flows can be drawn from this work.

Some of the numerical simulations that have been carried out have drawn attention to the difficulty of eliminating "numerical instabilities". The task of eliminating such instabilities was accomplished with some difficulty for a simulation carried out by this author that used an implicit numerical scheme which represented the solutions in terms of a Fourier series in the vertical direction, and used a finite difference scheme in the horizontal direction [*Kerr*, 1987]. A typical result is shown in Figure 2a. The scheme produced results that matched up with the flows which seemed to be present from the many descriptions of flows where the primary visualisation of the instabilities was made using the shadowgraph technique. They also seemed to agree with other numerical simulations. With the arrival of more powerful computers it was possible to enhance the resolution of this numerical simulation, and the "numerical instabilities" returned. This time instead of the instabilities being apparently grid related, they could be resolved by the use of a finer resolution grid (Figure 2b). Indeed these "numerical instabilities" turned out to resemble very closely the secondary instabilities that are frequently (if not universally) observed in the interior of cells in sidewall convection experiments where dye or tracer particles are used to visualise the flow. The apparent universality of these secondary instabilities in experiments, and their appearance in numerical simulations with sufficient resolution to resolve them [for example *Schladow et al.*, 1992] may lead one to view with some caution any numerical simulation that either has insufficient resolution to resolve such disturbances, or manages to suppress them by some artificial (numerical) means. However these secondary instabilities may not be present in all situations, in particular when the slot is relatively narrow.

The contrast between the apparent uniformity inside the sharply defined convection cells when viewed using the shadowgraph technique and the vigorous internal mixing that is observed when tracer particles or dyes are used for flow visualisation are made clearly by the comparison of these two approaches by *Tsinober et al.* [1983]. In their Figure 1 they showed simultaneous visualisations using both techniques of an experiment which

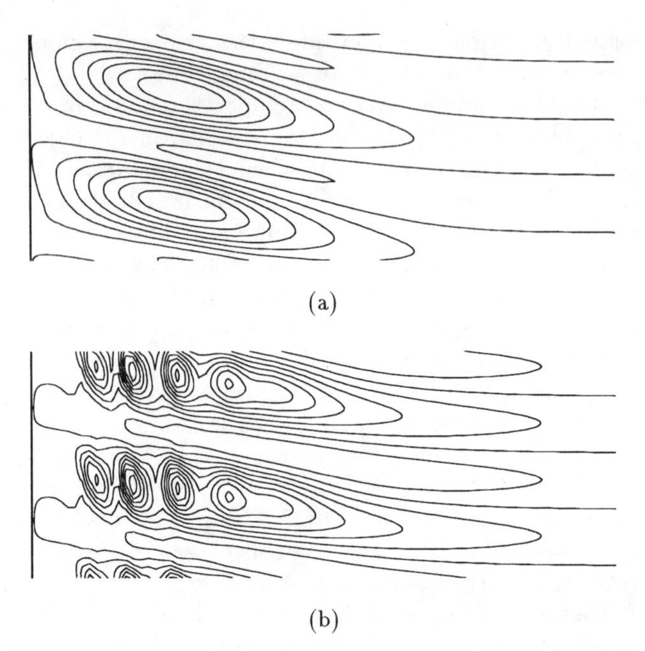

Fig. 2. Numerical calculations of the streamlines of instabilities due to the heating of a salinity gradient from a sidewall. Both have been calculated with $Q = 150\,000$, $H/L \approx 0.25$, $\sigma = 7$ and $\tau = 1/80$. These plots have been compressed in the horizontal direction, for the correct aspect ratio they should be stretched horizontally by a factor of about 3. The calculations in (a) successfully suppressed the "numerical instabilities" that are now resolved in (b) with a finer grid.

looked at the convection in a salinity gradient over a point source of heat. The convection cells observed grow horizontally from the plume above the heat source, and are closely related to the instabilities under consideration here. Ideally experiments would always be conducted using both methods of visualisation; the shadowgraph for showing the salinity structure, and tracers for showing the velocities.

The dynamics and driving mechanism behind these secondary instabilities has been attributed to shear. However this may well not be the case. In a study of the stability of interleaving intrusions in an unbounded fluid with linear compositional gradients and fluxes driven by molecular diffusivities [as in *Holyer*, 1983], *Kerr* [1992a] showed that the secondary instabilities can be driven by double-diffusive effects and not shear. For example, in the limit of very weak horizontal gradients the secondary instability starts when the local minimum Richardson number for the shear in the intrusions has values of 8·378 for the case where the background stable stratification is due only to temperature, and 42 590 when the stratification is due only to salinity. In neither case is the Richardson number remotely close to 1/4, the maximum possible value for shear instabilities to occur. Hence the instabilities in this case *cannot* be attributed to shear, but are driven by the altered temperature and salinity gradients caused by the growing intrusions. The typical instabilities found (Figure 3). are similar the secondary instabilities observed in numerical simulations or experiments.

An important aspect of the nonlinear dynamics of the convection cells at isolated boundaries is the process of cell merging. As the wall temperature increases, so too does the Chen scale, H. It has been observed in many experiments that, although the initial height of the convection cells corresponds closely to H, as this scale increases adjacent convection cells undergo a process of merging, and the *average* cell height settles down to approximately $\frac{2}{3}H$. This average height has been observed in a variety of experiments involving the heating of various substances stratifying a variety of fluids with differing viscosities [*Wirtz et al.*, 1972; *Huppert and Turner*, 1980; *Huppert and Josberger*, 1980; *Huppert et al.*, 1984; *Tanny and Tsinober*, 1988]. A detailed numerical investigation of cell merging is beyond the resources of the numerical investigations which have been conducted to date. However, a simple model of cell merging has been presented by *Kerr* [1992b]. This model assumes (1) cell interfaces do not move up or down the wall, but may vanish, and (2) cell interfaces vanish when the combined height of the adjacent pair of convection cells equals the instantaneous Chen height, H. When numerical simulations are carried out using this model it is found that the distribution of cell heights, nondimensionalised with respect to the instantaneous value of the Chen height, settles down to the steady distribution shown in Figure 4. This distribution has an average height of 0·6805, and a standard deviation of 0·1561. The former is in good agreement with the variety of experiments mentioned above, while the standard deviation shows reasonable agreement with the reported results of *Tanny and Tsinober* [1989]. That the above model gives such good results should not be accorded too much importance. It glosses over the physics of convection cell merging; it ignores the time dependency of the merging process, it assumes that the merging process is independent of the ratio of the adjacent cell heights, and it uses an arbitrary height H for the merging criterion (why not $0{\cdot}9H$ or $1{\cdot}1H$?). However, the reasonable predictions that this model makes would lead us to expect that the true underlying criteria for the merging process may not be too far from this simplistic model. What we can conclude from it is that real convection cells are likely to merge to give convection cells that are significantly larger that the observed *average*, and that this observed average height and size distribution is likely to be associated with the resistance of cell interfaces to migration and break down. What is needed to understand this whole process of cell merging is an improved understanding of the mechanisms which control the dynamics of the cell interfaces, when do they move and when and how do they break down.

There is one further aspect of convection initiated at

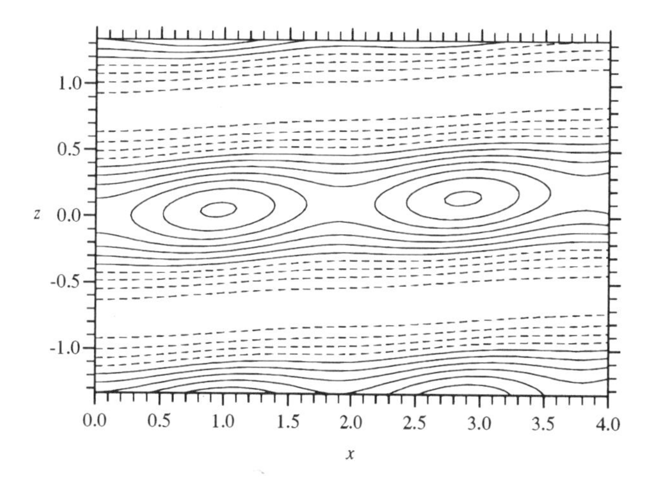

Fig. 3. Streamfunction for secondary instabilities in double-diffusive interleaving in an unbounded fluid with uniform vertical and horizontal gradients. [From *Kerr* 1992a.]

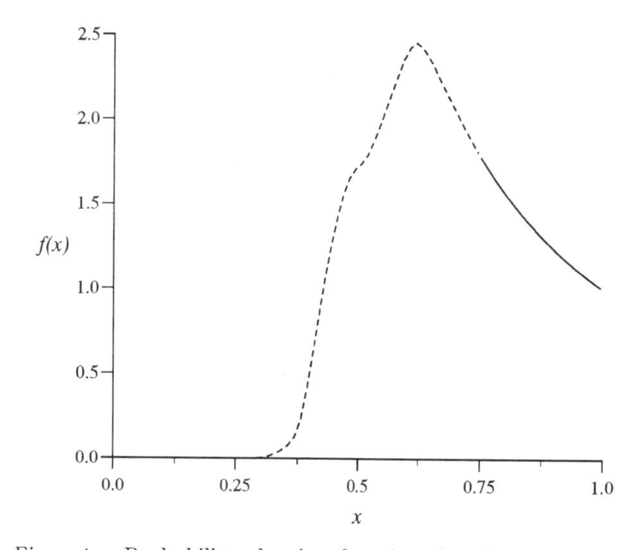

Fig. 4. Probability density function for the nondimensional convection cell heights derived from the model of *Kerr* [1992b]. The solid lines in the intervals $(0, \frac{1}{4})$ and $(\frac{3}{4}, 1)$ represent theoretical results, the dashed line in the interval $(\frac{1}{4}, \frac{3}{4})$ was found numerically.

isolated boundaries that has not been touched on here, principally because there has been little in the way of experimental study, and no theoretical nor numerical investigations. The convection cells which have been discussed so far are driven by the horizontal gradients near the boundary. However, if the bulk of the fluid has a destabilising temperature gradient and a stabilising salinity gradient, then the interior of the fluid may be stable to infinitesimal disturbances, but be unstable to finite amplitude subcritical layered convection. This leads to the possibility that convection cells initiated at the boundary may trigger self-propagating intrusions which grow into the interior of the stratified fluid which extract their driving energy from the vertical temperature gradient in the core of the fluid. This phenomenon has been observed in the experiments of *Schladow et al.* [1992]. When they heated a combined temperature and salinity gradient from a vertical sidewall they found that in some instances the instabilities at the wall did indeed propagate into the bulk of the fluid, but in other cases it did not, even though a destabilising temperature gradient was present. What governs when the sidewall instabilities can generate self-propagating disturbances is not understood. Such knowledge could have important implications for salt-gradient solar ponds where such convecting layers intruding into the bulk of the encapsulating salinity gradient would lead to a degradation of the thermal insulation provided by what is intended to be a static stable salinity gradient lying on the top of the reservoir. This mechanism could also be important in magma chambers where there are typically destabilising temperature gradients, and stabilising compositional gradients present. For any predictions to be made in such cases an understanding of the dimensionless parameters involved in governing when such intrusions grow and when they do not is needed. Currently this knowledge does not exist.

4. CONCLUSIONS

Some aspects of the formation and behaviour of double-diffusive convection cells at sidewalls are well understood. In particular the criterion for the onset of instability, and the behaviour of near critical nonlinear convection cells for the case when heating rates are low and the quasi-static assumption can be made are known. However there are many points concerning double-diffusive sidewall convection that need a better understanding. The onset of instabilities for faster wall heating rates is not well understood; there is currently no theory for the onset of instabilities at isolated boundaries in such cases. This theory would have to account for the transient behaviour of the background salinity and temperature gradients which evolve on time-scales comparable to the time-scale for the evolution of the instabilities themselves, and so there is no separation of scales that simplifies the analysis for the low heating rate case. Indeed these instabilities are not well understood experimentally – there is some disagreement amongst the experimentalists as to what is the most appropriate measure of the vertical length scale of these instabilities. It would aid the search for a better understanding of the processes involved if there was some agreement on this point.

Another area that needs to be understood more fully is the process of cell merging. The criteria which govern the behaviour of the interfaces between convection cells are not well understood. It would be helpful if the processes involved in the interface break-up and the time-scales involved were known, and how these depended on the rate of wall heating.

Lastly the self-propagating intrusions in a fluid with a destabilising temperature gradient which are triggered by convection at the wall are not well understood. A criterion that would clearly differentiate between regimes where this self-propagation would and would not occur would be most desirable. If such a criterion is to be applicable to diverse situations such as whether these intrusions could appear in magma chambers, it must have a sound theoretical basis.

Some of the important theoretical challenges in this area of double-diffusive convection lie in the main areas listed above. A proper understanding of these areas would help us to go beyond describing experimental observations and enable us to make quantitative predictions of the behaviour of double-diffusive instabilities at a sidewall for the large diversity of fluids, scales and geometries which experience these phenomena.

REFERENCES

Chen, C.F. and R.D. Sandford, Stability of time-dependent double-diffusive convection in an inclined slot. *J. Fluid Mech. 83*, 83–95, 1977.

Chen, C.F. and M.W. Skok, Cellular convection in a salinity gradient along a heated inclined wall. *Int. J. Heat Mass Transfer 17*, 51–60, 1974.

Chen, C.F., R.A. Briggs and D.G. Wirtz, Stability of thermal convection in a salinity gradient due to lateral heating. *Int. J. Heat Mass Transfer 14*, 57–65, 1971.

Chereskin, T.K. and P.F. Linden, The effect of rotation on intrusions produced by heating a salinity gradient. *Deep-Sea Res. 33*, 305–322, 1986.

Hart, J.E., On sideways diffusive instability. *J. Fluid Mech.*

49, 279–288, 1971.

Hart, J.E., Finite amplitude sideways diffusive convection. *J. Fluid Mech.* *59*, 47–64, 1973.

Holyer, J.Y., Double-diffusive interleaving due to horizontal gradients. *J. Fluid Mech.* *137*, 347–362, 1983.

Holyer, J.Y., T.J. Jones, M.G. Priestly and N.C. Williams, The effect of vertical temperature and salinity gradients on double-diffusive interleaving. *Deep-Sea Res.* *34*, 517–530, 1987.

Huppert, H.E. and E.G. Josberger, The melting of ice in cold stratified water. *J. Phys. Oceanogr.* *10*, 953–960, 1980.

Huppert, H.E. and J.S. Turner, Ice blocks melting into a salinity gradient. *J. Fluid Mech.* *100*, 367–384, 1980.

Huppert, H.E., R.C. Kerr and M.A. Hallworth, Heating or cooling a stable compositional gradient from the side. *Int. J. Heat Mass Transfer* *27*, 1395–1401, 1984.

Kerr, O.S. Horizontal effects in double-diffusive convection. Ph.D., Thesis 230 pp., University of Bristol, U. K., 1987..

Kerr, O.S., Heating a salinity gradient from a vertical sidewall: linear theory. *J. Fluid Mech.* *207*, 323–352, 1989.

Kerr, O.S., Heating a salinity gradient from a vertical sidewall: nonlinear theory. *J. Fluid Mech.* *217*, 529–546, 1990.

Kerr, O.S., Double-diffusive instabilities at a sloping boundary. *J. Fluid Mech.* *225*, 333–354, 1991.

Kerr, O.S, Two-dimensional instabilities of steady double-diffusive interleaving. *J. Fluid Mech.* *242*, 99–116, 1992a.

Kerr, O.S., On the merging of double-diffusive convection cells at a vertical boundary. *Phys. Fluids A* *4*, 2923–2926, 1992b.

Lee, J.W. and Hyun, J.M., Heating or cooling a stable compositional gradient from the side. *Int. J. Heat Mass Transfer* *34*, 2409–2421, 1991.

Linden, P.F. and J.E. Weber, The formation of layers in a double-diffusive system with a sloping boundary. *J. Fluid Mech.* *81*, 757–773, 1977.

McDougall, T.J., Double-diffusive interleaving I. Linear stability analysis. *J. Phys. Oceanogr.* *15*, 1532–1541, 1985.

Narusawa, U. and Y. Suzukawa, Experimental study of double-diffusive cellular convection due to a uniform lateral heat flux. *J. Fluid Mech.* *113*, 387–405, 1981.

Niino, H., A linear theory of double-diffusive horizontal intrusions in a temperature-salinity front. *J. Fluid Mech.* *171*, 71–100, 1986.

Paliwal, R.C. and C.F. Chen, Double-diffusive instability in an inclined fluid layer. Part 1. Experimental Investigation. *J. Fluid Mech.* *98*, 755–768, 1980a.

Paliwal, R.C. and C.F. Chen, Double-diffusive instability in an inclined fluid layer. Part 2. Stability analysis. *J. Fluid Mech.* *98*, 769–785, 1980b.

Ruddick, B.R., Intrusive mixing in a Mediterranean salt len–Intrusion slopes and dynamical mechanisms. *J. Phys. Oceanogr.* *22*, 1274–1285, 1992.

Ruddick, B.R. and T.G.L. Shirtcliffe, Data for double diffusers: Physical properties of aqueous salt-sugar solutions. *Deep-Sea Res.* *26*, 775–787, 1979.

Ruddick, B.R. and J.S. Turner, The vertical length scale of double-diffusive intrusions. *Deep-Sea Res.* *26*, 903–913, 1979.

Schladow, S.G., E. Thomas and J.R. Koseff, The dynamics of intrusions into a themohaline stratification. *J. Fluid Mech.* *236*, 127–165, 1992.

Stern, M.E., Lateral mixing of water masses. *Deep-Sea Res.* *14*, 747–753, 1967.

Tanny, J. and A.B. Tsinober, The dynamics and structure of double-diffusive layers in sidewall-heating experiments. *J. Fluid Mech.* *196*, 135–156, 1988.

Tanny, J. and A.B. Tsinober, On the behaviour of a system of double diffusive layers during its evolution. *Phys. Fluids A* *1*, 606–609, 1989.

Thangam, S., A. Zebib and C.F. Chen, Transition from shear to sideways diffusive instability in a vertical slot. *J. Fluid Mech.* *112*, 151–160, 1981.

Thangam, S., A. Zebib and C.F. Chen, Double-diffusive convection in an inclined fluid layer. *J. Fluid Mech.* *116*, 363–378, 1982.

Thorpe, S.A., P.K. Hutt and R. Soulsby, The effects of horizontal gradients on thermohaline convection. *J. Fluid Mech.* *38*, 375–400, 1969.

Toole, J.M. and D.T. Georgi, On the dynamics and effects of double-diffusively driven intrusions. *Prog. Oceanogr.* *10*, 123–145, 1981.

Tsinober, A.B., Y. Yahalom and D.J. Shlien, A point source of heat in a stable salinity gradient. *J. Fluid Mech.* *135*, 199–217, 1983.

Tsitverblit, N. and E. Kit, The multiplicity of steady flows in confined double-diffusive convection with lateral heating. *Phys. Fluids A* *5*, 1062-1064, 1993.

Turner, J.S., *Buoyancy effects in fluids*, 368pp., Cambridge University Press, 1973.

Walsh, D. and B. Ruddick, Double-diffusive interleaving: the influence of non-constant diffusivities, *J. Phys. Oceanogr.*, in press, 1994.

Wirtz, R.A., D.G. Briggs and C.F. Chen, Physical and numerical experiments on layered convection in a density-stratified fluid. *Geophys. Fluid Dyn.* *3*, 265–288, 1972.

Yoshida, J., H. Nagashima and H. Niino, The behaviour of double-diffusive intrusions in a rotating system. *J. Geophys. Res.* *94*, 4293–4937, 1989.

Oliver S. Kerr, Department of Mathematics, City University, Northampton Square, London EC1V 0HB, U.K.

Axisymmetric Double-Diffusive Convection in a Cylindrical Container: Linear Stability Analysis with Applications to Molten $CaO-Al_2O_3-SiO_2$

Yan Liang

Department of Geophysical Sciences, University of Chicago, Chicago, IL 60637

The onset of fingering instability in a multicomponent fluid confined between two horizontal plates or in a cylinder is discussed. Using a linear transformation of the compositional variables, linear stability analyses show that the critical Rayleigh number for the onset of fingering instability for a multicomponent fluid depends only on the boundary conditions and geometry of the container and is independent of the number of components in the system if the initial concentration distributions are linear and an effective Rayleigh number is introduced. The multicomponent nature of the system is contained in the definition of the effective Rayleigh number given in the matrix notion $\mathbf{Ra_e} \equiv gd^3\alpha^T[\mathbf{D}]^{-1}\Delta\mathbf{C}/\nu$. The result of linear stability analysis was applied to chemical diffusion and double-diffusive convection experiments in molten $CaO-Al_2O_3-SiO_2$ conducted at 1500 °C and 10 Kb using a cylindrical molybdenum capsule. Two approximate axisymmetric fingering instabilities were observed, one with upflow and the other with downflow along the center of the capsule. These "silica" fingers form under similar conditions as that for salt fingers when the fast diffusing component is stabilizing and the slow diffusing component is destabilizing. For diffusion across an initially sharp interface, there is a critical time for the onset of fingering instability consistent with that observed in the experiments.

1. INTRODUCTION

Among the most active areas of experimental petrology and geochemistry is the laboratory study of chemical diffusion in synthetic (e.g., $CaO-MgO-Al_2O_3-SiO_2$) and natural (e.g., basalt or rhyolite) molten silicates at high temperatures and pressures because diffusion plays an important role in such processes as magma mixing and convection and also crystal growth and dissolution [e.g., *Watson and Baker,* 1991 *and references therein*]. A typical isothermal chemical diffusion experiment involves juxtaposing two melts of different compositions in a cylindrical container using a piston cylinder apparatus (Figure 1). To avoid convection, melt of smaller density is always placed on top of melt of greater density. This configuration is not, however, guaranteed to be gravitational stable because of the possibility of double-diffusive convection [e.g., *Turner,* 1974; 1985].

Stability analyses of double-diffusive convection have been given in a number of studies all of which deals with a fluid confined between two horizontal plates [e.g., *Stern,* 1960; *Veronis,* 1965; *Baines and Gill,* 1969; *Griffiths,* 1979; *McDougall,* 1983; *Terrones,* 1993]. Linear stability studies of a fluid in a cylindrical container were mostly concerned with Rayleigh-Bénard convection [e.g., *Charlson and Sani,* 1970 *and* 1971; *Joseph,* 1971; *Jones and Moore,* 1979]. These studies have shown that the vertical sidewall has a significant effect on the stability of the fluid with the critical Rayleigh number increasing with the decreasing aspect ratio of the container. For axisymmetric Rayleigh-Bénard convection it is well know that two stable modes of convection exist, one with downflow and the other with upflow along the cylindrical axis [e.g., *Liang et al.,* 1969; *Joseph,* 1971; *Jones et al.,* 1976]. The actual flow direction is determined by the initial perturbation and remains unchanged once convection sets in [e.g., *Liang et al.,* 1969]. For a fluid with temperature dependent viscosity, *Liang et al.* [1969] have shown that the amplitudes of upflow and downflow are different. The flow with higher amplitude is in the direction of decreasing fluid

Double-Diffusive Convection
Geophysical Monograph 94

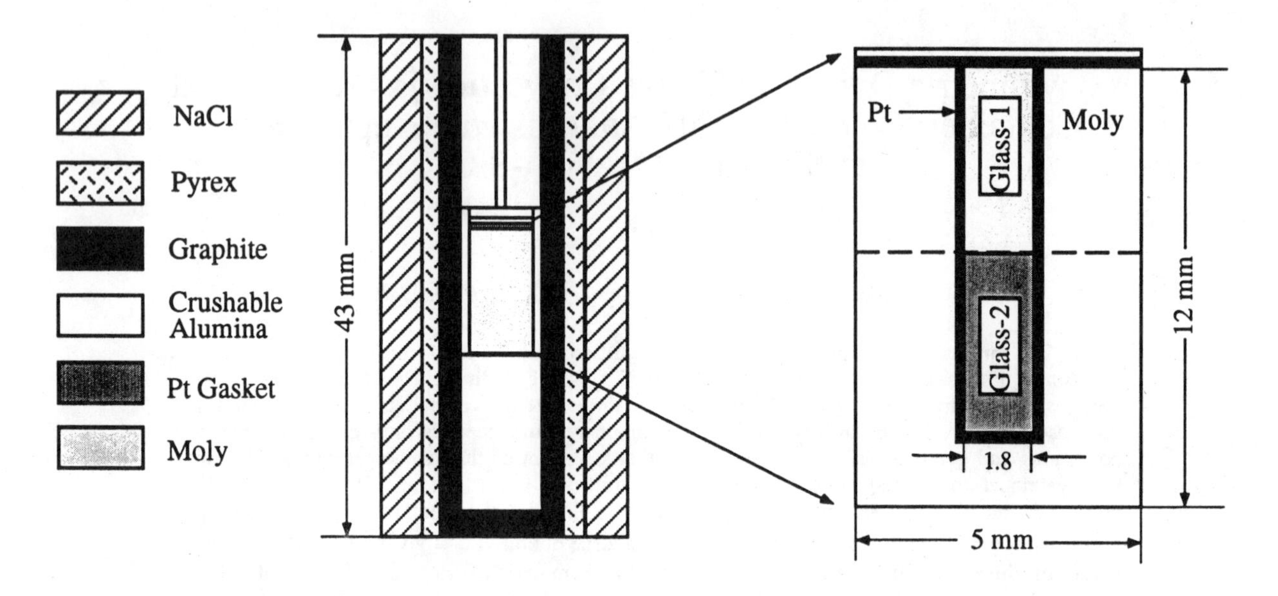

Fig. 1. Schematic cross section of the piston cylinder assembly and molybdenum capsule used for diffusion experiments. Diffusion or convection couple was formed either by loading two starting glass powders directly into the capsule (glasses 1 and 2, CVT-2 in Plate 1a) or by joining two halves of uniform glass rods from pre-synthesized runs (CD1/2 in Plate 1b). The diffusion or convection couple was then placed in the piston cylinder apparatus at 1500 °C and 10 Kb for the duration of the experiment. Upon quenching the charge was sectioned longitudinally, polished and mounted for electron microprobe analysis. Due to compaction and melting, the final capsule radius was reduced slightly to 0.7 ~ 0.9 mm. Previous experiments with similar capsules [*Ayers et al.*, 1992] show that the vertical temperature differences are small across the capsule (≤ 5-10 °C), and negligible in terms of their effect on the convection discussed in this paper.

viscosity.

The stability of double-diffusive convection in a cylindrical container has not been reported in the past. In this paper, I will extend the linear stability analysis of a simple fluid confined in a cylindrical container to a multicomponent fluid in which off-diagonal elements of the diffusion matrix play a key role in destabilizing the fluid system. The analysis is directly relevant to our recent laboratory experiments showing double-diffusive convection in a molten silicate system. In the next section, a linear transformation, which can be used to simplify the linear stability analysis of a multicomponent fluid with off-diagonal terms in the diffusion matrix, is introduced. This is followed by an example of a linear stability analysis of a ternary fluid confined in a cylindrical container. The results are then applied to our recent experimental study of chemical diffusion and double-diffusive convection in molten CaO-Al_2O_3-SiO_2 at 1500 °C and 10 Kb [*Liang et al.*, 1994].

2. MULTICOMPONENT DIFFUSION AND LINEAR TRANSFORMATION

The linear stability analysis of a multicomponent fluid composed of $n+1$ components involves solving a momentum together with conservation equations for each component and a statement about how density depends on composition. The special feature of multicomponent convection arises from the conservation equations of the form [e.g., *de Groot and Mazur*, 1962]

$$\frac{\partial C_i}{\partial t} + \mathbf{U}\cdot\nabla C_i = \sum_{j=1}^{n} D_{ij}\,\nabla^2 C_j \tag{1a}$$

or in the matrix notation

$$\frac{\partial \mathbf{C}}{\partial t} + \mathbf{U}\cdot\nabla\mathbf{C} = [\mathbf{D}]\nabla^2\mathbf{C} \tag{1b}$$

where $\mathbf{C}$ is the concentration vector whose element C_i (i = 1, n) is the concentration of component *i*, $\mathbf{U}$ is the velocity field. [$\mathbf{D}$] is a $n \times n$ diffusion matrix. C_1 can represent temperature if a thermal gradient exists. Component *n+1* is not considered an independent variable since it can always be calculated given the concentration of the other *n* components. Assuming a constant diffusion matrix [$\mathbf{D}$] is justified when temperature changes and concentration ranges are small. For many fluids, the off-diagonal terms of the diffusion matrix, D_{ij} ($i \neq j$), are not negligible and can play an important role in destabilizing a fluid system [*McDougall*, 1983; *McDougall and Turner*, 1983; *Liang et al.*, 1994]. Although this poses no real difficulties in carrying out a stability analysis, it does make the stability analysis very tedious, especially when the number of component in the system becomes large [e.g., $n \geq 3$, *McDougall*, 1983; *Terrones*, 1993]. The analysis can be simplified by noting that [$\mathbf{D}$] is positive definite [*de Groot and Mazur*, 1962]. Hence equation (1) can be diagonalized with the linear transformation [e.g., *Toor*, 1964; *Gupta and Cooper*, 1971]

$$\mathbf{T} = [\mathbf{B}]\mathbf{C} \tag{2a}$$

$$[\Lambda] = [\mathbf{B}][\mathbf{D}][\mathbf{B}]^{-1} \tag{2b}$$

where $\mathbf{T}$ is a column vector, $[\Lambda] = \text{diag}[\lambda_1, \lambda_2, ..., \lambda_n]$ and λ_i's are the eigenvalues of the diffusion matrix. The eigenvectors of [$\mathbf{D}$] form the columns of the $[\mathbf{B}]^{-1}$ matrix. Substituting (2) into (1), it follows that

$$\frac{\partial \mathbf{T}}{\partial t} + \mathbf{U} \cdot \nabla \mathbf{T} = [\Lambda]\nabla^2 \mathbf{T} \tag{3}$$

The boundary conditions for $\mathbf{C}$ can also be diagonalized in many cases (e.g., $\mathbf{C}$ = constants on the boundaries). Thus one has *n* uncoupled diffusion equations along with their boundary conditions expressed in $\mathbf{T}$ space. Stability analysis in $\mathbf{T}$ space is much simpler than in $\mathbf{C}$ space when the density-composition relation is linear. For example, consider a ternary fluid for which the diffusion-advection equations are

$$\frac{\partial C_1}{\partial t} + \mathbf{U} \cdot \nabla C_1 = D_{11}\nabla^2 C_1 + D_{12}\nabla^2 C_2 \tag{4a}$$

$$\frac{\partial C_2}{\partial t} + \mathbf{U} \cdot \nabla C_2 = D_{21}\nabla^2 C_1 + D_{22}\nabla^2 C_2 \tag{4b}$$

The density of the fluid can be expressed in a linear form in the neighborhood of C_1^0 and C_2^0 as

$$\rho = \rho_0\left[1.0 + \alpha_1(C_1 - C_1^0) + \alpha_2(C_2 - C_2^0)\right] \tag{4c}$$

To simplify the linear stability analysis, equations (4a)-(4c) are first diagonalized with the linear transformation

$$\begin{bmatrix} T_1 \\ T_2 \end{bmatrix} = \begin{bmatrix} b_{11} & b_{12} \\ b_{21} & b_{22} \end{bmatrix} \begin{bmatrix} C_1 \\ C_2 \end{bmatrix} \tag{5}$$

The uncoupled diffusion equations in $\mathbf{T}$ space are

$$\frac{\partial T_1}{\partial t} + \mathbf{U} \cdot \nabla T_1 = \lambda_1 \nabla^2 T_1 \tag{6a}$$

$$\frac{\partial T_2}{\partial t} + \mathbf{U} \cdot \nabla T_2 = \lambda_2 \nabla^2 T_2 \tag{6b}$$

And the density-composition relation becomes

$$\rho = \rho_0\left[1.0 + \beta_1(T_1 - T_1^0) + \beta_2(T_2 - T_2^0)\right] \tag{6c}$$

where $\lambda_{1,2} = \frac{1}{2}\left[D_{11} + D_{22} \pm \sqrt{(D_{11} - D_{22})^2 + 4D_{12}D_{21}}\right] > 0$, are the eigenvalues of [$\mathbf{D}$]. β_1 and β_2 are related to α_1 and α_2 through [$\mathbf{B}$]

$$\begin{bmatrix} \beta_1 \\ \beta_2 \end{bmatrix}^T = \begin{bmatrix} \alpha_1 \\ \alpha_2 \end{bmatrix}^T \begin{bmatrix} b_{11} & b_{12} \\ b_{21} & b_{22} \end{bmatrix}^{-1} \tag{7}$$

With this linear transformation, the stability analysis can be carried out using equations (6a)-(6c) which no longer have off-diagonal terms in the diffusion matrix.

In classical thermohaline convection, linear stability analysis based on constant concentration gradients shows that fingering instability occurs, in a horizontally infinite fluid layer of depth d with free upper and lower boundaries, when

$$R_1 + \frac{1}{\tau} R_2 > \frac{27}{4}\pi^4 \tag{8}$$

where $R_1 = g\beta_1 \Delta T_1 d^3/\nu\lambda_1$ and $R_2 = g\beta_2 \Delta T_2 d^3/\nu\lambda_1$ are Rayleigh numbers for component T_1 and T_2, respectively, g is the acceleration due to gravity, ΔT_i is the difference in component T_i between the upper and lower boundaries, ν is the kinematic viscosity of the fluid, and $\tau = \lambda_2/\lambda_1$ is the diffusivity ratio [e.g., *Stern*, 1960; *Veronis*, 1965; *Baines and Gill*, 1969]. Defining an effective Rayleigh number [*Turner*, 1974]

$$Ra_e = R_1 + \frac{1}{\tau} R_2 \tag{9a}$$

it can then be easily verified using relations (5) and (7) that

$$Ra_e = \frac{gd^3}{\nu}\begin{bmatrix}\beta_1\\ \beta_2\end{bmatrix}^T\begin{bmatrix}\lambda_1 & 0\\ 0 & \lambda_2\end{bmatrix}^{-1}\begin{bmatrix}\Delta T_1\\ \Delta T_2\end{bmatrix}$$

$$= \frac{gd^3}{\nu}\begin{bmatrix}\alpha_1\\ \alpha_2\end{bmatrix}^T\begin{bmatrix}D_{11} & D_{12}\\ D_{21} & D_{22}\end{bmatrix}^{-1}\begin{bmatrix}\Delta C_1\\ \Delta C_2\end{bmatrix} \quad (9b)$$

Hence fingering instability sets in when

$$Ra_e = \frac{gd^3}{\nu}\begin{bmatrix}\alpha_1\\ \alpha_2\end{bmatrix}^T\begin{bmatrix}D_{11} & D_{12}\\ D_{21} & D_{22}\end{bmatrix}^{-1}\begin{bmatrix}\Delta C_1\\ \Delta C_2\end{bmatrix} > \frac{27}{4}\pi^4 \quad (10)$$

Criterion (10) is equivalent to equation (17) of *McDougall* [1983] for the occurrence of fingering instability in a ternary fluid in which off-diagonal elements of the diffusion matrix are included. Similarly, the criterion for a four component system in which off-diagonal terms of the diffusion matrix are non-trivial [*Terrones,* 1993] can be readily recovered from the results of *Griffiths* [1979]. The general result is that for an *n+1* component fluid confined between two horizontally infinite plates with free upper and lower boundaries, fingering instability sets in when

$$Ra_e > \frac{27}{4}\pi^4 \quad (11a)$$

where Ra_e is defined by

$$Ra_e = \frac{gd^3}{\nu}\begin{bmatrix}\alpha_1\\ \alpha_2\\ \cdots\\ \alpha_n\end{bmatrix}^T\begin{bmatrix}D_{11} & D_{12} & \cdots & D_{1n}\\ D_{21} & D_{22} & \cdots & D_{2n}\\ \cdots & \cdots & \cdots & \cdots\\ D_{n1} & D_{n2} & \cdots & D_{nn}\end{bmatrix}^{-1}\begin{bmatrix}\Delta C_1\\ \Delta C_2\\ \cdots\\ \Delta C_n\end{bmatrix} \quad (11b)$$

Diagonalizing the diffusion equations along with their boundary conditions is clearly very useful when analyzing the stability of multicomponent systems with significant off-diagonal terms in their diffusion matrices. Not only is the algebra simplified and the connection to classical stability problems more obvious, but also it permits one to identify the relevant time and concentration scales as d^2/λ_1 and ΔT_i, as opposed to less relevant choices such as d^2/D_{11} and ΔC_i [e.g., *McDougall,* 1983; *Terrones,* 1993]. ΔC_i, for example, can be zero for a given component whereas ΔT_i is not. The advantage of working with the diagonalized diffusion equation can be further seen in the next section on the linear stability of axisymmetric double-diffusive convection in a cylindrical geometry.

3. DOUBLE-DIFFUSIVE CONVECTION IN A CYLINDRICAL CONTAINER

Consider a three component fluid confined in a vertical cylindrical container of radius r_o and height d with insulating sidewall and free upper and lower boundaries. The initial concentration distributions are linear in the z direction (z being positive upward) with gradients $\Delta C_1/d$ and $\Delta C_2/d$. ΔC_i is the concentration difference between the upper and lower boundaries where C_i remains constant. The stability of such a system subject to an infinitesimal perturbation is considered. As before, the original diffusion-advection equations (4a)-(4c) along with boundary conditions are first diagonalized through the linear transformation $\mathbf{T} = [\mathbf{B}]\mathbf{C}$. Because of the nature of the experiments discussed in the next section, I will restrict myself to the case of axisymmetric unstable fingers. Extensions to asymmetric situations and diffusive regimes are straightforward when the diffusion equations are diagonalized. The linearized perturbation equations, in non-dimensional form in **T** space are

$$\left(\frac{\partial}{\partial t} - \nabla^2\right)T_1 = -w \quad (12a)$$

$$\left(\frac{\partial}{\partial t} - \tau\nabla^2\right)T_2 = -w \quad (12b)$$

$$\left[\frac{1}{\sigma}\frac{\partial}{\partial t} - \left(\nabla^2 - \frac{1}{r^2}\right)\right]\left(\frac{\partial w}{\partial r} - \frac{\partial u}{\partial z}\right) = R_1\frac{\partial T_1}{\partial r} + R_2\frac{\partial T_2}{\partial r} \quad (12c)$$

$$\frac{\partial u}{\partial r} + \frac{u}{r} + \frac{\partial w}{\partial z} = 0 \quad (12d)$$

with boundary conditions

$$T_1 = T_2 = \frac{\partial u}{\partial z} = w = 0, \text{ at } z = 0, 1 \quad (12e)$$

$$\frac{\partial T_1}{\partial r} = \frac{\partial T_2}{\partial r} = u = w = 0, \text{ at } r = \Gamma \quad (12f)$$

where u and w are the radial and vertical components of velocity, respectively, $\sigma = \nu/\lambda_1$ is Prandtl number, and $\Gamma = r_o/d$ is the aspect ratio of the cylinder. d^2/λ_1, d, and ΔT_i were chosen to nondimensionalize time, length, and concentration T_i, respectively. ΔT_i relates to ΔC through (5). R_1, R_2, and τ are the same as in the previous section. The normal mode of solutions satisfying the upper and lower boundary conditions are of the form

$$u \sim e^{pt}\,U(r)\cos(n\pi z) \quad (13a)$$

$$w \sim e^{pt}\,W(r)\sin(n\pi z) \quad (13b)$$

$$T_1 \sim e^{pt}\,T_1(r)\sin(n\pi z) \quad (13c)$$

$$T_2 \sim e^{pt}\,T_2(r)\sin(n\pi z) \quad (13d)$$

where p is the growth rate. n is the vertical wave number.

Substituting (13a)-(13d) into (12a)-(12d), one can verify that $U(r) \sim J_1(ar)$, where J_1 is the Bessel function of the first order. The characteristic equation for the growth rate p is then

$$p^3+k^2(1+\tau+\sigma)p^2+\left[k^4(\tau+\sigma+\tau\sigma)-\frac{\sigma a^2}{k^2}(R_1+R_2)\right]p + \sigma\left[k^6\tau - a^2(\tau R_1+R_2)\right] = 0 \quad (14)$$

where $k^2 = n^2\pi^2 + a^2$. The state of marginal stability for the onset of fingers corresponds to $p = 0$, in which case equation (14) reduces to

$$R_1 + \frac{1}{\tau}R_2 = \frac{(n^2\pi^2 + a^2)^3}{a^2} \equiv R_c \quad (15)$$

where R_c is the critical Rayleigh number. a can be solved from (15) in terms of R_c and n. Since only three of the six roots of a are linearly independent, one can write the complete eigenfunctions for the radial components of velocity and concentration in terms of Bessel functions of the zeroth and first order

$$U(r) = A_1J_1(a_1r)+A_2J_1(a_2r)+A_3J_1(a_3r) \quad (16a)$$

$$W(r) = -\frac{1}{\pi}\left[a_1A_1J_0(a_1r)+a_2A_2J_0(a_2r)+a_3A_3J_0(a_3r)\right] \quad (16b)$$

$$T_1(r) = \frac{1}{\pi}\left[\frac{a_1A_1J_0(a_1r)}{\Gamma^2a_1^2+n^2\pi^2}+\frac{a_2A_2J_0(a_2r)}{\Gamma^2a_2^2+n^2\pi^2}+\frac{a_3A_3J_0(a_3r)}{\Gamma^2a_3^2+n^2\pi^2}\right] \quad (16c)$$

$$T_2(r) = \frac{1}{\tau}T_1(r) \quad (16d)$$

where A_i (i = 1, 3) are constants to be determined by the boundary conditions at $r = \Gamma$. For the least stable mode (n = 1), *Drazin and Reid* [1981] have shown that all three roots of a^2 from (15) are real and satisfy the relation

$$a_1^2 > a_2^2 > 0 > a_3^2 \quad (17)$$

when $R_c > 27/4\pi^4$. Applying boundary conditions for the case n = 1, one has

$$J_1(a_1\Gamma)A_1+J_1(a_2\Gamma)A_2+I_1(a_3\Gamma)A_3 = 0 \quad (18a)$$

$$a_1J_0(a_1\Gamma)A_1+a_2J_0(a_2\Gamma)A_2+a_3I_0(a_3\Gamma)A_3 = 0 \quad (18b)$$

$$\frac{a_1^2J_1(a_1\Gamma)}{\Gamma^2a_1^2+\pi^2}A_1+\frac{a_2^2J_1(a_2\Gamma)}{\Gamma^2a_2^2+\pi^2}A_2+\frac{a_3^2I_1(a_3\Gamma)}{\Gamma^2a_3^2-\pi^2}A_3 = 0 \quad (18c)$$

or in the matrix notion

$$[\mathbf{H}]\mathbf{A} = 0 \quad (18d)$$

Here the complementary Bessel functions I_0 and I_1 are used because $a_3^2 < 0$. It is understood that the a_i's in (18) are absolute values of the three linearly independent roots of (15). Equations (18a)-(18c) are linear homogeneous equations and their solutions exist only when the determinant of the coefficients is zero, that is

$$\det[\mathbf{H}] = 0 \quad (19)$$

Equation (19) is therefore an eigenvalue problem for R_c and Γ. The critical Rayleigh number R_c can be determined numerically from equations (15) and (19). One proceeds as follows: first select a trial Rayleigh number R_c, solve for the three roots of a_i^2 from (15) and then substitute the absolute values $|a_i|$ into (19) to solve for the aspect ratio Γ. A pair of (R_c, Γ) is obtained when $\Gamma > 0$. Given $R_c(\Gamma)$, one can rewrite (15) explicitly in C space. Thus fingering instability sets in when

$$Ra_e = \frac{gd^3}{\nu}\begin{bmatrix}\alpha_1\\ \alpha_2\end{bmatrix}^T\begin{bmatrix}D_{11} & D_{12}\\ D_{21} & D_{22}\end{bmatrix}^{-1}\begin{bmatrix}\Delta C_1\\ \Delta C_2\end{bmatrix} > R_c(\Gamma) \quad (20)$$

Figure 2 shows the computed critical Rayleigh number as a function of the aspect ratio of the cylinder for the least stable mode n = 1. $R_c(\Gamma)$ decreases with the increase of aspect ratio, identical to the result given by *Joseph* [1971] for Rayleigh-Bénard convection in a vertical cylinder. This is consistent with the earlier result in that the critical Rayleigh number for the onset of fingering instability in a multicomponent fluid is also the same as that in a simple fluid [$27/4\pi^4$, *Pellew and Southwell,* 1940; *Stern,* 1960; *Chandrasekhar,* 1961; *Veronis,* 1965; *Baines and Gill,* 1969; *Griffiths,* 1979; *McDougall,* 1983; *Terrones,* 1993].

The general result for an *n+1* component fluid confined in a vertical cylindrical container with insulating sidewalls and free upper and lower boundaries is that fingering instability sets in when

$$Ra_e = \frac{gd^3}{\nu}\begin{bmatrix}\alpha_1\\ \alpha_2\\ \cdots\\ \alpha_n\end{bmatrix}^T\begin{bmatrix}D_{11} & D_{12} & \cdots & D_{1n}\\ D_{21} & D_{22} & \cdots & D_{2n}\\ \cdots & \cdots & \cdots & \cdots\\ D_{n1} & D_{n2} & \cdots & D_{nn}\end{bmatrix}^{-1}\begin{bmatrix}\Delta C_1\\ \Delta C_2\\ \cdots\\ \Delta C_n\end{bmatrix} > R_c(\Gamma) \quad (21)$$

where $R_c(\Gamma)$ is given in Figure 2. Equations (11) and (21) show that the critical Rayleigh number for the onset of the fingering instability is independent of the number of components in the fluid when the initial concentration distributions are linear in z and an effective Rayleigh number Ra_e is introduced. In the next section, the result of the linear stability analysis is applied to our recent

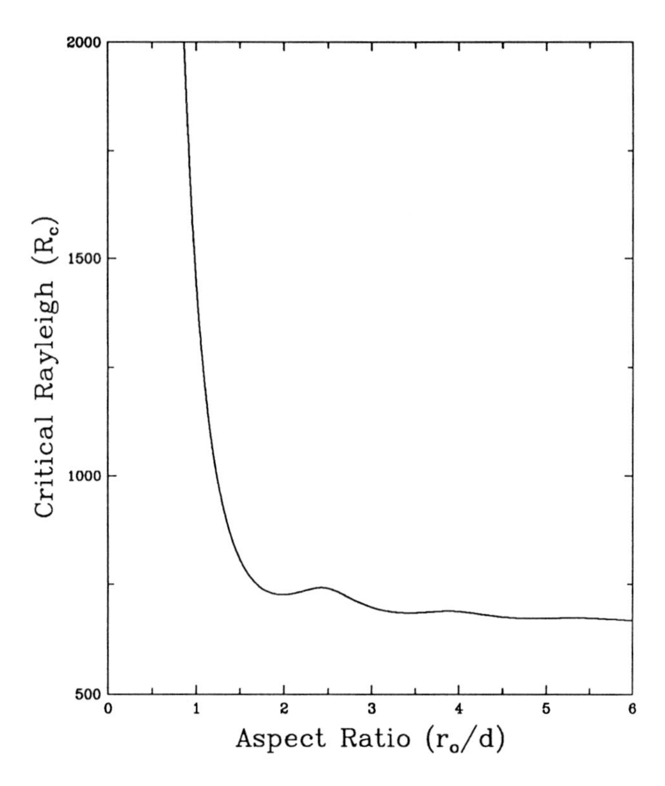

Fig. 2. Variations of the critical Rayleigh number R_c as a function of the aspect ratio of the cylinder.

laboratory studies of multicomponent chemical diffusion and double-diffusive convection in molten CaO-Al_2O_3-SiO_2 at 1500 °C and 10 Kb [*Liang et al.*, 1994].

4. APPLICATION TO MOLTEN CaO-Al_2O_3-SiO_2

Figure 1 shows a schematic diagram of the experimental setup [*Liang*, 1994; *Liang et al.*, 1994]. Table 1 lists the starting compositions of the seven glasses used to form diffusion and convection couples, along the direction of constant CaO (couple CT-4/CT-2), constant Al_2O_3 (couple CT-3/CT-5), and constant SiO_2 (couple CT-6/CT-1). These diffusion couples were run for various times (0.25, 0.5, 1.0, 2.0, and 6.0 hours) in the piston cylinder apparatus at 1500 °C and 10 Kb. Because the quenched charges from these runs are colorless, diffusion penetrations and convection patterns had to be resolved chemically through electron microprobe analysis. Instabilities were identified using concentration profiles and X-ray intensity maps of element Ca, Al, and Si from the sectioned charges [*Liang et al.*, 1994].

A series of diffusion/convection experiments were conducted using two types of setups. In the first setup a more dense melt was placed on top of a less dense melt (top/bottom, CT-1/CT-6, CT-2/CT-4, and CT-5/CT-3) and direct convective overturning was observed. The convection patterns are asymmetric about the cylindrical axis [*Liang et al.*, 1994]. This is an obviously unstable situation that one wants to avoid when conducting diffusion experiments. In the second setup the less dense melt was placed on top of the more dense one (CT-6/CT-1, CT-4/CT-2, CT-3/CT-5, and CT-7/CT-1) and now stable chemical diffusion was observed in experiments using couples CT-4/CT-2 and CT-3/CT-5 in runs lasting up to 6 hours and in CT-6/CT-1 for run durations of 30 minutes or less. These stable chemical diffusion runs were used to invert diffusion coefficients [Table 1, *Liang*, 1994; *Liang et al.*, 1994]. Fingering instabilities were observed for couple CT-6/CT-1 for run durations greater than 30 minutes. Fingering instability was also observed in a two hour run using couple CT-7/CT-1.

Plate 1a shows false-colored X-ray maps of element Ca, Al, and Si in a 2 hour run (CVT-2) using couple CT-6/CT-1. An approximate axial symmetric fingering instability is obvious with the less dense, SiO_2 rich finger rising along the center of the capsule and more dense melt sinking along the sidewalls. Plate 1b presents X-ray maps of Ca, Al, and Si from another two hour run (CD1/2) using CT-6/CT-1 which has a small offset at the interface. Again the fingering instability is apparent. Only now the flow direction is reversed with downflow (SiO_2 depleted) along the cylindrical axis and upflow along the capsule wall. The downgoing "silica" finger is clearly associated with a region of negative buoyancy as shown in the density field in Plate 1b. This convection pattern was confirmed by duplicate runs (1 and 2 hours) and was also observed in the 2 hour run using CT-7/CT-1. The depletion or enrichment in SiO_2 shown in Plates 1a and 1b is caused by the strong coupling between CaO and SiO_2 which produces a SiO_2 flux oriented opposite to that of CaO ($D_{SiO_2\text{-}CaO}/D_{SiO_2\text{-}SiO_2} = -2.7$, Table 1). It is also noted that the amplitude of downflow shown in Plate 1b is slightly higher than that of upflow shown in Plate 1a. These are all characteristic of axisymmetric convection with concentration-dependent viscosity [e.g., *Liang et al.*, 1969]. The viscosity of melt in couple CT-6/CT-1 decreases from 77 poise at the top to 27 poise at the bottom (Table 1), which explains the observed larger amplitude of the downflow [*Liang et al.*, 1969], even though convective instability may set in slightly earlier in upflow (r_o = 0.9 mm, no offset) than in downflow (r_o = 0.7 ~ 0.8 mm, with a slight offset).

A phenomenological explanation of these experiments is given by *Liang et al.* [1994]. Here the linear stability theory given in the previous section is used even though the stability analysis of free diffusion in a vertical cylinder is, strictly speaking, a time dependent problem which is much more complicated than the linear case discussed in this paper [e.g., *Foster*, 1968; *Gresho and Sani* 1971; *Jhaveri and Homsy*, 1982]. To proceed, it is assumed that the instability, once it sets in, evolves on a time scale

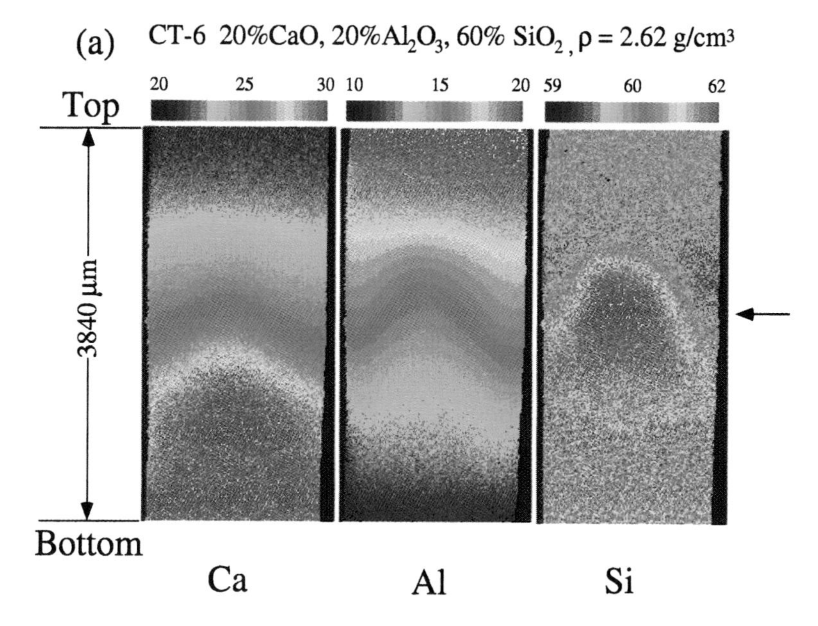

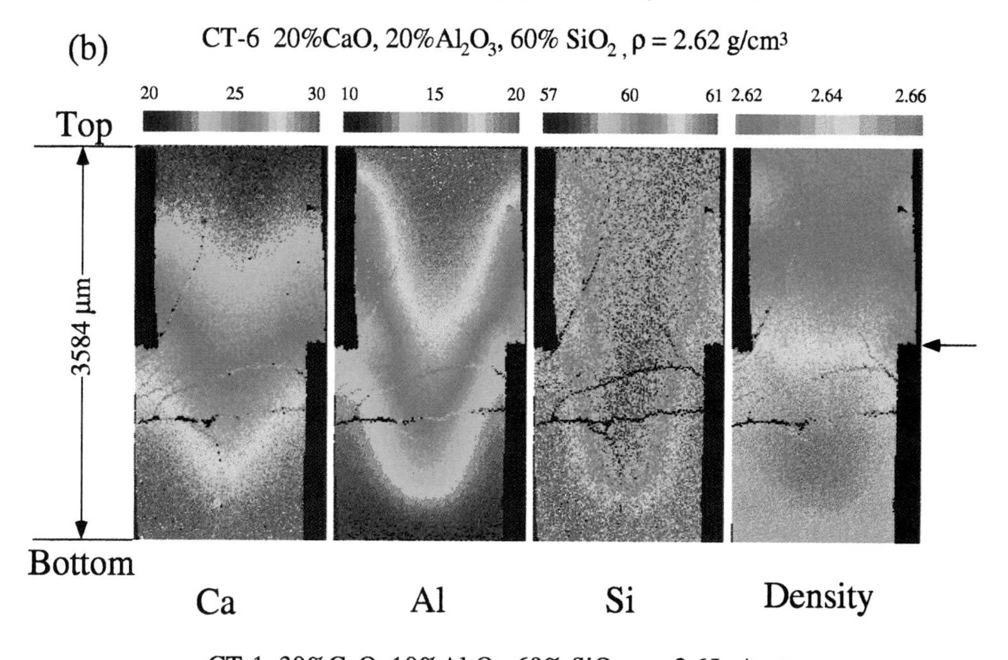

Plate 1. False-colored X-ray intensity maps of the elements Ca, Al, and Si showing convection in runs where the less dense melt (CT-6) was placed above the more dense melt (CT-1). The ranges of density (g/cm^3) and compositional (wt%) variations are represented by the horizontal scale bars on top of the images. The location of the original interface are marked by the arrows. The black strips on the sides of each image are the container walls. (a) X-ray maps of the central portion of the experimental charge showing an approximate axisymmetric upgoing fingering instability in a two hour run (CVT-2) using couple CT-6/CT-1. (b) X-ray maps of the central portion of the charge showing an approximate axisymmetric downgoing "silica" finger in another two hour run (CD1/2) using couple CT-6/CT-1. The false-colored density field was constructed from the intensity data. Thin irregular dark lines are dilation cracks produced when the sample was quenched at the end of the run. The step along the container wall is due to a small misalignment when joining capsules together. The X-ray intensities were collected using a Cameca SX-50 electron microprobe with an accelerating voltage of 15 Kv, beam current of 100 nA and beam spot of 5 mm.

TABLE 1. Starting Compositions, Density, Viscosity and Diffusion Matrix for Molten $CaO\text{-}Al_2O_3\text{-}SiO_2$ at 1500 °C and 10 Kb*

Sample	CaO	Al_2O_3	SiO_2	T_1	T_2	Density	Viscosity
CT-1	29.95	9.86	60.18	28.13	15.25	2.653	27
CT-2	25.15	19.97	54.87	21.84	24.30	2.660	36
CT-3	20.18	15.35	64.67	17.62	18.83	2.600	95
CT-4	25.04	9.93	65.03	23.26	14.40	2.614	56
CT-5	30.46	15.13	54.41	27.83	20.52	2.681	22
CT-6	20.01	19.85	60.14	16.78	23.23	2.618	77
CT-7	25.28	15.22	59.50	22.69	19.65	2.639	45
Diffusion Matrix	$D_{CaO\text{-}CaO}$ 4.66	$D_{CaO\text{-}SiO_2}$ 0.44	$D_{SiO_2\text{-}CaO}$ -2.72	$D_{SiO_2\text{-}SiO_2}$ 1.00		λ_1 4.303	λ_2 1.357
Density Relation	ρ_o 2.639	α_1 0.388		α_2 0.209		β_1 0.342	β_2 0.266

* Concentrations are in wt%. Density is in g/cm^3 computed using the partial molar value data of *Lange and Carmichael* [1987]. Viscosity is calculated using the method of *Bottinga and Weill* [1972, 1 atm]. Diffusivities are in units of 10^{-7} cm^2/s [*Liang*, 1994; *Liang et al.*, 1994]. T_1 and T_2 are concentrations after the transformation $\mathbf{T} = [\mathbf{B}]\mathbf{C}$. The linear density coefficients α and β are obtained using Taylor expansions of the density-composition relation around CT-7.

much faster than the diffusive time scale [e.g., *Huppert and Manins*, 1973; *McDougall*, 1983]. To a first approximation, I will use an averaged viscosity and neglect the effect of any offset at the interface.

Salt fingers form when hot salty water is place above cold fresh water and the fast diffusing component is stabilizing while the slow diffusing component is destabilizing [e.g., *Stommel et al.*, 1956; *Turner*, 1974]. The "silica" fingers observed in our experiments (Plates 1a and 1b) form under similar conditions with T_1 playing the role of temperature and T_2 that of salt. To compare directly with the formation of salt fingers I will, again, take the transformation $\mathbf{T} = [\mathbf{B}]\mathbf{C}$ and $\beta^T = \alpha^T[\mathbf{B}]^{-1}$ as in (5) and (7). The density coefficients α and β and diffusion matrix $[\mathbf{D}]$ along with its eigenvalues λ_1 and λ_2 are listed in Table 1. The diffusivity ratio between the slow and fast diffusing components (λ_2/λ_1) for the $CaO\text{-}Al_2O_3\text{-}SiO_2$ melt is 0.315. In contrast to that of temperature, the density coefficient for the fast diffusing component T_1 is positive for the silicate melt ($\beta_1 = 0.342$, Table 1). Thus a necessary condition for the formation of "silica" fingers is $\Delta T_1 < 0$ and $\Delta T_2 > 0$ so that $\beta_1\Delta T_1 < 0$ (stabilizing) and $\beta_2\Delta T_2 > 0$ (destabilizing). The concentration differences between the upper and lower end members for couple CT-6/CT-1 are $\Delta T_1 = -0.1135$ and $\Delta T_2 = 0.0789$, which satisfies the condition for the formation of "silica" fingers.

At large times when both R_1 and $R_2/\tau \gg R_c(\Gamma)$, it can be shown that a sufficient condition for the onset of fingers at the center of an initially sharp interface is the same as that given by *Huppert and Manins* [1973] for a fluid confined between two horizontal plates

$$-\frac{\beta_1\Delta T_1}{\beta_2\Delta T_2} < \tau^{-3/2} \qquad (22)$$

In (22), R.H.S. = 5.65, L.H.S. = 1.85 for couple CT-6/CT-1 and 1.59 for couple CT-7/CT-1. Thus fingering instability occurs along these directions. Analysis of couple CT-4/CT-2 and CT-3/CT-5 shows that both are stable with respect to fingering (and diffusive) instability in agreement with the stable chemical diffusion observed in these two directions.

At small times when both R_1 and R_2/τ are comparable to $R_c(\Gamma)$ one has to revolve the stability criterion (20). At a given time, the effective Rayleigh number Ra_e in the experiments can be estimated from the thickness of the diffusive boundary layer with $d \approx 2\sqrt{\lambda_1 t}$ in (20). The aspect ratio Γ is approximately $r_o/2\sqrt{\lambda_1 t}$. Thus both R_c and Ra_e are functions of experimental run durations. R_c is a also a function of capsule radius whereas Ra_e is not. Figure 3 displays R_c as a function of time for two choices of capsule radius r_o = 0.5 and 0.8 mm. Also plotted in Figure 3 are calculated Ra_e for couple CT-6/CT-1 with three choices of melt viscosity (μ = 27, 45, or 77 poise). At the onset of experiment, $Ra_e = 0$ because no diffusion had taken placed and $R_c = 27/4\pi^4$ ($\Gamma \to \infty$, Figure 2). Both Ra_e and R_c increase with the increment of time. The trajectory of R_c, however, depends on the radius of the capsule. Convective instability sets in earlier in a wider capsule than in a narrower one. Ra_e intersects R_c from below at $t_c \approx 0.5 \sim 1.1$ hours ($r_o \approx 0.7 \sim 0.9$ mm, Figure 3). This is quite consistent with our experimental observations for convection has not been observed in runs with durations of 30 minutes or less. This frozen time method, though crude, provides us with a very useful information for conducting diffusion experiments along these unstable directions. One can, for example, avoid the effect of convection by using a thinner capsule and

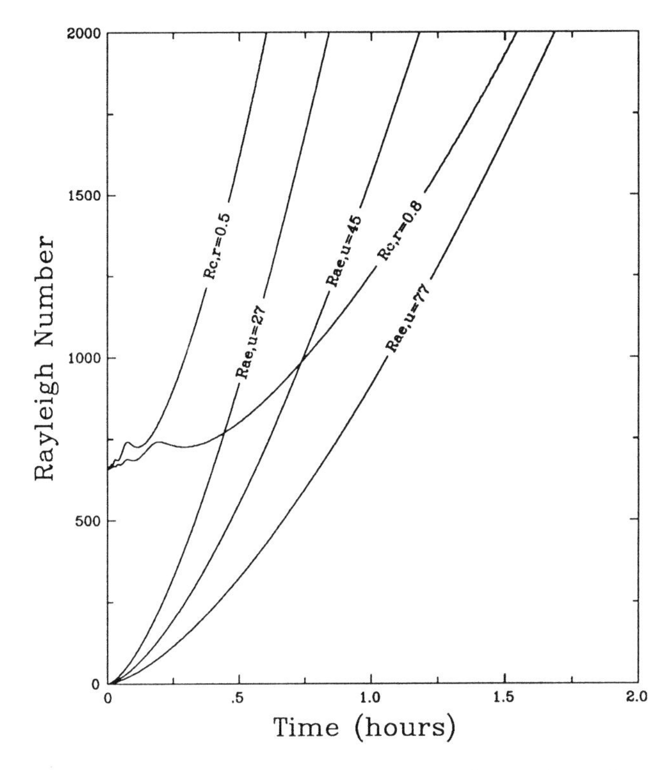

Fig. 3. Variation diagrams of the critical Rayleigh number and effective Rayleigh number as a function of time for two choices of tube diameter (R_c, r_o = 0.5 or 0.8 mm) and three choices of melt viscosity (Ra_e, μ = 27, 45, or 77 poise).

reducing the experimental duration.

4. SUMMARY AND DISCUSSION

Using a linear transformation of the compositional variables, it has been shown that the critical Rayleigh number for the onset of fingering instability for a multicomponent fluid is independent of the number of components in the system if the initial concentration distributions are linear, boundary conditions are similar for each component, and the effective Rayleigh number is introduced. The critical Rayleigh number is the same as for Rayleigh-Bénard convection and thus one can make use of numerous earlier studies that calculate the critical Rayleigh number as a function of container geometry and boundary conditions. The multicomponent nature of the system is contained in the definition of the effective Rayleigh number as given in its most general form by equation (11) or (21).

One of our initial objectives was to study chemical diffusion along various compositional directions in the system CaO-Al_2O_3-SiO_2 at 1500 °C and 10 Kb. It is obvious now that even in conventional diffusion setups where a lighter melt is placed on top of a heavier one the multicomponent system is not guaranteed to be stable against convection. Fingering or diffusive instabilities can occur along certain compositional directions. I have shown an experimental example of this in a case where "silica" fingers form in a molten silicate system in much the same way as salt fingers form in thermohaline convection. One of important applications of the stability theory of multicomponent systems is in identifying the cause of abnormal diffusion profiles in laboratory and natural systems which one might suspect are the result of convection. Clearly one needs to be cautious of this possibility when setting out to do purely diffusion experiments. The linear theory is a useful guide for identifying stable directions in compositional space. It is only a guide because the diffusion matrix is not known before hand and therefore one cannot carry out an exact analysis. In practice one can often estimate the diffusion matrix by comparing to other similar systems or use estimates based on the self diffusion coefficients of the individual components [e.g., *Cooper,* 1965; *Liang,* 1994]. The linear theory can also be used to estimated run durations such that diffusion experiments in unstable directions will not yet have been disturbed by convection. This is only a crude estimation, however, because of the assumption of frozen time. Given the increasing interests in measuring diffusion matrix in molten silicates, I hope the present results will stimulate future studies that include the time dependent nature of the diffusion problem.

Acknowledgments. I am indebted to Prof. Frank M. Richter for his invaluable suggestions and comments. I would like to thank Prof. Bruce E. Watson and his students John Hanchar and Janet Manchester for their advice and help in conducting diffusion/convection experiments, and Prof. Victor Barcilon for his comments. This work was supported by the NSF grand EAR-9206047 to Frank M. Richter.

REFERENCES

Ayers, J. C., Brenan, J. B., Watson, E. B., Wark, D. A., and Minarik, W. G., A new capsule technique for hydrothermal experiments using the piston-cylinder apparatus, *Am. Mineral.*, 77, 1080-1086, 1992.

Baines, P. G. and Gill, A. E., On thermohaline convection with linear gradients, *J. Fluid Mech.*, 37, 298-306, 1969.

Bottinga, Y. and Weill, D. F., The viscosity of magmatic silicate liquids: a model for calculation, *Amer. J. Sci.*, 272, 438-475, 1972.

Chandrasekhar, S., *Hydrodynamic and Hydromagnetic Stability*, Dover, New York, 1961.

Charlson, G. S., and Sani, R. L., Thermoconvective instability in a bounded cylindrical fluid layer, *Int. J. Heat Mass Transfer*, 13, 1479-1496, 1970.

Charlson, G. S., and Sani, R. L., On thermoconvective instability in a bounded cylindrical fluid layer, *Int. J. Heat Mass Transfer*, 14, 2157-2160, 1971.

Cooper, A. R. Jr., Model for multi-component diffusion, *Phys. Chem. Glasses*, 6, 55-61, 1965.

de Groot, S. R., and Mazur, P., *Non-Equilibrium Thermodynamics*, Dover, New York, 1962.

Drazin, P. G., and Reid, W. H., *Hydrodynamic Stability*, Cambridge University, 1981.

Foster, T. D., Effect of boundary conditions on the onset of convection, *Phys. Fluids*, 11, 1257-1262, 1968.

Gresho, P. M., and Sani, R. L., The stability of a fluid layer subjected to a step change in temperature: transient vs. frozen time analysis, *Int. J. Heat Mass Transfer*, 14, 207-221, 1971.

Griffiths, R. W., The influence of a third diffusing component upon the onset of convection, *J. Fluid Mech.*, 92, 659-670, 1979.

Gupta, P. K., and Cooper, A. R. Jr., The [D] matrix for multicomponent diffusion, *Physica*, 54, 39-59, 1971.

Huppert, H. E., and Manins, P. C., Limiting conditions fro salt-fingering at an interface, *Deep-Sea Res.*, 20, 315-323, 1973.

Jhaveri, B. S., and Homsy, G. M., The onset of convection in fluid layers heated rapidly in a time-dependent manner, *J. Fluid Mech.*, 114, 251-260, 1982.

Jones, C. A., Moore, D. R., and Weiss, N. O., Axisymmetric convection in a cylinder, *J. Fluid. Mech.*, 73, 353-388, 1976.

Jones, C. A., and Moore, D. R., The stability of axisymmetric convection, *Geophys. Astrophys. Fluid Dynamics*, 11, 245-270, 1979.

Joseph, D. D., Stability of convection in containers of arbitrary shape, *J. Fluid Mech.*, 47, 257-82, 1971.

Lange, R. A., and Carmichael, I. S. E., Density of Na_2O-K_2O-CaO-MgO-FeO-Fe_2O_3-Al_2O_3-TiO_2-SiO_2 liquids: New measure-ments and derived partial molar properties, *Geochim. Cosmochim. Acta*, 51, 2931-2946, 1987.

Liang, S. F., Vidal, A. and Acrivos, A., Buoyancy-driven convection in cylindrical geometries. *J. Fluid Mech.*, 36, 239-256, 1969.

Liang, Y., Models and experiments for multicomponent chemical diffusion in molten silicates, Ph.D. dissertation, The University of Chicago, 1994.

Liang, Y., Richter, F. M., and Watson, E. B., Convection in multicomponent silicate melts driven by coupled diffusion, *Nature*, 369, 390-392, 1994.

McDougall, T. J., Double-diffusive convection caused by coupled molecular diffusion, *J. Fluid. Mech.*, 126, 379-97, 1983.

McDougall, T. J., and Turner, J. S., Influence of cross-diffusion on 'finger' double-diffusive convection, *Nature*, 299, 812-814, 1983.

Pellew, A., and Southwell, R. V., On maintained convection motion in a fluid heated from below, *Proc. Roy. Soc.*, 176A, 312-343, 1940.

Stern, M. E., The "Salt Fountain" and thermohaline convection, *Tellus*, 12,172-175, 1960.

Stommel, H., Arons, A. B., and Blanchard, D., An oceanographical curiosity: the perpetual salt fountain, *Deep-Sea Res.*, 3, 152-153, 1956.

Terrones, G., Cross-diffusion effects on the stability criteria in a triple diffusive system, *Phys. Fluid A*, 5, 2172-2182, 1993.

Toor, H. L., Solution of the linearized equations of multicomponent mass transfer: I, *A.I.Ch.E. J.*, 10, 448-455, 1964.

Turner, J. S., Double-diffusive phenomena, *Ann. Rev. Fluid. Mech.*, 6, 37-56, 1974.

Turner, J. S., Multicomponent convection, *Ann. Rev. Fluid. Mech.*, 17, 11-44, 1985.

Veronis, G., On finite amplitude instability in thermohaline convection, *J. Mar. Res.*, 23, 1-17, 1965.

Watson, B. E., and Baker, D. R., Chemical diffusion in magmas: An overview of experimental results and geochemical applications, in *Physical Chemistry of Magmas*, edited by L. L. Perchuk and I. Kushiro, pp. 120-151, Springer-Verlag, New York.

Yan Liang, Department of Earth and Environmental Sciences, Rensselaer Polytechnic Institute, Troy, New York 12180-3590.

Dissipative Structure, Non-uniform Strain Fields, and Formation of Seismicity Patterns

Yijun Du

Department of Geological and Environmental Sciences, Stanford University, Stanford, California

The crustal deformation and seismicity along major plate boundaries display interesting spatial patterns. Some heterogeneous factors such as heterogeneity of strength on fault plane and discontinuous fault geometry have been proposed to explain the observed spatial patterns of deformation and seismicity. In this paper, it is demonstrated that nonuniform strain field patterns can arise under homogeneous conditions because of the nonlinear interaction between elastic strain accumulation and aseismic fault creep along a fault zone. It is found that in the presence of diffusion, it is possible that variation in the diffusivity ratio can bring about a breakdown of the steady-state and the formation of spatial heterogeneities (so-called dissipative structures) for the strain field at a certain well-defined critical wavelength. Certain conditions must necessarily hold for the occurrence of the above-described diffusive instability. Thus spatial seismicity patterns can form via dissipative structures of strain fields.

1. INTRODUCTION

The crustal deformation and seismicity along major plate boundaries display interesting patterns. For example, along the San Andreas fault, which forms the boundary between the Pacific and North American plates, the locked and freely-creeping sections form a spatial periodic pattern (Fig. 1). For example, from North to South there are the northern locked section from Point Delgado to Redwood City, the northern freely-creeping section from Redwood City to Cholame, the southern locked section from Cholame to San Bernardino and the southern freely-creeping section from San Bernardino further to the south. Recent seismicity data from 1980-1986 along the San Andreas fault (Fig.2) shows that most small to intermediate earthquakes have occurred along the creeping sections. However, the locked sections are the places where large earthquakes occurred. The California earthquake of April 18, 1906 (M=8.25) and the great Fort Tejon earthquake of January 9, 1857 (M=8.25) ruptured the northern and southern locked sections, respectively. The spatio-temporal variations of seismicity before major earthquakes have been studied by many investigators [for review, see *Karamori*, 1981] in an attempt to understand the physical mechanism of earthquakes and to use them as a tool for earthquake prediction. A natural way to explain the seismicity pattern is to find some sorts of heterogeneous factors which may be responsible for these patterns, such as the tectonic environment (strain rate, temperature or confining pressure) and the physical and geometrical heterogeneities of the fault plane. Two well-known models, one is called "asperity model" [*Kanamori*, 1981] and the another "barrier model " [*Aki*, 1979], have been suggested to explain these seismicity patterns. In this paper I will demonstrate that spatial seismicity patterns can form via dissipative structures of strain fields due to nonlinear interaction between two basic diffusive processes: elastic strain accumulation and aseismic fault creep along a fault zone.

Dissipative structures are temporal or spatial inhomogeneities which can arise in purely dissipative systems under suitable conditions. Such structures initially were proposed by *Prigogine et al.* [1969], and have been found in different systems such as biological [*Prigogine et al.*, 1969], ecological [*Segal and Jackson*, 1972] and chemical [*Glansdorff and Prigogine*, 1971] systems. *Ouchi et al.* [1985] were the first to point out that the local concentration of the elastic strain in certain regions can be caused by a "diffusive instability" and that spatial

Double-Diffusive Convection
Geophysical Monograph 94

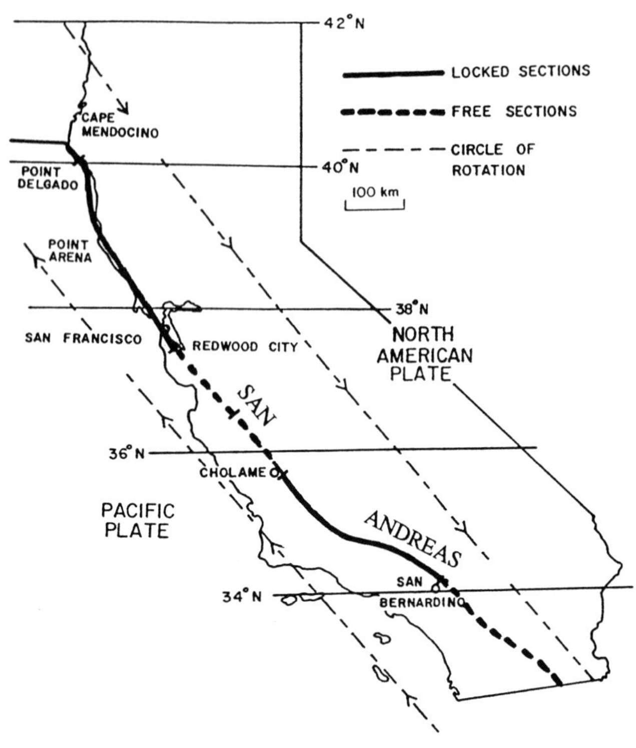

Fig. 1. Surface trace of locked and free sections of the San Andreas fault. From *Turcotte and Schubert* [1982].

and temporal patterns of the strain field are created by cooperative interactions of the differentiated diffusions. However they only performed the linear analysis and did not compute the spatial heterogeneous structures. They had to introduce asperities to model the seismicity patterns. Here I construct a nonlinear model based on *Ouchi et al.*'s model, and focus on the spatial dissipative structures. The temporal dissipative structures [e.g. see *Auchmuty and Nicolis*, 1976] will not be discussed here. My main purpose is to show that a stable uniform steady-state solution for elastic strain under homogeneous conditions can evolve to a stable non-uniform steady-state solution, i.e., a spatial dissipative structure which may explain the seismicity patterns.

2. MATHEMATICAL FORMULATION OF THE PROBLEM

Several attempts have been made by some investigators to formulate mathematically the stress-strain fields of the plate prior to large earthquakes [for review, see *Ouchi et al.*, 1985]. There are two models which are particularly important. First is the so-called Elasser model [*Elasser*, 1969; *Bott and Dean*, 1973; *Anderson*, 1975] in which one assumes that an elastic lithosphere rides over a viscous asthenosphere (Fig. 3). In this case the dynamic equation is given by

$$\frac{\partial u}{\partial t} = D_1 \frac{\partial^2 u}{\partial x^2} \tag{1}$$

where u is the elastic strain due to viscous-elastic effects of the elastic plate on the viscous layer, and D_1 is a parameter which includes the viscosity and the Young's or rigidity modulus [*Bolt and Dean*, 1973]. In the fault creep model, it is assumed that the physical properties of the stress-strain field are essentially attributed to the creep motion along the fault system [*Savage*, 1971; *Ida*, 1974]. *Savage* [1971] derived the equation for the dislocation flow or fault creep along a transform fault:

$$\frac{\partial v}{\partial t} = - A \frac{\partial v}{\partial x} + D_2 \frac{\partial^2 v}{\partial x^2} \tag{2}$$

where v denotes the dislocation flow and A, D_2 are advection and diffusion coefficients, respectively. These two models have been used to describe the migration phenomena of earthquakes, the propagation of the crustal movements and the occurrence of large earthquakes along the trench or the transform fault.

Ouchi et al. [1985] combined above two models and postulated that cooperation of elastic and inelastic effects produce the spatial-temporal variation of the strain patterns. They studied the stress-strain field of the fault zone under two geophysical situations: a transform and a subduction system of the elastic plate on the viscous layer of the asthenosphere. In their model, the dislocation flow or fault creep is the significant process and this dislocation flow is coupled with the elastic deformation of the plate. Furthermore, the viscous drag of the asthenosphere on the base of the lithosphere causes the elastic deformation to diffuse through the lithosphere. By coupling these two processes, *Ouchi et al.* [1985] postulated the following equations

$$\begin{aligned} \frac{\partial u}{\partial t} &= H(u,v) + D_1 \frac{\partial^2 u}{\partial x^2} \\ \frac{\partial v}{\partial t} &= I(u,v) - A \frac{\partial v}{\partial x} + D_2 \frac{\partial^2 v}{\partial x^2} \end{aligned} \tag{3}$$

Here the elastic strain (u) is considered to diffuse due to visco-elastic relaxation, while the dislocation flow (v) is mainly attained by diffusion and advection along the fault. A, D_1 and D_2 are constants. H and I show the nucleation and consumption of the elastic strain and dislocation flow.

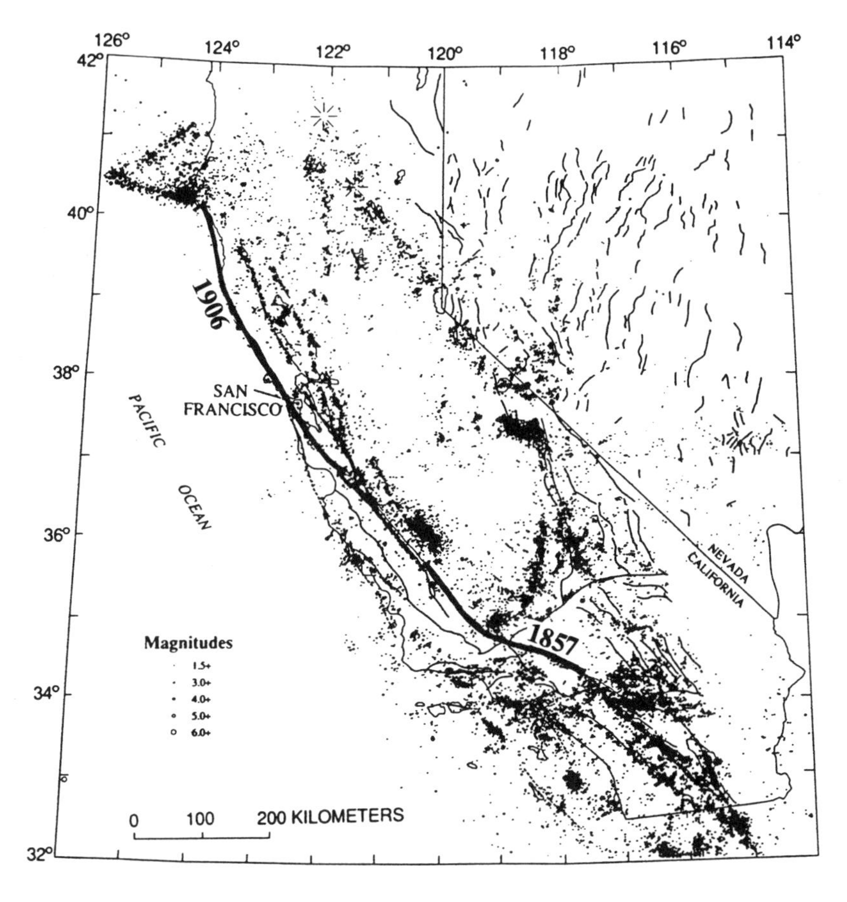

Fig. 2. Locations of 64,000 M≥1.5 earthquakes in California and western Nevada during 1980-86 and mapped Holocene faults. Heavy lines denote the segments ruptured during 1906 and 1857 big earthquakes.

They also include the driving force term due to plate motions and interacting or coupling effects of visco-elastic and plastic properties of the system [*Ouchi et al.*, 1985]. H and I will take rather complicated form and will be nonlinear. *Ouchi et al.* [1985] used the linear approximation for H and I to perform linear stability analysis. Here I will exploit effects of nonlinearity and accordingly consider a quadratic nonlinearity for the sake of simplicity. My purpose is to demonstrate the role of the nonlinearity in the formation of strain patterns. Therefore we assume the forms of H and I are following:

$$\begin{aligned} H(u,v) &= a_1 u + c_1 u^2 - b_1 uv \\ I(u,v) &= -a_2 v - c_2 v^2 + b_2 uv \end{aligned} \tag{4}$$

where a_i, b_i and c_i (i=1, 2) are positive constants. Here I have assumed that v does not contribute directly to $\partial u/\partial t$, but through a nonlinear interaction with u (uv). Similarly, u does not contribute directly to $\partial v/\partial t$, but through a nonlinear interaction with v (uv). Since the elastic strain should be self-enhancing and dislocation flow should be dissipative, we have negative terms $-b_1uv$, $-a_2v$ and $-c_2v^2$, and positive terms a_1u, c_1u^2 and b_2uv. In this paper, the influence of advection term will not be considered. Combining equations (3) and (4) and setting A=0, we have following system:

$$\begin{aligned} \frac{\partial u}{\partial t} &= a_1 u + c_1 u^2 - b_1 uv + D_1 \frac{\partial^2 u}{\partial x^2} \\ \frac{\partial v}{\partial t} &= -a_2 v - c_2 v^2 + b_2 uv + D_2 \frac{\partial^2 v}{\partial x^2} \end{aligned} \tag{5a}$$

subject to zero-flux boundary conditions which imply that there are no fluxes of the strain through the boundary:

$$\frac{\partial u}{\partial x} = \frac{\partial v}{\partial x} = 0 \qquad \text{at } x=0 \text{ and } x=1. \tag{5b}$$

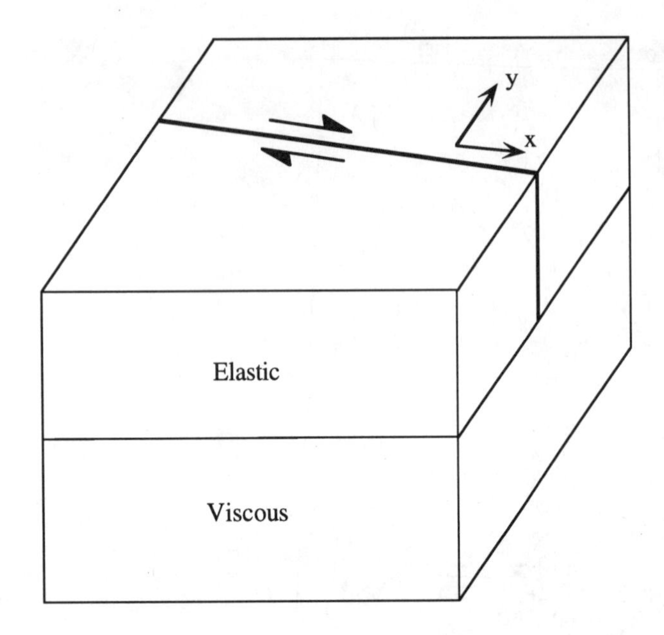

Fig. 3. A strike-slip fault system. The fault is situated in the elastic lithosphere riding over a viscous asthenosphere.

3. LINEAR STABILITY ANALYSIS IN THE CASE OF NO DIFFUSION

In the case of no diffusion, eqn(5) becomes

$$\frac{\partial u}{\partial t} = a_1 u + c_1 u^2 - b_1 uv \quad (6a)$$

$$\frac{\partial v}{\partial t} = -a_2 v - c_2 v^2 + b_2 uv \quad (6b)$$

This system admits a single non-zero uniform steady-state solution

$$\hat{u} = (a_1 c_2 + a_2 b_1)/(b_1 b_2 - c_1 c_2) \quad (7a)$$

$$\hat{v} = (a_1 b_2 + a_2 c_1)/(b_1 b_2 - c_1 c_2) \quad (7b)$$

Let

$$u = \hat{u} + \varepsilon_1, \qquad v = \hat{v} + \varepsilon_2 \quad (8)$$

where ε_1 and ε_2 are small perturbations around the steady-state u and $\hat{v}$. Then we obtain the linearized system for ε_1 and ε_2:

$$\dot{\varepsilon}_1 = c_1\hat{u}\varepsilon_1 - b_1\hat{u}\varepsilon_2 \quad (9a)$$

$$\dot{\varepsilon}_2 = b_2\hat{v}\varepsilon_1 - c_2\hat{v}\varepsilon_2 \quad (9b)$$

After a few algebraic manipulations, one finds the characteristic equation:

$$\lambda^2 + (c_2\hat{v} - c_1\hat{u})\lambda + (b_1 b_2 - c_1 c_2)\hat{u}\hat{v} = 0 \quad (10)$$

The solution of this equation is

$$\lambda_{1,2} = \frac{1}{2}\{-(c_2\hat{v}-c_1\hat{u}) \pm \sqrt{(c_2\hat{v}-c_1\hat{u})^2 - 4(b_1 b_2 - c_1 c_2)\hat{u}\hat{v}} \quad (11)$$

The condition for stability is that λ is negative if λ is real, or λ has a negative real part if λ is complex. It is found that the uniform steady-state solution $(\hat{u}, \hat{v})$ for the system (6) is stable if $b_1 b_2 > c_1 c_2$.

4. LINEAR STABILITY ANALYSIS IN THE CASE OF DIFFUSION

The uniform solution $(\hat{u}, \hat{v})$ is also an equilibrium point when D_1 and D_2 differ from zero, but the equilibrium is not necessary stable. In general, diffusion plays the role of increasing stability. However, there is an important exception known as "diffusion-induced instability" or "diffusive instability" [*Turing*, 1952]. This exception might not be a rare event especially in aquatic systems. Herein we will explore the Turing effect [*Turing*, 1952].

To examine the stability of the uniform state to perturbation, again we substitute eqn(8) into the original system (5) and then we can obtain the following linearized equations by assuming an infinitesimal perturbation:

$$\begin{aligned} \frac{\partial \varepsilon_1}{\partial t} &= c_1\hat{u}\varepsilon_1 - b_1\hat{u}\varepsilon_2 + D_1\frac{\partial^2 \varepsilon_1}{\partial x^2} \\ \frac{\partial \varepsilon_2}{\partial t} &= b_2\hat{v}\varepsilon_1 - c_2\hat{v}\varepsilon_2 + D_2\frac{\partial^2 \varepsilon_2}{\partial x^2} \end{aligned} \quad (12)$$

Here the advection term has been neglected. I should point out, however, that *Jorne* [1974] have shown that the inclusion of the advection effects increases the possibility of the diffusive instability.

To examine stability of the system, we assume that the perturbations have solutions in the form

$$\varepsilon_1 \sim \exp(\lambda t + ikx) \quad (13a)$$

$$\varepsilon_2 \sim \exp(\lambda t + ikx) \quad (13b)$$

where λ and k are the growth rate and wavenumber, respectively. The eigenvalue equation then reads

$$\begin{vmatrix} \lambda + D_1 k^2 - c_1\hat{u} & b_1\hat{u} \\ -b_2\hat{v} & \lambda + D_2 k^2 + c_2\hat{v} \end{vmatrix} = 0 \quad (14)$$

Solving for λ

$$\lambda = \frac{1}{2}\left\{(\hat{a}_{11}+\hat{a}_{22}) \pm \sqrt{(\hat{a}_{11}+\hat{a}_{22})^2 - 4(\hat{a}_{11}\hat{a}_{22}+b_1 b_2 \hat{u}\hat{v})}\right\} \quad (15)$$

where

$$\hat{a}_{11} = c_1\hat{u} - D_1 k^2$$

$$\hat{a}_{22} = -c_2\hat{v} - D_2 k^2$$

The condition k=0 corresponds to the absence of diffusion. Therefore, perturbations of zero wavenumber are stable when diffusive instability sets in provided $b_1 b_2 > c_1 c_2$ from the previous section. If $D_1=D_2=D$, i.e., fault creep and elastic strain diffuse with the same speed, then a uniform steady state is stable under small inhomogeneous perturbation for all values of k. This also means that for diffusive instability to occur, the diffusivity must not be the same for both fault creep and elastic strain.

Diffusive instability sets in when the following condition is violated

$$\hat{a}_{11}\hat{a}_{22}+b_1 b_2 \hat{u}\hat{v} > 0 \quad (16)$$

Reversal of the inequality (16) yields

$$H(k^2) \equiv D_1 D_2 k^4 - (c_1\hat{u}D_2 - c_2\hat{v}D_1)k^2 + (b_1 b_2 - c_1 c_2)\hat{u}\hat{v} < 0 \quad (17)$$

The minimum of $H(k^2)$ occurs at $k^2=k_m^2$, where

$$k_m^2 = (c_1\hat{u}D_2 - c_2\hat{v}D_1)/(2D_1 D_2) > 0 \quad (18)$$

Inserting (18) into (17) yields

$$(b_1 b_2 - c_1 c_2)\hat{u}\hat{v} - (c_1\hat{u}D_2 - c_2\hat{v}D_1)^2/(4D_1 D_2) < 0 \quad (19)$$

The final criterion for diffusive instability is then

$$c_1\hat{u}D_2 - c_2\hat{v}D_1 > 2\sqrt{(b_1 b_2 - c_1 c_2)\hat{u}\hat{v}D_1 D_2} > 0 \quad (20a)$$

or

$$c_1\hat{u}\theta^2 - c_2\hat{v} > 2\theta\sqrt{(b_1 b_2 - c_1 c_2)\hat{u}\hat{v}} > 0 \quad (20b)$$

where $\theta^2=D_2/D_1$. When inequality (20) is barely satisfied, diffusive instability is incipient. The critical conditions for the occurrence of the instability are obtained when the first inequality of (20) is an equality. The critical wavenumber k_c of the first perturbations to grow is found by evaluating k_m from (18).

The condition (20) requires that the diffusivity of creep flow D_2, i.e., the "stabilizer", must be larger than that of the elastic strain D_1, i.e., the "destabilizer". The same condition also requires that the ratio of the two diffusivities has a critical value beyond which diffusive instability sets in.

In summary, if θ_c^2 is the critical ratio of diffusivities, then the uniform steady-state solution is stable to a small perturbation when $\theta^2 < \theta_c^2$, but unstable when $\theta^2 > \theta_c^2$. This instability is expected to occur for the critical wavenumber k_c defined at $\theta^2=\theta_c^2$. In the following section, we will seek a steady-state solution beyond the critical point θ_c^2.

5. BIFURCATION, AND FORMATION OF DISSIPATIVE STRUCTURE

From the linear stability analysis one can infer that beyond the critical point θ_c^2, the uniform steady-state solution of eqn(12) evolves to a new steady-state solution. In this section we construct the analytical form of this new steady-state solution bifurcating beyond instability.

Mathematically, the transition to new steady state configurations can be understood as a phenomenon of branching of solutions of nonlinear partial differential equations [*Sattinger*, 1973]. *Auchmuty* and *Nicolis* [1975] developed a theoretical analysis for the nonlinear reaction-diffusion equation using bifurcation theory. A systematic perturbation scheme was introduced which allows the analytical calculation of the form of the bifurcating steady state in the neighborhood of the marginal stability point. The same perturbation technique is applied here to obtain the new steady state solution of (5) beyond θ_c^2.

Inserting (8) into original system (5), keeping this time the nonlinear contributions, we obtain:

$$\frac{\partial \varepsilon_1}{\partial t} = c_1\hat{u}\varepsilon_1 - b_1\hat{u}\varepsilon_2 + c_1\varepsilon_1^2 - b_1\varepsilon_1\varepsilon_2 + D_1\frac{\partial^2 \varepsilon_1}{\partial x^2} \quad (21a)$$

$$\frac{\partial \varepsilon_2}{\partial t} = b_2\hat{v}\varepsilon_1 - c_2\hat{v}\varepsilon_2 - c_2\varepsilon_2^2 + b_1\varepsilon_1\varepsilon_2 + D_2\frac{\partial^2 \varepsilon_2}{\partial x^2} \quad (21b)$$

To do rescaling, let

$$x = \eta\sqrt{\frac{D_1}{c_1\hat{u}}},\ \theta^2 = \frac{D_2}{D_1},\ t = \frac{\tau}{c_1\hat{u}},\ a = \frac{b_1}{c_1},\ b = \frac{1}{\hat{u}},$$

$$c = \frac{b_1}{c_1\hat{u}},\ d = \frac{b_2\hat{v}}{c_1\hat{u}},\ e = \frac{c_2\hat{v}}{c_1\hat{u}},\ f = \frac{c_2}{c_1\hat{u}},\ g = \frac{b_2}{c_1\hat{u}},$$

then we have

$$\frac{\partial \varepsilon_1}{\partial \tau} = \varepsilon_1 - a\varepsilon_2 + b\varepsilon_1^2 - c\varepsilon_1\varepsilon_2 + \frac{\partial^2 \varepsilon_1}{\partial \eta^2} \quad (22a)$$

$$\frac{\partial \varepsilon_2}{\partial \tau} = d\varepsilon_1 - e\varepsilon_2 - f\varepsilon_2^2 + g\varepsilon_1\varepsilon_2 + \theta^2\frac{\partial^2 \varepsilon_2}{\partial \eta^2} \quad (22b)$$

with boundary conditions:

$$\frac{\partial \varepsilon_1}{\partial \eta} = \frac{\partial \varepsilon_2}{\partial \eta} = 0 \text{ at } \eta = 0 \text{ and } \eta = \ell = \left(\frac{c_1\hat{u}}{D_1}\right)^{\frac{1}{2}} \quad (23)$$

To find a steady-state solution, let $\partial\varepsilon_1/\partial\tau = \partial\varepsilon_2/\partial\tau = 0$, then we have

$$\frac{\partial^2 \varepsilon_1}{\partial \eta^2} + \varepsilon_1 - a\varepsilon_2 = -b\varepsilon_1^2 + c\varepsilon_1\varepsilon_2 \quad (24a)$$

$$\theta^2\frac{\partial^2 \varepsilon_2}{\partial \eta^2} + d\varepsilon_1 - e\varepsilon_2 = f\varepsilon_2^2 - g\varepsilon_1\varepsilon_2 \quad (24b)$$

In vector form, it can be rewritten as

$$L\begin{pmatrix}\varepsilon_1\\ \varepsilon_2\end{pmatrix} = \begin{pmatrix}R(\varepsilon_1, \varepsilon_2)\\ Q(\varepsilon_1, \varepsilon_2)\end{pmatrix} \quad (25)$$

where

$$L = \begin{pmatrix}\frac{\partial^2}{\partial\eta^2} + 1 & -a\\ d & \theta^2\frac{\partial^2}{\partial\eta^2} - e\end{pmatrix},\quad R = -b\varepsilon_1^2 + c\varepsilon_1\varepsilon_2,\ Q = f\varepsilon_2^2 - g\varepsilon_1\varepsilon_2 \quad (26)$$

In order to calculate explicitly the bifurcating solution, we expand ε_1, ε_2 and θ^2 in the neighborhood of the critical point in terms of a small parameter ε:

$$\begin{pmatrix}\varepsilon_1\\ \varepsilon_2\end{pmatrix} = \varepsilon\begin{pmatrix}x_0\\ y_0\end{pmatrix} + \varepsilon^2\begin{pmatrix}x_1\\ y_1\end{pmatrix} + \cdots \quad (27a)$$

$$\theta^2 - \theta_c^2 = \gamma = \varepsilon\gamma_1 + \varepsilon^2\gamma_2 + \cdots \quad (27b)$$

Introducing these expansions into (25) and identifying equal powers of ε, we then obtain a set of equations of the form:

$$L_c\begin{pmatrix}x_k\\ y_k\end{pmatrix} = \begin{pmatrix}s_k\\ t_k\end{pmatrix},\quad k = 0, 1, \cdots \quad (28)$$

together with the boundary conditions

$$\frac{\partial x_k(0)}{\partial \eta} = \frac{\partial x_k(\ell)}{\partial \eta} = \frac{\partial y_k(0)}{\partial \eta} = \frac{\partial y_k(\ell)}{\partial \eta} = 0$$

where L_c is the operator L evaluated at the critical point of the first bifurcation. The first few coefficients of s_k and t_k are

$$s_0 = 0 \quad (29a)$$

$$t_0 = 0 \quad (29b)$$

$$s_1 = cx_0y_0 - bx_0^2 \quad (29c)$$

$$t_1 = -\gamma_1\frac{\partial^2 y_0}{\partial \eta^2} + fy_0^2 - gx_0y_0 \quad (29d)$$

$$s_2 = cx_0y_1 + cx_1y_0 - 2bx_0x_1 \quad (29e)$$

$$t_2 = -\gamma_1\frac{\partial^2 y_1}{\partial \eta^2} - \gamma_2\frac{\partial^2 y_0}{\partial \eta^2} + 2fy_0y_1 - gx_0y_1 - gy_0x_1 \quad (29f)$$

In order to solve eqn(28), we need to apply the theorem of the solvability condition [see, *Nicolis and Prigogine*, 1977]. We have the following theorem:

Theorem (Fredholm Alternative) The vector $(x_k, y_k)^T$ is a solution of eqn(28) provided the right-hand side $(s_k, t_k)^T$ is orthogonal to the null eigenvector of the adjoint operator L_c^*.

According to *Nicolis and Prigogine* [1977], the operator L has a one-dimensional null space spanned by

$$\begin{pmatrix}u_n\\ v_n\end{pmatrix} = \begin{pmatrix}h_1\\ h_2\end{pmatrix}\cos\frac{n\pi\eta}{\ell},\quad n = 0, 1, 2, \ldots\ldots \quad (30)$$

for the no-flux boundary conditions. The adjoint operator L^* of the operator L has the same eigenvalues as for L, whereas the eigenfunctions of L^* feature the same space dependence:

$$\begin{pmatrix}u_n^*\\ v_n^*\end{pmatrix} = \begin{pmatrix}l_1\\ l_2\end{pmatrix}\cos\frac{n\pi\eta}{\ell},\quad n = 0, 1, 2, \ldots\ldots \quad (31)$$

If we note that the eigenvalue problem

$$L\begin{pmatrix}u_n\\v_n\end{pmatrix}=\lambda\begin{pmatrix}u_n\\v_n\end{pmatrix} \quad (32)$$

$$L^*\begin{pmatrix}u_n^*\\v_n^*\end{pmatrix}=\lambda\begin{pmatrix}u_n^*\\v_n^*\end{pmatrix}, \quad (33)$$

then we can substitute (30) and (31) into (32) and (33) to obtain the following quantities

$$\delta_n=\frac{h_2}{h_1}, \text{ and } \Delta_n=\frac{l_2}{l_1}.$$

Now only h_1 and l_1 remain undetermined. We can fix it by requiring a specific normalization for $(u_n, v_n)^T$ and $(u_n^*, v_n^*)^T$ [*Auchmuty and Nicolis*, 1975; *Nicolis and Prigogine*, 1977]. Following *Auchmuty and Nicolis* [1975], here we require

$$h_1^2+h_2^2=2, \text{ and } l_1^2+l_2^2=2.$$

Now we can solve the equation (28). First we can immediately have

$$\begin{pmatrix}x_0(\eta)\\y_0(\eta)\end{pmatrix}=\begin{pmatrix}h_1\\h_2\end{pmatrix}\cos\frac{n_c\pi\eta}{\ell} \quad (34)$$

The solvability condition takes the explicit form:

$$\left\langle(u_n^*, v_n^*)\begin{pmatrix}s_k\\t_k\end{pmatrix}\right\rangle\equiv\int_0^\ell(s_ku_n^*+t_kv_n^*)d\eta$$

$$=\int_0^\ell(l_1s_k+l_2t_k)\cos\frac{n_c\pi\eta}{\ell}d\eta=0 \quad (35)$$

Now the procedure is as follows:

(a) Introducing (29) in (35) for $k=1$, we find

$$\gamma_1=0 \quad (36)$$

(b) To take the calculations to the next order, we need to solve $(x_1, y_1)^T$. They are given by the solution of (28) and (29) with k=1, i.e.,

$$L_c\begin{pmatrix}x_1\\y_1\end{pmatrix}=\begin{pmatrix}s_1\\t_1\end{pmatrix} \quad (37)$$

where

$$\begin{pmatrix}s_1\\t_1\end{pmatrix}=\begin{pmatrix}(ch_2-bh_1)h_1\\(fh_2-gh_1)h_2\end{pmatrix}\cos^2\frac{n_c\pi\eta}{\ell}. \quad (38)$$

(37) can be solved by using a standard method of Fourier series expansion as

$$\begin{pmatrix}x_1(\eta)\\x_1(\eta)\end{pmatrix}=\begin{pmatrix}p_0\\q_0\end{pmatrix}+\begin{pmatrix}p\\q\end{pmatrix}\cos\frac{2n_c\pi\eta}{\ell} \quad (39a)$$

where

$$p_0=\frac{(a\delta-\alpha e)}{2(ad-e)} \quad (39b)$$

$$q_0=\frac{(\delta-d\alpha)}{2(ad-e)} \quad (39c)$$

$$p=p_{2n_c}=\frac{a\delta+\alpha(\theta_c^2\beta-e)}{2[ad-(1-\beta)(\theta_c^2\beta-e)} \quad (39d)$$

$$q=q_{2n_c}=\frac{a[d\alpha-\delta(1-\beta)]-2\alpha(1-\beta)(\theta_c^2\beta-e)}{2a[ad-(1-\beta)(\theta_c^2\beta-e)]} \quad (39e)$$

$$\alpha=h_1(ch_2-bh_1),\quad \delta=h_2(fh_2-gh_1) \quad (39f)$$

$$\beta=(2n_c\pi/\ell)^2. \quad (39g)$$

(c) Using the expressions for (x_0, y_0) and (x_1, y_1), we may now determine γ_2 from (35) with k=2. This can be written as

$$\gamma_2\int_0^\ell l_2\frac{\partial^2y_0}{\partial\eta^2}\cos\frac{n_c\pi\eta}{\ell}d\eta=\int_0^\ell[l_1(cx_0y_1+cx_1y_0-2bx_0x_1)$$

$$+l_2h_2fy_0y_1-gx_0y_1-gy_0x_1]\cos\frac{n_c\pi\eta}{\ell}d\eta.$$

Substituting (34) and (39) into this equation gives

$$\gamma_2=\{p_0(2bh_1l_1+gh_2l_2-ch_2l_1)$$

$$+q_0(gh_1l_2-ch_1l_1-2fh_2l_2)\}[h_2(\frac{n_c\pi}{\ell})^2l_2]^{-1}. \quad (40)$$

If $\gamma_2=0$, the calculations should be continued to the next order. Since $\gamma_1=0$, ε is given by (27b) as

$$\varepsilon=\pm\left(\frac{\theta^2-\theta_c^2}{\gamma_2}\right)^{\frac{1}{2}}. \quad (41)$$

The bifurcating non-uniform steady-state solution near

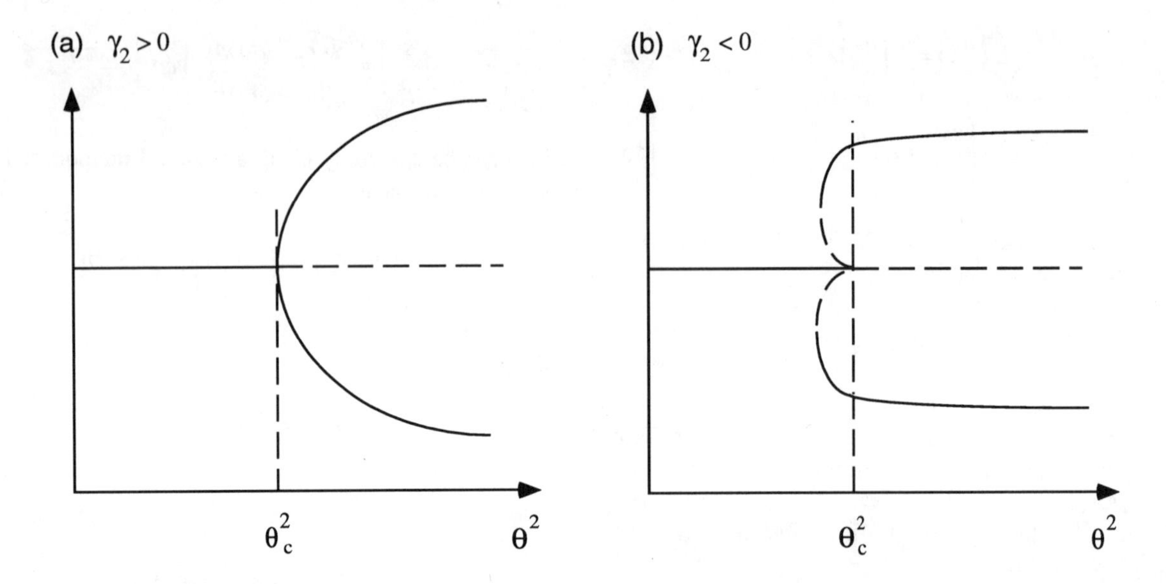

Fig. 4. Bifurcating structures near the critical point θ_c^2 for (a) $\gamma_2 > 0$ and (b) $\gamma_2 < 0$. The solid and dashed lines denote stable and unstable branches of the steady state solution, respectively.

$\theta^2 = \theta_c^2$ is thus approximated by

$$\varepsilon_1 = \pm\left(\frac{\theta^2 - \theta_c^2}{\gamma_2}\right)^{\frac{1}{2}} h_1 \cos\frac{n_c\pi\eta}{\ell} + \frac{\theta^2 - \theta_c^2}{\gamma_2}\left(p_0 + p\cos\frac{2n_c\pi\eta}{\ell}\right) + O(\varepsilon^3) \tag{42a}$$

$$\varepsilon_2 = \pm\left(\frac{\theta^2 - \theta_c^2}{\gamma_2}\right)^{\frac{1}{2}} h_2 \cos\frac{n_c\pi\eta}{\ell} + \frac{\theta^2 - \theta_c^2}{\gamma_2}\left(q_0 + q\cos\frac{2n_c\pi\eta}{\ell}\right) + O(\varepsilon^3) \tag{42b}$$

When $\gamma_2 > 0$, (42) holds for $\theta^2 > \theta_c^2$. If $\gamma_2 < 0$, (42) applies for $\theta^2 < \theta_c^2$. Moreover, we can have following conclusion according to *Sattinger* [1973] and *Nicolis and Prigogine* [1977]: In the vicinity of the critical point θ_c^2, the new bifurcating solutions are asymptotically stable in the supercritical region $\theta^2 > \theta_c^2$ if $\gamma_2 > 0$. However, the subcritical branches are unstable if $\gamma_2 < 0$. This conclusion suggests the bifurcation diagrams as shown in Fig.4. The bifurcating spatially non-uniform steady-state solution beyond the critical point θ_c^2 is called a dissipative structure, discovered first by *Prigogine et al.* [1969]. We will not discuss the detailed properties of the dissipative structures here. However, one can get some information about the general features of dissipative structures in *Nicolis and Prigogine*'s book [*Nicolis and Prigogine*, 1977].

6. SUMMARY

In this paper, we have shown that because of the nonlinear interaction between elastic strain accumulation and fault creep, it is possible that variation in the diffusivity ratio will bring about a breakdown of the steady state and the formation of spatial heterogeneities at a certain well-defined "critical wavelength" (so-called dissipative structures) which could explain the formation of non-uniform strain patterns and accordingly seismicity patterns. The necessary condition for the formation of spatially non-uniform strain field is that the diffusion speed of fault creep is large enough so that the ratio of the diffusivity for elastic strain over that for fault creep reaches a critical value. The seismicity pattern along the San Andreas fault clearly defines two wavelengths of creeping and locked sections, implying that the diffusity ratio for the system is already beyond the critical value. Thus, spatial seismicity pattern can form via dissipative structures of strain fields.

Acknowledgments. I would like to thank two anonymous reviewers for their helpful comments. This study was supported in part by the Stanford Rock Fracture Project.

REFERENCES

Aki, K., Characterization of barriers on an earthquake fault, *J. Geophys. Res., 84*, 6140-6148, 1979.

Anderson, D. L., Accelerated plate tectonics, *Science, 187*, 1077-1079, 1975.

Auchmuty, J. F. G., and G. Nicolis, Bifurcation analysis of nonlinear reaction-diffusion equations - I. evolution equations and the steady state solutions, *Bull. Math. Bio.*, *37*, 323-365, 1975.

Bott, M. H. P., and D. S. Dean, Stress diffusion from plate boundaries, *Nature, 243*, 339-341, 1973.

Elsasser, W. M., Convection and stress propagation in the upper mantle, in *The application of modern physics to the Earth and planetary interiors*, edited by S. K. Runcorn, Wiley-Interscience, New York, 223-246, 1969.

Glansdorff, P., and I. Prigogine, *Thermodynamics of structure, stability and fluctuations*, Wiley-Interscience, New York, 1971.

Hill, D. P., J. P. Eaton, and L. M. Jones, Seismicity, 1980-86, in *The San Andreas Fault System, California,* U.S. Geological Survey Professional Paper 1515, edited by R. E. Wallace, 115-151, 1990.

Ida, Y., Slow moving deformation pulses along tectonic faults, *Phys. Earth Planet. Inter., 9*, 328-337, 1974.

Jorne, J., The effect of ionic migration on oscillations and pattern formation in chemical systems, *J. Theor. Biol., 43*, 375-380, 1974.

Kanamori, H., The nature of seismicity patterns before earthquakes, In *Earthquake Prediction and International Review,* edited by D. W. Simpson and P. G. Richards, American Geophysical Union, Washington D. C., 1-19, 1981.

Nicolis, G. and I. Prigogine, *Self-organization in nonequilibrium systems*, John Wiley & Sons, 1977.

Ouchi, T., Goriki, S., and K. Ito, On the space-time pattern formation of the earthquake strain field, *Tectonophysics, 113*, 31-48, 1985.

Prigogine, I., R. Lefever, A. Goldbeter, and M. Herschkowitz-Kaufman, Symmetry breaking instabilities in biological systems, *Nature, 233*, 913-916, 1969.

Sattinger, D., *Lecture note in mathematics*, 309, Berlin, Springer, 1973.

Savage, J. C., A theory of creep wave propagating along a transform fault, *J. Geophys. Res., 76*, 1954-1966, 1971.

Segal, L. A., and J. L. Jackson, Dissipative structure: an explanation and an ecological example, *J. Theor. Biol., 37*, 545-559, 1972.

Turcotte, D. L., and G. Schubert, *Geodynamics; Applications of Continuum Physics to Geological Problems*, John Wiley & Sons, New York, 1982.

Turing, A. M., The chemical basis for morphogeneis, *Philos. Trans. R. Soc. London, 237*, 37-72, 1952.

Y. Du, Department of Geological and Environmental Sciences, Stanford University, Stanford, CA 94305.

Formation of Layered Structures in Double-Diffusive Convection as Applied to the Geosciences

Ulrich Hansen

Dept. Theoretical Geophysics, Earth Sciences Institute, Univ. Utrecht, Utrecht, Holland

David A. Yuen

Minnesota Supercomputer Institute and Dept. Geology and Geophysics, Univ. Minnesota, Minneapolis, Minnesota

We have studied the different scenarios for layered structures to form or be destroyed in time-dependent, double-diffusive convection for an infinite Prandtl number fluid. We have concentrated on the subcritical diffusive and finger regimes and examined the thermal-chemical evolution as applied to magma chambers. Both cooling from the top and side boundary conditions have also been examined. Subcritical double-diffusive convection yields the most favorable conditions for layering. A sufficiently high chemical buoyancy ratio keeps the layering intact and yields a small Nusselt number. Double-diffusive convection in the subcritical regime can lead to layering from an initially unlayered state. There are no evidences for layering in the finger regime. Finally we present a new result dealing with the potentially important role played by viscous dissipation in double-diffusive convection, as found in the Earth's outer core and crust.

1. INTRODUCTION

Double-diffusive convection (D.D.C.) is a fundamental fluid dynamical process involving the difference between thermal and chemical diffusivities in releasing gravitational potential energy. The major physical differences in the applications of D.D.C. in the geological sciences and oceanography are the Prandtl and Lewis numbers. The Prandtl number is effectively infinite for the mantle and silicic magmas, while for the oceans the Prandtl number is 0(10). The Lewis number for the mantle and magmas is many orders of magnitude greater than that for the oceans. The physical mechanism of double-diffusive convection has been introduced to the geological sciences for a long time [*Irvine*, 1980; *Spera et al.*, 1986; *Hansen and Yuen*, 1987]. First, this approach was employed for magma chamber studies [*Clark et al.*, 1987] and later the formalism of double-diffusive convection has been applied to study the thermal-chemical instabilities at the core-mantle boundary [*Hansen and Yuen,* 1988, 1989a and b, *Kellogg,* 1991, *Kellogg and King,* 1992, *Hansen and Yuen,* 1994]. A major question confronting geologists is the mechanism for the formation of layered structures in D.D.C. Yet these questions have not been addressed at all in these past numerical works. In this paper we will present examples for the development of layered structures and the destruction of these structures by boundary-layer instabilities. We will also investigate the issues of the generation of layered structures from an initially non-layered structure. Some of these issues have been studied already in the laboratory [*Tait and Jaupart*, 1992] but for Prandtl and Lewis numbers much lower than those associated with geological media, which have an effectively infinite Prandtl number and very large Lewis number [e.g. *Clark et al.* 1987]. Finally we will present results on the effects of viscous dissipation in

Double-Diffusive Convection
Geophysical Monograph 94

D.D.C. where there is a potential for generating large thermal anomalies from the conversion of gravitational potential energy to mechanical dissipation.

The paper is organized in the following manner. First, the governing equations and numerical methods will be presented. Then we will present the various examples of the formation or destruction of layered structures. We end with some interesting new results we have recently found for viscous dissipation in D.D.C. In the final section we present a discussion of the results and implications for magma chamber evolution.

2. MODEL AND EQUATIONS

In this section we will present the set of equations used to model D.D.C. in geological systems. We employ a linearized equation of state for the density ρ

$$\rho = \rho_o(1 - \alpha(T - To) + \beta(C - Co)) \tag{1}$$

where the coefficients of thermal expansion and its analogous compositional counterpart are respectively given by α and β. Reference values are denoted by the subscript 'o' . The temperature and compositional field are given respectively by T and C. We have considered double-diffusive convection in a two-dimensional cartesian configuration. We have employed both the Boussinesq and the extended Boussinesq (with viscous and adiabatic heatings) approximations for an infinite Prandtl number, constant property fluid. We have neglected the interaction between heat and mass transfer (i.e. Soret effects) or coupled diffusion [*Liang et. al.*, 1994]. Using the streamfunction ψ formulation for simplifying the conservation of mass equation, we can write down the non-dimensional partial-differential equations within the extended Boussinesq framework [*Christensen and Yuen*, 1985] for momentum, temperature and composition as:

$$\nabla^4\Psi = Ra\left(\frac{\partial T}{\partial x} - R\rho\frac{\partial C}{\partial x}\right) \tag{2}$$

$$\frac{\partial T}{\partial t} = \nabla^2 T + \frac{\partial \psi}{\partial x}\frac{\partial T}{\partial z} - \frac{\partial \Psi}{\partial z}\frac{\partial T}{\partial x} + D\frac{\partial \psi}{\partial x}(T + To) + \frac{D}{Ra}\phi \tag{3}$$

$$\frac{\partial C}{\partial t} = \left(\frac{1}{Le}\right)\nabla^2 C + \frac{\partial \Psi}{\partial x}\frac{\partial C}{\partial z} - \frac{\partial \Psi}{\partial z}\frac{\partial C}{\partial x} \tag{4}$$

where Ra and Ra_C are respectively the thermal and compositional Rayleigh numbers and Le is the Lewis number, given by Le = κ_T / κ_D Le for magma chamber lies between 10^4 and 10^8 [*Spera et al.*, 1986]. The non-dimensional surface temperature is given by To. The viscous dissipation function, consisting of the second invariant of the strain-rate tensor, [e.g. *Batchelor*, 1967] is given by ϕ. The dissipation number [*Jarvis and Mc Kenzie*, 1980] is given by D, where $D = \alpha gd/C_p$, α is the thermal expansivity and C_p is the heat capacity. The thermal and the chemical diffusivities are denoted by κ_T and κ_D respectively. The buoyancy ratio $R\rho$ which represents the relative importance between chemical and thermal buoyant forces, is given by Ra_c/Ra. The quantity ψ is the dimensionless streamfunction, and the dimensionless time is given by t, non-dimensionalized by the thermal diffusion time across the layer depth d for the box. The vertical and horizontal coordinates are denoted respectively by z and x, with gravity g aligned opposite to z. In characterizing D.D.C. flow with specified values for the temperature and compositional fields at the top and bottom boundaries, we have defined the two Rayleigh numbers by:

$$Ra = \frac{\alpha g \Delta T d^3}{\kappa_T \nu} \tag{5}$$

$$Ra_c = \frac{\beta g \Delta C d^3}{\kappa_T \nu} \tag{6}$$

where ν is the kinematic viscosity. We have taken a free-slip boundary condition all around the box. The boundary conditions are :

$$\Psi = \frac{\partial^2}{\partial z^2}\psi = 0 \tag{7}$$

on z = 0 , 1

$$\Psi = \frac{\partial^2}{\partial z^2}\Psi = \frac{\partial T}{\partial x} = \frac{\partial C}{\partial x} = 0 \tag{8}$$

on x = 0, λ, where λ corresponds to the width or aspect-ratio of the box. Boundary conditions for C along all the boundaries were of the Newmann type in all cases except for the finger case and the viscous dissipation. In these two cases Dirichlet conditions for C were employed at the top and bottom and Newmann conditions along the side walls. Dirichlet boundary conditions for T are used for z = 0 (bottom) and T = 0 at z = 1 (top) for the diffusive regime. Alternatively for the finger regime we have set T = C = 0 at the bottom and T = C = 1 at the top. The different situations depicting the various combinations of T and C boundary

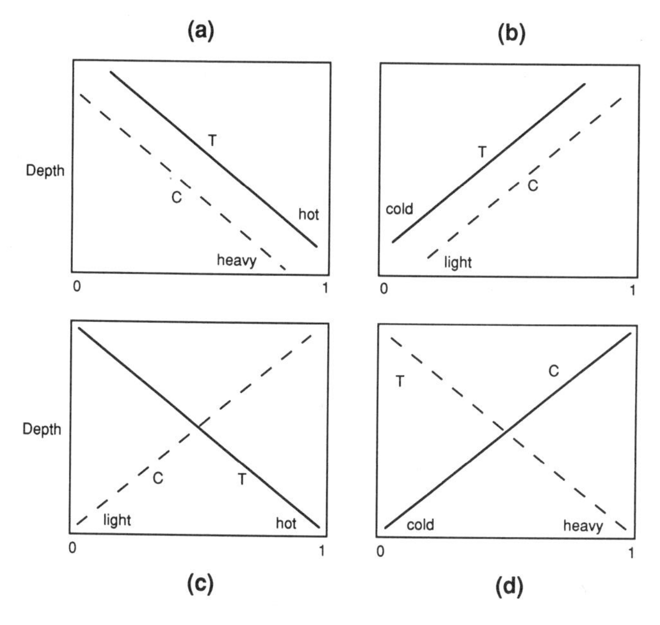

Fig.1 Depth profiles of the temperature T and composition C for the different regimes of double diffusive convection. Configurations (a) and (b) are respectively the diffusive and finger regimes. Within the diffusive and finger regimes one can further classify the cases into "supercritical" and "subcritical". Configuration (c) has both components acting in a destabilizing manner, while (d) represents a completely stabilized situation.

conditions are depicted in Figure 1.

More realistic boundary conditions of geological relevance can be found in Oldenburg et al. [1989], Oldenburg and Spera [1992 a and b]. Only in the last section will we discuss the effects of viscous dissipation. For all intents and purposes D is set to zero in most of this paper.

We have employed the finite-element techniques to solve the partial differential equations, given by equations (2) to (4). A description of the numerical techniques can be found in Hansen and Yuen [1989b] and Hansen et al. [1993]. This 2-D code has been benchmarked against other codes [*Blankenbach et al.*, 1989, *Hansen et al.*, 1992].

3. RESULTS

Double-diffusive convection takes place in fluids whose density is influenced by two components with different diffusivities. In this paper we will describe scenarios in which the temperature and a concentration of heavy component (C = 1) prescribe the density of the fluid. From a geological context the concentration of heavy material , such as MgO, in a magmatic melt is one example. Another example would be granite overlying a layer of basalt [*Marsh*, 1989]. We have outlined in Fig.1 four different scenarios : (a) hot material, enriched in a heavy component lies under a cooler material, which is relatively depleted in the heavy component. This situation in which the slowly diffusing component C acts as the restoring force, while the faster diffusing component , T, acts to drive the flow, has been coined the "diffusive" regime [*Turner,* 1973]. The opposite configuration in which cold material is compositionally light and underlies hot, compositionally heavy material is shown in Fig. 1b. Finger-like instabilities can develop under these circumstances and thus this regime is called the "finger" regime. Fig. 1c displays a configuration in which both components act as driving forces. This regime did not receive a special name and likewise did not receive much attention. Since both components act in the same destabilizing fashion, it has commonly been thought that both influences can simply be superimposed and the system can be treated as a usual one-component convection [*Turner,* 1973]. However, it is not clear to what extent one can assume a linear superposition of forces, since there is a vast difference in the diffusivities. We have also investigated this regime in view of the geological applications to be discussed below. For the sake of completeness, we will mention the configuration in which both forces stabilize the system (Fig. 1d), i.e. where the cold and compositionally heavy material underlies the hot and compositionally light material.

We can further subdivide the first two configurations, the diffusive and the finger regimes, into the "subcritical" and "supercritical" categories. In the "supercritical" category the driving force exceeds the restoring force. This means that in the diffusive regime the destabilizing thermal influence would overcome the stabilizing influence of compositional stratification, while in the finger regime a "supercritical" configuration would require the destabilizing compositional influence to surpass the stabilizing thermal distribution. *In sensu stricto*, the state of "supercriticality" demands that the driving force exceeds the restoring force by a certain amount [*Huppert and Moore*, 1976; *Hansen and Yuen,* 1989b]. On the other hand, in a "subcritical" configuration the restoring forces are dominant, i.e., the net density of the system is stably stratified. But it is certainly fascinating that in D.D.C. flow can in fact take place under statically stable circumstances because of the possibilities for "subcritical" instabilities. These "subcritical" instabilities can take place because of the phase lag between the compositional and thermal influences on the flow field due to the difference in the diffusivities. This "subcritical" mechanism has been described very nicely by Turner [1973, 1985].

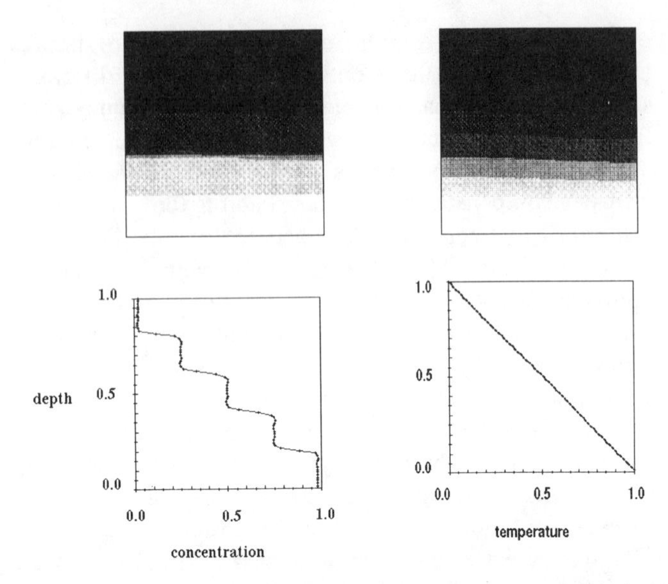

Fig.2. Initial model set-up for the experiments in the diffusive regime. Concentrational field consists of five distinct layers with the heavy material on the bottom. Newmann boundary conditions for C along all boundaries. For the temperature a linear stratification has been assumed. Contours at top panel range from light and cold material to heavy and hot material at the bottom. Contour intervals have the same value.

In the geological context D.D.C. has been proposed as a mechanism, which can give rise to the generation of strongly layered structures. Prominent examples are large intrusional complexes like the Skaergaard in Greenland and the Stillwater complex in Montana [*Turner and Campbell,* 1986]. They both display horizontal layering, which can not be explained at all by gravitational settling of crystals. Since layering has been observed in a wide range of laboratory experiments on D.D.C. [*Turner and Campbell,* 1986; *Fernando,* 1989], it has been proposed as a viable mechanism for inducing layered structures [*Huppert and Sparks,* 1984]. However, Hansen and Yuen [1989b] have shown that non-layered stationary states, in fact, do exist even in the subcritical regime. Here we wish to focus on the following questions dealing with the time-dependence of D.D.C.; (1) How do flows in D.D.C. evolve in a layered system ; (2) How can layers be generated by D.D.C. ?

In order to clarify the first question, we have conducted numerical experiments from initial conditions , which have a layered compositional structure (see Figure 2). A linear temperature field is then superimposed on this compositional field with a staircase structure consisting of five distinct layers. The depth profiles of C and T show that both temperature and composition increase with depth. This

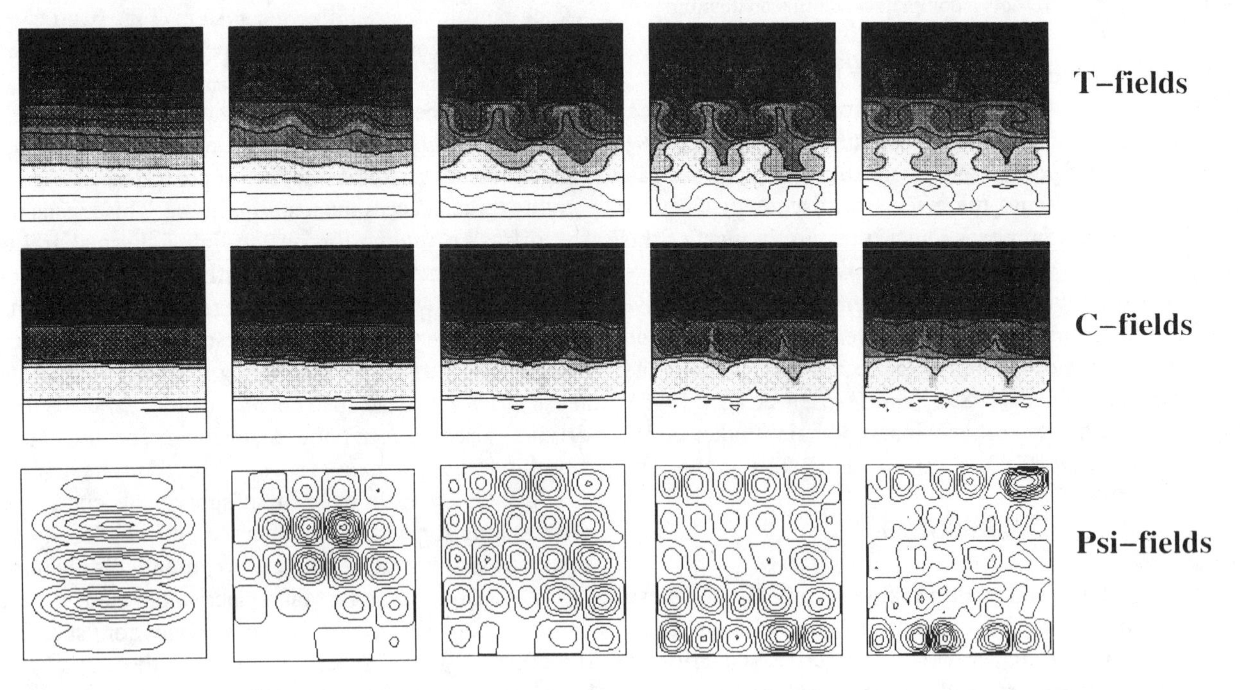

Fig.3 Snapshots of the evolution of C, T and Ψ fields for an experiment, started from an initial condition as described in Figure 2. For the T- and C-fields the greyscales are superimposed by contour lines. The parameters are Ra=10^7, Rρ=2 and the Le is 100. Grid resolution is 35 x 35 elements. Snapshots have been taken at t= (a) 0.0001 (b) 0.0011 (c) 0.00127 (d) 0.00147 and (e) 0.0020

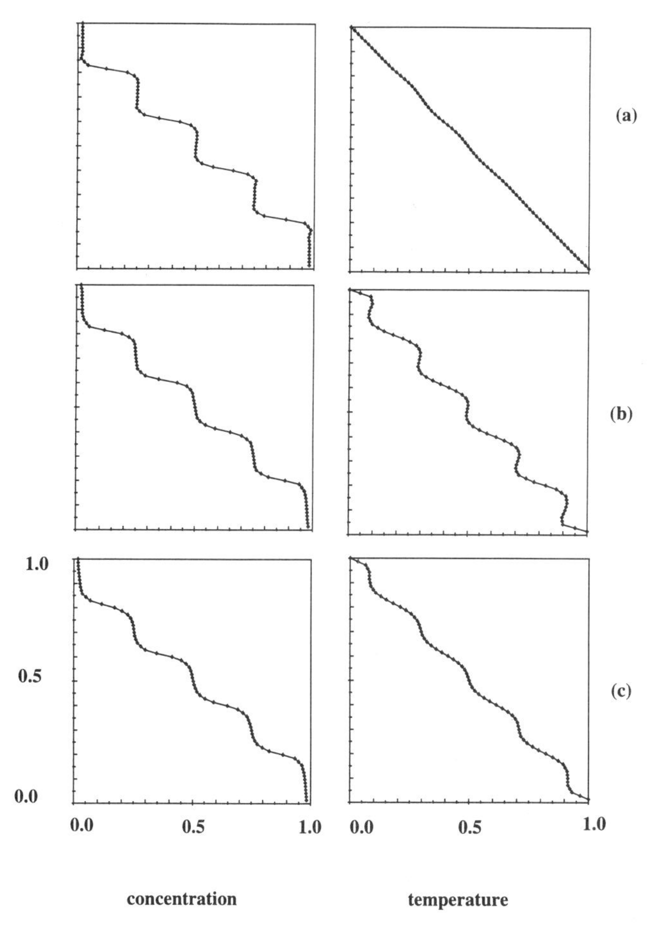

Fig.4 : Snapshots of the depths profiles of the concentration and the temperature for the experiment as described in Fig.4 The snapshots have been taken at the times (a) 0.0011 (b) 0.0020 and (c) 0.0025.

configuration would be then classified as being in the diffusive regime. Rayleigh numbers are based on the surface values. For thermal Rayleigh number $Ra=10^6$ and for compositional Rayleigh number $Ra_c=2x10^6$. Thus the buoyancy ratio $R\rho$ is 2 and the density stratification is subcritical. A Lewis number of Le=100 has been used throughout this paper. Hansen and Yuen [1988] have found that in the transient regime there are very little differences in the C and T fields between the solutions for Le=100 or larger values. In Fig. 3 we show the evolution of these configurations. From the snapshots of the T, C and Ψ fields we can identify the onset and development of the D.D.C. instabilities. Convection takes place separately in each of the five distinct compositional layers. Since the C field is essentially uniform with each individual layer, there is then no restoring force. The thermal stratification across each layer is sufficient for destabilizing the particular layer. Any sort of deflection of the compositional interface would result in a strong stabilizing restoring force, which would prevent the overturning of two or more of the layers. Snapshots of the depth profiles in both C and T (Fig. 4) reveal clearly the layered nature of the flow. The initially linear temperature profile (a) is replaced by a more step-like profile (b), once convection has ensued. Mixing within each layer leads to quite isothermal layers being separated from each other by diffusive interfaces with clearly distinguishable temperature drops across them.

In order to maintain a layered structure, one must maintain a density stratification, which is sufficiently subcritical. The snapshots in Figs 5a and 5b describe the evolution of an initially chemically layered system with a density stratification, being less subcritical than in the

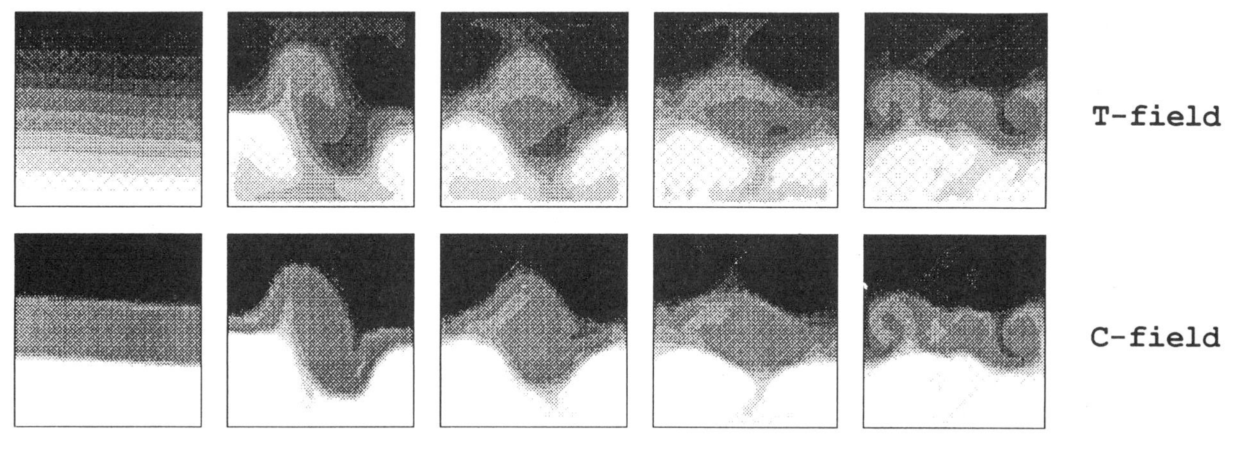

Fig. 5a: Evolution of the T- and C-field for an experiment started from an initial condition with a linear temperature profile and a compositional field consisting of 3 distinct layers (left panel Fig. 5a). Parameters are $Ra=10^7$, Rp=1 and Le = 100. Times are (a) 0.0 (b) 0.00036 (c) 0.00115 (d) 0.00194 and (e) 0.00336.

previous case. A linear temperature profile has been superimposed on a concentrational field consisting of three distinct layers (Fig.5a). We have employed a Rρ of 1 with $Ra=Ra_c=10^7$. As a result of strong convective motions in the upper- and lower-layers the light-grey middle layer is deformed significantly. This deformation creates a strong restoring force. The light-gray material has entered the white layer where it is compositionally lighter than the white (heavy) material. At the same time the light-gray material is relatively heavy in the topmost layer, as compared to the dark material. This strong horizontal compositional gradient almost restores the initial layered structure, as shown in the last snapshot of Fig. 5a. At this stage of the evolution convection takes place separately in each of the three layers. Entrainment leads to a further amalgamation of the layers. The first two snapshots of Fig. 5b show the disappearance of the middle layer, which has been entrained into the upper and lower layers. Only very little material with a pure white (heavy) concentration remains, because of the mixing with the initially light-gray material. The last three snapshots of Fig. 5b show the final overturning of the system, resulting in a single convective flow in a compositionally well-mixed fluid. The thermally driven flow is not influenced at all by compositional heterogeneities and the flow exhibits chaotically fluctuating boundary-layer instabilities (Fig 5b, T-field) for these high Rayleigh numbers [*Hansen et al.*, 1992; *Yuen et al.*, 1993].

Earlier Hansen and Yuen [1989] have demonstrated that non-layered flow fields can exist under subcritical conditions. In fact, these non-layered solutions are derived from the steady-state solutions under subcritical conditions. With these previous findings in mind, we can interpret that sub-critical conditions are a necessary, but not sufficient condition for a layered flow to persist for at least several overturns. Moreover, a sufficiently large Rρ is required. In other words, the compositional restoring force must be greater than the thermal driving force in order to maintain layered convection. We emphasize here that although layering is a phenomenon commonly observed in D.D.C. motions, one cannot make the foregone conclusion that the double-diffusive convection automatically implies layering.

The heat transport efficiency in convective flows has important ramifications for both natural and industrial applications. The heat transfer efficiency is commonly measured by the Nusselt number Nu, which is a horizontally averaged quantity. We have monitored the time history of Nu for a series of D.D.C. flows. In Fig. 6a we show the time history plots of Nu, measured at the surface, which have

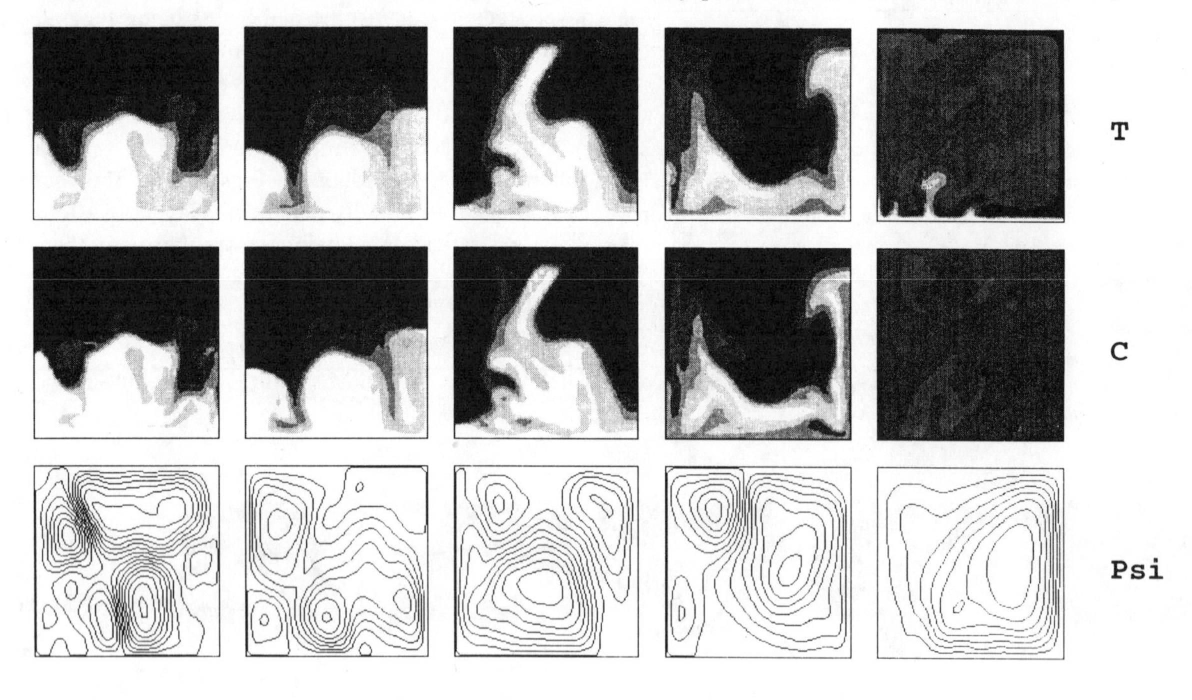

Fig 5b: Late stages of the experiment described in Fig.5a The temperature, composition and the streamfunction fields are displayed here. Times are (a) 0.00575 (b) 0.00800 (c) 0.0105 (d) 0.0245 (e) 0.0395

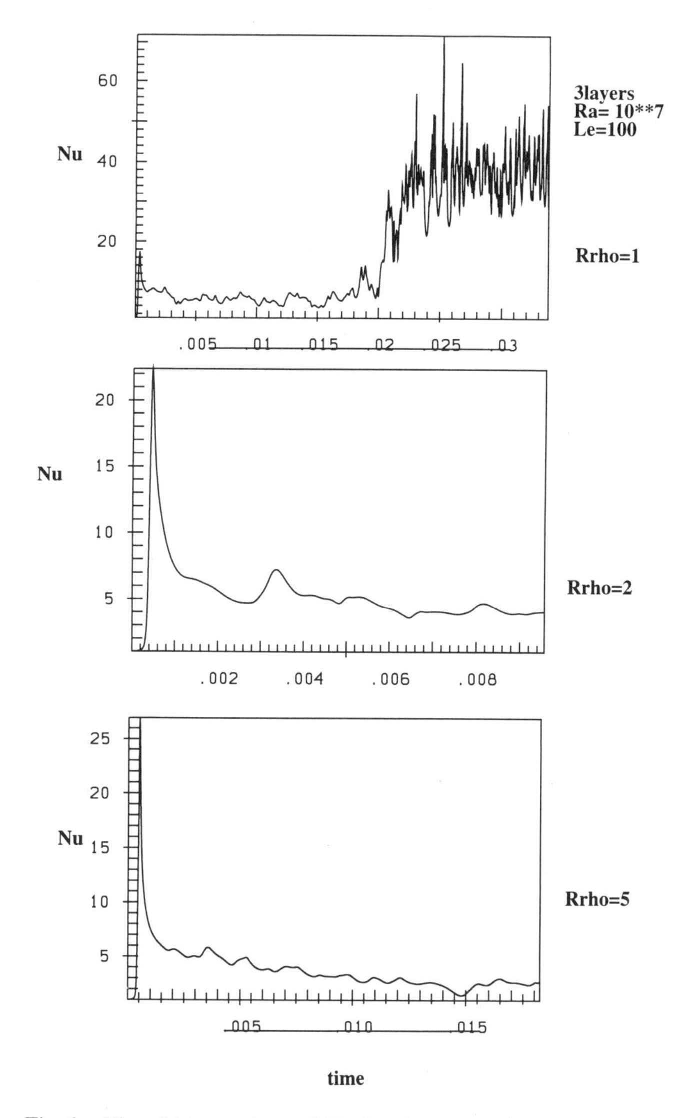

Fig 6a: Time history plots of Nu for three experiments with Rρ varying. All runs have been started from the same initial condition as described in Fig.5a. All cases have Ra=10^7, and Le =100 . Rρ is 1 (top) 2 (middle) and 5 (bottom).

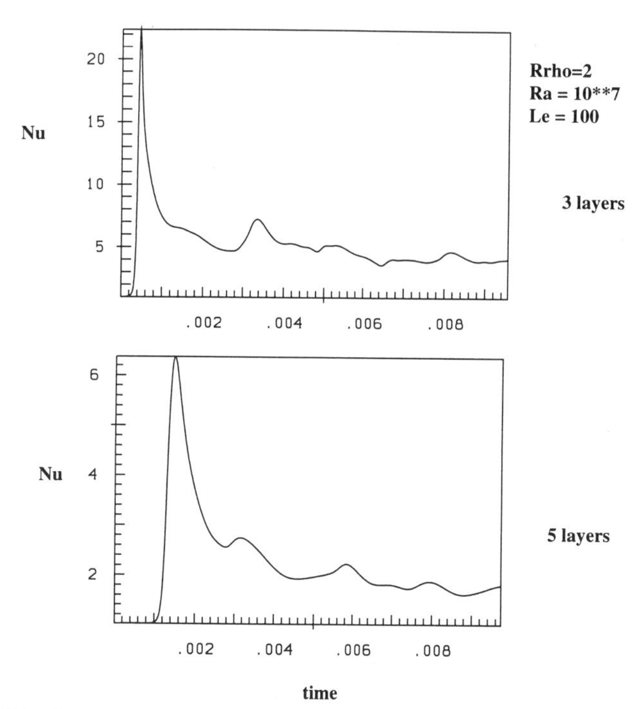

Fig 6b: Comparison of Nusselt time history for two experiments with the same control parameters (Ra=10^7, Rρ=2 and Le=100). One was started from the 3-layer initial condition (top) the other from the 5- layer initial condition as described in Fig.2 (bottom).

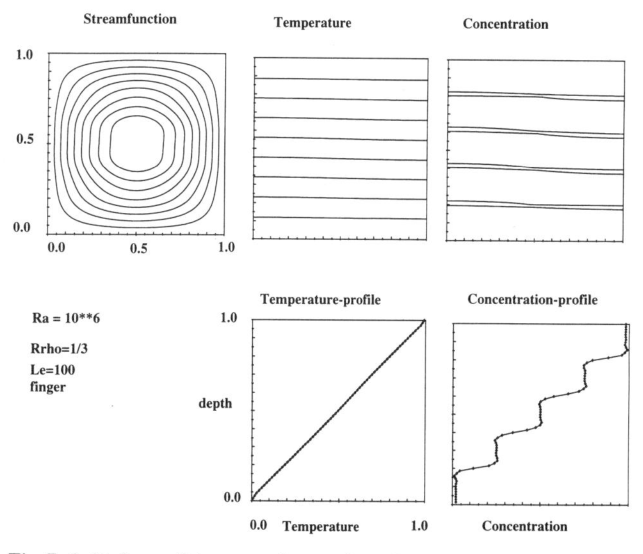

Fig.7: Initial conditions as chosen for a D.D.C. flow in the finger regime. An unstably stratified layered structure (5 layers) has been adopted for the concentrational field while a linear stably stratified temperature field has been assumed. Parameters for the experiments were Ra=10^6, Rρ=1/3 Le=100. The experiments has been carried out on a mesh of 45x45 equidistant elements. Boundary conditions were free-slip for ψ, no flux for T and C along the side walls and fixed T and C at the bottom and the top.

been taken from experiments carried out under the same conditions as those described in Fig. 5. We have considered Rρ =1 , 2 and 5. The top panel shows the case for Rρ =1, which is exactly the experiment discussed already in Fig. 5. As a comparison the value of Nu for a purely thermal convection case with Ra =10^7 is 45. Clearly, the period of layered convection can be distinguished from the fully overturning mode. Both phases are separated by a strong increase of Nu, upon the transition from the layered to non-layered mode. After an initial spin-up period the layered period persists until t=0.02 and is characterized by low values of Nu aorund 7 and 8. The single-layered flow produces values of Nu of around 45, which is very close to that for pure thermal convection.

An increase of Rρ to 2 changes the picture, as no overturn takes place (Fig. 6 middle row). We did not find any evidences for deformation of the interface, even though the flow has been run to t=0.01. We therefore conclude that a dynamically induced overturn cannot take place under these circumstances. Here we employ the terminology "dynamically induced overturn" for distinguishing cases in which the homogenization of the C field takes place by diffusion from those in which mixing is greatly facilitated by mechanical stirring.

A further increase of Rρ to 5 corroborates these conclusions (Fig. 6a bottom panel.) Even after more than 0.015 of the thermal diffusion time there was no deformation of the interfaces. For both Rρ = 2 and 5 the three compositonally distinct layers are maintained and vertical cross-layer mixing occurs only over the thermal diffusion timescale. Values of Nu are much lower than for the non-layered case but seem not to depend too much on Rρ. Values of Nu of around 5 are obtained for both Rρ=2 and 5. This seems quite understandable, since each layer is compostionally uniform so that thermal convection takes place within each layer without being disturbed by the presence of compositional heterogeneities. The Nusselt number is therefore determined by the number of layers.

Two experiments, performed under the same conditions as above except for the initial configuration confirm this result. We have used the same physical parameters as for the previous figure for both cases. One run has been started from an initial three layer state (Fig. 6b top), while for the other one an initial five layer state, as described previously for Fig. 2, has been employed. In both cases the layer structure remains intact but it is clear that the Nusselt number is further decreased, when we change from the 3-layer structure (Nu ~ 5) to the 5-layer structure (Nu ~ 2).

In the geosciences the diffusive regime has been the subject of most of the research efforts, because of the prime motivations to explain the prominent layered geological structures. There have been studies in oceanography on the finger regime, e.g. Schmitt [1994]. High-resolution numerical works have been carried out by Shen [1989, 1993], and Shen and Veronis [1991] following the earlier work by Piacsek and Toomre [1980]. Hansen and Yuen [1989b, 1990] treated D.D.C. in the finger regime within the context of magma chamber convection. Tait and Jaupart [1992] drew an analogy between D.D.C. in the finger regime and compositionally driven convection in evolving magma chambers. Numerical work on double-diffusive fingering is also under investigation within the context of flow in porous and fluid layer [*Chen and Chen*, 1988 ; *Chen and Lu,* 1991].

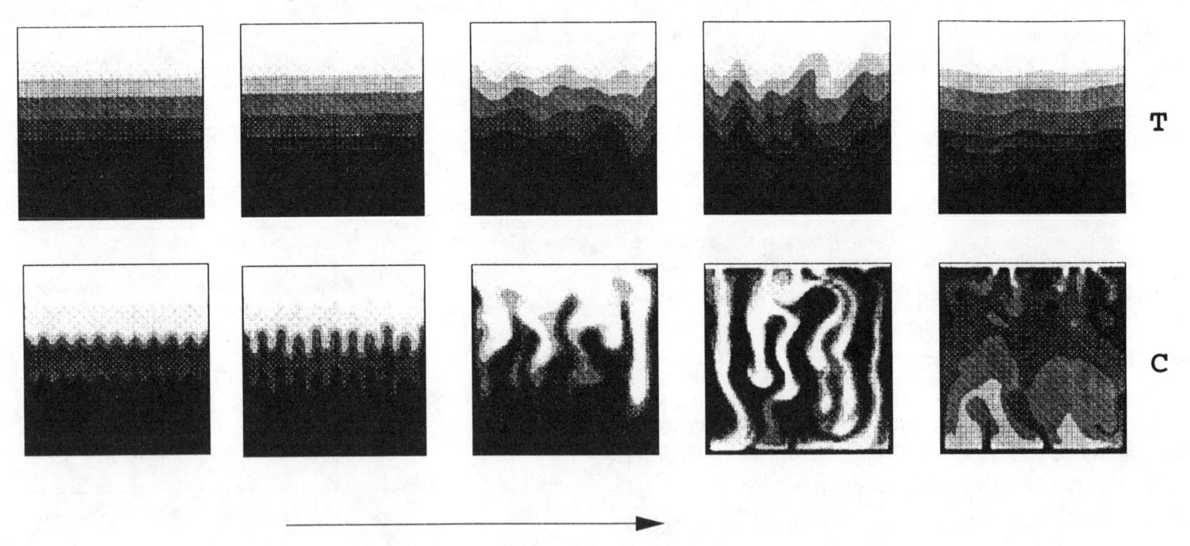

Fig. 8: Evolution of the T-and C-fields in the finger-regime. Times are (a) 0.0251 (b) 0.0383 (c) 0.0546 (d) 0.0705 and (e) 0.1308.

We now inquire what happens to a layered system, when it is subject to double-diffusive instabilities in the finger regime described above. Fig. 7 displays the set-up of the model and the initial conditions. As in the other cases, we have started from an initially compositionally layered configuration, which is subject to a stabilizing temperature gradient. The C-field is stratified in an unstable manner. The net density stratification is subcritical in that the stabilizing influence from the temperature prevails in the overall balance. According to our definition, Rρ, the buoyancy ratio is 1/3 and the thermal Rayleigh number Ra is set to 10^6. In contrast to the previous experiments, T and C have been kept fixed at the top and bottom boundaries.

Fig. 8 shows the development of the C and T fields. Initial perturbations with a high horizontal wavenumber, k greater than 300, have been employed for both fields. By choosing high wavenumber modes, we can avoid the excitation of a particular stage in the flow development. Initially about 12 instabilities develop, as shown by the greyscale plots of the C field. Because of the much higher diffusivity (Le =100) the temperature field does not display sharp features. As noted by Piacsek and Toomre [1980] for finite Prandtl number, the small perturbations merge and form a collective instability. The initially layered structure is destroyed immediately by the exponentially growing finger instabilities. Rapid mixing of the C field takes place, while the T-field remains relatively unscathed. We arrive at a situation in which the layering has been completely

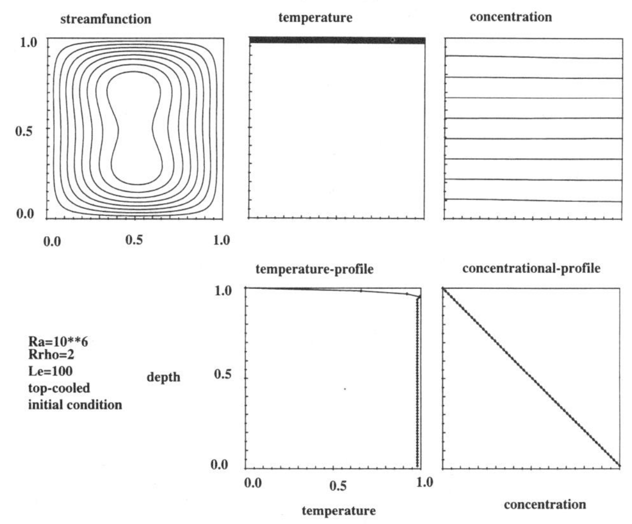

Fig.9 Cooling a stable compositional gradient from the top. Initial condition is a uniformly hot, compositionally stably stratified system. Boundary conditions are: free-slip for ψ along all boundaries, no-flux for C along all boundaries. For the temperature, no flux is assumed, except at the top where T is fixed to T=0. Parameters are Ra=10^6, Rρ=2, Le=100.

eradicated and the stabilizing linear temperature profile has nearly been restored. This numerical experiment has not revealed any evidence of the existence of layer formation in the finger regime. Instead we witness a rather efficient destruction of a pre-existing layered structure. The formation of distinct layers has been reported in laboratory experiments [e.g. *Ruddick and Turner*, 1979] but has, to our knowledge, never been observed in numerical simulations. The question of whether layer formation in the finger regime depends on the actual boundary condition [*Howard and Veronis*, 1987] or whether it may be a result of a specific initial condition must be clarified in the future. We can state here that, different from the diffusive regime, the finger regime does not seem to allow for the persistence of distinct layers for a sufficiently long timescales even for a sufficiently subcritical stratified state.

In the previous section we have investigated the fate of a compositionally layered system, subject to D.D.C. instabilities. A question of great geological importance is how can layers be generated from non-layered states ? This question has been discussed intensively [e.g. *Huppert and Worster*, 1992]. Here we wish to readdress this question within the framework of a cooling intrusional complex. The problem of cooling lava lakes has been recently studied by Davaille and Jaupart [1993], who have treated this within the framework of thermal convection. We will now turn our attention to the evolution of an initially hot intrusion with a stably stratified compositional field. We show the initial conditions in Fig. 9. The system cools only through the top. Adiabatic boundary conditions have been applied on all other boundaries. The zero-flux boundary condition for C is imposed along all boundaries. The net density stratification is stable. A buoyancy ratio of Rρ= 2 at a thermal Ra of 10^6 has been chosen. Initial cooling leads to the formation of the thermal boundary layer at the top, as seen from the contours of the T-field in Fig. 9. This boundary layer then grows diffusively until it reaches a critical thickness (Fig. 10). Then the instabilities begin to develop from this boundary layer and sink into the interior. However, the stably stratified compositional field prevents the cold instabilities from penetrating into great depths (Fig. 10 b). Mixing takes place efficiently only in the upper-part, while the lower part is left relatively unperturbed. In Fig.11 we show the evolution of the depth profiles for both C and T. Two distinct domains can be observed to develop. One at the top has a nearly uniform composition, while a second layer, situated between the middle to the bottom, remains internally stratified in the composition. Such a scenario can be used to explain the coexistence of compositionally homogeneous layers with stratified layers.

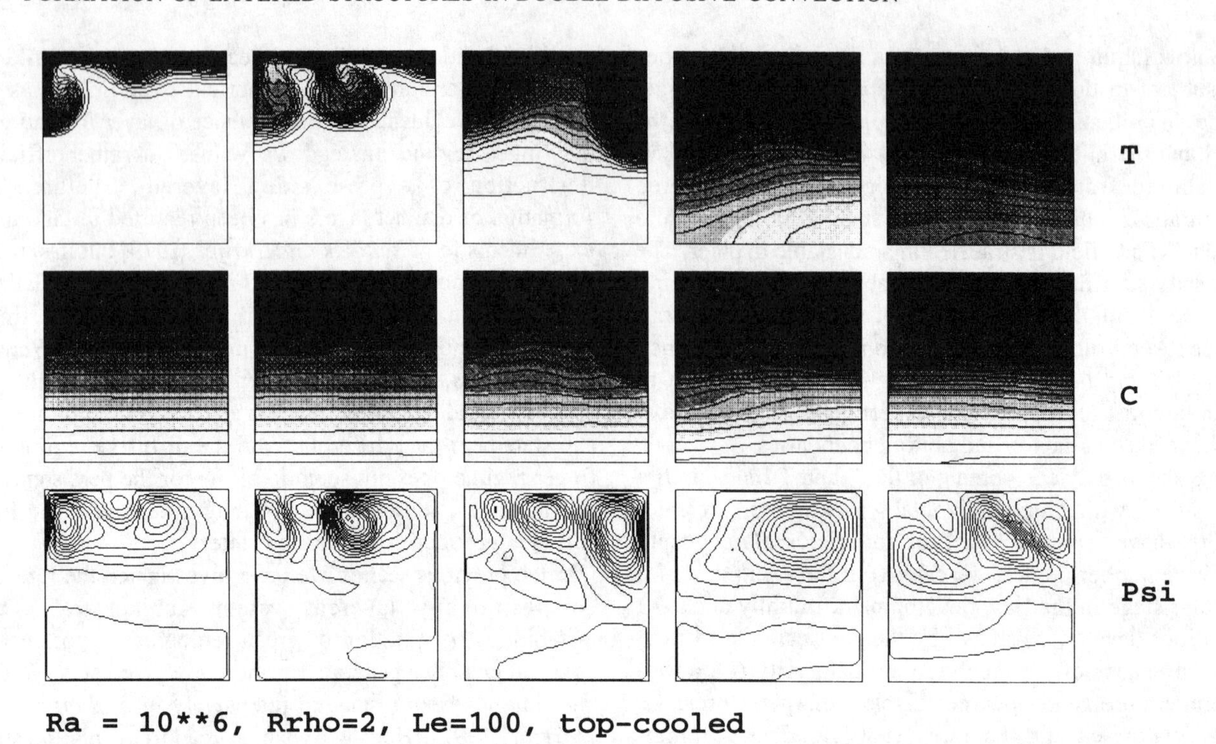

Fig.10: Evolution of the experiment as described in Fig.9 Times are (a) 0.0048 (b) 0.0067 (c) 0.02562 (d) 0.0989 and (e) 0.1766

Most of the work on the cooling of magma chambers has been conducted for the scenario of top-cooling [e.g. *Marsh*, 1989, *Worster et al.*, 1990]. However, it is also reasonable to assume that magma chamber cools also through the bottom [*Jaupart and Brandeis*, 1986] and through the sidewalls [*Bergantz*, 1988]. In order to investigate the effects due to different cooling mechanisms, we have carried out the same experiment as described just above, but now we allow for cooling to take place through all of the boundaries. For the C field we have assumed the no-flux boundary condition at all boundaries. Fig. 12 gives an overview of the evolution. Strong sinking instabilities develop along the sidewalls and they create a convecting layer at the top, which is gradually homogenized by the flow (see the C-fields in Fig. 12). Below this region another layer with a counter-flow pattern is formed. At the bottom one finds a stagnant region. Fig. 13a shows the evolution of the T and C profiles and Fig. 13b displays the situation close to the final state, when the system is cold and the convective vigor has vanished. The three resultant compositional layers reveal different degrees of internal stratification, ranging from a nearly uniform layer in the middle to a strongly stratified layer at the bottom. The thickness of the final layers is influenced by the ability of the instabilities to sink into the interior and therefore by the magnitude of $R\rho$ We have, in fact, performed the same experiment with a value of $R\rho = 1$ and in this case we have obtained a much thicker uniform layer at the top, due to the decrease in the restoring force.

We have considered, up to now, systems in which the temperature and composition fields exert opposite effects in the density. There are geological situations in which both the compositional and the thermal and chemical fields are unstably stratified. Crystallization at the inner-outer core boundary, for example, releases light material at the bottom of the outer core [e.g. *Buffett et al.*, 1992]. The thermal stratification is unstable only slightly as compared to the compositional stratification. Diapirs in the earth's crust may form by compositional density differences within a temperature field, which is only slightly unstable. A good example would be the emplacement of granites in the crust [e.g. *Stel et al.*, 1993]. There has been relatively little attention paid on this type of flow, driven in concert by both the destabilizing influences of temperature and composition. The reason may be due to the idea that the influences of composition and temperature can simply be superimposed and the situation then resembles essentially single-component convection. The situation changes when one takes into account the effects of adiabatic and viscous heatings (see equation 3 in section 2). In a flow driven

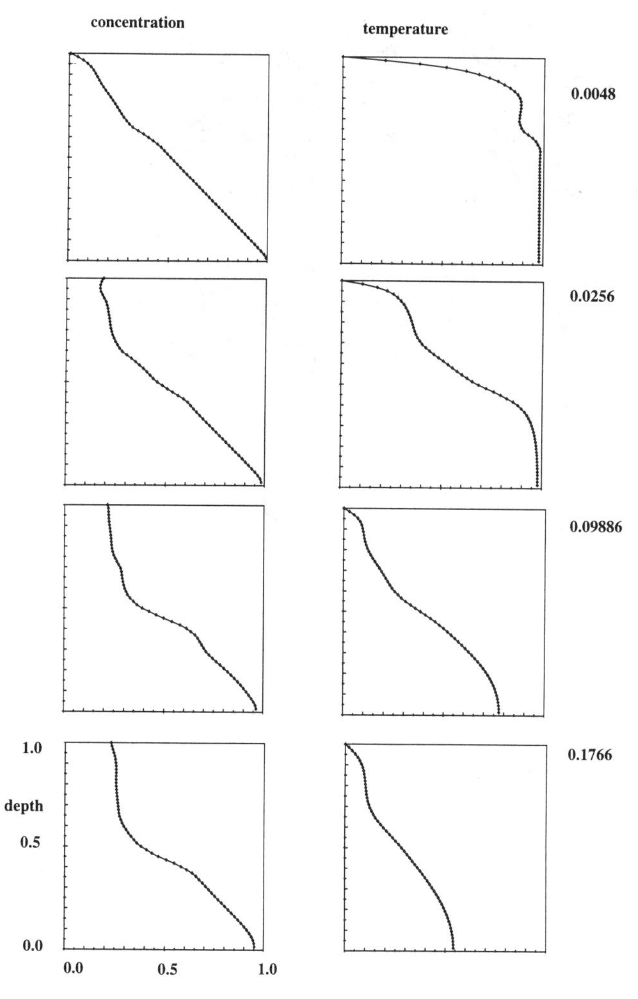

Fig.11: Snapshots of the T- and C- depth profiles at the corresponding times.

predominantly by compositional forces the heat produced by viscous dissipation can play a significant role. We have already examined the consequences of viscous heating in "subcritical" D.D.C. situations [*Hansen and Yuen*, 1990]. A significant heating effect was observed even there. For the "supercritical" situation more drastic effects can be expected. Gravitational potential energy, stored in the compositional field, is transformed into motion, which is converted to a mechanical source of heat by means of viscous dissipation. There is no other mechanism to compete with this effect so that a strong local build-up of heat can take place.

We have carried out a numerical experiment to demonstrate the effects of dissipative heating in a flow driven mainly by composition. We have employed a box with λ =3 and this box is filled with initially cold fluid being enriched in heavy components. At t =0, the lower boundary is set to C = 0 and T = 1 , i.e. hot and light material now enters through the bottom. The compositional Rayleigh number has been chosen to be 10^5, while the thermal Rayleigh number Ra is lower by a factor of 100 (Ra=10^3). The dissipation number D is taken to be 0.3 and the dimensionless surface temperature To is 0.2. Random perturbations on the C field have been imposed. As shown in Fig. 14, many small compositionally-driven diapirs develop initially. Because of the much higher diffusivity these instabilities are hard to be picked up in the temperature field. The small instabilities gather and erupt in two collective instabilities. In the middle panel of Fig. 14 there is a mark indicating the maximum of the temperature at that stage. We see that within the compositionally driven diapir viscous heating is sufficiently strong for producing a maximum temperature of 1.34 or 34 % hotter than the heated fluid entering the box! For an initially unstable situation, such as a linear stratification, an even stronger effect can be found. At the later stages new diapirs are regenerated , which lead again to a local build-up of a temperature maximum (T=1.06 in the bottom panel of Fig. 14). Finally in Fig. 15 we show the time history of the global maximum temperature. From this plot one can distinguish the different heating episodes, due to the production of the different diapirs. This heating mechanism, which is due to the conversion of chemical potential energy to mechanical heat dissipated, is novel and more work is needed for a better understanding because of its important geological implications for the crust [*Spera et al.*, 1994] and the outer core [*Hansen and Yuen*, 1995].

4. CONCLUDING REMARKS

In this work we have presented 2-D numerical results on some relevant issues pertaining to the physics of double-diffusive convection, as applied to magma chambers. Double-diffusive convection at infinite Prandtl number is a mechanism, which has important geological relevance for the stability regimes depicted in Fig. 1. We have focused our attention on matters such as what causes layering and the destruction of layering.

Subcritical D.D.C. in the diffusive regime provides the most favorable conditions for layering , for states sufficiently subcritical. This would depend mainly on Rρ. Heat-transfer out of the system depends on the type of layering (see Fig. 6). A sufficiently high Rρ keeps the layering intact and leads to a low Nusselt number. Magma chambers with this style of convection would cool relatively

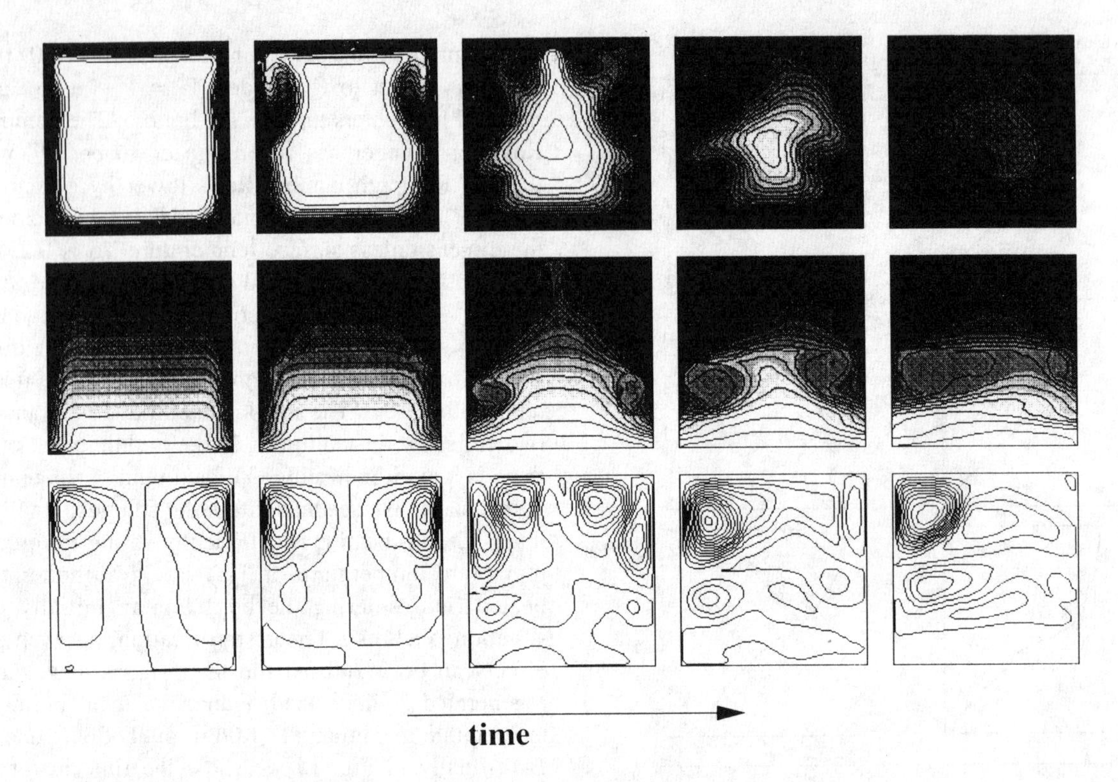

Fig.12. Cooling a stable concentrational gradient through all sides. Evolution of an experiment performed under the same condition as described in Fig. 9 however the temperature was kept fixed to 0 along all sides. Parameters are $Ra=10^6$, $\rho=2$ Le=100.

slowly. The cooling rate also decreases with the increased number of layers in the magmatic system. In the case of Rρ close to that needed for enforcing layering, a sudden overturn can take place after an initial transient layered period. By this mechanism a magma chamber can stay relatively hot during the layered state and then cools rapidly after the spontaneous overturn event.

In the subcritical regime D.D.C. can produce layering from an initially non-layered state. Magma chambers cooled from the top and from the side walls can develop a compositional layering from an initially unstratified condition. The thicknesses of the different layers depends strongly on Rρ. We have shown the possibilities for the formation of coexisting layers with almost uniform composition and layers with gradual gradations in the chemical composition.

In the finger regime we did not find any traces of layering. On the contrary, finger instabilities lead to the rapid destruction of pre-existing layers. Moreover, the exponential nature in the growth of the finger instabilities gives rise to much faster evolution. Magma chambers with compositionally lighter material underlying enriched material should develop much differently from the diffusive regime with respect to the layering and also to the thermal evolution. Discrepancies between laboratory observations of layered formation in the finger regime and numerical studies need further investigation.

Throughout this work we have not addressed the issues of solidification, which is complicated by the complex nature of the phase diagrams involved [*Oldenburg and Spera*, 1992a and b, *Emms and Fowler,* 1994]. But unless some of the basic issues of D.D.C. involving the modes of cooling and the interaction of instabilities with the interior are understood better, then it would be of no use to go into more complicated situations, where more fine-structure spatial

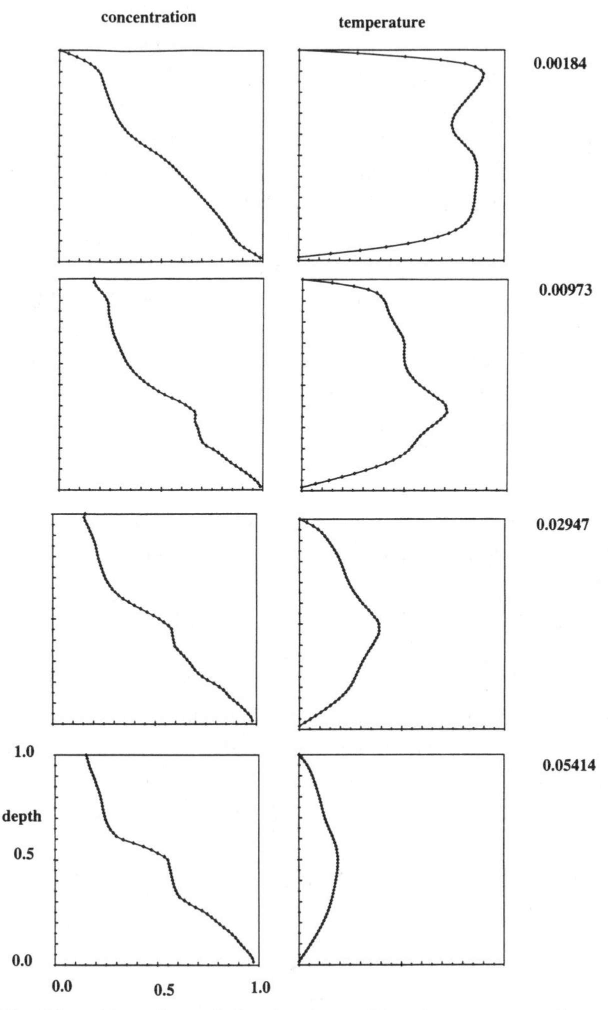

Fig.13 a: Snapshot of the depth profiles for concentration and temperature for the experiment described above. Times are (a) 0.00184 (b) 0.00973 (c) 0.02947 and (d) 0.5414.

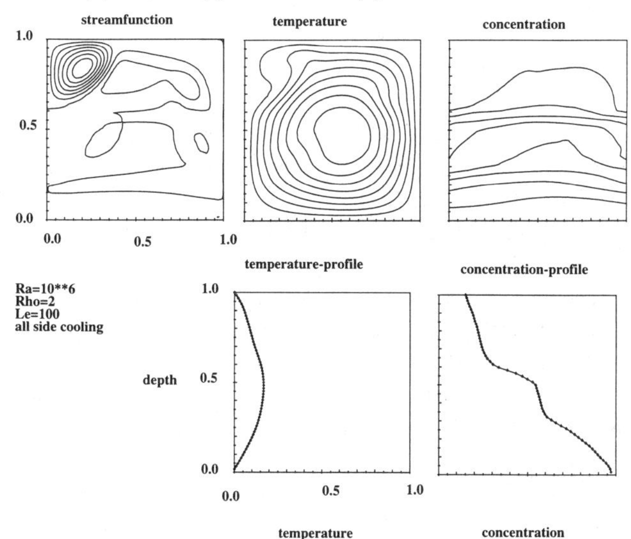

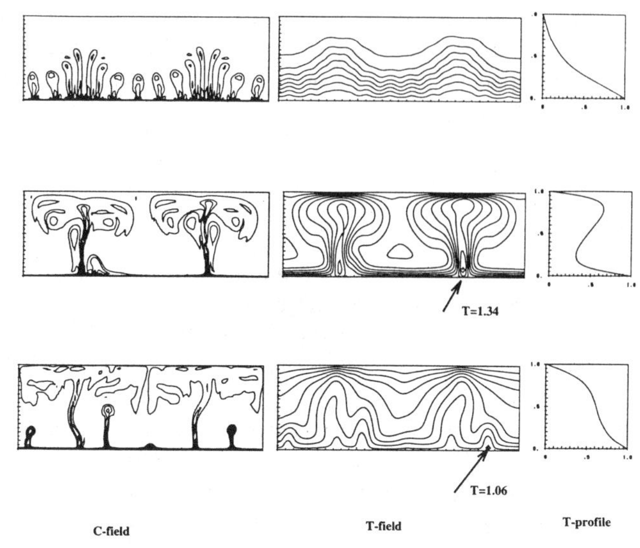

Fig. 14: The effects of viscous dissipation in a compositionally driven flow. Boundary conditions for C are Newmann along the side walls and Dirichlet at top and bottom. Snapshots of C- and T-fields and of the temperature-depth profiles. In the temperature plots, the arrows point to the locations of the maximum temperature and the values of maximum temperature Tmax are given. Times are (a) 0.0432 (b) 0.05579 and (c) 0.08943. Parameters are Ra=10^3 Ras=10, the dissipation number D=0.5 and Le=100. A grid of 35x145 points has been employed.

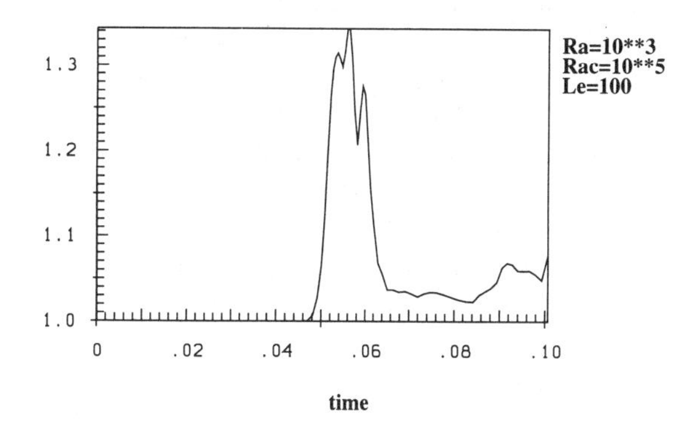

Fig. 15: Time history of the global maximum temperature for the experiment as described in the previous figure.

features, such as dendrites, are present. Finally, we have also pointed out a potentially very important mechanism for a concentrated heat source due to viscous heating in double-diffusive convection.

Fig 13b: Late stage of the experiment described in Fig.12.

Acknowledgments. We thank Dr. Frank J. Spera for discussions and for remarks by an anonymous reviewer. This research has been supported by the Geochemistry and the Ocean Sciences Programs of the N.S.F.

REFERENCES

Batchelor, G.K., *An Introduction to Fluid Dynamics,* chapter 3, Cambridge Univ. Press, Cambridge, 1967.

Bergantz, G.W., Convection and solidification in tall magma chambers, Ph.D. Thesis, Johns Hopkins Unv. Baltimore, MD, 127pp, 1988.

Blankenbach, B. et al., A benchmark comparison of mantle convection codes, *Geophys. J. Int., 98,* 23-38, 1989.

Buffett, B.A., Huppert, H.E., Lister, J.R. and A.W. Woods, Analytical model for solidfication of the Earth's core, *Nature, 356,* 329-331, 1992.

Chen F. and C.F. Chen, Onset of finger convection in a horizontal porous layer underlying a fluid layer, *J. Heat Transfer, vol 110,* 403-409, 1988.

Chen F. and J.W. Lu, Influence of viscosity variation on salt-finger instability in a fluid layer, a porous layer and their superposition, *J. Appl. Phys. 70,* 4121-4131, 1991 .

Christensen, U.R. and D.A. Yuen, Layered convection induced by phase transitions, *J. Geophys. Res., 90,* 10 291-10 300, 1985.

Clark, S., Spera, F.J. and D.A. Yuen, Steady-state double-diffusive convection in magma chambers heated from below, in Magmatic Processes: *Physicochemical Principles,* pp 289-305, ed. by B.O. Mysen, The Geochemical Society, 1987.

Davaille, A. and C. Jaupart, Thermal convection in lava lakes, *Geophys. Res. Lett., 20,* 1827-1830, 1993 .

Emms, P.W. and A.C. Fowler, Compositional convection in the soldification of binary alloys, *J. Fluid Mech., 262,* 111-139, 1994.

Fernando, H.J.S., Buoyancy transfer across a diffusive interface, *J. Fluid Mech., 209,* 1-34, 1989.

Hansen, U. and D.A. Yuen, Evolutionary structures in double-diffusive convection in magma chambers, *Geophys. Res. Lett., 14,* 1099-1102, 1987.

Hansen, U. and D.A. Yuen, Numerical simulations of thermal-chemical instabilities and lateral heterogeneities at the core-mantle boundary, *Nature,334*, 237-240, 1988.

Hansen, U. and D.A. Yuen, Dynamical influences from thermal-chemical instabilities at the core-mantle boundary, *Geophys. Res. Lett., 16,* 629-632, 1989a.

Hansen, U. and D.A. Yuen, Subcritical double-diffusive convection at infinite Prandtl number, *Geophys. Astrophys. Fluid Dyn., 47,* 199-224, 1989b.

Hansen, U. and D.A. Yuen, Nonlinear physics of double-diffusive convection in geological systems, *Earth Science Rev. 29,* 385-399, 1990.

Hansen, U., Yuen, D.A. and A.V. Malevsky, A comparison of steady-state and strongly chaotic thermal convection at high Rayleigh number, *Phys. Rev., A46,* 4742-4755,1992.

Hansen, U., Yuen, D.A., Kroening S.E. and T.B. Larsen, Dynamical consequences of depth-dependent thermal expansivity and viscosity on mantle circulations and thermal structure, *Phys. Earth. Planet. Inter., 77,* 205-223, 1993.

Hansen, U. and D.A. Yuen, Effects of depth-dependent thermal expansivity on the interaction of thermal-chemical plumes with a compositional boundary, *Phys. Earth Planet. Inter., 86,* 205-221 , 1994 .

Hansen, U. and D.A.Yuen, Potential role played by viscous heating in thermal-chemical convection in the outer core, submitted to *Geochimica Cosmochima Acta,* 1995.

Howard, L.N. and G. Veronis, The salt finger zone, *J. Fluid mech. 183,* 1-13, 1987

Huppert, H.E. and R.S.J. Sparks, Double-diffusive convection due to crystllization in magmas, *Ann. Rev. Earth Planet. Sci., 12,* 11-37, 1984.

Huppert, H.E. and M.G. Worster, Vigorous motions in magma chambers, *Chaotic Processes in the Geological Sciences* edited by D.A.Yuen, pp 141-173, Springer Verlag, New York, 1992.

Irvine, T.N., Magmatic infiltration metasomatism, double-diffusive fractional crystallization and adcumulus growth in the Muskox intrusion and other layered intrusions. In *Physics of Magmatic Processes,* ed. by R.B. Hargraves, Princeton Univ. Press, 585pp, 1980.

Jarvis, G.T. and D.P. McKenzie, Convection in a compressible fluid with infinite Prandtl number, *J. Fluid Mech., 96,* 515-583, 1980.

Jaupart , C. and G. Brandeis, The stagnant bottom layer of convecting magma chambers, *Earth Planet. Sci. Lett., 80,* 183-199, 1986.

Kellogg, L. H., Interaction of plumes with a compositional boundary at 670 km, *Geophys. Res. Lett., 18,* 865-868, 1991.

Kellogg, L.H. and S.D. King, Effects of mantle plumes on the growth of D" by reaction between the core and mantle, *Geophys. Res. Lett., 20,* 379-382, 1993.

Liang, Y., Richter, F.M. and E.B. Watson, Convection in multicomponent silicate melts driven by coupled diffusion, *Nature, 369,* 390-392, 1994.

Marsh, B.D., Magma chambers, *Annu. Rev. Earth Planet. Sci. 17,* 439-474, 1989.

Oldenburg, C.M., Spera, F.J. , D.A. Yuen and G. Sewell, Dynamic mixing in magma bodies: theory, simulations and implications, *J.Geophys. Res.94,* 9215-9236, 1989.

Oldenburg, C.M. and F.J. Spera , Hybrid model for solidification and convection, *Numer. Heat Transfer, part 2, Vol.21,* 217-229, 1992a.

Oldenburg, C.M. and F.J. Spera, Modelling transport processes in nonlinear systems in *Chaotic Processes in the Geological*

Sciences, edited by D.A.Yuen, pp 205-224, Springer Verlag, New York, 1992 b.

Piascek, S.A. and J. Toomre, Nonlinear evolution and structure of salt fingers in *Marine Turbulence,* edited by J.C.J. Nihoul, Elsevier, Amsterdam, pp 193-219, 1980.

Ruddick, B.R. and J.S. Turner, The vertical length scale of double-diffusive intrusions, *Deep-Sea Res. 26A,* 23-40, 1979.

Schmitt, R.W., Double diffusion in oceanography, *Ann Rev. Fluid Mech., 26,* 255-285, 1994 .

Shen, C. Y., The evolution of the double-diffusive instability: salt fingers. *Physics Fluids A1(5)*, 829-844, 1989.

Shen, C. Y. and G. Veronis, Scale transition of double-diffusive finger cells, *Physics Fluids A3(4)*, 58-69, 1991.

Shen , C.Y. , Heat-salt finger fluxes across a density interface, *Physics Fluids A5(11),* 2633 -2643, 1993 .

Spera, F.J., Yuen, D.A. and H.-J. Hong, Double-diffusive convection in magma chambers: single or multiple layers?, *Geophys. Res. Lett., 13,* 153-156, 1986.

Spera, F.J., Trial, A.F., Yuen, D.A. and U. Hansen, Viscous dissipation : role in magma generation and ascent, *E.O.S., American Geophys. Union Trans., Vol. 75,* no. 44, 703, 1994.

Stel, H., Cloetingh, S., Heeremans, M. and P. van der Beek, Anorogenic granites, magmatic underplating and the origin of intracratonic basins in a non-extensional setting, *Tectonophysics, 225,* 285-300, 1993 .

Tait, S. and C. Jaupart, Compositional convection in a reactive cyrstalline mush and melt differentiation, *J.Geophys. Res., 97,* 6735-6756, 1992 .

Turner, J.S., *Buoyancy Effects in Fluids,* Chapter 8, Cambridge Univ. Press, Cambridge, 1973 .

Turner, J.S., Multicomponent convection, *Annu. Rev. Fluid Mech., 17,* 11-44, 1985 .

Turner, J.S. and I.H. Campbell, Convection and mixing in magma chambers, *Earth Sci. Rev., 23,* 255-352, 1986.

Worster, M.G., Huppert, H.E.and R.S.J. Sparks, Convection and crystallization in magma cooled from above, *Earth Planet Sci. Lett. 101,* 78-89, 1990 .

Yuen, D.A., Hansen, U. , Zhao, W., Vincent, A.P. and A.V. Malevsky, Hard turbulent thermal convection and thermal evolution of the mantle, *J. Geophys. Res. , 98,* E3, 5355 -5373 , 1993.

Ulrich Hansen, Dept. Theoretical Geophysics, Earth Sciences Institute, Univerity Utrecht, TA3508, Utrecht, Holland.

David A. Yuen, Minnesota Supercomputer Institute and Dept. of Geology and Geophysics, University of Minnesota, Minneapolis, MN, 55415-1227, U.S.A.

Volume Separation in Double Diffusive Convection Systems

Anatoly S. Fradkov[1], Günter Nauheimer [2], Horst J. Neugebauer [2]

Geodydamik-Physik der Lithosphäre, Universität Bonn, Nußallee 8, 53115 Bonn, Germany

This is a numerical investigation of the influence of volume separation on the dynamic behavior of heat transfer properties as well as convective flow structures in weak nonlinear double diffusive systems. The mathematical model used here is an extension of double diffusive convection (DDC), where the separation of the heavy component is described by an additional advective flow in the direction of the volume force. This type of model was first used by Trubitsyn to investigate the convective stability of settling particles immersed in a fluid. The critical values of the thermal Rayleigh number, density number, separation velocity were determined numerically for different Lewis numbers with respect to the on-set of convection. A systematic investigation of the temporal and spatial system behavior after the first bifurcation to convection as a function of governing parameters is presen-ted. It is confirmed that separation promotes convection even under conditions where the corresponding DDC system is stable. Periodic oscillations, chaotic and steady states as well as pulsating time behavior, which is unknown in DDC, were investigated numericaly as a function of the governing parameters. In the pulsating state heat transfer remains quasi constant most of the time and is systematically interrupted by short pulsations. The dynamic layering of the flow structure, non-symmetrical concentration and temperature depth distributions, caused by the separation of the heavy chemical component, are flow structure features of the volume separation system which differ from that of DDC.

1. INTRODUCTION

Great progress has been made in understanding the dynamics of double-diffusive systems in the last few decades [*Huppert & Moore, 1976*], [*Knobloch et al. , 1986*], [*Turner, 1984*],etc... . Theoretical investigations were mostly limited to numerical considerations of a twodimensional, isoviscous, Newtonian fluid with stress free boundaries where the solute is the denser component. In this case the negative buoyancy of the solute acts as a stabilising force and the positive buoyancy of temperature is destabilising [*Huppert & Moore, 1976*], [*Knobloch et al. , 1986*], [*Veronis, 1968*].

Double-Diffusive Convection
Geophysical Monograph 94

A phase lag between temperature and solute distributions can appear if the diffusivities of temperature and solute are different. This results in phase lag time oscillation occuring after the first bifurcation to convection. Finite amplitude oscillations have been found in a series of experimental works [*Shirtclif, 1969*]. If the thermal Rayleigh number increases, oscillatory convection gives way to steady overturning motions.

In the above mentioned numerical works dynamic behavior was investigated systematically. Knobloch & et al. demonstrated that the transition to chaos takes place during a cascade of period doubling bifurcations by increasing the thermal Rayleigh number. As it was shown there "... the onset of chaos is apparently associated with homoclinic and heteroclinic bifurcation where the oscillatory branch meets the unstable portion of the steady state branch ..." [*Knobloch et al. , 1986*].

Crystal settling is likely the dominating factor defining temporary dynamics and flow structure in cooling magma chambers [*Marsh & Maxey, 1985*], [*Huppert & Sparks 1980*] , [*Martin & Nokes, 1988, 1989*] , [*Koyagushi et al. , 1990, 1993*].

In an earlier study, Marsh & Maxey (1985), Weinstein et al. (1988) investigated particle redistribution by integrating the paths of particles in a given one vortex, time independent circulation in a rectangular cell. In their model, the heavy particles did not influence convective flow structure, and only passive mixing was investigated.

A simple mathematical describtion of a convecting fluid with suspended particles which considers not only transport, but also the influence of particles on flow structure , is presented in the works of Trubitsyn & Kharybin (1987, 1988, 1991). These authors considered the stability problem of a fluid/particle mixture where the relative motions (sedimentation) of the particles is allowed. The transport of the concentration of particles is similar to that considered below except that no diffusive transport is considered.

We will briefly review the peculiar features of sedimentation in an infinite Prandtle number fluid for the isothermal case (called sedimentary convection SC) as explored by these authors. The stability of a transient, linear depth particle concentration was considered and they demonstrated that particle settling can give rise to macroscopic convective motion. The critical value of the sedimentary Rayleigh number $\mathrm{RaS} = \frac{\beta \Delta C g L^2}{k \mathrm{v}_{St}}$ for sedimentary convection (For symbols see symbol table) and free slip boundary conditions is, $\mathrm{RaS}^{crit} = -104.7$, with wave number $k = 3$. The minus sign indicates an upwardward directed gradient of particle concentration, i.e. an inverse density distribution. It follows from linear stability analysis of a transient state the suprising fact that the inverted density distribution is not always unstable when sedimentation of the heavy component is allowed. This means that the inverse density distribution can remain stabile, if RaS is below its critical value (e.g. sedimentation velocity is sufficiently high), keeping in mind that in SC there is no diffusive transport of the heavy material. This is in contrast to Rayleigh Taylor instability, where an inverted density distribution is unconditionally unstable.

This can understood intuitively if we consider an element of the fluid-particle mixture, which is initialy shifted downward in a surrounding with an upward directed particle concentration gradient . Due to sedimentation, the flux of particles into the top of the element is less than the loss through the bottom, and the fluid element becomes lighter. In other words sedimentation hinders the growth of a small disturbance and can stabilize inversed density distribution [*Trubitsyn & Kharybin, 1987*].

We are interested in the dynamics and flow structure of a fluid/particle mixture in the weak nonlinear case which can not characterized by linear stability analysis. For this purpose we present the results of numerical simulations of Thermo-Diffusive-Sedimentary Convection (TDSC) to clarify the role of governing parameters on the onset of convection and its dynamic features.

This paper is organised as follows. In section 2 we set up the mathematical problem. In section 3 we present the results of computations, where we show that increasing the settling velocity leads to convective instability. Subsequent topics covered are the influence of settling velocity, Rayleigh and density number on the dynamics of weak non-linear conditions. Lastly, we discuss possible applications and our conclusions.

2. SET UP OF THE MODEL

The equations we investigate differ from the standard double diffusion system (DDC) by the addition of flow to the concentration equation. This is why we have focused on this new form of "composition" transport.

2.1 Particle Transport Equation

We assume that particles suspended in a fluid can be described by a continuous distribution function $C = C(x, z)$, where C is the volume concentration of the particles. If there are no phase changes and the particles have constant density, the volume concentration C is a conservative quantity (For detailed discussion see appendix):

$$\begin{aligned} \frac{\partial C}{\partial t} + \vec{\nabla}\cdot(C\vec{v}_c) &= 0 \\ \vec{v}_c &= \vec{v}_m + \vec{v}_{pm} \end{aligned}$$

where $\vec{v}_c$ is the averaged particle velocity, $\vec{v}_m$ is the velocity of the center of mass of the fluid/particle mixture and $\vec{v}_{pm}$ is the relative diffusion velocity [*Ishii, 1975*], describing the motion of particles relative to the system's center of mass.

Flow, $C\vec{v}_{pm}$, caused by this relative velocity can be written as follows [*Landau & Lifschitz, 1991*], [*Ishii, 1975*]

$$C\vec{v}_{pm} = -D_{eff}\vec{\nabla}C + C\vec{v}_{rel} \quad (1)$$

The second term to the right of (1) describes the movement of particles relative to the center of mass of the fluid which is caused by gravity. In the simple case of settling spheres in dilute suspensions, we can set separation velocity $\vec{v}_{rel}$ equal to the Stokes velocity$\mathbf{v}_{St}$.

The first term (1) describes effective particle diffusion on a continuum level. Two of the contributions to this diffusion term require some explanation. In low viscosity fluids containing small particles, the effective diffusion is equal to the Brownian diffusion ($D_1 = \frac{kT}{6\pi\eta R}$) [*Landau & Lifschitz, 1991*] (here k is the Boltzmann constant). The second contribution to diffusion (D_2) is a model of dispersion in the velocity $\vec{v}_{pm}$, caused by the interactions with the other particles. According to Batchelor & van Rensburg (1986) the value of this effective diffusion is likely to be in the order of

$$D_2 \approx R \cdot |\vec{v_{rel}}|$$

where R is the particle size. D_2 is a model on continuum level in the same sense as D_1 is a model for averaged movement of Brownian particles. The validity of this approach depends on the timescale choosen. We are interested in a timescale τ_1, which is connected with the evolution of the macroscopic flow structure ($\tau_1 = \frac{L}{v}$, L=characteristic length, v = characteristic CMS velocity). The microscopic movement of the particles , movement caused by adjacent particles , occur on the timescale of $\tau_2 = \frac{R}{v_{rel}}$ (R=particle diameter , v_{rel}= separation velocity) , which is much smaler than τ_1 because $\frac{L}{R} \geq \frac{v}{v_{rel}}$ (see estimates below).

This is why we argue that particle concentration "diffuses" on the "slower" timescale τ_1.

We can use granitic melts as a reasonable quantitative estimate. The melt has a mean density of $\varrho_M \approx 2.5 \cdot 10^3 \frac{kg}{m^3}$ and a kinnematic viscosity of $\nu \approx 10^2 \frac{m^2}{sec}$. The particles have density $\varrho_C \approx 2.7 \cdot 10^3 \frac{kg}{m^3}$ and size of $R \approx 1cm$. At a temperature of 1000 ^oC the value of Brownian diffusion is in the order of $D_1 \approx 10^{-24} \frac{m^2}{sec}$ and "particle diffusion" is about $D_2 \approx 10^{-9} \frac{m^2}{sec}$. In basaltic melts "particle diffusion" can be comparable with temperature diffusivity.

For the numerical investigation presented below we use much greater values of particle diffusivity (lower Lewis number) than estimated above, to eliminate the influence of numerical diffusion on the results by long time integration of the field equations. The second reason for using lower values of the Lewis number is a property of the trivial solution which we use as a starting condition for calculations. This trivial solution has a lower boundary layer thickness proportional to 1/Le which has to be resolved properly by numerical calculation.

2.2 The whole System of Equation

The whole system of equations considers mass, momentum, energy conservation and transport of particles. In magmatic melts, the infinite Prandtle number case with free slip boundary conditions (BC) for the velocity is appropiate. Assuming Boussinesq' approximation, isoviscous Newtonian incompressible fluid, simple 2-D cartesian geometry a streamfunction approach and a standard dimensionalisation procedure we get the following system of equations (see symbol table for parameter definition)

$$\begin{aligned} \triangle \triangle \psi &= \mathrm{RaT}(\frac{\partial T}{\partial x} - \mathrm{Rp}\frac{\partial C}{\partial x}) \\ v_x = -\frac{\partial \psi}{\partial z} & \qquad v_z = \frac{\partial \psi}{\partial x} \\ \frac{\partial T}{\partial t} + \nabla \cdot T\vec{v}_m &= \triangle T \\ \frac{\partial C}{\partial t} + \nabla \cdot C\vec{v}_m &= \frac{1}{Le} \triangle C - \frac{\partial a\, C}{\partial z} \end{aligned}$$

Some limitations of the mathematical model should be mentioned. We have omitted the heat transport associated with additional particle flow $\vec{v}_{pm}C$. This term is proportianal to the concentration C and the separation velocity. Under our assumptions, these terms are small compared with other transport mechanisms.

The second type of limitations arise from using simple constant viscosity rheology for fluid-particle suspension. The rheological law for stress can be expanded if we allow the viscosity to depend on particle concentration [*Einstein, 1906*], [*Ronald et al., 1992*] and temperature.

To emphasize the differences between DDC and TDSC clearly we used Dirichlet BC at the top and bottom and v. Neumann's "no flux" BC at the side walls for the temperature and concentration. These equation were solved numerically on a staggered rectangular grid by a conservative finite-difference approach. For the elliptical part we used Fourier transformation in the x direction, and Gauss elimination "Pragonka" for the three band matrix for the Fourier coefficients. To decrease the influence of the numerical diffusion flux-corrected-transport method [*Zalesack, 1979*] was used for approximation of the advective terms. We have done several tests, to check our program:

For pure thermal convection simulation we compared the program with Benchmark [*Blankenbach et al. 1989*] The deviation between our calculations and the extrapolated value of the thermal Nusselt number for RaT $= 10^4$ on a mesh 32×32 was less than 0.6%. We have compared the results for double diffusion convection with the results in [*Hansen, 1987*]. For the case RaT $= 3 \cdot 10^4$, $R_c =$ Rp$\cdot$RaT $= 10^4$ and Le $= 3$ the deviation in the maximum of the streamfunction was less than the plot resolution (less than 3%). This support our assumption that our program is valid for simulations within the parameter range we have defined.

As mentioned previously, the numerical investigations presented below are limited to relative small values of the Lewis number, because of numerical constraints. Despite this limitation our investigation is important from a methodical point of view to illustrate which new features volume separation brings into convective dynamics of thermosolutal systems.

The time independent trivial solution of TDSC in the absence of convection ($\vec{v}_m = 0$) is:

$$T(x,z) = z$$

$$C(x,z) = \frac{\exp(a \cdot \mathrm{Le} \cdot z) - 1}{\exp(a \cdot \mathrm{Le}) - 1} \qquad (2)$$

The analytical stability analysis of TDSC is complicated because of the exponential concentration destribution and nonsymmetrical temperature and composition equations. For this reason we were obliged to perform numerical experiments studiing TDSC-system behavior in the weak non-linear case (i.e. parameters marginaly greater than critical value).

It is convenient to use the trivial solution, for this purpose, with a small (1%) harmonic disturbance in the temperature field as the initial condition. Other starting conditions would evolve to the trivial solution, if the parameters represented stable conditions for the system. By inspection it is evident that the trivial solution (2) changes when we vary parameters. Dispite this unusual property of TDSC it is still convenient to use the above trivial solution as an initial condition. But we expect that hysteresis effect, previous investigation have found in DDC [*Knobloch et al. 1986*], however occurs in TDSC also. Wether hysteresis occurs in TDSC and under which condition is beyond the scope of this paper.

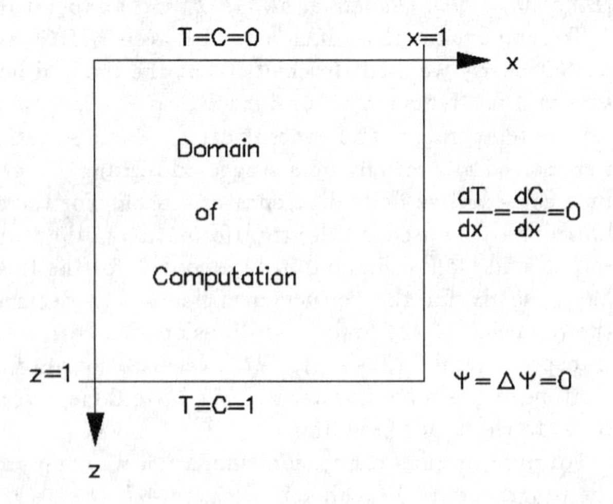

Fig. 1. Domain of computation and boundary condition

The parameters (Rayleigh-, density-, Lewis-number) for all numerical calculations were chosen, such that the DDC system (i.e. setting a to zero) is convective stable. Note that in the limit $a \to 0$ the TDSC-equation tends to the DDC-equation and that the trivial solution goes to a linear one for both temperature and concentration. A square cell was used for all numerical simulations.

3. NUMERICAL RESULTS

In the first part we describe the influence of settling velocity, thermal Rayleigh number and density number on time averaged heat transfer properties. In the second part we present temporary dynamics in detail.

3.1 The Influence of the Governing Parameters on the Time Averaged Heat Transfer

3.1.1 Influence of settling velocity. To clarify the influence of the dimensonless settling velocity, a, we progressively increased its value while maintaining the other parameters constant. If the system remains convectively stable, only from depth dependent trivial solution sets up in the cell, after the transition time of decay of the initial disturbance. If the system is convectively unstable, the initial disturbance does not decay and different types of flow structure are observed by numerical time integration.

Numerical results of the settling velocity influence on time and volume averaged thermal Nusselt are presented in figure 2.

Due to the exponential depth-dependent distribution of particle concentration in the trivial solution (2), the contribution of the particles to buoyancy forces in the upper part of the cell diminishes in the trivial solution with an increase in the product ($a \cdot$Le). That is why the critical value of settling velocity, corresponding to the onset of convection (Fig. 2), decreases with increasing Lewis number, if other parameters are fixed. In figure 2 it is obvious that the first bifurcation to convection (by constant Rayleigh and density numbers) is determined approximately by the product $a \cdot$ Le for different Rp values.

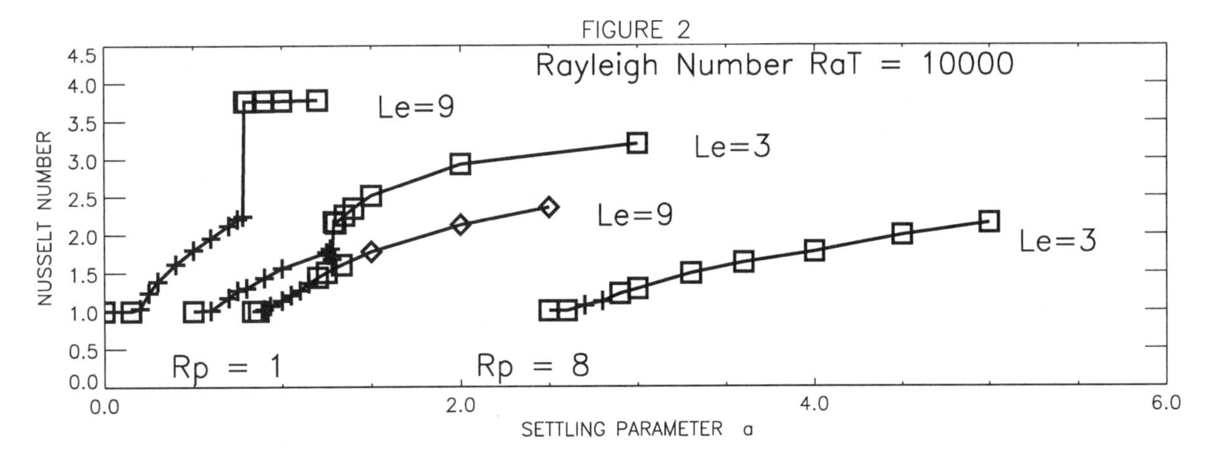

Fig. 2. The volume averaged thermal Nusselt number plotted against the settling parameter for two different Lewis Le = 3 , 9 and density numbers Rp = 1 , 8. The symbols denote numerical experiments: $\in$ steady state or time independent Nusselt number, + oscillations and $\ni$ pulsation state after first steady state. Oscillatoric and pulsating time behavior is characterized by the time averaged Nusselt number values: 1) - first oscillatory phase, 2) - transition zone, 3) - first steady state, 4) - pulsating phase

Damping by denser composition is the second factor we investigated, which strongly influences the onset convective instability, by changing the density number. If damping increases, the onset of convection begins with greater values of the settling velocity (Fig. 2), which is clear and does not require additional comments. Numerical simulations have shown that the onset of the TDSC at a thermal Rayleigh number of 10^4, Lewis numbers in the range of $3-9$ and density numbers of $\mathrm{Rp} = 2,4,8$ can be parametrized by a simple power law dependence (see table 1 where the product $a_{crit} \cdot \mathrm{Le}$ for different Lewis number are presented).

$$a_{crit} \cdot \mathrm{Le} = 3.4 \cdot \mathrm{Rp}^{0.41}$$

As shown in figure 2, the Nusselt number increases progressively by increasing a after the first bifurcation to convection. In the TDSC model the sedimentation is responsible for the removal of the heavy component from the inner part of the cell. We can expect that the Nusselt number will tend to the value of a pure thermal convection state (4.8 for $\mathrm{RaT} = 10^4$), if the settling velocity tends to infinity, which would remove the entire heavy component from the cell. However, we can not confirm this statement in terms of numerical simulations with high settling velocity, because one has to resolve the lower concentration boundary layer, which decreases linearly with increasing settling velocity.

Table 1

onset of convection ($a_{crit} \cdot \mathrm{Le}$), $\mathrm{RaT} = 10^4$

	Rp = 2	Rp = 4	Rp = 8
Le = 3	4.503	6.3	7.95
Le = 6	4.5	6.15	7.95
Le = 9	4.5	6.15	7.9499

Comparing the results for the two density number values presented on figure 1, you see that the route to the simple thermal convection state depends strongly on the damping influence of the heavy component. For Rp = 1 we observed a very narrow transition zone between the periodical and steady state. This zone appears for $a \approx 0.75(1.29)$ and Le = 9(3). In both cases the width of the transition zone is very narrow. The documented jump of the thermal Nusselt is 1.6(0.6) for Le = 9(3). For Le = 9 the thermal Nusselt number reaches almost 3.8, which is about 80% of the value for pure thermal convection.

After the large jump through the transition zone to steady state a flatter increase of the Nusselt number as a function of settling velocity is observed. This indicates that the Nusselt number vs settling velocity dependence is very sensitive to flow structure changes, but can have a relative weak gradient if the type of the flow structure does not changes.

We have not observed such a striking jump in the Nusselt number vs settling velocity dependence by increased density numbers Rp = 4, 8, where damping due to the

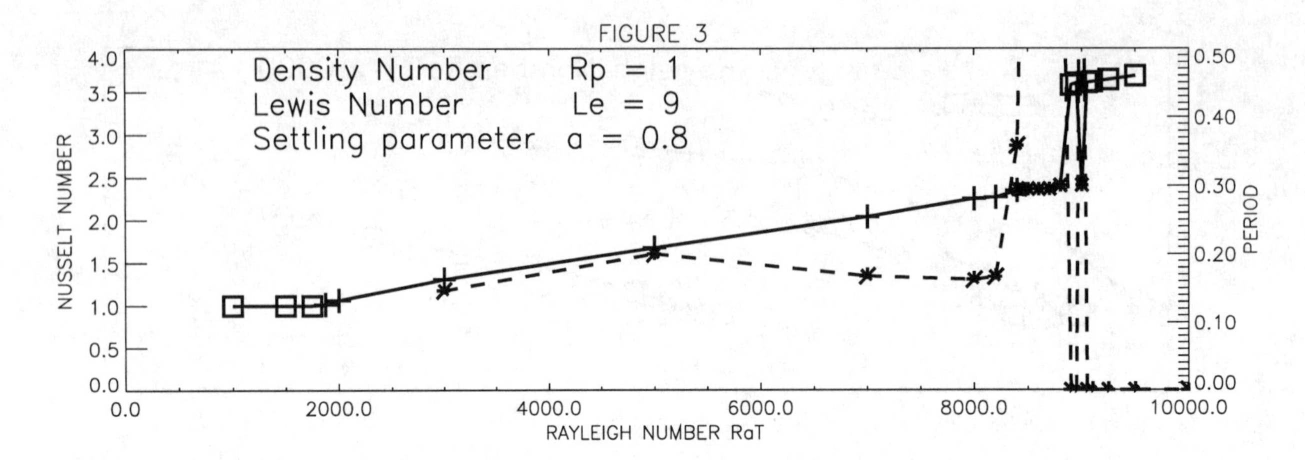

Fig. 3. The dependence of the time averaged thermal Nusselt number and characteristic time period of oscillation as a function of the thermal Rayleigh number. Symbols have the same meaning as in figure 1 . The dashed curve refers to the period of oscillation.

heavy component is increased. Thus strong damping by high density numbers has a dominant influence on the averaged heat transfer properties and hinders drastic change by a rebuilding of the flow structure. The numerical experiments have shown that the tendency to "jumping" becomes more pronounced by the an increase of the Lewis number, or if the diffusive mixing decrease (Fig. 2).

It is generaly known that the transition between different subsequent flow structures can be accompanied by changes of the dependent Nusselt number vs an independent parameter (thermal Rayleigh number, density number etc). It was shown by Huppert & Moore (1976) for DDC models that the thermal Nusselt number can increase significantly in the transition from an oscillatory to a steady state flow system. Our results confirm this behavior for the TDSC model.

3.1.2 Influence of heating. The influence of heating, as a function of the Rayleigh number, was investigated by starting with the disturbed trivial solution again. This differes from that of previous paragraph, as the initial condition is the same when investigating the influence of the Rayleigh number, because the trivial solution does not depend on Rayleigh and density numbers.

Figure 3 represents the Nusselt as a function of the Rayleigh number for numerical solutions with $\mathrm{Rp} = 1$, $\mathrm{Le} = 9$ and $a = 0.8$.

After the first bifurcation value ($\mathrm{RaT} \approx 1875$) simple oscillation with one local minimum (maximum) per period appears up to $\mathrm{RaT} \approx 8200$. At $\mathrm{RaT} = 8200$, a period doubling bifurcation occurs, which then changes pattern at $\mathrm{RaT} = 8400$ and additional low frequencies appear. (For details of the time behavior see figure 5 and corresponding description below). A transition window is then reached at $\mathrm{RaT} \approx 8410$ where temporary chaos occurs. We do not observe any periodical state in the transition window as was found in DDC [*Knobloch et al. 1986*] However, we did find two steady states within the transition zone at Rayleigh numbers ($\mathrm{RaT} = 8900$, 8950). For Rayleigh values greater than 9050 the oscillating chaotic state became unstable and steady (monotonic) state was observed. Similar to figure 2, the Nusselt curve shows a strong jump in the heat transfer during transition from oscillatory to steady state, resulting from rebuilding of the flow structure.

Our calculations show that the route to steady state from chaos differs from that observed by DDC. In contrast to Huppert & Moore (1976) where the cascade of period doubling bifurcations was observed, we found only one period doubling and the time period of the oscillations increases in our case also not monotonic.

3.1.3 Influence of the density number. It is obvious that the damping influence of the heavy component decreases with decreasing density number. Formally, the motions are determined by thermal effects only when Rp tends to zero. The influence of decreasing density number on time averaged heat transfer is summarized on figure 4 for two parameter sets of the Lewis number and settling velocity.

Due to a steeper initial concentration gradient in the case $\mathrm{Le} = 9$, $a = 1.5$ (Fig. 4) oscillating convective motions occurs at a greater value of Rp, (85 —30). The first steady state occurs after this when the time period of oscillations (dashed curve figure 4) becomes zero. For even lower density numbers the steady state becomes unstable and a new periodical, pulsating state

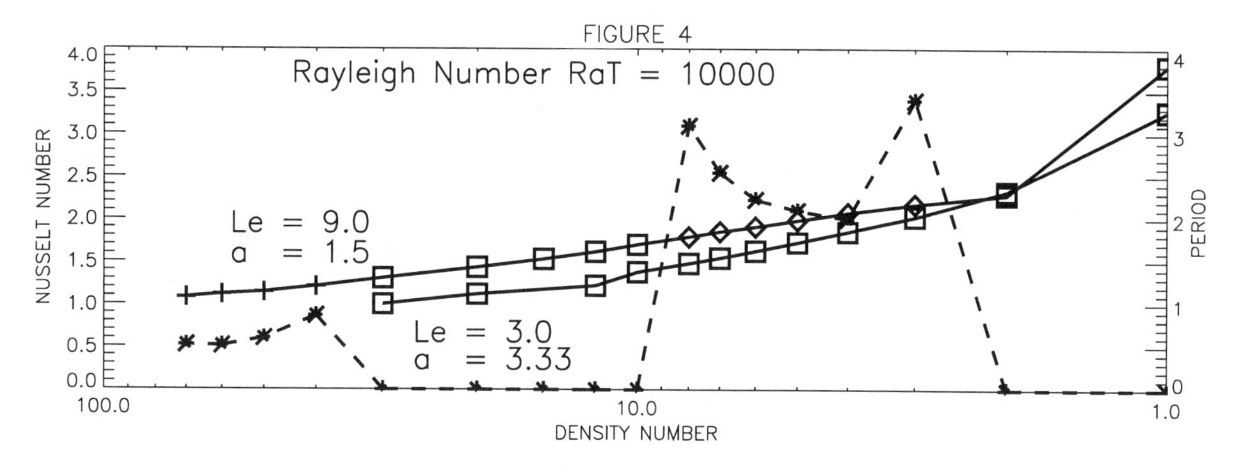

Fig. 4. The thermal Nusselt number as a function of density number for two parameter sets (solid lines) and the period of oscillations for the case Le $= 9$, $a = 1.5$ (dashed line).

is reached between Rp $\approx 8 - 3$. This periodical regime is characterized by a quasi steady state heat transfer, which is interrupted by short pulsations (Fig. 6). In quasi steady state, the Nusselt number does not change with time and the flow structure is characterised by one vortex. The single vortex flow structure is periodicaly disturbed by the apearence of a small vortex with opposite spin which grows and suppresses the previous one. The periodicity of flow pattern change is very stable and does not alter when the numerical scheme is improved by halving of the mesh spacing. A second steady state phase is reached with a decrease of the density number. The system now becomes sensitive to the change in density number indecated by the comparatively steep gradient in the Nu(Rp) curve in the Rp-range $1-2$. It should be emphasized that the transition from pulsating to steady behavior occurs within a very narrow zone of density numbers range.

The second solid curve in figure 4 corresponds to system parameters with greater diffusion. The two solid curves represent a factor of 3 in the difference of diffusion influence. This leads to the disapearence of any time oscillations for the second parameter set and steady state convection only.

Analysis of these overturning steady state flow structures show that for greater density number a two layer convection pattern appears (see figure 6). Relatively intensive convective motions exist in the upper part of the cell where the concentration of the heavy component is small.

Layering at the greater values of the density numbers can be rationalized physicaly. Due to the sedimentary flow in the concentration equation, the heavy component accumulates at the lower boundary. If the density number is high enough this lower layer has negative buoyancy and it does not take part in the overall convective motion. In this case, we have nearly pure thermal convection in the upper convective part of the cell, where the contribution of concentration to buoyancy is very small due to small particle content. In general, the Rayleigh number must exceed a specific value in order to ensure thermal convection in the upper layer and overcome the viscous friction with the lower stagnant layer. The numerical experiments have shown that the two layer convection appears first, while decreasing the density number. I.e. the thermal Rayleigh number first becomes sufficient for two layered convection. A further decrease of the density number also decreases

Table 2

averaged Nu , $vrms$ as a function of Rayleigh number

Rp $= 1$ Le $= 9$ $a = 0.8$

RaT	3000	8200	8400	8600	8800	8950	9000	9050
Nu	1.35	2.27	2.34	2.37	2.41	3.5	2.40	3.5
$vrms$	5.64	13.9	14.3	14.8	15.2	25.5	15.1	25.6

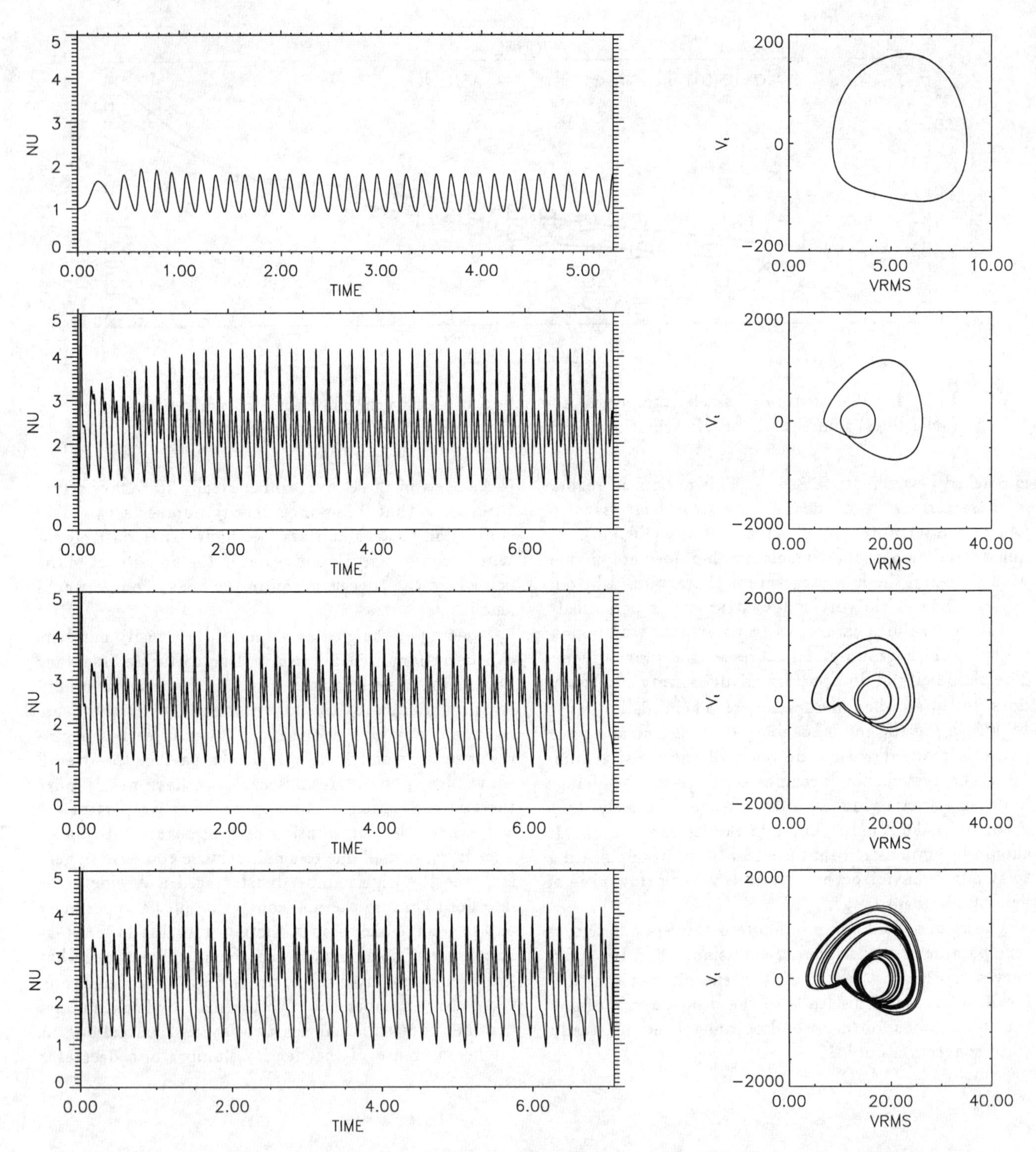

Fig. 5.a. Time plots of the volume averaged thermal Nusselt number for different density numbers (Rp = 1 − 60) and a phase map of the mean square velocity. The transition state connected with decay of the starting condition is not presented on the phase map as the starting conditions are "forgotten". Only the values corresponding to dimensionless time greater than 2 units are shown in phase map, mean square velocity (horizontal axes) and its time derivative (vertical axes). The other parameters are ($a = 1.5$, $\mathrm{Le} = 9$, $\mathrm{RaT} = 10^4$), for corresponding values of the density numbers, time averaged Nusselt numbers and the root mean square velocity see table 3.

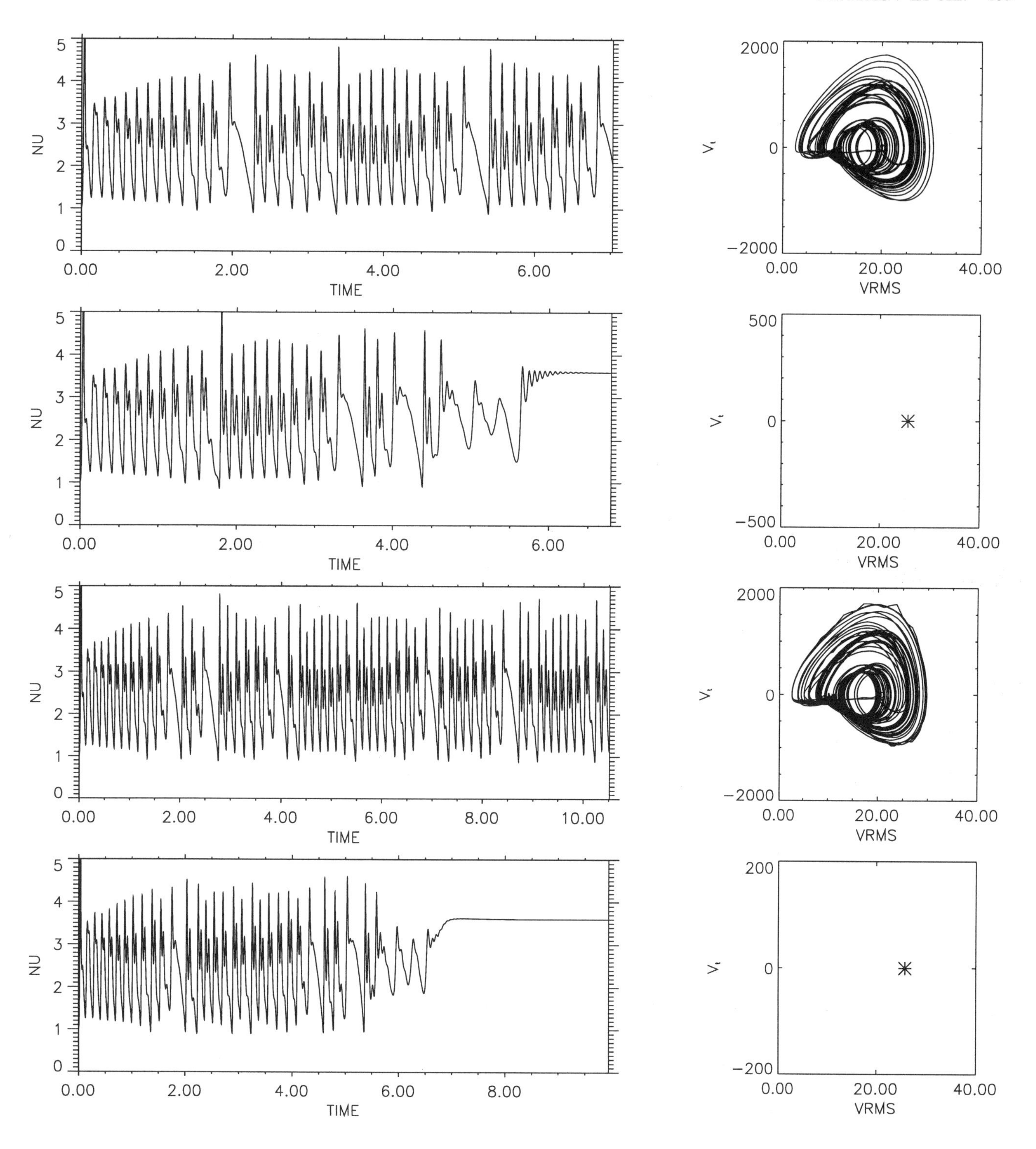

Fig. 5.b. Continuation of Fig. 5a.

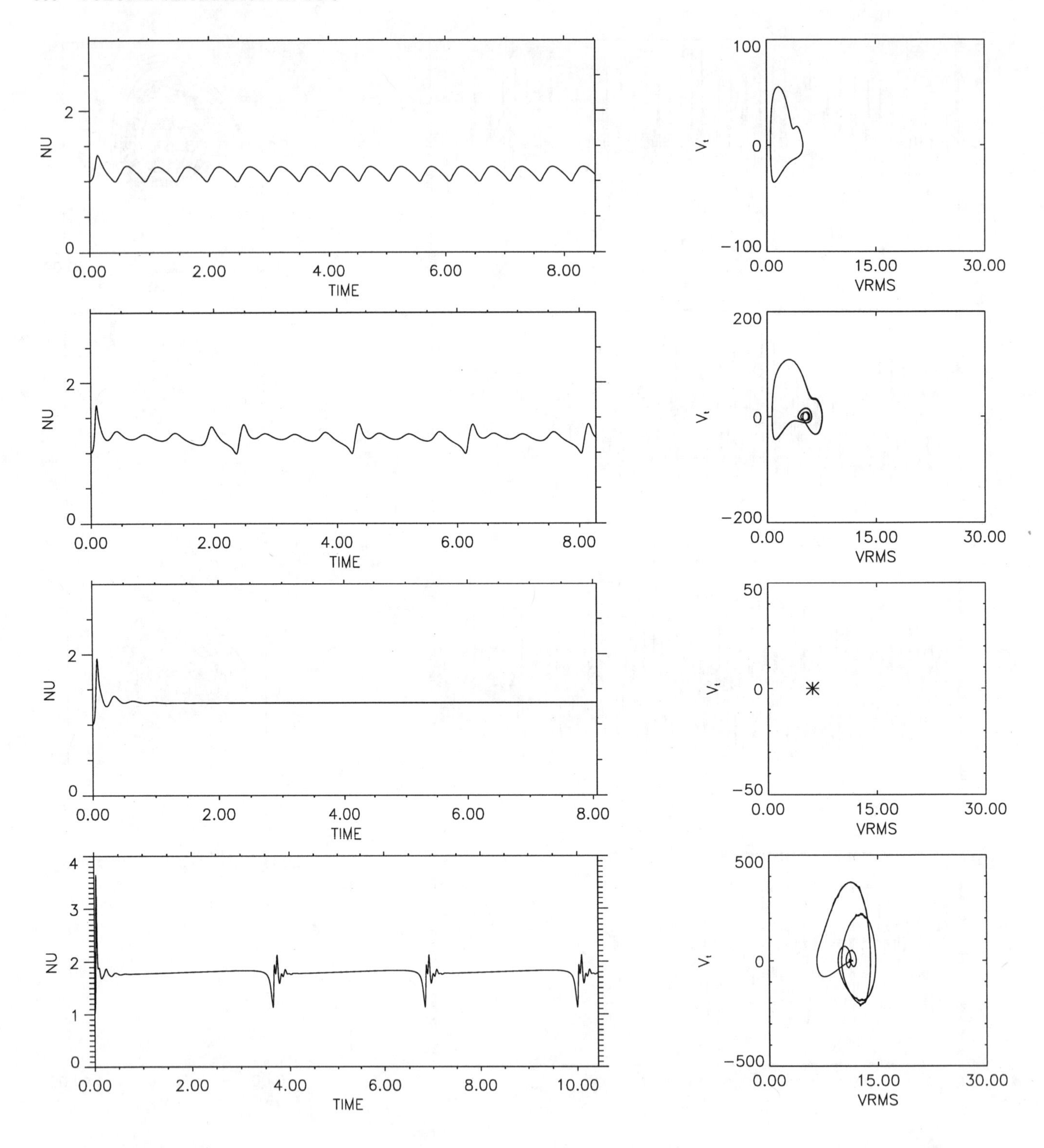

Figure 6.a. Time plots of the volume averaged thermal Nusselt number and phase space of the volume averaged mean square velocity presented as a function of time for different thermal Rayleigh numbers RaT. The other parameters for these calculations are(Rp=1,Le=9, a=0.8) and RaT increases progressively from top to bottom. The corresponding values of the Rayleigh, time averaged Nusselt numbers and the time averaged values of mean square velocity are presented in table 2.

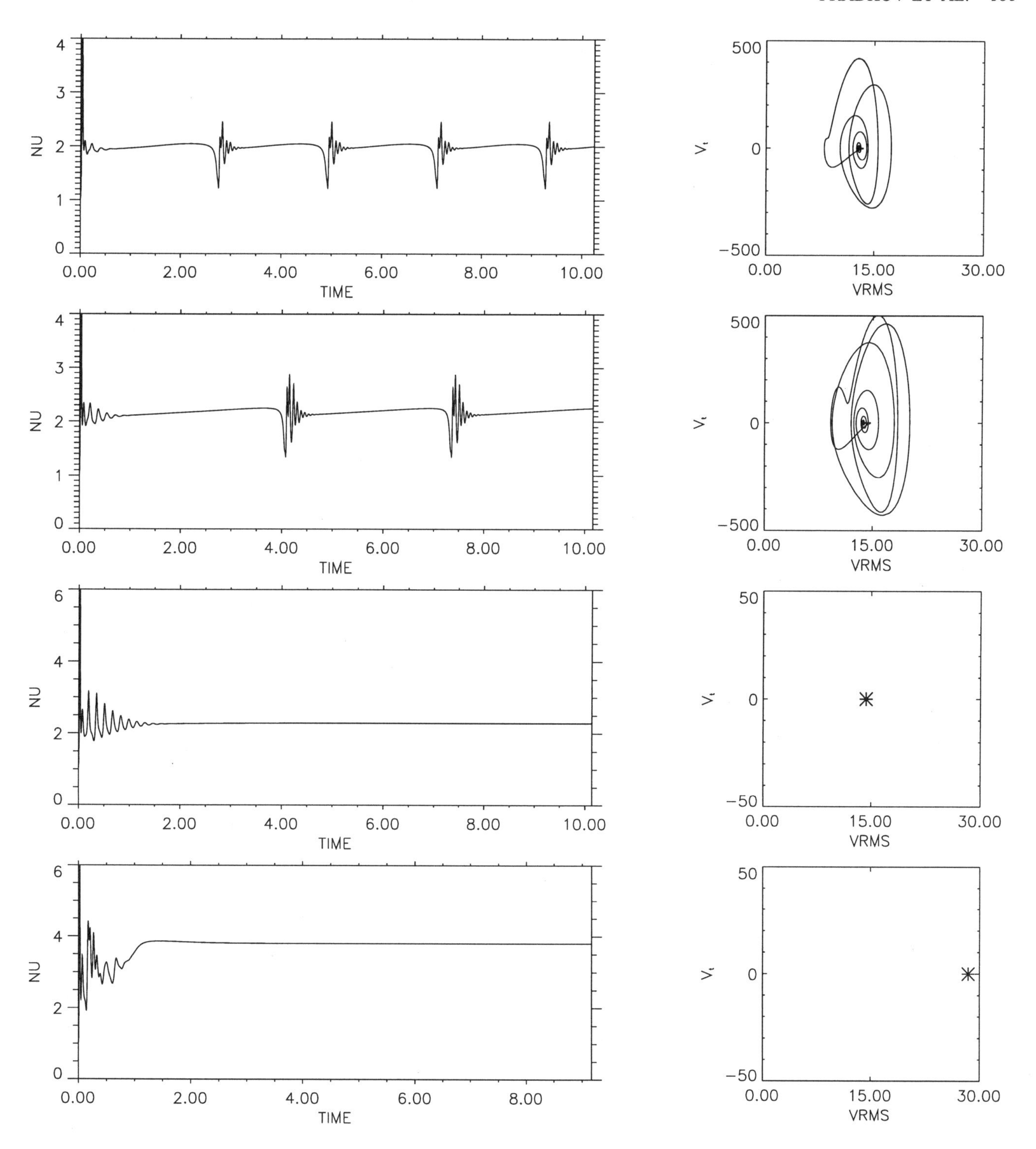

Fig. 6.b. Continuation of Fig. 6a

the damping due to the heavy component and the value of the thermal Rayleigh number becomes sufficient for single layer flow. For one layer flow, the influence of concentration on buoyancy is not negligible in the whole cell i.e. the motion are closer to DDC type, where symmetric boundary layers appear at the upper and lower boundary (see figure 7).

If the flow structure is layered, temperature and concentration distributions in the lower stagnant layer are more or less laterally homogeneous and convection does not penetrate into the lower boundary of the cell (figure 7). The horizontal temperature depth distribution in the two-layered case exhibits two distinct boundary layers where the upper one is essentially smaller than the lower one. In a thermal sense, the lower layer hinders convective heat transport and works as thermal screen. Calculations have shown that decreasing the density number also decreases the thickness of the lower layer, but the thermal Nusselt number increase is relatively weak if the flow structure remains layered due to this screening effect. Density numbers smaller than 3 cause the lower layer to disappear, thus the penetration depth of convective motion increases and the system becomes increasingly sensitive to changes in the density number. Obviously, there are two zones to distinguish. For Rp changes form 60 to 3 heat tranfer is not very sensitive to changes the density number. If Rp becomes less than 3, the heat transfer becomes more sensitive to a change in particle concentration. We admit that similar changes of the sensitivity by transition from two layers to one layer motion would also be possible for other parameter sets, nevertheless we presented numerical investigations on the influence of the density number only in a limited parameter range ($a \cdot Le$ about 10, RaT $= 10^4$, Rp $= 1 - 60$).

It should be mentioned that the horizontal-averaged concentration depth distribution in TDSC has a fundamentaly different nature than the temperature depth distribution. For a layered flow structure, the concentration is practical zero in the convective part of the cell. On the other side, the concentration increases strongly with depth in the lower layer. This asymmetry in depth concentration distribution is also a peculiarity of the TDSC model and is based on aditional sedimentary flow of the particles, which inturn makes the temperature and concentration transport equation nonsymmetrical. During the change of the flow structure from two to one layer, a concentration boundary layer also appears near the upper boundary. However, the concentration gradient near the the lower boundary is always stronger than at the upper boundary.

3.2 Dynamic Behavior of the TDSC

In the previous seccion we discussed time averaged heat transfer and the influence of system parameters. We now present some of the numerical results as a function of time.

3.2.1 Influence of heating. The following discusion illustrates the influence of change in the thermal Rayleigh number on the dynamics of heat transfer for (Rp $=$ 1 , Le $=$ 9 , $a = 0.8$), figure 5. For these parameters, the system is convectively unstable at a Rayleigh a slightly less than 1875 (Fig. 3). The transition from oscillating to steady state through chaos by increasing Rayleigh number is illustrated on figure 5

Periodic time behavior near the first bifurcation to convection is presented in panel 5a.1 . The periodic state is characterised by a limit cycle with characteristic period of about 0.17. The increase of RaT , (panels 5a.2,5a.3), leads to more complicated temporary dynamics (Fig. 3). The phase map shows internal loops but time behavior remains periodic. At RaT $=$ 8200 we found a state with two maximums per period, similar to DDC simulations [*Huppert & Moore, 1976*]. At RaT $=$ 8400, low frequency oscillations appear and the form splits (panel 5a.3). A chaotic state is reached at RaT $=$ 8410. Panels 5a.4- 5b.3 correspond to the transition zone between periodic oscillations and steady state. The transition to chaos is clearly seen in panel 5b.1 . For RaT less than 8600, we did not observe chaotic bursts and laminar states between oscillations. These bursts become more frequent with increasing Rayleigh number (panels 5b.1 and 5b.3). At RaT 8950, which lies within the transition zone, the oscillatory state becomes unstable and a steady state occurs (panel 5b.2).

The efficiency of heat transport in this steady state is about 45% larger than for the higher value of RaT $=$ 9000, where stochastic behavior is again observed (5b.3). The final panel (5b.4) corresponds to the stable steady state equivalent of the third section of the curve 1 in figure 3. The value RaT=9050 is close to the end of transition zone where stochastic behavior becomes unstable. By comparison, heat transfer is more efficient in the steady state than in the stochastic state. This is valid up to RaT $= 10^4$ which is the upper limit of our simulation.

3.2.2 Influence of density number. The phenomena connected with a change in density number between 1-60 are presented in figure 6 to clarify the temporary dynamics of the heat transfer, when others parameters were fixed. We refer to the case of figure 4 (RaT $=$ 10^4 , Le $=$ 9 , $a = 1.5$). The time averaged values

Fig. 7. Flow structures for steady state thermo-diffusive-sedimentary convection as a function of the density number. The parameters for these calculations are $a = 3.333$, $\mathrm{RaT} = 10^4$, $\mathrm{Le} = 3$ and variable density number values $\mathrm{Rp} = (1, 2, 4, 8, 16)$.

Table 3

averaged Nu, $vrms$ as a function of density number
RaT $= 10^4$ Le $= 9$ $a = 1.5$

Rp	1	2	3	5	8	30	40	60
Nu	3.8	2.28	2.17	1.98	1.79	1.30	1.21	1.18
$vrms$	28.45	14.34	14.15	13.05	11.22	6.14	4.96	3.6

of the Nusselt numbers and mean square velocities for the different density numbers corresponding to panels 6a.1- 6b.4 in figure 6 are listed in table 3. At Rap 60, harmonic oscillations occur after the transition state, close to the first bifurcation (panel 6a.1). When the density number decreases, the period of oscillation grows (panel 6a.2). Further decrease of the density number, the phase map becomes compressed to a single point and the steady state becomes stable (panel 6a.3). The first steady state phase remains stable until the density number approaches Rp $\approx$ 9. Decreasing Rp further, the steady state becomes unstable and a new oscillating state forms. Panels 6a.4-6b.2 show the pulsating time behavior which occurs after the first steady state phase. In this state, the thermal Nusselt number remains more or less constant with periodic interruptions by pulsation. The duration of these pulsations is essentially less than that of the quasi steady state. The minimum duration of the quasy steady state occurs at Rp $\approx$ 5 , panel 6b.1. The magnitude of the oscillations reaches its maximum at density number 3.

These periodic pulsating phases can be understood hydrodynamically by the weak imbalance between reentraintment of the heavy component in convective motions caused by diffusion and its removal through the lower boundary of the cell caused by sedimentation. We believe that this imbalance is relative small near the boundaries of the pulsating window, in figure 4, and that it lengthens the duration of quasi steady state phases. The imbalance attain a maximum in the middle of the pulsating window and the duration of quasy steady state phases reaches a minimum. At a density number of 2 (Fig. 4), the second "sensitive" branch of the steady state is reached, with a stronger Nu(Rp) dependence.

It is physically clear that the periodicity of the oscillations and the flow structure depends on geometric parameters. All our numerical experiments were done in square cell. We have not presented results on convective dynamics by greater aspect ratio. From the results of linear stability analysis [*Trubitsyn & Kharybin, 1987, 1988, 1990*] we know that TSC increases the tendency to horizontal spreading of the convective vortex, which is a characteristic property of thermal convection with sedimentation.

4. CONCLUSIONS

The main objective of our numerical investigations was to clarify the influence of governing parameters on TDSC and to emphasize the difference between TDSC and DDC.

Differences in the stability, temporary dynamics and flow structures of TDSC and DDC are obviously caused by the sedimentary flow of particle transport, which makes the temperature and composition equations asymetrical. It is important to note that in DDC it is possible to interchange temperature and composition and get the original equations after renaming the constants. This is not the case for TDSC which is why we investigated the weak non-linear case in considerable detail.

The additional separation term in the solute transport equation is responsible for removal of the heavy component from the core of the cell and leads to the formation of weak upper and stronger lower boundary layers. This first aspect makes TDSC convectively unstable, even for large damping caused by either the density number or by large compositional diffusion. All calculations presented above where done under conditions where the corresponding DDC system remains stable.

The following qualitative conclusions can be drawn from our numerical investigation of TDSC in the weak nonlinear case:

- The separation velocity a, is a critical parameter of convective stability. The onset of convection is determined by the product $a \cdot$ Le, not only by the usual parameters such as density number (Rp) and thermal Rayleigh number (RaT). Even in the case of the strong damping due to density and diffusion (figure 1), convective motion appears for relative small settling velocity values, which is only a small percent of the mean square values of thermal convection.
- Sedimentation flow of the heavy component , used in this work, promotes layering of the flow structure in vertical direction, where relatively intensive convective motions appear in the upper part of the cell while weak motions induced due to viscous frictions between two layers occurs in the lower part of the cell.

- Time averaged heat transfer depends strongly on the type of convective motion occurring in the cell: A considerable jump of the efficiency of the heat transfer was to be caused by rebuilding of the flow structure from one to two layer convection and from time dependent to steady state flows as a function of density, Rayleigh and settling parameter. Similar behavior was also observed in DDC [*Knobloch et al. 1986*].

- The numerically observed different dynamic states of TDSC are:
  - periodic oscillations;
  - periodic "laminar" pulsations;
  - stochastic time behavior;
  - steady state.

  The increase of the Lewis and decrease of density numbers amplify the oscillations and broaden the parameter ranges, where these oscillations occur.

- Dynamic time behavior of TDSC differs qualitatively from that of DDC, mainly through the presence of a pulsating state. Within this state, the quasi steady heat transfer is periodically interrupted by short pulsations. The maximum fraction between the duration of pulsations and following quasi steady state is about 0.25. This state was investigated as a function of Lewis and density numbers. Increasing these two numbers broadens the parameter range where pulsating state can be observed.

- A transition window was observed as the thermal Rayleigh number was increased between periodic time behavior and steady state, similar to DDC [*Knobloch et al. ,1986*]. Within this transition window we found only one period doubling bifurcation.

We used an infinite Prandtle number for our simulation and we feel it may be relevant for magmatic melts. The numerical experiments were mainly limited to a relative small Lewis number for numerical reasons. The applicability of TDSC to the real systems is of course limited by the constant parameters and simple boundary conditions we used.

Symbol Table

physical parameters	
thermal expansitivity of the mixture	α
magma density	ϱ_f
crystal density	ϱ_c
chemical expansitivity	$\beta = \frac{\varrho_c - \varrho_f}{\varrho_f}$
kinematic viscosity	ν
layer thickness	L
acceleration of gravity	g
thermal diffusivity	k
Stokes' velocity	$\mathbf{v}_{St} = \frac{\vec{g}}{g} \frac{2(\varrho_{cryst} - \varrho_o)R^2 g}{9\eta}$
time unity	$v_0 = \frac{L^2}{k}$
velocity unity	$t_0 = \frac{k}{L}$
effective diffusivity	$D = \gamma R \left\| v_{Stokes} \right\|$
dimensionless parameters in TDSC	
Rayleigh number	$\mathrm{RaT} = \frac{\alpha \Delta T g L^3}{k\nu}$
density number	$\mathrm{Rp} = \frac{\beta \Delta C}{\alpha \Delta T}$
Lewis number	$\mathrm{Le} = \frac{k}{D}$
dimensionless settling velocity	$a = \frac{v_{Stokes} L}{k}$
Nusselt number	$Nu = \int_0^1 dx [\frac{\partial T}{\partial z} - v_z \cdot T]$
root mean square velocity	$vrms = \sqrt{\int_0^1 \int_0^1 dx dz [v_x^2 + v_z^2]}$

APPENDIX

This is a more detailed derivation of the TDSC-model. For the more general context we use the phenomenological approach given in the appendix of McKenzie (1984), because this is more widely recognized in the geoscience community.

The considered system is a high viscous fluid with suspended settling particles. The volume concentration of particles is less than 1% and we assume that the particles do not grow or shrink. As we are interested in the evolution of the system as a whole, we restrict ourselfs to phenomena where the dynamics of fluid and particles can be adequately described by an effective medium. In this case, a concentration (one velocity) model is generaly accepted, which will now be derived using the more general (two velocity) model given in McKenzie (1984). In the following we use the notation given in McKenzie (1984).

A.1 Mass Conservation

In a concentration model, mass-distribution is described by the mixture mass density

$$\varrho_m = \phi\varrho_f + (1-\phi)\varrho_s$$

and the crystal-mass density $(1-\phi)\varrho_s$ of the mixture. In the text we defined $(1-\phi)$ as $C \equiv$ volume concentration of crystalls.

The evolution equation governing ϱ_m is obtained , by adding the mass equation of the fluid and particle phases.
Mass Conservation of fluid phase:

$$\frac{\partial \varrho_f \phi}{\partial t} + \nabla \cdot (\varrho_f \phi \vec{v}_f) = \frac{D_V M}{Dt}$$

Mass conservation of particle phase:

$$\frac{\partial \varrho_c(1-\phi)}{\partial t} + \nabla \cdot (\varrho_c(1-\phi)\vec{v}_c) = -\frac{D_V M}{Dt}$$

Introducing the center of mass-velocity $\vec{v}_m$ by the relation

$$\varrho_m\vec{v}_m = \varrho_f\phi\vec{v}_f + \varrho_c(1-\phi)\vec{v}_c \qquad \begin{cases} \vec{v}_f = \text{fluid velocity} \\ \vec{v}_c = \text{solid phase-velocity} \end{cases}$$

leads to the conservation equation of the mixture-mass.

$$\begin{aligned} \frac{\partial \varrho_m}{\partial t} + \nabla \cdot (\varrho_m \vec{v}_m) &= 0 \quad \leftrightarrow \\ \frac{\partial \varrho_m}{\partial t} + \vec{v}_m \cdot \nabla \varrho_m &= -\varrho_m \nabla \cdot \vec{v}_m \end{aligned}$$

The variation of mass ϱ_m from the mean value ϱ_o is assumed to be small and a linear function of temperature and the volume concentration of particles

$$\varrho_m(x,z) = \varrho_o \cdot (1 - \alpha(T(x,z) - T_o) + \beta C(x,z))$$

(for symbols see below). Because of the assumption $\beta \geq 0$ particle distribution is the heavy component of the system. From the equation for mixture mass conservation and assuming small mass variations it follows, that the velocity of the center of mass is solenoidal

$$\nabla \cdot \vec{v}_m \approx 0 \quad .$$

In the case of constant fluid-phase and particle-phase density the analysis can be done in terms of the "mixture volume flux" $\vec{j}_m = \phi\vec{v}_f + (1-\phi)\vec{v}_c$, which is solenoidal independendly of the density difference $\varrho_c - \varrho_f$. Doing so, the constitutive equations (see below) have to be reconsidered.

The conservation equation of the particle phase $(1-\phi)\varrho_s$ in the TDSC-model is obtained from the mass conservation of the particle phase, the definition of relative diffusion velocity $\vec{v}_{pm} = \vec{v}_c - \vec{v}_m$ and the assumption that no phase-changes occur ($\frac{D_v M}{Dt} = 0$)

$$\frac{\partial \varrho_c(1-\phi)}{\partial t} + \nabla \cdot (\varrho_c(1-\phi)\vec{v}_m) = -\nabla \cdot (\varrho_c(1-\phi)\vec{v}_{pm})$$

We further assume that ϱ_c is constant (i.e. a particle species of specific density) and in order to build up a concentration model we have to make model assumptions for the flow $(1-\phi)\vec{v}_{pm}$. [*Ishii, 1975*], [*Landau & Lifshitz, 1991*]

$$(1-\phi)\vec{v}_{pm} = -D\nabla(1-\phi) + (1-\phi)f(1-\phi)\mathbf{v}_{St}$$

where $\mathbf{v}_{St}$ is the Stokes settling velocity of a single particle in an unbounded fluid and f is a function (empirical determined), which takes into account the interaction with adjacent particles in not infinite dilute suspensions [*Davis & Acrivos, 1985*]. This function is set to unity, because we consider only very small particle content here.

In the notation given in the text we obtaine

$$\frac{\partial C}{\partial t} + \nabla \cdot (C\vec{v}_m) = \nabla \cdot D\nabla C - \nabla \cdot C\mathbf{v}_{St}$$

(For a detailed discussion on diffusivity see text.)

A.2 Momentum Equation

Summing up the momentum equation for the solid and

the fluid phase, we get the momentum equation for the mixture $\varrho_m \vec{v}_m$.

Momentum equation for the solid phase:

$$-(1-\phi)\varrho_c g\delta_{i3} + I_i + \frac{\partial}{\partial x_j}\left[(1-\phi)\sigma^s_{ij}\right] = 0$$

Momentum equation for the fluid phase:

$$-\phi\varrho_f g\delta_{i3} - I_i + \frac{\partial}{\partial x_j}(\phi\sigma^f_{ij}) = 0$$

(see also discussion in McKenzie (1984) leading to (16,17), which also fits our situation)

The momentum equation for the mixture is now:

$$\begin{aligned} -\left[(1-\phi)\varrho_c + \phi\varrho_f\right]g\delta_{i3} + \frac{\partial}{\partial x_j}\left[(1-\phi)\sigma^s_{ij} + \phi\sigma^f_{ij}\right] &= 0 \\ -\varrho_m g\delta_{i3} + \frac{\partial}{\partial x_j}\sigma^m_{ij} &= 0 \end{aligned}$$

The use of this last equation means that we assume that the mixture's behavior can be described by an effective material. This is a usual accepted approach, if the particle and fluid phase are closely coupled.

In order to make a model for the effective mixture stress tensor σ^m_{ij}, we assume that "The behavior in shear is likely to be linear if the stress is sufficiently small, and hence the stress can be written as (Landau & Lifshitz, 1959)" [*McKenzie, 1984*]

$$\sigma^m_{ij} = -P\delta_{ij} + \zeta\delta_{ij}\frac{\partial v^l_m}{\partial x_l} + \eta_{eff}\left[\frac{\partial v^i_m}{\partial x_j} + \frac{\partial v^j_m}{\partial x_i} - \frac{2}{3}\delta_{ij}\frac{\partial v^l_m}{\partial x_l}\right]$$

Because $\vec{v}_m$ is solenoidal, the second and the last contribution to σ^m_{ij} vanish. The effective viscosity for the case of dilute suspension of spherical particles can be calculated theoreticaly [*Einstein, 1906*], [*Batchelor, 1970*]

$$\eta_{eff} = \eta_f\left(1 + 2.5(1-\phi)\right)$$

For suspensions with greater volume fraction of particles ($(1-\phi) = C \leq 0.5$) an empirical law for the $\eta(C)$ dependence is given by $\eta(C) = \eta_f \cdot (1 - \frac{C}{C_m})^{-1.81}$ [*Ronald et al. 1992*].

We neglect the dependence of effective viscosity on particle content in the following, because $(1-\phi) \ll 1$ so that we obtain the simple Stokes equation with constant viscosity (in vector notation).

$$0 = -\nabla p + \eta \triangle \vec{v}_m + \varrho_m \mathbf{g}$$

A.3 Energy Equation

For energy conservation we adopted the integral balance law for the energy given in McKenzie (1984):

$$\begin{aligned} \frac{d}{dt}\int_V \left(\varrho_c E_s(1-\phi) + \varrho_f E_f\phi\right) d^3r &= \int_S k_T \nabla T \cdot d\vec{f} + \\ \int_V H\, d^3r + \int_S \sigma^s \cdot \vec{v}_c(1-\phi)\cdot d\vec{f} &+ \int_S \sigma^f \cdot \vec{v}_f \phi \cdot d\vec{f} \\ - \int_S \varrho_c E_s(1-\phi)\vec{v}_c \cdot df &- \int_S \varrho_f E_f \phi \vec{v}_f\, d\vec{f} \\ + \int_V (1-\phi)\varrho_c \vec{v}_c \cdot \mathbf{g}\, d^3r &+ \int_V \phi\varrho_f \vec{v}_f \cdot \mathbf{g}\, d^3r \end{aligned}$$

In our case, the kinetic energy is negligible so that $E_s \approx C^s_p T$ and $E_f \approx C^f_p T$ are the internal energy per unit mass of the solid and the fluid phase respectively (we assumed that both phases are incompressible so that the internal energy is proportional to temperature T). We do not consider internal heat generation $H = 0$, and neglect the work done by the surface stresses and the body force. If we introduce the mixture heat capacity by the relation

$$\varrho_m C^m_p = \varrho_c(1-\phi)C^s_p + \varrho_f\phi C^f_p$$

we obtain the conservation law

$$\begin{aligned} \frac{d}{dt}\int_V \varrho_m C^m_p T\, d^3r &= \int_V \nabla\cdot k_T \nabla T\, d^3r \\ -\int_S \varrho_m C^m_p T\vec{v}_m \cdot d\vec{f} &+ \int_S (1-\phi)\varrho_c(E_s - E_f)\vec{v}_{pm}\cdot d\vec{f} \end{aligned}$$

From this integral balance law you can go easily to the differential equation describing the thermal energy.

$$\begin{aligned} \frac{\partial}{\partial t}\varrho_m C^m_p T &+ \nabla\cdot\left(\varrho_m C^m_p T\vec{v}_m\right) \\ +\nabla\cdot\left(\varrho_c(1\right. &- \left.\phi)(C^s_p - C^f_p)T\vec{v}_{pm}\right) = \nabla\cdot k_T\nabla T \end{aligned}$$

For the systematic calculation presented in the text, we set $C^s_p = C^f_p$ and assumed that the mixture mass ϱ_m is constant (Boussinesq approximation). So finaly we obtain:

$$\varrho_m C^m_p\left(\frac{\partial T}{\partial t} + \nabla\cdot(T\vec{v}_m)\right) = \nabla\cdot k_T\nabla T$$

For numerical integration we assumed that all material parameters are constant and by a standart scaling procedure one gets the dimensionless equations given in the text.

Acknowledgments A.S.F. would like to thank V. Trubitsyn and E. Kharybin for introduction in the sedimentation problems and many discussion in the early stage of research. A.S.F. and G.N. would like to thank N.C. Laube for critical discussions during the elaboration of this work and George Hartley for the help with the English text, which greatly improved the reading of our paper. A.S.F. thanks the Deutsche Forschungsgemeinschaft for financial support during the elaboration of this work. We thank an anonymous reviewer for the friendly criticism.

REFERENCES

Batchelor, G.K., *The Stress System in a Suspension of Force free Particles* J. Fluid Mech. , vol 41, part 3 , pp 545 - 570, 1970.

Batchelor, G.K. and R.W. Janse van Rensburg, *Structure formation in bidisperse sedimentation* J. Fluid Mech. Vol 166 pp 379-407, 1986.

Blankenbach et al., *A Benchmark Comparisson for Mantle Convection Codes* Geophysical Journal International, Vol 98 , No 1, pp 32 - 38, 1989 .

Davis, R.H. and Acrivos, A., Sedimentation of noncollodial Particles at low Reynolds number Ann. Rev. Fluid Mech. Vol 17, pp 91-118, 1985

Drew, D. A., Mathematical Modeling of two-phase flow Ann. Rev. Fluid Mech., Vol 15, pp 261-291, 1983

Einstein, A.,Eine neue Bestimmung der Moleküldimensionen Annln. Phys., 19, 289, 1906

Gibb, F. G. F. and C.M.B. Henderson Convection and Crystal Settling in Sills Contrib. Mineral. Petrol., Vol 109, pp 538-545, 1992.

Hansen,U.,Zur Zeitabhaengigkeit der thermischen Konvektion im Erdmantel und der doppelt- diffusiven Konvektion in Magmakammern Mitteilungen aus dem Institut fuer Geophysik und Meteorologie, Koeln, 52, 155p, 1987 .

Huppert, H.E., D.R.Moore, *Non-linear double-diffusive convection.*, J. Fluid. Mech., 78, 821-854, 1976

Huppert, H.E., R.S.I Sparks, The fluid dynamics of a basaltic magma chamber replenished by influx of hot, dense ultrabasic magma., Contrib. Mineral. Petrol., 75, 279-289, 1980

Huppert, H.E., R.S.I Sparks, Double-Diffusive Convection Due to Crystallization in Magma., Annual Review of Earth and Planetary Sciences, Volume 12, 1984

Ishii, M., *Thermo-fluid Dynamic Theory of two-phase flow* Paris: Eyrolles, 1975

Knobloch, E., D.R.Moore, J. Toomre, N.O. Weiss, *Transitions to the chaos in two dimensional double diffusive convection.*, J. Fluid. Mech., 166, 409-448, 1986

Koyagushi, T., M.A. Hallworth, H.E. Huppert, R.S.I. Sparks, Sedimentation of particles in convecting fluid,Nature, 343, 447-450, 1990

Koyagushi, T., M.A. Hallworth, H.E. Huppert, *An experimental study on the effects of phenocrysts on convection in magmas.* Journ. of Volcanology and geothermal Research, 55, pp 15-32, 1993

Landau, L., E.M. Lifschitz, *Lehrbuch der theoretischen Physik, Band VI* , Akademie Verlag Berlin 1991

Marsh, B.D., M.R.Maxey, *On the distribution and the separation of crystals in convecting magma.* J. Volcanol. Geotherm. Res, 24,95-150, 1985

Martin, D., R.I. Nokes, Crystal settling in vigorously convecting magma chambers. Nature, 332, 534-536, 1988

Martin, D., R.I. Nokes, *A fluid dynamical study of crystal settling in convecting magmas.* J. Petrology, 30, 1471-1500, 1989

McKenzie, D., *The Generation and Compaction of Partially Molten Rock* J, Petrology, Vol 25, part 3, pp 713-765, 1984

Nauheimer, G., Untersuchungen des Modelles der Thermo - Diffusiven - Sedimentären Konvektion im schwach nichtlinearen Bereich und mögliche Anwendungen auf Konvektion in Magmakammern
Dipl. Phys. Thesis , 1994

Ronald, J.P., Armstrong R.C., Brown R.A., Graham A.L., Abbott J.R. A constitutive equation for concentrated suspension that accounts for shear-induced particle migration Phys. Fluids A 4 (1), January 1992

Shirtcliffe, T.G.L, *An experimental investigation of thermosolutal convection at marginal stability* J. Fluid. Mech.,35, pp 677-688, 1969

Trubitzyn, V.P., E.V. Kharybin, Convective instability of the sedimentation state in the mantle, Physics of the solids Earth (Fizika zemli), N8, pp 21-30, 1987 (in russian)

Trubitzyn, V.P., E.V. Kharybin, Hydrodynamical model of the differentiation process in the Earth's interior, Physics of the solid Earth (Fizika zemli), N4,pp 83-86, 1988 (in russian)

Trubitzyn, V.P., E.V. Kharybin, Thermoconvective instability of the two components viscous fluid, Physics of the solids Earth (Fizika zemli), N2,pp 3-17, 1991 (in russian)

Turner, J.S., Multicomponent convection, Annual Review of Fluid Mechanics,Vol 16, pp 11-45, 1984

Veronis, G. , *Effect of a stabilizing gradient of solute on thermal convection*, J. Fluid. Mech., 34, pp 315-336, 1968

Weinstein, S.A., D.A. Yuen, P.L. Olson, Evolution of crystal-settling in magma chamber convection, Earth Planet Sci. Let., 87, pp 237-248, 1988

Zalesak, S.T., *Fully multidimensional Flux- Corrected Transport Algorithms for Fluids* J. Comp. Phys., Vol 31, pp 335-362, 1979

[1]Earth & Planetary Sciences, McGill University, 3450 University Street, Montreal, QC H3A 2A7, Canada
[2]Geodynamik - Physik der Lithosphäre, Universität Bonn, Nußallee 8, D-53115 Bonn, Germany

The Shearing Instability in Magnetoconvection

A.M. Rucklidge and P.C. Matthews[1]

Department of Applied Mathematics and Theoretical Physics,
University of Cambridge, Cambridge, UK

Magnetoconvection is an example of double-diffusive convection where the stabilizing secondary component is an imposed magnetic field. Numerical experiments on two- and three-dimensional convection in the presence of a vertical magnetic field reveal a bewildering variety of periodic and aperiodic oscillations. Steady two-dimensional rolls can develop a shearing instability, in which rolls turning over in one direction grow at the expense of rolls turning over in the other, resulting in a net shear across the layer. As the temperature difference across the fluid is increased, two-dimensional pulsating waves occur, in which the direction of shear alternates. In three dimensions, more complicated alternating pulsating waves are observed: the fluid develops rolls with their axes aligned along the x-axis, which are unstable to shear and a strong streaming motion in the y-direction is generated, suppressing the x-rolls and stretching out the magnetic field in the y-direction. Rolls with their axes aligned along the y-axis are not suppressed by this field; these rolls grow, and in turn are suppressed by streaming in the x-direction. This pattern repeats periodically, with the streaming rotating by 90° during each quarter cycle. The numerical experiments are interpreted in terms of low-order models, which confirm that pulsating waves appear in a global bifurcation.

1. MOTIVATION

Convection in a horizontal layer heated from below typically sets in in a cellular pattern (for example, rolls or hexagons). As the temperature difference across the layer is increased, cellular convection may develop an instability that generates a large-scale flow [*Krishnamurti and Howard*, 1981], which has important consequences for the transport properties of the flow. The new mode of convection breaks the reflection symmetry that separates the original cells: fluid moves in one direction at the top of the layer and in the opposite direction at the bottom, resulting in a net shear across the layer.

The onset of the shearing instability can be understood by the following feedback mechanism [*Howard and Krishnamurti*, 1986]. Convection rolls that are tilted to the left (for example) carry left-moving fluid to the top of the layer (and right-moving fluid to the bottom), leading to a net shear flow across the layer, which enhances the leftward tilt of the rolls. Since shear damps the vigour of the original cellular flow, an oscillation may develop: if convection is unstable to shear, the shear will grow, damping the convection; with nothing to drive it, the shear decays viscously, which allows the original convection cells to reform. This shearing instability is a general phenomenon, and may occur in a wide variety of flows.

One of the interesting questions that arises is whether the direction of the shear can reverse over the course of such an oscillation. A reversing oscillation is called a pulsating wave (PW): after half a period, the oscillation is in a state that is the mirror image of its initial state

[1]Present address: Department of Theoretical Mechanics, University of Nottingham, Nottingham, NG7 2RD, UK

Double-Diffusive Convection
Geophysical Monograph 94

[*Proctor and Weiss*, 1993]. Whether or not pulsating waves arise hinges on a delicate balance between the temperature difference that forces the system and the rates of decay of the shear and the mode that causes the rolls to tilt. The imposition of a vertical magnetic field across the layer brings another dimension to the problem: the field will be pulled out by a shearing flow, but this will be opposed by magnetic forces that get larger as the field is stretched. In a strong field, this interaction can lead to an oscillating shear, but even a weak field can cause the shear to change direction. Again, there is a delicate balance between the decay rates of shear and the horizontally stretched field.

Two-dimensional pulsating waves have been observed numerically in several convective systems [for example, *Matthews et al.*, 1993; *Proctor et al.*, 1994; *Prat, Massaguer and Mercader*, 1994; D.P. Brownjohn, N.E. Hurlburt, M.R.E. Proctor and N.O. Weiss, unpublished, 1994; N. Brummell and K.A. Julien, unpublished, 1994]. Landsberg and Knobloch [1991] deduced using invariant bifurcation theory that they could be formed in a local bifurcation. In this paper, we focus on the situation where pulsating waves are created in global bifurcations, using low-order models and maps, and extend these techniques to three-dimensional convection.

Magnetoconvection is an example of double-diffusive convection: here, the stabilizing secondary component is the imposed vertical magnetic field. Like in convection with a stabilizing solute gradient, an imposed magnetic field can impede convection and can lead to oscillations at onset if the field is strong enough. Magnetoconvection has two additional complications: first, the field reacts back on the flow through the nonlinear Lorentz force, rather than through a linear buoyancy force, and second, the vector nature of the magnetic field (as opposed to the scalar solute concentration) greatly complicates the dynamics when the flow is three dimensional. The instabilities discussed in this paper may also be present in thermosolutal convection.

This paper presents work, done in collaboration with N.O. Weiss and M.R.E. Proctor, on the analysis of the shearing instability in magnetoconvection; more details and comparisons with the the behaviour of the partial differential equations (PDEs) are to be found in Rucklidge and Matthews (unpublished, 1994). In section 2, we derive a low-order model that describes the shearing oscillation in two-dimensional convection in the presence of a vertical magnetic field. We address the question of the reversal of the shear in section 3 using the techniques of nonlinear dynamics, and compare the predictions of a low-order model with solutions of the PDEs. In section 4, we describe some preliminary three-dimensional calculations that show an alternating pulsating wave, in which the direction of the shear rotates by 90° over each quarter of the oscillation. This behaviour is interpreted in terms of a more elaborate low-order model. We summarise in section 5.

2. TWO-DIMENSIONAL BEHAVIOUR

The PDEs for two-dimensional Boussinesq convection in a vertical magnetic field are:

$$\frac{\partial \omega}{\partial t} + \mathrm{J}(\Psi, \omega) = \sigma \nabla^2 \omega - \sigma R \frac{\partial \theta}{\partial x} - \sigma \zeta Q \left(\frac{\partial \nabla^2 A}{\partial z} + \mathrm{J}(A, \nabla^2 A) \right), \quad (1)$$

$$\frac{\partial \theta}{\partial t} + \mathrm{J}(\Psi, \theta) = \nabla^2 \theta + \frac{\partial \Psi}{\partial x}, \quad (2)$$

$$\frac{\partial A}{\partial t} + \mathrm{J}(\Psi, A) = \zeta \nabla^2 A + \frac{\partial \Psi}{\partial z}, \quad (3)$$

where $\omega = -\nabla^2 \Psi$ is the vorticity, Ψ is the streamfunction, θ is the deviation from the conducting temperature profile, A is the deviation of the flux function from a uniform vertical magnetic field, and x, z and t are the horizontal, vertical and time coordinates respectively [*Knobloch, Weiss and Da Costa*, 1981]. The nonlinearities in the equations are in the Jacobian operator $\mathrm{J}(f,g) = (\partial f/\partial x)(\partial g/\partial z) - (\partial g/\partial x)(\partial f/\partial z)$. The physical parameters are the Prandtl number σ and magnetic diffusivity ratio ζ, the Rayleigh number R (proportional to the temperature difference across the layer) and the Chandrasekhar number Q (proportional to the square of the imposed magnetic field). The boundary conditions are chosen for convenience: $\Psi = \omega = \theta = \partial A/\partial z = 0$ on the top and bottom walls ($z = 0, 1$). We impose periodic horizontal boundary conditions in a box of length $2L$ and define k, the spatial wave number, by $k = \pi/L$.

We construct a low-order model by truncating the PDEs (1)–(3): ordinary untilted convection is represented by terms like $\Psi \propto \sin kx \sin \pi z$. Horizontal shear is described by a term proportional to $\sin \pi z$, and nonlinear interactions between these two modes generate a $\cos kx \sin 2\pi z$ term, which gives the rolls a tilted appearance. We therefore pose the eleven-mode truncation:

$$\Psi = \Psi_{11} \sin kx \sin \pi z + \Psi_{01} \sin \pi z + \Psi_{12} \cos kx \sin 2\pi z, \quad (4)$$

$$\theta = \theta_{11} \cos kx \sin \pi z + \theta_{02} \sin 2\pi z + \theta_{12} \sin kx \sin 2\pi z, \quad (5)$$

$$A = A_{11} \sin kx \cos \pi z + A_{20} \sin 2kx + A_{01} \cos \pi z + A_{12} \cos kx \cos 2\pi z + A_{10} \cos kx, \quad (6)$$

where the mode amplitudes are functions only of time. Here, we have fixed the phase of the rolls in the periodic box, so we have ruled out solutions that travel. This truncation yields an eleventh-order set of ordinary differential equations (ODEs):

$$\begin{aligned}
\dot{\Psi}_{11} &= -\sigma k_{11}^2 \Psi_{11} + \frac{\sigma k R}{k_{11}^2} \theta_{11} - \sigma \zeta Q \pi A_{11} \\
&\quad - \frac{k\pi}{2k_{11}^2} (k^2 + 3\pi^2) \Psi_{01} \Psi_{12} \\
&\quad + \sigma \zeta Q \frac{k\pi}{2k_{11}^2} \{ 2(\pi^2 - 3k^2) A_{11} A_{20} \\
&\quad + 2(\pi^2 - k^2) A_{01} A_{10} + (k^2 + 3\pi^2) A_{01} A_{12} \}, \\
\dot{\theta}_{11} &= k\Psi_{11} - k_{11}^2 \theta_{11} + \tfrac{1}{2} k\pi \{ 2\Psi_{11} \theta_{02} + \Psi_{01} \theta_{12} \}, \\
\dot{A}_{11} &= \pi \Psi_{11} - \zeta k_{11}^2 A_{11} - \tfrac{1}{2} k\pi \{ 2\Psi_{11} A_{20} + \Psi_{01} A_{12} \\
&\quad + \Psi_{12} A_{01} + 2\Psi_{01} A_{10} \}, \\
\dot{\theta}_{02} &= -4\pi^2 \theta_{02} - \tfrac{1}{2} k\pi \Psi_{11} \theta_{11}, \\
\dot{A}_{20} &= -4\zeta \pi^2 A_{20} + \tfrac{1}{2} k\pi \{ \Psi_{11} A_{11} - 2\Psi_{12} A_{12} \}, \\
\dot{\Psi}_{12} &= -\sigma k_{12}^2 \Psi_{12} - \frac{\sigma k R}{k_{12}^2} \theta_{12} - 2\sigma \zeta Q \pi A_{12} \\
&\quad + \frac{k\pi}{2k_{12}^2} k^2 \Psi_{11} \Psi_{01} + \sigma \zeta Q \frac{k\pi}{2k_{12}^2} \{ k^2 A_{01} A_{11} \\
&\quad + (12k^2 - 16\pi^2) A_{12} A_{20} \}, \\
\dot{\theta}_{12} &= -k\Psi_{12} - k_{12}^2 \theta_{12} - \tfrac{1}{2} k\pi \Psi_{01} \theta_{11}, \\
\dot{A}_{12} &= 2\pi \Psi_{12} - \zeta k_{12}^2 A_{12} \\
&\quad + \tfrac{1}{2} k\pi \{ \Psi_{01} A_{11} - \Psi_{11} A_{01} + 4\Psi_{12} A_{20} \}, \\
\dot{A}_{10} &= -\zeta k^2 A_{10} + \tfrac{1}{2} k\pi \{ \Psi_{01} A_{11} + \Psi_{11} A_{01} \}, \\
\dot{\Psi}_{01} &= -\sigma \pi^2 \Psi_{01} - \sigma \zeta Q \pi A_{01} + \tfrac{3}{4} k\pi \Psi_{11} \Psi_{12} \\
&\quad + \sigma \zeta Q \tfrac{1}{2} k\pi \{ A_{11} A_{10} - \tfrac{3}{2} A_{11} A_{12} \}, \\
\dot{A}_{01} &= \pi \Psi_{01} - \zeta \pi^2 A_{01} + \tfrac{1}{2} k\pi \{ -\Psi_{11} A_{10} \\
&\quad + \tfrac{1}{2} \Psi_{12} A_{11} + \tfrac{1}{2} \Psi_{11} A_{12} \},
\end{aligned} \quad (7)$$

where $k_{1n}^2 = k^2 + n^2\pi^2$. This system includes as subsystems the Lorenz [1963] equations and the equations of Howard and Krishnamurti [1986] for sheared Bénard convection (no magnetic field); in addition, the fifth-order truncated model of magnetoconvection without shear of Knobloch, Weiss and Da Costa [1981] is an invariant subsystem. Lantz [1995] has studied the analogous truncated model for sheared convection in a horizontal field.

The number of parameters is reduced by considering the limit of narrow rolls: $L = \pi/k \to 0$, using the approach of Hughes and Proctor [1990], who studied the analogous non-magnetic problem. This is justified by numerical experiments that show that the instability to shearing behaviour occurs most readily in narrow rolls. Taking this limit also has the advantage of reducing the order of the model from eleven to five while retaining the essential dynamics. The following scalings lead to an appropriate balance between the linear and nonlinear terms:

$$\begin{aligned}
&\Psi_{11} \sim \Psi_{01} \sim \Psi_{12} \sim A_{01} \sim L, \\
&\theta_{11} \sim \theta_{02} \sim \theta_{12} \sim L^2, \\
&A_{11} \sim A_{12} \sim A_{10} \sim L^3,
\end{aligned} \quad (8)$$

with

$$R = R_C(1 + L^2\mu), \quad (9)$$

where

$$R_C = \frac{(1 + L^2)^3 \pi^4}{L^4} + \pi^2(1 + L^2)Q. \quad (10)$$

Here, R_C is the Rayleigh number at which the trivial solution (with no motion) first becomes unstable to steady convection, and μ is the parameter that plays the role of the Rayleigh number. With these scalings, the six variables θ_{11}, A_{11}, θ_{12}, A_{12}, A_{10} and A_{20} are slaved to the other five variables. Finally, time, the variables and the parameters are rescaled:

$$\begin{aligned}
&t \to t/4\pi^2, \quad \mu \to \frac{4(1+\sigma)}{\sigma} \mu, \\
&\Psi_{11} \to \sqrt{\frac{32(1+\sigma)}{\sigma}} \Psi_{11}, \\
&\Psi_{01} \to 8\Psi_{01}, \quad \theta_{02} \to \frac{4(1+\sigma)}{\sigma\pi} \theta_{02}, \\
&\Psi_{12} \to \sqrt{\frac{32(1+\sigma)}{\sigma}} \Psi_{12}, \quad A_{01} \to \frac{8}{\zeta\pi} A_{01};
\end{aligned} \quad (11)$$

this scaling yields the set of model equations [*Rucklidge and Matthews*, 1993]:

$$\begin{aligned}
\dot{\Psi}_{11} &= \mu \Psi_{11} + \Psi_{11} \theta_{02} - \Psi_{01} \Psi_{12}, \\
\dot{\theta}_{02} &= -\theta_{02} - \Psi_{11}^2, \\
\dot{\Psi}_{12} &= -\nu \Psi_{12} + \Psi_{11} \Psi_{01}, \\
\dot{\Psi}_{01} &= -\frac{\sigma}{4} \Psi_{01} - \frac{\sigma Q}{4\pi^2} A_{01} + \frac{3(1+\sigma)}{4\sigma} \Psi_{11} \Psi_{12}, \\
\dot{A}_{01} &= \frac{\zeta}{4} \Psi_{01} - \frac{\zeta}{4} A_{01},
\end{aligned} \quad (12)$$

where $\nu = (9\sigma/4(1+\sigma)) - \mu$, and we require that $\nu > 0$. This system has a two-dimensional invariant subspace when $\Psi_{12} = \Psi_{01} = A_{01} = 0$: trajectories that start in this subspace remain in it; thus the untilted solutions are represented by Ψ_{11} and θ_{02} alone.

The symmetries relevant to this problem (using the notation of Proctor and Weiss [1993]) are given in Table 1: these are the identity e, reflections in the vertical planes $x = L$ and $x = \frac{1}{2}L$ and their product, and, since we will be considering time-periodic solutions with period P, the advance of half a period in time. Along with the products of these elements, they form an eight element group. The reason for considering the advance of one-half period in time is that operating on a periodic orbit with a reflection (which is its own inverse) may (or may not) map the orbit to itself, but shifted by half a period [see *Golubitsky, Stewart and Shaeffer*, 1988]. In addition, the system has an up-down symmetry (reflection in the horizontal mid-plane) due to the incompressibility; this symmetry is not broken in the bifurcations that will be discussed in this paper. As a result, the leftward motion at the top of the layer will always be equal to the rightward motion at the bottom of the layer, so the rolls will not travel. In the analogous problem of compressible convection, there is no up-down symmetry, so solutions that break the vertical reflection symmetries will be free to travel to the left or to the right.

When $\mu < 0$ ($R < R_C$), the trivial solution is stable in (12); as μ increases, there is first a pitchfork bifurcation to symmetric steady (SS) convection: $\Psi_{11}^2 = -\theta_{02} = \mu$. This state is invariant under the symmetry operations m, t_e and t_m, and the two SS fixed points are mapped to each other by m'. The reflection symmetry m is broken in one of two ways: either in a Hopf bifurcation directly to pulsating waves (PW) with symmetry t_m, or in a pitchfork bifurcation when fixed points corresponding to steady tilted convection (STC) are created, with symmetry t_e. In the PDEs for non-magnetic convection, we only find the Hopf bifurcation at high Prandtl numbers (as do Prat, Massaguer and Mercader [1994] in the case of rigid boundaries at the top and bottom of the layer). In this paper, we focus on the case with a pitchfork bifurcation to STC fixed points. As μ is increased further, these lose stability in a Hopf bifurcation to oscillatory tilted convection (OTC), breaking the t_e symmetry. This sequence is readily found in numerical solutions of the PDEs: Figure 1a,b show streamlines corresponding to steady untilted and tilted convection; oscillatory tilted convection is shown in Figure 1c,d. Finally, there is a gluing

TABLE 1. Symmetries of magnetoconvection[a]

	$(x,z,t) \rightarrow$	$(\Psi,\theta,A) \rightarrow$
e:	(x,z,t)	(Ψ,θ,A)
m:	$(-x,z,t)$	$(-\Psi,\theta,-A)$
l:	$(L+x,z,t)$	(Ψ,θ,A)
m':	$(L-x,z,t)$	$(-\Psi,\theta,-A)$
t_e:	$(x,z,t+\frac{1}{2}P)$	(Ψ,θ,A)
t_m:	$(-x,z,t+\frac{1}{2}P)$	$(-\Psi,\theta,-A)$
t_l:	$(L+x,z,t+\frac{1}{2}P)$	(Ψ,θ,A)
$t_{m'}$:	$(L-x,z,t+\frac{1}{2}P)$	$(-\Psi,\theta,-A)$

[a]After Proctor and Weiss [1993].

bifurcation, followed by pulsating waves shown in Figure 2. Note that the gluing bifurcation restores the spatio-temporal symmetry t_m: the pulsating wave is invariant under the combined symmetry operation of advance of half a period in time followed by a reflection in the vertical mid-plane.

The dynamics of the problem is best understood in terms of phase portraits (see Figure 3): the horizontal coordinate represents the amplitude of the untilted rolls (Ψ_{11}) and the vertical coordinate represents the shear (Ψ_{01}). In Figure 3a, the four OTC periodic orbits have been created in a Hopf bifurcation from the four STC fixed points. As the Rayleigh number (or μ) increases, the four orbits grow until they collide with the two SS fixed points (b). Beyond the gluing bifurcation (c), there are a pair of pulsating waves, invariant under the symmetry t_m. This sequence contains an example of a homoclinic, or global, bifurcation (called a gluing bifurcation in this case, as it involves gluing a pair of asymmetric orbits together to form a larger symmetric orbit). Here, there are two simultaneous gluing bifurcations.

The dynamics of a system with a reflection symmetry near a gluing bifurcation is well understood [see, for example, *Lyubimov and Zaks*, 1983; *Glendinning*, 1984; *Rucklidge*, 1993]. There are a number of cases (the simplest of which was described above) characterised by three properties: whether, at the gluing bifurcation, the trajectories approach (and leave) the fixed point (SS in this case) in directions that are invariant or that are reversed under the reflection symmetry, whether the eigenvalues with real parts closest to zero are real or complex, and whether the ratio of the real parts of these eigenvalues is greater than or less than one, in absolute value. Several of the possible combinations of these criteria occur in the fifth-order model (12), but

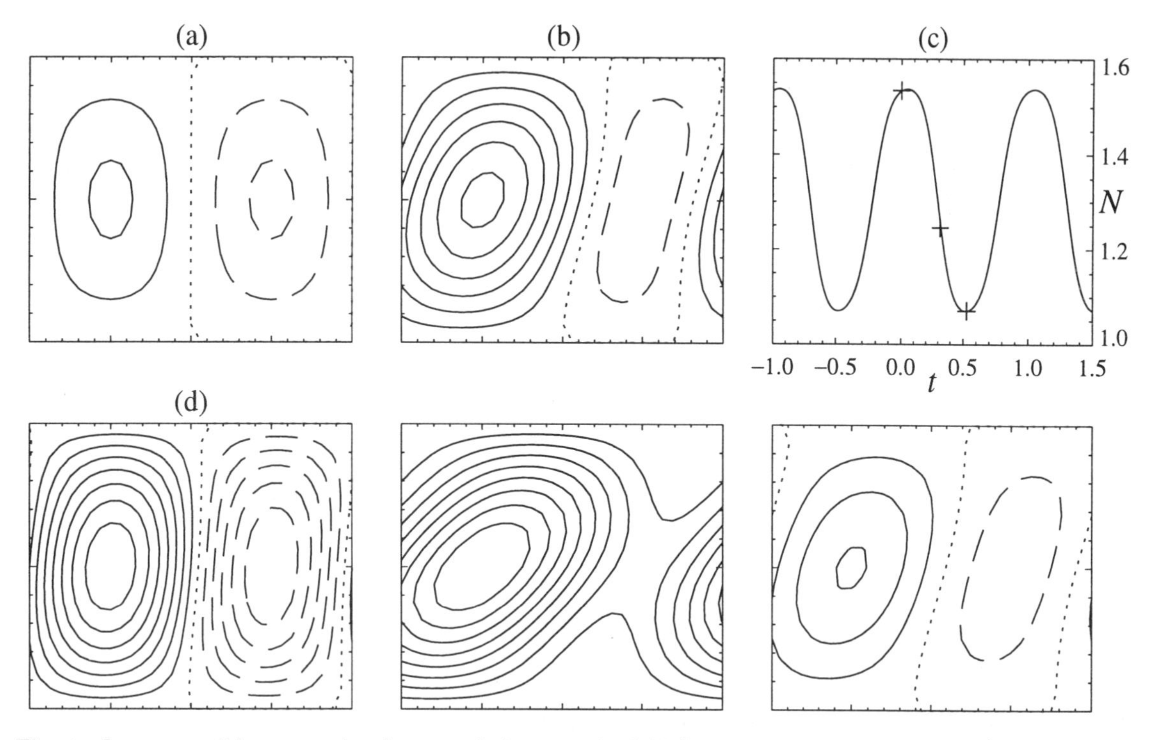

Fig. 1. Incompressible convection in a vertical magnetic field [after *Matthews et al.*, 1993]. The spacing between the streamlines is uniform and is the same in each figure. The zero streamline is dotted and negative streamlines are dashed. Note that the system retains point symmetry. The parameters are $\sigma = 0.5$, $\zeta = 0.2$, $L = 0.378$ and $Q = 63.2$. (a) Symmetric steady (SS) convection: $R = 8193$. (b) Steady tilted convection (STC): $R = 9262$. Oscillatory tilted convection (OTC) at $R = 10{,}331$: (c) periodic variation of the Nusselt number N, a dimensionless measure of the heat transport across the layer, with time in units of the period P; (d) streamlines at times $t = 0, 0.32, 0.53P$, which are represented by crosses in (c).

rather than discuss each one in detail, we focus instead on the interesting behaviour that is observed when the magnetic field parameter Q is small or zero.

In the case of no magnetic field, the global bifurcation that occurs as μ increases is more complicated (Figure 4): the OTC periodic orbits (a) collide with the SS and trivial fixed points simultaneously (b); beyond this heteroclinic bifurcation, there are a pair of orbits of the type described by Howard and Krishnamurti [1986], which we will denote by HK. These differ from the pulsating waves in the magnetic case (Figure 3) in that the shear does not change sign over the course of the oscillation and they do not have the symmetry t_m; they do, however, have the symmetry t_l.

With a small magnetic field, periodic orbits do not collide with the trivial and SS fixed points simultaneously: the presence of even a small magnetic field breaks this degeneracy for a reason that will be described in the next section, and allows the periodic orbits to collide first with the SS fixed point (Figure 3) and subsequently with the trivial fixed point. In fact, there is a sequence of global bifurcations that has the same net effect as the global bifurcation with $Q = 0$. This sequence is depicted in Figure 5: the four OTC periodic orbits (a) glue together to form a pair of PW (b). These collide with the origin; there is an interval of chaotic trajectories (c), from which a different type of pulsating wave emerges (d). These orbits are invariant under the symmetry $t_{m'}$, and so will be named PW′: they are pulsating waves, in that they are invariant under the combination of a vertical reflection and an advance of half a period in time, but the plane of reflection is $x = \frac{1}{2}L$, rather than $x = L$ for PW. The two PW′ orbits collide with the two SS fixed points simultaneously to form (e) a pair of four-looped orbits that are invariant under t_m; finally, each one of these two collides with one of the SS fixed points and unglues to form (f) a pair of HK orbits, invariant under t_l.

3. GLOBAL BIFURCATIONS

Global bifurcations occur when trajectories that start

Fig. 2. Incompressible convection in a vertical magnetic field: as in Figure 1 but with $R = 10{,}687$ [after *Matthews et al.*, 1993]. Pulsating wave (PW) at times $t = 0, 0.08, 0.14, 0.26, 0.41, 0.50P$.

very close to a fixed point end up back at that fixed point (homoclinic bifurcation) or very close to another fixed point (heteroclinic bifurcation). Near global bifurcations, trajectories spend most of their time near the fixed points, and move rapidly between them. Using this separation of time-scales, the flow can be approximated by low-dimensional maps, constructed with standard techniques [see, for example, *Guckenheimer and Holmes*, 1983]. We begin by considering the case with no magnetic field. With smallish σ, trajectories approach the origin tangent to the Ψ_{01}-axis. We consider a trajectory, depicted in Figure 6, that starts near the origin with $\Psi_{01} = h_0$, where h_0 is a small positive constant, and follow this trajectory as it passes the origin, travels out to one of the two SS fixed points, and returns to a neighbourhood of the origin, until it intersects the

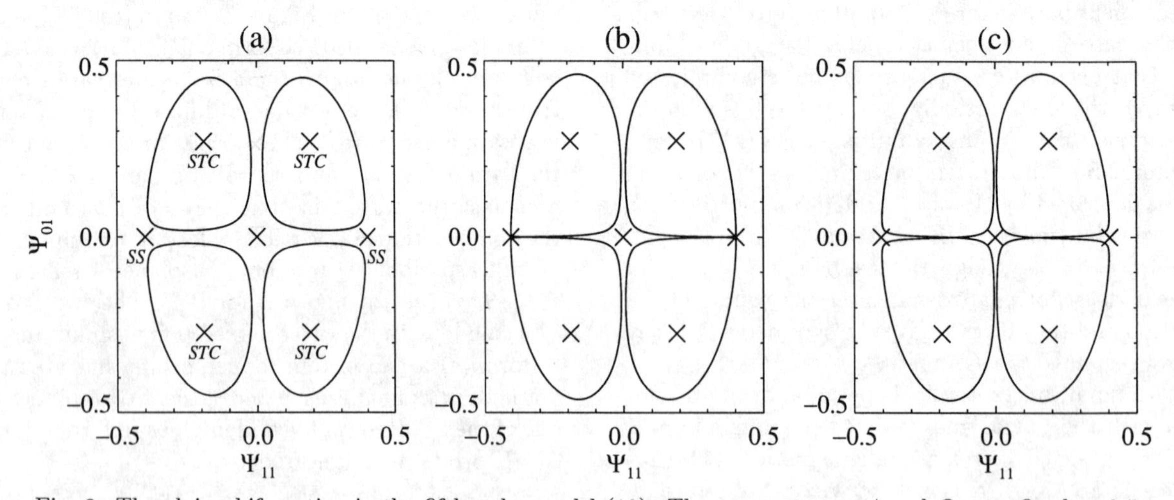

Fig. 3. The gluing bifurcation in the fifth-order model (12). The parameters σ, ζ and Q were fixed at 0.5, 0.2 and 1.0 respectively. As μ increases, the four periodic orbits (OTC) in (a) $\mu = 0.160$ glue together in pairs near (b) $\mu = 0.163$, with two periodic orbits (PW) emerging in (c) $\mu = 0.164$. The fixed points are represented by crosses: SS indicates steady symmetric convection and STC indicates steady tilted convection.

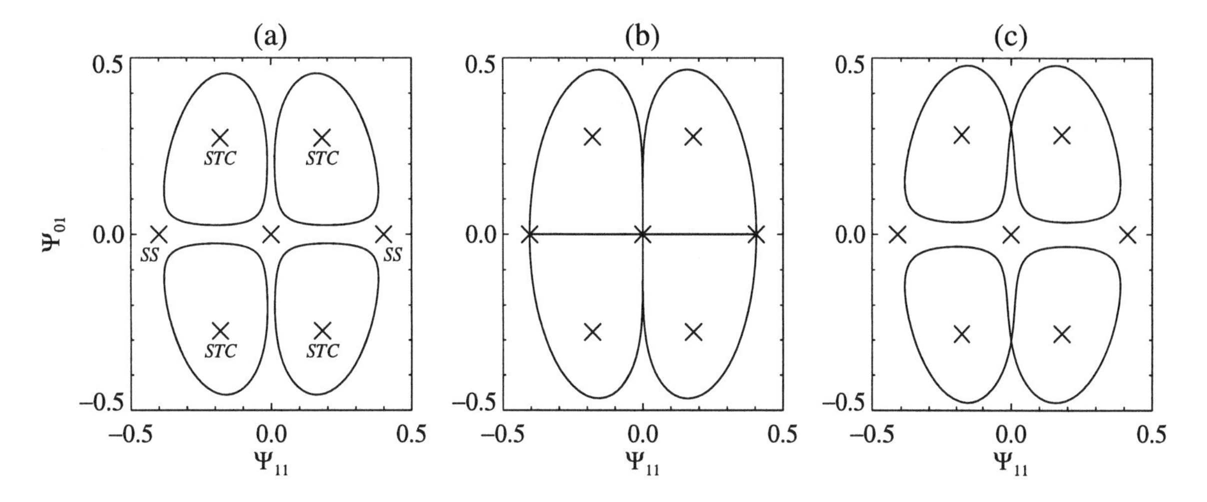

Fig. 4. The global bifurcation in (12) with no magnetic field ($Q = 0$). The parameter σ was fixed at 0.5. As is μ increased, the four periodic orbits (OTC) in (a) $\mu = 0.16$ collide with the trivial and nontrivial (SS) fixed points near (b) $\mu = 0.163875$, with two periodic orbits (HK) emerging in (c) $\mu = 0.17$.

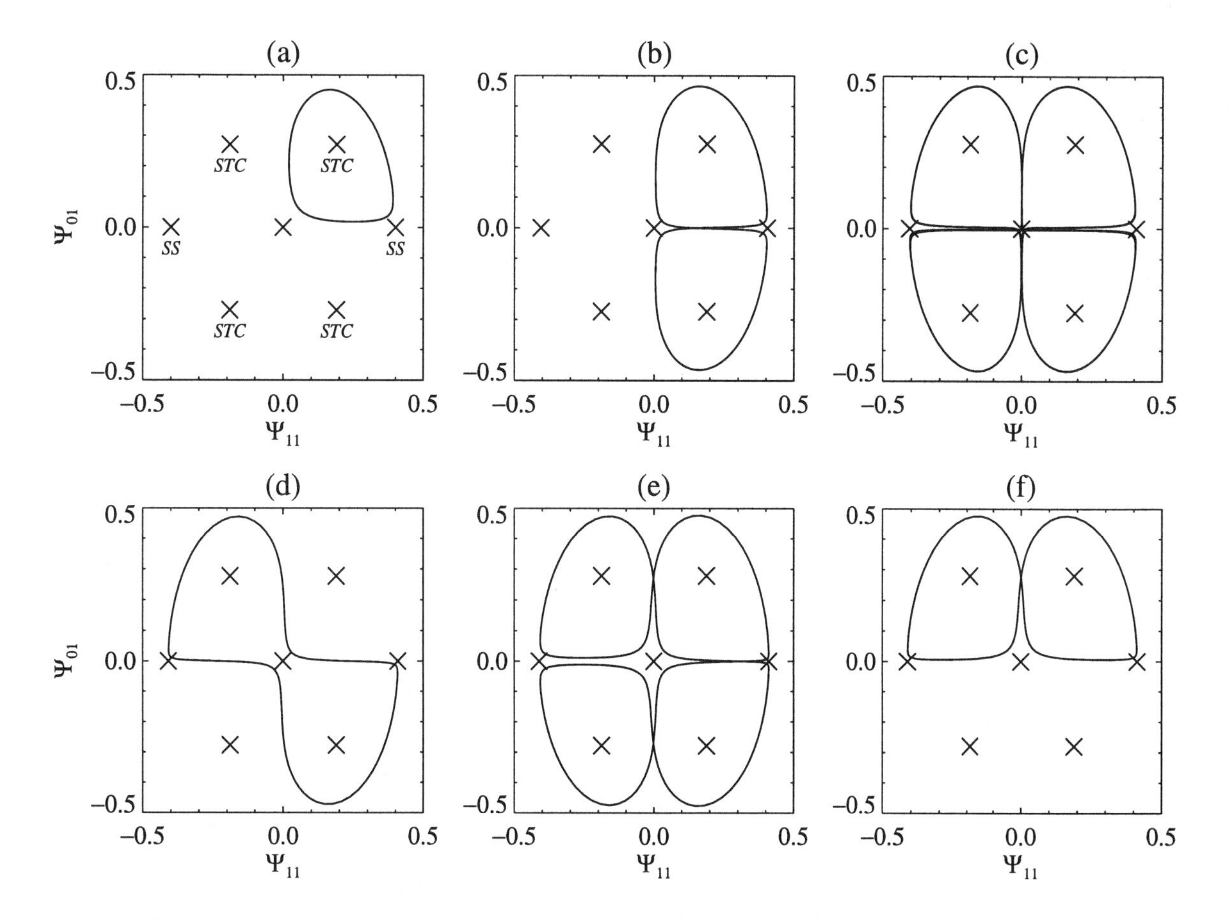

Fig. 5. The full sequence of global bifurcations with $Q = 1.0$, $\sigma = 0.5$ and $\zeta = 0.2$. (a) OTC: $\mu = 0.160$; (b) PW: $\mu = 0.164$; (c) chaos: $\mu = 0.1655$; (d) PW′: $\mu = 0.167$; (e) four-looped: $\mu = 0.169$; (f) HK: $\mu = 0.170$. Only one of several periodic orbits that map to each other are shown in each case.

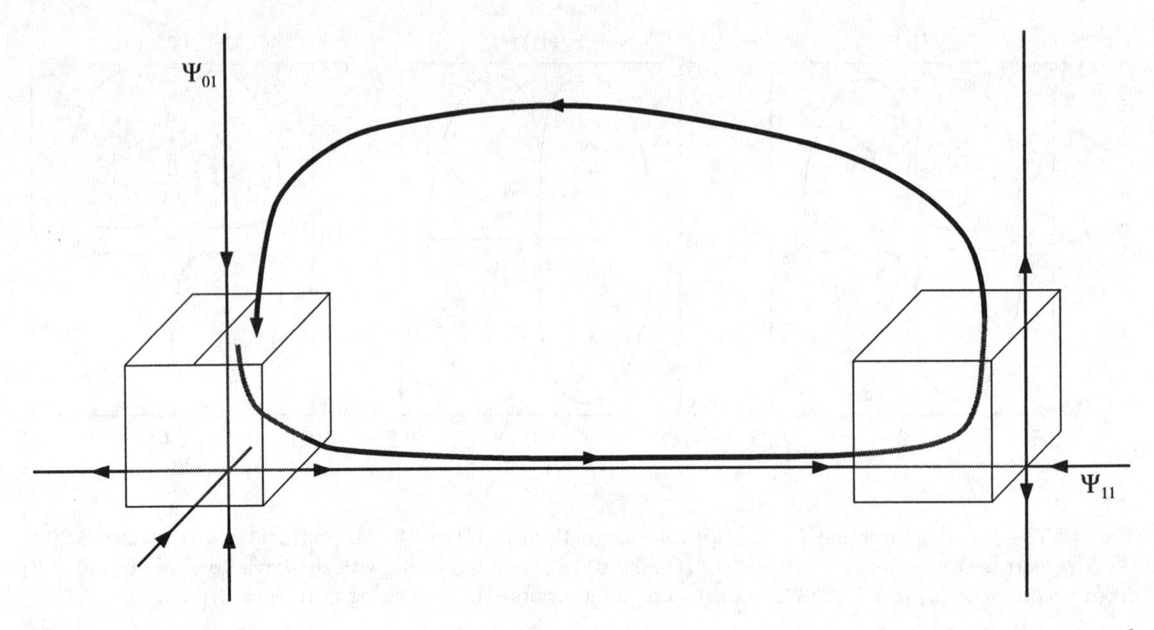

Fig. 6. The trajectory starts (on the left) near the origin with $\Psi_{01} = h_0$, travels through the box around the origin to one of the SS fixed points (right), through the box around that fixed point then back to the plane $\Psi_{01} = h_0$.

plane $\Psi_{01} = h_0$. Thus the flow defines a map from this plane back to itself. This map may be calculated approximately by considering small boxes around each of the fixed points: within these boxes, the flow is approximately linear and is dominated by the eigenvalues of the fixed points, and between the boxes, the flow is linearised about the one-dimensional unstable manifolds of the fixed points. The resulting map can be simplified to a one-dimensional map if the contraction in the Ψ_{12}-direction is strong; this yields the well-known Lorenz map to relate successive values of the Ψ_{11}-coordinate of intersections with the plane $\Psi_{01} = h_0$:

$$\Psi_{11} \to \text{sgn}(\Psi_{11})\left(-\kappa + a|\Psi_{11}|^{\delta}\right), \qquad (13)$$

where $-\kappa$ is the Ψ_{11}-coordinate of the intersection of the unstable manifold of the SS fixed point with the plane $\Psi_{01} = h_0$, a is a constant that depends on the global properties of the flow, and δ is the product of the ratio of the dominant negative to positive eigenvalues of the trivial and SS fixed point. In this context, the dominant eigenvalues are those with smallest positive real part and largest (closest to zero) negative real part. At $\kappa = 0$, there are heteroclinic connections from the SS fixed points to the origin, which form a heteroclinic cycle with the structurally stable connections from the origin back to the SS fixed points. This map (13) is well understood [*Guckenheimer and Holmes*, 1983]: with $1 > a > 0$ and $\delta > 1$, there is a simple gluing bifurcation when $\kappa = 0$, while with $\delta < 1$, there is a sequence of bifurcations with the same net effect as a gluing bifurcation, but involving an interval with a chaotic attractor. In the ODEs (12), the heteroclinic bifurcation occurs with $\delta < 1$ when $\sigma \lesssim 0.37$, so we expect chaotic trajectories for small σ. For $0.37 \lesssim \sigma \lesssim 0.43$, we have $\delta > 1$ and the bifurcation is a simple gluing bifurcation. For larger σ, the dominant stable eigenvalues of the SS fixed point become complex, and a more complicated map is required [Rucklidge and Matthews, unpublished, 1994].

The derivation of the map (13) above assumes that the contraction in the Ψ_{12} direction is strong at the global bifurcation, that is, the Ψ_{01} eigenvalue $-\frac{1}{4}\sigma$ dominates (is closer to zero than) the Ψ_{12} eigenvalue $-\nu$. This condition holds in the ODEs (12) for the values of σ where the dominant eigenvalues of the trivial and SS fixed points are real; for these values of σ, it is Ψ_{01} that does not change sign in the periodic orbit created in the global bifurcation, while Ψ_{12} does change sign. In this case, an HK orbit is created in the global bifurcation. The Ψ_{12}-direction dominates for larger σ, and it will be the shear Ψ_{01} that changes sign after the global bifurcation. These oscillations with alternating shear are of the type PW′, invariant under the symmetry $t_{m'}$. Thus we expect that the PDEs will have HK oscillations, without reversals of the shear, whenever the shear (Ψ_{01}) eigenvalue dominates the tilt (Ψ_{12}) eigenvalue at the global bifurcation, and that this will occur for smallish σ. Conversely, with larger σ, we expect that the PDEs will have shear reversals beyond the global bifurcation. The transition value of σ occurs when the eigenvalue of Ψ_{01} ($-\sigma\pi^2$) in (7) is equal to the dominant eigenvalue in the (Ψ_{12}, θ_{12})-plane.

With non-zero magnetic field, the situation becomes more complicated, and the question of whether or not the shear reverses depends on the magnetic diffusivity ratio ζ as well. With Q just greater than zero, the eigenvalues of the origin will still be real, provided that $\sigma \neq \zeta$. If the Ψ_{12} eigenvalue $-\nu$ dominates, then we should expect reversals of shear as in the non-magnetic case. On the other hand, suppose that σ is small; then trajectories will approach the origin in the (Ψ_{01}, A_{01})-plane along the eigenvector with eigenvalue closest to zero. With small Q, the two eigenvalues in this plane are close to $-\frac{1}{4}\sigma$ and $-\frac{1}{4}\zeta$, with eigenvectors slightly rotated from the Ψ_{01} and A_{01} axes. If $\sigma < \zeta$, then Ψ_{01} essentially continues to dominate, and the shear will not reverse in the HK oscillations after the global bifurcation. If $\sigma > \zeta$ (as in Figure 5), then a trajectory that approaches the origin with Ψ_{01} positive may change sign if it comes in along the eigenvector that points into the $\Psi_{01} < 0$ half-space, as illustrated in Figure 7a. It is this change of orientation of the eigenvectors that permits the gluing bifurcation from OTC to PW to take place: as the system approaches the heteroclinic bifurcation at $\kappa = 0$, trajectories will spend longer and longer in the box around the origin, and there must come a point when the shear changes sign before the trajectory leaves the box. At that point, the trajectory enters the stable manifold of an SS fixed point, so there will be a gluing bifurcation to form pulsating waves. The condition $\sigma > \zeta$ for reversals of shear can be understood physically in the following way: shear decays more rapidly than horizontally stretched magnetic field, so when the shear has decayed to small levels and convection is just about to restart, there is enough residual magnetic field to flip the shear over to the other direction. However, this condition is not strict: if σ and ζ are nearly equal, then additional effects will need to be included.

Once Q exceeds $\pi^2(\sigma - \zeta)^2/4\sigma\zeta$, the eigenvalues of the origin become complex. In this case, as the system approaches the heteroclinic bifurcation at $\kappa = 0$, the shear will change sign several times before leaving the box around the origin – see Figure 7b. The trajectory will first acquire a half-twist, so the shear will change sign once, forming a pulsating wave; closer to the heteroclinic bifurcation, the trajectory will acquire a full twist, changing sign twice before leaving the box, forming a twisted periodic orbit with shear that changes sign briefly near the origin. Thus we expect an infinite

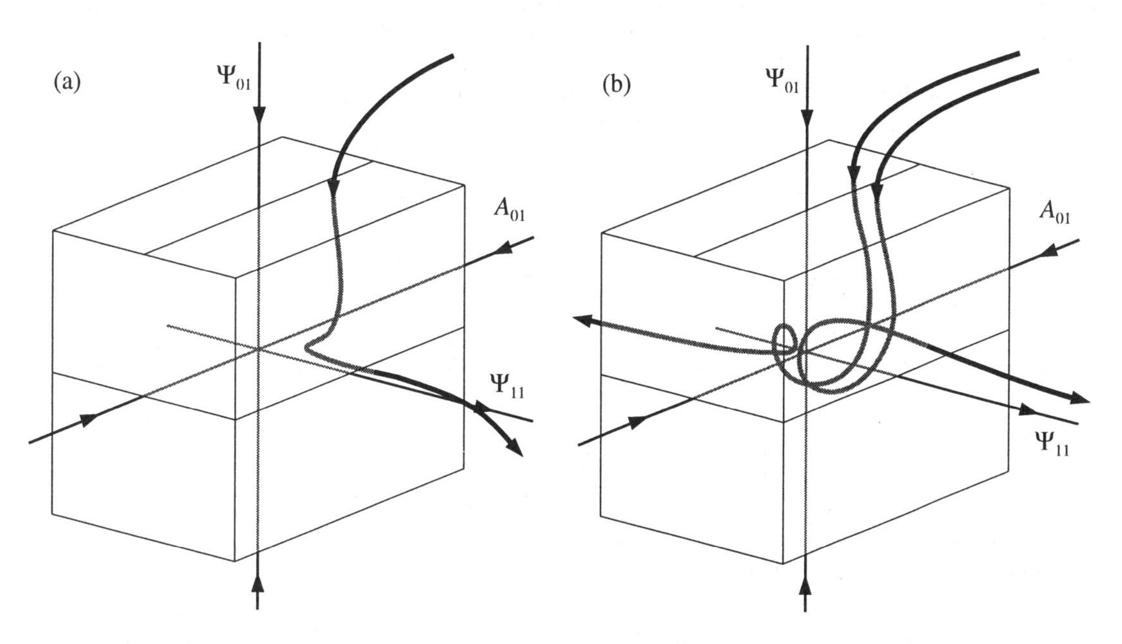

Fig. 7. As Q increases, first the eigenvectors in the (Ψ_{01}, A_{01})-plane no longer point along the axes (a), so trajectories that approach the origin with $\Psi_{01} > 0$ can leave with $\Psi_{01} < 0$. If Q is large enough, the eigenvalues become complex (b); the two trajectories illustrated twist once before leaving to the right and one and a half times before leaving to the left.

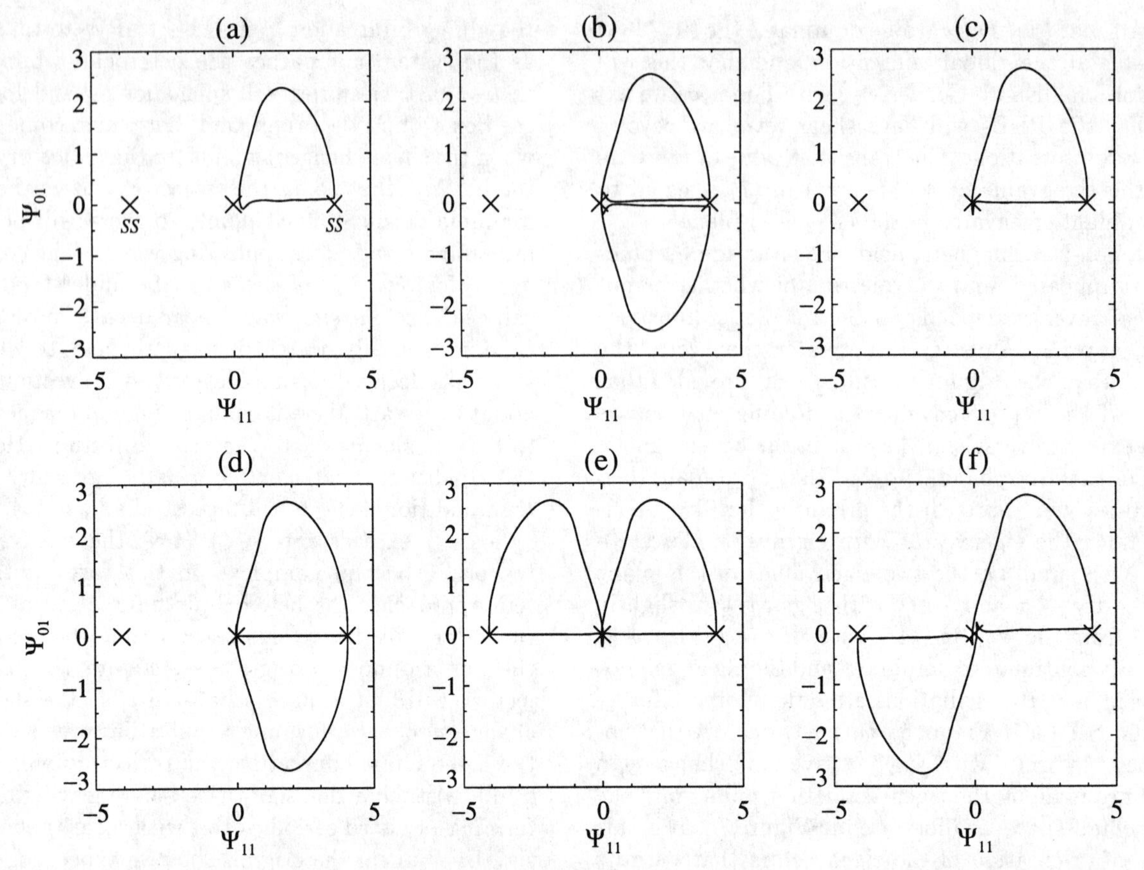

Fig. 8. A sequence of global bifurcations in the PDEs, with $Q = 19.74$, $L = 1.0$, $\sigma = 0.1$ and $\zeta = 0.1$, showing how the orbits wind up around the origin and then unwind. (a) OTC: $R = 2.000R_S$; (b) PW: $R = 2.100R_S$; (c) OTC: $R = 2.150R_S$; (d) PW: $R = 2.156R_S$; (e) HK: $R = 2.160R_S$; (f) PW′: $R = 2.200R_S$, where $R_S = 8\pi^4$. At the heteroclinic bifurcation ($R \approx 2.156R_S$), the relevant ratio of eigenvalues is 18.2. The trivial and SS fixed points are indicated by crosses.

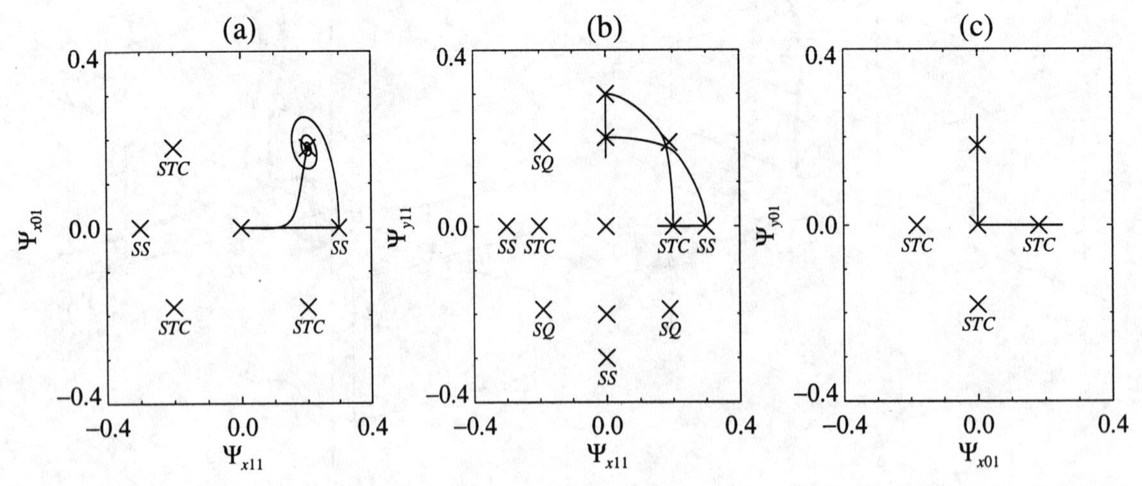

Fig. 9. An attracting and structurally stable heteroclinic cycle in the model for three-dimensional convection (14). The parameters are $\mu = 0.09$, $Q = 1.0$, $\sigma = 0.5$, $\zeta = 0.2$ and $\beta = 0.5$. Three projections are shown: (a) Ψ_{x01} against Ψ_{x11}, (b) Ψ_{y11} against Ψ_{x11} and (c) Ψ_{y01} against Ψ_{x01}. The crosses indicate some of the many fixed points: SS indicate pure rolls, STC indicate rolls plus corresponding shear, and SQ represent pure squares. Only a few of the many possible heteroclinic connections are shown.

sequence of gluing bifurcations leading up to the heteroclinic bifurcation as trajectories acquire additional half-twists in the box around the origin, alternating between the OTC and PW type of oscillation. After the heteroclinic bifurcation, when the unstable manifold of the SS fixed point hits the origin, there will be another infinite sequence of untwisting bifurcations, alternating between the HK and PW′ type of oscillation. Each gluing bifurcation in the sequence may or may not involve an interval of chaotic trajectories, depending on the eigenvalues at the SS fixed point. Examples of twisted periodic orbits from the sequence in the PDEs are shown in Figure 8.

4. THREE-DIMENSIONAL BEHAVIOUR

Three-dimensional convection will obviously generate a much wider variety of solutions than in two dimensions. Physically, an instability to three-dimensional convection should be expected, as magnetic field and shear stretched out in one direction (say, along the x-axis) will suppress x-rolls (with their axis in the y-direction), but will not suppress y-rolls.

The initial bifurcation to steady convection can give rolls, squares or hexagons, but with a small magnetic field or with a Boussinesq fluid, rolls are typically preferred for the parameter values of interest [*Clune and Knobloch*, 1994]. Near the onset of convection in narrow rolls, the reduction to a low-order model described in section 2 can be repeated for three-dimensional convection. The initial truncation requires 44 modes (including modes with vertical vorticity), and a similar scaling and elimination procedure yields the following ninth-order model, which turns out to be two copies of the fifth-order model (12) coupled in the obvious way:

$$
\begin{aligned}
\dot{\Psi}_{x11} &= \mu\Psi_{x11} + \Psi_{x11}\theta_{02} - \Psi_{x01}\Psi_{x12} - \beta\Psi_{y11}^2\Psi_{x11}, \\
\dot{\Psi}_{x12} &= -\nu\Psi_{x12} + \Psi_{x11}\Psi_{x01}, \\
\dot{\Psi}_{x01} &= -\frac{\sigma}{4}\Psi_{x01} - \frac{\sigma Q}{4\pi^2}A_{x01} + \frac{3(1+\sigma)}{4\sigma}\Psi_{x11}\Psi_{x12}, \\
\dot{A}_{x01} &= \frac{\zeta}{4}\Psi_{x01} - \frac{\zeta}{4}A_{x01}, \\
\dot{\theta}_{02} &= -\theta_{02} - \Psi_{x11}^2 - \Psi_{y11}^2, \qquad (14) \\
\dot{\Psi}_{y11} &= \mu\Psi_{y11} + \Psi_{y11}\theta_{02} - \Psi_{y01}\Psi_{y12} - \beta\Psi_{x11}^2\Psi_{y11}, \\
\dot{\Psi}_{y12} &= -\nu\Psi_{y12} + \Psi_{y11}\Psi_{y01}, \\
\dot{\Psi}_{y01} &= -\frac{\sigma}{4}\Psi_{y01} - \frac{\sigma Q}{4\pi^2}A_{y01} + \frac{3(1+\sigma)}{4\sigma}\Psi_{y11}\Psi_{y12}, \\
\dot{A}_{y01} &= \frac{\zeta}{4}\Psi_{y01} - \frac{\zeta}{4}A_{y01}.
\end{aligned}
$$

Here, the parameters are the same as in (12); β is a parameter that is formally zero in the limit of narrow rolls, but that is required to break the roll/square degeneracy that would otherwise occur ($\beta > 0$ as rolls are preferred at onset). Ψ_{x11} and Ψ_{y11} represent rolls with their axes in the y- and x-directions; Ψ_{x12} and Ψ_{y12} represent the tilt of these rolls, and Ψ_{x01} and Ψ_{y01} their shear. Similarly, A_{x01} and A_{y01} represent magnetic field stretched out along the x- and y-axes, and θ_{02} is the horizontally averaged temperature.

The ODEs (14), and the PDEs from which they were derived, have many invariant subspaces; for example,

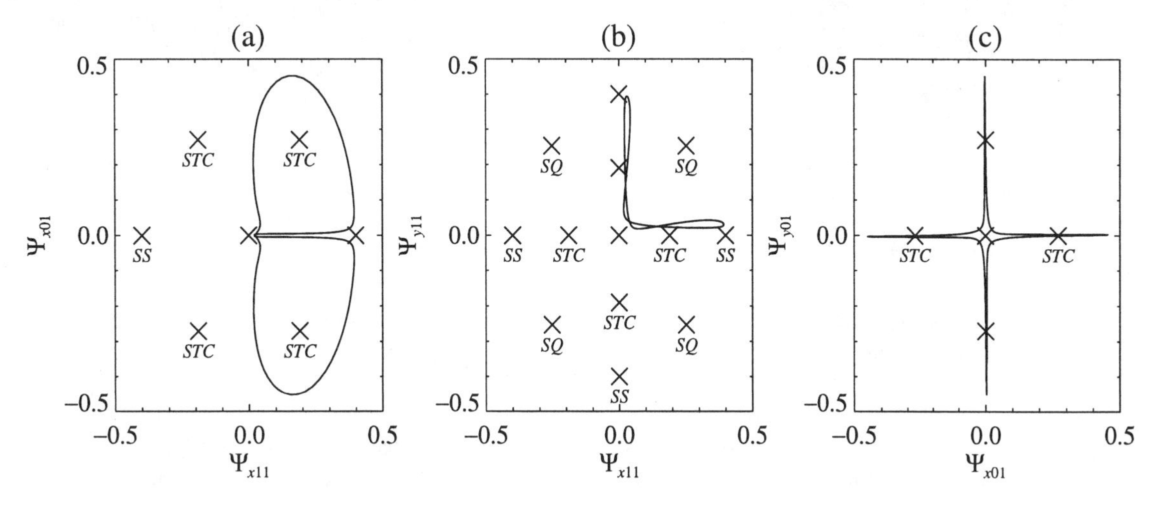

Fig. 10. An alternating pulsating wave in (14). The parameters are as in Figure 9, but with $\mu = 0.16$. (a) Ψ_{x01} against Ψ_{x11}; (b) Ψ_{y11} against Ψ_{x11}; (c) Ψ_{y01} against Ψ_{x01}. This periodic orbit shears in the positive x direction, then the positive y direction, then the negative x direction, then the negative y direction.

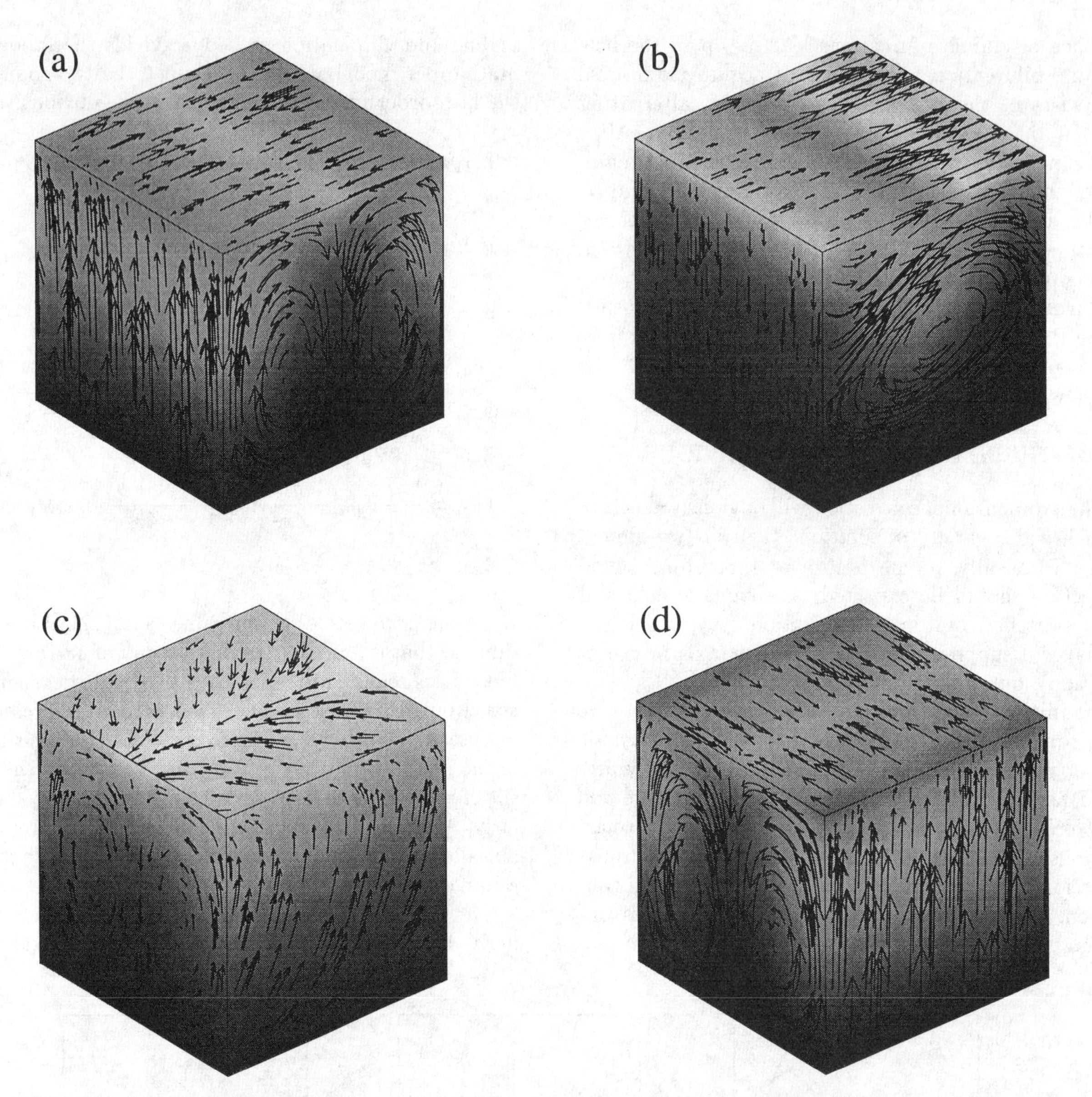

Fig. 11. An alternating pulsating wave in compressible magnetoconvection [after *Matthews et al.*, 1994]. The three-dimensional shearing instability of rolls in a magnetic field. The frames show streaklines and the gas density (black represents dense fluid and white represents light fluid) at times (a) $t = 0$, (b) $t = 0.08P$, (c) $t = 0.16P$ and (d) $t = 0.25P$, where P is the period of the oscillation.

pure x-rolls and pure y-rolls are each two-dimensional subspaces, and the mixture of these, which will include squares, is three dimensional. Additional subspaces are: x-rolls plus x-shear (five dimensional), y-rolls plus y-shear (five dimensional) and x-rolls plus y-rolls plus x- or y-shear (six dimensional). This proliferation of invariant subspaces leads to the formation of a network of structurally stable heteroclinic cycles, which are observed in the ODEs (14) (see Figure 9) as well as the PDEs. This cycle begins with pure x-rolls (SS), which are unstable to x-shear, so the trajectory goes to STC. In this five-dimensional subspace STC is stable, but it is unstable to y-rolls, in a six-dimensional subspace. As y-rolls develop, they suppress x-rolls, x-shear decays, and the trajectory reaches pure y-rolls. The cycle continues with the development of y-shear, then of x-rolls when the system returns to its original state. The cycle is structurally stable because at each step, the system is travelling in an invariant subspace from an unstable fixed point to a fixed point that is stable within the subspace (but unstable in a direction out of the subspace). In the ODEs and in the PDEs, when the heteroclinic cycle is attracting, the amplitudes of the modes vary over many orders of magnitude and can descend to the levels of numerical noise before growing again as the system moves from one invariant subspace to another. Thus we expect the dynamics in this regime to be noise-sensitive: tiny differences will determine whether the shear will develop one way or the other.

At higher values of μ, the heteroclinic cycle loses stability and is replaced by an alternating pulsating wave (Figure 10), which is strictly periodic and is not noise-sensitive. In this example, the direction of the shearing motion rotates by 90° over each quarter-cycle. An example of this oscillation in the PDEs for compressible magnetoconvection is shown in Figure 11. The system begins near pure x-rolls, then x-shear develops, which suppresses x-rolls but not y-rolls. The y-rolls grow (completing a quarter-cycle), y-shear develops, suppressing y-rolls. The x-rolls grow again and x-shear develops again, but this time in the opposite direction from before, and the cycle is completed with y-shear, again in the opposite direction from the first time. Over the course of this oscillation, the trajectory comes close to the fixed points but does not hit them.

The next step in the analysis of the three-dimensional behaviour will be the construction of a map corresponding to (13), but this is beyond the scope of this paper.

5. CONCLUSIONS

Using ideas from nonlinear dynamics, we have unravelled some of the complicated shearing behaviour observed in numerical simulations for the PDEs for convection. The low-order sets of ODEs have greatly aided the interpretation of the numerical results, and an understanding of the precise nature of the global bifurcations requires the use of low-dimensional maps. We have shown under what circumstances pulsating waves, with reversals of shear flow, should be expected. With no magnetic field, we expect PW′ to be created in the global bifurcation with moderate to large Prandtl number σ (when Ψ_{12} dominates Ψ_{01}). With a weak magnetic field, we expect reversals if Ψ_{12} dominates Ψ_{01} and A_{01}; otherwise, we need $\sigma > \zeta$ for reversals, or Q large enough for there to be complex eigenvalues at the origin. These predictions will be compared against numerical solutions of the PDEs in a future paper. The transitions in three-dimensional convection are a great deal more complicated, and await further analysis.

Acknowledgements. We are grateful for comments and suggestions from P. Glendinning, A. Shil'nikov, M. Silber and C. Sparrow. We are also grateful for a travel grant from the U.S. Office of Naval Research, and we thank Profs. Brandt and Fernando for organising the AGU Chapman Conference on Double-Diffusive Convection. This research was supported by the Science and Engineering Research Council, U.K., and by Peterhouse, Cambridge.

REFERENCES

Clune, T. and Knobloch, E., Pattern selection in three-dimensional magnetoconvection, *Physica*, *74D*, 151–176, 1994.

Glendinning, P., Bifurcations near homoclinic orbits with symmetry, *Phys. Lett.*, *103A*, 163–166, 1984.

Golubitsky, M., Stewart, I. and Shaeffer, D.G., *Singularities and Groups in Bifurcation Theory. Volume II*, Springer, New York, 1988.

Guckenheimer, J. and Holmes, P., *Nonlinear Oscillations, Dynamical Systems and Bifurcations of Vector Fields*, Springer, New York, 1983.

Howard, L.N. and Krishnamurti, R., Large-scale flow in turbulent convection: a mathematical model, *J. Fluid Mech.*, *170*, 385–410, 1986.

Hughes, D.W. and Proctor, M.R.E., A low-order model of the shear instability of convection: chaos and the effect of noise, *Nonlinearity*, *3*, 127–153, 1990.

Knobloch, E., Weiss, N.O. and Da Costa, L.N., Oscillatory and steady convection in a magnetic field, *J. Fluid Mech.*, *113*, 153–186, 1981.

Krishnamurti, R. and Howard, L.N., Large-scale flow generation in turbulent convection, *Proc. Nat. Acad. Sci. U.S.A.*, *78*, 1981–1985, 1981.

Landsberg, A.S. and Knobloch, E., Direction-reversing traveling waves, *Phys. Lett.*, *159A*, 17–20, 1991.

Lantz, S.R., Magnetoconvection dynamics in a stratified layer. II. A low-order model of the tilting instability, *Astrophys. J.*, in press, 1995.

Lorenz, E.N., Deterministic nonperiodic flow, *J. Atmos. Sci.*, *20*, 130–141, 1963.

Lyubimov, D.V. and Zaks, M.A., Two mechanisms of the transition to chaos in finite-dimensional models of convection, *Physica*, *9D*, 52–64, 1983.

Matthews, P.C., Proctor, M.R.E., Rucklidge, A.M. and Weiss, N.O., Pulsating waves in nonlinear magnetoconvection, *Phys. Lett. A*, *183*, 69–75, 1993.

Prat, J., Massaguer, J.M. and Mercader, I., Mean flow in 2D thermal convection, in *Mixing in Geophysics*, edited by Redondo, J.M. and Metais, O., pp. 208–215, Universitat Politècnica de Catalunya, Barcelona, 1994.

Proctor, M.R.E. and Weiss, N.O., Symmetries of time-dependent magnetoconvection, *Geophys. Astrophys. Fluid Dynamics*, *70*, 137–160, 1993.

Proctor, M.R.E., Weiss, N.O., Brownjohn, D.P. and Hurlburt, N.E., Nonlinear compressible magnetoconvection. Part 2. Instabilities of steady convection, *J. Fluid Mech.*, *280*, 227–253, 1994.

Rucklidge, A.M., Chaos in a low-order model of magnetoconvection, *Physica*, *62D*, 323–337, 1993.

Rucklidge, A.M. and Matthews, P.C., Shearing instabilities in magnetoconvection, in *Solar and Planetary Dynamos*, edited by Proctor, M.R.E., Matthews, P.C. and Rucklidge, A.M., pp. 257–264, Cambridge University Press, Cambridge, 1993.

A.M. Rucklidge and P.C. Matthews, Department of Applied Mathematics and Theoretical Physics, University of Cambridge, Cambridge CB3 9EW, UK

Diffusion and Double Diffusive Convection in Isothermal Liquid Boundaries

Vincenzo Vitagliano

Dipartimento di Chimica, Università di Napoli Federico II, Via Mezzocannone 4, 80134 Napoli, Italy. FAX (3981) 5527771

Gravitational instabilities in multicomponent liquid diffusion boundaries are briefly discussed. Two types of convective motions have been found in a diffusion boundary: (1) Dynamic instability arises at the center of the boundary and develops with a finger-like mechanism. (2) Diffusional instability arises at the borders of the boundary promoting a layering of the boundary that keeps itself apparently stable and sharp for long time. The experimental techniques that allow to detect quite accurately the conditions for onset of convection are briefly illustrated.

1. DIFFUSION

The aim of this paper is to illustrate briefly the behavior of isothermal free diffusion boundaries, in connection with the possible growth of gravitational instabilities inside them (*double diffusive convection* [*Chen and Johnson*, 1984]).

In fact, diffusion layers of fluid systems, where transport is promoted by at least two independent driving forces, may develop convective motions so that the transport process is completely modified.

This is the case of isothermal diffusion in three or more component systems, or diffusion in binary systems coupled to a temperature gradient. In the first case the independent driving forces are the concentration gradients of two, or more, components (among three, or more), in the second case, that has largely interested ocenographers, the independent driving forces are the concentration gradient of one component (salt in sea water) and the temperature gradient.

Isothermal diffusion in a n-component system can be described by the extended Fick's law [*Kirkaldy and Young*, 1987; *Tyrrell and Harris*, 1984; *Vitagliano*, 1991]:

$$J_i = -\sum_j D_{ij} \operatorname{grad} C_j \qquad (i, j = 1, \ldots, n-1) \qquad (1)$$

where J_i is the flow of component i, C_j the concentration of component j, and D_{ij} are the diffusion coefficients.

Double-Diffusive Convection
Geophysical Monograph 94

The concentration gradient of the n-th component (generally chosen as the solvent) and its flow, J_n, are not independent quantities; the following relation holds among concentrations

$$\sum C_i \bar{V}_i = 1 \qquad (2)$$

where $\bar{V}_i$ is the partial specific volume of component i (or its partial molar volume, if concentration is expressed in terms of mol L^{-1}).

The relation among flows arises from the continuity law and depends on the reference frame chosen for describing the motion of components:

$$\sum J_i = 0 \quad \text{(in terms of mass fixed reference frame)} \qquad (3)$$

$$\sum J_i \bar{V}_i = 0 \quad \text{(in terms of volume fixed ref. frame)} \qquad (4)$$

$$J_i = C_i (v_i - v_n) \quad \text{(in terms of relative motion of component i with respect to component n)} \qquad (5)$$

v_i being the velocity of component i.

For concentration differences not too large, the diffusion coefficients, in Eq.(1), may be assumed as constant through the diffusion boundary (differential diffusion). This assumption does not mean that the D_{ij} are independent of concentration, but only that in actual experimental conditions they may be assumed independent of concentration gradients. On this assumption, if diffusion occurs only in one direction (say y) eq.(1) may be written as

$$\partial C_i / \partial t = \sum_j D_{ij} (\partial^2 C_j / \partial y^2) \qquad (i, j = 1, \ldots, n-1) \qquad (6)$$

Eqs. (1) and (6) show that diffusion in a n-component system is described by $(n-1)^2$ coefficients: 4 diffusion coefficients in a three component system, 9 in a four component system.

Experiments have shown that the cross diffusion terms (D_{ij} $i \neq j$) cannot be ignored [*Vitagliano*, 1991]. In multi-component systems they may be large and even larger than the main terms, although the condition

$$\lim_{c_i \to 0} D_{ij} = 0 \qquad (7)$$

applies. This condition has a simple physical explanation. In absence of component i, its flow, J_i, must vanish. However, the concentration gradient of component j can assume any arbitrary value. Consequently, in order to keep $J_i = 0$, it is necessary that D_{ij} goes to zero.

For the stability of the diffusion process in ternary systems two conditions are required:

$$\mathrm{Trace}(\mathbf{D}) > 0 \qquad (8)$$

and

$$|\mathbf{D}| \geq 0 \qquad (9)$$

where the equality sign applies at the plait point of a phase separation region (or along a spinodal curve).

It is worth while to stress here that the presence of large cross terms in Eqs. (1) or (6) has nothing of peculiar. These equations describe the interchange motion of all n components in terms of $(n-1)^2$ coefficients. Their values depend on the arbitrary choice of the n-th component. A set of transform equations [*Vitagliano* , *et al.*, 1986; *Miller, et al.*, 1986] allows to change the set of diffusion coefficients in connection with the choice of component n. Although the determinant of **D** and the trace of **D** are invariant [*Miller, et al.*, 1986; *Vitagliano, et al.*, 1978].

Acrtally, a suitable choice of component n may allow a better insight on the transport mechanism. It can be remembered that the cross terms coefficients of the system *sucrose - sodium chloride - water* are very small if water is chosen as component 3, while they become large if NaCl is chosen as component 3, showing that the water transport is largely driven by sucrose [*Vitagliano, et al.*, 1986].

The four D_{ij} have beeen obtained for a variety of systems during the last decades [*Kirkaldy and Young*, 1987; *Tyrrell and Harris*, 1984; *Vitagliano*, 1991], and data are available also for some four component system [*Paduano, et al.*, 1992].

2. GRAVITATIONAL INSTABILITIES

The stability analysis of diffusion boundaries was widely discussed in the literature [*Veronis*, 1965; *Baines and Gill*, 1969; *Sartory*, 1969; *Turner*, 1973; *Huppert and Manins*, 1973], including the particular case of isothermal multi-component boundaries [*McDoughall*, 1983; *Vitagliano, et al.*, 1984, 1986, 1992; *Ambrosone, et al..*, 1993] and various experimental cases were examined [*Vitagliano, et al.*, 1972, 1984, 1986, 1988; *Miller and Vitagliano*, 1986].

In a two component system only a static condition is required for the boundary stability, namely the absence of density inversions inside the boundary. Assuming that the y axis be positive in the upward direction:

$$\partial \rho / \partial y \leq 0 \qquad (10)$$

where ρ is the density of the solution. This condition is generally fulfilled when the top solution has a lower density than the bottom one.

In a three component system an additional condition is required, it may be called a dynamic condition:

$$(1/y)(\partial^2 \rho / \partial y^2) \geq 0 \qquad (11)$$

When this condition is not fulfilled, convection grows even in absence of density inversions inside the boundary.

Any further component imposes an additional condition for the boundary stability. However, it has been proved that, at least in a four component system, this additional condition is redundant (*Vitagliano, et al.*, 1992, Eq. 44a), in fact it always fails when conditions (10) and (11) had already failed.

It has not been proved, yet, whether conditions (10) and (11) are the only necessary and sufficient conditions for stability in systems with an arbitrary number of components.

3. DIFFUSION BOUNDARIES

The initial and boundary conditions of free diffusion in one dimension are:

$$C_i = C_i^o - \frac{1}{2} \Delta C_i \quad \text{for } y > 0 \text{ and } t = 0$$
$$C_i = C_i^o + \frac{1}{2} \Delta C_i \quad \text{for } y < 0 \text{ and } t = 0 \qquad (12)$$

$$\lim_{y \to +\infty} C_i = C_i^o - \frac{1}{2} \Delta C_i \quad \text{for } t > 0$$
$$\lim_{y \to -\infty} C_i = C_i^o + \frac{1}{2} \Delta C_i \quad \text{for } t < 0 \qquad (13)$$

Under these conditions Eqs. (6) can be integrated leading, in a three-component system, to the sum of two error functions for any property which is a linear function of the concentration. In eqs. (12) and (13) ΔC_i is the concentration difference between bottom and top solutions (bottom - top), and the origin of axes was put at the center of the diffusion boundary. In terms of solution density the integration of Eqs.(6) gives:

$$\rho = \rho^{o} + \frac{\Delta n}{2}\left[\Gamma_{+}\Phi\left(z\sqrt{\sigma_{+}}\right) + \Gamma_{-}\Phi\left(z\sqrt{\sigma_{-}}\right)\right] \qquad (14)$$

$$\frac{1}{\Delta\rho}\frac{d\rho}{dz} = \frac{1}{\sqrt{\pi}}\left[\Gamma_{+}\sqrt{\sigma_{+}}\,e^{-\sigma_{+}z^{2}} + \Gamma_{-}\sqrt{\sigma_{-}}\,e^{-\sigma_{-}z^{2}}\right] \qquad (15)$$

where:

$$z = \frac{y}{\sqrt{4t}} \qquad (16)$$

and

$$\Phi(q) = \frac{2}{\sqrt{\pi}}\int_{0}^{q} \exp(-q^{2})\, dq \qquad (17)$$

Table 1 A collects all the expressions required in Eqs. (14) and (15).

Obviously, actual boundaries cannot correspond to a step function at t=0 [initial conditions (12)], however they approach the density distribution predicted by Eqs. (14)-(15) as $t \to \infty$.

Figure 1 shows a schematic representation of a possible initial diffusion boundary in a three-component system, including some possible evolution of the density, according to eq.(14). It can be seen that gravitational instabilities due to the failing of condition (10) or (11) may arise either at the center of the boundary (graphs **2** and **3**, C) or at its borders (graphs **4**, C).

In the case of failure of condition (11) [and eventually also condition (10)] at the center of the boundary a *finger-like* convection grows up, comparable to that found in sea water when diluted cold water stratifies below salty hot water. In this case the diffusion boundary is generally destroyed in a time much shorter than that required by the simple diffusional mixing.

When density inversions appear at the borders of the boundary, because of the failing of condition (10), the gravitational instability promotes the growing of large convective cells that *wash* the upper and lower borders of the diffusion boundary keeping it sharp and narrow for much longer time than that required by simple diffusion. The process is similar to the stratifications appearing in sea water when hotter concentrated water is stratified below a cooler and more diluted water (we called it *overstability*).

The evolution of a three-component boundary [Eqs. (14) and (15)] can be discussed in terms of the ratio $\Delta C_2/\Delta C_1$ between the concentration differences of bottom and top solutions. Let us represent the composition of top and bottom solutions in a *clock-like* diagram where the concentration of component 1 is plotted on the abscissa and that of component 2 on the ordinate (we assume that component 2 is the faster diffusing one, namely, $D_{22} > D_{11}$). This diagram is shown in Fig.2 where some characteristic $\Delta C_2/\Delta C_1$ conditions are drawn. Fig.3 shows, in a three-dimensional graph, the evolution of the density gradient (Eq.15) by changing the ratio $\Delta C_2/\Delta C_1$. The mathematical expressions for some characteristic $\Delta C_2/\Delta C_1$ are collected in Table 1 B.

Some features of the expressions given in Table 1 are interesting to be pointed out [*Miller and Vitagliano,* 1986].

(1) The slope of line G is independent of the concentration coefficients of density, H_1 and H_2, depending only on the four diffusion coefficients.

(2) In absence of cross terms the slope of line G is zero, while that of line F becomes:

$$\Delta C_2/\Delta C_1 = -(H_1/H_2)/(D_{22}/D_{11})^{3/2} \qquad (18)$$

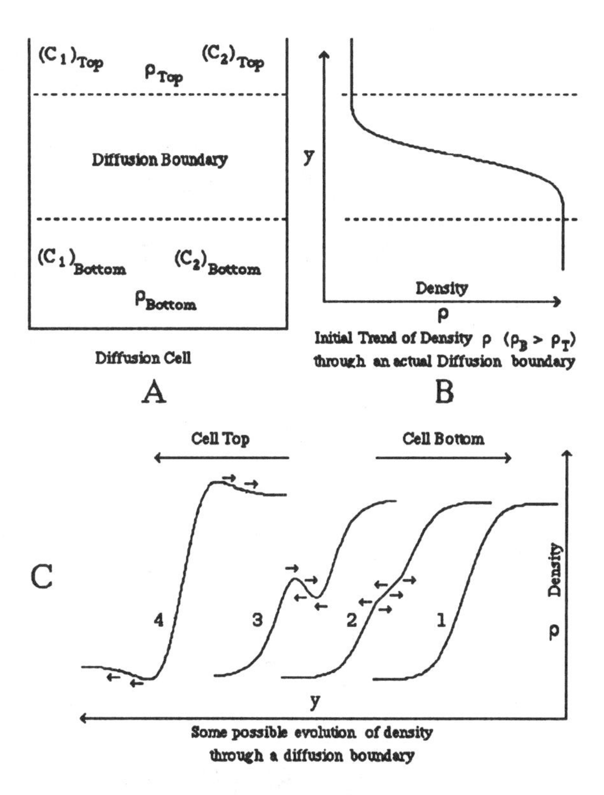

Fig. 1. **A** and **B**, schematic representation of an actual initial diffusion boundary in a 3-component system. **C**, four possible evolutions of a diffusion boundary according to Eq.(14): **1**, stable boundary, conditions (10) and (11) are fulfilled; **2**, boundary unstable at the center where condition (11) fails, although no density inversions evolved due to the diffusion process (dynamic instability); **3**, boundary unstable at the center, both conditions (10) and (11) fail, the diffusion process promoted a density inversion at the center of the boundary; **4**, boundary unstable at the borders where condition (10) fails and density inversions appear due to diffusion.

TABLE 1. Useful Mathematical Expressions

$$\Gamma_+ = 2\frac{H_1K_1^+ + H_2K_2^+}{H_1\Delta C_1 + H_2\Delta C_2} \quad \text{and} \quad \Gamma_- = 2\frac{H_1K_1^- + H_2K_2^-}{H_1\Delta C_1 + H_2\Delta C_2} \qquad (*)$$

$$\sigma_+ = \frac{1}{2}\frac{(D_{22}+D_{11}) + [(D_{22}-D_{11})^2 + 4D_{12}D_{21}]^{1/2}}{D_{22}D_{11} - D_{12}D_{21}} \qquad (*)$$

$$\sigma_- = \frac{1}{2}\frac{(D_{22}+D_{11}) - [(D_{22}-D_{11})^2 + 4D_{12}D_{21}]^{1/2}}{D_{22}D_{11} - D_{12}D_{21}} \qquad (*)$$

$$K_1^+ = \frac{1}{2}\frac{\{(D_{22}-D_{11}) + [(D_{22}-D_{11})^2 + 4D_{12}D_{21}]^{1/2}\}\Delta C_1 - 2D_{12}\Delta C_2}{4[(D_{22}-D_{11})^2 + 4D_{12}D_{21}]^{1/2}} \qquad (*)$$

$$K_1^- = -\frac{1}{2}\frac{\{(D_{22}-D_{11}) - [(D_{22}-D_{11})^2 + 4D_{12}D_{21}]^{1/2}\}\Delta C_1 - 2D_{12}\Delta C_2}{4[(D_{22}-D_{11})^2 + 4D_{12}D_{21}]^{1/2}} \qquad (*)$$

where:

$H_1 = \partial\rho/\partial C_1$, $H_2 = \partial\rho/\partial C_2$, and K_2^+ and K_2^- are obtained by interchanging indices 1 and 2 throughout. (*)

$$\Delta\rho = H_1\,\Delta C_1 + H_2\,\Delta C_2 \qquad (**)$$

$$\frac{\Delta C_2}{\Delta C_1} = \frac{\left(\frac{H_2}{H_1}\frac{D_{21}}{D_{11}} - \frac{D_{22}}{D_{11}}\right)(A - BZ) - 2D_{21}\left(\frac{D_{12}}{D_{11}} - \frac{H_2}{H_1}\right)}{\left(\frac{D_{12}}{D_{11}} - \frac{H_2}{H_1}\right)(A - BZ) + 2D_{12}\left(\frac{H_2}{H_1}\frac{D_{21}}{D_{11}} - \frac{D_{22}}{D_{11}}\right)} \qquad \text{Line F} \qquad (**)$$

$$\frac{\Delta C_2}{\Delta C_1} = -\left(\frac{H_1}{H_2}\right)\frac{(D_{22}D_{11} - D_{12}D_{21})^{1/2} + D_{22} - (H_2/H_1)D_{21}}{(D_{22}D_{11} - D_{12}D_{21})^{1/2} + D_{11} - (H_1/H_2)D_{12}} \qquad \text{Line D} \qquad (**)$$

$$\frac{\Delta C_2}{\Delta C_1} = -\left(\frac{H_1}{H_2}\right) \qquad \text{Line N} \qquad (**)$$

$$\frac{\Delta C_2}{\Delta C_1} = -2D_{21}/(A + BZ) \qquad \text{Line G} \qquad (**)$$

where:

$A = (D_{22} - D_{11})$; $B = (1 + w)/(1 - w)$; $w = (\sigma_+/\sigma_-)^{1/2}\exp[-z^2(\sigma_+ - \sigma_-)]$;
and $Z = (A^2 + 4\,D_{12}\,D_{21})^{1/2}$ (**)

(*) Expressions used in Eqs. (14) and (15)

(**) Limiting conditions for the onset of gravitational instabilities in free diffusion boundaries, for $D_{22} > D_{11}$

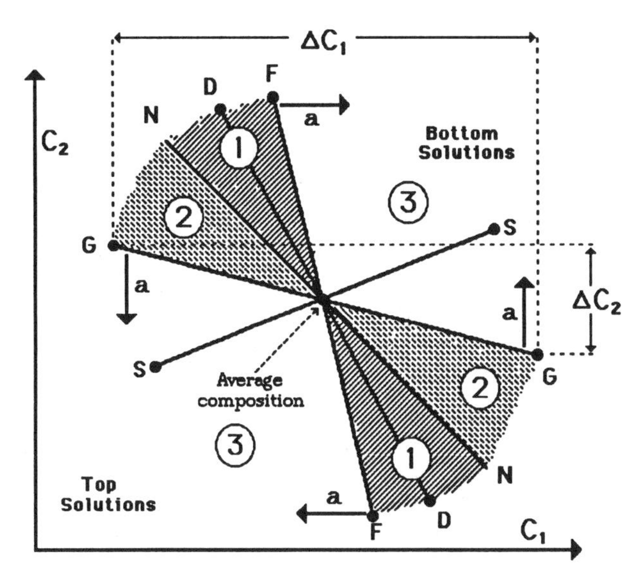

Fig. 2. Clock-like diagram showing the compositions of top and bottom solutions in a diffusion run (●). **(1)** Field of instabilities at the center of the boundary. **(2)** Field of instabilities at the borders of the boundary. **(3)** Field of gravitationally stable boundaries. **S**, example of $\Delta C_2/\Delta C_1$ conditions for a stable boundary. **F**, limit $\Delta C_2/\Delta C_1$ for the failing of condition (11) at the center of the boundary (dynamic instability). **D**, limit $\Delta C_2/\Delta C_1$ for the failing of condition (10) at the center of the boundary. **N**, neutral density line (ρ = constant). **G**, limit $\Delta C_2/\Delta C_1$ for the failing of condition (10) at the borders of the boundary. All boundaries having $\Delta C_2/\Delta C$ included between **F** and **N** should promote finger-like convection while those having $\Delta C_2/\Delta C_1$ included between **N** and **G** should give rise to stratification. Arrows **a** show the effect of increasing D_{12} and /or decreasing D_{21} on the slope of Lines **F** and **G**.

(3) On the assumption that $H_2 > H_1 > 0$ and

$$P_2 = D_{22} + (H_2/H_1)D_{21} > \\ > D_{11} + (H_1/H_2)D_{12} = P_1 \qquad (19)$$

increasing D_{12} and/or decreasing D_{21} turns line F clockwise and line G contreclockwise (see arrows **a** in Fig.2).

It is possible to have line F or line G, or both with a positive slope, that is to have instability conditions even with both components 1 and 2 (heavier than the third component) more concentrated at the bottom. Systems like these were found experimentally [*Miller, et al.*, 1988; *Costantino, et al.*, 1992]. In particular line G has always a positive slope if $D_{21} < 0$, this, generally happens when attractive interactions are present between component 1 and component 2.

(4) If $P_1 = P_2$ [Eq. (19)] lines F, D, and G collapse on line N, diffusion boundaries are always gravitationally stable for any $\Delta C_2/\Delta C_1$ values.This peculiar case has also been proved experimentally for the system KBr - tetrabuthylammonium bromide - water [*Vitagliano, et al.*, 1988].

4. EXPERIMENTAL TECHNIQUES

Various experimental techniques allow the measurements of diffusion coefficients in multicomponent systems [*Kirkaldy and Young*, 1987; *Tyrrell and Harris*, 1984]. Presently, interferometric techniques, such as Gouy or Rayleigh techniques, yield quite accurate values of the D_{ij}. Fig.4 **A** is a representation of the Gouy diffusiometer optics. Fig.4 **B** is a scheme of the diffusiometer set up allowing also the Rayleigh fringes registration (which shows the refractive index as a function of the diffusion direction y through the diffusion boundary), and that of the schlieren optics. This last technique allows the registration of the gradient of refractive index through the diffusion boundary. Details on these optical techniques can be found elsewhere [*Kirkaldy and Young*, 1987; *Tyrrell and Harris*, 1984].

The boundary thickness, in a diffusion run, is quite narrow, it never exceeds 10-12 mm, so that the direct observation of convective motions inside it is very difficult. However, the optical techniques used in diffusion measurements allow to observe easily the boundary distorsions due to the onset of convection, so they are useful to test accurately the fluid-dynamic theory describing the process of double diffusive convection [*Miller and Vitagliano*, 1986].

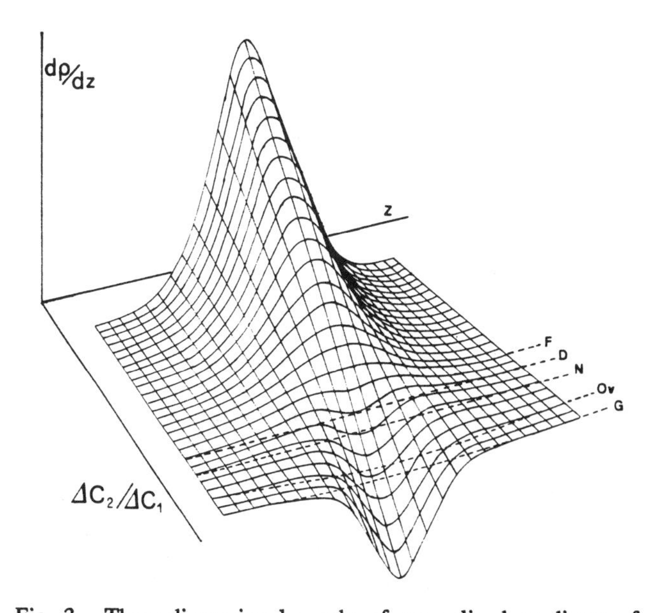

Fig. 3. Three-dimensional graphs of normalized gradients of density (Eq.15). Curves corresponding to the $\Delta C_2/\Delta C_1$ plotted in Fig. 2 are indicated with the corresponding letters. **Ov** is a density gradient curve corresponding to conditions favouring the boundary stratifications.

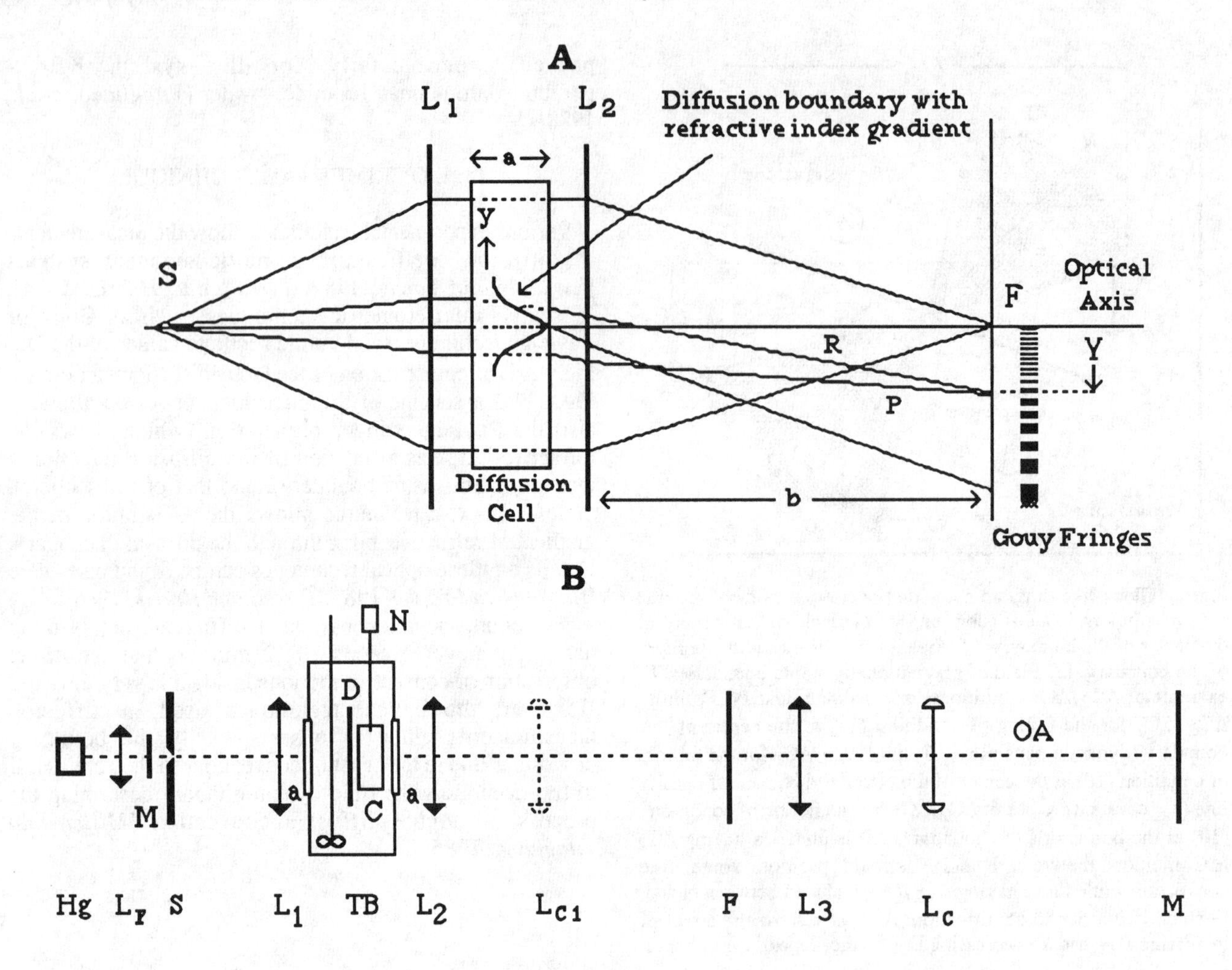

Fig. 4. **A.** Scheme of the Gouy Diffusiometer showing the formation of interference fringes: S is the light source, L_1 and L_2 are two lenses that give the source image in F. Light passing through the diffusion boundary, where a vertical gradient of refractive index is present, deviates to form the source image at different levels in F. Light deviation Y, in F, is given by Y = ab (dn/dy), where dn/dy is the refractive index gradient at level y into the diffusion cell. Light wave fronts R and P, passing through the diffusion boundary at different levels, but joining in F, have different optical paths and interfere on the focal plane F to form a fringe pattern. **B.** Hg, high pressure mercury vapor lampe (or laser lamp). L_F, focusing lens (focuses the lamp image on the source S). M, filter or interference monochromator (not necessary with laser lamp). S, light source (horizontal slit for Gouy or Philpot arrangement, vertical slit for Rayleigh arrangement). L_1, first lens (gives a parallel beam passing through the diffusion cell C). (a), optical windows. TB, thermostatic bath. D, screen with suitable slits put in front of the diffusion cell. C, diffusion cell. N, siphoning needle for the formation of the initial diffusion boundary. L_2, second lens (gives in F the image of the source slit). L_{C1}, cylindrical lens (focuses the cell image along the vertical direction in F, for the Rayleigh arrangement; it is used alternatively to the arrangement using lenses L_3 and L_C). F, plane where the image of the source S is focused (the camera with the photografic plate is set in this plane for Gouy arrangement, or Rayleigh with L_{C1} lens arrangement; a diagonal phase plate is set on this plane for the Philpot arrangement). L_3, third lens (gives in M the image of the diffusion cell). L_C, cylindrical lens, focuses the source image S in M (vertical axis for the Rayleigh arrangement and for the Philpot arrangement, with a horizontal axis it focuses again the Gouy fringes in M). OA, optical axis.

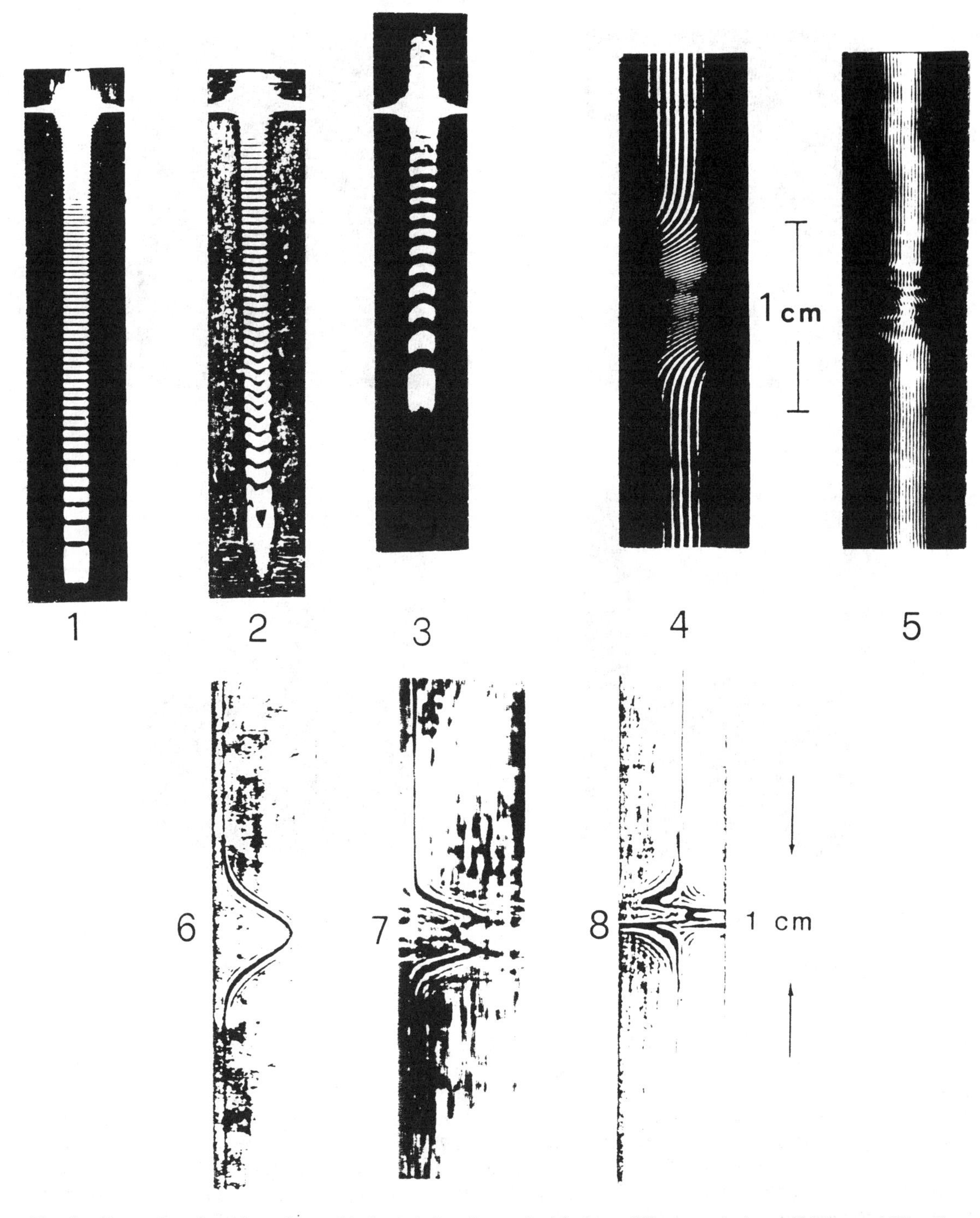

Fig. 5. Examples of stable and unstable boundaries detected with three different techniques [*Miller and Vitagliano*, 1986]. Gouy technique: **1**, stable boundary; **2**, boundary unstable at the center, perturbed fringes correspond to the level where instabilities are growing into the diffusion boundary; **3**, boundary unstable at the borders, perturbed fringes can be seen on both sides of the undeviated light. Rayleigh technique, the refractive index into the cell is registered: **4**, stable boundary; **5**, unstable boundary. Schlieren optics technique, the gradient of refractive index within the cell is registered: **6**, stable boundary; **7**, boundary unstable at the center; **8**, boundary unstable at the borders.

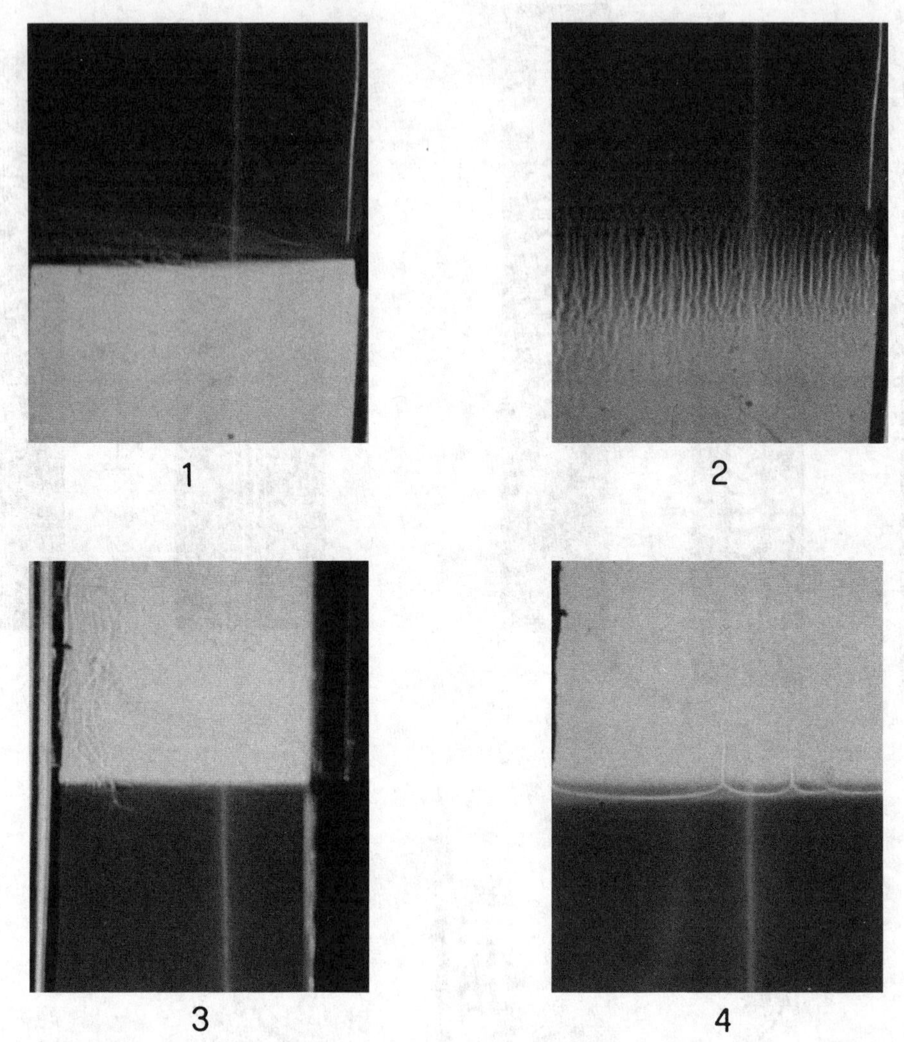

Fig. 6. Pictures of gravitationally unstable boundaries in the ternary system: red wine (10% alcohol) - sucrose - water. **1** and **2**: upper solution wine + 0.3 % sucrose (ρ = 0.993 g/ml), lower solution distilled water (ρ = 0.997 g/ml). **1**, boundary formation. **2**, *fingers* after 6 hours of run. **3** and **4**: upper solution distilled water, lower solution wine + 1.5% sucrose (ρ = 0.998 g/ml). **3**, boundary formation (large convection is present). **4**, *overstability* after 24 hours (the boundary is still sharp and large convection cells are present). Width of the diffusion cell 5 cm.

Fig.5 shows a set of photographs taken with different optical techniques, where stable and unstable boundaries are compared.

The drastic change on the convection mechanism in a diffusion boundary, by slightly changing the upper and lower solutions composition can be seen in Fig. 6 where the boundaries obtained by stratifying red wine (10% alcohol) containing a small amount of sucrose and distilled water are shown.

In pictures **1** and **2** red wine + 0.3% (density 0.993 g/ml) is stratified over distilled water (density 0.997 g/ml). Picture **1** is taken during the formation of the diffusion boundary by the siphoning technique, upper and lower solutions are clear and stable. Picture **2** is taken after about 6 hours and the evolution of fingers can be seen.

In pictures **3** and **4** distilled water is stratified over red wine + 1.5% sucrose (density 0.998 g/ml). Picture **3** is taken during the formation of the diffusion boundary. During the siphoning and soon after the starting of the run, large and turbolent convection involves all the system. It can be seen in the upper side of the cell. Picture **4**, taken after about 24 hours shows the boundary still sharp and narrow, a few convection channels, evident in the upper side of the diffusion cell, show the presence of large convective cells grown up into the system (*overstability*).

Acknowledgement. The author wishes to thank all his collaborators whose names appear in the references list. This work was supported by the Italian M.U.R.S.T. and the Italian C.N.R.

REFERENCES

Ambrosone, L., Costantino, L., Paduano, L., and Vitagliano V., Ecological problems and diffusion in *Trends in Ecological Physical Chemistry*, Bonati, L., Cosentino,U., Lasagni, M., Moro, G., Pitea, D., Schiraldi, A., Eds., Elsevier, Amsterdam 1993, pp. 65-78.

Baines, P. G. and Gill, A.E., On thermohaline convection with linear gradients, *J. Fluid Mech.*, *37*, 289-306 (1969).

Chen, C.F., and Johnson, D.H., Double - diffusive convection: A report on on an engineering foundation conference, *J. Fluid Mech.*, *138*, 405-416 (1984).

Costantino, L., Della Volpe, C., Ortona, O., and Vitagliano, V., Isothermal diffusion in a peculiar ternary system: The micro-emulsion AOT - Water - Heptane, *J. Chem. Soc. Faraday Trans.*, *88*, 61-63 (1992).

Huppert, H.E., Manins, P.C., Limiting conditions for salt-fingering at an interface, *Deep Sea Res.*, *20*, 315-323 (1973).

Kirkaldy, J.S. and Young, D. J., *Diffusion in Condensed State*, The Institute of Metals, London 1987.

McDoughall ,T. J., Double- diffusive Convection caused by coupl ed molecular diffusion, *J. Fluid Mech.*, *126*, 379- 397 (1983).

Miller, D.G., Ting, A.W., Rard, J.A., Mutual diffusion coef ficients of various $ZnCl_2$ (0.5 M) - KCl - H_2O mixtures at 298.15 K by Rayleygh i nterferometry, *J. Electrochem. Soc.*, *135*, 896 (1988).

Miller, D.G., and Vitagliano, V., Experimental test of McDoughall theory for the onset of convective instabilities in isothermal ternary systems, *J.Phys. Chem.*, *90*, 1706-1717 (1986).

Miller, D.G., Vitagliano, V., and Sartorio, R., Some comments on multicomponent diffusion. Negative main term diffusion coefficients, second law constraints, solventc hoice, and refer ence frame transformations, *J. Phys. Chem.*, *90*, 509 - 519 (1986).

Paduano, L., Sartorio, R.,Vitagliano V., Albright, J.G., and Miller, D.G., Measurement of the mutual diffusion coefficients at one composition of the four - component system: α–cyclodextrin - L- phenylalanine - monobuthyl-urea - H_2O at 25°C, *J. Phys. Chem.*,*96*, 7478-7483 (1992).

Sartory,W.K., Instability in diffusing fluid layers, *Biopolyymers*, *7*, 251-263 (1969).

Turner, J.S. , *Buoyancy Effects in Fluids*, Cambridge University Press, New York 1973

Tyrrell, H.J.V. and Harris, K.R., *Diffusion in Liquids*, Butter worths, London 1984

Veronis, G., On finite amplitude instability in hermohaline conv ection, *J.Marine Res.*, *23*, 1-17 (1965)

Vitagliano, V., Some phenomenological and thermodynamic aspects of diffusion in multicomponents systems, *Pure & Appl. Chem.*, *63*, 1441- 1448 (1991).

Vitagliano, P.L.,Ambrosone, L., and Vitagliano, V., Gravitational Instabilities in Multicomponent free diffusion boundaries, *J. Phys. Chem.*, *96*, 1431-1437 (1992).

Vitagliano, V., Borriello, G., Della Volpe, C., and Ortona, O., Instabilities in free diffusion boundaries of NaCl - sucrose-water solutions at 25°C, *J. Solution Chem.*, *15*, 811-826 (1986).

Vitagliano, P.L., Della Volpe, C., and Vitagliano, V., *J. Solution Chem..*, *13*, 549 - 562 (1984).

Vitagliano, V., Della Volpe, C., Ambrosone, L., and Costantino, L., Diffusion and double diffusive convection in isothermal ternary systems, *Phys. Chem. Hydrodynamics*, *10*, 239 - 247 (1988).

Vitagliano, V., Sartorio, R., Scala, S., and Spaduzzi,D., Diffusion in a ternary system and the critical mixing point, *J. Solution Chem.*, *7*, 605 - 621 (1978).

Vitagliano, V., Zagari, A., Sartorio, R., and Corcione, M., Dissipative structures and diffusion in ternary systems, *J. Phys. Chem.*, *76*, 2050 -2055 (1972).

Double Diffusively Unstable Intrusions Near an Oceanic Front: Observations from R/P FLIP

Steven P. Anderson
Woods Hole Oceanographic Institution, Woods Hole, Massachusetts

Robert Pinkel
Marine Physical Laboratory, Scripps Institution of Oceanography, La Jolla, California

An oceanic front was advected past the experiment site while the R/P FLIP was moored 500 km off the coast of central California during the Surface Waves Processes Program (SWAPP). A pair of CTD's and the Marine Physical Laboratory Coded-Pulse Doppler Current Profiler were deployed from FLIP to simultaneously profile from the surface to 420. The CTD's profiled every 130 seconds for 16 days obtaining nearly 9,200 profiles of potential temperature, salinity and potential density. The front is characterized below 200 m by warm, salty water encountered at the start of the experiment and a transition in the middle of the time series to colder, fresher water of nearly the same density. The temperature and salinity changes on an isopycnal located at 300 m are 0.425°C and 0.0825 ppt. To study the time and space scales of the intrusive interleaving associated with the front, a depth-timeseries of temperature and salinity anomalies is presented. The anomalies are calculated by removing the mean temperature and salinity at constant density for each CTD profile. Small scale intrusions with vertical lengths of 5 to 15 m are observed to be double diffusively unstable on the top and bottom and slope upward from the warm side of the front to the cold side. The intrusions are consistent with scales predicted by the linear instability theory of *Toole and Georgi* [1981].

1. INTRODUCTION

In this paper the fine-scale vertical structure of temperature and salinity associated with an oceanic frontal zone is examined. An oceanic front is a boundary between two water masses with different thermohaline characteristics. Fronts play an important roll in the transformation of mesoscale inhomogeneities into thermohaline finestructure. Intrusions (a layer of one water type surrounded by water with different thermohaline characteristics), temperature inversions and salinity inversions are commonly observed in thermohaline finestructure near oceanic fronts [*Horne*, 1978; *Georgi*, 1978; *Joyce et al.*, 1978; *Posmentier and Houghton*, 1978; *Fedorov*, 1986]. The formation of finestructure is an important process since the interleaving of the two water masses acts as stirring which enlarges the surface area between the two water masses and thus enhances cross frontal fluxes [*Joyce*, 1977].

There are two driving mechanisms for interleaving near a frontal zone. Horizontal advection by internal waves may move a water parcel from one side of the front to the other. If horizontal temperature gradients are strong enough, this may lead to a salinity or temperature inversion and double diffusive convection. *Turner* [1978] observed a second driving mechanism in laboratory models. He found that double diffusive convection alone can drive interleaving at the interface of two water masses with different thermohaline characteristics.

The observations of vertical interleaving presented here were obtained as part of the Surface Waves Processes Program (SWAPP). SWAPP was conceived as an experiment to study mixed layer dynamics, Lang-

Double-Diffusive Convection
Geophysical Monograph 94

muir circulation and surface wave growth. The SWAPP experiment took place off the coast of central California from February 22 to March 18, 1990. During this time, the R/P FLIP was moored on station at latitude 35°8.2′N and longitude 126°59.0′W. The R/P FLIP is a research platform which is essentially a large spar buoy [*Rudnick*, 1964]. It provides a stable laboratory and instrument platform in the open ocean. Deployed from FLIP was a rapid profiling conductivity, temperature and depth (CTD) system developed by the Marine Physical Laboratory (MPL) of the Scripps Institution of Oceanography (Figure 1). The MPL rapid profiling CTD system profiled from the surface to 420 m every 130 seconds for 16 continuous days and collected over 9000 temperature and salinity profiles.

The SWAPP experiment was not originally designed to be a study of local hydrographic features. The data set consists of a timeseries of velocity, temperature and salinity profiles from a fixed position. The frontal structures are observed as the water masses are advected past the experiment site. An XBT survey over the waters surrounding the SWAPP experiment site, taken on yearday 55, 1990 (Figure 2), reveals a front in the 200 to 400 m depth range. The front is initially located to the northwest of FLIP's location and is characterized by warm water to the southeast and cold water to the northwest. This is the only spatial survey available of the deep frontal structure. The high temporal resolution of the temperature and salinity profiles obtained during SWAPP provides a unique opportunity to track intrusions. The sample rate is above the local buoyancy frequency maximum which is 7.0 cpm. Intrusions are tracked in density coordinates. This allows the removal of internal wave straining and vertical advection and isolates the thermohaline finestructure associated with intrusions. This provides a clear picture of interleaving at the frontal interface. A summary of the CTD instrumentation and the observations of the front are presented in section 2. The thermohaline finestructure and interleaving is investigated in section 3. The observations are tested for consistency with the hypothesis that the interleaving near the front is driven by double diffusive convection rather than internal wave advection. Section 4 is a discussion of the interleaving observations at the front and other possible interleave driving mechanisms.

2. THE EXPERIMENT

2.1. The CTD Instrumentation

Finestructure temperature and salinity estimates are made using conductivity, temperature and pressure data collected with a repeat profiling CTD. Sea-Bird Electronics (SBE) instruments were deployed during the SWAPP experiment in a similar arrangement as has been used in past upper ocean experiments on FLIP [*Williams,* 1985; *Sherman,* 1989]. The conductivity sensors are model SBE 4-01 (5×10^{-5} S m^{-1} resolution at 12 Hz), and the temperature sensors are model SBE 3-01/F (5×10^{-4} °C resolution at 12 Hz). The CTD sensors and a pressure case containing sampling electronics are mounted in a streamlined cage with a 55 kg lead weight in front to allow for high fall rates. The shape of the cage allows profiling only in one direction. The cycle period was 130 seconds from the start of one profile to the beginning of the next and fall rates were 3.65 m s^{-1}. A lower CTD profiled from 200 to 420 m. The temperature, pressure and conductivity sensors are sampled at 24 Hz. The lower CTD was operated from March 1 to March 16, 1990, collecting ~9200 profiles.

A pitot tube equipped pressure sensor, used to measure the static pressure during profiling, is used to estimate depths, z. The raw pressure data contains a high frequency fluctuation associated with water turbulence and does not reflect real fluctuations in fall rate. The pressure signal is low pass filtered below 0.14 cpm to remove the turbulent noise. The pressure data was then resampled at half the original sample rate.

Following the procedure developed by *Williams* [1985], the time response of the temperature (T) and conductivity (C) sensors are matched. The method assumes that the conductivity fluctuations at scales smaller than 10 m are due mainly to temperature fluctuations. By computing the cross spectrum between temperature and conductivity, the phase and amplitude of the transfer function can be estimated. Here this transfer function is estimated using 100 profiles from the CTD when there was minimal haline finestructure. The profiles are first differenced then Fourier transformed to compute temperature and cross spectra. The transfer function is modeled by fitting a fifth order polynomial to the estimated amplitude and phase. The transfer function is applied to match the spectral response of T to that of C. Both T and C are then low pass filtered with a cut off of 1.3 cpm and resampled at half the data rate to form corrected profiles. Salinity, potential temperature and potential density are estimated using the equation of state [*Fofonoff and Millard,* 1983].

2.2. The Front

The CTD depth-timeseries clearly indicate two different water masses. A scatter plot of temperature and salinity (Figure 3) from 200 to 420 m reveals a clustering of data points along two different characteristic T–S lines. Few data points are found to lie near the cruise averaged T–S curve. A picture of the frontal structure

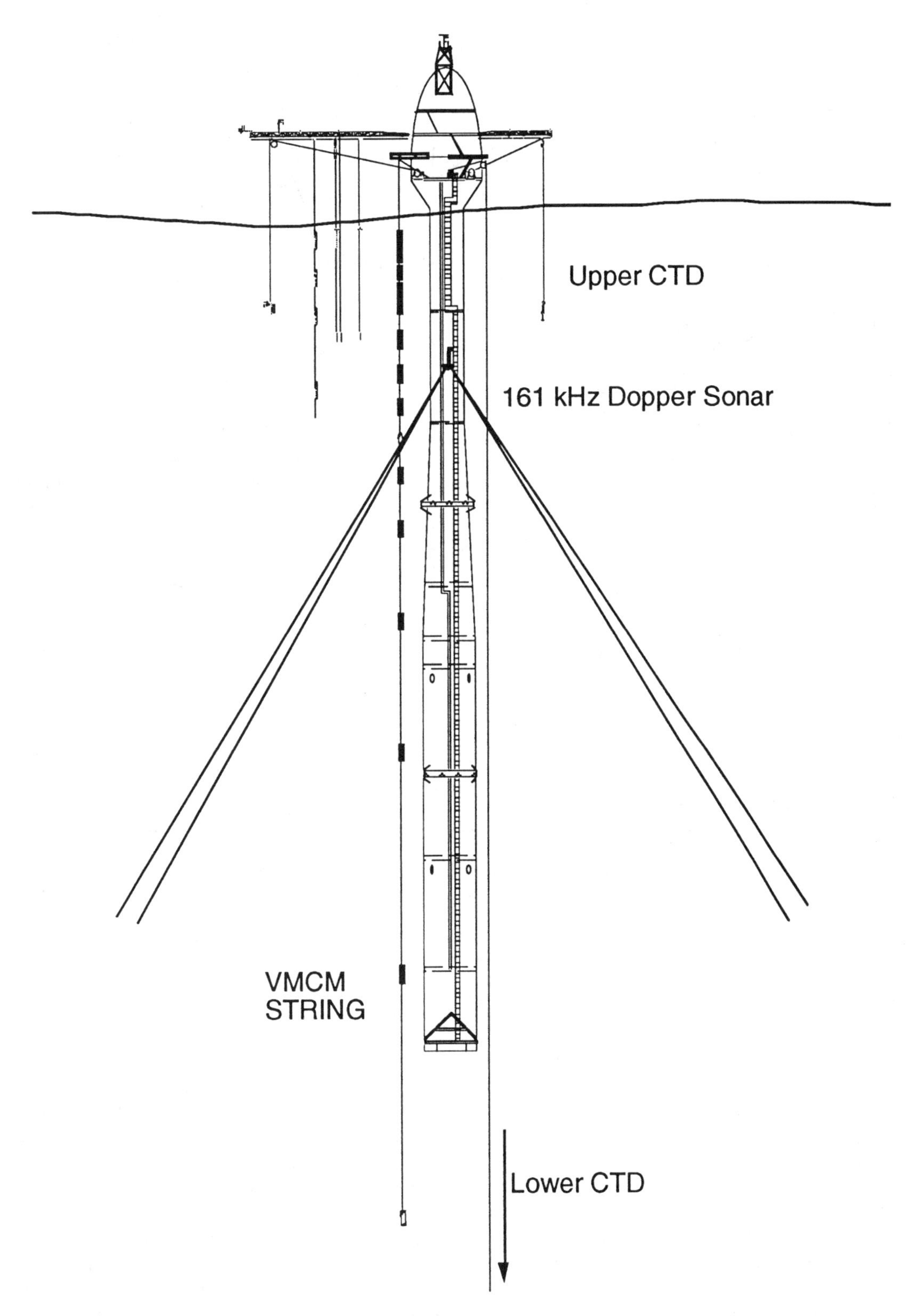

Fig. 1. R/P FLIP Instrumentation Diagram. Upper ocean observations are made with the Marine Physical Laboratory 161 kHz Doppler sonar system and a pair of CTD's. This diagram shows the deployment configuration used during SWAPP.

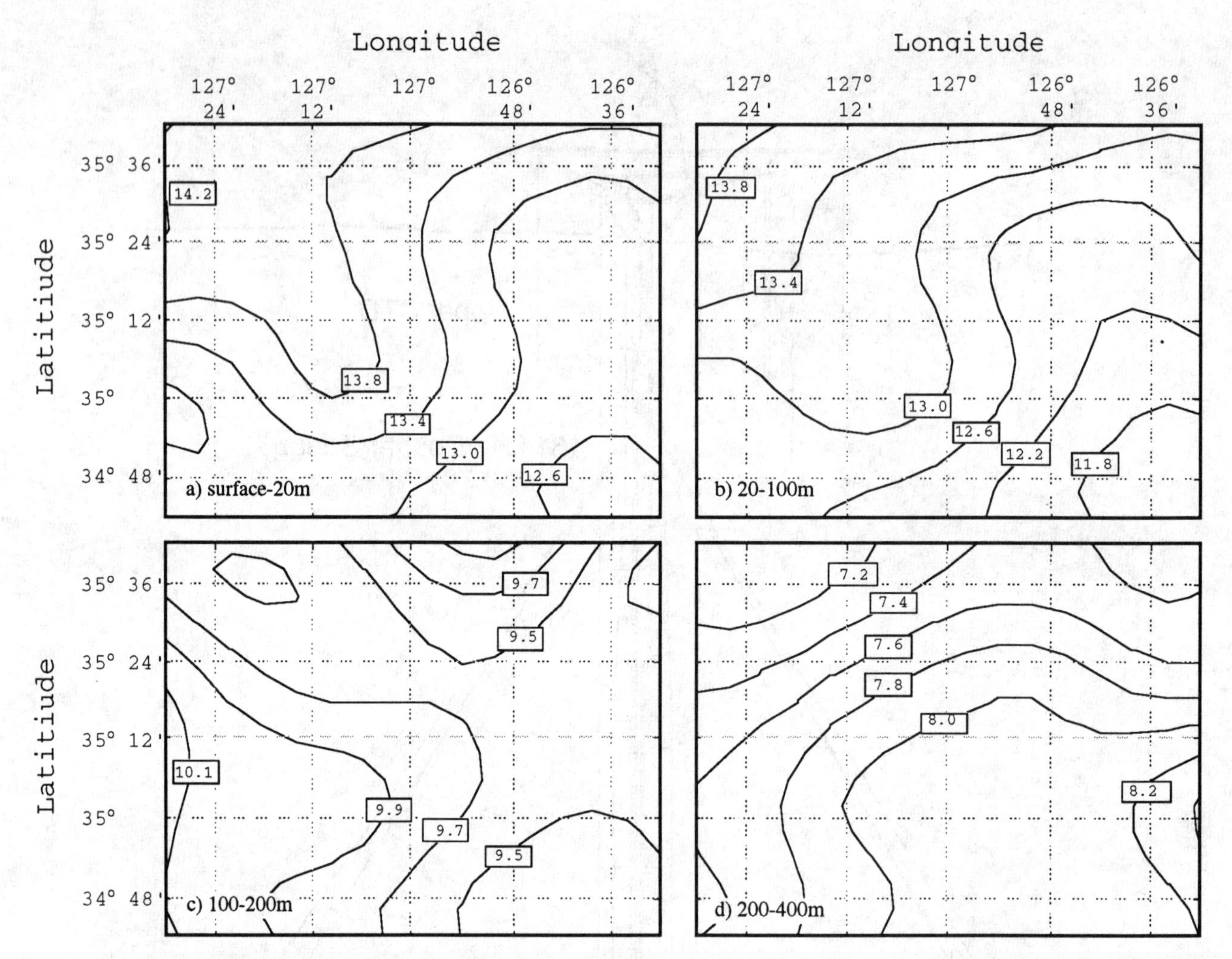

Fig. 2. a–d. Depth Averaged Temperature Contours. From February 20 to February 25, 1990, R. Weller's Group from W.H.O.I. performed an XBT survey from the *Navajo*. From 57 temperature profiles taken, 30 are selected here which appear reliable down to 400 m. For the contour plots, temperature profiles from the XBT survey are averaged in four depth ranges: (a) surface to 20 m, (b) 20 to 200 m, (c) 100 to 200 m and (d) 200 to 400 m. These depth averaged temperatures are contoured with labels in degrees Celsius. The survey reveals a front. Near surface water temperature increases by 2°C from southeast to northwest, while the water temperature below 200 m decreases by 1°C.

can be formed by tracking the temperature and salinity along an isopycnal in depth and time (Figure 4). Isopyncals are labeled by their mean depth, $\bar{\zeta}$. It is useful at this point to divide the timeseries into three segments. The first segment contains yeardays 60 to 66. It is characterized by warmer and saltier water than average. The water mass is relatively free from intrusions. The temperature and salinity along all the isopycnals slowly decreases in time. The second segment contains yeardays 66 to 70. It is characterized by a rapid drop in temperature and salinity along isopycnals. The deepest isopycnals ($420 > \bar{\zeta} > 300$ m) continually decrease in temperature during this time while the shallower isopycnals ($300 > \bar{\zeta} > 200$ m) go from cold to warm to cold. Many temperature inversions occur at all depth ranges. The third segment contains yeardays 70 to 75 and is characterized by colder and fresher water than average at all depth ranges. Many temperature inversions can still be seen. The temperature change along an isopycnal increases with depth and thus the average vertical gradient of temperature and salinity is larger at the end of the experiment than at the beginning.

Estimating the horizontal gradients of temperature and salinity across the front is done in two ways. As the SWAPP experiment was being set up, an XBT survey of the area surrounding the SWAPP site was conducted.

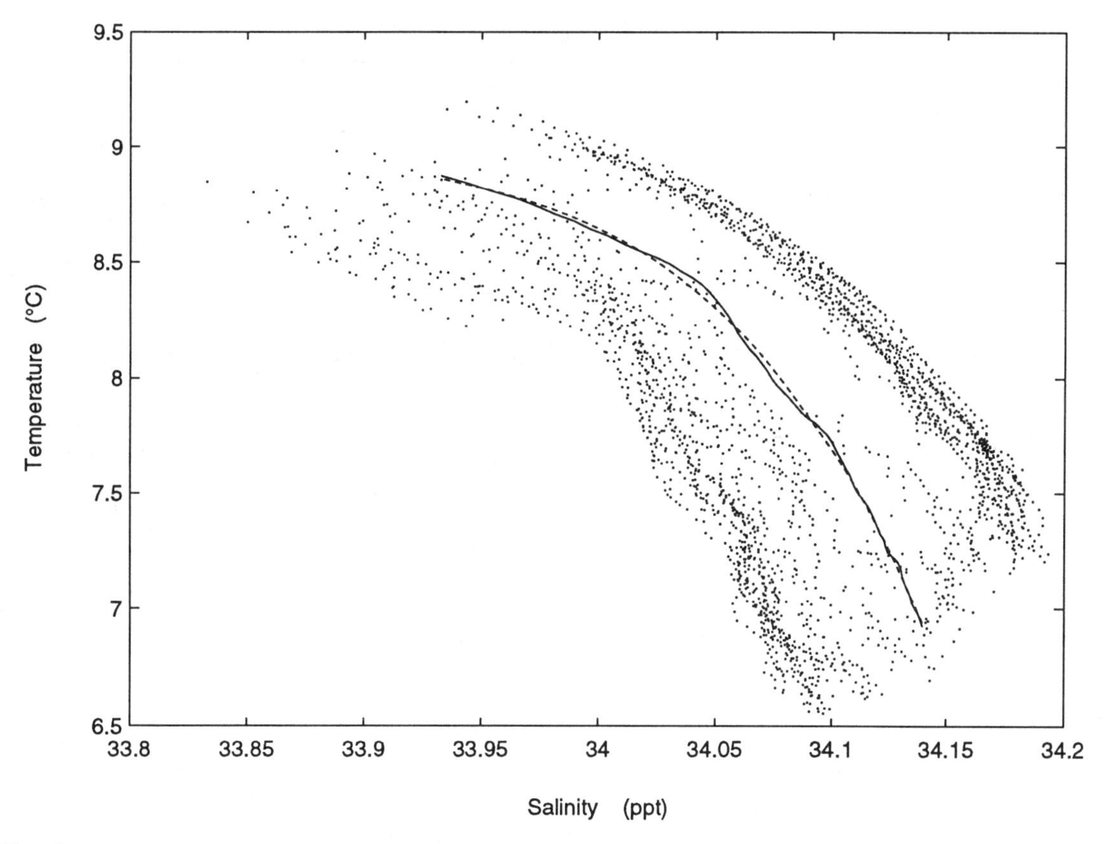

Fig. 3. T–S Diagram from 200 to 420 m. Scatter plot of salinity plotted versus potential temperature. A point is plotted for every 3 m from 200 to 420 m. One profile from the lower CTD is plotted every 6 hours. Two distinct water masses can be seen. The cruise averaged profile (solid line) and a cubic curve fit (dashed line) are also plotted.

This survey revealed a horizontal temperature gradient in the 200 to 400 m depth range (Figure 2). The estimated temperature gradient is 1.85×10^{-5}°C/m. Assuming that the density gradient is small, the salinity gradient can be estimated by

$$\frac{d\bar{S}}{dy} = \frac{\alpha}{\beta}\,\frac{d\bar{T}}{dy} \tag{1}$$

where $\beta = \rho_0^{-1}\frac{\partial\rho}{\partial S}$; $\alpha = \rho_0^{-1}\frac{\partial\rho}{\partial T}$ and y is the across front coordinate. α and β are nearly constant from 200 to 420 m and are equal to $\alpha = 1.508 \times 10^{-4}$°C^{-1} and $\beta = 7.712 \times 10^{-4}$ ppt^{-1}. Using (1) yields a salinity gradient of 3.62×10^{-6} ppt m^{-1}. The second method is to compare a timeseries of temperature on an isopycnal with the time integrated velocity field. The temperature and salinity on three isopycnals with average depths of 250, 300 and 350 m are tracked (Figure 5). Between yeardays 66 and 68, the temperature and salinity decrease by approximately 0.425°C and .0825 ppt on the 300 m isopycnal. The largest temperature change occurs on the deepest isopycnal which continues to decrease until yearday 70. The vertical temperature gradient becomes larger on the cold side of the front. The horizontal range of temperatures measured is still not as large as that seen in depth averaged temperatures from the XBT survey due to the temperature gradient change.

The horizontal displacement is found by time integrating the depth averaged velocity. The velocity data is obtained from the Marine Physical Laboratory 161 kHz Coded-pulse Doppler Sonar. The sonar was mounted on the hull of FLIP and profiled continuously during the experiment. Velocity profiles were taken every minute from 50 to 350 m with a vertical resolution of 5.5 m. A complete description of this instrument is provided by *Anderson* [1992]. A trajectory is estimated by time integrating the depth-average velocity from 100 to 350 m (Figure 6). The resulting trajectory assumes the velocities result from FLIP moving through a 'frozen' ocean. The depth averaged velocities are dominated by a current of nearly 9 cm s^{-1} flowing towards the east. Using the displacement from yearday 66.7 to 67.2 (~7.1 km) yields temperature and salinity gradients of

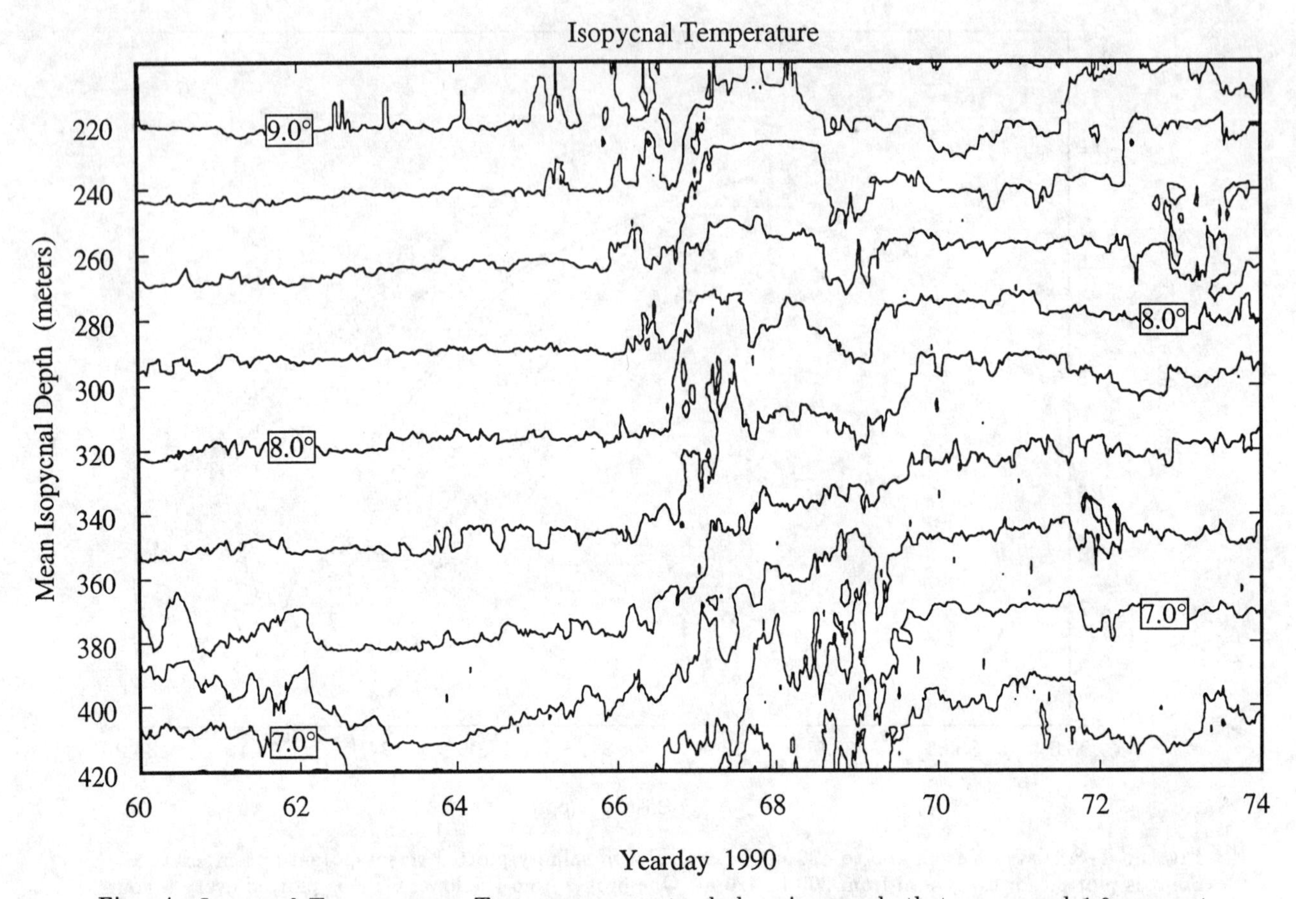

Fig. 4. Isopycnal Temperatures. Temperatures are tracked on isopycnals that are spaced 1.0 m apart on average. The anomalies are contour plotted versus mean isopycnal depth and CTD drop number. The contour interval is 0.25°C.

5.27×10^{-5}°C m^{-1} and 1.03×10^{-5} ppt m^{-1}. It is unlikely that the velocity field is perpendicular to the front, so the width of the front is overestimated and the horizontal gradients are underestimated.

3. VERTICAL INTERLEAVING

Typical temperature and salinity profiles from the lower CTD reveal many temperature and salinity inversions on scales of 3 to 20 m (Figure 7). Temperature and salinity inversions are matched such that the resulting density profile is hydrostatically stable. The potential energy stored in the inversions will drive double diffusive convection leading to enhanced vertical mixing rates [*Schmitt*, 1994]. The difference in the molecular diffusivity between salt and heat in the ocean has a significant effect on vertical fluxes when salinity and temperature both increase or decrease with depth.

The potential energy driving these fluxes comes from the thermohaline property that has a destabilizing effect on the density gradient. The diffusive case occurs when the temperature gradient is unstable (temperature and salinity increase with depth). It is characterized by convective layers and by more heat flux than salt flux through the layer. The salt fingering case occurs when the salinity gradient is unstable and salt fluxes are larger than heat fluxes. The potential for one of these processes to take place is quantified by the density ratio

$$R_\rho = \frac{\alpha \frac{\partial T}{\partial z}}{\beta \frac{\partial S}{\partial z}} \tag{2}$$

or Turner angle [*Ruddick*, 1983]

$$Tu = \tan^{-1}\left(\frac{\alpha T_z - \beta S_z}{\alpha T_z + \beta S_z}\right) = \tan^{-1}(R_\rho) - 45°. \tag{3}$$

The two cases are then described by

$$\left.\begin{array}{c} 0 \le R_\rho \le 1 \\ -90° \le Tu \le -45° \end{array}\right\} \rightarrow \textit{Diffusive}$$

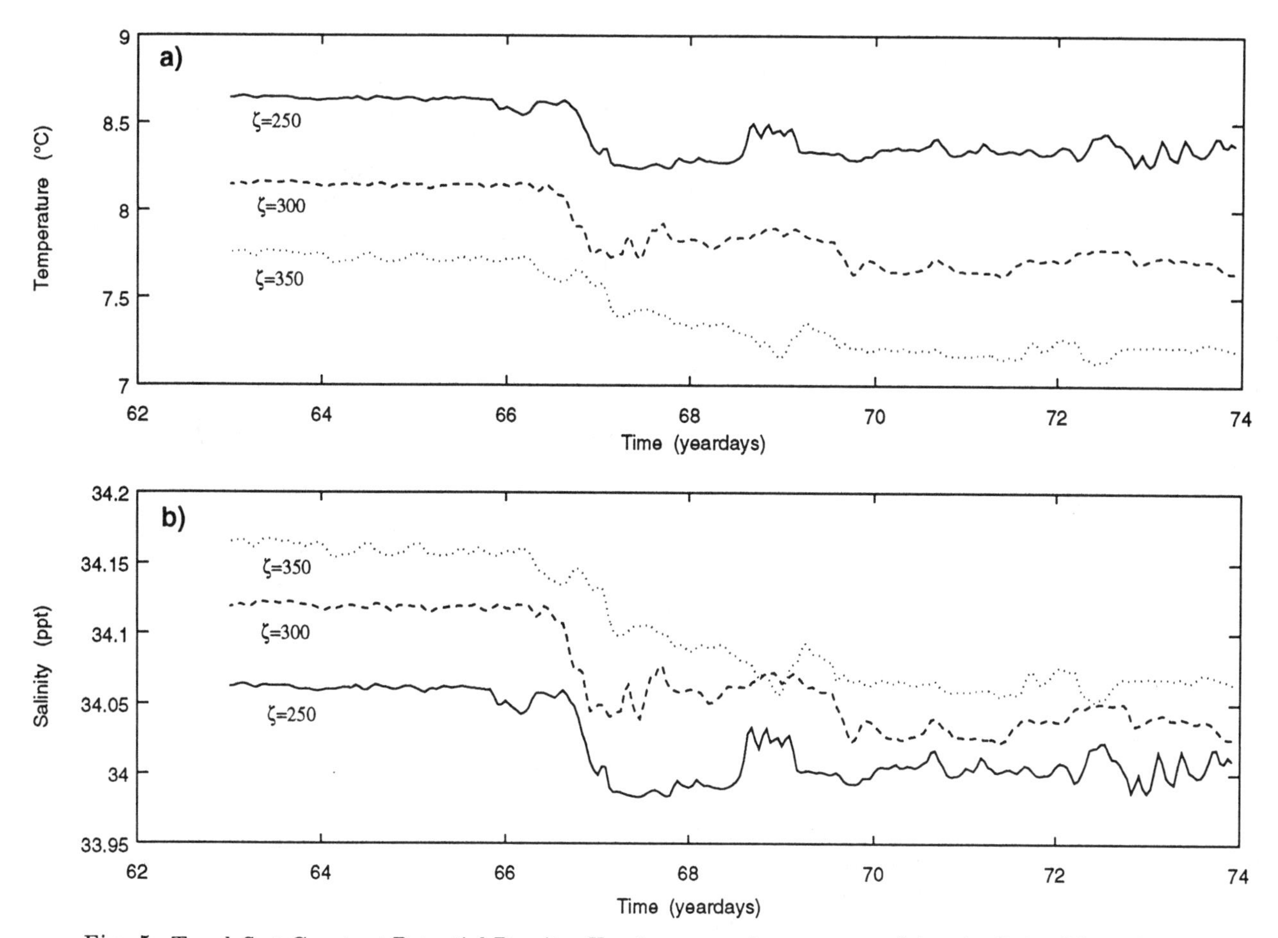

Fig. 5. T and S at Constant Potential Density. Hourly averaged temperature (a) and salinity (b) are found along three isopycnals (z = 250, 300 and 350 m) and plotted versus time.

$$\left.\begin{array}{c} 1 \leq R_\rho \\ 45^\circ \leq Tu \leq 90^\circ \end{array}\right\} \rightarrow \text{Fingers.}$$

A typical profile of Tu reveals that much of the water column (200 to 420 m) is subject to double diffusive processes (Figure 8).

The percent of the water column that is subject to salt fingering and diffusive convection is plotted versus time (Figure 9). Consistently, the percent of the Tu profile that is favorable to salt fingering ($Tu > 45^\circ$) is twice as large as the percent of the profile that is favorable to diffusive convection ($Tu < -45^\circ$). During the first part of the timeseries (yeardays 60 to 65), diffusively unstable water occurs less than 3% of the time, whereas fingering occurs 10 to 20% of the time. On yearday 65, occurrences of temperature and salinity inversion increases dramatically. This increase takes place before the large isopycnal temperature drop that takes place at yeardays 66 to 67. The incidents of temperature and salinity inversions reach a maximum at the time of the large temperature drop (yeardays 66–67) and again at yearday 69 when the isopycnals in the 200 to 300 m depth range warm up for a half a day. During these maximums, as much as 50% of the water column between 200 to 420 m is double diffusively unstable. After yearday 70, the percentage of temperature inversions decreases to 5% then slowly rises to 10% at yearday 75 and the percentage of salinity inversions stays large at 30% to 35%.

Schmitt [1990] notes the oceans avoidance of density ratio, R_ρ, values too close to unity. *Schmitt* [1990] explains that the mixing rate for salt finger interfaces increases as R_ρ approaches one. When advection works to sharpen interfaces and gradients, driving R_ρ towards one, the enhanced mixing weakens the interface and drives R_ρ back to larger values. Similarly with diffusive interfaces, the mixing rate increases as R_ρ approaches one which works to weaken gradients and drive R_ρ back towards zero. A histogram of density ratio from SWAPP reveals a slight dip in the distribution of R_ρ near one (Figure 10a) which is consistent with the

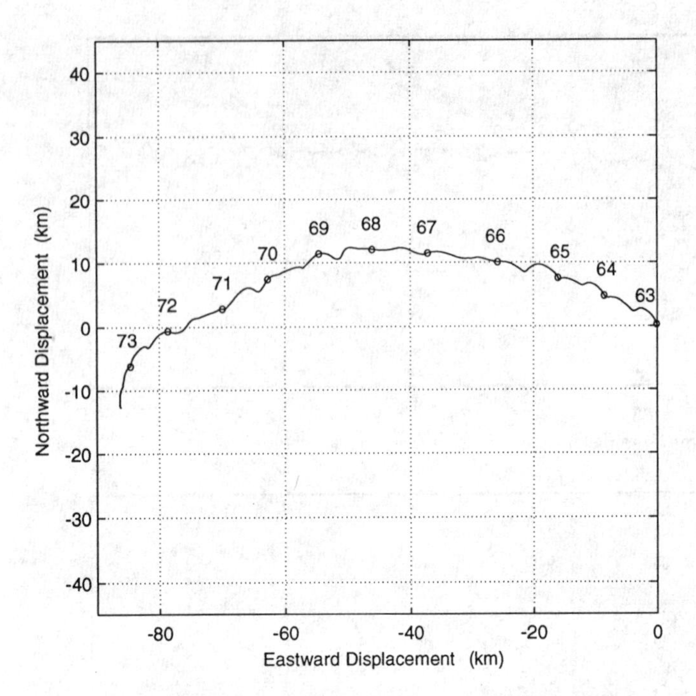

Fig. 6. Integrated Current Trajectory. Hourly averaged velocity profiles are depth averaged from 200 to 350 m depths and integrated in time.

findings of *Schmitt* [1990]. A histogram using a subset consisting of half a day of data from the beginning of the timeseries show a lack of values of R_ρ between zero and 2.0. While a histogram taken at the frontal interface has increased occurrences of R_ρ between zero and 2.0 with only a very slight dip near one.

The likely cause for the large number of temperature and salinity inversions is cross frontal advection. Water from one side is advected across the front and encounters a water mass with different thermohaline characteristics. With the small vertical mixings rates typically found in the ocean interior, these intrusions will carry some of the thermohaline characteristics of their source waters for a long time. This may be why the number of temperature and salinity inversions rises before the actual front is encountered. There are two kinematic mechanisms which would cause the horizontal displacement of water parcels across a front. The first is horizontal velocities from the internal wave field. The second is horizontal intrusions or noses that are driven by the vertical flux of heat and salt by double diffusion. To study the intrusions and possible driving mechanisms, the salinity and temperature signals associated with mesoscale as well as internal wave advection and straining must be removed leaving just the finestructure thermohaline anomalies associated with the intrusions.

Three scales are considered to resolve the finestructure thermohaline anomalies in a similar manner as developed by *Joyce* [1977] and *Fedorov* [1986]. The overbar denotes the mean field and is associated with the large (mesoscale) features. The tilde marks are those finescale fluctuations associated with vertical interleaving which are anisotropic. The accent signifies turbulent micro-fluctuations associated with dissipation and is isotropic in nature. Thus any scalar quantity, ϕ, is described by

$$\phi = \bar{\phi} + \tilde{\phi} + \phi' \tag{4}$$

where

$$\bar{\tilde{\phi}} = \tilde{\phi}' = 0$$

Temperature and salinity anomalies, $\tilde{T}$ and $\tilde{S}$, are deter-

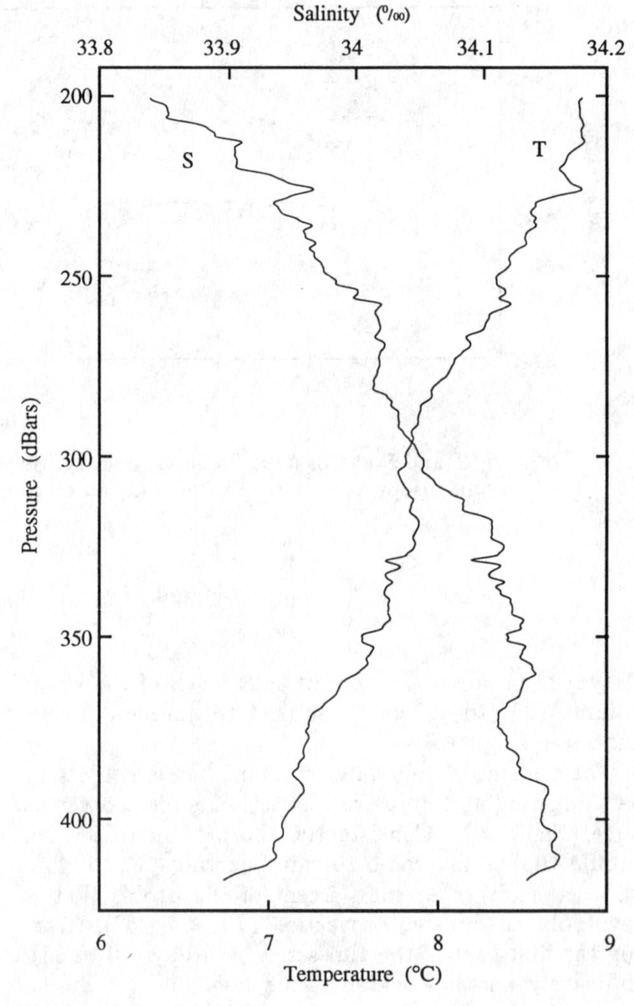

Fig. 7. Typical Temperature and Salinity Profile from Yearday 67. A typical temperature and corresponding salinity profile are plotted versus pressure. Temperature inversions are matched by salinity inversions to yield a stable density profile. Typical scales of temperature inversions are 5 to 15 m.

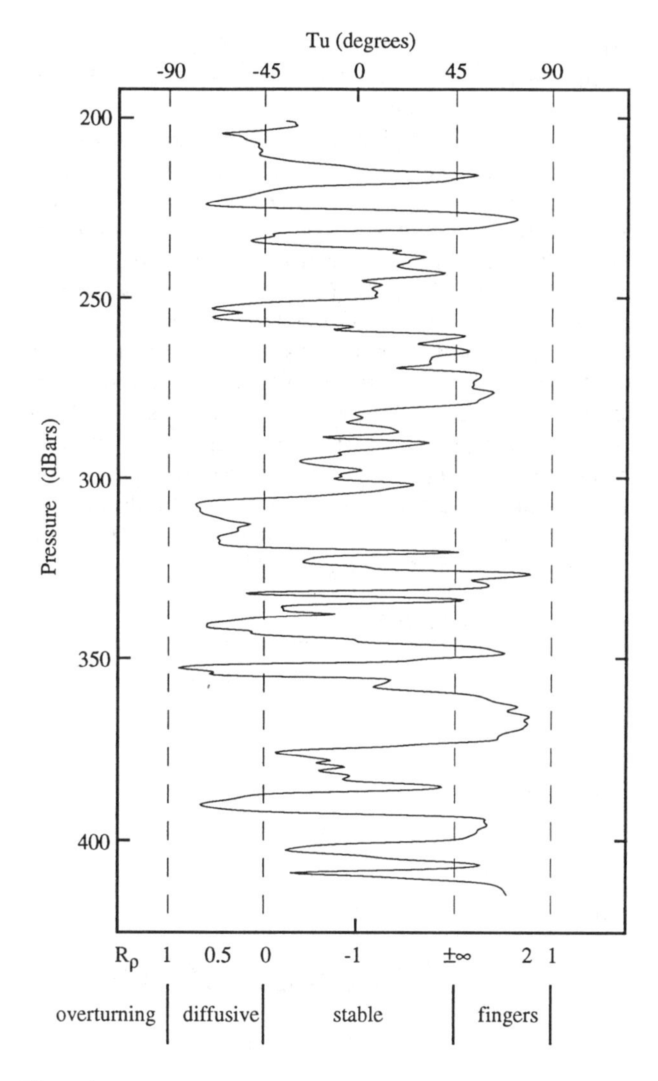

Fig. 8. Typical Turner Angle Profile from Yearday 67. Same as Figure 7 except Turner angle, Tu, is plotted. Tu is computed from temperature and salinity differences taken over 5 m in depth.

mined by removing the cruise averaged mean temperatures and salinity, $\bar{T}$ and $\bar{S}$. To eliminate the effects of vertical advection due to internal waves, $\bar{T}$ and $\bar{S}$ are found as functions of potential density using a cubic fit to the cruise average. The cruise averaged functions used are

$$\begin{aligned}\bar{T}(\sigma_t) &= a_3(26-\sigma_t)^3 + a_2(26-\sigma_t)^2 \\ &\quad + a_1(26-\sigma_t)^1 + a_0 \qquad (5)\\ \bar{S}(\sigma_t) &= b_3(26-\sigma_t)^3 + b_2(26-\sigma_t)^2 \\ &\quad + b_1(26-\sigma_t)^1 + b_0 \qquad (6)\end{aligned}$$

where

$$\begin{aligned}a_0 &= 8.5861;\ a_1 = 4.285;\ a_2 = -12.72;\\ &\quad a_3 = 5.438;\\ b_0 &= 33.491;\ b_1 = 2.143;\ b_2 = -2.651;\\ &\quad \text{and } b_3 = 1.264.\end{aligned}$$

The scale of the turbulent fluctuations is not resolved by the CTD and are assumed to be zero. The temperature and salinity anomalies, $\tilde{T}$ and $\tilde{S}$, are then determined for each hydrographic profile by

$$\tilde{T}_n(z) = T_n(z) - \bar{T}(\sigma_{t_n}(z)) \qquad (7)$$

and

$$\tilde{S}_n(z) = S_n(z) - \bar{S}(\sigma_{t_n}(z)). \qquad (8)$$

T_n, S_n, and σ_{t_n} are the temperature, salinity and density profiles for profile number n. This technique uses density as a semi-Lagrangian coordinate to remove the effects of internal wave vertical advection and straining leaving only the temperature and salinity signals associated with intrusions of different water masses.

3.1. Internal Wave Driven Interleaving

Horizontal displacement, due to internal waves, of water parcels from two different water masses is a possible cause of vertical interleaving near a front. *Joyce et al.* [1978] and *Georgi* [1978] relate the intrusion temperature signal, $\tilde{T}$, to the horizontal displacement across the front by internal waves and the horizontal temperature gradient. This relationship can be written in vertical wavenumber frequency space as

$$\tilde{T}(\omega, m)\frac{\partial \bar{T}}{\partial y}\frac{\hat{V}(\omega, m)}{i\omega} \qquad (9)$$

where y is the cross frontal coordinate and $\hat{V}$ is the cross frontal velocity. An upper bound on the vertical wavenumber spectrum of $\tilde{T}$ is then

$$\begin{aligned}\Phi_{\tilde{T}}(m) &\leq f^{-2}\left(\frac{\partial \bar{T}}{\partial y}\right)^2 \Phi_{\hat{V}}(m)\\ &\leq (4.9 \times 10^{-2}\,^\circ\mathrm{C}^2/(\mathrm{m/s})^2)\Phi_{\hat{V}}(m)\end{aligned} \qquad (10)$$

where $\Phi_{\tilde{T}}$ is the spectrum of $\tilde{T}$ and $\Phi_{\hat{V}}$ is the spectrum of $\hat{V}$. Since the internal wave field is nearly isotropic, the velocity spectrum is calculated from one component of velocity from the MPL sonar. The velocity spectrum is determined using the East component of velocity, from depths 50 to 306 m and ensemble averaged from yeardays 63 to 78. For comparison, the temperature anomaly (instantaneous temperature minus the strain

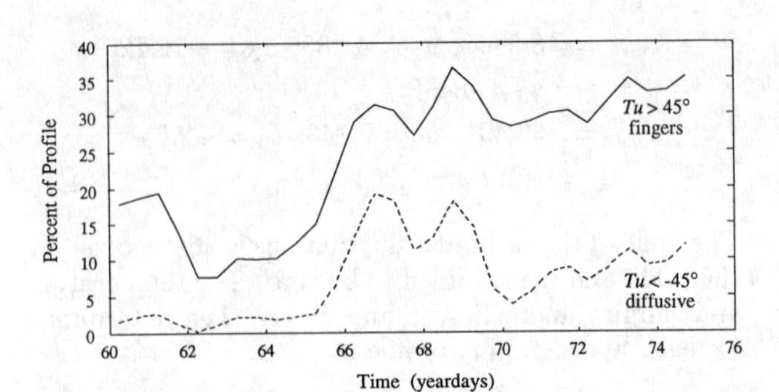

Fig. 9. Percent of Profile that is Subject to Double Diffusive Convection. Turner angle, Tu, is computed from temperature and salinity differences taken over 5 m in depth. The percent of each profile that is subject to diffusive instabilities ($Tu < -45°$; dashed line) and salt fingering ($Tu > 45°$; solid line) is averaged over half day intervals and plotted versus time. During strong interleaving on yeardays 66 to 70, as much as 50% of the water column is double diffusively unstable.

corrected, mean temperature profile) from depths 200 to 420 m is ensemble averaged from yeardays 60 to 75. The temperature and velocity data overlap by 12 days. The full records are used to add confidence to the spectra. Since internal wave velocities will scale with buoyancy frequency, calculating a velocity spectrum from shallower depths will over estimate the velocity variance at 200 to 420 m. The spectra are converted to gradient spectra by multiplying by $(2\pi m)^2$ where m is the vertical wavenumber. Comparing the two spectra reveals that the internal wave horizontal advection can only account for up to 5% of the total observed temperature signal (Figure 11). (This estimation is sensitive to the horizontal temperature gradient estimate. If the true gradient is twice as large as the estimated gradient, displacements from internal waves could still only account for 20% of the temperature variance.) Thus, the internal wave field is not strong enough to account for the

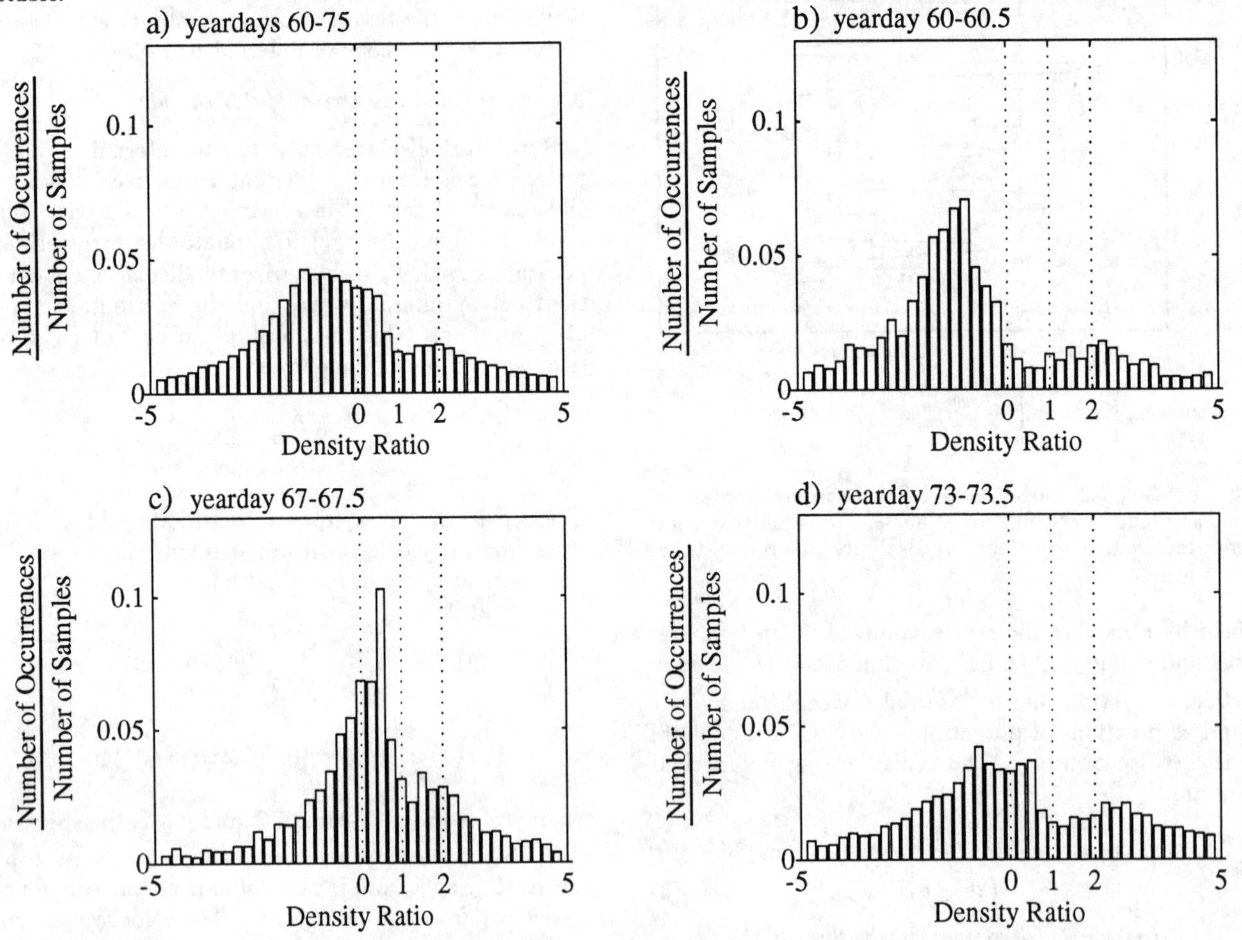

Fig. 10. Histograms of Density Ratio. Histograms of Density Ratio calculated over 3 m in depth from depths 200 to 420 m. (a) Histogram using all the available profiles taken during SWAPP. Histograms using a subset of profiles each spanning half a day representing data from (b) the warm side of the front (yearday 60–60.5), (c) near the interface (yearday 67–67.5) and (d) the cold side of the front (yearday 73–73.5). The histograms values are normalized by the total number of samples used.

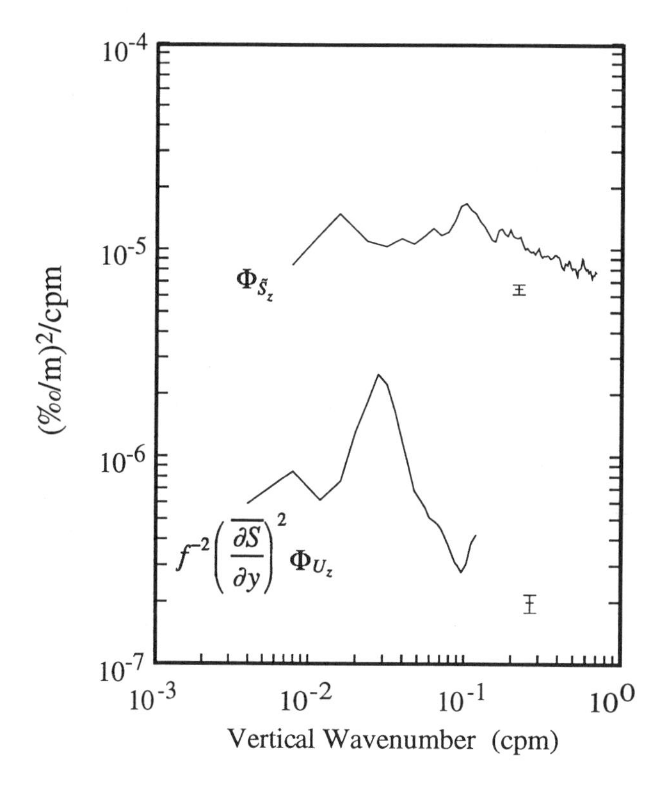

Fig. 11. Internal Wave Temperature Anomalies. Temperature gradient anomaly spectrum is calculated using temperature fluctuations from depth ranges 220 to 420 m and ensemble averaged over yeardays [60–75]. Maximum internal wave temperature anomaly spectrum is calculated using vertical shear spectrum (from 6.5 minute averaged velocity profiles, over depth ranges 50 to 306 m and ensemble averaged over yeardays [63–77]) and scaling by the cross frontal temperature gradient squared. Internal waves can only account for 5% of observed vertical interleaving.

observed interleaving unless the horizontal gradients are much larger then estimated. This is the same conclusion reached by *Joyce et al.* [1978] and *Georgi* [1978].

3.2. Double Diffusively Driven Interleaving

Since the vertical interleaving is not driven by the internal waves, it is likely that double diffusive processes are driving them. The potential for the double diffusive processes to occur has been shown, but no direct measurements of salt fingering or diffusive convection were made. Two typical finestructure features of double diffusively active intrusions are compared to observations for consistency. The features are the expected vertical scale of the intrusions and the slope of the intrusions across isopycnals.

3.2.1. Interleaving scales. *Ruddick and Turner* [1979] used a laboratory model to study mixing at an oceanic front across which there are large T–S anomalies but a small net horizontal density difference. They found that the vertical scales of the intrusions formed is directly proportional to the horizontal concentration gradient and inversely proportional to the vertical density gradient. They conclude that the vertical length scale of cross-frontal intrusions, H, can be determined to within a factor of two by

$$H = \frac{3}{2}(1-n)\frac{\beta\Delta\bar{S}}{\frac{1}{\rho}\frac{d\rho}{dz}} \tag{11}$$

where n is the density flux ratio (0.56 for heat/salt fingers),

$\Delta\bar{S}$ is the salinity contrast measured across the front along an isopycnal,

$\frac{1}{\rho}\ \frac{d\rho}{dz}$ is the vertical density gradient and $\beta = \frac{1}{\rho}\ \frac{\partial\rho}{\partial S}$.

Ruddick and Turner [1979] compare their estimated length scales with near frontal ocean observations. Using the mean density gradient from 200 to 400 m (2.36×10^{-6} m^{-1}), the expected vertical length scale of the intrusions from (11) is 17.8 m.

Toole and Georgi [1981] contend that it is incorrect to extrapolate the laboratory results of *Ruddick and Turner* [1979]. Their argument is that in the laboratory, the interleaving takes on the same vertical scales as the salt fingers suggesting the intrusion scale is determined by the energy in the salt fingering. In the ocean, the expected scales of the interleaving is much larger than the vertical scale of the salt fingers (which may be on the order of 1 m [*Williams*, 1974]) and thus are unimportant to the behavior of large intrusions.

Toole and Georgi [1981] use linear stability analysis of intrusions in a frontal zone and look for the fastest growing mode. Their model predicts the fastest growing mode as a function of lateral salinity gradient, the vertical density gradient of the mean field, the vertical viscosity, haline diffusivity and flux and the local value of the Coriolis acceleration. The interleaving scale is given by

$$H = 2\pi\sigma^{-1/4}\left[\frac{4K_v N}{g(1-\gamma_f)\beta S_x}\right]^{1/2} \tag{12}$$

where γ_f is the flux ratio of heat to salt, σ is the eddy diffusivity ratio of buoyancy to salt, and S_x is the cross frontal salinity gradient. Unlike the model of *Ruddick and Turner* [1979] which assumes a step function in salinity across the front, *Toole and Georgi* [1981] use the horizontal salinity gradient which is not well known for the SWAPP experiment. Only the lower bound on the horizontal salinity gradient is known. The gradients are

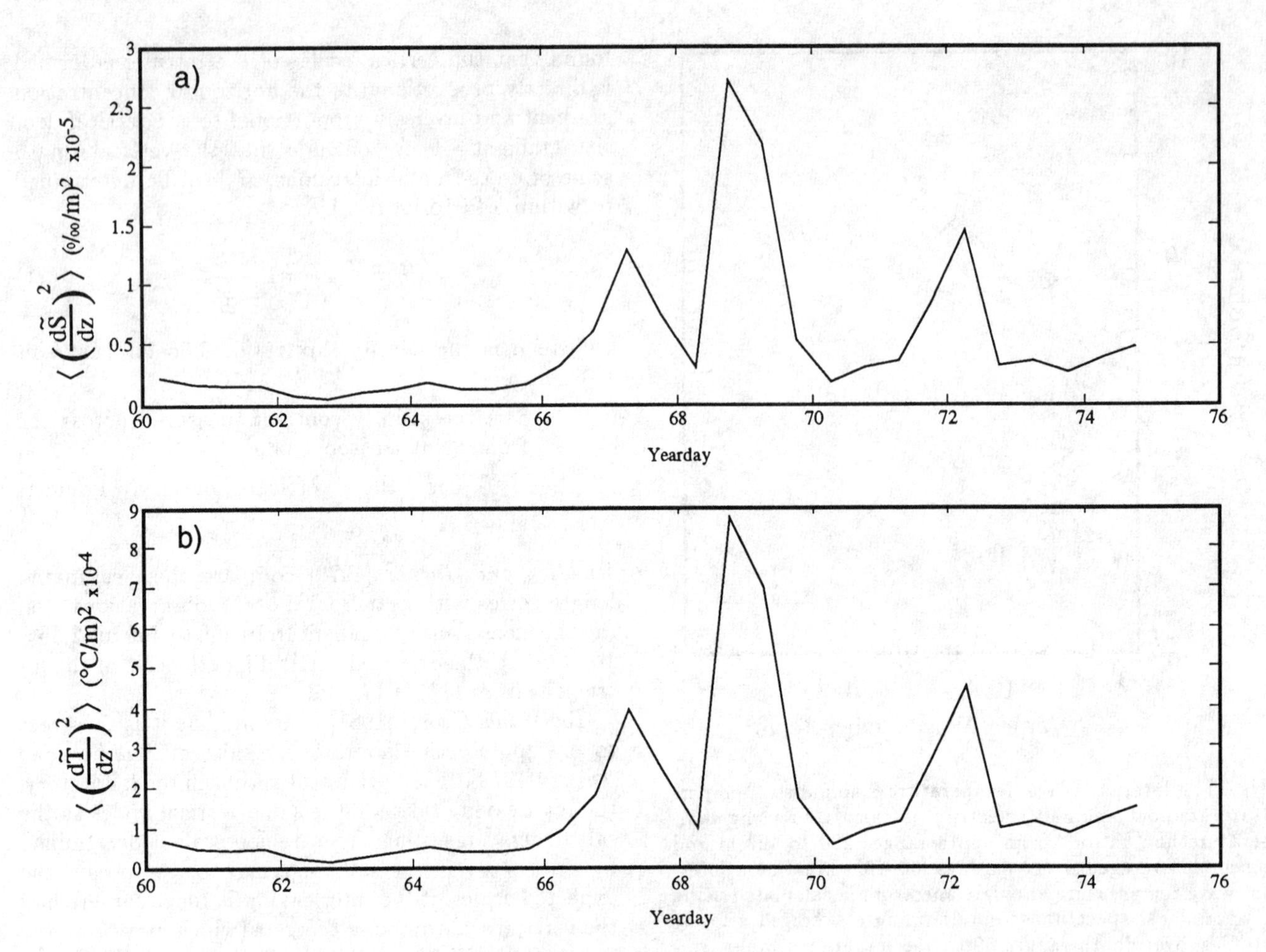

Fig. 12. Variance of Vertical Gradients of Salinity and Temperature Anomalies. Salinity (a) and temperature (b) gradient variances are calculated from half day ensemble averaged spectra over the wavenumber band [0.0167–0.7] cpm and plotted versus time.

$$\beta\bar{S}_x > 7.94 \times 10^{-9}\ \mathrm{m}^{-1}, \text{ and}$$
$$N = 5.24 \times 10^{-3}\ \mathrm{s}^{-1}$$

The constants suggested by *Toole and Georgi* [1981] are

$$\sigma = 0.1$$
$$\gamma_f = 0.6$$

A vertical eddy diffusivity, K_v, of 5×10^{-6} m^2 s^{-1} is chosen. This is consistent with mixing rates estimated by *Gregg* [1989] for this region. Using (12), the estimated vertical scale of the intrusions is H<38 m. This provides an upper bound on the size of the intrusions. The vertical scales estimated by *Ruddick and Turner* [1979] are consistent with this bound.

The observed intrusion scales are estimated using vertical wavenumber spectra. Half day ensemble averaged spectra are calculated for $\tilde{T}$ and $\tilde{S}$ over the depth range 220 to 420 m using triangle windowing on two overlapping 128 m depth segments. These are converted to gradient spectra by multiplying by $(2\pi m)^2$, where m is the vertical wavenumber and corrected for high wavenumber smoothing (Figure 12). The temperature and salinity anomaly variance varies throughout the cruise (Figure 13) reaching a maximum on yearday 68. The spectra from the half day averages corresponding to the times of maximum (yearday 62–62.5) and minimum (yearday 68.5–69) variance are plotted along with the 15 day averaged spectrum for comparison (Figure 12). The cruise average spectra are nearly white out to 0.1 cpm and then fall off slightly out to 0.7 cpm. The minimum variance spectra are red at all wavenumbers except for a slight rise at 0.2 cpm. The maximum spectra rises out to 0.1 cpm then falls off at higher wavenumber. The maximum spectra have excess variance in the wavenumber band [0.05–0.2] cpm

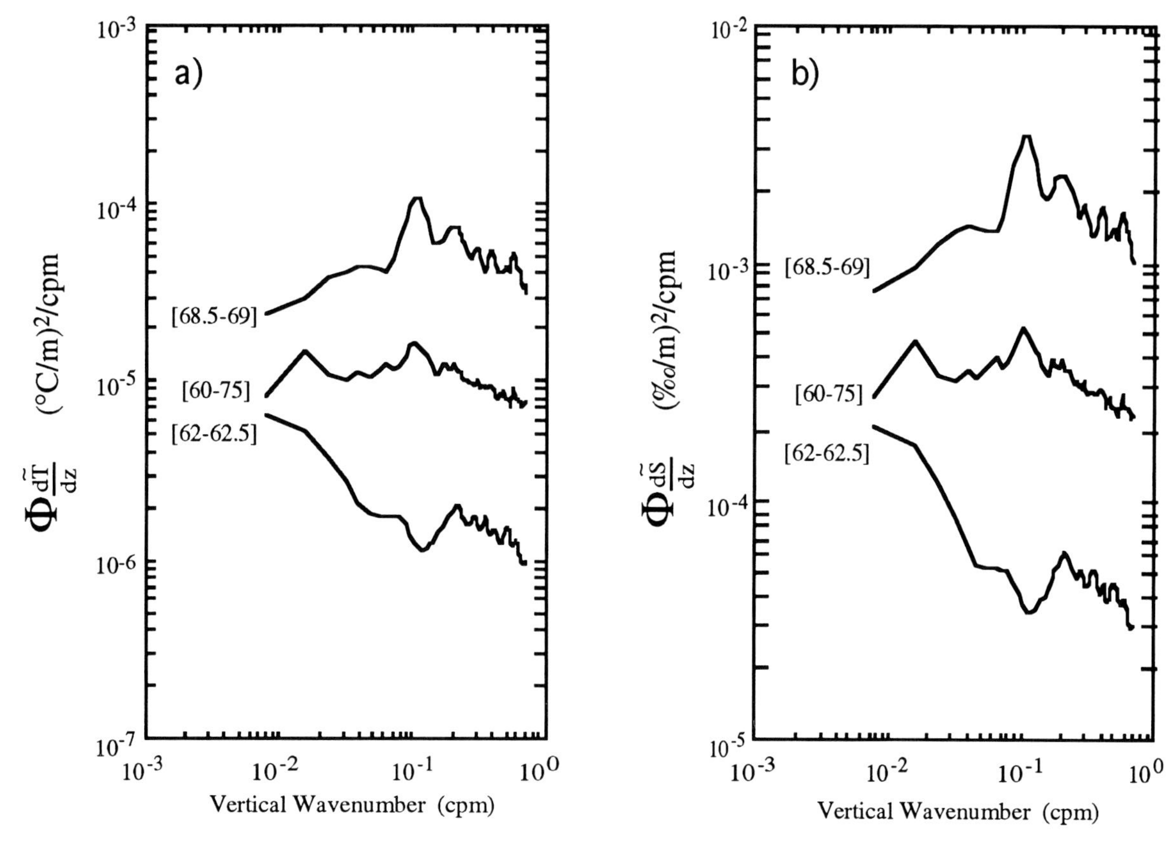

Fig. 13. Vertical Wavenumber Spectra of Gradient Temperature and Salinity Anomalies. Spectra are calculated over the depth ranges 220 to 420 m. Three ensemble averaged spectra are shown: yeardays [62–62.5], [68.5–69] and a 15 day average. Half day ensembles are smoothed in frequency.

while the minimum spectra have a lack of variance in the same band. The variance found in the wavenumber band [0.05–0.2] cpm grows more rapidly than variance found at other wave lengths approaching the front. The expected wavelength of the intrusions, as predicted by *Ruddick and Turner* [1979] and *Toole and Georgi* [1981], falls within this band. Thus the observed excess variance in the expected wavenumber bands associated with the intrusions is consistent with double diffusive driving of the intrusions.

3.2.2. Intrusion slopes. The slope of the intrusions across isopycnals can also be an indication of active double diffusive processes. *Turner* [1978] used laboratory experiments to show that cross frontal intrusions will slope across isopycnals surfaces. His laboratory experiments demonstrate that for diffusively unstable intrusions, the increase in density below a salt fingering region dominates the decrease above the diffusive regime. Thus warm salty intrusions will tend to rise when surrounded by cold, less salty waters. The opposite is true as well; cold, fresh intrusions will tend to sink when surrounded by warm, salty waters. Similar conclusions were reached theoretically by *Stern* [1967]. Ocean observations by *Joyce et al.,* [1978], *Gregg and McKenzie* [1979] and *Ruddick* [1992] have shown intrusions crossing isopycnals in the predicted manner.

Tracking intrusions along isopycnals is done here by sampling temperature anomalies on isopycnal surfaces. This is the semi-Lagrangian, or isopycnal following, coordinate system and in this case isopycnals that are spaced 0.5 m apart on average are tracked. The temperature anomaly on an isopycnal is then just the difference between the average temperature on that isopycnal and the instantaneous temperature. A contour plot of temperature anomalies reveals many intrusions which propagate across isopycnals (Plate 1a). Without a cross section of profiles directly across the front, it is difficult to say whether the intrusions are sloping across the front in the correct sense. The horizontal advection of intrusions below FLIP confuses the orientation of intrusion slopes. Assuming the intrusions slope up from the warm water encountered at the beginning of the cruise to the cold water encountered at the end of the cruise, any mean water velocity reversal will change the direction of the slope in time. For this reason, only a subset of the data where the temperature anomalies change most rapidly during yeardays [66.7–67.5] is investigated for isopycnal slope (Plate 1a).

By eye, there appears to be equal energy in upward and downward propagating intrusions so Fourier analysis is used to investigate the slopes. Only the sense of the slope can be determined from the timeseries. The depth range 300 to 464 m is used since temperatures decrease steadily during this time. A depth–timeseries consisting of 128 points in depth and 512 points in time is windowed with a triangle taper in both directions before two-dimensional Fourier transformation. Upward and downward sloping phase contributions are separated and summed over all frequency bands to make vertical wavenumber temperature anomaly spectra. Gradient spectra are found by multiplying the anomaly spectra by $(2\pi m)^2$ (Figure 14). Upward sloping corresponds to intrusions sloping upward from the warm, salty water to the cold, fresh water. The upward sloping intrusions contain two thirds of the total variance, $< \tilde{T}^2 >$. The upward sloping vertical wavenumber spectrum shows more variance than the downward sloping spectrum at all wavenumbers below 0.225 cpm. This is the wavenumber band predicted to contain the scales of the interleaving intrusions. These results are consistent with the hypothesis that the intrusions are actively double diffusive.

4. DISCUSSION

It is worthwhile to take a closer look at the structure of the intrusions during the time when the temperature and salinity characteristics change most rapidly. There appears to be two vertical scales associated with the temperature anomalies on isopycnals (Plate 1a). The larger scale is 40 to 60 m and takes the form of noses of warm water protruding into cold water. Two of the warm noses can be seen at depths 220 to 270 m, day 66, hour 18:00 to 22:00 and at depths 310 to 360 m, day 67; hour 2:00 to 4:00. The cold noses lie above and below the warm noses. The smaller scale comprises the structure on the noses which has vertical scales of 5 to 15 m.

The potential for double diffusion to occur is mapped using Turner angle, Tu, calculated during yeardays [66.7–67.5]. Following the same isopycnals (separated by 0.5 m in depth on average), Tu is calculated for each isopycnal from the temperature and salinity differences across isopycnals that are separated by 3.0 m in depth on average. This yields a depth-timeseries of Tu (Plate 1b). The characteristic scale for the diffusive and fingering interfaces is 5 to 15 m. Alternating diffusive and fingering interfaces are observed at the edges of the larger scales noses which appear at nearly constant density or rise slightly with time. Diffusive and fingering layers are absent at depths 200 to 270 m from 5:00 to 12:00; day 67, where the temperature anomalies suggest only one water mass is present ($\tilde{T} = -0.3°$C). Away from the noses, the alternating layers are not as apparent and diffusive and fingering interfaces tend to rise and fall across isopycnals (depths 300 to 420 m, 2:00 hours; day 67). The smaller layers seen in the temperature anomalies (5 to 15 m) are double diffusively active above and below and their scales are likely set by the processes as described by *Ruddick and Turner* [1979] and *Toole and Georgi* [1981].

The physics governing the larger noses (40 to 60 m) is not as clear. One possibility is that the noses are driven by mesoscale dynamics. The dynamics that created the front in the first place may have led to this larger scale distortion on its interface. A second possibility is that strong shears for near inertial internal waves are distorting the front. *Kunze* [1990] shows that salt fingers still form in the presence of strong vertical

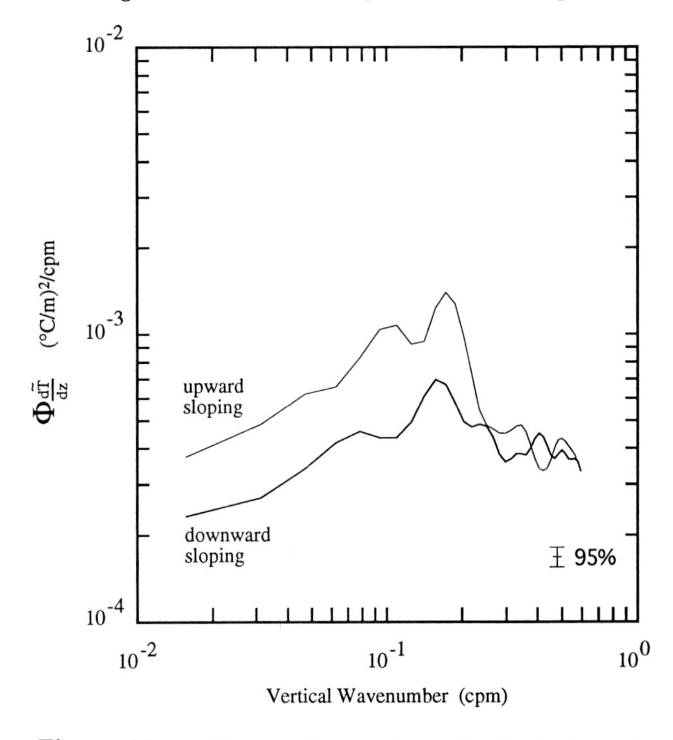

Fig. 14. Sloping Temperature Gradient Anomaly Wavenumber Spectrum. Temperature anomalies on isopycnals are tracked. Upward and downward phase wavenumber spectra are extracted from vertical wavenumber/frequency spectra. Wavenumber temperature gradient anomaly spectrum corresponding to upward (heavy line) and downward (light line) propagating phases are plotted. Data is from yeardays [66.7–67.5] and depths 300 to 364 m. Upward sloping intrusions correspond to warm, salty (cold, fresh) water intrusions rising (sinking) across isopycnals and they move into cold, fresh (warm, salty) water.

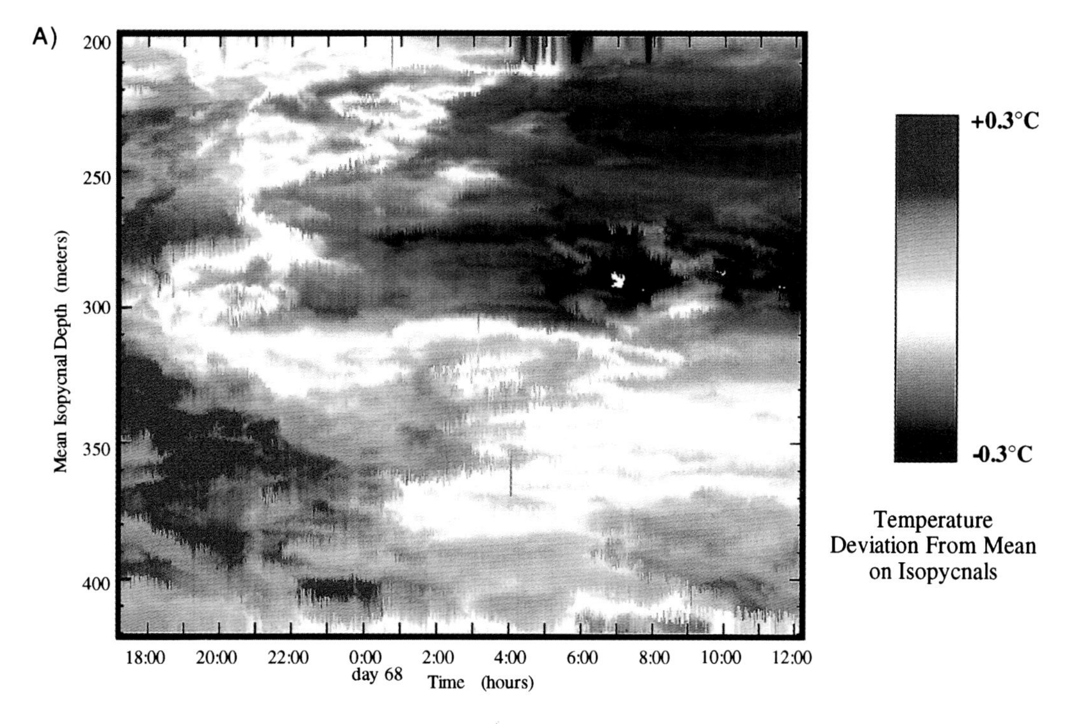

Plate 1a. Temperature Anomalies on Isopycnals During Yeardays 66.7 to 67.5. Temperature anomalies are tracked on isopycnals that are spaced in depth 0.5 m. apart on average. The anomalies are contour plotted versus mean isopycnal depth and time from yeardays 66.7 to 67.5.

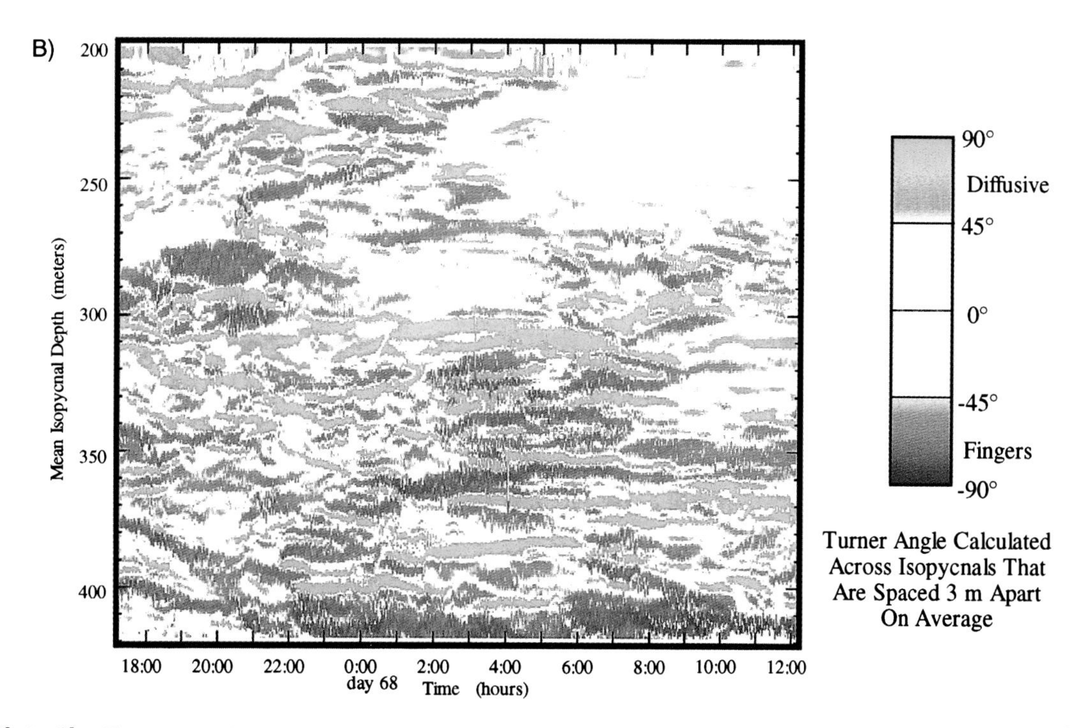

Plate 1b. Turner Angles on Isopycnals During Yeardays 66.7 to 67.5. Turner Angles, Tu, are tracked on isopycnals that are spaced in depth 0.5 m apart on average. Tu is contour plotted versus mean isopycnal depth and time from yeardays 66.7 to 67.5. Tu is calculated by taking the temperature and salinity differences across isopycnals separated by 3 m on average.

shear so the smaller layers are still double diffusively active in the presence of larger scale inertial shears. The shear analysis performed by *Anderson* [1992] shows that the vertical shear variance during SWAPP is dominated by near inertial waves with vertical wavelengths of 35 to 45 m which is nearly consistent with the scale of the noses.

A third possibility is that larger scales are linked in some way to the double diffusive processes. *Fedorov* [1986] performed an interesting experiment. He placed a single (warm and salty) intrusive wedge into a surrounding body of cold water. Salt fingers began to form at the bottom interface of the wedge and, in time a counterflow (opposite to the direction of the moving wedge) was observed. This led to the formation of a second wedge below the original. Eventually the intrusion split into several laminae. The resulting structure was characterized by the uppermost layer protruding the farthest from the source and deeper layers protruding successively less. This has many of the characteristics of the two scales observed here. Two larger scale noses (located at 220 to 280 m and 310 to 370 m) are warm and salty, the top of the nose protrudes the deepest into the cold, fresh side of the front. Beneath, several smaller layers are found with the general shape of the nose sloping away from the cold water. This process still needs an outside force to define the larger scale noses but may explain the generation of the smaller scale layers and their link to the larger scales.

Newly formed intrusive lenses, referred to as 'calving' by *Fedorov* [1986], are another feature seen during yeardays [66.7–67.5]. Calving is the separation of an intrusive tongue from the frontal interface and its transformation into an isolated lens. Several lenses can be seen in Plate 1. One such feature occurs at 300 m at 20:00; day 66; where a lens of warm water is surrounded by colder water. Without a three dimensional survey, it is impossible to tell whether this lens is completely separated form its origin. Separated lenses near frontal zones have been observed before [*Fedorov,* 1986]. These lenses are likely formed from motions in the horizontal plane. The calving of interleaving layers will lead to increased variability in isopycnal temperatures near a front and to enhanced horizontal and vertical mixing.

5. SUMMARY

Observations of increased variance in thermohaline finestructure during the SWAPP experiment is linked to physical processes associated with a frontal zone. The front is characterized in the depth range 200 to 420 m by warm water to the southeast and warm water to the northwest. Horizontal salinity and temperature gradients across the front are at least 1.0×10^{-5} ppt m^{-1} and 5.3×10^{-5}°C m^{-1}.

The fine scale thermohaline structure is sampled as a CTD depth–timeseries taken from a moored FLIP while the frontal zone was advected past the sampling location. The timeseries is characterized by warm, salty water starting at yearday 60, 1990, then a rapid drop in isopycnal temperatures near yearday 67 as the water characteristics become colder and fresher. Individual temperature and salinity profiles are characterized by many salinity and temperature inversions. Profiles of Turner angle reveal indirectly the presence of double diffusive convection. The percentage of the profile that favors salt fingering ($Tu > 45°$) is larger than the percent that favors diffusive convection ($Tu < -45°$). The number of temperature and salinity inversions rises before yearday 67 and precedes the front.

Temperature and salinity anomalies are calculated by removing the mean temperature and salinity on isopycnals and allows the tracking of intrusions. The intrusions have two dominant vertical scales near the front. The small scale intrusions have vertical lengths of 5 to 15 m, are driven by double diffusive processes, slope upward from the warm side to the cold side of the front and are consistent with scales predicted by the linear instability theory of *Ruddick and Turner* [1979]. The larger frontal disturbances have vertical scales of 40 to 60 m and may be determined by mesoscale or internal wave forcing.

Acknowledgments. The authors thank Michael Goldin, Lloyd Green and Eric Slater for the design and construction of the CTD and sonar systems used in this effort and for their at-sea operation. Discussions with Raymond Schmitt provided insight and encouragement. This research was supported by the Office of Naval Research under grants N00014-90-J-1099, N00014-90-J-1495 and by the Woods Hole Oceanographic Institution Postdoctoral Scholar program. This is contribution 8726 from the Woods Hole Oceanographic Institution.

REFERENCES

Anderson, S. A., Shear, strain and thermohaline vertical finestructure in the upper ocean, Ph.D. thesis, University of California, San Diego, 1992.

Fedorov, K. N., *The Physical Nature and Structure of Oceanic Fronts*, C. Garrett (editor), Springer-Verlag, Berlin, 333 pp., 1986.

Fofonoff, N. P., and R. C. Millard, Jr., Algorithms for computation of fundamental properties of seawater. *UNESCO Tech. Pap. in Mar. SciI.*, No. 44, 53 pp, 1983.

Georgi, D., Fine structure in the Antarctic Polar Front Zone: its characteristics and possible relationship to internal waves, *J. Geophys. Res.*, *83*, 4579–3671, 1978.

Gregg, M. G., Small-scale mixing: a first-order process?

Parameterization of Small-Scale Processes, Proceedings of the 'Aha Huliko'a Hawaiian Winter Workshop, Honolulu, P. Muller and D. Henderson, editors, 117–126, 1989.

Gregg, M. G., and J. H. McKenzie, Thermohaline instrusions lie across isopycnals, *Nature, 280*, 310–311, 1979.

Horne, E. P., Interleaving at the subsurface front in the slope water off Nova Scotia, *J. Geophys. Res., 83*, 3659–3671, 1978.

Joyce, T. M., A note on the lateral mixing of water masses, *J. Phys. Oceanogr., 7*, 626–629, 1977.

Joyce, T. M., W. Zenk, and J. M. Toole, The anatomy of the Antarctic Polar Front in the Drake Passage, *J. Geophys. Res., 83*, 6093–6113, 1978.

Kunze, E., The evolution of salt fingers in inertial wave shear, *J. Mar. Res., 48*, 471–504, 1990.

Posmentier, E. S., and R. W. Houghton, Fine structure instabilities induced by double diffusion in the shelf/slope water front, *J. Geophys. Res., 83*, 5135–5138, 1978.

Rudnick, P., FLIP: An oceanographic buoy, *Science, 146*, 1268–1273, 1964.

Ruddick, B. R., A practical indicator of the stability of the water column to double-diffusive activity, *Deep-Sea Res., 30*, 1105, 1983.

Ruddick, B. R., Intrusive mixing in a Mediterranean salt lens: intrusion slopes and dynamical mechanisms, *J. Phys. Oceanogr., 22*, 1274–1285, 1992.

Ruddick, B. R., and J. S. Turner, The vertical length scale of double-diffusive intrusions, *Deep-Sea Res., 25A*, 903–913, 1979.

Schmitt, R. W., Double diffusion in oceanography, *Ann. Rev. Fluid Mech., 26*, 255–285, 1994.

Schmitt, R. W., On the density ratio balance in the Central Water, *J. Phys. Oceanogr., 20*(6), 900–906, 1990.

Sherman, J. T., Observations of fine-scale vertical shear and strain, Ph.D. dissertation, University of California, San Diego, 1989.

Stern, M. E., Lateral mixing of water masses, *Deep-Sea Res., 14*, 747–753, 1967.

Toole, J. M., and D. T. Georgi, On the dynamics and effects of double-diffusively driven intrusions, *Prog. Oceanog., 10*, 123–145, 1981.

Turner, J. S., Double-diffusive intrusions into a density gradient, *J. Geophys. Res., 83*, 2887–2901, 1978.

Williams, A. J., Salt finger observations in the Mediterranean Outflow, *Science, 185*, 941–943, 1974.

Williams, R. G., The internal tide off southern California, Ph.D. dissertation, University of California, San Diego, 1985.

Steven P. Anderson, Department of Physical Oceanography, Woods Hole Oceanographic Institution, Woods Hole, Massachusetts 02543.

Robert Pinkel, Marine Physical Laboratory, Scripps Institution of Oceanography, La Jolla, California 92093.

Salt Fingering and Turbulence-Induced Microstructure Measured by a Towed Temperature–Conductivity Chain

Stephen A. Mack and Howard C. Schoeberlein

The Johns Hopkins University Applied Physics Laboratory, Laurel, Maryland

Towed measurements from an array of temperature and conductivity sensors are used to investigate the relative importance of salt fingering and turbulence-induced microstructure for long horizontal tows in the upper thermocline of the Sargasso Sea. The small-scale signature of the microstructure regions is measured with high-frequency conductivity sensors and the discrimination technique of Mack and Schoeberlein using the conductivity gradient spectral slope, and conductivity gradient kurtosis is applied to separate salt fingering from turbulence-induced microstructure. The discrimination relies on the fact that salt fingering exhibits higher spectral slopes and lower kurtosis values than those of turbulence and uses a two-dimensional likelihood ratio approach to separate the two mechanisms while minimizing the unavoidable errors due to overlap of the empirical probability density functions. Tows in a relatively homogeneous region with a nearly constant temperature–salinity relationship and across a portion of the Subtropical Convergence Zone front with significant temperature–salinity variability are compared to determine the relative contribution of salt fingering and turbulence within and between the two varying environments. The approach is shown to be fairly robust over the varying regions of the 260 km of tow investigated. Bimodal distributions of spectral slope and likelihood ratio are apparent in each environment, suggesting the presence of microstructure of at least two types. The incidence of elevated microstructure for the Frontal region (12.5% of data exhibit microstructure, 2.1% due to salt fingering and 10.4% due to turbulence) is slightly greater than that of the Homogeneous region (8.4% exhibit microstructure, 1.7% due to salt fingering and 6.7% due to turbulence). The occurrence of salt fingering is rather random between the 14-km tow segments used to examine the variability within either environment. It appears that the weak salinity inversions necessary to support salt fingering are abundantly present in both environments. Preliminary assessment of the geometry of each type of microstructure region over only a 10-km test region implies some difference between the geometry of turbulence-generated microstructure and salt fingering regions, but this analysis has not yet been applied to the same long horizontal tows of the major part of the analysis.

1. INTRODUCTION

Although the evidence of double-diffusive processes in the ocean is substantial [*Schmitt*, 1994] and the presence of salt fingering and diffusive convection is accepted, few measurements can detect the small-scale features of salt fingering, distinguish them from turbulence-induced microstructure, and sample large ocean regions to determine the relative importance of each physical process. We report here measurements with an array of towed conductivity sensors used to detect the small-scale signatures of salt fingering and turbulence-generated microstructure and distinguish between each microstructure type over significant tow regions. The observations are an extension of the work of *Mack and Schoeberlein* [1993] and *Mack* [1989], who showed that the conductivity (or tempera-

Double-Diffusive Convection
Geophysical Monograph 94

ture) gradient spectral slope of salt fingering differed from that of turbulence-generated microstructure and that the conductivity gradient kurtosis for salt fingering also differed from that of turbulence-generated microstructure. Their analysis of limited regions of towed chain data was consistent with suggestions by *Holloway and Gargett* [1987] on the potential use of temperature gradient kurtosis, as well as with the results of *Gargett and Schmitt* [1982], who suggested the use of spectral slope as a potential discriminant.

Mack and Schoeberlein [1993] showed that for their tows in the thermocline of the Sargasso Sea, both conductivity gradient kurtosis and spectral slope could be useful discriminants between salt fingering and turbulence-induced microstructure. They applied a two-dimensional discrimination approach (called the log likelihood ratio technique) that uses both the kurtosis and slope information in a joint likelihood formalism. They showed that the unavoidable errors in discrimination due to overlaps of the probability density functions are minimized when the two-dimensional density functions are used instead of either of the one-dimensional density functions. The technique was developed on the basis of "control" or "training" patches and was then applied to a 10-km test tow as a limited demonstration of its potential utility.

This paper extends that work, which suggested that the log likelihood ratio technique be applied to large tow regions to determine the relative importance of salt fingering and turbulence-induced microstructure. We have analyzed two long tows in the Sargasso Sea using the two-dimensional log likelihood ratio technique to determine the relative distribution of microstructure sources in two different environments, one in a relatively homogeneous environment and the other across the Subtropical Convergence Zone front.

An understanding of the source mechanisms of ocean microstructure regions and their physical distribution in space has implications for large-scale ocean models that parameterize such small-scale processes. In particular, because salt fingering is more efficient than turbulence in transporting salt, model predictions for the vertical salt diffusivity can vary by a factor of 10 for the same measured turbulent dissipation rate, depending upon whether the mixing is caused by turbulence or by salt fingering [*Osborn*, 1980; *McDougall*, 1988; *Schmitt*, 1988; and *Hamilton et al.*, 1989]. A knowledge of the source of microstructure is paramount in knowing which model to use for an estimate of the diffusivity over an extended ocean region.

One of our long-term objectives is to determine the importance of double-diffusive processes and turbulence-induced microstructure in regions where both are possible. We ultimately envision a towed system that can map out the microstructure field, identify the type of microstructure (double-diffusive or turbulence), and relate the microstructure to the source mechanisms described by the larger scale temperature, salinity, shear field, and internal-wave field. The present discussion is limited to the log likelihood ratio discrimination technique applied to long tows in the Sargasso Sea. Some preliminary results to quantify the size and shape of the microstructure regions are also presented and are based upon the same short 10-km tow segment used in *Mack and Schoeberlein* [1993].

The remainder of the paper is organized as follows: The towed chain is described in section 2, followed by a description of the analysis approach using the spectral slope, kurtosis, and log likelihood ratio technique in section 3. A brief description of the environmental setting for the measurements is given in section 4. The major results for tows in two environments are described in section 5. The preliminary patch geometry results are found in section 6, followed by a discussion and summary in section 7.

2. INSTRUMENT DESCRIPTION

The towed chain consists of a vertical array of 30 co-located temperature (Thermometrics P20 thermistors) and conductivity cells with about 50 cm vertical spacing. Detailed descriptions of the conductivity sensors and general chain characteristics have been presented in *Farruggia and Fraser* [1984], *Mack* [1989] (henceforth M89), and *Mack and Schoeberlein* [1993] (henceforth MS93). The major part of the analysis makes use of the data from the four-electrode planar conductivity cells [*Farruggia and Fraser*, 1984], which are sampled at 320 samples per second and have a spatial response (3 dB down at 8 cpm] able to detect a portion of the small-scale signatures of salt fingering or turbulence-generated microstructure while not fully resolving the signatures. The conductivity spectra can be corrected for sensor spatial response to 30 cpm, and the analysis makes use of the limited wavenumber regime to 20 cpm for the discrimination. The conductivity output has been pre-emphasized with an analog filter that has unity gain below 0.7 Hz and increases at 6 dB per octave from 0.7 Hz to 100 Hz. In addition to the conductivity sensors used for the discrimination analysis, the co-located temperature and conductivity sensors are used to determine the larger scale features in salinity and for the determination of the density ratio [MS93].

3. DATA ANALYSIS

3.1. *Spectral Slope*

We briefly review the analytical techniques used in the discrimination developed in MS93. Although the conductivity gradient spectral slopes can be computed by several techniques, including straight-line fits of the spectrum computed over various wavenumbers and with varying types of smoothing, any discrimination approach simply requires that the slope distributions of salt fingering and turbulence differ sufficiently to minimize the errors caused by overlaps in the distributions. An efficient approach, described in MS93, called the power ratio technique, uses the ratio of the conductivity gradient variance in two wavenumber bands (high band: 3–20 cpm, and low band: 0.3–1 cpm) to estimate the spectral slope necessary to achieve the variance ratio. Spectral slopes will differ slightly, depending upon the exact wavenumber band chosen for either the direct spectral technique or the power ratio technique, but for our purpose here, the power ratio technique offers an efficient method and in fact provides a slightly greater separation of the resulting probability density functions for the salt fingering and turbulence "control" patches. MS93 found the mean and standard deviation for the slope estimates using the power ratio approach to be 1.5 and 0.26, respectively, for salt fingering and 0.71 and 0.35, respectively, for turbulence-generated control patches.

3.2. *Kurtosis*

We compute the kurtosis of the pre-emphasized conductivity c'

$$K = \langle c'^4 \rangle / \langle c'^2 \rangle^2 \tag{1}$$

on 1-s intervals after the conductivity has been wild-point edited and de-meaned. The results obtained by MS93 for the control patches were consistent with the results of *Holloway and Gargett* [1987]. MS93 obtained a mean of 3.18 and a standard deviation of 0.56 for the salt fingering control patch and a mean of 6.11 and a standard deviation of 3.4 for the turbulence control patch. *Holloway and Gargett* [1987] quoted a kurtosis of 3.04 for salt fingering and 6.75 for turbulence. The distributions of the kurtosis values obtained by MS93 (their Figure 6) were not Gaussian but rather had tails at high values. Taking the log of the kurtosis made the distributions more Gaussian. The slope distributions were more nearly Gaussian. Such Gaussian distributions are useful in modeling the probability density functions (PDFs) for a likelihood ratio discrimination technique that computes the ratio of two PDFs. Hence, MS93 used log kurtosis rather than kurtosis as the variable in the likelihood ratio technique to be described.

3.3. *Log Likelihood Ratio Discrimination Technique*

The likelihood ratio technique [*Whalen*, 1971] is a general formalism for decision making that minimizes the errors due to overlap of the PDFs of the observables that have different distributions. For a random variable x, the log likelihood ratio (LLR) for our oceanographic example, defined as

$$\lambda = \ln\left[\frac{p(x \mid SF)}{p(x \mid Turb)}\right], \tag{2}$$

is the optimal discriminator, where $p(x|SF)$ and $p(x|Turb)$ are the PDFs of x, given salt fingering and turbulence, respectively. An event is called salt fingering for all values of x such that the salt fingering PDF is greater than the PDF for turbulence. This can be used for more than one variable, and for our case where we have the variables slope (s) and log kurtosis (k), we can write the LLR as

$$\lambda(s,k) = \ln\left[\frac{p(s,k \mid SF)}{p(s,k \mid Turb)}\right]. \tag{3}$$

Since the slope and log kurtosis PDFs are nearly Gaussian, we use the joint Gaussian probability density functions as our model PDFs. The joint PDFs for salt fingering and turbulence based upon the mean, standard deviation, and correlation coefficients of s and k from the MS93 control patches are shown in Figure 1. Figure 1a shows the PDFs, and Figure 1b shows several contours for each PDF as well as the contour where the PDFs are equal ($\lambda = 0$). For values of slope and log kurtosis such that $\lambda > 0$, the sample is called salt fingering, and for $\lambda \le 0$ the sample is called turbulence. MS93 showed that the errors of such a two-dimensional approach (portion of the salt fingering PDF under the turbulence PDF, hence $\lambda \le 0$, or portion of the turbulence PDF under the salt fingering PDF, $\lambda > 0$) are less than the equivalent errors for a one-dimensional approach using either slope or log kurtosis alone. Hence, we use the two-dimensional approach to assess the occurrence of salt fingering and turbulence for our long tows.

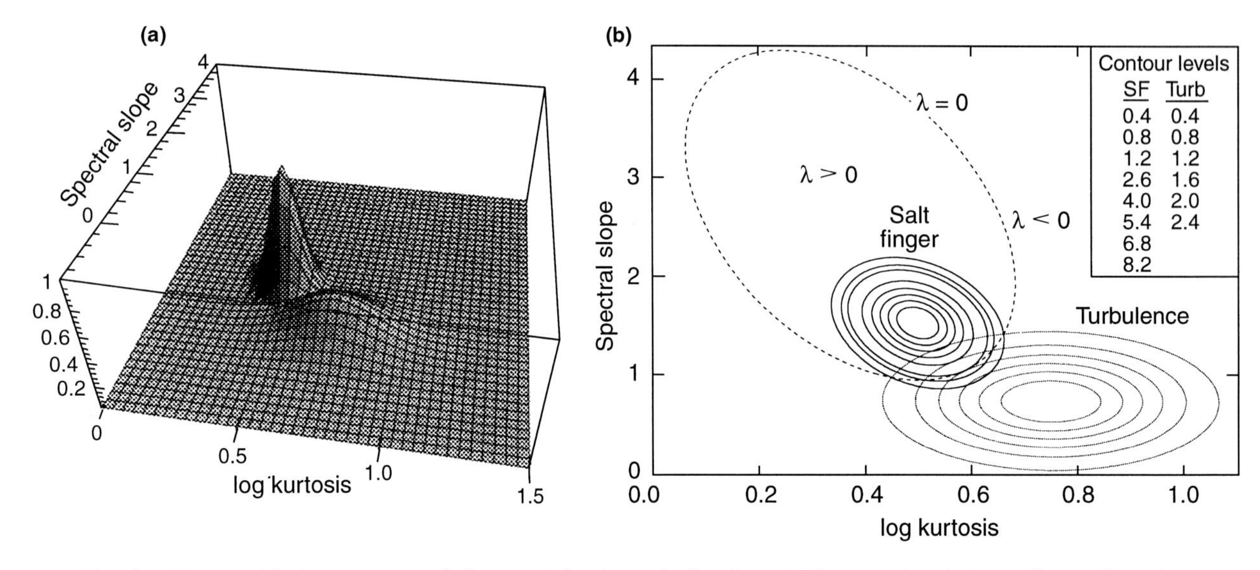

Fig. 1. The empirical parameters of slope and log kurtosis for the salt finger and turbulence "control" regions were used to form (a) a three-dimensional plot of the salt finger and turbulence two-dimensional Gaussian models, and (b) elliptic contours of the two-dimensional Gaussians and elliptic contour for $\lambda = 0$ (where the PDFs are equal).

TABLE 1. Statistics for Salt Fingering and Turbulence

	Salt Fingering	Turbulence
Mean slope	1.5	0.71
Standard deviation of slope	0.26	0.35
Mean log kurtosis	0.50	0.75
Standard deviation log kurtosis	0.064	0.16
Correlation coefficient	−0.28	−0.03

The algorithm is implemented as follows: The slope and log kurtosis are computed for every 1 s of data (approximately 3 m at the 3 m/s tow speed), and the LLR is computed on the basis of the control statistics given in Table 1. If $\lambda > 0$, the 3-m segment is called salt fingering, and if $\lambda \leq 0$, it is called turbulence-induced microstructure.

A sample of the spatial distribution of microstructure labeled by a generation mechanism based upon such a discrimination approach is shown in Figure 2. Pre-emphasized conductivity time series are plotted at their appropriate depth for a 12-min tow segment (approximately 2.2-km distance), and a gray scale of the LLR shows whether the region is due to salt fingering or turbulence. The LLR is computed only if the region has enhanced microstructure (using a conductivity gradient zero-crossing algorithm described in M89). The region near 0800 for sensors 3–17 labeled as a is identified as due to salt fingering. The patches labeled as b, c, d, e, and f are identified as turbulence-induced.

4. ENVIRONMENT

The data were collected during west to east tows obtained in the Sargasso Sea (Figure 3) during October and November 1984. During one tow near 30°N, with the chain array aperture between 100 and 115 m, a relatively homogeneous environment persisted over the length of the tow. The average buoyancy frequency was approximately 4.7 cph (obtained from accompanying CTD profiles). During the other tow, along 29°N, the Subtropical Convergence Zone front was crossed. It should be noted that this was the general location of the FASINEX experiment [see *Weller*, 1991 for an overview of FASINEX], although these measurements were not part of that experiment. The location of the most significant frontal crossing shown in Figure 3 agrees with the frontal location obtained from Advanced Very High Resolution Radiometer (AVHRR) images and shown by *Weller* [1991, Figure 2] for November 1984. The chain aperture during this frontal crossing was between 115 and 130 m (buoyancy frequency of approximately 5.1 cph). Although the tows were generally from west to east, the ship also executed some turns that have been excluded from the analysis. The straight tow regions have been divided into segments about 14 km in length (average for Homogeneous region tow was 15 km and average for Frontal region was 13 km for each segment). The segments are not contiguous in time or space and have an average separation between segments on the order of 2 km. There were

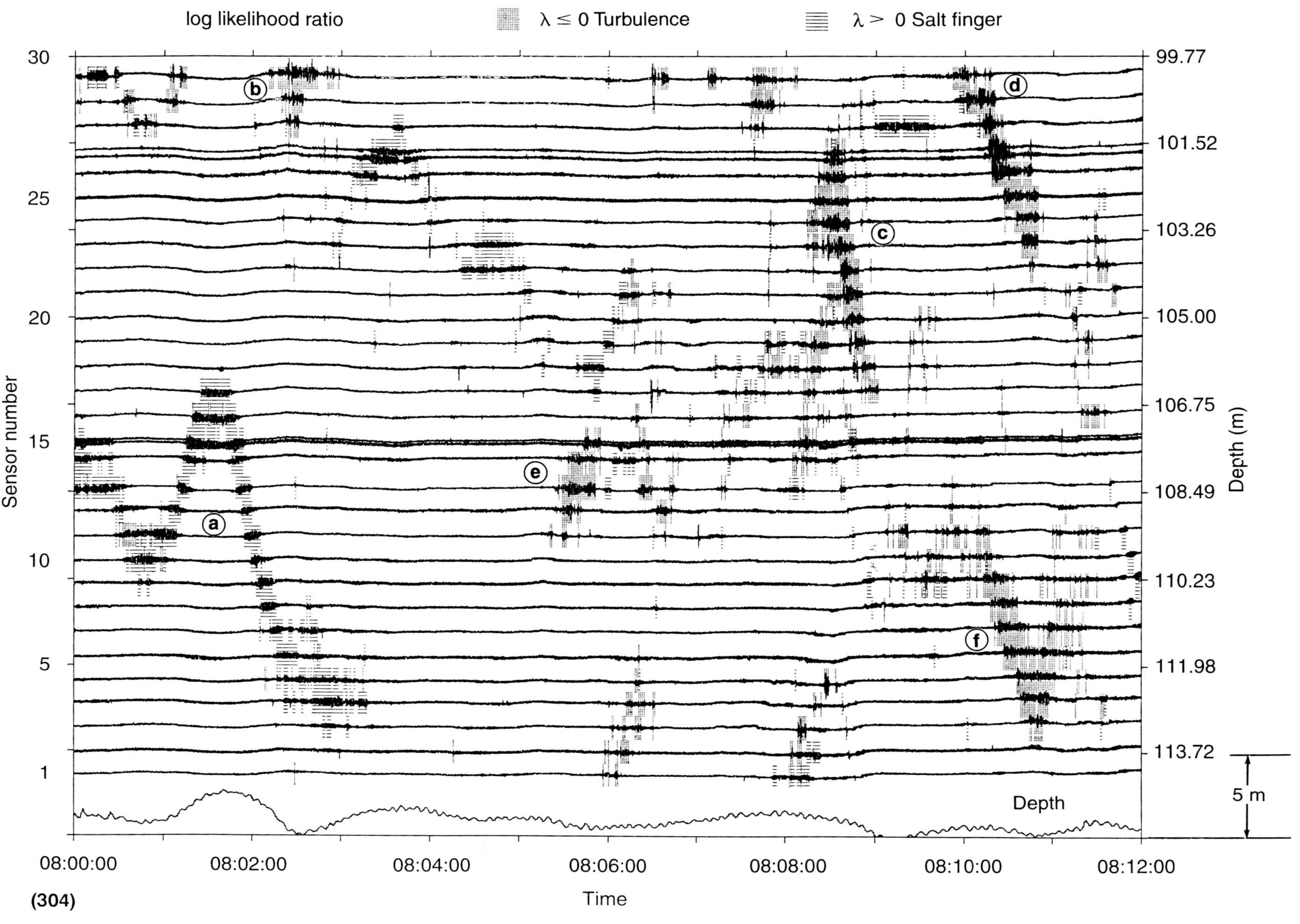

Fig. 2. Pre-emphasized conductivity, in min-max format, over a 12-min tow period (~2.2 km at typical tow speeds of 3 ms^{-1}) showing patches of both salt fingering and turbulence-generated microstructure. Gray scale is based on the LLR using slope and log kurtosis. $\lambda > 0$ implies salt fingering, and $\lambda \leq 0$ implies turbulence. The depth, obtained from a pressure transducer, is shown at the bottom with its own relative vertical scale at the right (increasing upward). The vertical motion should be noted before estimating the extent of the microstructure, such as for patch (a) where a 2-m depth excursion occurred.

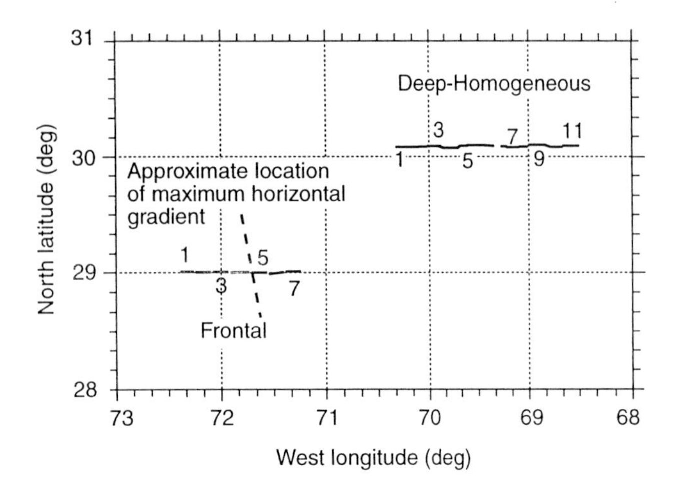

Fig. 3. Locations of the tow segments for two tows (Deep-Homogeneous and Frontal) in the Sargasso Sea. Alternate tow segment numbers are labeled. The average tow segment length is approximately 14 km.

11 tow segments in the Homogeneous region and 7 tow segments in the Frontal region. The total length of data segments was 168 km during the Homogeneous tows and 92 km for the Frontal tows. The results of our microstructure assessment using the LLR analysis have been separately compiled for each segment in order to assess the variability of microstructure type within a particular environment and to determine the overall differences between frontal and nonfrontal regions.

The manifestation of the frontal and nonfrontal behavior is reflected in the general temperature–salinity (T–S) diagrams for each of the tow segments (Plate 1). These were obtained from a single temperature–conductivity[1] sensor pair. The Homogeneous region shows a very tight T–S structure (some variability is in part due to residual vertical chain motion of about ±2 m as well as relative displacement due to internal waves) over the 168 km of tow. The Frontal tow shows a variable T–S behavior over the first several tow segments. The strongest horizontal gradient occurred between tow segment 4 and tow segment 5.[2] During segments 5, 6, and 7, the T–S characteristics very nearly match those of the Homogeneous tow. Note that the front is a strong temperature front and also a density front as indicated by the sigma-t contours.

5. OBSERVATIONS

5.1. *Homogeneous Region Tows*

The spectral slope and the kurtosis estimates over 3-m intervals have been computed for each conductivity sensor, and the LLR (λ) has been computed according to Equation (3) for each 3 m of data that exceeds 2 conductivity gradient zero-crossings per second [MS93]. The resulting histograms of λ for each of the eleven 14-km tow segments for the Homogeneous environment are shown in Figure 4. Several observations are noteworthy. Many of the histograms have clearly bimodal distributions, such as for tow segments 1, 2, 3, 5, and 7. Since the distributions are bimodal rather than more continuous over all λ values, this suggests the presence of two distinct sources of mixing (turbulence-generated microstructure and salt fingering) and supports the contention that these processes generally can be distinguished by their slope and kurtosis characteristics. Tow segments 7 and 2 show the largest percentage of salt fingering microstructure, 49.1 and 41.9%, respectively. Tow segments 6, 8, 9, and 10 show the largest percentage of turbulence-induced microstructure. The unavoidable errors predicted by the overlap of the two-dimensional Gaussian distributions [MS93, Table 3] imply that 3.1% of the turbulence-induced microstructure will have $\lambda > 0$ and hence will be falsely called salt fingering. This suggests that the small percentages of salt fingering identified in tows 6, 9, and 10 are probably due to such errors. Several of the tows show salt fingering contributions of about 13 to 30%.

The relative contributions of salt fingering and turbulence-generated microstructure over all tows in the Homogeneous environment are 19.9% for salt fingering and 80.1% for turbulence (Figure 5). A clear bimodal distribution for λ (Figure 5a) is contrasted with the weaker indication of a bimodal distribution for spectral slope (Figure 5b). The histogram for log kurtosis shows little indication of a bimodal distribution (Figure 5c). It is noteworthy that the bimodal peaks in the spectral slope appear to be at about 0.7 and 1.5. Hence, the mean spectral slopes computed over nearly 168 km of tow with about 30 conductivity sensors appear to be consistent with the spectral slopes obtained over the few kilometers of the control or training regions in MS93.

In addition to the relative distribution of salt fingering and turbulence-generated microstructure, it is useful to determine the contribution of microstructure type relative to all ocean samples. This is shown in the form of histograms for the $\log_{10}$ of the conductivity gradient variance in the two bands used for determination of the spectral slope, the 0.3–1 cpm band (Figure 6a), the 3–20 cpm band (Figure 6b). The unconditional histograms represent all of the data for the 11 tow segments. The dark shaded histogram represents the subset of the data with $\lambda > 0$ (salt fingering), and the lighter shaded histogram is for

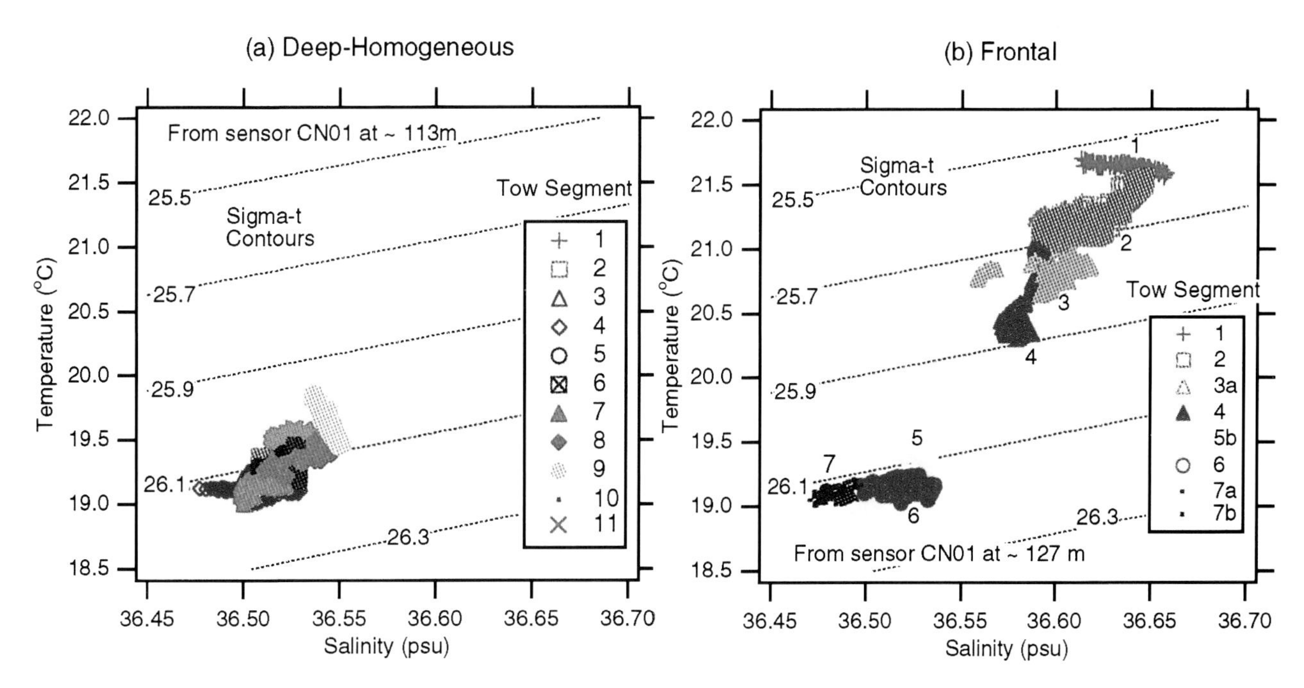

Plate 1. The temperature–salinity (T–S) diagrams from one temperature-conductivity sensor pair show (a) a nearly constant T–S relationship for the Deep-Homogeneous tow and (b) a strongly varying T–S behavior across the Subtropical Convergence Zone front.

$\lambda \leq 0$ (turbulence). Salt fingering is suggested for 1.7% of the data, and turbulence is suggested for 6.7%. Note that in the 0.3–1 cpm band (Figure 6a), the variance for the turbulence-generated microstructure (mean of $\log_{10}$= –3.85 or mean variance = 1.41×10^{-4} ((mmho/cm)/s)^2) is greater than that for the salt fingering (mean of $\log_{10}$= –4.66 or mean variance = 2.19×10^{-5} ((mmho/cm)/s)^2). This is consistent with the relatively broadband nature of turbulence with significant energy over a wide spectral range compared to the narrowband salt fingering process with little energy in this relatively low-wavenumber band (scales of about 1–3 m). In contrast, for the higher wavenumber band, 3–20 cpm, salt fingering and turbulence appear to have about the same conductivity gradient variance, –1.96 and –2.06, respectively, for $\log_{10}$. It is interesting to note that these results are very similar to the results of MS93 (see their Figure 17) for another 10-km sample tow in the Homogeneous region used for preliminary testing of the discrimination approach. Those limited results appear to be consistent with the longer tows and considerably larger data set over the 168 km of tow in this region.

5.2. *Frontal Region Tows*

The same analysis was applied to the Frontal tows, and the results are shown in the next three figures. For the seven Frontal tows, the percentage of microstructure exhibiting the characteristics of salt fingering ranged from 5.9 to 33.7% (Figure 7). The tow with the highest percentage of salt fingering is tow 3 on the west side of the strongest frontal gradient region. Only tow 2 and tow 3 show a strong bimodal distribution of LLR. Tows 4 and 6 show only 7.5 and 5.9% salt fingering, respectively, just slightly above the amount predicted from the errors due to overlap of the underlying parent distributions. The ensemble results over all of the tows of the Frontal region are shown in Figure 8 in terms of the histograms of LLR, slope, and log kurtosis. The histograms of LLR indicate a stronger bimodal appearance than that for spectral slope or log kurtosis, thus suggesting that the two-dimensional discrimination should outperform either one-dimensional approach. However, LLR is not as strongly bimodal as that observed during the Homogeneous tows. The distribution of LLR implies that 16.5% of the microstructure over the entire Frontal tow region is due to salt fingering, and 83.5% is due to turbulence-generated microstructure.

To determine the occurrence of these microstructure sources as a percentage of the total tow region, we compare the total number of samples with LLR > 0 or LLR $\leq$ 0 with the unconditional samples. This is shown in Figure 9 in terms of histograms for the log conductiv-

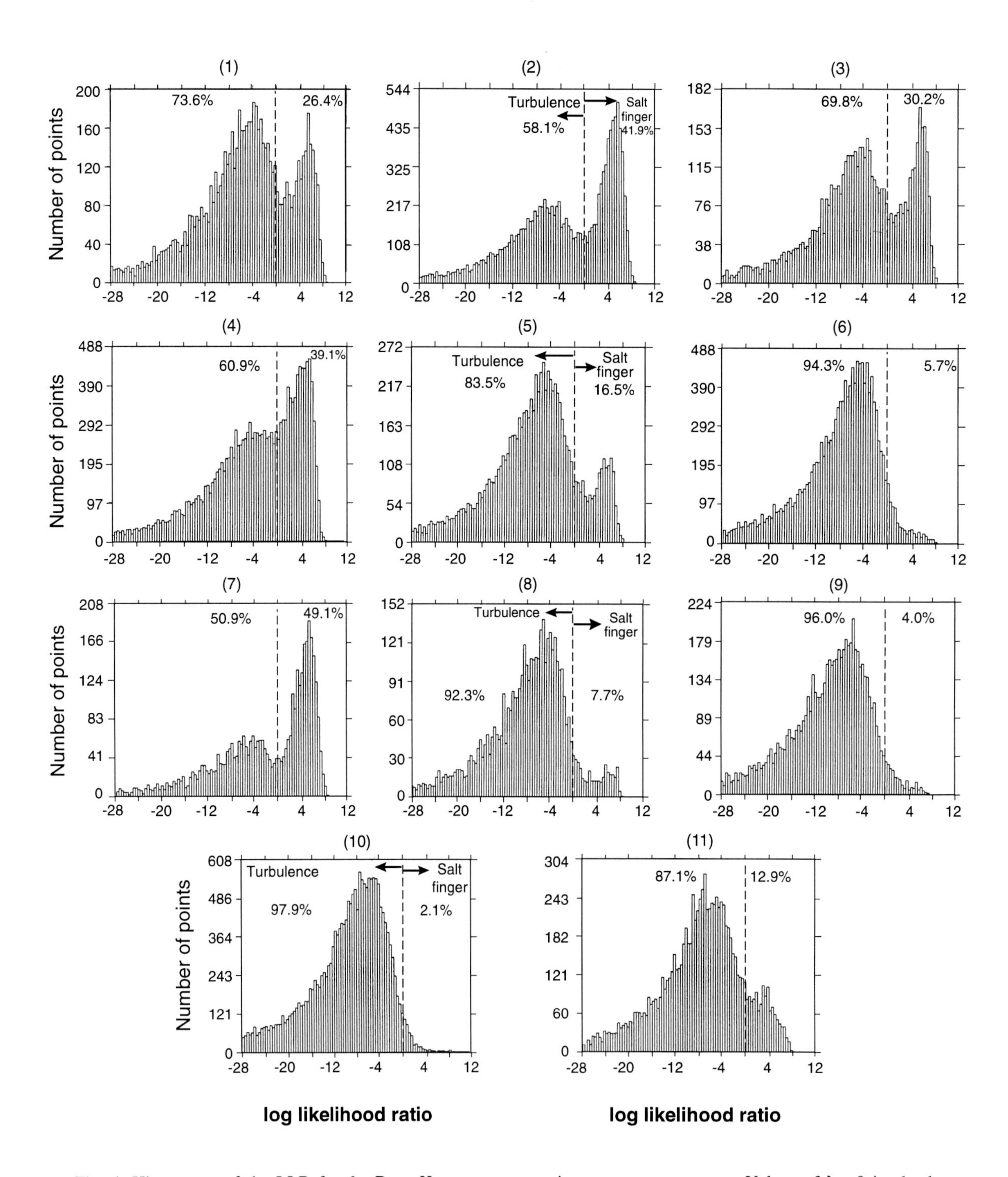

Fig. 4. Histograms of the LLR for the Deep-Homogeneous environment tow segments. Values of $\lambda > 0$ imply that the combination of slope and kurtosis is consistent with salt fingering, and those values of $\lambda \leq 0$ imply that the slope and kurtosis values are consistent with turbulence-generated microstructure.

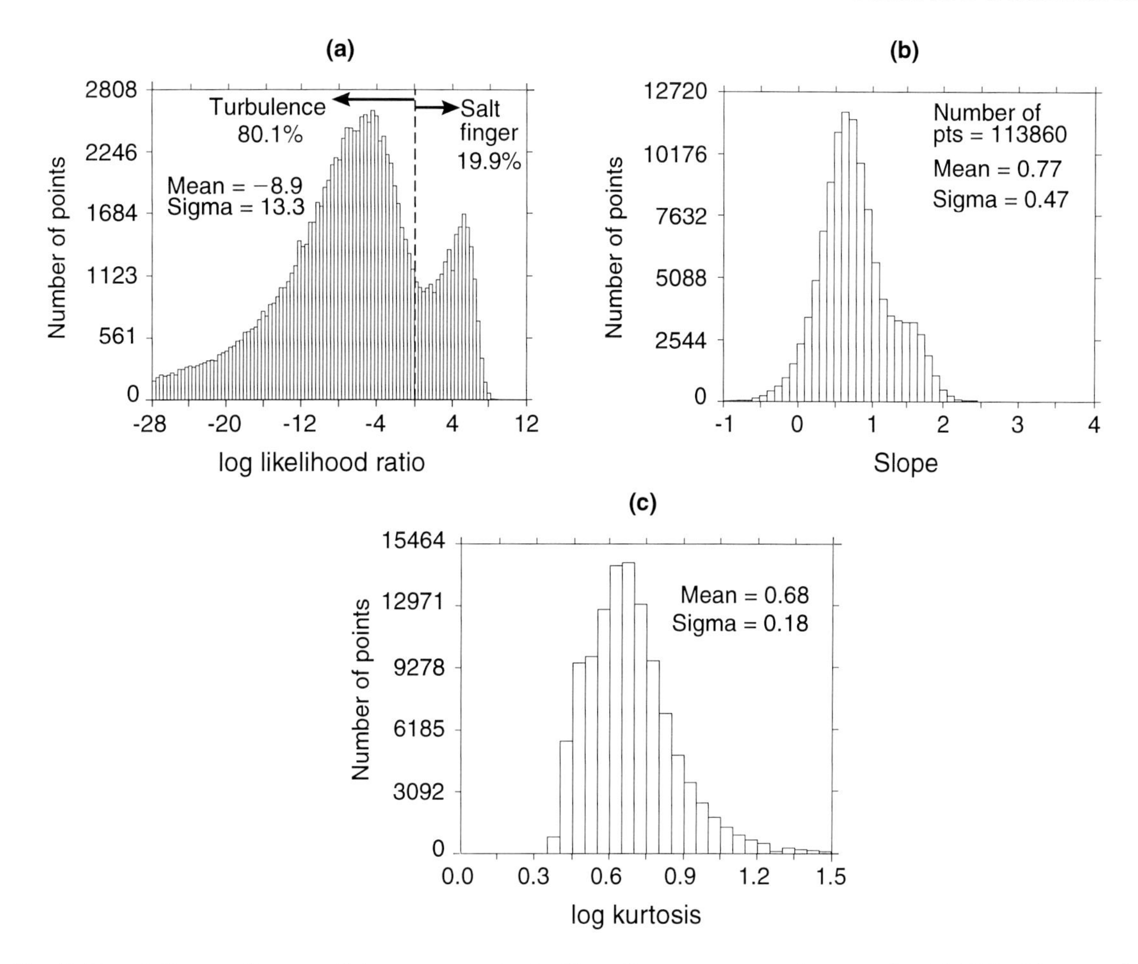

Fig. 5. Cumulative results over all tow segments in the Deep-Homogeneous environment for (a) LLR, (b) spectral slope, and (c) log kurtosis.

ity gradient power for the low-wavenumber band (Figure 9a) and the high-wavenumber band (Figure 9b). Salt fingering constitutes 2.1% of the data, and turbulence constitutes 10.4%. Distribution of the variance is similar to that observed for the Homogeneous tows (Figure 6); the turbulence-generated microstructure strength for the low band (0.3–1 cpm) has a mean of $\log_{10} = -3.69$, compared with $\log_{10} = -4.46$ for the salt fingering. In the 3–20 cpm band, the salt fingering shows slightly greater variance than the turbulence ($\log_{10} = -1.85$ and $\log_{10} = -2.0$, respectively).

5.3. *Comparison of Homogeneous and Frontal Results*

It is useful to summarize the relative percentage of microstructure type, the absolute contribution of microstructure type, and the spatial distribution over tow segments for both environments. Such a summary is shown in Figure 10 where each shaded histogram represents the fraction of data identified by the LLR approach as due to salt fingering (diagonal bars) and turbulence (dotted areas). Each histogram represents the tow segments from west to east (see Figure 3). The cumulative results for each environment are shown to the right of each display. This clearly shows the variation of the relative microstructure contribution from segment to segment. Salt fingering is observed in both environments but not in every tow segment. As indicated previously, the small percentage of salt fingering identified in tow segments 8, 9, and 10 in the Homogeneous environment could be true salt fingering or simply represent the errors due to overlap of the underlying distributions. The Frontal environment shows only slightly greater microstructure (12.5%) than the Homogeneous environment (8.4%), and

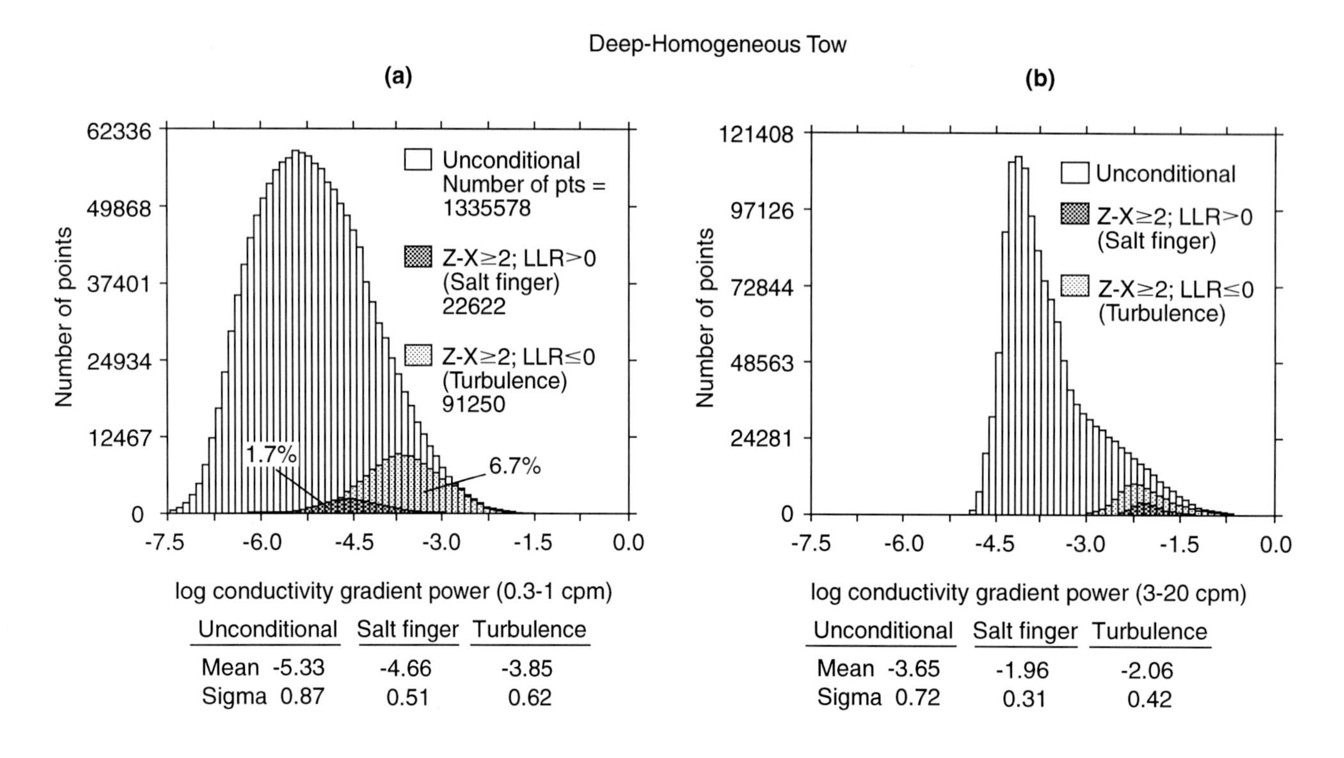

Fig. 6. Histograms of the log conductivity gradient variance for the Deep-Homogeneous environment in two wavenumber bands: (a) 0.3–1 cpm and (b) 3–20 cpm.

slightly greater salt fingering (2.1%) and turbulence (10.4%) than the Homogeneous environment (1.7 and 6.7%, respectively). The most significant events appear to be the large contribution of turbulence-generated microstructure observed during the westernmost tow segments (1 and 2) of the Frontal tows. During the first tow segment, nearly 26% of the data shows excess microstructure, for which 22.3% is due to turbulence and 3.2% is due to salt fingering. The strong contribution of turbulence-generated microstructure observed in tow segment 1 (Frontal) does not seem to affect the occurrence of salt fingering (3.2%). It is interesting to note that although the greatest incidence of microstructure occurs in the Frontal tows, it is not observed at the region of strongest horizontal temperature gradient, which occurs within tow segment 5, but rather during the first two segments on the west side of the maximum gradient region.

5.4. *Comparison of LLR Discrimination with Density Ratio*

It should be instructive to compare the results of the LLR discrimination with the density ratio (R_ρ) information available from the horizontal tows. The advantages and limitations of the R_ρ computation based upon single-sensor T–C pairs as they cross isopycnals have been thoroughly discussed in M89 and MS93. Within the limitations of R_ρ values so determined, we can examine the consistency of the LLR discrimination with this additional information. For an ocean that follows our notions about salt fingering and turbulence, we would expect that the assessment of microstructure type based upon the LLR also should be consistent with regimes of the density ratio (R_ρ) for a particular instability type. Salt fingering is permitted only for $1 < R_\rho < K_T/K_S \approx 100$ (where K_T and K_S are the molecular diffusivities of heat and salt, respectively) and is more likely or supercritical as R_ρ approaches 1 [*Schmitt*, 1979; *Mack*, 1985; M89]. Figure 11 shows R_ρ histograms[3] and conditional probabilities for both the Deep and Frontal environments. Similar histograms over a 10-km test tow were shown in MS93 (their Figure 15). Consider first the results for the Deep-Homogeneous environment (Figures 11a and 11b). The upper panel in Figure 11a shows the unconditional histogram and those (light shading) conditioned on the presence of microstructure (with conductivity gradient zero-crossings ≥ 2 [see M89]) and further conditioned (dark shading) upon $\lambda > 0$ (salt fingering). The associated conditional probabilities are shown in the lower panel along

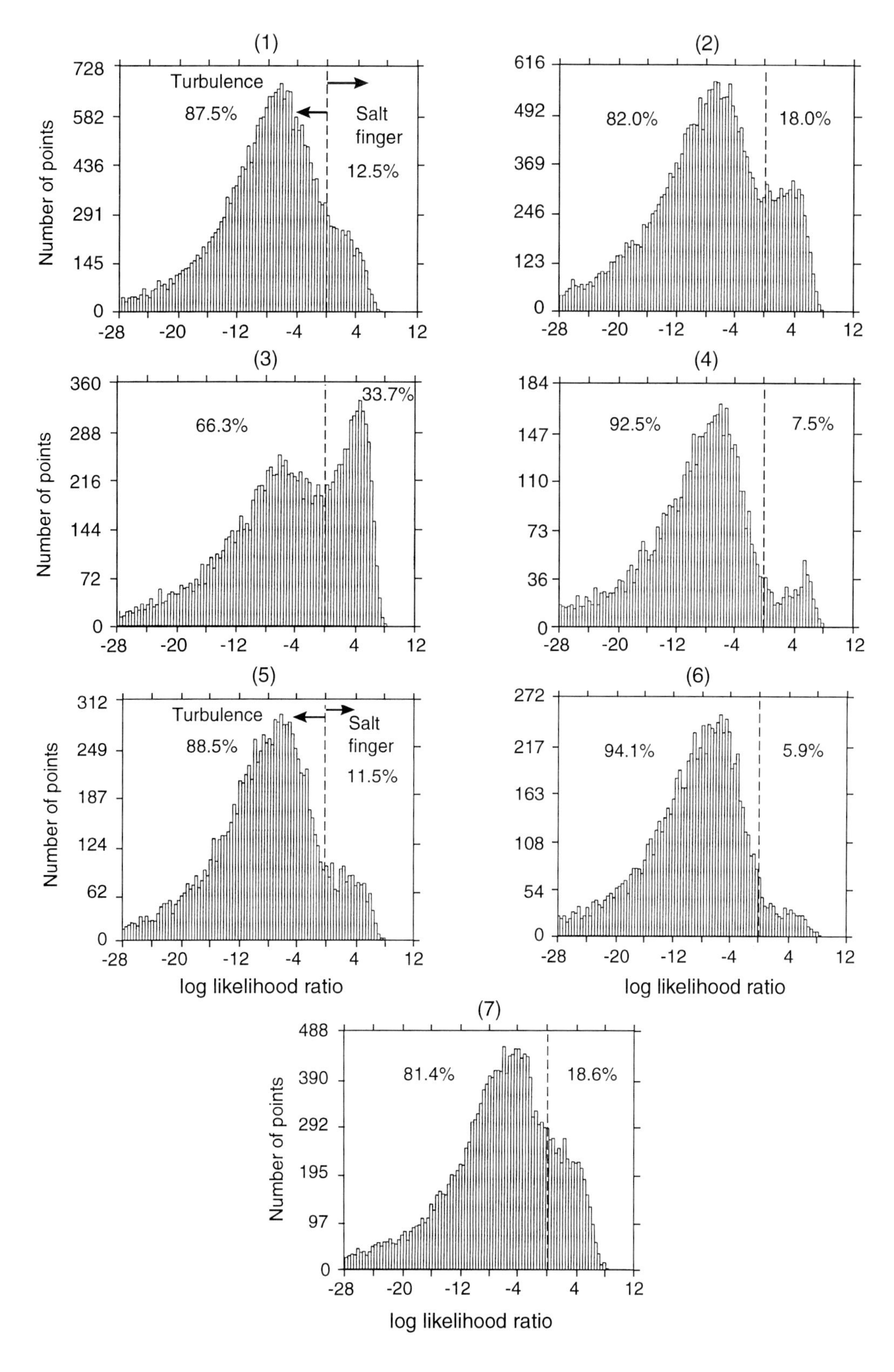

Fig. 7. Histograms of the LLR for the Frontal environment tow segments.

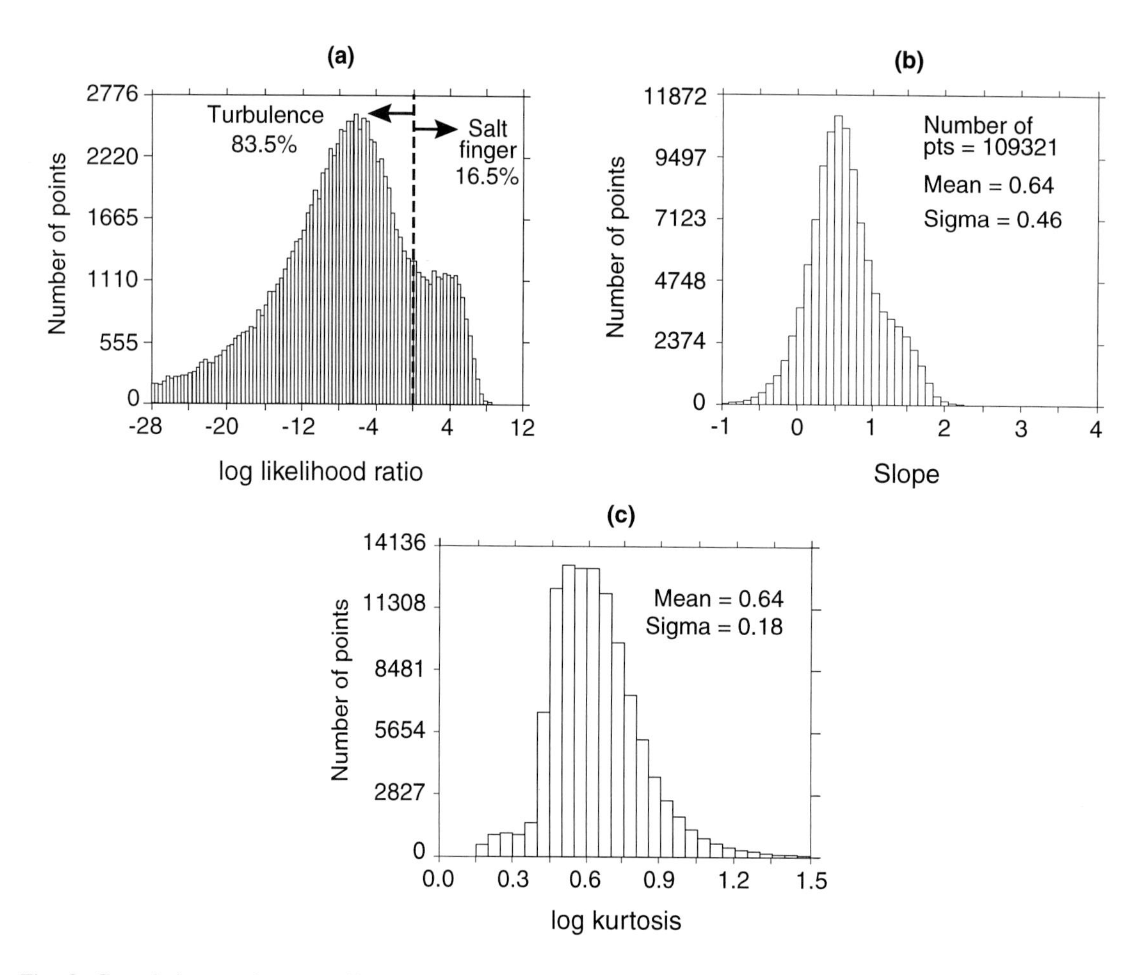

Fig. 8. Cumulative results over all tow segments in the Frontal environment for (a) LLR, (b) spectral slope, and (c) log kurtosis.

with the levels of the null hypotheses (upper, 0.083 for "microstructure is independent of R_ρ," which fails, and lower, 0.017 for "microstructure with $\lambda > 0$ is independent of R_ρ," which also fails). Both probabilities peak in the supercritical salt finger regime for R_ρ between 1 and 2. The light curve peaks in the salt finger regime but also has elevated probability over all R_ρ values. M89 and MS93 argued that since turbulence should be independent of R_ρ, it should produce elevated microstructure over all R_ρ. Conditioning on only zero-crossings ≥ 2 should include both sources, and this is reflected in the light shaded histogram and probability. Further conditioning upon the LLR discrimination with $\lambda > 0$ should limit the microstructure to salt fingering thus reducing the turbulence contribution. Indeed, by conditioning on $\lambda > 0$ (dark shaded probability curve in Figure 11a), the probability outside the supercritical salt finger regime is reduced to nearly zero. A peak for R_ρ between 1 and 2.5 consistent with salt fingering remains. On the other hand, by conditioning on turbulence only ($\lambda \leq 0$), the probability in the salt finger R_ρ regime should be reduced, and microstructure probability in the stable regime ($R_\rho < 0$) should remain unchanged. Results approaching this are observed (Figure 11b).

It should also be noted in Figure 11b that even for $\lambda \leq 0$, a small residual peak in the probability still exists for R_ρ between about 0.5 and 2.5. To understand this, first consider the salt finger regime ($1 < R_\rho < 2.5$). A residual peak is to be expected for any discrimination technique using empirical PDFs as a result of the overlap for the salt finger and turbulence distributions (such as our slope and kurtosis PDFs of Figure 1; also see MS93). Because of this overlap, a portion of salt fingering will be misidentified as turbulence and vice versa. These two types of errors are expected to manifest themselves differently on an R_ρ histogram. When true salt fingering is misidentified as turbulence, the erroneous values should be grouped within the range of R_ρ values mostly between

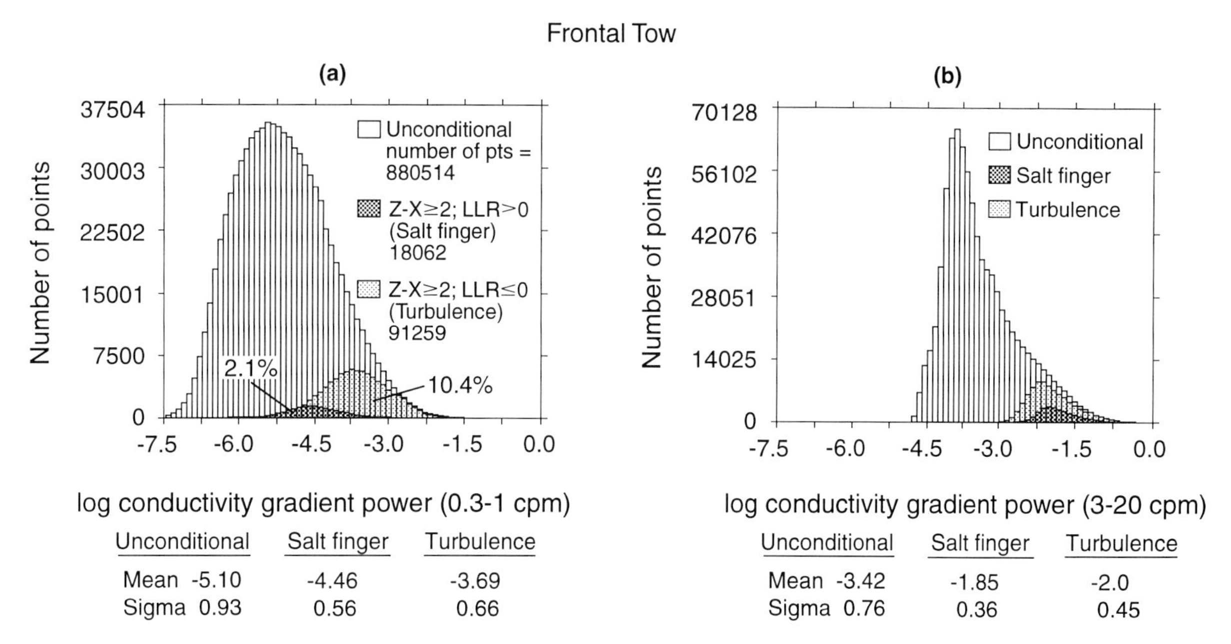

Fig. 9. Histograms of the log conductivity gradient variance for the Frontal environment in two wavenumber bands: (a) 0.3–1 cpm and (b) 3–20 cpm.

1 and 2.5, as observed in Figure 11b. On the other hand, when true turbulence is misidentified as salt fingering, we expect that the erroneous values should be spread over all R_ρ, as observed in Figure 11a, which shows a small residual probability even after applying the salt finger ($\lambda > 0$) condition. We can estimate the magnitude of such errors by using the errors associated with the two-dimensional slope and kurtosis PDF overlap discussed in MS93 and summarized here in Table 2. In particular, consider the error of incorrectly labeling salt fingering as due to turbulence, $P(\lambda \leq 0 \mid SF)$, stated as the probability that $\lambda \leq 0$ given salt fingering. The PDF overlap implies that 3.1% of the slope and kurtosis values that lie in the turbulence regime, $\lambda \leq 0$, are actually due to salt fingering.

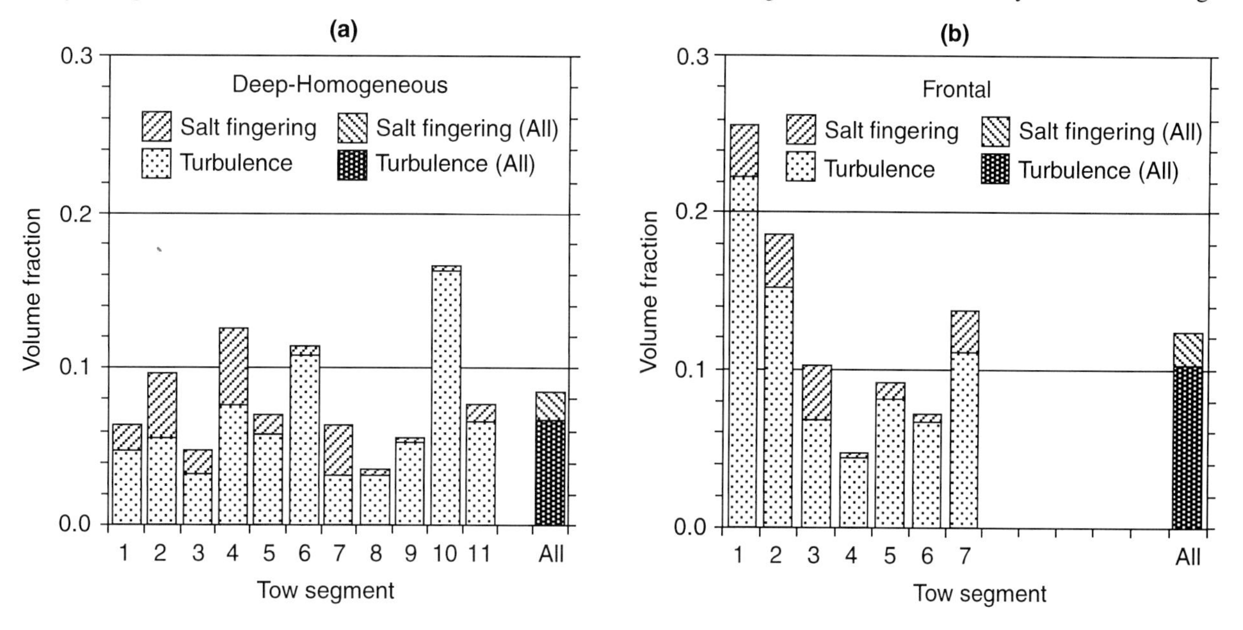

Fig. 10. Fraction of data showing salt fingering and turbulence over the tow segments of (a) Deep-Homogeneous and (b) Frontal environments. The cumulative results over each environment are shown as the histograms at the right labeled as "All".

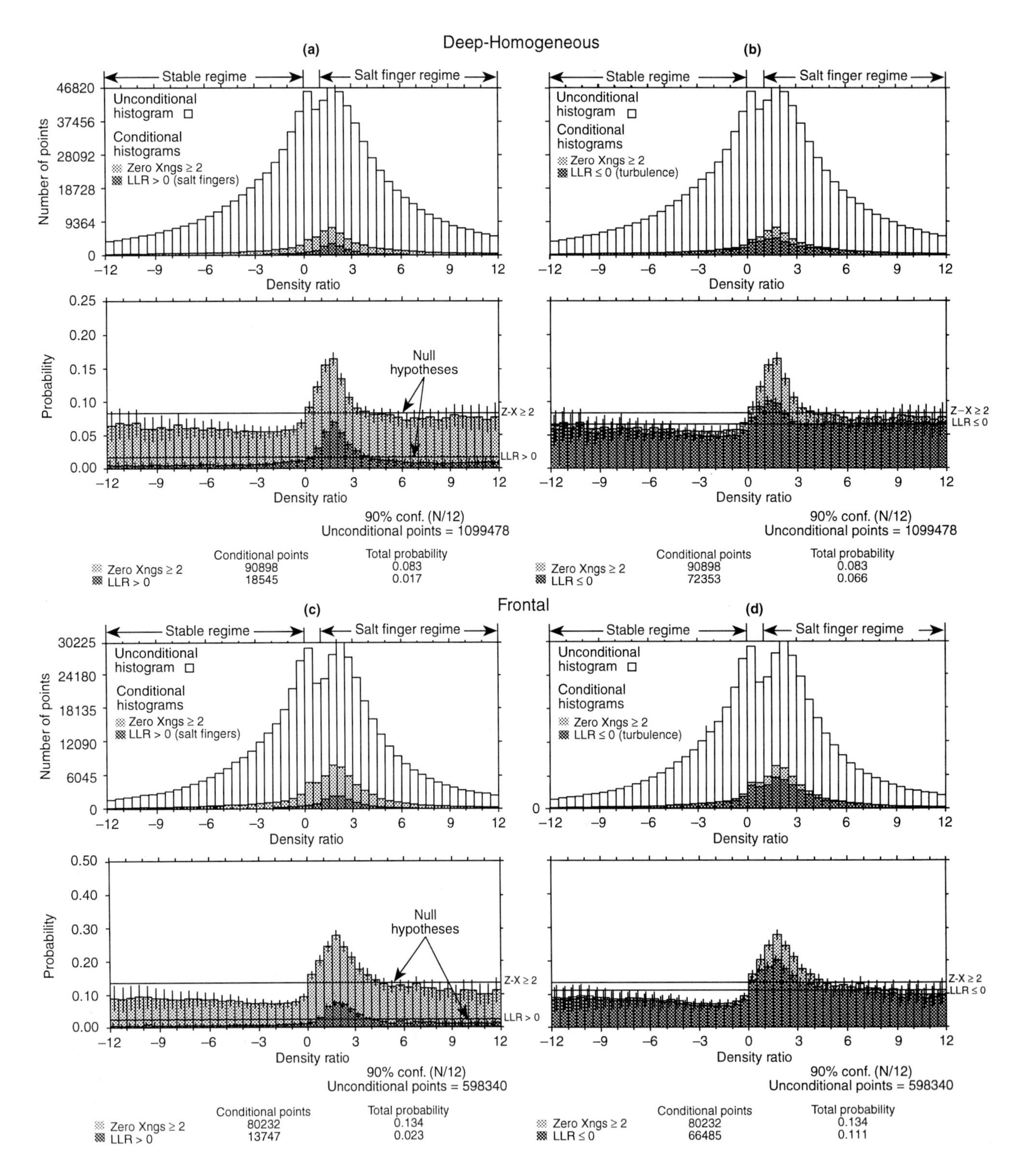

Fig. 11. Histograms of the density ratio and probability of microstructure for the Deep-Homogeneous environment: (a) conditioned for salt fingering, $\lambda > 0$; (b) conditioned for turbulence, $\lambda \leq 0$; and for the Frontal environment: (c) conditioned for salt fingering, $\lambda > 0$; (d) conditioned for turbulence, $\lambda \leq 0$.

TABLE 2. Errors in Discrimination

Discrimination Type	$P(\lambda \leq 0 \mid SF)$	$P(\lambda > 0 \mid turb)$	Average Error
LLR using slope and kurtosis	0.031	0.053	0.042

We approximate the effect on the histogram by taking 3.1% of the microstructure identified as turbulence (0.031×72353 = number of points with $\lambda \leq 0$), 2243, and redistribute these erroneous values to the salt finger R_ρ population between 1 and 2.5 (add 2243 to the salt finger population obtained by multiplying the uniform null hypothesis probability, 0.066, and the unconditional population in the bins $1 < R_\rho < 2.5$, 135889; hence $0.066 \times 135889 = 8969$). The new population between R_ρ of 1 and 2.5 incorporating these errors due to PDF overlap is $2243 + 8969 = 11212$, and the new probability based upon this new population is 0.083. The measured probability within these same R_ρ bins is 0.093 (Figure 11b). It is clear that the basic features of the probability, namely, an enhancement in the salt finger regime and a very small reduction for other R_ρ values, is consistent with the errors due to overlap of slope and kurtosis PDFs used in the LLR discrimination approach.

The analogous results for the Frontal environment are shown in Figure 11c and 11d. Note that the unconditional histograms are bimodal (as are those for Deep-Homogeneous) with a minimum near $R_\rho = 1$. This is consistent with double diffusion, which acts to redistribute the temperature and salinity gradients away from $R_\rho = 1$. Conditioning on salt fingering, $\lambda > 0$ (Figure 11c), reduced the probability in the stable regime and weak salt finger regime while leaving an enhanced peak for R_ρ between 1 and 2.5, consistent with salt fingering. Conditioning on turbulence, $\lambda \leq 0$ (Figure 11d) leaves the probabilities in the stable regime and weak salt finger regime nearly the same as the probabilities before applying the LLR discrimination and reduces the peak in the supercritical salt finger regime. Note, however, that this peak is not fully removed by conditioning on $\lambda \leq 0$, and it appears more significant than the residual peak for the Deep environment (Figure 11b). We can use the same arguments as described above for the Deep environment to show that a peak should remain due to overlap in the two-dimensional (slope and kurtosis) PDFs for salt fingering and turbulence. By using the same procedures described above for the Deep environment, we find that the probability within the R_ρ regime of 1 to 2.5 is increased from 0.11 (null hypothesis level) to 0.14. This elevated value is still less than the measured value of 0.18. It may be that the overlap errors of Table 2, which were determined from the "control" or "training" patches of MS93 in the Deep environment, underestimate the PDF overlap for the Frontal environment. Additional errors due to unreliable values in R_ρ, computed when the sensors do not cut through isopycnals, may also contribute. In part, it was due to such limitations of computing R_ρ from horizontal tows that led us to examine the discrimination approaches that only rely on the small-scale internal characteristics of the microstructure. The main point is that these histograms of R_ρ are generally consistent with the physics and models of salt fingering and turbulence and with an error assessment of our LLR discrimination approach.

A further observation should be noted for R_ρ in the diffusive regime ($0 < R_\rho < 1$). In both environments, but particularly in the Frontal environment, there appears to be an enhanced probability in this regime that remains after the condition $\lambda \leq 0$ is applied. That is, although the probability in the salt finger regime is reduced, less reduction is observed in the diffusive regime. Some of the enhanced microstructure could be due to active diffusive convection in the process of forming the step features normally attributed to the diffusive regime on temperature inversions. Conditioning on $\lambda > 0$ (salt fingering) does produce a probability reduction in this regime (Figures 11a and 11c). The observations of microstructure within the diffusive regime are more limited, but they appear to have slope and kurtosis characteristics similar to the turbulence-generated microstructure found in the stable regime.

6. PRELIMINARY PATCH GEOMETRY ANALYSIS

Up to this point, we have assessed the total volume fraction of microstructure and examined how this microstructure was divided between the two generating mechanisms: salt fingering and turbulence. At this point, we wish to examine the spatial characteristics of the microstructure patches and determine if their geometric characteristics differ between the two generation mechanisms.

Accordingly, the overall objective of the patch geometry analysis is to characterize the two-dimensional geometry of conductivity microstructure patches in the Sargasso Sea thermocline as a function of environment (Homogeneous and Frontal). The characterization will focus on the height, width, area, slope, and aspect ratio (width divided by height) of microstructure patches, and will emphasize the statistics of the patches as a function of the generation mechanism as indicated by the LLR. We wish to determine if there is a statistically significant difference in the geometry of patches generated by salt fingering and patches generated by turbulence. In addition, we seek to determine if the patches are relatively uniform, being generated by one mechanism or the other, or are they frequently mixtures of the two generating mechanisms. Accomplishing these objectives requires automated processing procedures be implemented, tested, and applied to the same data sets described previously.

This section presents an approach and some preliminary results for detecting microstructure patches on single sensors, grouping the single sensor patches into patch groups with vertical extent, and calculating their geometrical parameters. The approach will be illustrated using the same short segment of data (hereafter called Tow A) presented in Figure 2 and thoroughly discussed in M89 and MS93. Patch (a) in this data segment was used in those references to derive the moments of the spectral slope and gradient kurtosis for salt fingering necessary in constructing the LLR function. Following the example of M89 and MS93, we will then examine a specific 10-km tow segment in the same environment (hereafter called Tow B) to further illustrate the analysis techniques we intend to use to quantify differences between patches generated by salt fingering and those generated by turbulence.

The basic approach is to first detect single-sensor microstructure patches (SSPs) in the conductivity data. Then group these patches together into multiple-sensor patch (MSP) groups, and calculate the geometrical characteristics while preserving information regarding the generating mechanism as quantified by the LLR. The approach discussed here is by no means unique but was developed with the following goals in mind: First, the single-sensor processing should be simple and significantly reduce the data volume so that the patch statistics of a large data set can be assessed. The single-sensor processing should also be consistent with calculation of the LLR; specifically, the definition of a patch must be consistent with the definition of microstructure used in M89 and MS93, and a value of the LLR must exist at every point in the interior of each patch (the LLR is defined only in microstructure, not in ambient data). Sufficient information regarding the LLR statistics within the patch must be preserved so that statistics of the patch geometries as a function of the LLR can be calculated. Second, the algorithm that groups (or clusters) the SSPs into MSPs needs to be efficient and produce results consistent with a visual assessment of the number of significant patches in the data. The clustering algorithm should take into account the fundamental intermittency of the microstructure patches; "dropouts" within a microstructure patch are common. Although many clustering techniques are available, most of them require the multiple recalculation of a "distance matrix" that, in the application considered here, is a matrix of the physical distance between every SSP and every other one. Unfortunately, even a relatively small piece of data (say ~2 km of 30 sensors, similar to Tow A described later) can have ~500 patches (9 SSPs/km), making such an analysis extremely unwieldy. In addition, optimality criteria are common only when the number of clusters (MSPs) are known *a priori*, which is not the case in our application. A much simpler approach will be discussed later. Finally, the geometrical characteristics of an MSP can be defined in many ways, particularly when one considers the possibility of weighting the observations on the basis of the spatial distribution of activity (e.g., number of zero-crossings) within the patch. Our approach, discussed later, emphasizes simplicity and is easy to interpret.

An SSP will be defined as a contiguous run of zero-crossings greater than a threshold; the threshold chosen for this analysis, 2 zero-crossings in each second of data, is consistent with the analysis presented previously and in M89 and MS93. This criterion was used to determine if there was sufficient signal on the conductivity sensors to calculate the spectral slope and thus the LLR. Once the boundaries of the SSP are determined, the average value of the LLR in the patch is calculated and saved, in addition to the fraction of the patch with $LLR > 0$, indicative of salt fingering. Thus, information regarding the likely generation mechanism (mean LLR) and uniformity (fraction with $LLR > 0$) of that generating mechanism is saved with each patch. Patch lengths and arrival statistics can then be characterized by the likely generating mechanism. Depth variations are represented by preserving the sensor depth at the beginning and ending of the patch, assuming rigid body motion of the towed chain as measured by the depth sensor. The patch detections can then be plotted as parallelograms over the time and depth region spanned by the patch, as illustrated in Figure 12.

Plate 2 presents an example patch plot from Tow A, which can be directly compared with Figure 2. Patches whose mean $LLR > 0$, indicative of salt fingering microstructure, are plotted in red, whereas patches with $LLR \leq 0$, indicative of turbulence-generated microstructure, are plotted in blue. The red salt fingering regions, such as

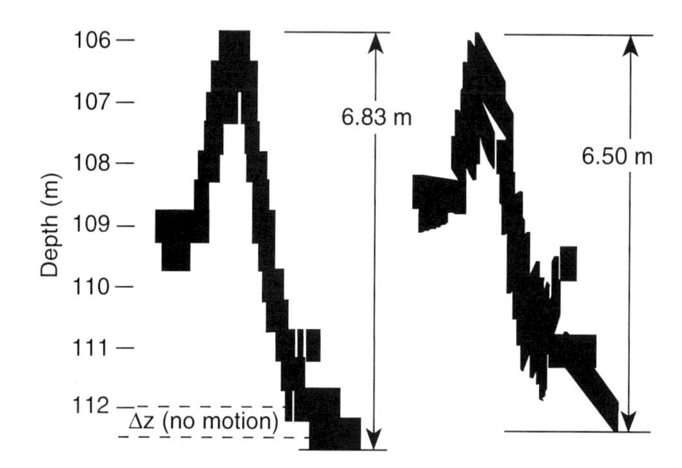

Fig. 12. Example of the effects of chain motion on the height of an MSP. The same segment from Tow A (patch labeled a in Figure 2) is represented in two ways. The left representation has not been compensated for chain motion; each SSP is represented as a rectangle, whose height is the sensor spacing Δz and whose width is determined by the horizontal extent of the patch (zero-crossings greater than a threshold). The same segment is plotted on the right using the actual sensor depth at the beginning and ending of each SSP, forming a set of parallelograms. This reduces the total height of the patch from 6.83 to 6.50 m.

that between 107 and 113 m during the first three minutes of the data, correspond to regions with horizontal bars in Figure 2, whereas the blue turbulence regions correspond to the gray shading. The high correspondence between these two plots is a consequence of the fact that the same criteria were used in defining patches (the number of zero-crossings in a 3-m sample must be greater than 2), and both are based on the LLR. Only two differences are found between these two representations. First, the patch plot assigns a single LLR value to each SSP on the basis of the average LLR over the SSP, whereas in Figure 2, each individual 3-m data point can have a separate gray scale. The other difference is that the patch plot in Plate 2 includes depth changes of the patches by plotting the patches in parallelogram form, whereas this information in Figure 2 is represented as a single depth sensor time series.

Two important observations should be noted from Plate 2. First, most of the visually apparent MSPs in the data appear relatively uniform; the LLR of each SSP in the MSP is usually either salt fingering or turbulence. Relatively few of the MSPs appear to be mixtures of the two generation mechanisms. Second, note the large number of very small, isolated patches of one-second dura-

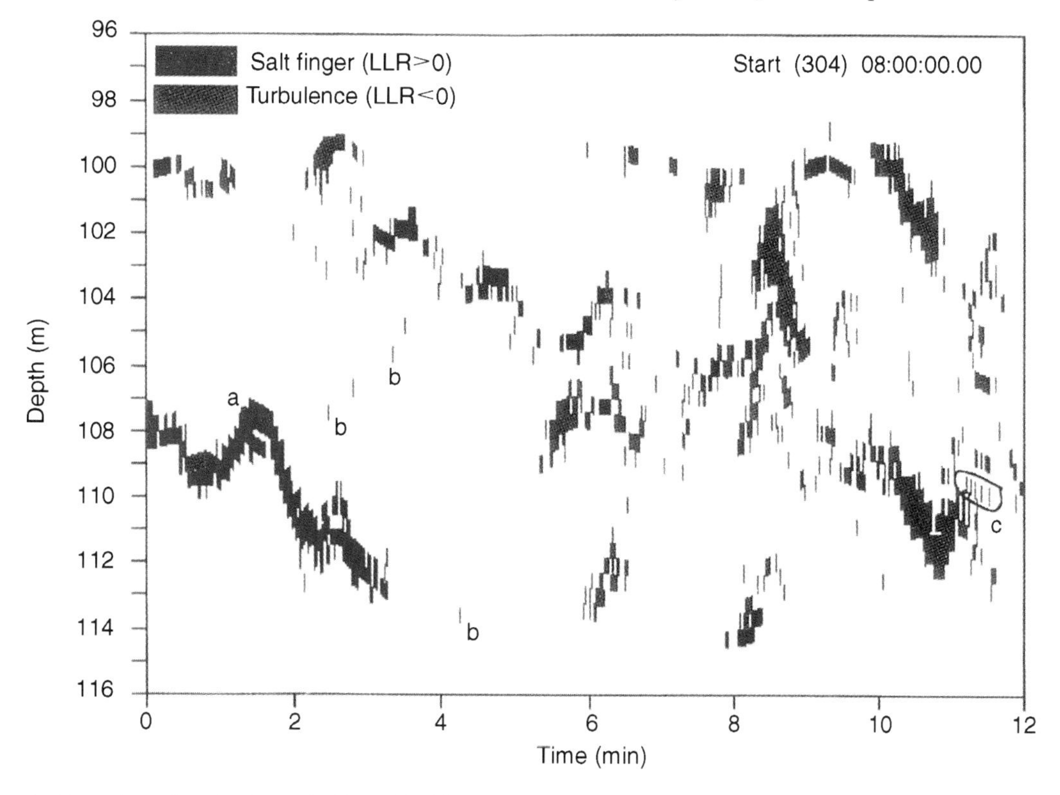

Plate 2. Plot of patches for Tow A (day 304, 08:00–08:12). Patches are plotted as parallelograms using the depth at the beginning and ending of the patch. The color of the patch is determined by the sign of the mean LLR within the patch.

tion (examples are labeled "b" in the figure). These events could be due to several different phenomena other than short episodes of microstructure in the data. First, they may be indicative of very weak events in the data, such that the threshold is exceeded for only very brief periods. However, a weak but wide patch of microstructure on a single sensor, with zero-crossings nearly equal to the threshold, would produce a series of short, disconnected patches extending over the length of the weak patch; the patch labeled "c" in Plate 2 is an example of this situation, but it appears to be connected with a larger MSP. Second, relatively weak biologic spikes slipping though the conductivity wild point editor can trigger events not truly associated with conductivity microstructure. Whatever the explanation of these small isolated patches, it is clear that the fraction of them that represents microstructure events is relatively insignificant in comparison with the larger patches. However, because of their relative abundance, if they are included in the statistics, these events would completely dominate the geometry results described later. In addition, the actual geometry of these patches, consisting of only a single sample, is not well defined.

The simplest definition of an MSP, and the one adopted here, is a spatially contiguous collection of SSPs. Intuitively, such a definition is obvious, but it may be too stringent. For example, in the sample patch illustrated in Figure 12, there is an SSP that is clearly part of the main patch but is not connected to it. The strict definition just given would identify the example as two MSPs instead of one. Several *ad hoc* procedures can be added to the basic algorithm to address the problem, such as requiring a minimum separation between patches on a single sensor, or adjacent sensors, in order to identify them as members of different MSPs. However, in the analysis presented here, we will adhere to our strict definition (we are currently exploring alternatives). Each MSP identified in the data has an associated height, width, slope, and aspect ratio, as well as mean LLR and fraction of 3-m sample points within the MSP with LLR > 0. Figure 13 illustrates an MSP (in the absence of vertical chain motion) and defines its geometric parameters.

In the case of Tow A, the grouping algorithm identifies 298 MSPs consisting of one or more patches (recall the definition of an MSP includes single isolated patches), and exactly half of these consist of a single sample (3 m wide). As discussed previously, several circumstances having nothing to do with microstructure can trigger these small patches, and the relatively large number of these implies that they will adversely influence

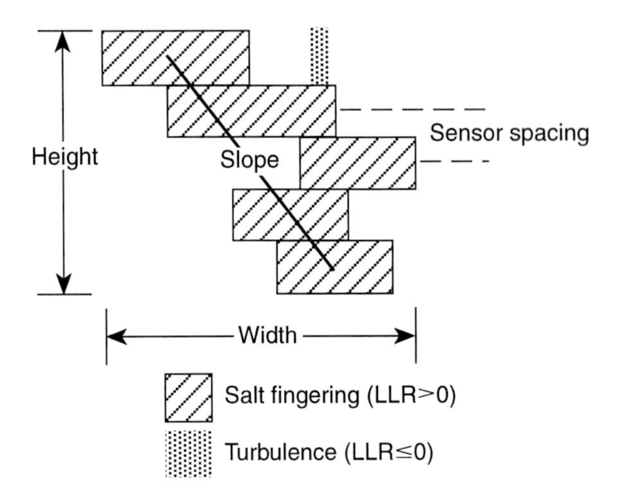

Fig. 13. Illustration of geometric parameters for an MSP consisting of five salt finger patches (the mean LLR in these patches is greater than zero) and one turbulence patch (mean LLR ≤ 0) in the absence of chain motion. The top and bottom of an SSP is half way between the sensor above and the one below (in this case, the number of zero-crossings must be greater than two). The length is defined by all the points above threshold. Slope is defined as shown from the top left patch to the bottom right patch.

the patch geometry statistics. Therefore, when we compare the statistics of salt fingering and turbulence MSPs, we will delete SSPs from consideration. Despite this factor of 2 reduction in the number of MSPs, it could be argued that in Tow A, the eye is naturally drawn to the larger patches evident in the data and generally identifies many fewer MSPs than 149.

The preliminary results described in the following paragraphs are based on Tow B, a 10-km segment of data in the same deep, homogeneous environment as Tow A, and the same segment discussed in MS93.[4] In Tow B, 1466 MSPs were detected, and 677 of those were retained after single-sample editing, thus providing a better statistical base for comparing the geometry of salt fingering and turbulence microstructure patches than Tow A (149 MSPs).

Figure 14 presents the heights and widths of the MSPs as a function of the likely generation mechanism, as indicated by the sign of the LLR. One-third of the MSPs have mean LLR > 0, indicative of salt fingering, whereas two-thirds of the MSPs have mean LLR ≤ 0, indicative of turbulence generation. Aspect ratios of 10:1 and 100:1 are indicated on the figure. We see that the patch width and height are correlated; the correlation coefficient is 0.69 (the 95% confidence ratio for the correlation coefficient is [0.65,0.73]). Additionally, the aspect ratio of most of the patches is between 10:1 and 100:1.

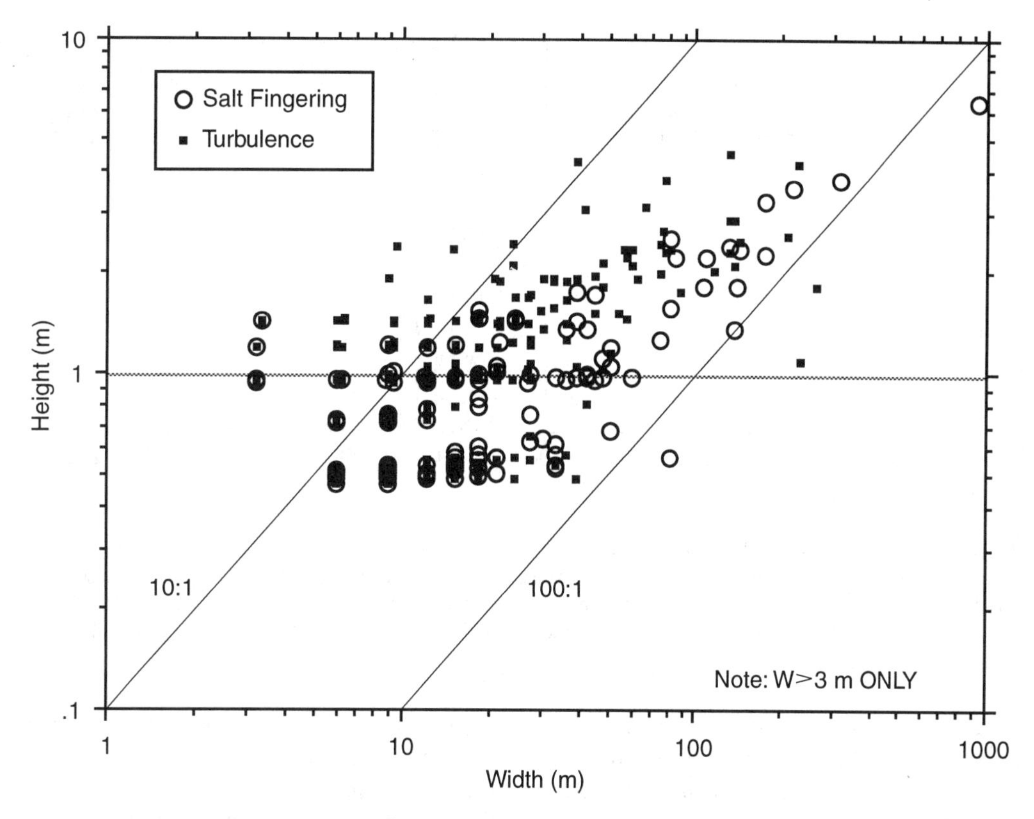

Fig. 14. Plot of MSP heights vs. widths for the 10-km segment designated Tow B (day 304, 01:07–02:05) in the Deep-Homogeneous environment. It excludes SSPs consisting of only one sample (widths of 3 m).

Although the heights and widths of the patches are similar in the salt fingering and turbulence populations, there is some preliminary indication that MSPs with LLR > 0 (salt fingering) tend to be wider and (thus) have greater aspect ratios that MSPs with LLR ≤ 0 (turbulence).

Figure 15 presents histograms of the MSP width, split into two populations on the basis of the sign of the LLR. Considerable overlap occurs between the two histograms, and both exhibit a roughly exponential shape. The distributions of height and aspect ratio also have this basic shape. However, the details of the distributions are different for salt fingering and turbulence generation mechanisms; the salt fingering patches tend to be wider (by 8.5 m in this case) than turbulence-generated patches, and the difference appears to be (marginally) statistically significant. A simple *t*-test for the difference in the mean of two populations indicates that the difference in mean patch width is significant at the 2.6% *p*-level; this is the significance level at which the null hypothesis (both means are equal) would just be rejected, given the observed data.

Intuitively, it could be argued that patches of salt finger microstructure should be longer than patches of turbulence-generated microstructure on the basis of the greater intermittency of turbulence-generated microstructure in comparison with salt finger microstructure. As discussed in M89 and MS93, the kurtosis of the horizontal conductivity gradient is greater in turbulence-generated microstructure than in salt fingering; in fact, the LLR discriminator uses this difference to distinguish between the two generation mechanisms. Higher kurtosis is indicative of greater intermittency within the patch, which means that the amplitude of the fluctuations varies more than in salt fingering. This is dramatically illustrated in Figure 2 and discussed in M89. The increased intermittency in turbulent patches causes dropouts (or separation), which are interpreted as separate MSPs by the grouping algorithm. Thus, at least a portion of the reduced width in turbulence-generated microstructure may be due to the simplicity of the grouping algorithm, breaking a single patch into two because of intermittency within the patch. However, turbulence patches also tend to be higher than salt finger patches; if intermittency within patches is causing their separation, the effect should be just as valid in the vertical as in the horizontal.

Table 3 summarizes the statistics of MSP geometry

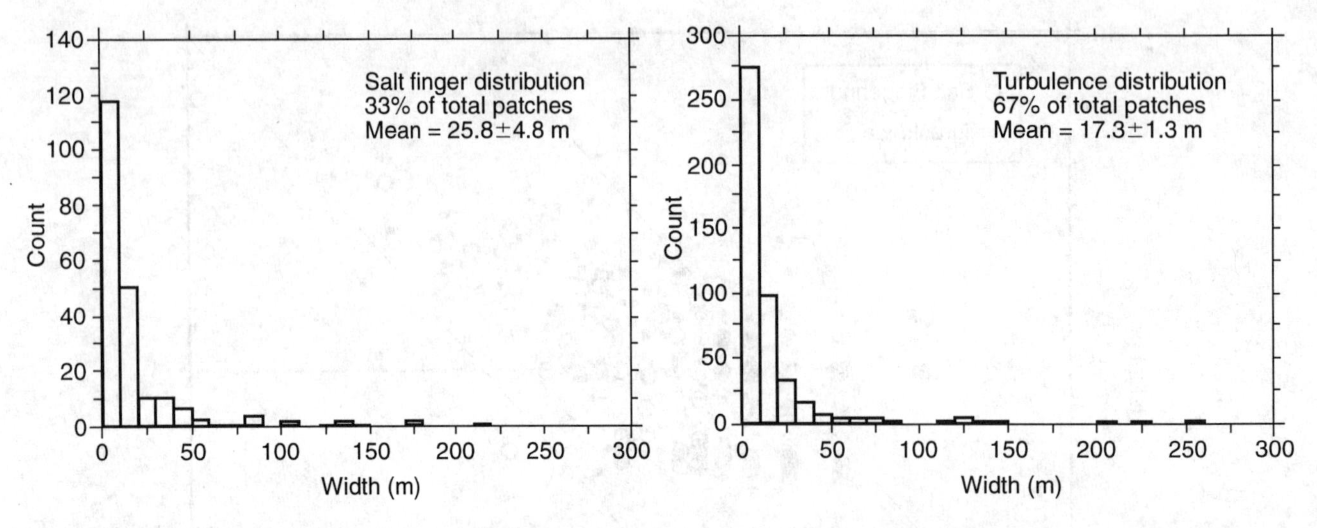

Fig. 15. Histograms of MSP width, conditioned on the sign of the LLR. The left plot is a histogram of MSPs with mean LLR greater than zero, indicative of salt fingering, whereas the right histogram is for LLR ≤ 0, indicative of turbulence. Both have a roughly exponential shape, and they overlap considerably; however, the mean salt finger patch is 8.5 m wider than the turbulence patches. This difference is moderately significant at the 2.6% level.

observations in Tow B. The table lists the sample mean and standard deviation of the listed parameters for salt finger and turbulence-generated microstructure, the difference in the mean and associated unpaired t-value, and finally the p-value, based on the Student t-distribution of the mean difference [see *Srivastava and Carter*, 1983, for an example]. These results could be preliminary indications of a difference in the geometry of turbulence-generated and salt finger patches. Patches generated by salt fingering tend to be wider and shorter than patches generated by turbulence. These observations imply that the aspect ratio should be very different in the two populations, which is, in fact, observed; the aspect ratio exhibits the largest degree of statistical significance (smallest p-value) of the geometrical parameters presented here. In contrast, one would expect less difference in the area of the patches because the larger width of salt finger patches relative to turbulent patches is offset somewhat by their shorter height. Table 3 indicates that the difference in area is not significant, even at the 10% level.

For this limited tow segment (~10 km), the aspect ratio for turbulence (~16) is less than the aspect ratio sometimes quoted for microstructure patches (~100), although few published observations are based on statistically significant databases, and investigators use different grouping techniques. *Dugan et al.* [1992], using towed thermistor data, observed an aspect ratio of about 70 for patches of activity with wavelengths between 10 and 30 cm and smaller aspect ratios for longer wavelengths. They used the correlation scales for log scalar variance as a measure of "patch" height and width. *Schoeberlein* [1985] reported aspect ratios of about 200 for wavelengths of 0.6 to 100 m in his towed thermistor data. *Gregg et al.* [1986] was able to track 4.5- to 10-m meter tall patches with a vertical profiler over distances greater than 1.4 km (implied aspect ratio > 100). It is not known whether our shorter aspect ratios (the heights of ~1 m are consistent with those generally quoted, but the widths appear shorter) are simply due to the limited sampling region (10-km test tow) or due to differences in the patch grouping algorithms used by various investigators

TABLE 3. Statistics of MSPs from Tow B

Parameter	Salt Finger	Turbulence	Difference (t-value)	P-Value
Width	25.8 ± 4.8 m	17.3 ± 1.3	8.5 m (2.2)	2.6%
Height	0.84 ± 0.044 m	0.94 ± 0.029 m	−0.1 m (−2.0)	5.0%
Aspect Ratio	22.4 ± 1.3	16.4 ± 0.7	6.0 (4.2)	<0.01%
Area	28 ± 142 m²	17 ± 44 m²	11 m² (1.5)	13.6%

for estimating geometry. This needs to be further examined. We expect to apply the geometry algorithms to the hundreds of kilometers in the Deep and Frontal regimes for better statistical reliability and to determine if the preliminary results are unique to this first sample tow.

Figure 16 presents a statistic that attempts to quantify the uniformity of MSPs: a histogram of the fraction of each MSP with LLR > 0 (salt fingering). This statistic is calculated by counting all the 3-m samples within the patch with LLR > 0 and dividing by the total number of 3-m samples in the patch. If all the observed MSP patches were either 100% salt finger or turbulence, the data would have only two values, 0 and 1, and the histogram would indicate that 445 observations out of the 667 (two-thirds) are equal to 0 (turbulence) and 222 are equal to 1 (one-third). In fact, as shown in Figure 16, a fraction (~20%) of the patches are mixtures of the two characteristics, but most of the patches are associated either entirely with salt fingering or entirely with turbulence-generated microstructure.

Although the analysis described in this section indicates geometrical differences for the two mixing mechanisms, we caution that the results should be considered only suggestive at this point. The results are based on a single 10-km test tow. Before applying the analysis to the larger data set discussed earlier, other grouping algorithms should be examined to verify that the geometrical differences are independent of the exact nature of the algorithm used to combine SSPs into MSPs with vertical extent.

7. SUMMARY

Salt fingering and turbulence-generated microstructure have been distinguished by the discrimination approach of MS93, which uses the conductivity gradient spectral slope and kurtosis of microstructure regions in a two-dimensional likelihood ratio formalism. MS93 demonstrated the technique on a limited 10-km tow in the Sargasso Sea and showed how the errors due to overlap of the slope and kurtosis distributions for salt fingering and turbulence-generated microstructure were minimized when both spectral slope and log kurtosis were used in the two-dimensional likelihood ratio approach. We have examined long tows in two environments in the Sargasso Sea, namely, in a homogeneous environment with relatively constant T–S structure (~168 km), and across a manifestation of the Subtropical Convergence Zone front (~92 km) with significant T–S variability. Our objectives

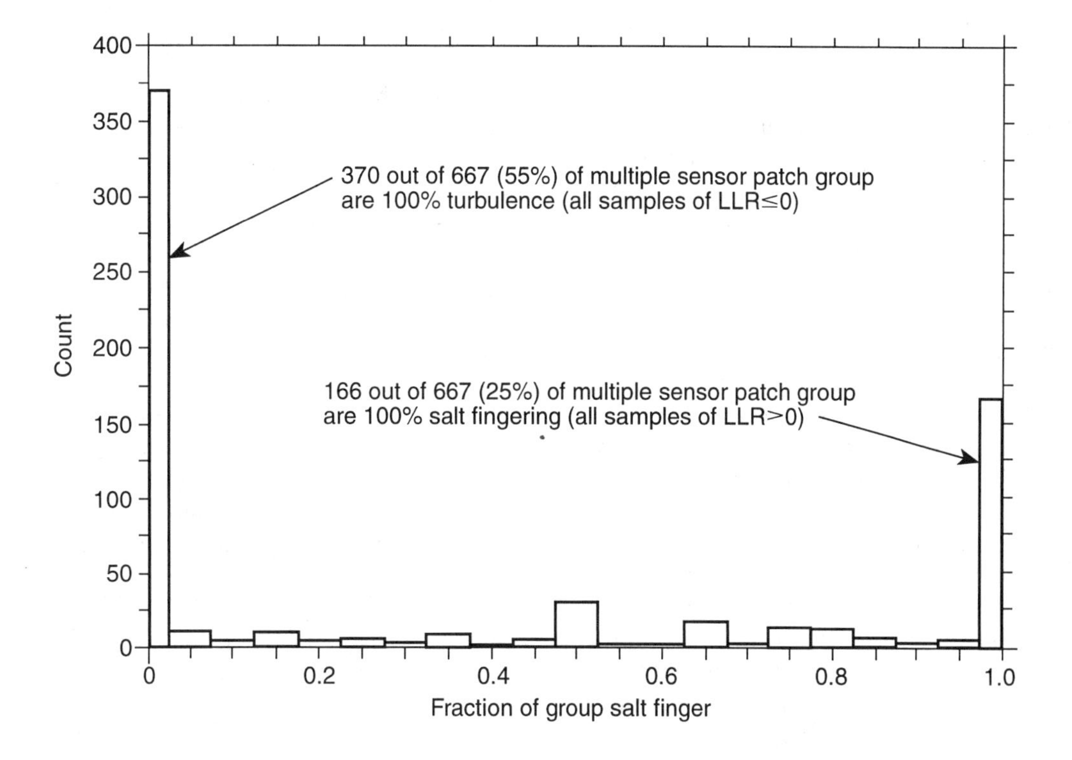

Fig. 16. Histogram of the fraction of MSPs with LLR > 0 (indicative of salt fingering based on Tow B (Deep-Homogeneous). If all MSPs were either uniformly salt fingering or uniformly turbulence, then only the cells containing 0 or 1 would be nonzero. These results indicate that 80% are either all salt fingering or all turbulence.

were to examine the usefulness of the approach over significantly longer tows than the initial 10-km test tow and to determine the relative importance of salt fingering and turbulence in two different environments.

The presence of salt fingering and turbulence in both environments is suggested by the bimodal shape of the spectral slope distributions. The LLR distributions show a stronger bimodal appearance than the spectral slope distributions, suggesting that the two-dimensional approach better separates the two source mechanisms. The density ratio histograms conditioned on LLR further suggest that LLR is a useful discriminant. Probabilities obtained from density ratio histograms conditioned first on zero-crossings only and then on $LLR > 0$ (salt fingering) show a clearer peak in the supercritical salt fingering regime and a reduction in the probability over the diffusively stable R_ρ regime after applying the LLR criteria. Furthermore, when the R_ρ histograms are conditioned on $LLR \leq 0$ (turbulence), the probability in the diffusively stable regime is relatively unaffected, and a reduction in the salt fingering R_ρ regime is observed. These observations suggest that the LLR is relatively robust over this extended data set. It has been used to determine the strength and distribution of microstructure sources in the two environments.

Salt fingering and turbulence-generated microstructure are observed in both environments; however, slightly more microstructure of both types is observed during the Frontal crossings (12.5% of data has microstructure exceeding our threshold, 2.1% exhibits salt fingering characteristics, and 10.4% exhibits turbulence characteristics) than during the Homogeneous tows (8.4% microstructure, 1.7% salt fingering, and 6.7% turbulence). Both the amount of microstructure and type appear to be rather random between the 14-km tow segments used to break up the tows in each environment. In some tow segments, as much as 4.9% of the observed data is identified as salt fingering. In other tow segments, salt fingering represents less than 1% of the data. Many of the features are consistent with salt fingering on weak salinity inversions, or turbulence caused by shear or internal-wave overturning. Even though there should be a greater likelihood for interleaving within the Frontal tows, we observe only a modest increase of salt fingering within the Frontal environment relative to the Homogeneous tows. It appears that the weak salinity inversions necessary to support the correct density ratios for salt fingering are abundantly present in both environments. In this sense, the two environments may be too similar to observe significant differences; however, the rapidly changing T–S diagrams during the Frontal (Plate 1) crossing are distinctly different from the tight T–S diagrams of the Homogeneous environment. In the absence of co-located shear measurements, we cannot relate turbulence-induced microstructure with Richardson number. This should be an objective of future field tests that could combine the LLR technique with shipborne acoustic Doppler current profiler (ADCP) measurements. Some observations of microstructure within the diffusive regime at temperature inversions have the slope and kurtosis characteristics of turbulence-generated microstructure.

We have also attempted to describe the geometry of the patches of microstructure. An approach has been developed and applied to the limited 10-km tow segment used as a test case in MS93. These preliminary results suggest that microstructure widths and heights follow an approximate exponential distribution. Salt fingering regions are wider and shorter than turbulence regions. However, the geometry assessment as a function of microstructure type has not yet been applied to the same hundreds of kilometers of tows in the Deep and Frontal environments. Such an analysis is necessary to build the statistical footing to determine whether the patch geometries truly differ, and if so, whether differences in geometry are consistent with salt fingering on weak salinity intrusions, turbulence microstructure at shear instabilities, or turbulence microstructure due to internal-wave overturning. This is planned as a future assessment.

It would be useful to obtain towed temperature and high-frequency conductivity data from a tow-yo mode to better provide R_ρ values in addition to spectral slope to combine R_ρ, slope, and kurtosis information. Such data should be coupled with shipborne shear information from ADCP instruments to test the correlation of turbulence-induced microstructure with low Richardson numbers.

Since we now have some ability to distinguish salt fingering and turbulence (with the caveats and errors discussed previously), it would be useful to compute internal-wave parameters such as temperature displacements from the towed temperature array to correlate the turbulence microstructure with the internal-wave features and better understand the relationship between turbulence and internal waves. This is planned for a future analysis.

Acknowledgments. We are grateful to Mel Hennessy for providing software development and for his help and diligence in crunching through a rather massive data set. We thank Jack Calman and Mark Baker for providing helpful comments on the manuscript. This work has been supported by the Office of Naval Research under Contract N00039-94-C-0001.

REFERENCES

Dugan, J. P., B. W. Stalcup, and R. L. DiMarco, Statistics of small-scale activity in the upper ocean, *J. Geophys. Res., 97(C4),* 5665–5675, 1992.

Farruggia, G. J., and A. B. Fraser, Miniature towed oceanographic conductivity apparatus, in *Proc. IEEE Conf. Oceans '84*, pp. 1010–1014, IEEE, 1984.

Gargett, A. E., and R. W. Schmitt, Observations of salt fingers in the central waters of the Eastern North Pacific, *J. Geophys. Res., 87*, 8017–8092, 1982.

Gregg, M. C., E. A. D'Asaro, T. J. Shay, and N. Larson, Observations of persistent mixing and near-inertial internal waves, *Phys. Oceanogr., 16*, 856–885, 1986.

Hamilton, J. M., M. R. Lewis, and B. R. Ruddick, Vertical fluxes of nitrate associated with salt fingers in the world's oceans, *J. Geophys. Res., 94*, 2137–2145, 1989.

Holloway, G., and A. E. Gargett, The inference of salt fingering from towed microstructure observations. *J. Geophys. Res., 92*, 1963–1965, 1987.

Mack, S. A., Two-dimensional measurements of ocean microstructure: The role of double diffusion, *J. Phys. Oceanogr., 15*, 1581–1604, 1985.

Mack, S. A., Towed-chain measurements of ocean microstructure, *J. Phys. Oceanogr., 19*, 1108–1129, 1989.

Mack, S. A., and H. C. Schoeberlein, Discriminating salt fingering from turbulence-induced microstructure: Analysis of towed temperature–conductivity chain data. *J. Phys. Oceanogr., 23*, 2073–2106, 1993.

McDougall, T. J., Some implications of ocean mixing for ocean modeling, in *Small-Scale Turbulence and Mixing in the Ocean*, edited by J. Nihoul and B. Jamart, pp. 21–35, Elsevier Oceanography Series, 46, Elsevier, New York, 1988.

Osborn, T. R., Estimates of the local rate of vertical diffusion from dissipation measurements. *J. Phys. Oceanogr., 10*, 83–89, 1980.

Schmitt, R. W., The growth rate of super-critical salt fingers, *Deep-Sea Res., 26A*, 23–40, 1979.

____, Mixing in a thermohaline staircase, in *Small-Scale Turbulence and Mixing in the Ocean*, edited by J. Nihoul and B. Jamart, pp. 435–452, Elsevier Oceanography Series, 46, Elsevier, New York, 1988.

____, Double diffusion in oceanography, *Annu. Rev. Fluid Mech., 26*, 255–285, 1994.

Schoeberlein, H. C, Statistical analysis of patches of oceanic small-scale activity, *Johns Hopkins APL Tech. Dig., 6(3),* 194–202, 1985.

Srivastava, M. S., and E. M. Carter, *An Introduction to Applied Multivariate Statistics*, p. 39, North-Holland, 1983.

Weller, R. A., Overview of the frontal air–sea interaction experiment (FASINEX): A study of air–sea interaction in a region of strong oceanic gradients, *J. Geophys. Res., 96*, 8501–8516, 1991.

Whalen, A. D., *Detection of Signals in Noise*, pp. 124–153, Academic Press, 1971.

S. A. Mack and H. C. Schoeberlein, The Johns Hopkins University Applied Physics Laboratory, Johns Hopkins Rd., Laurel, MD 20723-6099.

---

[1]This display uses a stable Neil Brown conductivity sensor that was also a part of the sensor suite [M89] for the computation of salinity.

[2]The apparent gap in the continuity of the T–S diagram is due to an 8-km gap between the regions of straight tow for segment 4 and the portion of segment 5 shown here.

[3]These data exclude six conductivity sensors that were not exactly co-located with thermistors and hence gave slightly degraded salinity.

[4]MS93 called this tow 5, but we rename it Tow B so as not to confuse it with either tow segment 5 shown in Figure 3. See MS93 section 4b and Figures 13 and 14 therein.

Sheet Splitting and Hierarchy of "Convective Plumes" in the North-Western Tropical Atlantic Salt Finger Staircase

Iossif D. Lozovatsky

P.P.Shirshov Institute of Oceanology Russian Academy of Sciences, Moscow 117851, Russia

Long distance tow-yo-yo CTD transects were carried out in January 1987 east of Barbados. Comparison with C-SALT data [*Schmitt et al.*,1987] leads to conclusion that step like structure parameters of basic quasi homogeneous layers preserve almost constant values approximately two years at least. Generation of pronounced narrow thermohaline fronts appear to be caused by vertical motions in convergence and divergence zones between synoptic eddies. Vertical shift of the staircase at the distances of a few kilometers achieves 100 m. The horizontal temperature and salinity oscillations were found in the quasi homogeneous layers. These "mesoscale plumes" were significant differ from the convective plumes observed by *Marmorino et al.*[1987], that have been characterized by temperature amplitudes of 0.005 K and aspect ratio, m ≈ 0.3 - 0.5. Average wave length of the revealed oscillations was about 6 km, aspect ratio m ≈ 10^{-2}, and typical amplitudes of temperature and salinity variations were 0.015 K and 0.005 psu. Lateral density ratio, R_x, varied remarkably along the homogeneous layer from R_x = 0.63, that is equal to value of the flux ratio for salt fingers [*Kunze*, 1987], to R_x = 0.92, as a local frontal zone is approached. The lowest R_x values were observed in the "plume area" and the highest, R_x = 1.20, - in the region of local upwelling at the opposite side of the front. Sheet splitting was also observed just above the mesoscale plumes. Probability distribution function, $F(R_\rho)$, of vertical density ratio, R_ρ, at different sheets were estimated. High probability of small R_ρ values, $F(R_\rho < 1.6) = 0.9$, was revealed at the splitted parts of the interface, whereas for nonsplitted interfaces probability of $R_\rho < 1.6$ was equal to 0.6 only. Mesoscale eddies steering within the homogeneous layers might be partly responsible to the plume origin. However it seems more likely these "mesoscale plumes" were caused by internal wave forcing of salt finger flux at the interfaces. A hypothesis is suggested concerning a hierarchy of convective plume scales. In salt finger favorable stratification a set of such multi scale plumes may provide the creation and sustenance of the staircases with the layers in various sizes and shapes, notably the homogeneous layers achieving 50 -70 meters thick.

1. INTRODUCTION

Step like structures generated by salt fingering convection in the oceanic thermocline were found in many regions at different depths. Stable averaged geometric and thermohaline characteristics of such structures are varied very slowly keeping almost constant values along several tens or even several hundreds of kilometers [*Schmitt et al.*, 1987; *Lueck*, 1987; *Marmorino et al.*, 1987; *Zhurbas et al.*, 1988]. However, the long distance transects, section 2, that had been carried out with high horizontal resolution in the north western tropical Atlantic, allowed to reveal the local deformations of the thermohaline staircase, section 3, related to upwelling and downwelling processes at the narrow fronts. Interface splitting, section 4, and mesoscale horizontal inhomogeneities within the quasi homogeneous layers were also found, section 5. Mesoscale eddies, turbulent mixing, salt finger convection and internal waves forcing are considered as processes that might produce these deformations. Interrelation between the variance of salt finger flux at the different interfaces and sheet deformations caused by internal waves is presented in Section 5 as the result of spectral and correlation analysis. In each section the interpretation is separated into two parts:

Double-Diffusive Convection
Geophysical Monograph 94

"Analysis and Inferences" that follow directly from observation analysis, and *"Discussion"* that contains additional more speculative consideration. A hypothesis on a hierarchy of "convective plumes" with various horizontal scales that might be responsible for the layer mixing in salt finger favorable stratification is suggested. The main conclusions are concentrated in the Summary.

2. DATA

Data were obtained during 13-th cruise of R/V AKADEMIK MSTISLAV KELDYSH in January 1987 at the C-SALT area using CTD-profiler NEIL BROWN MARK-III mounted into the towed streamlined hull. Observation were carried out in the scanning mode with a mean towing speed about 6 knots. Towed body was winched up and down front the stern. Location of the sections, depth ranges and directions of the transects are shown at the right lower panel in Figure 5,a. Mean horizontal distances between the consequent casts were 1.5 km approximately along the whole A transect as well as at the first part of B transect (the casts Nos'. 1-20). The last 47 casts (Nos'.21-68) at the B transect were separated by 0,9 km each other because the scanning depth range had became more narrow.

3. FRONTS AND STEP LIKE STRUCTURE

3.1 *Observations*

Three dimensional temperature image along A transect, T(z,L), is presented in Figure 1. "Temperature staircase" with almost dark plates (homogeneous layers) separated by steep light ledges (high gradient sheets) cuts down sharply at the central part of the transect (frontal zone). At the right hand of this front the temperature field in the depth range of 250-550 m looks like "hilly rising valley" (interleaving intrusions with relatively warm and cold water). At the upper right part of the section, depth range of 210-340 m, the vertical staircase appears again, although the steps are not so pronounced. At the frontal wall near the ledges between the steps one can see the transversal "canyons". Intermittent layers and sheets in the frontal zone might due to transfrontal heat and salt advective fluxes, if the revealed effects are not caused by insufficient horizontal resolution between two types of the vertical structures (staircase and inversions). This transition zone with the slope of 0,05 is narrower than 2 km.

The same frontal zone was found at the B transect between 12B and 13B casts. Temperature and salinity vertical sections in the depth range of 330-450 m at this transect are plotted in Figure 2 and Figure 3. Uplifting and downlifting of isolines look like intermittent plumes of warmer saltier and of cooler fresher water in the vicinity of the front. Typical horizontal sizes of these structures are about 5-6 km, the vertical displacements of isolines achieve 40-50 m. The vertical plumes or "thermohaline jets" near the frontal wall are characterized by temperature and salinity amplitudes run to 0.3-0.5 K and to 0.08-0.1 psu correspondingly. Density fluctuations are not so impressive, however remarkable uplifting and downlifting of the isolines are shown in the vicinity of the thermohaline front (Figure 4).

3.2 *Analysis and Inferences*

The genetic relation between the particular homogeneous layers located at the different depths NW and SE off the narrow local front (Figure 1) are confirmed by θ-S analysis, where θ is a potential temperature. This approach was successfully used by *Schmitt et al.* [1987] for discussion of staircase origin in the C-SALT area. The θ-S indexes of homogeneous layers for the casts 25A (upper step like structure, USS,) and 7A (lower step like structure, LSS,) are plotted in Figure 5,a against background θ-S diagram, taken from[*Schmitt et al.*,1987]. Mean θ-S values for each layer are almost the same for the 25A and 1B casts and for the 7A and 20B casts accordingly, so they are repeated at both transects. In Figure 5,a the indexes of six LSS layers (big light circles) are lying at the straight lines formed by θ-S indexes of the homogeneous layers observed in 1985 during C-SALT. The indexes of four USS layers are lying exactly at the prolongations of these straight lines which cross the θ-S curve of the first cast at the B transect. This curve coincides in the depth range of 250-550 m with the familiar θ-S curve of the South Atlantic Central Water. Consecutive shifts of four USS layer indexes in the transition zone (15A-21A casts) are revealed in Figure 5,a along the corresponding straight lines: locations of θ-S indexes for 19A cast layers are shown, for example, in Figure 5,a by crests. It is felt that we obtained the evidence of genetic relation not only for the particular homogeneous layers at the different sides of that front but of these layers and basic ones as well, that were found during C-SALT in 1985 more than a year before our observations. It leads to conclusion that the basic staircase is a quasi permanent phenomena. Permanent mean

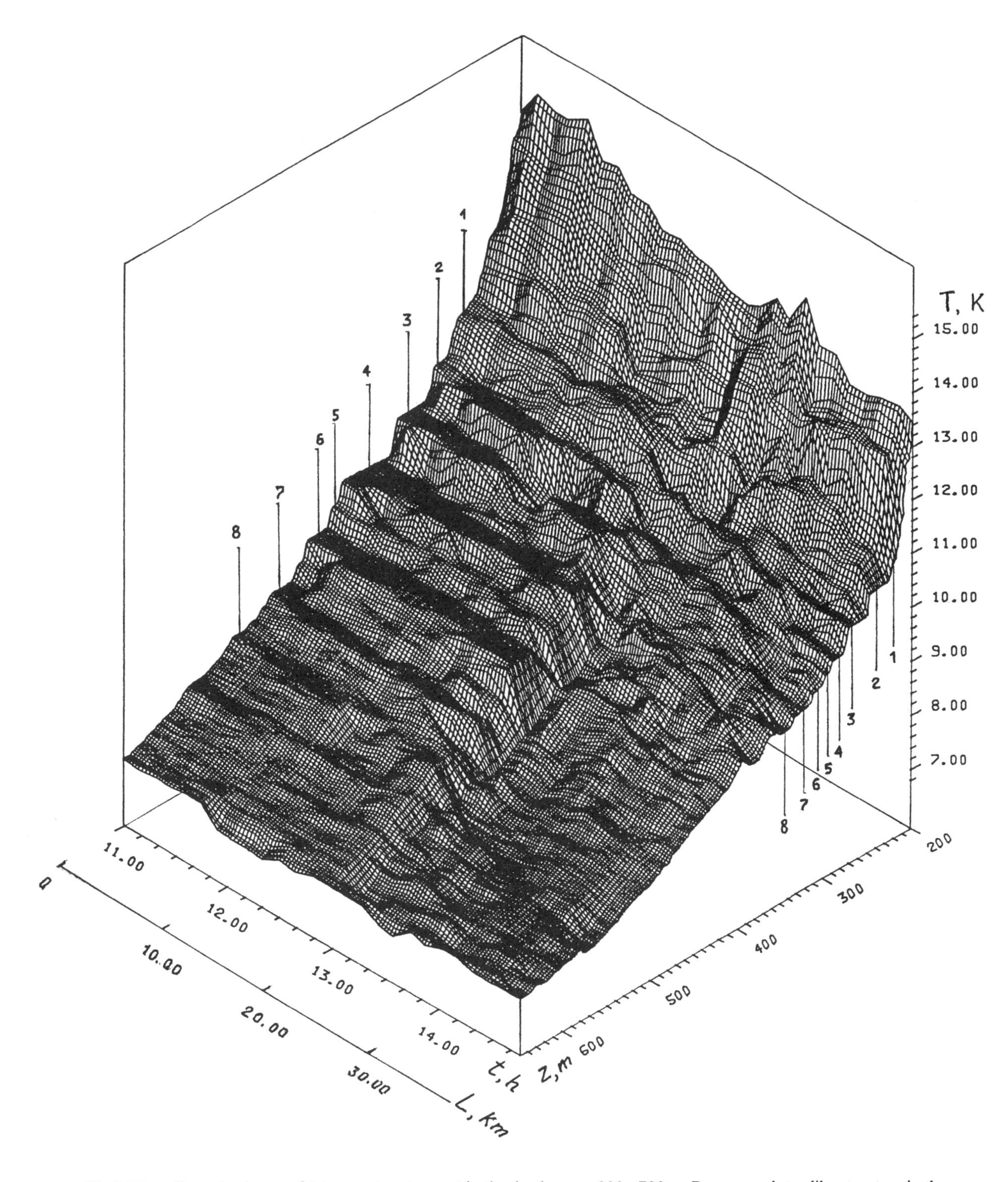

Fig.1. Three dimension image of A temperature transect in the depth range 200 - 700 m. Pronounced step like structure in the depth range 250 -550 m at the left side of the plot is represented as almost plate dark steps (homogeneous layers) and steep light ledges (high gradient interfaces). At the central part of the transect these "terraces" cut down sharply near the narrow front. At the upper right part of the plot in the depth range of 210-340 m, the vertical staircase appears again. The layer numbers are marked by figures 1,...8.

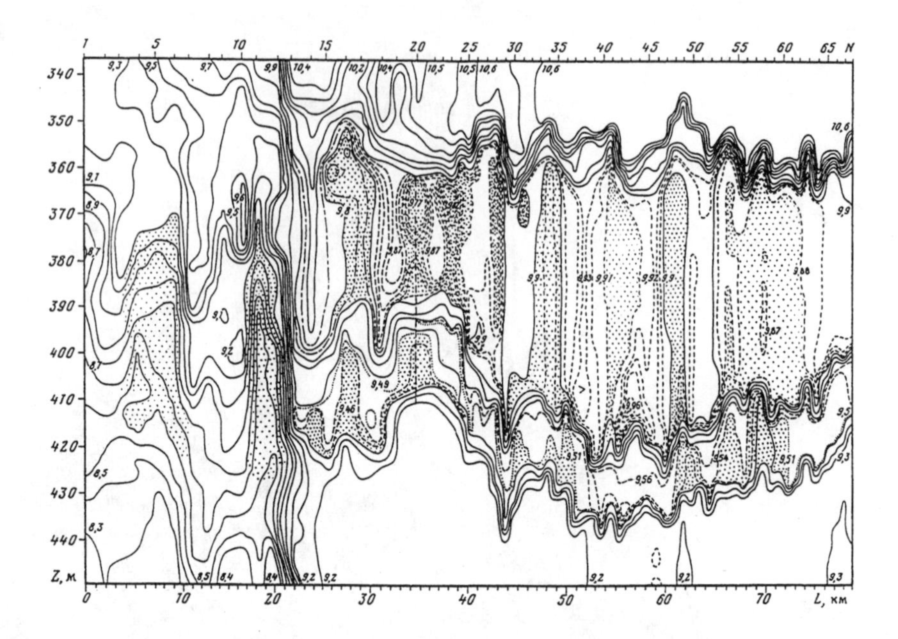

Fig.2. The vertical temperature section, T(z,L), at B transect in the depth range of 330 - 450 m. The solid isotherm lines are separated by step of $\Delta T = 0.1$ K; for the dashed lines $\Delta T = 0.01$ K in the central (middle) homogeneous layer and $\Delta T = 0.02$ K in the lower (intermediate) layer; additional isotherm lines T = 9.49 K and T = 9.51 K are shown by dotes. Intermittent areas of isotherm uplifting in the homogeneous layers and near the frontal zone are marked by the small dotes, whereas altered downlifting areas are light. The cast numbers, N, are at the upper horizontal axis.

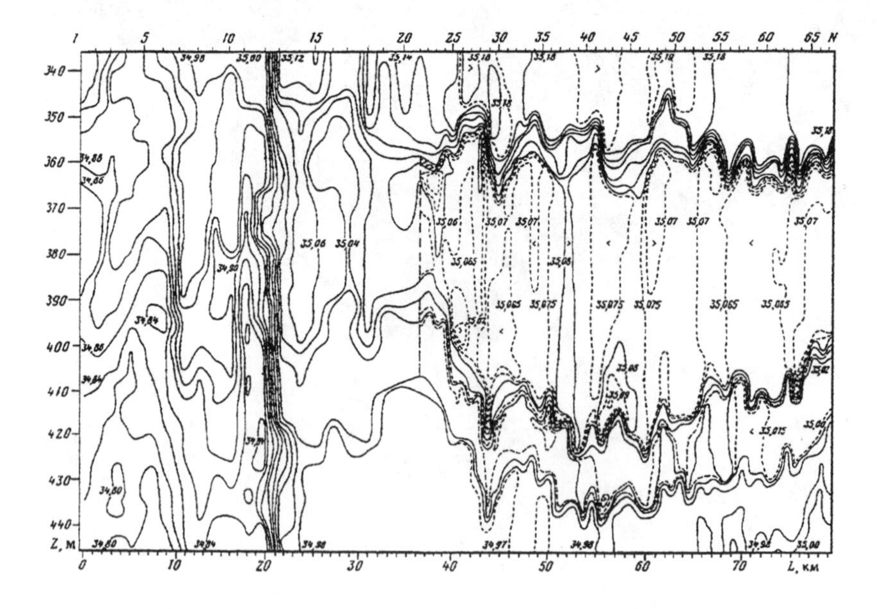

Fig 3. The vertical salinity section, S(z,L), at B transect in the depth range of 330 - 450 m. The solid isohaline lines are separated by step $\Delta S = 0.02$ psu, for the dashed lines in the homogeneous layers $\Delta S = 0.005$ psu.

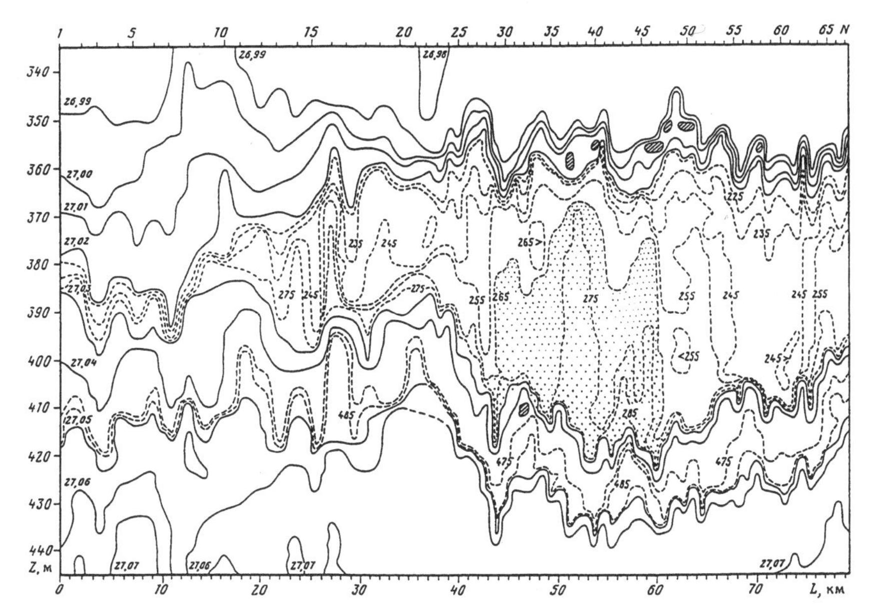

Fig 4. The vertical section of specific density, $\sigma_t(z,L)$, at B transect in the depth range of 330 - 450 m. The solid isopicnic lines are separated by step $\Delta\sigma_t = 0.01$; for the dashed lines in the homogeneous layers $\Delta\sigma_t = 0.001$. Figures near the dashed lines signify specific density values $(\Delta\sigma_t - 27.000)\cdot10^3$, but it is well to bear in mind that accuracy of $\Delta\sigma_t$ calculation is of 10^{-3}. So the boundaries between samples varying by value of $10^{-3}\sigma_t$ are marked arbitrarily with accuracy of $5\cdot10^{-4}\sigma_t$. In the center of the middle thickest layer an area of prominent density increasing is marked by the small dotes ($\sigma_t \geq 27.026$).

water mass parameters provide generation and maintaining of one the same basic homogeneous layers. However the basic staircase are affected by local deformation and distortion.

3.3 *Discussion*

It is interesting that main parameters of mesoscale thermohaline perturbations in the C-SALT area and of the mesoscale structures in the Canary coastal upwelling [*Erofeev and Lozovatsky*, 1990] are similar in appearance.

Perhaps one can concede that the thermohaline front in the salt finger staircase might be formed by vertical motions associated with upwelling or downwelling in the divergence or convergence zones between the mesoscale eddies or local current jets. The geostrophic velocities between casts 1B and 19B in the depth range of 200-500 m with the horizontal resolution of 3 km and vertical resolution of 10 m were estimated to check this conjecture. A pronounced SW flow are revealed in the depth range of 200-380 m between the casts 11B and 13B (Figure 6), i.e. exactly above the thermohaline front. At the right and left hands of this flow the narrow jets of the counter currents appear to be exist. The horizontal differences of geostrophic velocities may amount to as much as 25, 45 and 50 cm/s at the depths of 380, 300 and 200 m in the NW direction, and as much as 20, 40 and 50 cm/s at the same depths in the SE direction. In the depth range of 380 - 420 m the values of horizontal gradients of geostrophic velocity, dv/dx, was found to be about $(3 - 8)\cdot10^{-5}$ s^{-1} and the value of vertical gradient, dv/dz, proved to be equal $(1 - 2)\cdot10^{-3}$ s^{-1}. *Woods* [1988] has mentioned that such type of current structure leads to generation of narrow fronts with high thermoclinicity owing to the vertical circulation at the boundaries between the eddies or current jets. The values of background vertical temperature and salinity gradients were equal to $\langle dT/dz \rangle = -1.6\cdot10^{-2}$ K/m, and $\langle dS/dz \rangle = -2.2\cdot10^{-3}$ psu/m. If typical amplitudes of temperature, $A_T = 0.5$ K, and salinity, $A_S = 0.1$ psu, fluctuations are due to vertical and isopicnal advection principally, the influence of each process can be estimated by a parameter $\gamma = (A_S \langle dT/dz \rangle / A_T \langle dS/dz \rangle - 1)/(\langle R_\rho \rangle - 1)$, [*Fedorov*, 1976; *Lozovatsky and Lilover*, 1986], where $\langle R_\rho \rangle$ is the vertical density ratio, $\langle R_\rho \rangle = \alpha \langle dT/dz \rangle / \beta \langle dS/dz \rangle$, α, β are the thermal expansion and salinity contraction coefficients. Vertical motions are characterized by $\gamma = 0$, whereas for isopicnic motions $\gamma = 1$. In our case $\alpha/\beta = 0.225$, $\langle R_\rho \rangle = 1.64$, so $\gamma = 0.71$. That is why the vertical advection may be responsible for about of 30% of the amplitudes of temperature and salinity oscillations in the vicinity of the frontal zone. While more than

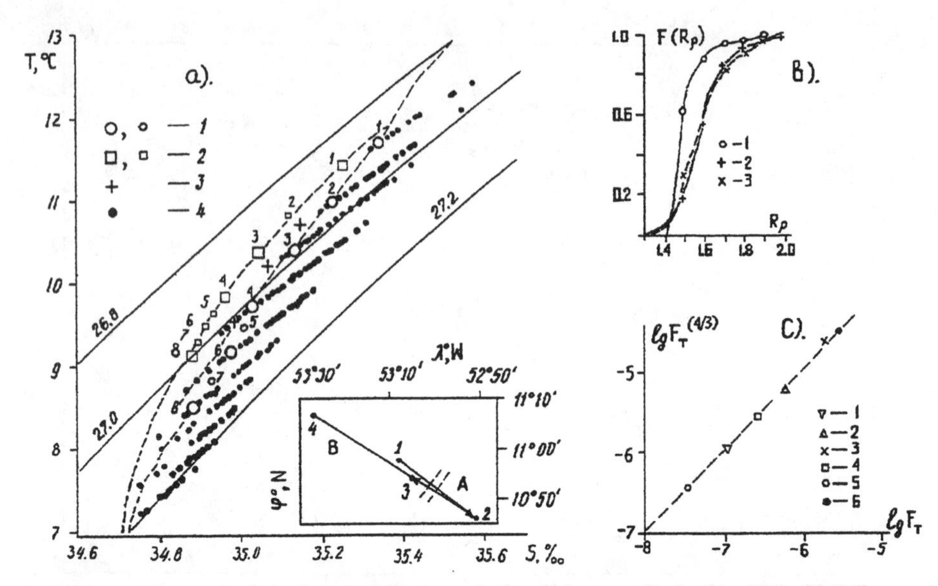

Fig.5. a) - θ,S indexes for the homogeneous layers of the thermohaline staircase at the background of θ,S diagram taken from [*Schmitt et al.*, 1987]: 1 - cast No.7A, 2 - 25A, 3 - 19A, 4 - [*Schmitt et al.*, 1987] data. Figures near the circles and squares are the layer numbers, denoted in Figure 1. The map with A and B transect locations is at the right low panel: measurements were carried out in the depth range ~ 200 ÷ 650 m between points 1 - 2 and 2 - 3, and in the range of ~ 300 ÷ 450 m between point 3 - 4. Dashed lines are approximate boundaries of the local front.

b) - cumulative distribution functions, $F(R_\rho)$, for the upper (1), middle (2), and the lowest (3) interfaces. Conditions at the upper interface ($R_\rho > 1.6$ is 0.1 only) are more favorable for salt fingers appearance in compare with two lower sheets, where the probability of $R_\rho > 1.6$ is in excess of 0.4.

c) - interrelation between thermal convective fluxes obtained from the dissipation measurements, F_T, and calculated by use of the "4/3 low", $F_T^{(3/4)}$. The regression line $\lg F_T^{(3/4)} = \lg F_T + 1$ approximates data from: 1 - [*Gargett and Schmitt*, 1982], 2 - [*Lueck*, 1987], 3 - [*Gregg and Sanford*, 1987], 4 - [*Marmorino*, 1987], 5 - [*Osborn*, 1988], 6 - [*Lozovatsky and Nabatov*, 1990].

70% of these amplitudes are governed by isopicnic processes. However the influence of horizontal or isopicnic mixing is decreased far enough from the frontal zone. Therefore the salt fingering convection becomes a basic factor in forming of the vertical and horizontal thermohaline structure owing to a favorable stable mean hydrological conditions in the North Western tropical Atlantic. Mesoscale deformations of the step like structure caused by upwelling and downwelling between the eddies and/or current jets lead to formation of pronounced thermohaline fronts.

4. INTERFACE SPLITTING

4.1 *Observations*

Besides of 10 basic layers generated under the action of the permanent mean hydrological factors, additional secondary layers may locally appear. It is easy to find out such layers in Figure 5,a.: namely the layers No.5 and No.7 at 1A-11A profiles, and No's. 2, 5, 7 at 21A - 26A profiles. The local secondary layers were observed at the 34B-41B, 44B-51B and 58B casts also, resulting from the main interfaces splitting. Interface splitting was revealed and analyzed recently by *Marmorino* [1991], *Fleury and Lueck* [1991].

Almost periodic oscillations at three observed interfaces (Figures 2 - 4) indicate internal waves in the entire depth range. Since high gradient sheets are nothing else than the local pycnoclines, internal waves dynamics should be consider as one of the governed factors. The average amplitudes of the detrended interface displacements are 3.8 - 3.9 m. The prevalent Doppler contaminated wave lengths are 2.5 and 6.0 km approximately. Let us examine some statistics of sheet thicknesses, h, and of the density ratio $R_\rho = \alpha\Delta T/\beta\Delta S$, where ΔT and ΔS are the temperature and salinity differences across the sheets. The set of 42 casts, Nos'. 27B - 68B, was chosen for such calculations. These profiles are located far enough of the front and influence of frontal processes have proved to be negligible. The lowest sheet

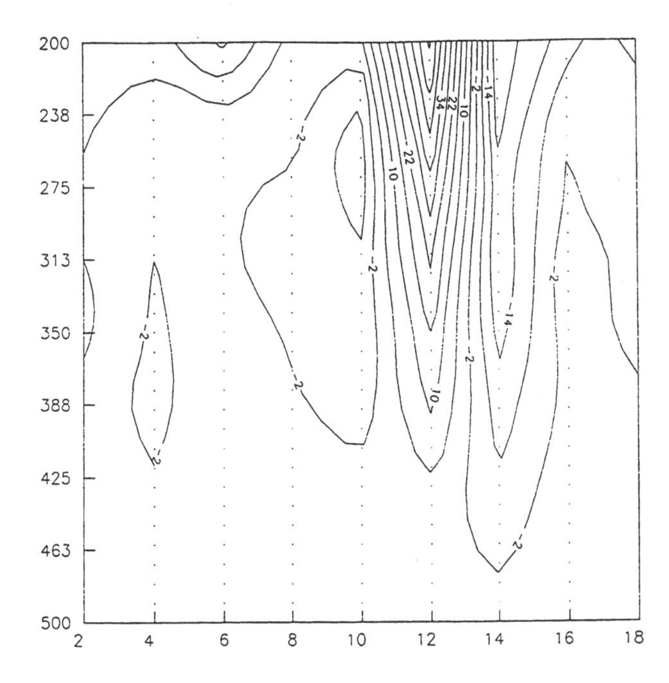

Fig. 6. The vertical section of geostrophic velocity in the depth range 200-500 m at B transect between casts 1B - 19B. Isolines are marked in cm/s, the reference level is 500 m. Positive values are related to the SW current, negative - to the NE current. The cast numbers (up-runs) are at the lower axis.

in Figure 2 is the thinnest: the mean thickness $<h> = 2.9$ m and the rms, $\sigma_{hl} = 0.9$ m, while the mean value of $<R_{\rho l}> = 1.59$ and the rms, $\sigma_{Rl} = 0.13$. The values of $<R_{\rho m}>$ and σ_{hm} for the middle sheet are almost the same: the mean thickness $<h_m>$ achieves 3.5 m and $\sigma_{hm} = 1.4$ m. The upper sheet is the thickest: $<h_u> = 5.3$ m, $\sigma_{hu} = 2.0$ m, and density ratio is significantly lower: $<R_{\rho u}> = 1.48$, $\sigma_{Ru} = 0.05$.

The main question is, which of the physical processes may be responsible for the local homogeneous layer generation?

4.1 *Analysis and Inferences*

In spite of the oscillation parameters were not differ significantly for various interfaces the splitting was observed at the one of them only. Why so?

Because of the origin of salt finger convection is strongly dependent on the values of the vertical density ratio, R_ρ, let us consider the probability distribution functions, $F(R_\rho)$, for the different sheets. According to the empirical cumulative distribution functions presented in Figure 5,b, the probability of the low R_ρ values at the upper sheet, $F(R_\rho < 1.6) = 0.9$, is much higher than at the second and third sheets, where $F(R_\rho < 1.6) =$ 0.6 only. What this means is the upper interface is more favorable for salt finger appearance in compare with two lower sheets. Variations of the rms values σ_h were found to be about 30 - 40 % of the mean value, $<h>$, whereas variability of R_ρ is much less: 4 - 8%. These estimates are probably typical for the step like structure in the North-West tropical Atlantic.

If we evaluate salt finger part of buoyancy flux through the staircase, $F_S^{(4/3)}$,using the "4/3 law" [*Fedorov,* 1976], $\beta F_S^{(4/3)} \sim (\beta \Delta S)^{4/3}$, it turns out that the relative fluctuations of $\beta F_S^{(4/3)}$ exceed of 20-25% for small R_ρ values. Since buoyancy flux fluctuations are so high, finger convective mixing may start in some particular parts of the sheet to generate additional internal homogeneous layers.

According to [*Gregg and Sanford,* 1987; *Lueck,* 1987; *Marmorino,* 1987; *Osborn,* 1988; *Lozovatsky and Nabatov,* 1990] the diffusive fluxes in the interfaces, calculated from the stationary turbulent energy balance equation by use of direct viscous dissipation rate measurements, are always lower than was generally appreciated started from the "4/3 law". Inasmuch as our observations were not accompanied by turbulent measurements in the depth range under consideration, an effort was made to recalculate $F_S^{(4/3)}$ and $F_T^{(4/3)}$ estimates to more realistic values taking into account all appropriate data. The regression function lg $F_T^{(4/3)}(\lg F_T)$ is presented in Figure 5,b, where F_T is a "microscale measured" heat flux and $F_T^{(4/3)}$ is a heat flux calculated by the "4/3 law" using the following relation: $F_T^{(4/3)} = n(\beta/\alpha) \cdot F_S^{(4/3)}$ with n dependent upon R_ρ [*Kunze,* 1987]. The heat flux estimates from [*Gargett and Schmitt,* 1982] and from the above mentioned five papers were used for the comparative analysis. The linear regression of $F_T = 0.1\ F_T^{(4/3)}$ with high correlation between F_T and $F_T^{(4/3)}$ results from these data undoubtedly. That is a reason why such empirical formula was used to calculate a set of the mixing parameters presented in the Table 1. There are the mean values of the Vaisala frequency $<N^2>$, specific heat $<F_T>$, salt $<F_S>$, and buoyancy $<B> = g(\beta<F_S> - \alpha<F_T>)$ fluxes through two upper interfaces as well as the thermal $<K_T> = - <F_T>/<dT/dz>$ and salt $<K_S> = - <F_S>/<dS/dz>$ diffusivities. For the upper interface 17 splitted profiles were considered separately from the other 25 casts without intermediate homogeneous layer, i.e. likely before the splitting. These estimates confirm our assumption that interface splitting significantly depends on the fingering convection. Indeed, the values of diffusivities at the segments of the upper interface without intermediate homogeneous layers are almost two times higher than $<K_T>$ and $<K_S>$ at the other two sheets, as well as at the new sheets after splitting.

TABLE 1: Fluxes, diffusivities and Vaisala frequencies through the upper and middle sheets

Parameters	Upper sheet before splitting	Upper sheet after splitting	Middle sheet
$N^2 \cdot 10^5, s^{-1}$	7.2	5.8	5.7
$F_S \cdot 10^7$, psu·m s^{-1}	8.0	3.2	2.7
$F_T \cdot 10^7$, K·m s^{-1}	20.0	7.8	6.7
$B \cdot 10^{10}, m^2\ s^{-3}$	21.6	9.1	7.4
$\langle K_S \rangle \cdot 10^5, m^2\ s^{-1}$	4.0	1.7	0.6
$\langle K_T \rangle \cdot 10^5, m^2\ s^{-1}$	1.6	0.6	0.8

4.3. *Discussion*

Convective mixing at some segments of interfaces may locally produce separate homogeneous layers, leading finally to complete sheet splitting and appearing of the secondary staircase steps. Fingering convection begins when salt finger length, l, becomes equal to about 10 cm during the Vaisala period, $\tau_N = 2\pi/N$, [*Kunze*, 1987]. An estimate of the effective exchange coefficient, K_N, may be expressed as $K_N = l^2N$. In accordance with the Table 1, $K_N = 8.5 \cdot 10^{-5}$ m^2s^{-1} for $N^2 = 7.8 \cdot 10^{-5}$ s^{-2}, i.e. at the splitting interface mass exchange coefficient achieves 10^{-4} m^2s^{-1} approximately and buoyancy flux, $B = - K_N N^2$, increases to $(6\text{-}8) \cdot 10^{-9}$ m^2s^{-3}. Complete mixing of an unit base layer with thickness h increases the potential energy of this layer by $\Delta E_p = h^2N^2/12$. Corresponding changing of the layer splitting kinetic energy, $\Delta E_k = \Delta E_p$, is reasonably spent equally to dissipation, ε, and mixing during a splitting time τ_h. In the other words $\Delta E_k / \tau_h = B - \varepsilon = 0.5 \cdot B = B_0$. Hence, the buoyancy flux $B_0 = 0.5 \cdot B = (3\text{-}4) \cdot 10^{-9}$ m^2s^{-3} can be used to estimate a typical splitting time, $\tau_h = h^2N^2/12B_0$. Consequently, if the thickness of a new layer is 1 m, it is being formed during 30 minutes, formation of a layer of 5 m thick takes 12 hours. Salt finger convection inside the sheet might produce the intermediate homogeneous layer with thickness about 5 m, because R_p values at the interfaces surrounded this layer increase to 1.6 while K_S and K_T decrease to the values typical for the stationary conditions, as at the lower sheets. It is reasonable that layer thickness variations are affected by constriction and expansion of the internal wave modes. Buoyancy flux B in quasi stationary state is approximately equal to the kinetic energy dissipation rate ε. Therefore, at the nonsplitted interfaces (the middle sheet in the Table 1) the energy dissipation rate is close to $\varepsilon = 7 \cdot 10^{-10}$ m^2s^{-3} that proved to be in a good agreement with the turbulent measurements presented in [*Gregg and Sanford*, 1987; *Lueck*, 1987; *Fleury and Lueck*, 1991]. Hence, the homogeneous layers with θ-S indexes do not coinciding with the basic regression lines in Figure 5,a seems to be generated by salt finger mixing inside the basic sheets, providing the mean exchange coefficient increasing up to 10^{-4} m^2s^{-1}. Interaction between the water masses of Antarctic and subtropical origin supports the existence of the basic layers themselves.

5. HORIZONTAL STRUCTURE IN THE VERTICALLY QUASI HOMOGENEOUS LAYERS

5a. *Observations*

In this section we consider temperature, salinity and density spatial variability within the vertically quasi homogeneous layers (QHL). Fine horizontal temperature structure in such layers was revealed by *Marmorino et al.*, [1987] (will be referred as MBM87). Towed thermistor chain containing of 180 sensors was used in that measurements along the ten miles transects. Wave length of quasi periodic isotherm oscillations in 8 meters QHL was about 20 m, and mean amplitudes of temperature fluctuations were close to 0.005 K. Such structures were called the convective plumes.

Spatial resolution of our measurements was significantly worse, but the transect was much longer. Horizontal structure inside the QHL looks like the structure observed in MBM87, however spatial scales of our "plumes" are quite different. Isotherm lines in the thickest middle layer are plotted in Figure 2. One can take notice the evident intermittence of uplifting of cold less salted water (the dark dotted areas) and downlifting of warm more-salted water (the light areas). Certain of the temperature isolines were observed at the neighborhood interfaces simultaneously, connecting these interfaces across whole layer. The amplitudes of quasi periodic lateral fluctuations of T and S achieve 0.015 - 0.02 K, and 0.01 psu; horizontal wavelength is about of 6 km. These oscillations are well defined in Figure 7, where temperature and salinity values from the middle QHL are presented at the reference density level $\sigma_t^0 = 27.025$ for the casts

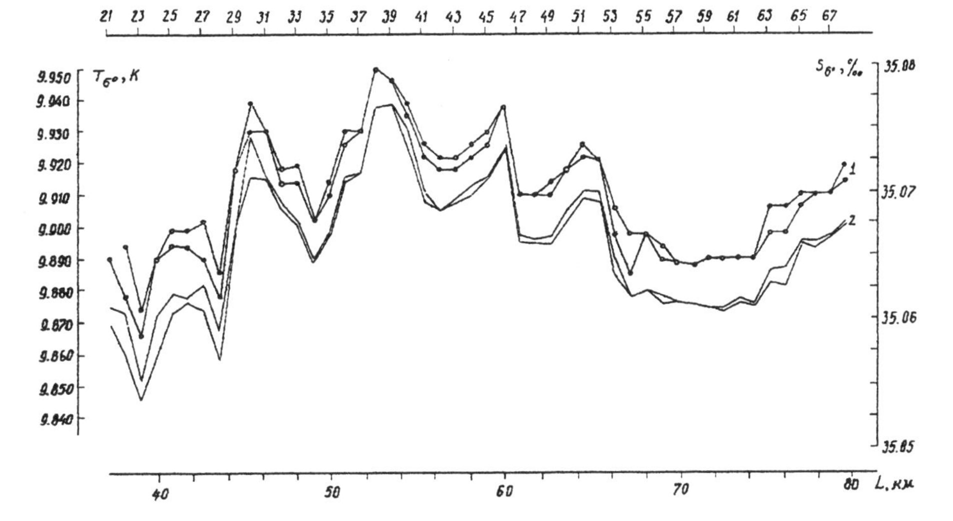

Fig.7. Potential temperature, $\theta(\sigma_t^0)$ - 1, and salinity, $S(\sigma_t^0)$ - 2, variations at the upper and lower boundaries of the isopicnic area, where $\sigma_t^0 = 27.025$, within the middle quasi homogeneous layer.

Nos'. 21 - 68. The isopicnic line σ_t^0 is located at the different depths in that layer, Figure 4. Temperature and salinity variability at the fixed depths in the central part of the layer is quite similar for $T(\sigma_t^0)$ and $S(\sigma_t^0)$.

5b. *Analysis and Inferences*

A comparative role of the different hydrophysical processes to be responsible for QHL spatial structure generation can be evaluated in terms of variability of the lateral density ratio, $\langle R_x \rangle$, along the layer. For pure isopicnal advection $\langle R_x \rangle \equiv \alpha \langle dT/dx \rangle / \beta \langle dS/dx \rangle = 1$, for double-diffusive convection $\langle R_x \rangle = n \equiv \alpha F_T / \beta F_S$, and for vertical advection of heat and salt $\langle R_x \rangle = \langle R_\rho \rangle \equiv \alpha \langle dT/dz \rangle / \beta \langle dS/dz \rangle$. According to [*Kunze*, 1987], n = 0.63 for $\langle R_\rho \rangle = 1.54$. Joint action of different processes leads to wide range of $\langle R_x \rangle$ values, namely between n and R_ρ. The values of θ-S indexes are shown in Figure 8 for the depth level z = 381 m, that crosses the frontal zone and locates at the center of the QHL. Five separate groups are clearly identified by various slopes of the regression lines. Hence, $\langle R_x \rangle$ is still constant along 10 - 12 km approximately: for the first group $\langle R_x^{(1)} \rangle = 1.20$, for the second group $\langle R_x^{(2)} \rangle = 1.13$ and for the third one $\langle R_x^{(3)} \rangle = 0.92$. It is reasonable that the greatest variation in $\langle R_x \rangle$ values was found between the profiles located at the different sides of the front (profiles 12 and 13). It should be underlined that $\langle R_x^{(1)} \rangle > 1$ and $\langle R_x^{(2)} \rangle > 1$ whereas $\langle R_x^{(3)} \rangle < 1$. By this is meant that at the left of the front isopicnic motion are affected by vertical advection, while at the right of the front by salt fingering convection. As horizontal thermohaline gradients become smaller and smaller, corresponding θ-S indexes in Figure 8 are concentrated near the point No.20. However detailed analysis of $\langle R_x \rangle$ variability between profiles 21 and 68 revealed evident increasing of double diffusion mixing in NNW direction from the local front. If we plot $\langle R_x \rangle$ values in the vicinity of the point No.20 with 10 times better θ-S resolution, these indexes locate at five straight lines. Regression lines for two groups of profiles are shown in Figure 8 as an example. The value of $\langle R_x^{(3)} \rangle = 0.92$ is almost constant up to 32nd profile. As soon as pronounced quasi periodic structure appears (Figures 2 and 7), the values of $\langle R_x \rangle$ fall down: $\langle R_x^{(4)} \rangle = 0.82$ (33-39 profiles) and $\langle R_x^{(5)} \rangle = 0.70$ (40-47 profiles). The value of $\langle R_x^{(5)} \rangle$ is close to n = 0.63. Therefore, θ-S horizontal fluctuations in that part of the QHL might be related to salt fingering convection. Horizontal fluctuations of T and S between 47 and 58 profiles are weak, $\langle R_x^{(6)} \rangle$ increases again to 0,81. Following increasing of temperature and salinity horizontal gradients between 59 and 68 sounding is accompanied by decreasing of the slope of regression line. The value of R_x here coincides with the flux ratio n ($\langle R_x^{(7)} \rangle = 0.63$), i.e. the role of double - diffusion becomes prevalent.

5c. *Discussion*

Temperature and salinity inhomogeneities are accompanied by weak, but appreciable fluctuations of the horizontal density gradient. Weak density inhomogeneities are typical for any QHL. However, only in the local zones of relatively sharp horizontal

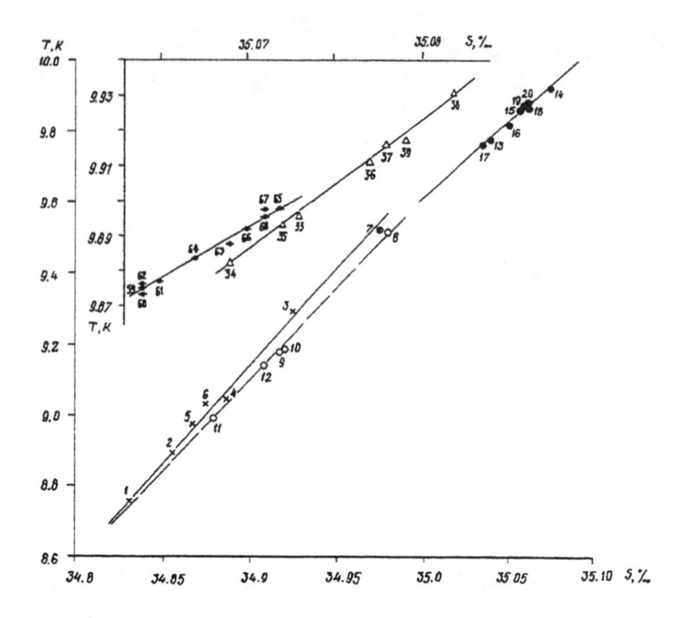

Fig. 8. Five groups of Θ-S indexes at the fixed depth $z_2 = 381$ m along the middle quasi homogeneous layer. Figures near the points are the profile numbers (see Fig. 2, 3). Regression lines correspond to several constant values of the lateral density ratio R_x in accordance with the different processes of the quasi homogeneous layer formation. Profiles 1-7, $R_x = 1.20$, were subjected to the local frontal upwelling region. Profiles 8-12, $R_x = 1.13$, demonstrate a mixture of vertical and isopicnal motions. Quasi isopicnal advection with a little convection contamination affect to profiles 13-20, $R_x = 0.92$. Double diffusion with a variable divergence of convective flux [McDougall, 1991] responsible for $R_x = 0.81$ (profiles 33-39). The ordinary salt finger convection ($R_x = 0.63$) was revealed between profiles 59-68.

gradients, $\rho_x \approx 10^{-6}$ kg m^{-4}, the geostrophic velocity shear, $V_{ZG} \approx 4 \cdot 10^{-4} s^{-1}$, becomes comparable with the shear measured by *Gregg and Sanford* [1987]. According to[*Lozovatsky and Nabatov*, 1990], the vertical momentum turbulent flux, $|\tau^{(z)}|$, at the layer borders can be characterized by values of $\sim 2 \cdot 10^{-8}$ m^{2}s^{-2}. Supposing vertical variability of $|\tau^{(z)}|$ is negligible, we can estimate the value of the momentum diffusivity, $K_M = |\tau^{(z)}| / V_{ZG} \approx 0.5 \cdot 10^{-4}$ m^{2}s^{-1}. This value seems to be quite typical for the main thermocline. However, in the QHL such values of K_M lead to negligible heat and salt turbulent fluxes in comparison with the same fluxes produced by salt fingering convection. Comparative role of turbulent and double- diffusive mixing through the staircase has been analyzed recently by *Fleury and Lueck* [1991].

Temperature and salinity fluctuations with horizontal length scales comparable with layer thickness are undoubtedly generated by convective processes. However the origin of fluctuations with aspect ratio $m = H_L/\lambda_p \approx 10^{-2}$, where λ_p is a typical wave length of plume like structures and H is a layer thickness, is not so obvious. If the "mesoscale plumes" are produced by upwelling and downwelling vertical motions arising between the mesoscale eddies rotating in the opposite directions inside the layers, the typical size of such eddies, D_E, should be equal to $\lambda_p/2$ and for $\lambda_p \approx (6\text{-}7) \cdot 10^3$ m, $D_E \approx (3.0\text{-}3.6) \cdot 10^3$ m. On the other hand, if one suppose the vertical size of an eddy is equal to the layer thickness, H_L, the eddy radius, R, can be estimated as the Rossby radius $Ro = H_L N_L / f$. So, for the typical values of the Vaisala frequency within the layers, $N_L \approx 10^{-3}$ s^{-1}, the layer thicknesses $H_L \approx 30\text{-}50$ m, and the Coriolis parameter, f, equal to $2.8 \cdot 10^{-5}$ s^{-1}, the Rossby radius $Ro \approx (1.1\text{-}1.8) \cdot 10^3$ m. This estimation for the eddy radius is in a good agreement with previous one, D_E, taking into account that $D_E = 2Ro$. Probably, motions in the convergence and divergence zones, that alternate approximately through 2Ro km, may produce plume like structures with the horizontal wave length λ_p of order of $2D_E \approx 4Ro$.

6. INTERNAL WAVES AND CONVECTIVE FLUXES.

6a. *Observations*

In this section we consider an assumption concerned with the internal waves propagation and periodic increasing of buoyancy flux at the interfaces that might be related to "mesoscale plume" formation. A statistical analysis of certain parameters for the homogeneous layers and interfaces was in particular carried out for checking up such hypothesis. Mean values of layer thicknesses, $\langle H_L \rangle$, and corresponding rms values, σ_H, are presented in the Table 2. The same parameters, $\langle T_L \rangle$ and σ_{TL}, were calculated for the temperature series at the fixed depths, zf, in the first, second and fourth layers : $zf_1 = 337$ m, $zf_2 = 381$ m, $zf_4 = 445$ m. Inasmuch as the third layer was apparently nonhorizontal, the temperature values have been chosen from the different depths, $zf_3 \equiv zf_c$, assigning to the center of the layer at each cast. Computations were carried out by use of 42 profiles: Nos'. 27B - 68B. As mentioned previously these casts were chosen because they had been obtained far enough from the frontal zone, where the influence of the frontal processes seems to be negligible. The values of σ^*_{TL} in the Table 2 represent typical rms amplitudes of horizontal temperature oscillations with the wave lengths less than 7 km approximately, because they have been obtained as a result of long wave trend

TABLE 2: Statistics of the homogeneous layer parameters (Figure 2)

Parameters	Upper Layer	Middle Layer	Intermediate Layer	Lower Layer
$<H_L>$, m	34.4	48.8	2.5	33.9
σ_{HL}, m	5.1	6.6	3.1	7.8
zf, m	337	381	zf_c	454
$<T_L>\|_{zf}$, K	10.631	9.903	9.531	9.197
$\sigma_{TL}\|_{zf}$, K	0.0274	0.0167	0.0187	0.0165
$\sigma^*_{TL}\|_{zf}$K	0.0142	0.0106	0.0125	0.0087

subtraction from the temperature series. This trend (Figure 7) is characterized by wave length about of 30 km and temperature amplitude of 0.03-0.04 K.

6b. *Analysis and Inferences*

An internal waves influence on the QHL convective mixing (or vise versa) can be consider by a statistical correlation analysis between the interface oscillations and the horizontal temperature variations within the homogeneous layers. Spectral calculations were carried out for a few of individual wave numbers taking into account short lengths of the temperature rows, $T_i(zf_j)$, and of the rows of isotherm displacements, $h_i(T_{1,2})$, where $i = 23 \div 68$ are the profile numbers and $j = 1 \div 4$ are the layer numbers. Displacement rows h_i of the isotherm line T_1 = 10.3 K at the upper interface and of the isotherm line T_2 = 9.5 K at the intermediate interface are considered. The maximum of spectral density for the upper interface displacements, $Eh^{(1)}(\varkappa)$, as well as for the spectra of horizontal temperature fluctuations in the upper, $ET^{(1)}(\varkappa)$, and in the middle, $ET^{(2)}(\varkappa)$, layers was found at the wave number $\varkappa_m = 10^{-3}$ m^{-1}. Besides, significant correlation between the temperature oscillations, $T_i^{(1)}$ and $T_i^{(2)}$, in the upper and in the middle layers was also found. Correlation coefficient, r, proved to be equal to -0.76 for the minimal spatial shift, $\Delta\varkappa$ = 0.9 km. The correlation is even closer for the individual harmonics with wave number $\varkappa_m = 10^{-3}$ m^{-1}. The coherence coefficient $C_{T1,T2}(\varkappa_m)$ = 0.97 and the correspondent phase shift, $\varphi_{T1,T2}(\varkappa_m)$, is equal to 160^0, i.e. 2.8 km. Correlation between the isotherm oscillations at the upper interface, $h^{(1)}$, and the temperature fluctuations in this layer, $T^{(1)}$, are characterized by the coherence coefficient $C_{h1,T1}(\varkappa_m)$ = 0.96,and $\varphi_{h1,T1}(\varkappa_m)$ = 130^0 (2.3 km). The same high coherence was found between the fluctuations at the same (No.1) interface and in the underlying (No.2) layer: $C_{h1T2}(\varkappa_m)$ = 0.93, $\varphi_{h1T2}(\varkappa_m)$ = 27^0 (0.47 km).

On the other hand, the correlation between the oscillations at the upper and at the middle interfaces is relatively weak: $r_{h1,h2}$ = 0.5, Δx = 0.9 km, as well as rather weak correlation was revealed for horizontal temperature inhomogeneities in the middle and intermediate layers: T_2 and T_3.

On the contrary, the middle and the lowest interfaces oscillate almost without any phase shift. It leads to high correlation between the temperature fluctuations in the middle layer and in the lowest layer. The oscillations at these two interfaces and in the lowest layer are high correlated also. The values of the correlation parameters for such series are follows: $r_{T3,T4}$ = 0.79 for Δx = 0, and $r_{T3,T4}$ = - 0.88 for Δx = 6.3 km. In the case, when the main displacement wave length was approximately equal to 10 km, $r_{T3,h3}$ = 0.96 for Δx = 2.7 km, and $r_{T3,h2}$ = - 0.7 for Δx = 5.4 km.

If we suppose that interface oscillations are caused by internal waves, the simplest dispersion equation for such waves at the interface between two homogeneous layers is follows,[*Phillips*, 1966]: $\omega^2 = g(\Delta\rho/\rho_0)\varkappa_{iw}$, where ω and $\varkappa_{iw}$ are intrinsic frequency and wave number, $\Delta\rho$ is the density jump and ρ_0 is a reference density. If we accept for raw estimation $\lambda_{iw}^{(1)} \equiv 2\pi/\varkappa_{iw} = 2\pi/\varkappa_m \approx 6\cdot10^3$ m^{-1} and $(\Delta\rho^{(1)}/\rho_0)$ for upper interfaces equal to $3.5\cdot10^{-5}$, thus the wave frequency $\omega_{iw}^{(1)} \approx 5.9\cdot10^{-4}$ s^{-1} and the wave period $\tau_{iw}^{(1)} \equiv 2\pi/\omega_{iw}^{(1)} \approx 1.1\cdot10^4$ s can be obtained. For the other interface $\lambda_{iw}^{(2)} \approx 10^4$ m, $(\Delta\rho^{(2)}/\rho_0) \approx 1.5\cdot10^{-5}$ that lead to $\omega_{iw}^{(2)} \approx 3\cdot10^{-4}$ s^{-1} and $\tau_{iw}^{(2)} \approx 2.1\cdot10^4$ s.

6c. *Discussion*

This set of statistic parameters, reflecting considerable correlation between the wave like interface oscillations and the horizontal thermohaline fluctuations in the vertical quasi homogeneous layers, can be used to check an idea that origin of plume like structure is associated with modulation of convective - diffusive fluxes at the interfaces owing to the internal wave forcing. If so, characteristic generation time of periodic structure should be depended on wave period. Assuming the temperature variation, δT, is caused by divergence of convective heat flux, F_T, during a typical internal wave period, τ_0, a time integrated equation of heat transfer across the interface can be obtained in the following form: $\delta T \approx \tau_0\cdot(\delta F_T/\delta h)$, where δh is a characteristic internal wave amplitude, it seems to be equal to rms of isotherm displacement at the interface, i.e. $\delta h \approx \sigma_h$. As was mentioned in section 4, $\sigma_{hu} = 2$ m for the upper interface and $\sigma_{hm} = 1.4$ m for the middle sheet. According to the calculations presented in the Table 1 and Table 2, $F_T^{(1)} = 2\cdot10^{-6}$ Km/s, $F_T^{(2)} = 6.7\cdot10^{-7}$ Km/s, and if equivalence of $\delta T = \sigma_T^*$ can be accepted, i.e. $\delta T^{(1)} \approx 0.011$ K, and $\delta T^{(2)} \approx 0.012$ K, the following estimates of typical generation time are obtained: $\tau_0^{(1)} \approx 1.1\cdot10^4$ s, and $\tau_0^{(2)} \approx 2.5\cdot10^4$ s for the middle and intermediate layers correspondingly. Closeness of τ_0 and τ_{iw} values lead to preliminary conclusion that a hypothesis of observed plume like structure origin by periodic forcing of salt finger convective fluxes under influence of internal waves should not be rejected.

Statistic analysis of the different parameters of homogeneous layers presented in section 5 and 6, as well as the data published in MBM87, allow to assume a hierarchy of horizontal scales of the convective structures in QHL. Indeed, if the relation between the layer thickness, H_L, and the horizontal size of the smallest convective plumes, $\lambda_p^{(s)}$, would be the same as reported in MBM87, i.e. $H_L/\lambda_p^{(s)} \approx 0.3 - 0.5$, than the wave length of plume like structure $\lambda_p^{(s)}$ in the middle homogeneous layer with $\langle H_L \rangle \approx 50$ m would be equal to $\approx 100 - 150$ m. The amplitudes of such temperature fluctuations should be the same as mentioned in MBM87, i.e. about 0.005K. It is to be underlined that observed T and S oscillations in the QHL have not been connected with aliasing caused by rare casts, because the amplitudes of these fluctuations are of 2-3, or even 10 times higher compare with the convective plume amplitudes offered in MBM87. Apparent confirmation of the "mesoscale plume" existence is presented in Figures 2 and 3 between 18 and 25 profiles especially, as well as in Figure 7. A long wave trend which is clearly defined in the same figures appear to be almost periodic with $\lambda_p^{(l)} \approx 30$ km and temperature amplitude of 0.03 - 0.04 K. Why would not suggest that such oscillation in the QHL was related to large scale convective motions also. Therefore a hypothesis on a hierarchy of plume like structure scales in the quasi homogeneous layers of salt finger staircase may be extended to the scales of several tens kilometers.

7. SUMMARY

7a. *Conclusions*

Detailed analysis of the lateral variability of thermohaline structure at the long distance transects in the north western tropical Atlantic allowed to formulate the following conclusions.

1. Vertical motions associated with local upwelling and downwelling between the mesoscale eddies or separated current jets may generate the narrow thermohaline fronts at the background of salt finger staircase. Vertical shift of the staircase at the distances of a few kilometers achieves 100 m.

2. A genetic relationship was revealed not only for the particular homogeneous layers observed at the opposite sides of the front, but for the basic layers found in the C-SALT area previously. However, this basic staircase is affected by deformation and distortion owing to local dynamic processes.

3. Homogeneous layers with θ-S indexes that do not coincide with the main regression lines for the "basic layers" at the θ-S diagram result from the interface splitting due to increased convective flux. It is reasonable that layer thickness variations are affected by constriction and expansion of internal wave modes.

4. Horizontal fluctuations with wave length of 6 km, aspect ratio 0.01, typical amplitudes 0.015 K and 0.005 psu were found in the quasi homogeneous layers. Mesoscale eddies steering within the layers might be partly responsible for the "mesoscale plume" origin as well as internal wave forcing of salt finger flux at the interfaces. "Large scale plumes" with wave length about 30 km probably also exist.

5. A hypothesis is suggested concerning a hierarchy of various scale convective plumes in the quasi homogeneous layers of salt finger staircase.

7b. *Discussion*

According to an alternative popular hypothesis the step like structure is caused by salt finger convection at the vertical boundaries of the frontal intrusions. Such mechanism is certain produces intrusions (and the following steps sometimes) at the low-baroclinic fronts with high thermoclinity (high values of T, S isopycnal gradients). However it is difficult to answer for a series of important questions accepting this hypothesis. For example, why the thermohaline staircases in the north western tropical Atlantic occupy a space of millions square kilometers, persisting the main parameters over the years, if there are no any permanent mesoscale fronts in that area? Why the staircase to the left of the front is shifted more than 100 m vertically compare with *the same* staircase to the right of the front, if they both have to be found at the same depths as is obvious from the classic laboratory experiments? And how the "intrusive hypothesis" can explain the origin of rather thick homogeneous layers that were found at the splitted interfaces far enough off the front? The most reasonable answer on these questions is follows: the step like structure in the area under consideration is governed mainly by free and forced convection rather than by isopycnal advection. Frontal intrusions and upwelling should be considered as the secondary phenomena. Vertical mixing and transport of heat, salt, oxygen, and nutrients across the staircase pycnocline are provided by a set of "convective plums". The plumes likely represent the main convective elements forced by internal waves and mesoscale eddies at the background of permanent salt finger favorable stratification.

7c. *Reflections*

One of the comments of an unknown reviewer prompted me to add this subsection however I understand that these reflections have no sufficient justification. On the other hand they seems to be useful for the future discussions.

The question was: "You suggest that plumes create layers. If so, then how did plumes exist before the layers were created?" In my opinion it is a classic problem from the series on the chicken and the egg. Plume like structures may certainly appear in the quasi homogeneous layers that have been created previously. However it is more likely the plumes as typical elements of forced convection (in the hydrothermal vents, for example) mix a layer which thickness is governed by background stratification and energy of the external forcing. Why we can not think that small plumes are self-organizing elements of free double-diffusion convection while mesoscale and large scale "plumes" are caused by forced convection? We can even guess the large scale plumes represent coherent vortices in the almost two-dimensional weak turbulent flow, taking into account that the aspect ratio for such structures is in the range of 10^{-2} - 10^{-3}. Step like thermohaline structure in salt finger favorable stratification may reflect a synergetic nature of mesoscale layer dynamics. A hierarchy of "convective plumes" may be considered probably as a manifestation of negative eddy viscosity - a reverse eddy cascade. If so, they may produce and survive sharp horizontal (fronts) and vertical (sheets) thermohaline gradients in conditions of unstable thermodynamic equivalence. On the other hand it is an effective mechanism of convective mixing and lateral stirring providing vertical and horizontal transport of salt, heat, nutrients. Thus, the plume like eddies and salt fingering convection can be considered as the interrelated processes leading to self-maintaining of the thermohaline staircase. Self control in such open thermodynamic system is governed by the interaction between Antarctic and subtropical water in the western tropical Atlantic. Double diffusive and hydrodynamic instability in that area are related to that permanent climatic source of high available potential energy.

Acknowledgments. The author gratefully acknowledges Vadim Paka who headed towing CTD measurements in the cruise. Preliminary data processing has been made by Igor Morozov. I am extremely thankful for the traveling grant from American Geophysical Union and Office of Naval Research providing my participation in the Chapman AGU Conference on Double-Diffusive Convection. This work was partially supported by International Science Foundation, grant MEV00, ONR grant N00014-9410325and RFFI grant No.94-05-16371.

REFERENCES

Lozovatsky, I. D., and V. N. Nabatov, Spectral structure of convective layers and high gradient sheets in the thermocline, *Izvestiya of USSR Acad. of Sci., ser. Atmosph. and Ocean Phys.*, 26, 412-420, 1990 (in Russian).

Lozovatsky, I. D., and M-J. H. Lilover, Spectral structure of thermohaline inhomogeneities of cold core cyclonic ring in the ocean, *Doklady USSR Acad. of Sci.*, 291, 703-708, 1986 (in Russian).

Erofeev, A. Yu., and I. D. Lozovatsky, Comparative analysis of lateral fine structure parameters in the Lomonosov current and Canary upwelling, *Oceanological Res.*, vol.42, 50-61, 1990 (in Russian).

Fedorov, K. N., *Thermohaline ocean fine structure.*, 184 pp., Hydrometeorological Publisher, Leningrad, 1976 (in Russian)

Fleury, M., and R. G. Lueck, Fluxes across a thermohaline interface, *Deep-Sea Res.*, 38, 745-769, 1991.

Gargett, A. E., and R. W. Schmitt, Observations of salt fingers in the central water of the eastern North Pacific., *J.Geophys.Res.*, 87, 8017-8029, 1982.

Gregg, M. C., and T. B. Sanford, Shear and turbulence in thermohaline staircases, *Deep-Sea Res.*, 34, No.10A., 1698-1696, 1987.

Kunze, E., Limits on growing finite-length salt fingers: A Richardson number constraint, *J. Mar. Res.*, 45, 533-556, 1987.

Lueck, R. G., Microstructure measurements in a thermohaline staircase., *Deep-Sea Res.*, 34, No. 10A, 1677-1688, 1987.

Marmorino, G. O., Intrusions and diffusive interfaces in a salt finger staircase, *Deep-Sea Res.*, 38, 1431-1454, 1991.

Marmorino, G. O., W. K. Brown, and W. D. Morris, Two-dimensional temperature structure in the C-SALT thermohaline staircase, *Deep-Sea Res.*, 34, No.10A, 1667-1676, 1987.

Marmorino, G. O., Observation of small-scale mixing processes in the seasonal thermocline., Part 1: Salt fingering, *J. Phys. Ocenogr.*, 17, 1339-1347, 1987.

McDougall, T. J., Interfacial advection in the thermohaline staircase east of Barbados, *Deep-Sea Res.*, 38, 357-370, 1991.

Osborn, T. R.,Signatures of double diffusive convection and turbulence in an intrusive regime, *J. Phys. Ocenogr.*, 18, 145-155, 1988.

Phillips, O. M., *The dynamics of upper ocean.*, 266 pp., Cambridge Press, 1966.

Schmitt, R. W., H. Perkins, J. D. Boyd, and M. C. Stalcup, C-SALT: an investigation of the thermohaline staircase in the western tropical North Atlantic, *Deep-Sea Res.*, 34, No.10A, 1655-1665, 1987.

Woods, J., Scale upwelling and primary production: Toward a theory on biological-physical interactions in the world ocean, in *Proc. NATO Adv. Res. Workshop*, edited by B.J. Rothschild, pp. 7-38, Kluwer Acad. Publ., 1988.

Zhurbas, V. M., J. Laanemets, R. V. Ozmidov, and V. T. Paka, Horizontal variability of thermohaline fields with step layering in the ocean, *Oceanology*, 33, 903-909, 1988 (in Russian).

I. D. Lozovatsky, The Program of Environmental Fluid Dynamics, Department of Mechanical and Aerospace Engineering, Arizona State University, Tempe, AZ 85287-6106.

Salt Fingering in the Cyprus Eddy

Tal Berman[1], Stephen Brenner[2] and Nathan Paldor[1]

(1) Institute of Earth Sciences, The Hebrew University of Jerusalem, Jerusalem, Israel.
(2) Israel Oceanographic & Limnological Research, Haifa, Israel.

The Cyprus Eddy, is a warm core eddy wedged between the seasonal and permanent thermoclines. Density Ratio ($R\rho$) and Turner Angle (Tu) indicate salt-fingering favorable conditions in both thermoclines (Tu > 45^0). Buoyancy fluxes due to heat and salt estimation using Kunze's (1987) model for thick interfaces indicate that during summer, the buoyancy fluxes in the seasonal thermocline are one order of magnitude larger than in the permanent thermocline. During all other seasons when the seasonal thermocline exists the buoyancy flux through it is of the same order of magnitude as that of the permanent thermocline. These fluxes are significant in determining the T and S signals of the eddy. The flux of salt into the eddy by salt fingering through the seasonal thermocline during the 9 summer months accounts for the excess of salt observed in the eddy (0.25-0.35 ptt). These suggested fluxes are further substantiated by the following observed changes in the chemical composition of the water in the eddy. High nitrate and low oxygen (>1.5 μM/kg; <175 mM/kg, respectively) are found in the eddy during winter when the seasonal thermocline dissapears, so the eddy is exposed to the atmosphere its primary source of oxygen and the opposite is expected.This discrepancy can be resolved by assuming a flux of high nitrate/low oxygen water into the eddy from below the permanent thermocline. During summer when the eddy is sealed off from the atmoshere by the seasonal thermocline the appearance of high oxygen and low nitrate (>220 μM/kg; <1μM/kg respectively) in the eddy indicates the exchange of water with the euphotic zone, overlying the seasonal thermocline.

1. INTRODUCTION

The Cyprus Eddy located south of the island of Cyprus in the Eastern Mediterranean (see Fig. 1 for exact location) has a 50 km radius and is best characterized as a lens of warm and saline water wedged between the seasonal and permanent thermoclines. The eddy has no apparent surface signal in summer and only a weak signal (less than 0.5^0C) in winter (Fig. 2). The strongest signal of the eddy in all seasons is subsurface and it reaches maximum values of 2-2.4 ^{0}C temperature difference and 0.25-0.34 ppt salinity difference in the 200-400 m layer (Fig. 2). In winter, there is no seasonal thermocline and the eddy extends from the surface to a depth of 400-600 m, the depth of the permanent thermocline in the center of the eddy. In other seasons the overlying water (between the surface and the seasonal thermocline) is warmer and saltier than the eddy's by as much as 10^0C and 0.5 ppt, respectively. Temperature within the lens varies at most by 0.03^0C and salinity is constant within the observation error of 0.003 ppt. The diameter of the eddy was found to be a constant, 100 km, (within the sampling resolution) as defined by the intersection of the 16^0C isotherm with the 300 m level. For a more detailed description of the observed synoptic features of the eddy see Brenner (1993).

Profiles of oxygen, nitrate and phosphate during the years 1989 - 1992 (Fig. 3) are similar to those given in Krom et al. (1992) (the latter describes the observations for the year 1989 only) and resemble the T and S profiles from those years. During winter, mixing occurs in the water column from the surface to the permanent thermocline with high concentration of oxygen (>220 μM/kg) and low

Double-Diffusive Convection
Geophysical Monograph 94

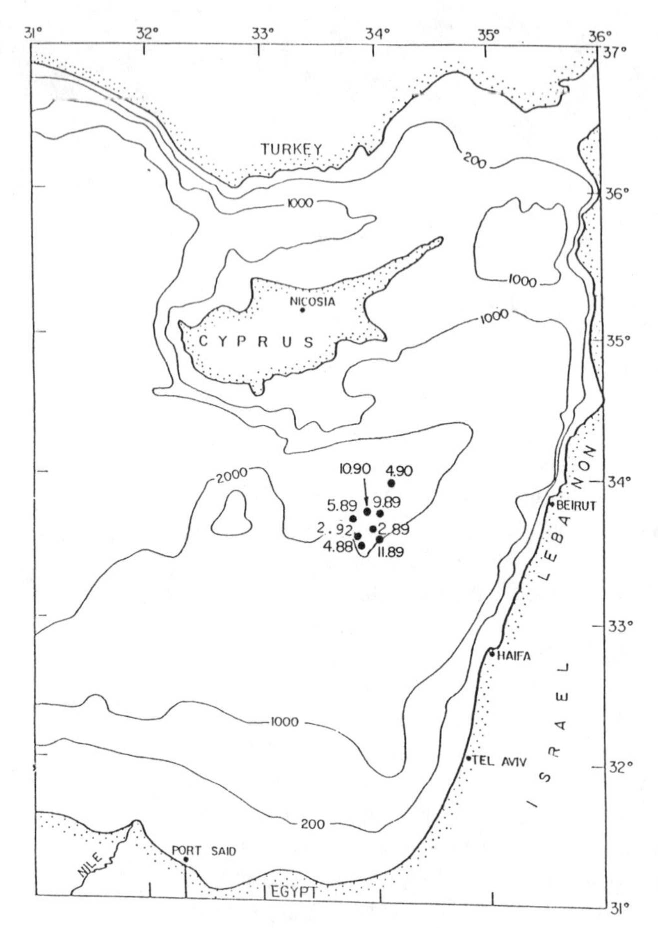

Figure 1: Bathymetric map of the eastern Mediterranean sea and the location of the Cyprus eddy from May 1982 to March 1992.

concentration of nitrate and phosphate (less than 2 and 0.1 m/kg respectively). Below the permanent thermocline, relatively low concentrations of oxygen (< 180 μM/kg) and high concentration of nitrate and phosphate (more then 6 μM/kg and 0.2 μM/kg respectively) are observed. In summer, the core of the eddy (200- 400 m) has those same characteristics, high concentrations of oxygen and low concentrations of nitrate and phosphate. Above the seasonal thermocline, concentrations of nitrate and phosphate are negligible during summer as expected but the oxygen profile has an unexplained local minimum at 70 m during this season.

It is possible to recognize within the eddy all the major water masses of the Mediterranean [*Hecht et al.,* 1988; *Brenner,* 1993]. During summer, the Levantine Surface Water with its high salinity (>39.15 ppt) extends from the surface to a depth of 40 m. Below the Levantine Surface Water underlies the Atlantic Water with its low salinity (38.6 ppt) which can be observed to a depth of 100 m. The main eddy water in all seasons consists of Levantine Intermediate Water with its high salinity(39.10 -39.15 ppt). Below 600 m, the base of the eddy is the Deep Water (DW) which is fresher (38.7 ppt) and colder (13.6^{0}C) than the overlying Levantine Intermediate Water .

Two main hypotheses concerning the location and formation of the eddy have been proposed. One suggestion is that the eddy is formed locally as a result of the flow of the Mid Mediterranean Jet over the Erastothenes Seamount (33^{0}42' N 32^{0}52'E). According to this suggestion, the eddy is formed in the region [*Brenner et al.,* 1991] as either a Taylor column or a detached meander of the Mid Mediterranean Jet. On the other hand, Feliks (1987; 1990) suggests that the eddy may not have been formed in the region but rather along the coast of Turkey at the North-Eastern corner of the Mediterranean where a downwelling front develops as a result of winter storms. The sinking water creates geostrophic currents which in turn cause eddies to develop. The eddies separate from the front and drift to the south where at some point they join the Mid Mediterranean Jet and are advected eastward.

Due to its thermal and salinty structure the eddy is suspected of having salt fingering instability at both its seasonal and permanent thermoclines. The controlling parameter in the growth of salt fingering and the resulting buoyancy fluxes across the interface is the well known density ratio Rρ defined as $\alpha T_z / \beta S_z$ where α is the coefficent of thermal expansion and β is the coefficent of halin contraction, T_Z and S_Z are the vertical gradients of temperture and salinity respectively. More recently alternative - the Turner angle [*Turner,* 1978; *Schmitt,* 1981], Tu. The latter commonly replaces Rρ for following reasons: (1) The infinite scale of Rρ is replaced by a finite one for Tu (running from +180^0 to -180^0) and (2) The poorly defined Rρ value obtained when S_Z=0 is well-defined in terms of Tu [*Ruddick, 1983*]. An additional reason for using Tu rather than Rρ is that gravitationally unstable water and water unstable to salt fingering have very close Rρ values [*Ruddick,* 1983] while the corresponding Tu values are clealy distinct. According to Ruddick (1983) Tu is defined as:

$$\tan(Tu) = \frac{}{\alpha T_Z - \beta S_Z} = \frac{}{R_\rho - 1} \quad (1)$$

Measurements of fluxes of salt due to salt-fingering between well-mixed layers show that these fluxes depend on both the salinity difference between the layers and on Rρ. The flux of salt across a salt fingering interface increases with increasing salinity difference and with decreasing Rρ. Strong salt fingering (SSF) are

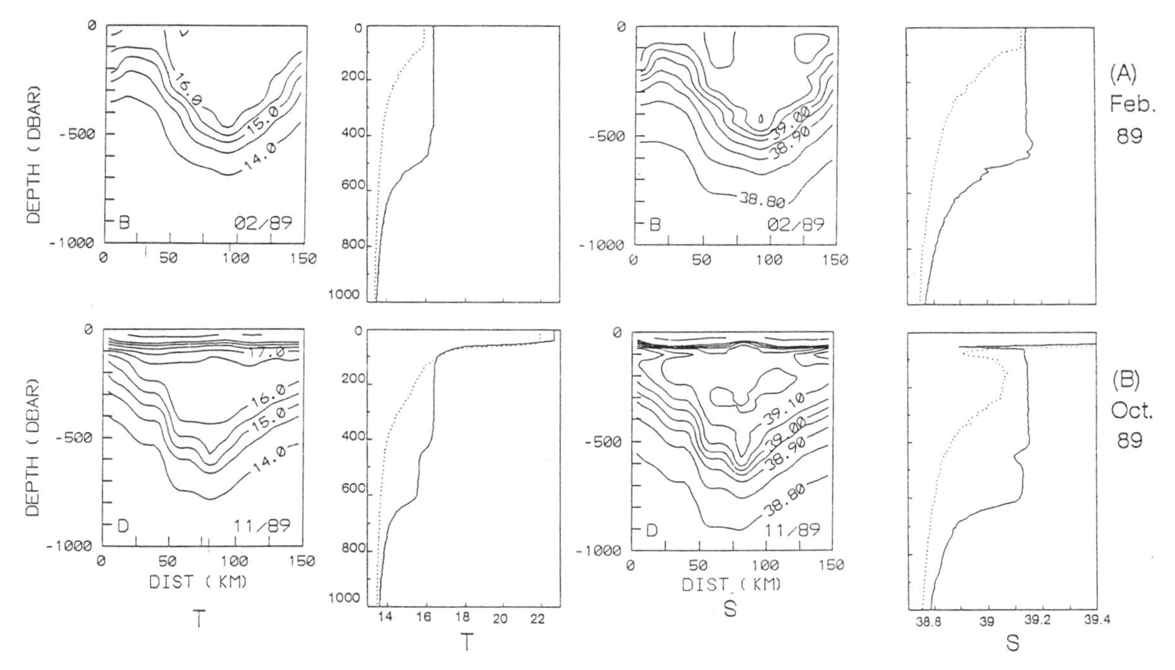

Figure 2: Cross sections and T and S profiles for the center (solid line) and the edge (dotted line) of the eddy: (a) Winter (Feb. 1989) there is no seasonal thermocline, and the eddy's warm and saline water reaches the surface while the surrounding water is cooler and fresher. (b) Summer (Oct. 1989), the water above the seasonal thermocline is very warm and very salty, while the water within the eddy is warmer and saltier then the surrounding water.

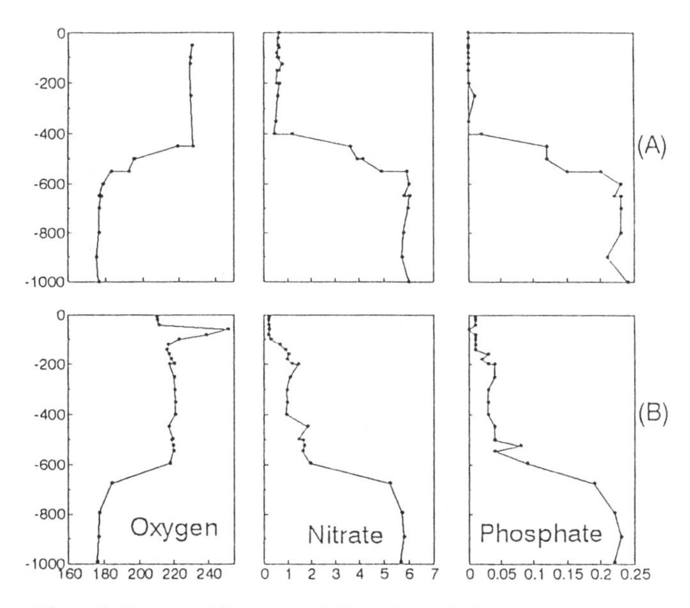

Figer 3: Oxygen, Nitrate, and Phosphate during winter- (a) and summer- (b) in the center of the eddy. Low nitrate and phosphate (negligible) and high amounts of oxygen (>230 μM) above the seasonal thermocline. In the eddy, small amounts of nitrate and phosphate (less than 2 and 0.1 μM/kg receptively) and high amounts of oxygen (220-230 μM/kg). Below the permanent thermocline high amounts of nitrate and phosphate (more then 6 and 0.2 receptively) and less oxygen (<180 μM/kg), (1989).

characterized by $2 > R\rho > 1$ ($71^0 < Tu < 90^0$) and weak salt fingering (WSF) by $\infty > R\rho > 2$ ($45^0 < Tu < 71^0$).

Several theoretical models for calculating the fluxes of both water and boyancy (i.e. heat and salt) from one layer to the other were proposed in the past. One of those model is Kunze's (1987) model for both thin and thick interfaces. It is well established [*Kunze*, 1987; *Hebert*, 1988] that since the thickness of any interface separating observed steps in the ocean (tens of centimeters) is generally thicker than the maximum vertical length of salt fingers, these interfaces consist of several layers of salt fingers. Thus, the laboratory flux laws which apply to interface with only one layer of salt fingers give the incorrect fluxes for many of the interfaces observed in the ocean. When the interface thickness is the same as the vertical scale of the salt fingers, i.e., when the interface is thinner than h^*_{max} defined by:

$$(k_T \beta \Delta S)^{1/3} \qquad (2)$$

then the thin interface model is applicable which is relevent in laboratory flux measurements [e.g. *Turner*, 1967; *Schmitt*, 1979 *and Kelly*, 1986].

In oceanic observations, h_{max} is given by :

$$_{max} \qquad (\kappa_T g \langle \delta S / \delta z \rangle)^{1/4} \qquad (3)$$

where ν is the molecular viscosity (10^{-6} m/s) of seawater; $_T$ ($1.4x10^{-7}$ m/s) is the molecular diffusivty of temperature; g is the gravitational constant (9.8 m/s^2) and $<\delta S/\delta z>$ is the average of all calculated salinity gradients through the layer [*Kunze*, 1987; *Hebert*, 1988]. In our observation all interfaces (steps) are thicker than this distinguishing height scale so we need to use the thick interface model in which the equations used for calculating the buoyancy fluxes due to heat and salt are:

$$g\alpha F_T = \gamma g \beta F_S \tag{4}$$

where:

$$\gamma = R_\rho^{1/2} \lfloor R_\rho^{1/2} - (R_\rho - 1)^{1/2} \rfloor \tag{5}$$

A brief introduction, intended for the physical oceanographer, is in order before going into the details of the seasonal changes in the chemical composition of the eddy's water. The ratio of oxygen to nutrients (O_2/NO_3 and O_2/PO_4) changes with depth. It is high near the surface where photosynthesis and atmospheric fluxes are responsible for the high oxygen values and it decreases with depth due to oxygen consumption by biological activity [*Broecker and Peng,* 1982]. Therefore, in the absence of vertical exchange of water the expected ratio should decrease with depth while exchange of water with the surface layer (where oxygen is abundant) is expected to increase the ratio and an exchange with the underlying deep water (rich in nutrients and depleted in oxygen) causes the ratio within the eddy to decrease. Seasonal changes in oxygen and nutrients concentration across the eddy can be used to trace the exchange of water and its source.

The purpose of this study is to establish the existence of salt fingering in the Cyprus eddy as a first step towards determining the processes responsible for its genesis and subsequent long life span. A secondary goal of this study is to establish the existence of cross- isopycnal mixing using two different indicates: Physical parameters- (temperature and salinity) and Chemical concentrations of (oxygen and nutrient) composition. Both indicates to the same pattern in the annual cycle of vertical fluxes into the eddy.

2. DATA

The data presented here consists of two simultaneous experiments which took place between 1988 and 1992. The ED (Eddy) experiment consisted of seven cruises to the eddy conducted between May 1988 and Feb. 1992 and the LBDS (Levantine Basin Dynamics Study) experiment of two cruises between 1989 and 1992. The ED data set is the most detailed on the eddy and these data provide a fairly dense coverage of its temporal and spatial variabilities. In these cruises, physical (T,S) as well as chemical (oxygen, nitrate, phosphate, silica) data were collected with the specific goal to study and trace the evolution of the Cyprus eddy. The data on the eddy was collected at stations 10 - 20 km apart. At each station, the cast went down to a depth of 1000 m at least. The number and locations of the stations varied from one cruise to the other depending upon the location of the eddy. In the LBDS experiment, the stations were located on a fixed grid of 0.5^0. In Table 1 we provide a summary of the cruises to the eddy, the observed location of the eddy's core and the physical properties of the eddy's core during these cruises. The depth chosen in Table 1 is that where the maximal horizontal difference in T and S, between the core and the edge of the eddy, were observed (see also Fig. 2).

The T, S raw data were collected aboard the R/V Shikmona with a Neil Brown CTD. The CTD was lowered at a speed of 0.7 - 1 m/sec and the sampling frequency was 30 hertz, The data used in the study are 1 m running averages of the raw data so that each data point used in this study represents an average of about 30 measurments. The raw data was corrected for the mismatch between the T and S measurments to avoid spurious spiking. Water samples both for chemical analysis and for salinity calibration were taken in Niskin bottles. Reversing thermometers were used for temperature and pressure calibrations. Additional details on the sampling and analysis of both types of data are given in Hecht et al. (1988), Ozsoy et al. (1989) and Krom et al. (1992).

TABLE 1. Data from the center of the Cyprus Eddy.

Cruise	Date	P(db)	T(^{0}C)	S(ptt)	σ_t	Location
ED02	May 88	200	16.42	39.08	28.79	33^{0}34'N, 33^{0}52'E
ED03	Feb. 89	200	16.41	39.15	28.84	33^{0}40'N, 33^{0}55'E
ED04	Apr. 89	200	16.39	39.14	28.85	33^{0}42'N, 33^{0}45'E
ED06	Oct. 89	200	16.39	39.15	28.85	33^{0}38'N, 34^{0}06'E
ED07	May 90	200	16.64	39.27	28.88	33^{0}57'N, 34^{0}08'E
ED08	Nov. 90	200	16.64	39.27	28.88	33^{0}45'N, 33^{0}52'E
ED09	Feb. 92	200	16.63	39.33	28.93	33^{0}38'N, 33^{0}50'E

In each of the LBDS cruises, at least one station was located within the eddy. While this is sufficient to indicate the presence of the eddy, it can not be expected to yield a statistically significant mean eddy profile and is insufficient to determine its center and thickness. The sampling during ED cruises was done on a denser grid and therefore provides more precise location and depth of the eddy center. In each of the ED cruises, at least one cross-section through the 100 km diameter of the eddy was obtained. In all ED cruises, the diameter of the eddy was found to be a constant 100 km as defined by the intersection of the 16^0C isotherm with the 300 m level. Within the sampling resolution, on any given cruise, temperature within the lens varied at most by 0.03^0C and salinity was constant within the observation error of 0.003 ppt. For a more detailed description of the observed synoptic features of the eddy see Brenner (1993) and Berman (1992).

3. RESULTS

3.1 Salt Fingering and Buoyancy Fluxes

Due to large differences in temperature and salinity across the seasonal and the permanent thermoclines, these interfaces are conducive for salt-fingering instability. Salt-fingering favorable stratification is often characterized by staircases in the profiles of salinity and temperature. Such staircases were indeed found in all profiles taken inside the eddy. An example of such observed staircase structure is shown in Fig. 4. The irregularity in the S profile is not due to a mismatch between the T and S sensors as this was removed in the initial processing of the raw data. Rather the spike indicates a true behaviour of the profile (If it were due to the sensor difference this would have been observed in all profiles taken, which is not the case). Furthermore, spiking due to sensor mismatch would accur everywhere in the profile and not in one isolated depth.

Calculation of Tu becomes problematic whenever T and S are nearly homogeneous in some parts of the water column. In this case the difference (αT_z- βS_z),(RHS Eqn. (2)) is very small and the calculated value of Tu fluctuates vigorously between positive and negative values, as measurement errors determine this value. To reduce this problem in the stable area of the profile, all $|\alpha T_z-\beta S_z|$ values smaller than the cutoff value of 8×10^{-3} were arbitrarily assigned a value of Tu = 0 (corresponding to a density ratio $R\rho$ = -1) .

The profiles of Tu and density ratio from the core of the eddy shown in Fig 5, clearly indicate the existence of salt fingering whenever Tu>45^0 is encountered. For clarity, all data with zero (as described above) Tu values were omitted.

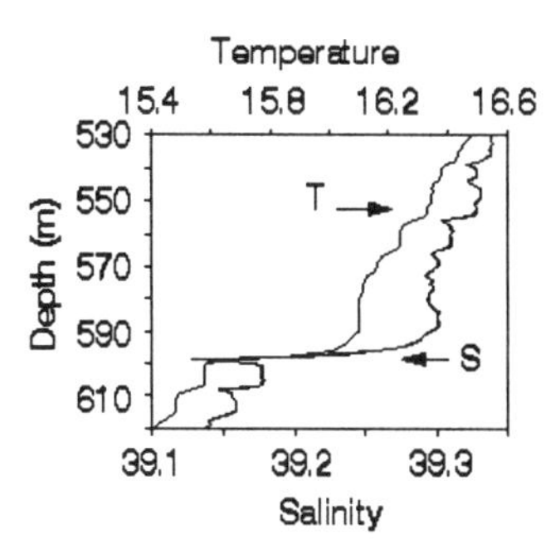

Figure 4: Example of Staircases at base of the eddy (Feb. 1992).

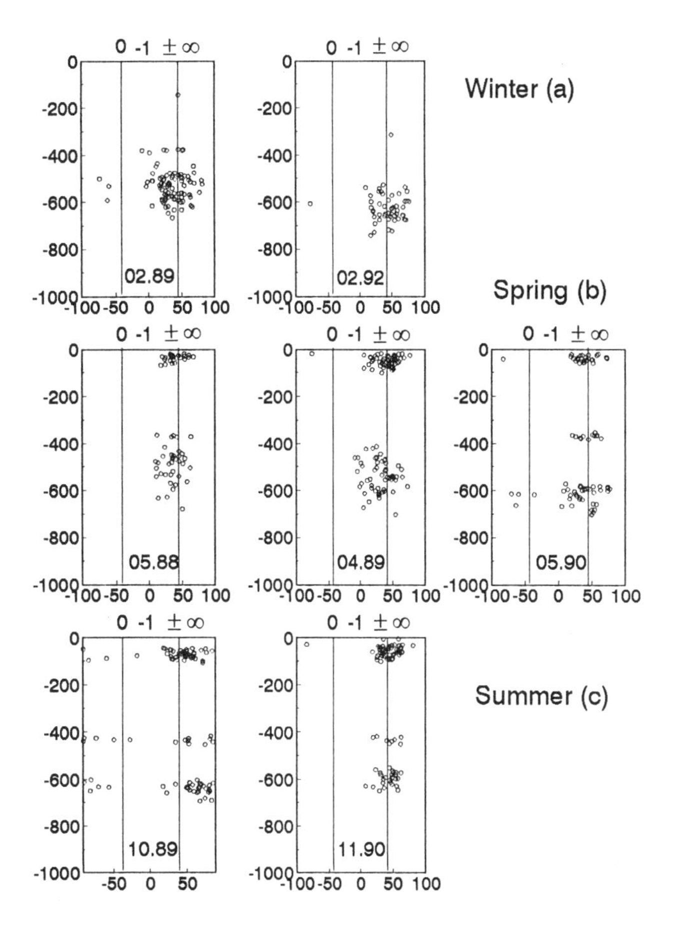

Figure 5: Tu values on the lower X axis and R on the upper X axis indicating instability regimes. The vertical lines 45^0 and -45^0 separate between different types of instability: -90^0 to -45^0 diffusive regime; -45^0 to 45^0 stable; 45^0 to 90^0 salt fingering regime. (a) Winter, Tu values indicate WSF in the permanent thermocline. (b) Spring, Tu values show WSF in both thermocline. (c) Summer, Tu values show SSF in seasonal thermocline and WSF in permanent thermocline. See text for a more detailed discussion.

TABLE 2. Observed interface with maximal fluxes (according to Kunze's thick interface model) measured in each cruise at the center of the eddy in the seasonal and permanent thermocline. h is the observed interface thickness, DS, DT are the salinity and temperature jumps across the observed interface, Rρ is the density ratio, h_{max} calculated from Eqn. (2), $g\beta F_s$ and $g\alpha F_T$ are the heat and salt fluxes calculated according to Eqn. (3) and Eqn. (4).

Note a well developed seasonal thermocline in April 1989, in other years the seasonal thermocline reaches a similar stage of development about two month later. The winter 1991/92 was exceptional in the number and intensive of storms which caused both the low temperature and the high rainfall in that winter.

Date	Press. (db)	h (m)	DS (ppt)	DT (^{0}C)	Rρ	Season	h_{max} m)	$g\beta F_s$ (10^{-10}kg^{-1})	$g\alpha F_T$ (10^{-10}kg^{-1})
May 88	21	4	0.046	0.15	2.42	Spring	0.623	12.7	7.21
	364	2	0.025	0.05	1.49		0.355	6.73	4.28
Feb. 89	509	1	0.032	0.06	1.39	Winter	0.257	15.3	10
Apr. 89	25	2	0.161	0.61	2.82	Spring	0.442	108	59.8
	540	4	0.027	0.123	3.39		0.972	11.3	6.16
Oct. 89	54	1	0.147	0.42	2.12	Summer	0.289	136	79
	634	1	0.02	0.063	2.33		0.522	21	12
May 90	36	1	0.019	0.05	1.96	Spring	0.441	15.7	9.25
	590	2	0.069	0.128	1.38		0.249	16.2	10.6
Nov. 90	34	1	0.221	1.136	3.82	Summer	0.452	427	230
	571	3	0.015	0.069	3.42		1.057	8.48	4.61
Feb. 92	599	4	0.15	0.35	1.73	Winter	0.326	25.9	15.7

The results of the buoyancy flux calculations are given in Table 2 from which it can be easily seen that there is a constant buoyancy flux due to heat and salt yearlong through the permanent thermocline (7 to 21×10^{-10} Wkg^{-1}, except for Feb. 1992) while a seasonal pattern can be seen in the buoyancy fluxes through the seasonal thermocline. The Tu profiles and the buoyancy fluxes of the different cruises in Table 2 are divided into three seasons: winter (a), spring (b) and summer (c). The following seasonality observed in the eddy was also found during POEM (Physical Oceanography of the Eastern Mediterranean) experiment (e.g. Oct. 1985, Apr. 1986, Sep.1987) and during the LBDS (Levantine Basin Dynamics Study) experiment [see *Berman*, 1992]:

A: Winter - Dec. to March; Fig. 5a. The water column from the surface to 400 m is homogeneous so Tu was assigned the value of 0 there. Below 400 m and through to the base of the eddy (600 m) Tu values are indicative of salt fingering instability (Tu>45^0), and at some depths even of SSF (Tu>71^0). Below the permanent thermocline (the base of the eddy) low values of Tu (<45^0) are encountered which indicates stability. Buoyancy fluxes due to heat and salt through the permanent thermocline (Table 2) are significant (i.e. not negligible), nearly equal one another and are fairly uniform throughout the year.

Winter 1991/92 was the coldest and most stormy winter recorded in the region in the last century, which caused a larger difference in temperature and salinity between the water above and below the permanent thermocline than in all other years. This increased gradient manifests itself in exceptionally high activity of salt fingering and relatively high buoyancy fluxes due salt and heat (25.9 and 15.7 10^{-10}Wkg^{-1} respectively).

B: Spring - April to July; Fig. 5b. During these months the seasonal thermocline is gradually building up but it is not yet fully developed. The values of Tu are indicative of WSF at the surface layer (i.e. above 150 m). Homogeneity of the water column above 300 m within the eddy gives Tu value of 0 while Tu values increase below 300 m. Note that in April 1989 there is an exceptionally well developed seasonal thermocline (see DT and DS in Table 2) while in other years the seasonal thermocline reached a similar stage of development only about two months later. This is the reason for the high values of Tu observed in the seasonal thermocline. Fluxes both in the seasonal and permanent thermoclines are of the same order of magnitude.

C: Summer - August to November 91; Fig 5c. The seasonal thermocline is at it's peak. The eddy is well defined as a homogeneous lens with a stable value of Tu = 0 throughout. SSF can be seen in the upper part of the eddy (near the seasonal thermocline) where the temperature gradient is the strongest. The buoyancy fluxes through the well developed seasonal thermocline are an order of magnitude larger than those in the permanent thermocline.

As expected, the mixed layer, from the surface to the seasonal thermocline is thicker and the area of high buoyancy fluxes penetrates deeper.

These Tu profiles and buoyancy fluxes indicate the presence of salt fingering during the entire year. However, during spring and winter, the strong salt fingering and buoyancy fluxes are located in the lower part of the eddy so the dominant buoyancy flux by double diffusion into the eddy is from the underlying deep water. During summer, on the other hand, the strong salt fingering is located in the upper part of the eddy where the differences in temperature and salinity are highest. Therefore the dominant buoyancy flux by double diffusion into the eddy during this season is from the warm and salty surface layer lying above the seasonal thermocline.

3.2 Chemical Evidence for Cross Isopycnal Mixing

The sources of oxygen in the eddy are the atmosphere and photosynthesis in the euphotic zone while the source of nutrients (nitrate and phosphate) is the deep water where no boilogical consumption of these nutrients takes place [*Broecker and Peng*, 1982]. We therefore expect that during winter, when the eddy is exposed to the atmosphere and the water is well mixed, high concentration of oxygen and low concentration of nutrients will be encountered. During summer on the other hand, when the eddy is underlying the euphotic zone and cut off from any contact with the atmosphere, a decrease in the concentration of oxygen and an increase in nutrients is expected due to respiration and mixing with the surrounding water.

The concentration of oxygen vs. nitrate (phosphate concentrations are too low and noisy to work with) below the euphotic zone (120 to 550 m) is shown in Fig. 6 for the month of February 1989 and in Fig. 7 for November 1989 along with a schematic representation of the density structures during these seasons. In the former the eddy has been directly exposed to the atmosphere for about 2 months while the latter depicts the relationship after 9 month of complete insulation of the eddy from its atmospheric and biological (below the euphotic zone) oxygen source.

Both Figs. 6a and 7a show the geometric schematic description of the eddy and the location of the samples inside it. All data reported here are those collected between the two dashed line. The eddy itself (given by the depth of the 29.09 isopycnal) was not necessarily above the lower dashed line. The points marked E are taken from the eddy (defined as the water between the 28.8 and 29.09 isopycnals and between 150 - 450 m depths). Those marked O are from the surrounding water (depth: 150 - 550 m, σ_t>29.09) and those marked B are from within the eddy near its lower boundary (450 - 550 m, σ_t<29.09).

During winter (Fig. 6b) the deep water (points marked O) have the expected composition - low oxygen due to the permanent thermocline insulating the water from the atmosphere and high nitrate due to lack of biological consumption. These points are clustered in the box at the lower right corner of Fig. 6b. By comparison the water in the main body of the eddy (points marked E) have the typical composition of surface water - high oxygen and low nitrate. These points clustered together in the circle located at the upper left corner of Fig. 6b. All the points taken in the lower boundary of the eddy (points marked B) have an intermediate composition as expected in a diffusive boundary layer. It is important to note that quite a few points from the main body of the eddy have this intermediate composition (points marked E in the center of Fig. 6b). The presence of this intermediate composition at sufficiently shallow levels within the eddy (at least 100 m above the lower boundary of the eddy) indicates the intensive fluxes through the permanent thermocline into the eddy. These fluxes from the deep water are strong enough not to be completey masked by downward convection of the surface water. In fact, as many as 30% of the samples taken in the eddy (both E and B points in the center of Fig. 6b) have a deep-water signature in their composition.

In summer the eddy is insulated from the surface water by the seasonal thermocline (Fig. 7a) so that an oxygen depletion and nitrate increase is expected to occurs in the eddy. This expected pattern is indeed observed in the circled cluster of data in the upper left corner of Figs. 6b, 7b (i.e. the data with the strongest surface signal) which have shifted down (lower oxygen) and to the right (increasing nitrate).

By contrast the data with intermediate composition show an entirely opposite pattern: Nearly all E, B points in the center of Fig. 6b are altogether absent from Fig. 7b; The highest nitrate value within the eddy is 3 μM/kg in summer (Fig. 7b) as compered with 4-5 μM/kg in winter (E points in the lower right squer are of Fig. 6b). The only way for this nitrate depletion (and increasing in oxygen) to have occurred well below the euphotic zone (i.e.) at depth where no photosynthesis can consumes the nitrate) is by downward mixing of high nitrate/low oxygen surface water through the seasonal thermocline. The permanent thermocline is a yearlong permanent feature so that the flux of high nitrate, deep water into the eddy through it does not vanish in summer. Therefore the downward mixing of high oxygen surface water through the seasonal thermocline should more than compensate for the flux of nitrate rich water through the peramanent thermocline. The points marked C (depth 120 - 150 m) in Fig. 7b have very high oxygen and

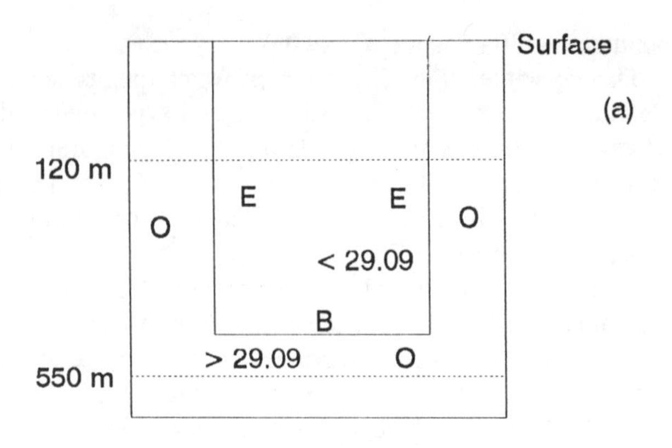

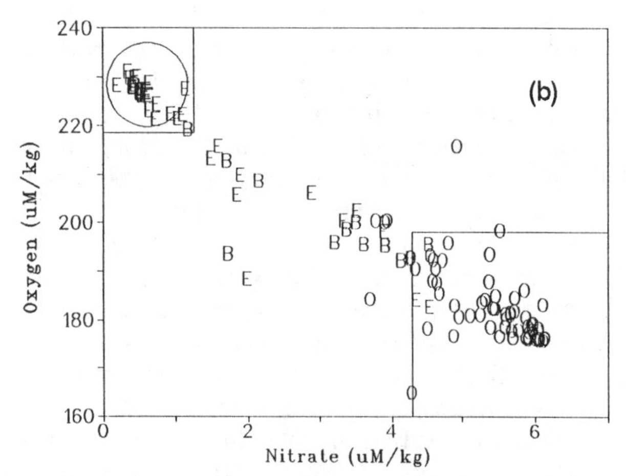

Figure 6: (a) A schematic sketch of the eddy in the winter. The two dotted lines indicate depth of sampling, letter E indicates samples taken from the main eddy water B samples from the lower boundary O samples taken from the surrounding and deep waters (which have identical chemical composition). (b) 30% of the samples taken within the eddy (E + B) fall in the area between the two squares and have deep water signature, indicating mixing with the underlying water. The large square on the right bottom contains deep water samples- high NO_3 , low O_2 . The upper left square contains typical eddy samples- high O_2 low NO_3 . The circle inside is described is for comparison with the summer cluster in Fig. 7b.

low nitrate values which characterizes the surface water further indicating downward fluxes through the seasonal thermocline.

Naturally a quantitative comparison between the fluxes calculated from all physical and chemical data (heat, salt oxygen and nitrate) would be of interest at this point. However, this is an ambitious undertaking beyond the scope the present work for several reasons; First the physical fluxes represent an instantaneous value for the particular time of observation while the fluxes based on changes in chemical concentration represent an average value for the time lapse between the observations. Second, flux estimation from changes in chemical composition demands a large number of assumptions and estimations: The rate of production/consumption of nitrate and oxygen by biological processes; constancy of the eddy's volume; charge balance due to diffusion, to name a few.

To summarize the implication of the chemical concentration data: Fluxes of high nitrate, low oxygen water into the eddy through the permanent thermocline at its bottom persist yearlong while oxygen rich, nitrate depleted surface water is mixed into the eddy through the seasonal thermocline in summer.

4. DISCUSSION

Regardless of the origin of the Cyprus Eddy, its persistence seems to pose an additional problem. During

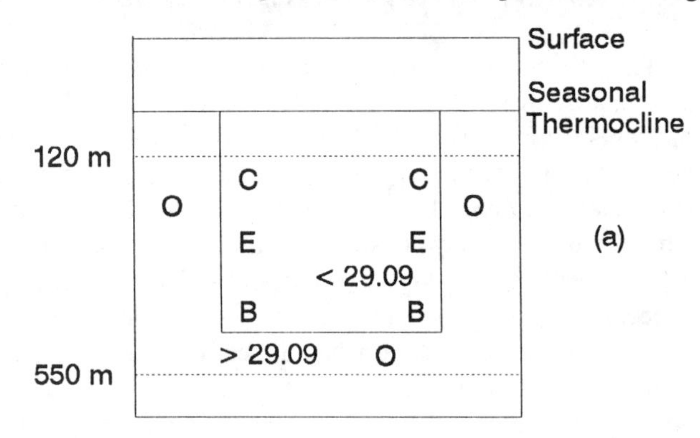

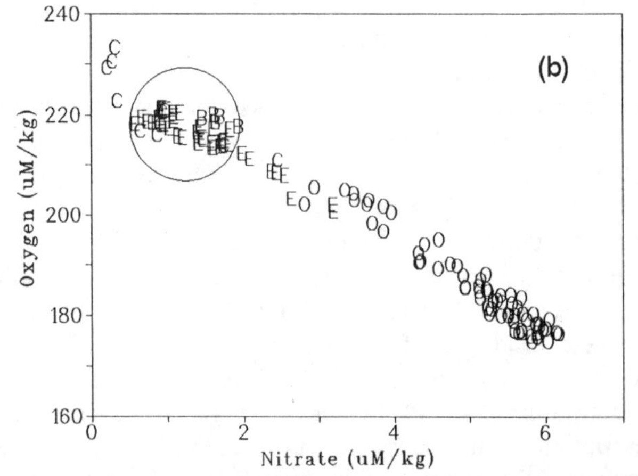

Fig. 7: (a) As figure 6a just for summer and with the additional C points indicating the upper part of the eddy near the boundary with the seasonal thermocline. (b) The samples with circle show typical behavior of a closed dark cell with respiration only and no photosynthesis- decreasing O_2 and increasing NO_3. The disappearance of points with intermediate values those between the squares of Fig. 7b. can only be explained by exchange with water in the euphotic zone.

summer, when the eddy is isolated between the seasonal and permanent thermoclines, we would expect the eddy to slowly exchange properties with the surrounding waters by turbulent mixing so its thermal, salinity and velocity signatures should weaken. This is not at all the case in our observations. To further substantiate this pattern we have also examined data from summers 1982 - 88 *[Berman, 1992]* with the same result - the distinctive temperature and salinity signatures of the eddy remained unchanged. This problem of "no-aging" of the eddy during summer is accentuated by the following observation of Brenner (1993); During summer only, potential vorticity (PV) surfaces are parallel to the density surfaces (σ_θ) which is a favorable condition for lateral mixing. Thus, if lateral mixing is to account for some of the fluxes into the eddy then we expect its effect to be most pronounced during summer- contrary to nearly perfect isolation of the eddy during this season. The oxygen- nitrate balance also supports the notion of the eddy not aging during the summer months. During winter there is a large fraction (30%) of eddy samples with intermediate values of oxygen and nitrate (between the squares in Fig. 6b). These points nearly vanish in summer (Fig.7b), whereas mixing with the surrounding water should increase this fraction. Convection is ruled out as a possible source of new water since the entire water column is gravitationally stable throughout the summer. One possible mechanism which might balance the loss of T and S signals by turbulent mixing is fluxes into the eddy of salt and heat by salt fingering.

Let us try and determine whether SF through the seasonal thermocline can account for the salt fluxes necessary to maintain the eddy's salinity signal. The volume of the eddy is:

$$\pi \times 50{,}000^2{}_{[m]} \times 400_{[m]} \approx 3.14 \times 10^{12}{}_{[m^3]} = 3.14 \times 10^{15}{}_{[lit]} \approx 3.14 \times 10^{15}{}_{[kg]}$$

Taking eddy's salinity to be 39 ppt its total salt content should be 1.2×10^{14} kg. From the data given in Table 2 and choosing a typical value of flux during summer through the seasonal thermocline, $g\beta F_s = 250 \times 10^{-10}$ W/kg, $g \approx 10$ m/s and $\beta \approx 1/1000$, the amount of salt entering into the eddy during the 9 month long summer (April to December) is:

$$250 \times 10^{-10}(1000/10) \times \pi \times 50{,}000^2 \times 3600 \times 24 \times 30 \times 9 \approx 4.6 \times 10^{11} \text{ kg/(9 month)}$$

This implies that about 0.4% ($4.6 \times 10^{11}/1.2 \times 10^{14}$) of the amount of salt stored within the eddy is transported into it through the seasonal thermocline in summer. Thus, the annual flux of salt due to SF is comparable to the salinity excess signal of the eddy (i.e. its access salinity over the surrounding water, 0.2ppt/40ppt 0.5%) and can easily explain the observed "no aging" of the eddy during the summer months!

One simple minded hypothesis for explaning the origin of the eddy is that it results from the collapse of the seasonal thermocline in early winter. While this collapse will indeed convect saline and warm surface water into greater depths it will take place everywhere in the Eastern Mediteranean and hence can not account for the observation of a permanent 100 km feature. In addition, this process can not convect lower density water into the eddy and will take place when the density of the surface water equals that of the eddy.

Although a definite determination of the origin of the eddy is beyond the scope of this work, it is appropriate to add a note regarding its location. The eddy is located above the deepest point in the basin where the water depth exceeds 2000 m (Fig. 1) while the eddy's own depth is about 500 m. This ratio is in line with the theory of Paldor and Nof (1990) who showed that when the ocean is more than four times deeper than an eddy, the eddy will be linearly stable. According to this theory, a shallower ocean will render the eddy unstable so that any small perturbation will grow exponentially with time causing dissipation of the eddy. The proximity of the eddy to the seamount suggests that the eddy might originate as a lee cyclone [*Pedlosky*, 1979] when the thermohaline, zonal, flow passes over the seamount. The problem with this scenario is that a lee cyclone in the northern hemisphere as its name indicates spins cyclonically while the Cyprus eddy spins anticyclonically. Thus, the region of the eddy is far from clear while its location might be related to the bottom topography.

5. CONCLUSION

1. The values of Tu and $R\rho$ are favorable to the existence of salt fingering fluxes into the eddy from both the overlying seasonal thermocline and from the underlying permanent thermocline . This is the first time that Salt Fingering has been observed in the Eastern Mediterranean. We expect salt fingering to occur in other places in the Mediterranean sea as well, especially during summer when large temperature differences are encountered.
2. Additional evidence for convection through the seasonal and the permanent thermoclines is provided by the changes in chemical composition. The values of nitrate and oxygen inside the eddy point to the fact that there must be an exchange of water between the eddy and the layers above and below.

3. Lateral mixing is found to have a negligible effect in diminishing the eddy's salinity signal even in summer when the coincidence of PV and σ_θ surface enables such mixing.

4. Salt fingering is a significant process in supplying salt into the eddy during summer by SF flux from the overlying salty surface water. We conjecture that this flux balances the loss of salt to the surrounding water by turbulent mixing.

Acknowledgments. We would like to acknowledge the continuing interest and advice of Dr. B. Lazar throughout all the period of this work. Thanks to Dr. N. Kress for providing the raw data used in Fig. 3, 6 and 7 .

REFERENCES

Berman, T., *Salt finger in a warm core eddy in the Eastern Mediterranean* (in Hebrew). M.Sc. Thesis, Hebrew University,Jerusalem, Israel, 1992.

Brenner, S., Z. Rozentrub, J. Bishop, M. Krom, The mixed layer/thermocline cycle of a persistent warm core eddy in the Eastern Mediterranean. *Dynam. Atmos. Oceans., 15,* 457- 476,1991.

Brenner, S., Long term evolution and dynamics of a persistent warm core eddy in the Eastern Mediterranean Sea. *Deep-Sea Res.* 40, 1193-1206, 1993.

Broecker, W.S., T.S.Peng, *Tracers in the sea.* 690 pp., A publication of Lamont-Doherty Observatory, Columbia University, Palisades, NY, 1982.

Feliks, Y., S. Itzikowitz, Movement and geographic distribution of anticyclonic eddies in the Eastern Levantine Basine. *Deep Sea Res.*, 34, 1499-1508, 1987.

Feliks, Y., Downwelling along the norhteast coast of the Eastern Mediterranean. *J. Phys. Oceanogr.*, 511-526, 1990.

Hebert, D., Estimates of salt-finger fluxes. *Deep-Sea Res.* 35, 1887-1901, 1988.

Hecht, A., N. Pinardi, A. R. Robinson, Currents, water masses, eddies and jets in the mediterranean levantine basin. *J. Phys. Ocean.,*18, 1320-1353, 1988.

Kelley, D., *Oceanic thermohaline staircases.* Ph.D. Thesis, Dalhousie University, Halifax, NS, 1986.

Krom, M. D., S. Brenner, N. Kress, A. Neori, L. I. Gordon, Nutrient dynamics and new production in a warm-core eddy from the Eastern Mediterranean Sea. *Deep Sea Res.*, 39, 467-480.

Kunze, E., Limits on growing, finite-length salt fingers: A Richardson number constraint. *J. Mar. Res.*, 45, 533 -556, 1987.

Ozsoy, E, A. Hecht, U. Unluata, Circulation and hydrography of the Levantine Basin. *Result of POEM coordinated experiments 1985-1986. Prog. Oceanography,* 22, 125 - 170, 1989.

Paldor, N., D. Nof, Linear Instability of an anticyclonic vortex in a two-layer ocean. *J. Geophys. Res.*, 95(c10), 18075-18079, 1990.

Pedlosky, J., *Geophysical fluid dynamics.* 624 pp., Springer - Verlag, N.Y., 1979.

Ruddick, B., A practical indicator of the water column to double diffusive activity. *Deep Sea Res.* , 30, 1105 - 1107, 1983.

Schmitt, R.W., Flux measurements at an interface. *J. Mar. Res.*, 37, 419-436, 1979.

Schmitt, R.W., Form the temperature-salinity relationship in the central water: Evidence for double-diffusive mixing. *J. Phys. Oceanogr.*, 11, 1015- 1026, 1981.

Turner, J. S., Salt fingers across a density interface. *Deep -Sea Res.*, 14, 599-611, 1967.

Turner, J. S., *Buoyancy effects in fluids.* 368 pp. Cambridge University Press, 1981.

Tal Berman and Nathan Paldor, Institute of Earth Sciences, The Hebrew University of Jerusalem, Jerusalem, Israel, 91904.

Stephen Brenner, Israel Oceanographic & Limnological Research, P.O. Box 8030, Haifa, Israel,.

Sources of Double Diffusive Convection and Impacts on Mixing in the Black Sea

Emin Özsoy and Şükrü Beşiktepe

Institute of Marine Sciences, Middle East Technical University, Erdemli, Icel, Turkey

The Black Sea is a uniquely stratified environment supporting double diffusive convection in most of its interior. Originating from the Bosphorus, and modified by entrainment of Cold Intermediate Water on the continental shelf, the inflowing Mediterranean Water drives double diffusive intrusions penetrating horizontally into the interior of the Black Sea from the continental slope. Intrusions aided by the inherent double diffusive instability of the ambient stratification drive a vertical circulation, creating ventilation across the halocline. The intrusions are also significant in transporting shelf-derived materials into the interior of the basin. Various scales of convection are indicated, with increasing amplitude near the basin boundaries. The relatively smaller scale disturbances are confined above the halocline, possibly driven by atmospheric forcing and buoyancy inputs. Double diffusive intrusions occur at or below the halocline, up to a depth of $\sim 500m$, where a peculiar zone of vanishing temperature gradient exists. A bottom convection layer occupying several hundred meters near the bottom is driven by geothermal heat fluxes. The slow but efficient convection in this layer homogenizes the water properties across the basin. The transports of heat and salt upwards from the bottom convective layer are most likely determined by double diffusive fluxes.

1. INTRODUCTION

With a maximum depth of $\sim 2200m$, a surface area of $4.2 \times 10^5\ km^2$ and a volume of $5.3 \times 10^5\ km^3$, the Black Sea is a unique marine environment, representing the largest land-locked basin in the world (Figure 1). Its positive water budget, with total freshwater input largely in excess of evaporation [*Ünlüata et al.*, 1990], is balanced by the two-way exchange through the Turkish Straits System (the Bosphorus, Dardanelles Straits and the Sea of Marmara). Almost complete isolation from the world ocean, except for the exchanges controlled by a shallow (60 m) sill, has resulted in a strong density stratification to be developed in the Black Sea.

As a result, the Black Sea deep waters have become stagnant, with renewal time scales of a few thousand years [*e.g., Boudreau and Leblond*, 1989; *Östlund*, 1974; *Grasshoff*, 1975; *Tolmazin*, 1985]. The basin is almost completely anoxic, containing oxygen in the upper $\sim 150m$ depth (13% of the sea volume) and hydrogen sulphide in its deep waters. A permanent halocline separates the oxic and anoxic waters. On the other hand, the ventilation of the deeper layers and the structure of the halocline essentially depend on the inflow of Mediterranean water through the Bosphorus.

It is important to understand processes contributing to pycnocline and deep mixing, and the three-dimensional transport of materials in the Black Sea, because it is being seriously threatened by environmental degradation [*e.g., Mee*, 1992] in recent years, as a result of man's activities around this enclosed region. Very little is known of the relative roles of the various trans-

Double-Diffusive Convection
Geophysical Monograph 94

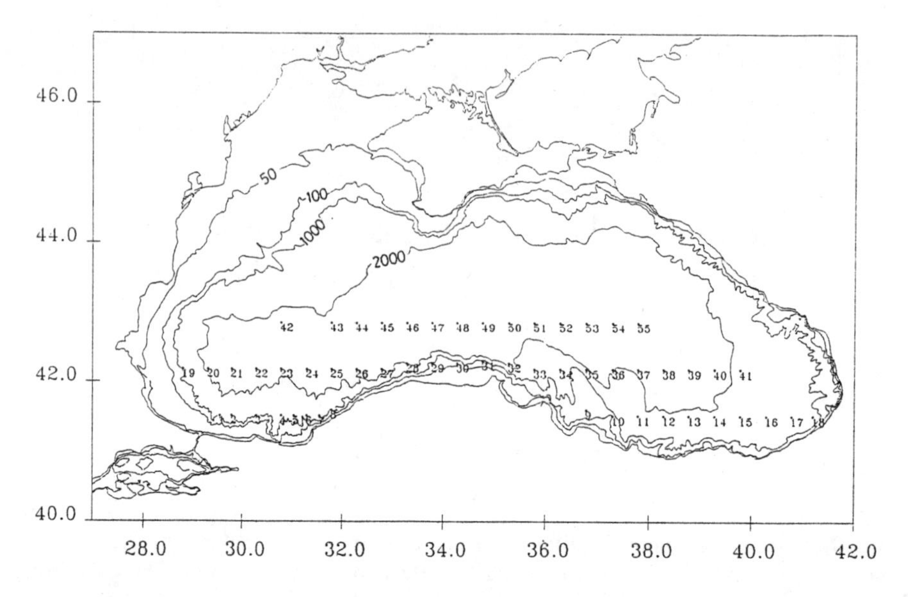

Fig. 1. The regional map and bottom topography of the Black Sea. Also shown are the locations of the three east-west transects for which station data are displayed in Figures 6, 7, 8 and 10.

port and ventilation mechanisms. On the other hand, the Black Sea is one of the primary areas of the world ocean where small-scale processes such as double diffusive convection can have a significant impact on the large scale properties of a system.

Although the present review derives largely from our earlier publications [*Özsoy et al.*, 1991; *Murray et al.*, 1991; *Latif et al.*, 1991; *Özsoy et al.*, 1993]), further discussion has been added, based on some recent observations.

2. HYDROGRAPHIC DATA

Our analyses are largely based on hydrographic data of high resolution and accuracy, collected by RV BİLİM of the Institute of Marine Sciences, Middle East Technical University during 1986-1993, and by RV KNORR during 1988 [*Murray et al.*, 1991; *Özsoy et al.*, 1993]. The temperature and salinity measurements were obtained with Seabird 9/11 CTD profilers in all cruises. In 1988 the KNORR, and in July 1992 the BİLİM collected light transmission data with a Sea-Tech transmissometer with a $25 cm$ optical path, interfaced with the CTD systems. After 1990, the data have been obtained collaboratively with the riparian countries of the Black Sea, in an international, interdisciplinary program presently known as CoMSBlack (Cooperative Marine Science Program for the Black Sea) [*Ünlüata et al.*, 1993]. Although these data cover the entire basin with a nominal horizontal resolution of 30 km, we only use data from BİLİM, because only these data were suitable for studies of fine structure. However, the analysis showing the depth distribution of the bottom diffusive interface in Figure 13 have been constructed using the entire data set.

3. DOUBLE DIFFUSIVE INTRUSIONS

3.1. *Shelf Modification of the Mediterranean Water*

The cross-shelf spreading of the Mediterranean Water into the Black Sea has been described to some extent [*Tolmazin*, 1985; *Yüce*, 1990], including a more complete, recent description [*Latif et al.*, 1991]. The warm, saline Mediterranean Water enters the Black Sea through the lower layer of the Bosphorus, overflowing a sill and following a bottom channel till the middle of the wide shelf region. From then on, the Mediterranean Water spreads out to form a thin sheet of warm and saline anomalous bottom water on the shelf [*Latif et al.*, 1991]. The overlying Cold Intermediate Water (CIW) is a product of winter convection (temperature less than $\sim 8°C$) in the Black Sea. The intruding Mediterranean Water is rapidly mixed (Figures 2 and 3), and diluted by a factor of 3-6 by entraining the CIW on the shelf. As a result, the shelf-mixed water becomes relatively colder, yet remains more saline than the Black Sea waters off the shelf edge [*Özsoy et al.*, 1993]. The cold anomaly of the shelf-mixed water (relative to the interior water

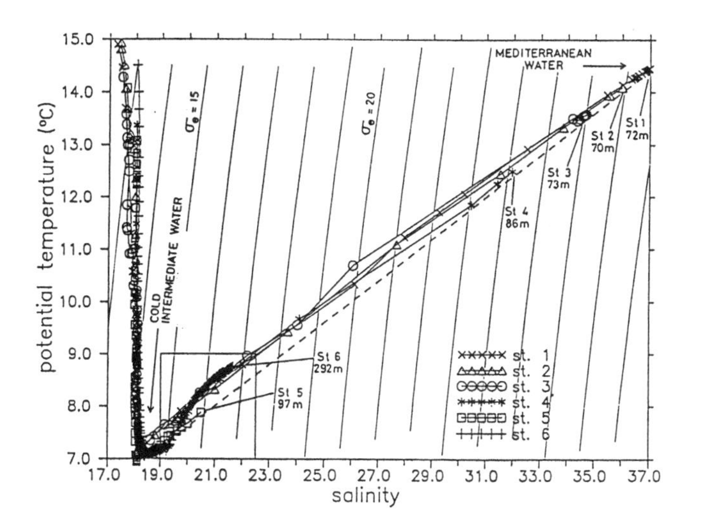

Fig. 2. Evolution of temperature - salinity across the shelf. Stations 1-5 extend from the Bosphorus to the shelf break. Station 6 is a deeper station immediately offshore. The dashed line models the changes in the 'Mediterranean effluent' at the bottom. At the shelf break (station 5), the modified bottom waters are colder than the waters at comparable depths of the continental slope (station 6). After Özsoy *et al.* (1993).

mass at the same depths) is then used to identify its intrusion into the interior at intermediate depth.

3.2. *Intermediate Depth Intrusions and Double Diffusive Instability*

The introduction at the shelf edge of shelf-mixed (cold, saline) water into a doubly stratified interior immediately leads to convection along the southwest margins of the Black Sea [*Özsoy et al.*, 1993]. The shelf-mixed water sinking along the continental slope drives double diffusive convection with a series of intermediate depth, horizontally spreading intrusions, as schematized in Figure 3. In the potential temperature versus salinity ($\theta - S$) diagram (Figure 4), the intrusions are identified first in the form of a cold sheet of water on the continental slope (dashed lines), then as discrete layers of anomalous characteristics (solid lines) spreading into the interior. The intrusions are characterised by a series of cold anomalies in the 100-500m depth range.

Two-dimensional effects created by a buoyancy source located on the lateral boundary of a stratified or a doubly stratified fluid environment often generates similar

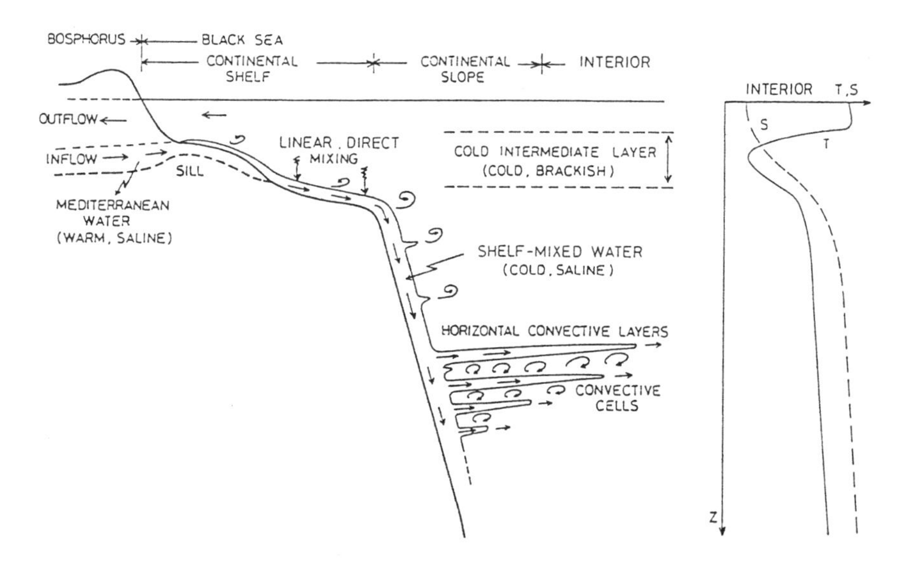

Fig. 3. Schematization of the boundary mixing processes driven by the Mediterranean Water issuing from the Bosphorus. Linear, direct mixing occurs on the shelf region and on part of the slope. At intermediate depths, double diffusive instabilities are generated due to the temperature and salinity contrasts of the intrusions and the potential instability of the interior. After Özsoy *et al.* (1993).

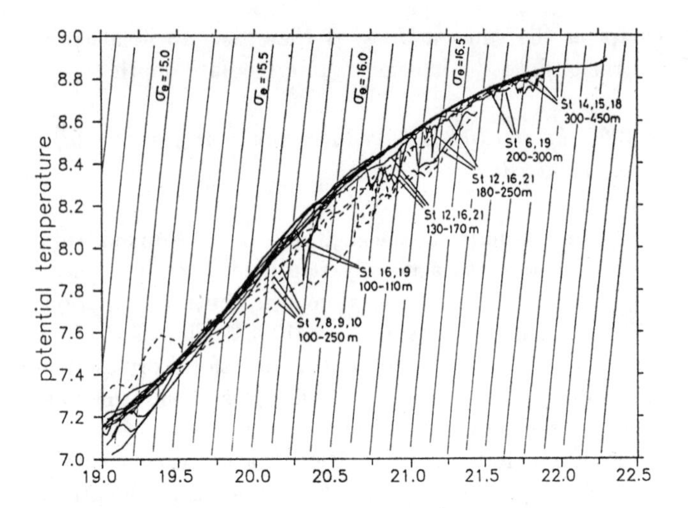

Fig. 4. The potential temperature - salinity relationship for stations near the southwestern shelf of the Black Sea. Dashed lines represent stations closest to the continental slope, *i.e.* within the boundary layer. The intrusive features at other stations offshore of the shelf region occur in the form of discrete layers spreading into the interior. After Özsoy *et al.* (1993).

convection patterns [*e.g.*, *Turner*, 1973, 1978; *Huppert and Turner*, 1980].

The essential element in the case of the Black Sea is a source of buoyancy with two diffusive components introduced into doubly stratified water which is potentially unstable. The vertical profile of Turner angle $Tu = tan^{-1}\{(1 + R_\rho)/(1 - R_\rho)\}$ [*Ruddick*, 1983], and density ratio $R_\rho = (\beta\partial S/\partial z)/(\alpha\partial T/\partial z)$ (where z is depth and α and β are the respective expansion coefficients for temperature, T, and salinity, S) plotted in Figure 5 indicates that almost the entire depth of the Black Sea is stratified in the double diffusive sense and in the diffusive range, a fact which can considerably increase the mixing efficiency of the intrusions [*Turner*, 1978]. (Note that the density ratio is defined following the convention used in the diffusive regime. In some references [*e.g.*, *Ruddick*, 1983], as well as in our earlier work [*Özsoy et al.*, 1993], it has been defined in the salt fingering sense, which is the inverse of the present definition. The inverse definition is easier to plot in the case of the Black Sea, because $R_\rho \to \infty$ at mid-depth in Figure 5a.)

It is expected that a lateral buoyancy source located in a stratified medium (characterized with two diffusive properties of either the source or the ambient fluids) can create complicated convection patterns. These effects are best illustrated in experiments on sidewall heating or cooling applied to a salinity gradient [*e.g.*, *Turner*, 1978; *Huppert and Turner*, 1980; *Tanny and Tsinober*, 1988; *Jeevaraj and Imberger*, 1991]. Accordingly, the critical Rayleigh number for layered convection is $Ra_c = 1.5 \times 10^4$. Computation of the Rayleigh number $Ra \equiv g\alpha\Delta T\eta^3/\nu\kappa_h$ (g being the gravity, ΔT the temperature difference at the sidewall, ν the kinematic viscosity, κ_h the thermal diffusivity, and $\eta \equiv \alpha\Delta T/\phi_0$, with $\phi_0 = -\beta dS/dz$, a length scale proportional to the vertical spacing of the convecting layers) for typical parameters of the Black Sea yields $Ra \simeq 10^{11} - 10^{13}$, exceeding the critical values of $Ra = Ra_c$ for convection. The estimated layer thicknesses of 20 - 40 m [*Özsoy et al.*, 1993] were also consistent with the observations.

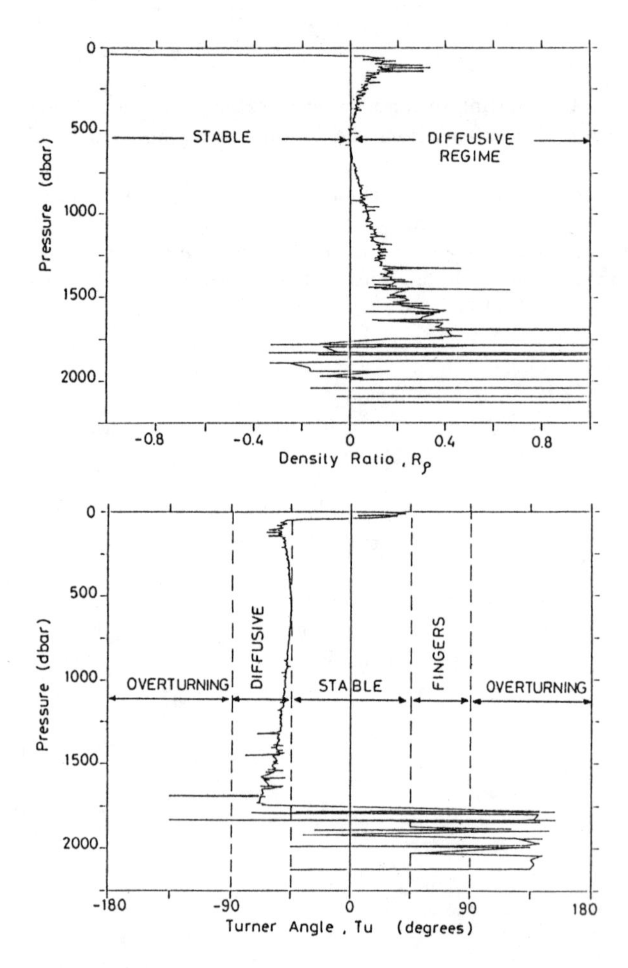

Fig. 5. The average stratification parameters computed from an ensemble of Black Sea deep water profiles: (a) the inverse density ratio $R_\rho^{-1} = \{(\beta dS/dz)/(\alpha dT/dz)\}^{-1}$, and (b) the Turner angle $Tu = tan^{-1}\{(1+R_\rho)/(1-R_\rho)\}$, where α and β are coefficients of expansion for temperature and salinity. The ranges for stable, statically unstable, and double diffusively unstable regimes are indicated. After Özsoy *et al.* (1993).

Short term variability and intermittency were found to be basic features of the intrusions. Part of the intermittency and filamentation was interpreted to be a result of the interaction of the sinking, dense, shelf-mixed water with ambient currents and the major topographic features in the region, such as Sakarya Canyon [*Özsoy et al.*, 1993; *Sur et al.*, 1994].

Based on data sets from a number of different cruises, it was shown [*Özsoy et al.*, 1993] that the anomalies could be traced back to a main source region in the southwest margin of the Black Sea, although much diluted imprints, in the form of smaller amplitude cold temperature anomalies and suspended material concentrations, can be found further east along the Anatolian coast [*Kempe et al.*, 1991].

More recent surveys covering the entire basin consistently indicate the southwestern shelf region to be the source for the intermediate depth intrusions. Figure 6 shows a series of temperature profiles in the southern Black Sea along 41°30'N, for repeated surveys of September 1990, September 1991 and July 1992. Note that in all of the cases, the intrusions with relatively larger cold anomalies are detected in the southwest, in the vicinity of the Bosphorus. The Turner Angle computed for the July 1992 profiles (Figure 7) show rapid fluctuations in the southwestern Black Sea, caused by alternating diffusive and fingering interfaces of the intrusions.

Along 42°10'N (Figure 8), only slightly north of the section in Figure 6, intermediate depth cold anomalies (at depths of 100-500m) are less prominent, but it can be verified that the cold anomalies have moved further east following the coast along the main direction of cyclonic circulation.

3.3. *Material Transport from the Continental Shelf and Slope Regions*

Intermediate depth inorganic particulate maxima near the basin boundaries are well known in the Black Sea [*e.g., Brewer and Spencer*, 1974; *Spencer et al.*, 1972]. Earlier sediment-trap measurements in the open waters of southwestern Black Sea have also identified large quantities of shelf-derived materials reaching their locations [*İzdar et al.*, 1986; *Hay, 1987; Honjo et al.*, 1987; and *Kempe et al.*, 1991].

A significant part of the transport of materials from the continental shelf and slope region into the interior is driven by the horizontally spreading intrusions. The most direct evidence of such transport is given by light transmission measurements [*Özsoy et al.*, 1993]. The perfect coincidence of seawater, particulate and nutrient anomalies [*Codispoti et al.*, 1992; *Özsoy et al.*, 1993], such as shown in Figure 9, indicate a common source of the materials. This pattern of transport, derived from the shelf and occurring across the halocline, is verified by independent measurements of Iron, Manganese and Chernobyl radiotracers [*Buesseler et al.*, 1991].

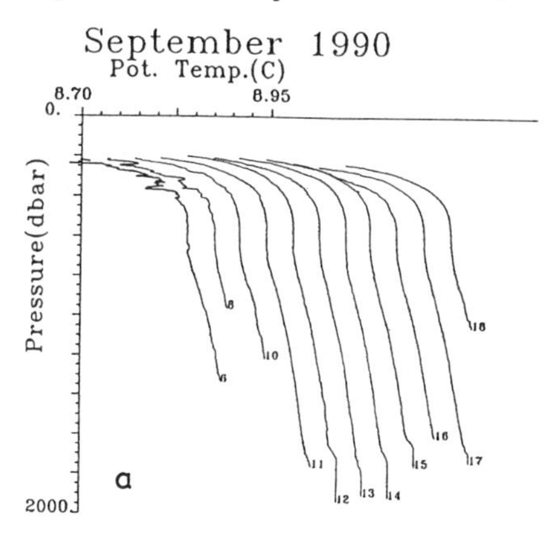

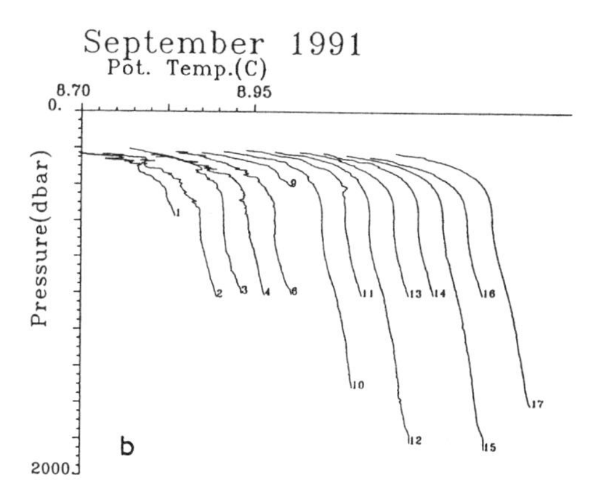

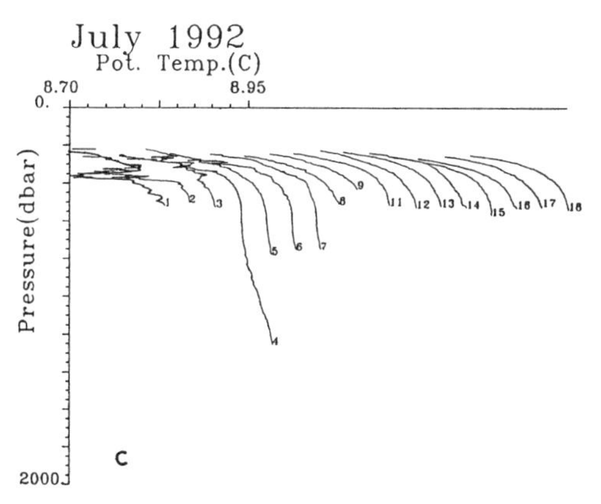

Fig. 6. Temperature profiles along 41° 30' N (Figure 1) in (a) September 1990, (b) September 1991, and (c) July 1992. Consecutive profiles are offset by 0.03°C in temperature units.

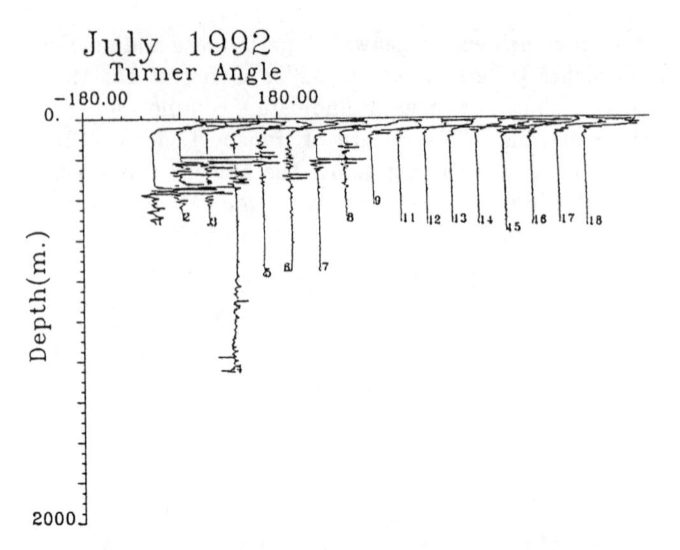

Fig. 7. Computed Turner angle (Tu) profiles along 41° 30' N (Figure 1) in July 1992. Consecutive profiles are offset by 90° in Turner angle units. In the western part of the transect, intrusions with alternating interfaces are identified by Tu fluctuating between the ranges of $-90° < Tu < -45°$ (diffusive regime) and $45° < Tu < 90°$ (salt finger regime) of double diffusive instabilities (Ruddick, 1983).

Light transmission measurements, repeated during the July 1992 survey (Figure 10), showed intermediate depth particulate maxima covering a large area of the basin. The peaks were prominent in the southwest Black Sea (Figure 10a), but also reached the central parts of the southern coast (Figure 10b) where the transect of profiles approaches the coast.

4. BOTTOM CONVECTION

Deep CTD casts in the Black Sea indicate constant temperature and salinity in a 300-400m thick layer above the bottom (Figures 6, 8 and 11). The perfect vertical and horizontal homogeneity of the properties across the basin within this layer suggest homogenization by convectively driven motions, believed to be formed by geothermal heating working against the existing gravitationally stable gradient of salinity [*Özsoy et al.*, 1991; *Murray et al.*, 1991]. Even the low level of geothermal heat fluxes in the Black Sea ($H \simeq 40\ mW\ m^{-2} = 0.9\ \mu cal\ cm^{-2}s^{-1}$), as compared with the relatively higher values in some neighboring seas [*Zolotarev et al.*, 1979; *Haenel*, 1979], are sufficient to drive convective motions in the otherwise stagnant waters of the deep Black Sea.

The thickness of the convective layer varies across the basin (Figure 12), possibly in response to the variable geothermal heat flux, modalities of the convection or other interior balances.

Laboratory convective layers under similar conditions of a salinity gradient heated from below have been studied extensively [*e.g., Turner*, 1968; *Huppert and Linden*; 1979, *Fernando*, 1987], and many oceanographic and

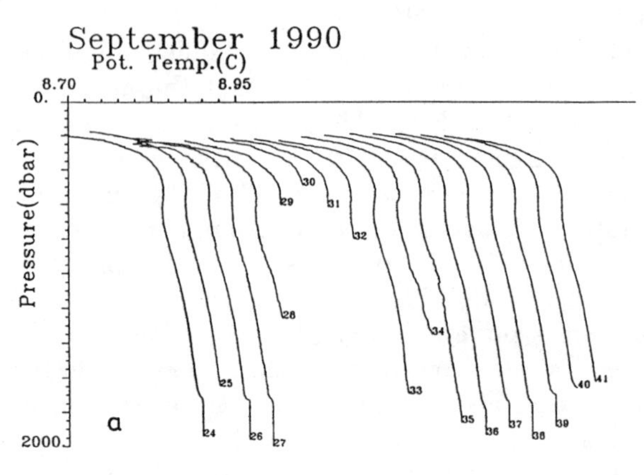

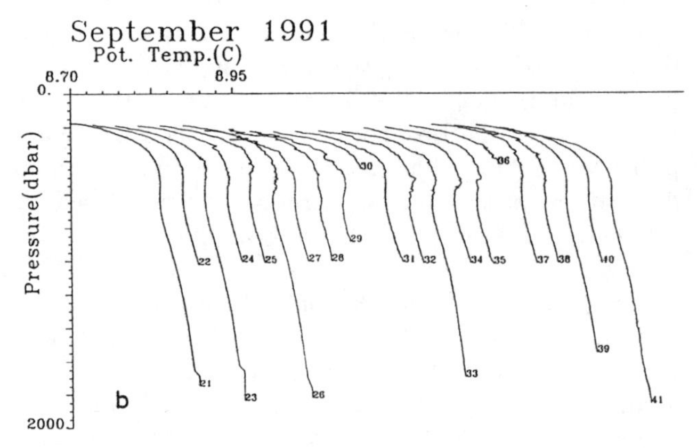

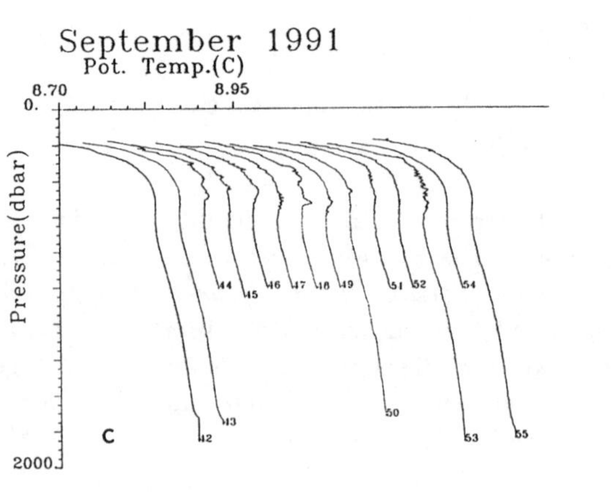

Fig. 8. Temperature profiles (a) along 42° 10' N in September 1990 and (b) along 42° 10' N in September 1991, and (c) along 42° 50' N in September 1991. Consecutive profiles are offset by 0.03°C temperature units.

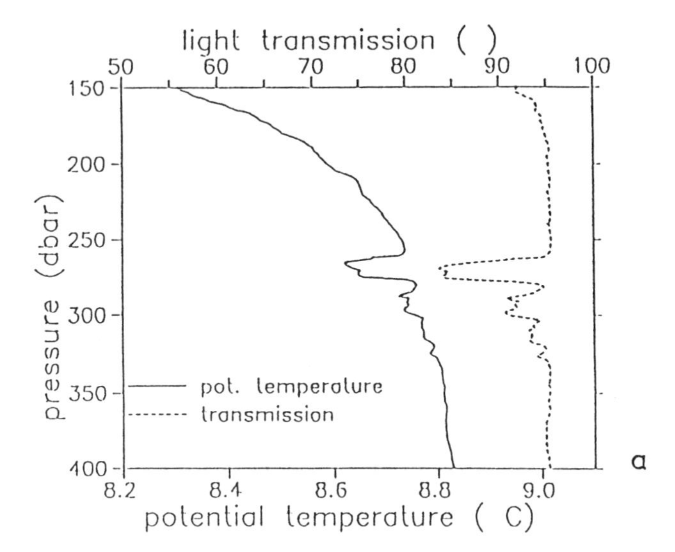

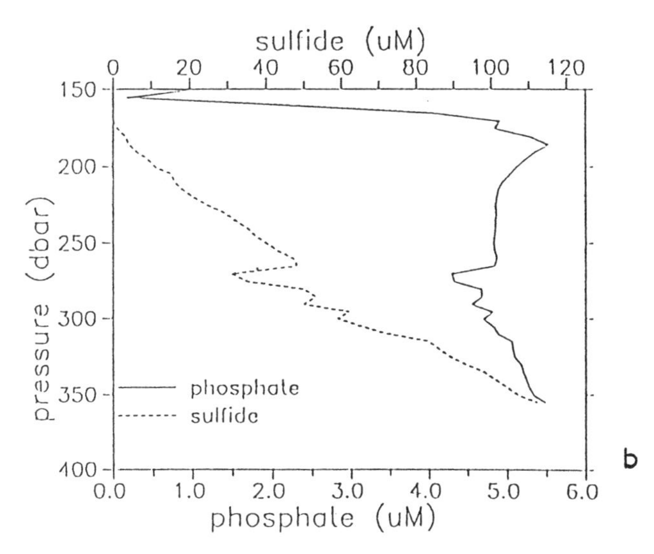

Fig. 9. (a) Potential temperature (solid line), light transmission (dashed line), (b) phosphate (solid line) and sulfide (dashed line) profiles in the southwestern Black Sea. The intrusions (like the discrete layers between 260-330 m depths) advect the water properties modified on the shelf and the continental slope, into the interior. Because the intrusions penetrate below the pycnocline (or the oxycline), they contribute to the mixing across the halocline. After Özsoy *et al.* (1993).

limnological examples are known [*Turner*, 1969, *Fernando*, 1989]; yet the available theory is far from fully explaining the time evolution of the convective layer in the Black Sea.

Based on the earlier results [*Turner*, 1968; *Huppert and Linden*, 1979], and for given values of geothermal heating, the time required for the formation of a layer with the present thickness is estimated to be on the order of 34 *years* or less. On the other hand, the maximum thickness at which the diffusive interface would split into a second convective layer is estimated to be 43 m, that would occur in less than a year after the initiation of the heat flux [Özsoy *et al.*, 1991]. Since there is no reason to suspect such rapid changes in the system, the age of the convective layer is expected to be on the same order as the deep water itself. (The convective layer is the bottom layer with uniform properties below $\sim 1700m$; the deep water is the water below $\sim 500m$). On the other hand, these comparisons indicate that there are large discrepancies between the predictions and the observations.

The observed structure can not be scaled with laboratory experiments, except with those in a limiting case [*Fernando*, 1987]. Recent experiments and a survey of oceanographic applications [*Fernando*, 1987, 1989] have indicated the possibility of different regimes in the development of the convective layer. In the 'low stability' regime, the eddies from the convection region initially penetrate into the interfacial layer, preventing its splitting into many layers; however, the growth of the layer slows down in the long time limit, when the kinetic energy of the eddies can no longer overcome the potential energy of the buoyancy jump across the interface. In particular, the layer thickness initially increases as $h \sim t^{1/2}$, until a critical thickness $h_c = C(HN^{-3})^{1/2} = cH_*^{1/2}S_*^{-3/4}$ is reached at time $t_c \simeq 860N^{-1}$ (where $N^2 = 2S_*$, $S_* = -(1/2)g\beta(dS/dz)$, $H_* = -g\alpha H/\rho c_p$, where H is the bottom heat flux, ρ and c_p are respectively the density and specific heat of sea water). It is found that $h_c = 20m$ and $t_c \simeq 1$ month in the case of the Black Sea [*Özsoy et al.*, 1991; *Murray et al.*, 1991].

The existence of a single mixed layer much thicker than h_c suggests that the Black Sea bottom convection is in the long time limit of the 'low stability' regime. By comparing the calculated value of the Rayleigh Number $Ra_h = g\alpha\Delta T h_s^3/\kappa_h\nu = 4\times10^{15}$ (ΔT is the temperature difference across the interface, and h_s is the convective layer thickness) versus the density ratio $R_\rho \simeq 2$ at the interface, and making use of the classification of various regimes [*Fernando*, 1989], we verify that the Black Sea interface is potentially in the 'low stability' regime. The fact that the effects of eddy entrainment have become vanishingly small in the long time limit can be verified by comparing the interface Richardson Number, $Ri_\star = \Delta b h w_\star^{-2}$ (where $\Delta b \equiv g\alpha\Delta T - g\beta\Delta S$ is the buoyancy jump defined by the temperature and salinity differences ΔT and ΔS at the interface, $w_\star$ is the *r.m.s.*

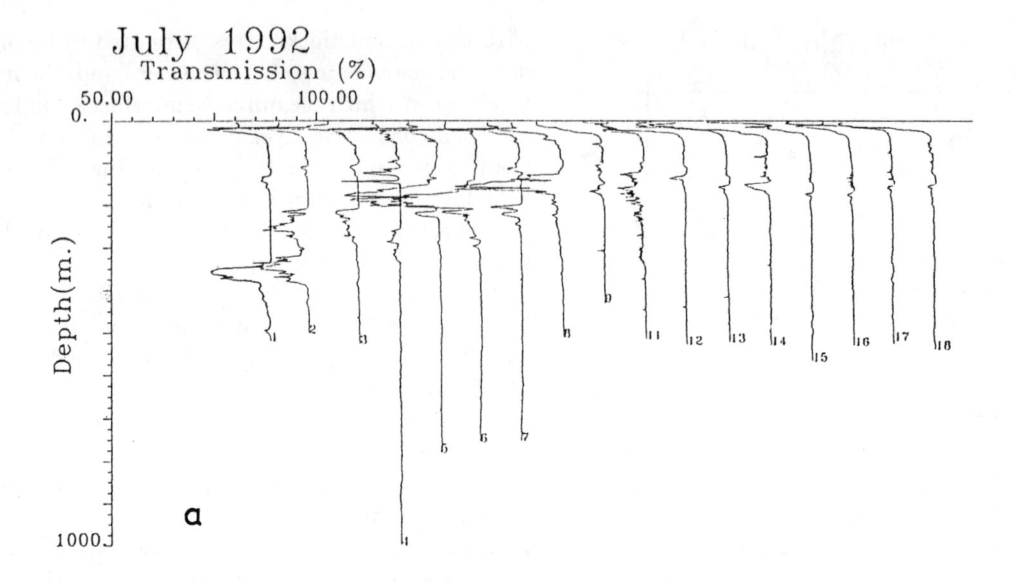

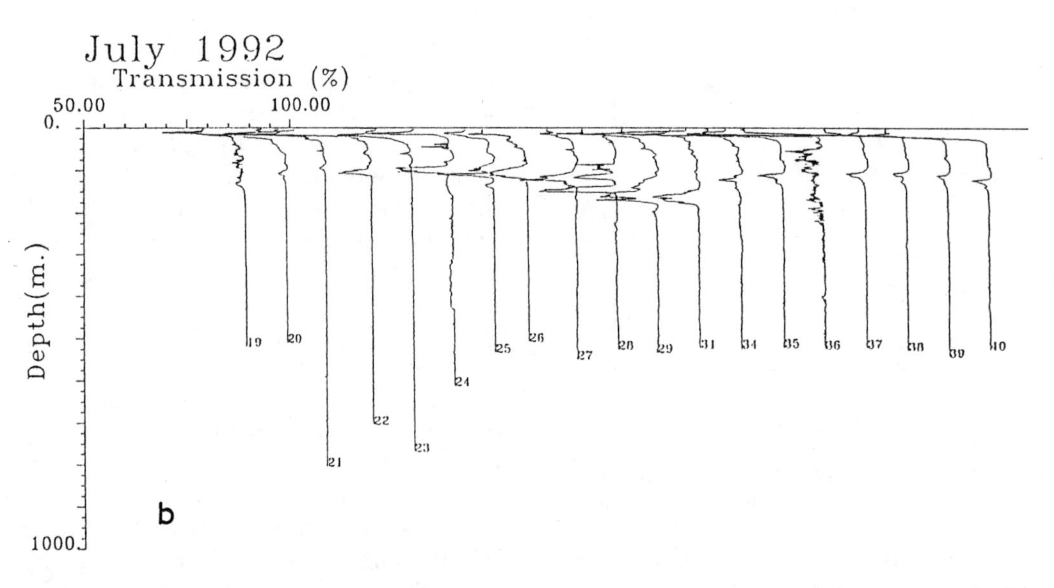

Fig. 10. Light transmission profiles along (a) 41° 30' N and (b) 42° 10' N (Figure 1) in September 1990. Consecutive profiles are offset by 10 % in light transmission units.

convective velocity), with the values reported for other cases [*Fernando*, 1989]. For $Ri_\star > 240$, the entrainment is expected to become vanishingly small; this is certainly true in the Black Sea, since the interface Richardson Number has a value of $Ri_\star \simeq 600$, the largest reported so far. This is not surprising, because the Black Sea also has the largest convective layer thickness among known examples in the world.

On the other hand, another criterion was found [*Fernando and Ching*, 1991] to differentiate between the cases with multiple splitting of the diffusive interface and a single layer development. According to this scheme, if $R_\rho > \tau^{1/2}$ ($\tau = \kappa_s/\kappa_h$ is the ratio of salt / heat diffusivities), the convection would continue to grow as a single layer because the interface would be a non-entraining (or detraining) type. With τ=0.01, and $R_\rho = 2$ at the diffusive interface, this criterion is satisfied in the Black Sea. Experiments indicate that, in this regime, the convective layer growth almost comes to a stop after the depth h_c is reached; *i.e.* the thickness of

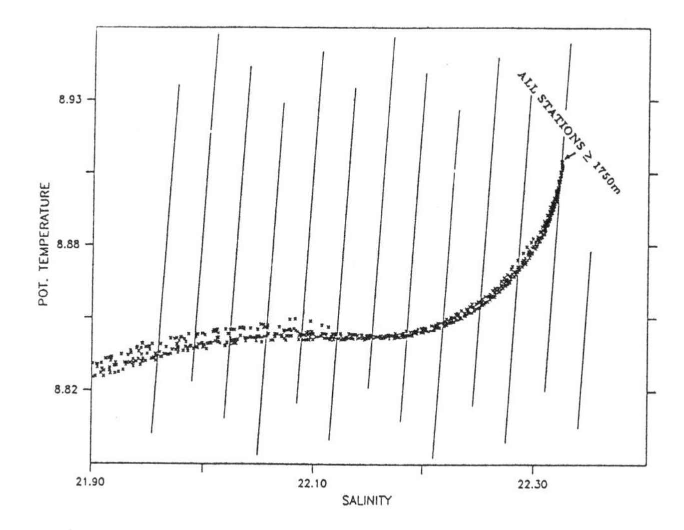

Fig. 11. The potential temperature - salinity relationship for the deep waters of the Black Sea. All the data with uniform properties within the bottom convective layer converge to a single point in potential temperature - salinity space.

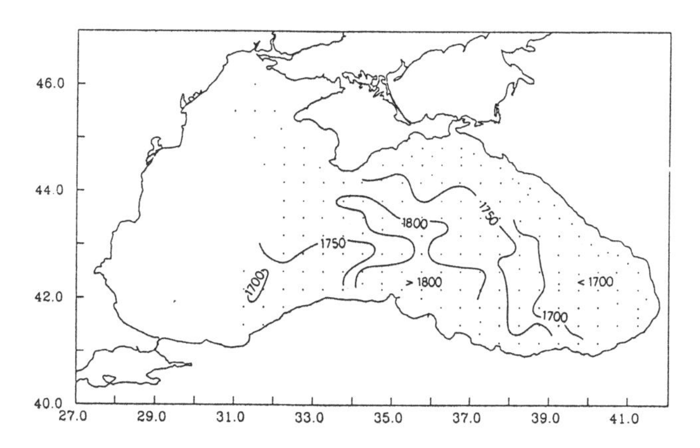

Fig. 12. The depth (m) of the diffusive interface bounding the bottom convective layer in September 1990.

the convective layer increases at a much slower rate as compared to the earlier regime. This is why we believe that the age of the Black Sea convective layer (time for the layer to reach the present thickness) must be comparable to the mean residence time of 1000 - 2000 years reported for the deep waters of the Black Sea below a depth of $\sim 500m$.

As for the thermal part of the buoyancy flux, q_h, across the diffusive interface, the [*Huppert and Linden*, 1971] formula $q_h = c_6(\kappa_h^2\nu^{-1})^{1/3}(g\alpha\Delta T)^{4/3}R_\rho^2$ yields $q_h = 0.61 \times 10^{-6}cm^2s^{-3}$. Similarly, the [*Fernando*, 1989] formula for the 'low stability' regime, $q_h = c_{11}(g\alpha\Delta T)w_\star$, is found to yield $q_h = 0.9 \times 10^{-6}cm^2s^{-3}$. (Both c_6 and c_{11} are empirical constants). Both estimates of the thermal buoyancy flux are greater than the measured thermal buoyancy flux of $H_* = -g\alpha H/\rho c_p = -0.12 \times 10^{-6}cm^2s^{-3}$ corresponding to the geothermal heat flux H. The disparity between calculated and measured fluxes is most probably associated with the long time limit of convection, for which none of the models are appropriate. We note that this result is consistent with [*Fernando*, 1989], who suggests an overestimating tendency in the flux computations in the long time limit, *i.e.* with large values of $Ri_\star$.

5. IMPACT ON MIXING

The identification of various mechanisms of mixing in the Black Sea has not been straightforward. Because of its rather unusual setting as an enclosed basin with a specific form and history of stratification, it may have very little in common with oceanographic settings in other regions of the world.

The various possible mechanisms of mixing in the Black Sea are schematized in Figure 13. A rather unique combination of factors exists in the Black Sea: a double diffusive environment, wind stress forcing in a closed basin geometry and the resulting Ekman pumping, buoyancy-driven lateral boundary layers, and exchanges through the Bosphorus Strait. All these mechanisms lead to mass fluxes, which simultaneously have to be balanced by conservation arguments.

Typical values of effective vertical diffusivities in the deep ocean are often found to be much larger than the molecular values, in regions where direct turbulent mixing is not expected. This fact has been explained, with some success, by boundary mixing theories [*Garrett*, 1979, 1990; *Ivey and Corcos*, 1982; *Phillips et al.*, 1986; *Woods*, 1991; *Salmun et al.*, 1991], in which variable turbulent boundary layers drive a vertical circulation forced at the pycnocline. Boundary mixing, driven directly by the Bosphorus buoyancy source, as well as other sources appears to be an important component of the Black Sea vertical circulation. It has been hypothesized [*Özsoy et al.*, 1994] that the boundary layer structure along the continental slope region, and its subsequent disintegration, resulting in the intrusions of water from the boundaries towards the interior, could partially be responsible for driving a recirculation in the upper part of the Black Sea (Figure 13).

The lateral intrusions of shelf-mixed inflowing water from the continental slope region into the interior constitute a source of mass introduced into the interior region. The time and depth variability of the corre-

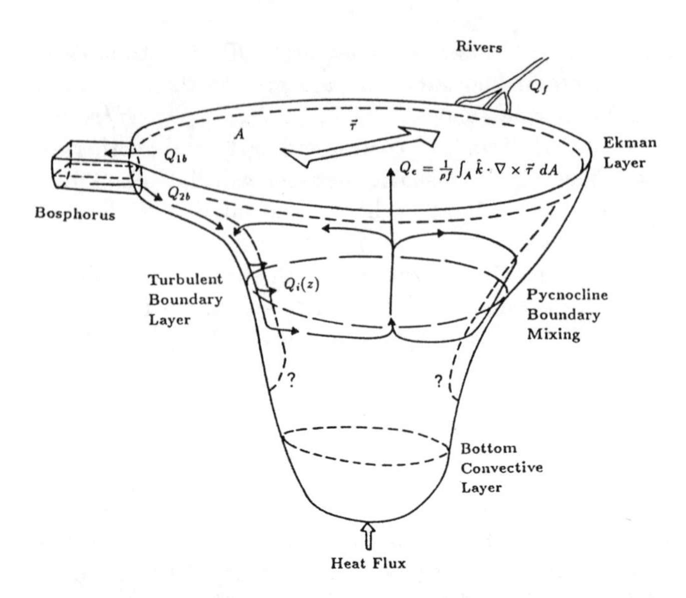

Fig. 13. Schematization of the recirculation driven by boundary mixing processes in the Black Sea. Mechanisms capable of driving a recirculation between boundary layers and the interior are emphasized. After Özsoy *et al.* (1993).

sponding source function at the halocline has important implications for the interior stratification in the upper part of the Black Sea, in which the vertical circulation may be dominated by the above mechanism of boundary transports. This conceptual model describing the upper ocean vertical recirculation in the semi-enclosed Black Sea basin is consistent with models of ventilation proposed earlier, based on the random termination depths of cascading water: *e.g.* the [*Rooth*, 1986] model for the Black Sea, and the Baltic model of [*Stigebrandt*, 1987].

It is possible that the zone of vanishing temperature gradient near a depth of $500m$ (Figures 4, 6 and 8) is linked with the observed ultimate depth of penetration of Bosphorus intrusions; the double diffusive fluxes likewise tend to vanish at the same depths corresponding to double diffusive marginal stability, $R_\rho \rightarrow \infty$ [*Özsoy et al.*, 1993]. This depth limit for efficient vertical mixing is consistent with other tracers, *e.g.* ^{14}C, showing smaller mean residence times of intermediate waters (depth $\leq$ 500m) compared to the more uniformly aged deep waters [*Östlund, 1974*, 1986]. Likewise, the Tritium penetration reaches similar depths [*Top et al.*, 1991]. Interestingly, the ventilation of the upper water column has been attributed [*Grasshoff*, 1975] to mixing along the Anatolian coastal margin; we suggest that a major source for the ventilation is the effects of the Bosphorus outflow.

Anomalous temperature fine structure is observed at all depths in the water column, and appears to be amplified near the basin lateral boundaries. For example, in Figure 8b, a variety of fine scale features appear at deeper depths, near the zone of vanishing temperature gradient at $\sim 500m$. Note, however, that these features are not characterised by cold anomalies as in the case of Bosphorus intrusions, but rather have temperature anomalies of both signs, with positive ones apparently dominating. Note also that these features were more common in September 1991 as compared to other surveys, at which time they also covered a larger area in the central parts of the basin. The origin of these features are therefore not clear, though they too are probably linked with double diffusive instabilities in the regime of high density ratio, possibly in response to cooling of the overlying CIW in the central regions. A number of other fine scale features are observed at the deeper depths. Deep features (500-2000m) are indicated in Figure 8a, especially near the central part, where the profiles were close to the basin boundary. These deep features provide clear evidence of boundary mixing in the deeper waters adjacent to the continental slope.

The transport between the bottom convective layer and the overlying waters occur through a diffusive interface between the two regions, where the fluxes are appropriate for a diffusive regime of double diffusion. In the deep water region above the interface, different types of assumptions could be made with regard to the fluxes. The deep water fluxes could either be related to double diffusion (with different diffusivities of salt and heat) or turbulent diffusion (with equal diffusivities of salt and heat). On the other hand, if the vertical transports of heat and salt were only related to one-dimensional diffusion (*i.e.* if lateral effects and variations in horizontal area were to be negligible), then we would expect the fluxes, and hence the flux ratio, to be constant throughout the deep waters.

One way to assess the regime of vertical transport in the deep water is to compare the computed flux ratios $R_f = q_s/q_h = \beta F_s/\alpha F_h$ (F_s and F_h respectively are the salt and heat fluxes) corresponding to the assumptions with regard to the form of vertical mixing.

The interfacial flux ratio in the 'low stability' regime is estimated as $R_f = 0.15 R_\rho = 0.3$ by the Fernando (1989) model, with a value of $R_\rho = 2$ at the interface. It should be reasonable to expect similar values of the flux ratio in deep water above the interface. If the vertical transport were to be by double diffusion across some staircase interfaces, then we could use either $R_f = \tau^{1/2} R_\rho$

[*Turner*, 1973], or $R_f = \tau^{1/2}$ [*Linden and Shirtcliffe*, 1978], to compute flux ratios of $R_f = 0.1 - 0.3$, immediately above the interface, and $R_f \rightarrow \infty$ near the zone of vanishing temperature gradient at about $500m$ depth. On the other hand, for $2 < R_\rho < 7$ (such as in most part of the Black Sea), it is often assumed that the flux ratio is constant, $R_f = 0.15$; at high values of density ratio, for $R_\rho > 7$, the dependence becomes $R_f = \tau R_\rho$ [*Newell*, 1984, cited in *Fernando*, 1989]. In contrast to the above, if we were to assume turbulent transport with equal eddy diffusivities for heat and salt, we should require $R_f = \{\beta(dS/dz)\}/\{\alpha(dT/dz)\} = R_\rho$, resulting in much higher values, increasing from $R_f = 2$ at the convective layer interface to $R_f \rightarrow \infty$ at about a depth of $500m$. Comparison of turbulent versus molecular diffusive fluxes therefore indicates that the double diffusive transport would be the more likely mechanism of vertical transport in deep water [*Özsoy et al.*, 1991; *Murray et al.*, 1991].

If the vertical transport in deep water was dominated by double diffusive fluxes we would expect to see staircase structures. However, our calculations show that even if layers were present, they could hardly be detected because of their size, and noise levels in the CTD measurements. For example, if the 'stairs' are formed by repeated splitting of the convective layer interface, we would expect the average thickness of staircases to be about $1m$ [*Özsoy et al.*, 1991; *Murray et al.*, 1991]. Another estimate, using the parameterization of Fedorov (1989), yields $h = 32.9\nu\kappa_h^{-1/2}(g\alpha\partial T/\partial z)^{-1/4} = 4m$ (taking an average temperature gradient of $\partial T/\partial z = 0.05°C/1000m$ in deep water).

A separate argument in favour of the role of double diffusive vertical fluxes in deep water derives directly from the typical 'S' pattern of the potential temperature versus salinity relationship. Curvilinear temperature - salinity relationships are typical of double diffusive regimes in other parts of the world (*e.g.* the salt-finger regime of [*Schmitt*, 1981]. In the case of the Black Sea, the different effective diffusivities of temperature and salinity could explain a curvilinear $\theta - S$ relationship, if two water masses separated by an initial discontunity were allowed to evolve by diffusion. In fact, if we assume that the Mediterranean Water reaching the bottom of the Black Sea several millenia before the present would initially form a discontinuous layer, the time evolution of the temperature and salinity structure would be analogous to the classical case studied in [*Mamayev*, 1975]. In his study, Mamayev (p. 189) showed that the resulting $\theta - S$ relationship would be 'S-shaped', as it happens to be the case in the Black Sea. Furthermore, for the infinite depth case studied, the shape of the $\theta - S$ curve is shown to be a function of the diffusivities and depth alone, and independent of time. This pattern of double diffusion could potentially explain the observed inflexion in the $\theta - S$ curve.

Turbulent convection seems to have an important impact on the homogenization of water properties across the basin within the bottom convective layer. The *r.m.s.* turbulent velocity, $w_\star$, within the convective layer is estimated to be $w_\star = c_2^{1/2}(H_\star h)^{1/3} = 0.25\ cm\ s^{-1}$ [*Özsoy et al.*, 1993] (with c_2 defined as an empirical constant) [*Fernando*, 1989]. The characteristic time scale of overturning within the layer with thickness $h = 400m$ is then calculated to be $T_\star \equiv \pi h/w_\star \simeq 6\ days$. Since the east-west axis of the basin has a length of about $L = 1000\ km$, an assumption of exchange between eddies occurring on every cycle implies a basin-wide transport time scale, T_b, of about $T_b \equiv T_\star L/h = 40\ years$ for homogenization of properties across the basin. We should caution that this idealised estimate is based on an average velocity scale which may not be representative of the various length and time scales of turbulent convection modes in an enclosed domain [*e.g., Knobloch et al.*, this volume].

The turbulent mixing within the bottom convective layer could have other important implications. In addition to its role in the observed homogeneity of water properties, we could hypothesize turbulent motions to have a similar effect on homogenization of sediment properties across the basin, although the faster rate of apparent mixing (in comparison to our estimate), yielding well correlated seasonal or annual varve structures across the basin, must be accounted for. A layer of organically rich, unbioturbated, high porosity (up to 99%) material, appropriately called the 'fluff layer', occupies the upper few centimeters of the Black Sea bottom sediments [*e.g. Moore and O'Neill*, 1991; *Lyons*, 1991] in the form of a fluid bed load. This unique layer contains freshly settled sediments preceeeding their conversion to varved bottom sediments by diagenesis, dissolution, and compaction processes. The origin of this floating material could in large be related to the bed load carrying capacity of the turbulent motions within the bottom convective layer. Despite large geographic differences in plankton blooms [*Sur et al.*, 1994], the sediment laminae (varves) resulting from the seasonal / interannual productivity cycles of the upper ocean display exceptional continuity across the basin [*Hay*, 1987, 1991; *Lyons*, 1991], suggesting that the homogenization could take place only after settling of the material near the bottom. These scientific questions with regard to the redistribution of bottom sediments are much worthy of further examination, because of the potential for sediment records in evaluating the history of hydrological, ecological and climatic changes in the region.

6. CONCLUSIONS

The peculiar temperature and salinity stratification of the Black Sea has great impact on its mixing characteristics. Double diffusion has primary importance in determining the mixing and transformation of its water masses. This setting of the Black Sea offers unique opportunities for observing double diffusive convection and for testing hypotheses in relation with such processes.

There are many areas of the world (*e.g.* the Mediterranean outflow and 'meddies' in the Atlantic Ocean, the Caribbean Sea and the Arctic Ocean staircases *etc.*) where small scale processes can determine ocean mixing in meso-scale and basin-scale dimensions. The Black Sea appears to be one of the unique areas of the world where double diffusive convection occurs on a large scale under restrictive conditions of a landlocked geometry.

The inflow of Mediterranean Water excites a large scale double diffusive convective pattern in this environment of potentially unstable double stratification. As a result, the inflow itself breaks up at intermediate depths, partially driving a basin-scale recirculation in the upper ocean, and transporting materials on the same scale.

Similarly, a relatively small geothermal heat flux at the bottom is sufficient to drive a bottom convective layer whose age is on the same order as the average age of the existing basin stratification. The convective layer homogenizes the bottom water mass, and could similarly play a major role in distributing bottom sediments across the basin. Features of this convective layer suggest a limiting case of convective layers studied in the laboratory, and could guide further studies on the possible regimes of convection.

Acknowledgements. The present paper has benefited from the combination of results from a number of studies: a project carried out in the Bosphorus and adjoining regions for the City of İstanbul, Water and Sewerage Administration (İSKİ), a study of the Black Sea fisheries supported by the Science for Stability program of NATO, and a national monitoring program supported by the Turkish Scientific and Technical Research Council (TÜBİTAK).

REFERENCES

Boudreau, B. P., and P. H. Leblond, A Simple Evolutionary Model for Water and Salt in the Black Sea, *Paleoceanography*, *4*, 157-166, 1989.

Brewer, P. G., and D. W. Spencer, Distribution of Some Trace Elements in Black Sea and Their Flux Between Dissolved and Particulate Phases. In: *The Black Sea - Geology, Chemistry and Biology*, E. T. Degens and D. A. Ross (editors), Am. Assoc. Pet. Geol. Memoir 20, Tulsa, Oklahoma, 137-143, 1974.

Buesseler, K. O., Livingstone, H. D., and S. Casso, Mixing Between Oxic and Anoxic Waters of the Black Sea as Traced by Chernobyl Cesium Isotopes, *Deep-Sea Res.*, *38*, S725 - S745, 1991.

Codispoti, L. A., Friederich, G. E., Murray, J. W., and C. Sakamoto, Chemical Variability in the Black Sea: Implications of Data Obtained with a Continuous Vertical Profiling System that penetrated the Oxic/Anoxic Interface, *Deep-Sea Res.*, *38*, S691-S710, 1991.

Fedorov, K. N., Layer Thicknesses and Effective Diffusivities in "Diffusive" Thermohaline Convection in the Ocean, In: *Small Scale Turbulence and Mixing in the Ocean*, J. C. J. Nihoul and B. M. Jamart, (editors), Elsevier, 471-480, 1989.

Fernando, H. J. S., The Formation of Layered Structure when a Stable Salinity Gradient is Heated from Below, *J. Fluid Mech.*, *182*, 525-541, 1987.

Fernando, H. J. S., Oceanographic Implications of Laboratory Experiments on Diffusive Interfaces, *J. Phys. Oceanogr.*, *19*, 1707-1715, 1989.

Fernando, H. J. S., and C. Y. Ching, An Experimental Study on Thermohaline Staircases, In: *Double Diffusion in Oceanography: Proceedings of a Meeting, September 26-29, 1989*, R. W. Schmitt (editor), Technical Report, Woods Hole Oceanographic Institution, 141-149, 1991.

Garrett, C., Mixing in the Ocean Interior, *Dyn. Atmos. Oceans*, *3*, 239-265, 1979.

Garrett, C., The Role of Secondary Circulation in Boundary Mixing, *J. Geophys. Res.*, *95*, 3181-3188, 1990.

Gertman, I. F., Ovchinnikov, I. M., and Y. I. Popov, Deep Convection in the Levantine Sea, *Rapp. Comm. Mer Medit.*, *32*, 172, 1990.

Grasshoff, K., The Hydrochemistry of Landlocked Basins and Fjords, In: *Chemical Oceanography*, J. P. Riley and G. Skirrow (editors), Academic Press, New York, *2*, 647 pp., 1975.

Hay, B. J., Particle Flux in the Western Black Sea in the Present and over the Last 5000 Years: Temporal Variability, Sources, Transport Mechanism, Ph.D. Thesis, Joint Program M.I.T./Woods Hole Oceanographic Institution, 202 pp., 1987.

Hay, B. J., Arthur, M. A., Dean, W. E., Neff, E. D., and S. Honjo, Sediment Deposition in the Late Holocene Abyssal Black Sea with Climatic and Chronological Implications, *Deep Sea Research*, *38*, Suppl. 2, S1211-S1236, 1991.

Haenel, R., A Critical Review of Heat Flow Measurements in Sea and Lake Bottom Sediments, *In:* V. Cermak and L. Rybach (editors), *Terrestrial Heat Flow in Europe*, Springer-Verlag, Berlin, 1979.

Huppert, H. E., and P. F. Linden, On Heating of a Stable Salinity Gradient from Below, *J. Fluid Mech.*, *95*, 431-464, 1979.

Huppert, H. E. and J. S. Turner, Ice Blocks Melting Into a Salinity Gradient, *J. Fluid Mech.*, *100*, 367-384, 1980.

Ivey, G. N., and G. M. Corcos, Boundary Mixing in a Stratified Fluid, *J. Fluid Mech.*, *121*, 1-26, 1982.

İzdar, E., Konuk, T., Ittekott, V., Kempe, S., and E. T. Degens, Particle Flux in the Black Sea: Nature of Organic Matter in the Shelf Waters of the Black Sea, In: *Particle Flux in the Ocean*, E. T. Degens, E. İzdar and S. Honjo (editors), Mitt. Geol.-Paleont. Inst. Univ. Hamburg, SCOPE/UNEP Sonderband, *62*, 1-18, 1987.

Jeevaraj, C. G., and J. Imberger, Experimental Study of Double Diffusive Instability in Sidewall Heating, *J. Fluid Mech.*, *222*, 565-586, 1991.

Kempe, S., Diercks, A. R., Liebezeit, G., and A. Prange, Geochemical and Structural Aspects of the Pycnocline in the Black Sea (R/V Knorr 134-8 Leg 1, 1988), In: *The Black Sea Oceanography*, E. İzdar and J. M. Murray (editors), NATO/ASI Series, Dordrecht, Kluwer Academic Publishers, 89-110, 1991.

Latif, M. A., Özsoy, E., Oğuz, T., and Ü. Ünlüata, Observations of the Mediterranean inflow into the Black Sea, *Deep Sea Research*, *38*, Suppl. 2, S711-S723, 1991.

Linden, P. F., and T. G. L. Shirtcliffe, The Diffusive Interface in Double Diffusive Convection, *J. Fluid Mech.*, *87*, 417-432, 1978.

Lyons, T., Upper Holocene Sediments of the Black Dea: Summary of Leg 4 Box Cores (1988 Black Sea Oceanographic Expedition), in: *The Black Sea Oceanography*, E. İzdar and J. M. Murray (editors), NATO/ASI Series, Dordrecht, Kluwer Academic Publishers, 401-441, 1991.

Mamayev, O. I., *Temperature - Salinity Analysis of World Ocean Waters*, Elsevier, Amsterdam, 374 pp. 1975.

Mee, L. D., The Black Sea in Crisis: the Need for Concerted International Action, *Ambio*, *24*, 278-286, 1992.

Moore, W. S., and D. J. O'Neill, Radionuclide Distributions in recent Sea Sediments, in: *The Black Sea Oceanography*, E. İzdar and J. M. Murray (editors), NATO/ASI Series, Dordrecht, Kluwer Academic Publishers, 257-270, 1991.

Murray, J. W., Top, Z., and E. Özsoy, Hydrographic Properties and Ventilation of the Black Sea, *Deep Sea Research*, *38*, Suppl. 2, S663-S689, 1991.

Östlund, H. G., Expedition Odysseus 65: Radiocarbon Age of Black Sea Water, In: *The Black Sea - Geology, Chemistry and Biology*, E. T. Degens and D. Ross (editors), The American Association of Petroleum Geologists, Memoir No. 20., Tulsa, Oklahoma, 127-131, 1974.

Östlund, H. G., and D. Dyrssen, Renewal Rates of the Black Sea Deep Water, In: *The Chemical and Physical Oceanography of the Black Sea*, Univ. of Göteborg, Rep. on the Chemistry of the Sea XXXIII. Presented in the Meeting on the Chemical and Physical Oceanography of the Black Sea, Göteborg, Sweden, 1986.

Özsoy, E., Top, Z., White, G., and J. W. Murray, Double Diffusive Intrusions, Mixing and Deep Sea Convection Processes in the Black Sea, In: *The Black Sea Oceanography*, E. İzdar and J. M. Murray (editors), NATO/ASI Series, Dordrecht, Kluwer Academic Publishers, 17-42, 1991.

Özsoy, E., Ünlüata, Ü., and Z. Top, The Mediterranean Water Evolution, Material Transport by Double Diffusive Intrusions, and Interior Mixing in the Black Sea, *Prog. Oceanog.*, *31*, 275-320, 1993.

Phillips, O. M., Shyu, J-H., and H. Salmun, An Experiment on Boundary Mixing: Mean Circulation and Transport Rates, *J. Fluid Mech.*, *173*, 473-499, 1986.

Rooth, C. G. H., Comments on Circulation Diagnostics and Implications for Chemical Studies of the Black Sea, In: *The Chemical and Physical Oceanography of the Black Sea*, Univ. of Göteborg, Rep. on the Chemistry of the Sea XXXIII. Also presented in the Meeting on the Chemical and Physical Oceanography of the Black Sea, Göteborg, Sweden, 1986.

Ruddick, B., A Practical Indicator of the Stability of the Water Column to Double-Diffusive Activity, *Deep-Sea Res.*, *30*, 1105-1107, 1983.

Salmun, H., Killworth, P. D., and J. R. Blundell, A Two-Dimensional Model of Boundary Mixing, *J. Geophys. Res.*, *96*, 18447-18474, 1991.

Schmitt, R. W., Form of the Temperature-Salinity Relationship in the Central Water: Evidence for Double-Diffusive Mixing, *J. Phys. Oceanogr.*, *11*, 1015-1026, 1981.

Spencer D. P., Brewer, P. G., and P. L. Sachs, Aspects of the Distribution and Trace Element Composition of Suspended Matter in the Black Sea. *Geochim. et Cosmochim. Acta*, *36*, 71-86, 1971.

Stanley, D. J., and C. Blanpied, Late Quaternary Water Exchange between the Eastern Mediterranean and the Black Sea', *Nature*, *285*, 537-541, 1980.

Stigebrandt, A., A Model for the Vertical Circulation of the Baltic Deep Water, *J. Phys. Oceanog.*, *17*, 1772-1785, 1987.

Sur, H. İ., Özsoy, E., and Ü. Ünlüata, Boundary Current Instabilities, Upwelling, Shelf Mixing and Eutrophication Processes In The Black Sea, *Prog. Oceanog.*, *33*, 249-302, 1994.

Swart, P. K., The Oxygen and Hydrogen Isotopic Composition of the Black Sea, *Deep-Sea Res*, *38*, Suppl. 2, S761-S772, 1991.

Tanny, J., and A. B. Tsinober, The Dynamics and Structure of Double-Diffusive Layers in Sidewall Heating Experiments, *J. Fluid Mech.*, *196*, 135-156, 1988.

Tolmazin, D., Changing Coastal Oceanography of the Black Sea, II: Mediterranean Effluent, *Prog. Oceanog.*, *15*, 277-316, 1985.

Top, Z., Östlund, H. G., Pope, L., and C. Grall, Helium and Tritium in the Black Sea: A Comparison with the 1975 Observations, *Deep-Sea Res.*, *38*, Suppl. 2, S747-S760, 1991.

Tsinober, A. B., Yahalom, Y., and Shlien, D. J., A Point Source of Heat in a Stable Salinity Gradient, *J. Fluid Mech.*, *135*, 199-217, 1983.

Turner, J. S., The Behaviour of a Stable Salinity Gradient Heated from Below, *J. Fluid Mech.*, *33*, (1), 183-200, 1968.

Turner, J. S., A Physical Interpretation of Hot Brine Layers in the Red Sea, In: *Hot Brines and Recent Heavy Metal Deposits in the Red Sea*, E. T. Degens and D. A. Ross (editors), Springer, New York, 164-172, 1969.

Turner, J. S., *Buoyancy Effects in Fluids*, Cambridge University Press, 367 pp., 1973.

Turner, J. S., Double-Diffusive Intrusions into a Density Gradient, *J. Geophys. Res.*, *83*, 2887-2901, 1978.

Ünlüata, Ü., Oğuz, T., Latif, M. A., and E. Özsoy, On the Physical Oceanography of the Turkish Straits, In: *The Physical Oceanography of Sea Straits*, L. J. Pratt (editor), NATO/ASI Series, Kluwer, Dordrecht, 25-60, 1990.

Ünlüata, Ü., Aubrey, D. G., Belberov, Z., Bologa, A., Eremeev, V. and M. Vinogradov, International Program Investigates the Black Sea, *EOS*, *74(36)*, 408-412, 1993.

Woods, A. W., Boundary-Driven Mixing, *J. Fluid Mech.*, *226*, 625-654, 1991.

Zolotarev, V. G., Sochel'nikov, V. V., and Y. P. Malovitskiy, Results of Heat-Flow Measurements in the Black and Mediterranean Sea Basins, *Oceanology*, *19*, 701-705, 1979.

E. Özsoy and Ş. Beşiktepe, Institute of Marine Sciences, Middle East Technical University, P. K. 28, Erdemli, İçel 33731 Turkey

Observations of Mixing Processes Downstream from the Confluence of the Mississippi and St. Croix Rivers

John A. Moody

U.S. Geological Survey, Lakewood, Colorado

The initial lateral mixing downstream from a confluence of two rivers with different water properties and velocities has a near-field mixing zone that is typically 5-10 river widths long and visually similar to the free shear-flow zone in an infinitely deep fluid. Vertical and horizontal profiles, and time-series measurements of temperature and specific conductance were collected in the near-field mixing zone at three distances downstream from the confluence of the Mississippi and St. Croix Rivers on April 11, 1992. The vertical interface between the two fluids was characterized by step-like gradients in color, temperature, specific conductance, and velocity. Mixing in the near-field zone consisted of stirring and diffusion processes. The difference in velocity between the two fluids (0.18 m/s) created lateral shear instabilities that were visible at the surface and appeared to have 3-dimensional structure. These instabilities stirred the water, increasing the surface area of the interface, and created horizontal intrusions of water with temperature and specific-conductance properties that contrasted with the surrounding water. Twenty meters downstream from the confluence (about 0.05 river widths), the downstream size of these intrusions was 10-80 m and the vertical size was 0.2-0.5 m. There were sharp gradients of temperature and specific conductance across the horizontal interface between the intrusions and surrounding fluid (differences of 3.4°C and 500 μS/cm), and measurements made inside some intrusions suggest that mixing at the edges of these intrusions was double diffusive. Double diffusion may have coupled with the shear (estimated gradient Richardson number was 0.2) to produce triple diffusion. Turbulent instabilities continued to stir the water causing the intrusions to decrease in size downstream such that the final measurements of temperature and specific conductance about 8 river widths downstream from the confluence were relatively smooth, monotonic functions of the cross-channel distance.

1. INTRODUCTION

The lateral mixing process of two rivers downstream from the confluence can be spectacular if visual, chemical, and hydraulic properties of the two rivers are different (Figure 1). Mixing processes consist of advective or stirring processes resulting from lateral shear and diffusion processes resulting from molecular and turbulent diffusion. The region of lateral mixing downstream from the confluence of two rivers can be divided into two zones depending upon the importance of the lateral shear. The near-field mixing zone begins at the confluence point and extends downstream for about 5-10 river widths and is a region of strong lateral shear and sharp visual boundaries between the two water masses. The far-field mixing zone begins 5-10 river widths downstream from the confluence point and extends downstream 10-100's of river widths and is characterized by weak lateral shear, weak gradients, and indistinct boundaries between the two water masses. Field studies of mixing in rivers [*Mackay*, 1970; *Yotsukura et al.*, 1970; *Matsui et al.*, 1976; *Stallard*, 1987; and *Weibezahn et al.*, 1989] have been 2-dimensional and focused on the far-field mixing zone where the first sampling cross section in these field studies was located about 4-200 river widths downstream from the confluence. Laboratory studies of lateral mixing [*Sayre and Chang*, 1968; *Fischer*, 1969; *Krishnappan and Lau*, 1977; *Webel and Schatzmann*, 1984; and *Bruno et al.*, 1990] have also been primarily 2-

Double-Diffusive Convection
Geophysical Monograph 94

Published in 1995 by the American Geophysical Union

Fig. 1. Aerial view of the confluence of the Mississippi and St. Croix Rivers, July 20, 1992. The view is looking upstream; the Mississippi River is on the left and St. Croix River is on the right. Estimated water discharge and cross-sectional velocity were 600 m^3/s and 0.50 m/s for the Mississippi River and 236 m^3/s and 0.24 m/s for the St. Croix River. (Photograph by Bob Meade, U. S. Geological Survey.)

dimensional and have focused on the far-field mixing zone with the first sampling cross section in these laboratory studies located 4-10 flume widths downstream from the initial mixing point.

The near-field mixing zone in rivers is often a zone of convergence between two vertically homogenous water masses that differ in conductivity, temperature, and suspended-sediment concentration. The water masses meet at an angle and with different velocities such that streamwise components of velocity parallel to the vertical interface create strong lateral shear along the entire depth of the interface, and the components perpendicular to the interface create vertical circulation near the interface. The lateral shear is often strong enough to deform the interface and generate horizontal vortices with strong gradients that appear as sharp boundaries when the merging rivers differ in color. These horizontal vortices and vertical circulation near the interface may be independent stirring processes, or they may combine to produce a complex stirring process. Stirring processes were described by Eckart [1948] in his discussion of the three stages of mixing, which consisted of: (1) sharp gradients at the interface with the average gradient over the entire volume being small, (2) the surface area of the interface increasing with a corresponding increase in the average gradient, and (3) gradients disappearing. If stirring processes create vertical gradients of conductivity or temperature by the horizontal interleaving of two different water masses then double diffusion might play an important part in mixing. Large-velocity gradients might also create unstable conditions (gradient Richardson number near 0.25; *Turner*, 1973) favorable for triple diffusion [*Williams*, 1981; *Ruddick*, 1992] where Kelvin-Helmholtz type instabilities create rapidly diffusing velocity gradients that combine with double diffusion to enhance mixing.

Three-dimensional phenomena associated with the mixing of two parallel fluids in the near-field zone have been studied in the laboratory [*Browand and Troutt*, 1985; *Jimenez et al.*, 1985; and *Lasheras et al.*, 1986] for free shear flows having a horizontal interface in an essentially infinitely deep fluid with no boundaries. Pairing of spanwise vortices across the horizontal interface of the fluid [*Winant and Browand*, 1974] is one possible type of stirring mechanism that has been observed, and additional streamwise vortices [*Bernal and Roshko*, 1986] are another possible stirring mechanism. Streamwise vortices are perpendicular to the spanwise vortices, project as counter-rotating vortices into the adjacent fluid with a spacing equal to approximately 0.7 times the wavelength of the initial spanwise instabilities, and add a third dimension to the original 2-dimensional mixing layer. Recently some laboratory investigators have begun to address the effects of finite depth and bottom friction. Laboratory models of tributaries entering a deeper river [*Best and Roy*, 1991] have been shown to produce enhanced mixing and vertical upwelling just downstream from the mouth of the tributary. Shear instabilities in shallow water depths were shown to be stabilized by bottom friction [*Alavian and Chu*, 1985] although the mechanism was not clearly understood. Lateral mixing rates in the near-field mixing zone were found [in laboratory experiments where width-to-depth ratios were about 10; *Chu and Babarutsi*, 1988; *Babarutsi and Chu*, 1991] to be about twice the lateral mixing rate for free shear flows because the finite depth increased the large-scale horizontal turbulence by reducing the transfer of large-scale horizontal turbulence to small-scale turbulence. The bottom friction has been related [*Chu et al.*, 1983] to a bottom-friction stability parameter,

$$S_f = \frac{c_f U}{2hU_y} \tag{1}$$

where turbulent motions of a scale smaller than the water depth were ignored and where c_f is the bottom-friction coefficient, h the water depth, U the average velocity, and U_y the lateral velocity gradient across the shear layer. The bottom-friction coefficient is given by

$$c_f = 2\left(\frac{u_*}{\bar{u}}\right)^2 \tag{2}$$

where u_* is the bottom-shear velocity and $\bar{u}$ the depth-average velocity. Indirect measurements [*Chu and Babarutsi*, 1988] indicate that the large-scale horizontal turbulence decreases as S_f increases up to a critical value of approximately 0.09. Thus, contrary to the effect produced by confinement to a finite depth, bottom friction is reported to decrease the lateral mixing rate.

Three-dimensional field measurements within the near-field mixing zone are needed to verify laboratory results, and to identify physical phenomena that may not have been present in laboratory experiments and which require further research. This paper describes some 3-dimensional measurements (time series, and vertical and horizontal profiles of temperature and specific conductance) collected in the near-field mixing zone downstream from the confluence of the Mississippi and St. Croix Rivers. Shear instabilities were observed visually at the surface in this relatively shallow (~6 m) channel, so these field measurements should provide some insight into the mixing processes of naturally occurring free shear flows in shallow water when the width-to-depth ratio is much greater than 10. The purpose of the investigation was to measure the 3-dimensional spatial structure in the near-field mixing zone as a function of distance downstream from the confluence point and to identify the importance of different components of the mixing process.

2. FIELD SITE AND INSTRUMENTATION

The confluence of the Mississippi and St. Croix Rivers is located at Prescott,Wis., about 50 km downstream from St. Paul, Minn. Average discharge of the Mississippi River at the confluence is about 317 m^3/s and the average discharge of the St. Croix River is about 131 m^3/s. The Mississippi River is often muddy and the St. Croix River is frequently clear, so there is usually a visible contrast between the two water masses. Besides the visible differences, the two rivers often have very different specific conductances--the Mississippi River's median specific conductance is 450 μS/cm (based on 41 measurements at St. Paul, Minn., 45 km upstream from the confluence, from October 1973 to September 1977) and the St. Croix River's median specific conductance is 175 μS/cm (based on 105 measurements at St. Croix Falls, Wis., 84 km upstream from the confluence from October 1974 to February 1986).

Conductivity, temperature, instrument depth, and the time were recorded every 0.5 s within a conductivity-temperature-depth profiler (CTD profiler, Sea Bird, Model SeaCat 19, any use of trade, product, or firm names is for descriptive purposes only and does not imply endorsement by the U.S. Government). The profiler was deployed from a small research boat with the axis of the cylindrical profiler rotated horizontal and facing into the flow, rather than in its normal vertical position. This permitted measurements to be made close to the bottom and oriented the conductivity sensor so that water would flow through it. The 95-percent response time was about 1.0 s for the conductivity sensor when it was immersed in water at about 1 m/s (which approximated the flow rate through the sensor as it was used in the field). The 95-percent response time was about 2.5 s for the temperature sensor under the same conditions. Conductivity was not lagged relative to the temperature because an examination of the data collected for a horizontal profile through a step-like change in temperature and conductivity indicated the lag between the sensors was less than 1 s. The data were averaged over 2.0 s (4 samples), which was a compromise that retained reasonably sharp gradients but improved the precision of the measurements. The temperature dependence of conductivity was removed by converting all conductivity measurements to the conductivity at a reference temperature of 25°C by using the following formula (*Clesceri et al.*, 1989):

$$S = \frac{C \times 10,000}{1 + 0.02 \times (T - 25)} \quad (3)$$

where S is the specific conductance in microsiemens per centimeter (μS/cm), C is the conductivity in Siemens per meter at temperature T in degrees Celsius.

The depth-averaged velocity was measured by using a handwinch to lower and raise a Price AA current meter (mounted above a 30-pound sounding weight) at a constant vertical transit rate. The transit rate varied from 0.04 to 0.09 m/s. The water depth was measured with an acoustic depth recorder (Lowrance, Model X16), which produced a continuous analog output. A mark was made manually on the analog record every 10 s. Horizontal distances between a microwave transmitter/receiver in the boat and a microwave transmitter located near the left edge of water were measured by a distance-measuring unit (Del Norte Technology Trisponder system) located in the boat. The distances and corresponding times were logged on a portable computer. Time was later used to merge all the digital and analog data (specific conductance, temperature, instrument depth, water depth, and horizontal distance).

3. SAMPLING STRATEGY

The initial specific conductance, temperature, and velocity conditions of both rivers at the confluence on April 11, 1992 were determined by measuring vertical profiles at five locations across the St. Croix River but at only two locations across the Mississippi River. The Mississippi River was sampled at fewer locations because the depth-averaged velocity, surface temperature, and surface specific conductance had been measured at nine locations across the

Mississippi River just 1 km upstream from the confluence on April 10, 1992, as part of a related research project.

The width of the Mississippi River downstream from the confluence is about 400 m and three measurement cross sections were selected at 20 m, 900 m, and 3400 m downstream from the confluence corresponding to about 0.05, 2, and 8 river widths (Figure 2). At these measurement cross-sections, three types of data were collected: (1) horizontal profiles across the river, (2) vertical profiles, and (3) a time series. The horizontal profiles were repeated 3-5 times at the same cross section at a boat speed of about 1 m/s and with the CTD profiler located about 0.3 m below the water surface. The vertical profiles were made by using a handwinch to lower and raise the CTD profiler at about 0.1 m/s while the boat was anchored. One vertical profile was located in the St. Croix River water mass, one in the Mississippi River water mass, and one in the mixing zone. The time series was recorded while the boat was anchored in the mixing zone and while the CTD profiler was held fixed at a depth of 1 m. These time series data were converted to an equivalent longitudinal profile by multiplying the time by an estimate of the water velocity at a depth of 1 m (~ 0.6 m/s). This velocity was obtained by dividing the total discharge by the cross-sectional area 3400 m downstream from the confluence point.

4. RESULTS

4.1. *Initial Conditions*

The initial water depth, temperature, specific conductance, and depth-averaged water velocity were measured about 20 m upstream from the confluence where the St. Croix River was deeper (mean depth about 7 m) than the Mississippi River (mean depth about 4 m). Vertical profiles of temperature and specific conductance indicate that both rivers were vertically homogenous with maximum ranges of 0.1 °C and 6 μS/cm in the St. Croix River and <0.1 °C and 5 μS/cm in the Mississippi River. The depth-averaged velocities varied from 0.13 to 0.40 m/s in the St. Croix River (20 m upstream from the confluence point) and from 0.3 to 0.6 m/s in the Mississippi River (~1 km upstream from the confluence point). Cross-sectional averaged velocity of the Mississippi River was greater (about 0.50 m/s) than that of the St. Croix River (0. 32 m/s). The corresponding water discharges were 570 m^3/s in the Mississippi River and 320 m^3/s in the St. Croix River. Estimates of the bottom-shear velocity, u_*, based on the depth-slope product, range from 0.03 to 0.06 m/s for the St. Croix River and from 0.04 to 0.06 m/s for the Mississippi River. Using the depth-averaged velocity, the estimate of the local bottom-friction coefficient, c_f, was about 0.05 for the St. Croix River and 0.02 for the Mississippi River.

The horizontal cross-channel profiles of temperature and specific conductance (Figure 3) measured a few meters downstream from the confluence point show similar step-like gradients with the greater temperature and specific conductance (about 7.8°C and 650 μS/cm) in the Mississippi River water and the lower temperature and specific conductance (about 4.4°C and 150 μS/cm) in the St. Croix River water. The densities of the water from each of the two rivers, calculated using the equation of state for lake water given by Chen and Millero [1986], were 1.000091 kg/m^3 for the St. Croix and 1.000245 kg/m^3 for the Mississippi River. Typical suspended-sediment concentrations are 5 mg/L and 36 mg/L which increases the approximate densities to 1.000096 and 1.000281 kg/m^3. Initially the difference in density occurred across a very thin vertical interface between the two fluids; approximating this thickness by 0.01 m and estimating the vertical velocity shear by the ratio of u_* and the height above the bottom gives a minimum estimate of the gradient Richardson number of approximately 0.2.

4.2. *20 Meters Downstream*

This measurement cross section was about 0.05 river widths downstream from the confluence and the width of the mixing zone was estimated visually to be about 10 m. A vertical profile made in about 4 m of water recorded specific conductances on the downcast that were nearly constant with depth (143-160 μS/cm) and were typical of pure St. Croix River water (Figure 4A). However, the corresponding

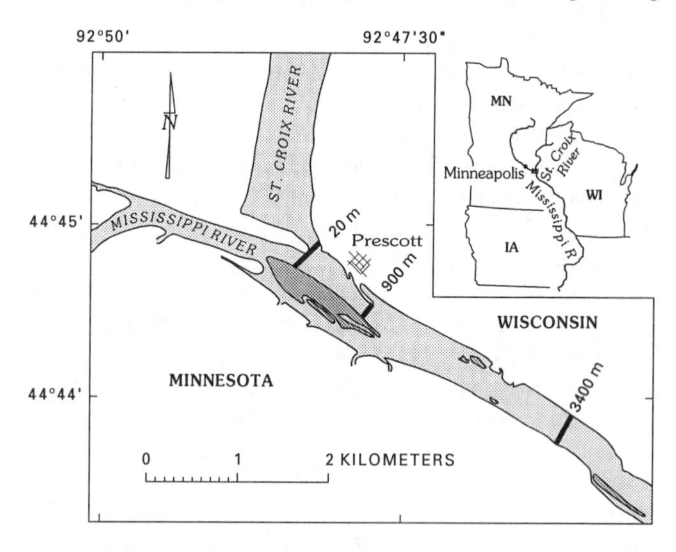

Fig. 2. Map showing the measurement cross sections downstream from the confluence of the Mississippi and St. Croix Rivers.

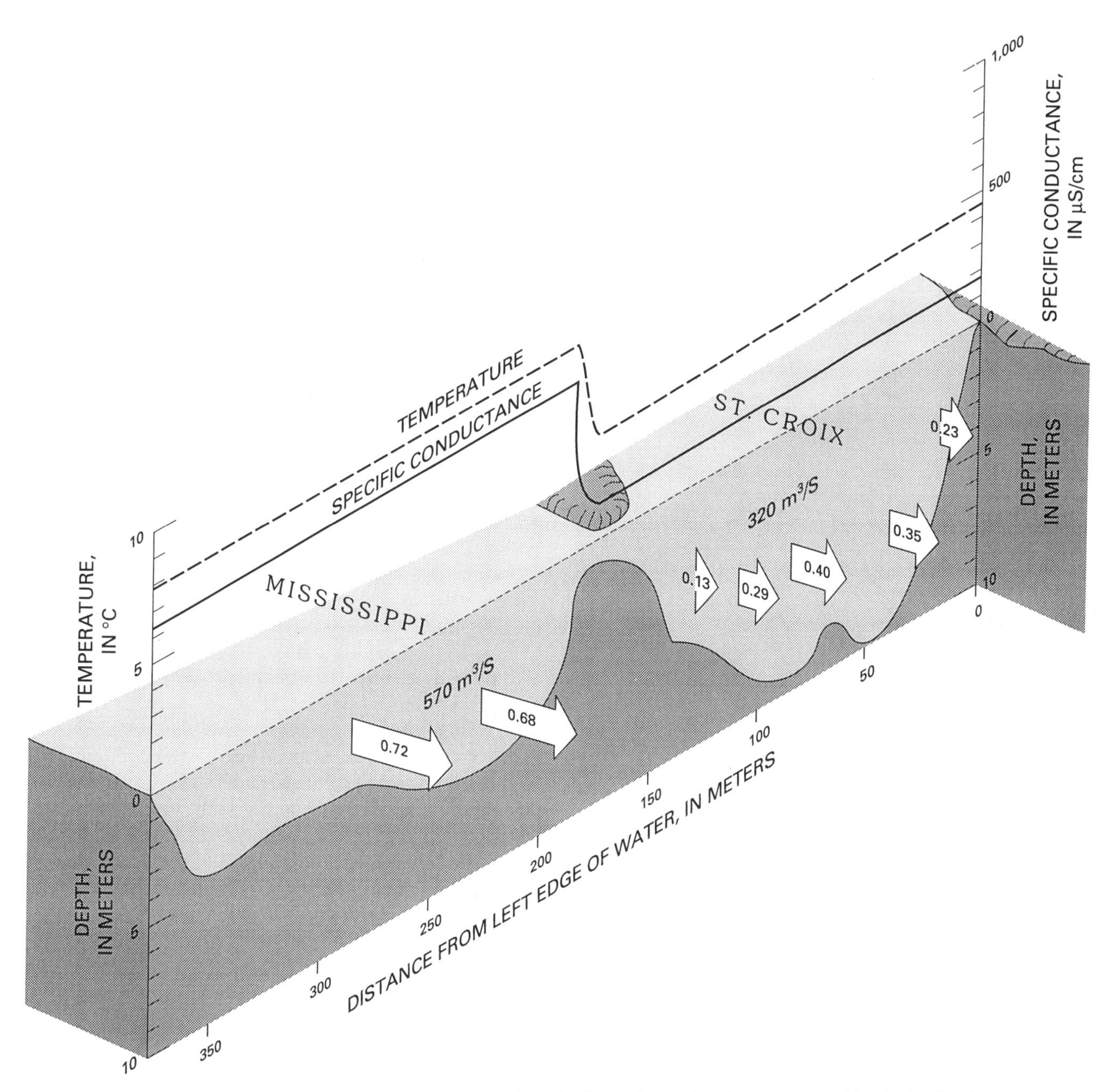

Fig. 3. Initial conditions at the confluence of the Mississippi and St. Croix Rivers on April 11, 1992. The depth-averaged velocities are shown as arrows with the velocity in meters per second shown inside the arrow. These depth-averaged velocities were measured about 20 m upstream from the confluence point. Discharge of the Mississippi River was 570 m^3/s, measured on April 10, 1992, approximately 1 km upstream from the confluence, based on 9 water depths and depth-averaged velocities. Discharge of the St. Croix River was 320 m^3/s, measured on April 11, 1992, based on 5 water depths and depth-averaged velocities. The difference in the temperature (dashed line) and specific conductance (solid line) across the interface between the two rivers was 3.4°C and 500 µS/cm.

temperatures measured during the downcast were not typical of the pure St. Croix River water but ranged from 4.4 to 5.9 °C (Figure 4B). There were indications of some intrusions of water masses (0.2-0.5 m thick) near the bottom of the upcast for specific conductance and near the top of the upcast for temperature. The upcast of temperature and specific conductance also recorded water that was not just a simple uniform mixture of Mississippi and St. Croix River waters.

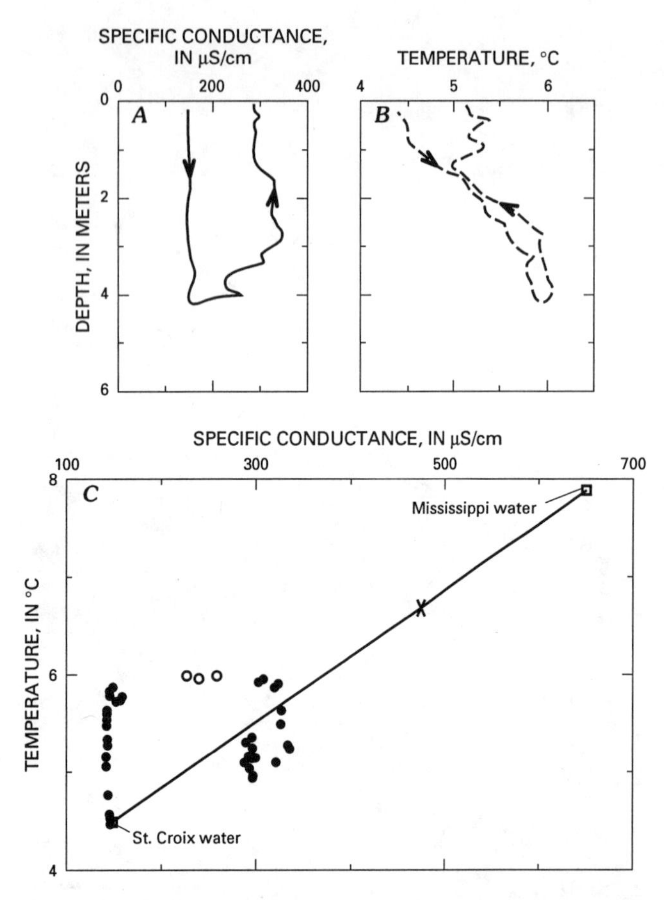

Fig. 4. Vertical profile in the near-field mixing zone 20 meters downstream from the confluence. A. Specific conductance--the arrow indicates the direction of travel of the CTD profiler. B. Temperature--the arrow indicates the direction of travel of the CTD profiler. C. TS-diagram for the data shown in 4A and 4B. The square boxes at either end of the uniform mixing line indicates the temperature and specific-conductance properties of pure Mississippi River water and of pure St. Croix River water. The X indicates the final temperature and specific-conductance values if the two water masses were completely mixed. Measurements near sharp gradients where response times may be important are shown as open circles.

This can best be seen in a plot of temperature versus specific conductance or TS-diagram (Figure 4C). On the TS-diagram, the pure Mississippi and St. Croix River water masses plot within the corresponding labeled square boxes. If mixing is only a mechanical process and not a double-diffusive process, then the temperature and specific conductance of any mixture of pure Mississippi and pure St. Croix River water masses should fall somewhere on the straight line of uniform mixing drawn between the two water masses; thus, the location on the uniform mixing line depends only on the ratio of the volumes of the two water masses. This, however, is not the case for the temperature and specific conductance measured in this vertical profile because several points do not plot on the uniform mixing line. The TS-diagram suggests that an intrusion of St. Croix River water was gaining heat from a Mississippi River water mass faster than it gained dissolved solids or specific conductance. The pattern in the TS-diagram (Figure 4C) indicates relatively long periods of time (several points in a group off the line of uniform mixing) when the CTD profiler was within intrusions of water in which the temperature and specific conductance had mixed at different rates. Some of the points off the uniform mixing line in the TS-diagram are the result of crossing a discontinuous gradient of temperature or specific conductance and the corresponding different response times of the temperature and specific-conductance sensors (for example, the three points between 200 and 300 μS/cm at about 6 °C in Figure 4C). The root-mean-square (rms) deviation of the temperature from the uniform mixing line was 0.46°C for the vertical profile (Figure 4C) and 0.50°C for the time series (Figure 5B). For comparison, the rms deviation from the uniform mixing line for temperature and specific-conductance profiles measured in pure Mississippi River water and in pure St. Croix River water was 0.01°C.

The longitudinal size of the intrusions of water, which varied from about 10 to 80 m (Figure 5A), was estimated from the time-series data by multiplying the length of the intrusion in seconds by the estimated water velocity (~0.60 m/s). Some of these intrusions were large enough that the CTD profiler recorded many measurements inside the intrusion. The length of time that the CTD profiler was inside an intrusion ranged from 8 to 121 s. The vertical gradients of temperature and specific conductances across the interface between intrusion are proportional to the temperature and specific-conductance differences, which are about 2°C and 400 μS/cm. Some intrusions shown in the time series (Figure 5A) are labeled with a letter and the corresponding data are enclosed and labeled in the TS-diagram for the time series (Figure 5B). Most temperature and specific conductance for the water intrusions again plot off the uniform mixing line.

4.3. 900 Meters Downstream

This measurement cross section was about 2 river widths downstream from the confluence. The width of the mixing zone was about 120 m, the cross-channel size of the intrusions was about 10 m (Figure 6A), and the vertical thickness of these intrusions averaged about 1 m (Figure 6B). The time series or longitudinal profile of specific conductance (Figure 6C) shows more numerous but smaller intrusions (3-60 m) than were recorded 20 m downstream. The large gradients of temperature and specific conductance that were observed 20 m downstream still exist, but there is

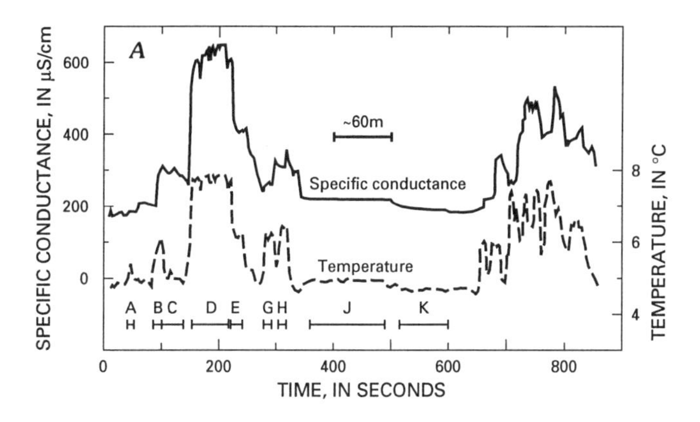

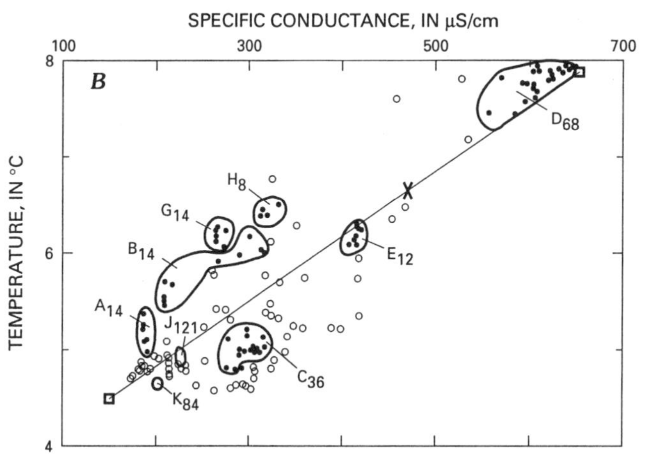

Fig. 5. Time series in the mixing zone at 1-m depth and 20 meters downstream from the confluence. A. Temperature (dashed line) and specific conductance (solid line) with letters identifying different intrusions of water. The length scale is based on an estimated velocity of 0.60 m/s at a depth of 1 m. B. TS-diagram for the data in 5A. The letters refer to the intrusions of water in Figure 5A, and the corresponding data points are enclosed. The subscript associated with each letter is equal to the time (in seconds) that the CTD profiler was inside the intrusion. Most of the data points outside the enclosed points represent measurements recorded while the CTD was crossing a sharp gradient and are shown as open circles.

an increase in the number of smaller gradients corresponding to the increase in smaller intrusions. Differences in temperature and specific conductance between intrusions were typically about 1 °C and 100 μS/cm. The TS-diagram (Figure 7A) shows most of the data points plotting closer to the uniform mixing line so that the rms deviation of temperature from the line had decreased to 0.18 °C.

4.4. 3400 Meters Downstream.

This measurement cross section was about 8 river widths downstream from the confluence. The width of the mixing zone was almost 200 m and the width-to-depth ratio of the river was about 100. The cross-channel size of the intrusions was about 5-10 m, but the differences in temperature and specific conductance between intrusions had decreased to about 0.5°C and less than 50 μS/cm (Figure 8A) so that the temperature and specific conductance were relatively smooth, monotonic functions of the cross-channel distance. Vertical profiles within the mixing zone showed nearly uniformly mixed water from surface to bottom with only faint evidence of interleaving intrusions and with variations of temperature ranging from 0.1-0.2 °C and with variations of specific conductance (Figure 8B) ranging from 5-30 μS/cm. The time series or longitudinal profile showed even more numerous but smaller intrusions (about 4 m long) with similar differences in temperature and specific conductance (Figure 8C) between the intrusions of 0.1-0.3 °C and 5-30 μS/cm. The TS-diagram for one horizontal profile (Figure 7B) shows almost all the data plotting on the uniform mixing line. The rms deviation of temperature from the uniform mixing line was only 0.03°C, which is very close to the rms value for the initial well-mixed pure Mississippi and St. Croix River water masses.

5. DISCUSSION

Mixing downstream from the confluence of two rivers seems to consist of two processes similar to those described by Eckart [1948]. The fluids are stirred by advective shear instabilities then mixed by diffusion. The instabilities deform the interface between the fluids, increasing the surface area of sharp gradients. Visual observations show that the interface is convoluted laterally into vortex-like filaments. Evidence from vertical profiles of temperature and specific conductance collected 20 m and 900 m downstream from the confluence indicated that there is interleaving of water masses, producing fully 3-dimensional structures that further increase the surface area of sharp gradients. The spacing of the interleaving (0.2-0.5 m) relative to the initial size of the shear instabilities (10-80 m) is much less than the ratio of 0.7 given for laboratory spacing of streamwise vortices in unbounded free shear flow and may be a result of the mixing layer being confined to a finite depth. The estimated gradient Richardson number is near the critical value of 0.25 for stability, so that the velocity shear may produce microscale turbulent instabilities near the interfaces that enhance the mixing by triple diffusion. The velocity shear also transforms the original vertical interface into numerous horizontal interfaces along which either finger or diffusive regimes could exist [*Turner*, 1973]. Assuming no turbulent diffusion, the final temperature and specific conductance that would result from double diffusion between two well-mixed intrusive water masses can be estimated by using the initial

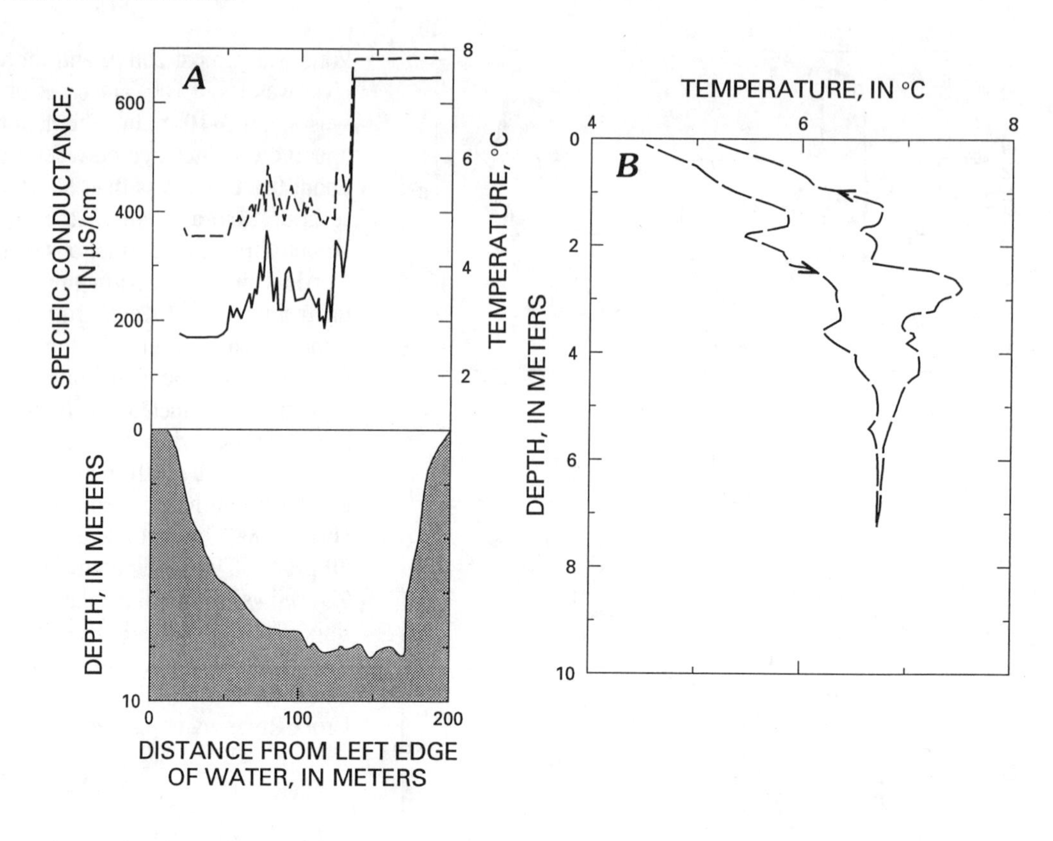

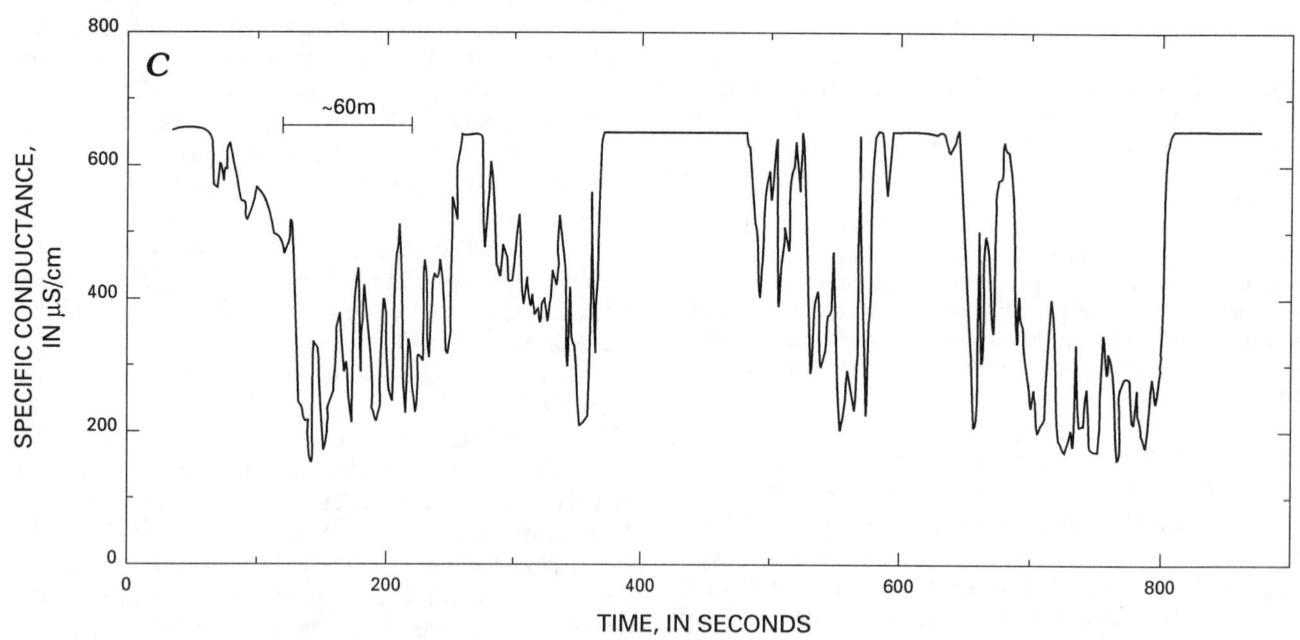

Fig. 6. Horizontal cross-channel profiles, vertical profiles, and time series measured 900 meters downstream from the confluence. A. Temperature (dashed line) and specific conductance (solid line) plotted above the shaded bottom profile. B. Vertical profile of temperature in the mixing zone (arrows indicate the direction of travel of the CTD profiler). C. Time series at 1-m depth in the mixing zone. The length scale is based on an estimated velocity of 0.60 m/s.

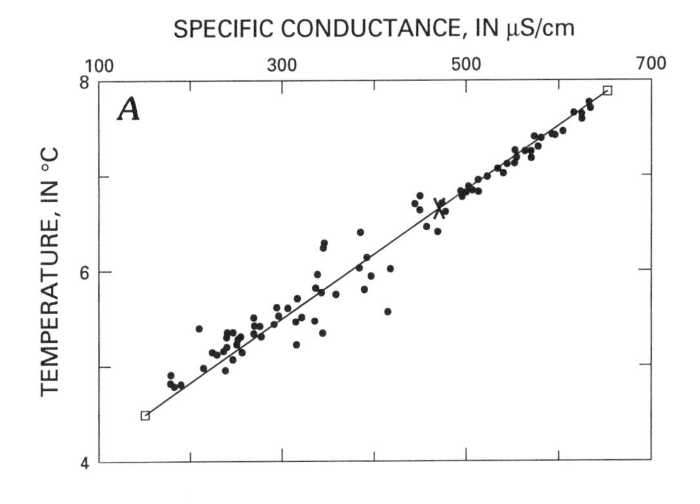

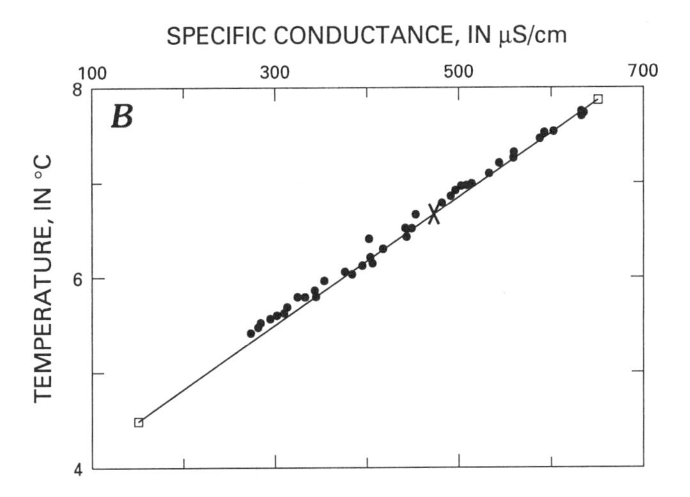

Fig. 7. TS-diagrams. A. Data are the first 200 seconds of a time series measured at 1-m depth and 900 meters downstream from the confluence (Figure 6C) B. Data are from a horizontal profile that was measured 3400 meters downstream from the confluence (Figure 8C).

conditions of temperature and specific conductance and equations for the temperature, T(z,t), and the specific conductance, S(z,t), which are error functions of the distance away from the interface, z, and the time, t [*Williams*, 1981]:

$$T(z,t) = \frac{T_2 + T_1}{2} + \frac{T_2 - T_1}{2} erf\left(\frac{z}{\sqrt{4K_T t}}\right) \tag{4}$$

$$S(z,t) = \frac{S_2 + S_1}{2} + \frac{S_2 - S_1}{2} erf\left(\frac{z}{\sqrt{4K_S t}}\right) \tag{5}$$

where the diffusivity for temperature and specific conductance $K_T = 1.4 \times 10^{-7}$ $m^2 \cdot s^{-1}$ and $K_S = 1.5 \times 10^{-9}$ $m^2 \cdot s^{-1}$. At distances of 0.1 m from the interface (within a typical intrusion) and after 10 seconds, the increase in temperature of pure St. Croix water (4.4°C, 150 µS/cm) would be about 1°C while the increase in specific conductance would be much less than 1 µS/cm. After 100 seconds the increase in temperature and specific conductance would be about 1.5°C and 20 µS/cm, resulting in final values of about 5.9°C and 170 µS/cm. These final temperature and specific-conductance values predicted by double diffusion are very similar to the measured values that plot off the uniform mixing line in the TS-diagram for profiles measured 20 m downstream from the confluence (Figure 4C).

Because the mixing zone is also a zone of convergence, there is a component of velocity normal to the interface. This component may act independently or in concert with the lateral shear to deform the interface and to produce interleaving intrusions with a corresponding increase in surface area. The finite depth may cause complex 3-dimensional structures to evolve from the initially 2-dimensional vortices along the interface as a result of the vertical shearing of vortices by the streamwise velocity. Regardless of the cause of this deformation, the interface is characterized by intrusions that were initially large enough that the CTD profiler could log many data records of temperature and specific conductance inside each intrusion. These measurements inside the intrusion suggest that mixing occurs in two steps. Heat diffuses first because the diffusivity for heat is approximately 100 times the diffusivity for dissolved solids [*Thorpe, et al.*, 1969; *Williams*, 1981]. This diffusion decreases the temperature gradients, reduces the heat flux, and allows the flux of dissolved solids to catch up to the flux of heat so that the intrusions eventually becomes uniformly mixed.

Concurrently with this double-diffusion process, the water is being advected downstream by the mean velocity and intrusions are broken up into smaller intrusions by perhaps the lateral and vertical shear. The breaking up of large intrusions into smaller ones involves turbulent motions smaller than the water depth. These turbulent motions may be as important a mechanism for extracting energy from the lateral shear and hence stabilizing the growth of the shear instabilities as is the bottom-friction mechanism suggested by Alavian and Chu [1985]. For most flows, the cross-sectionally averaged value of the bottom-friction coefficient will probably be constant as a function of downstream distance unless there are some radical changes in the roughness of the channel. Therefore, a downstream increase in the bottom friction stability parameter, S_f, is primarily a result of the decrease in the lateral shear and not due to an increase in the bottom-friction. The smaller intrusions limit the time the CTD profiler can log data inside the intrusions

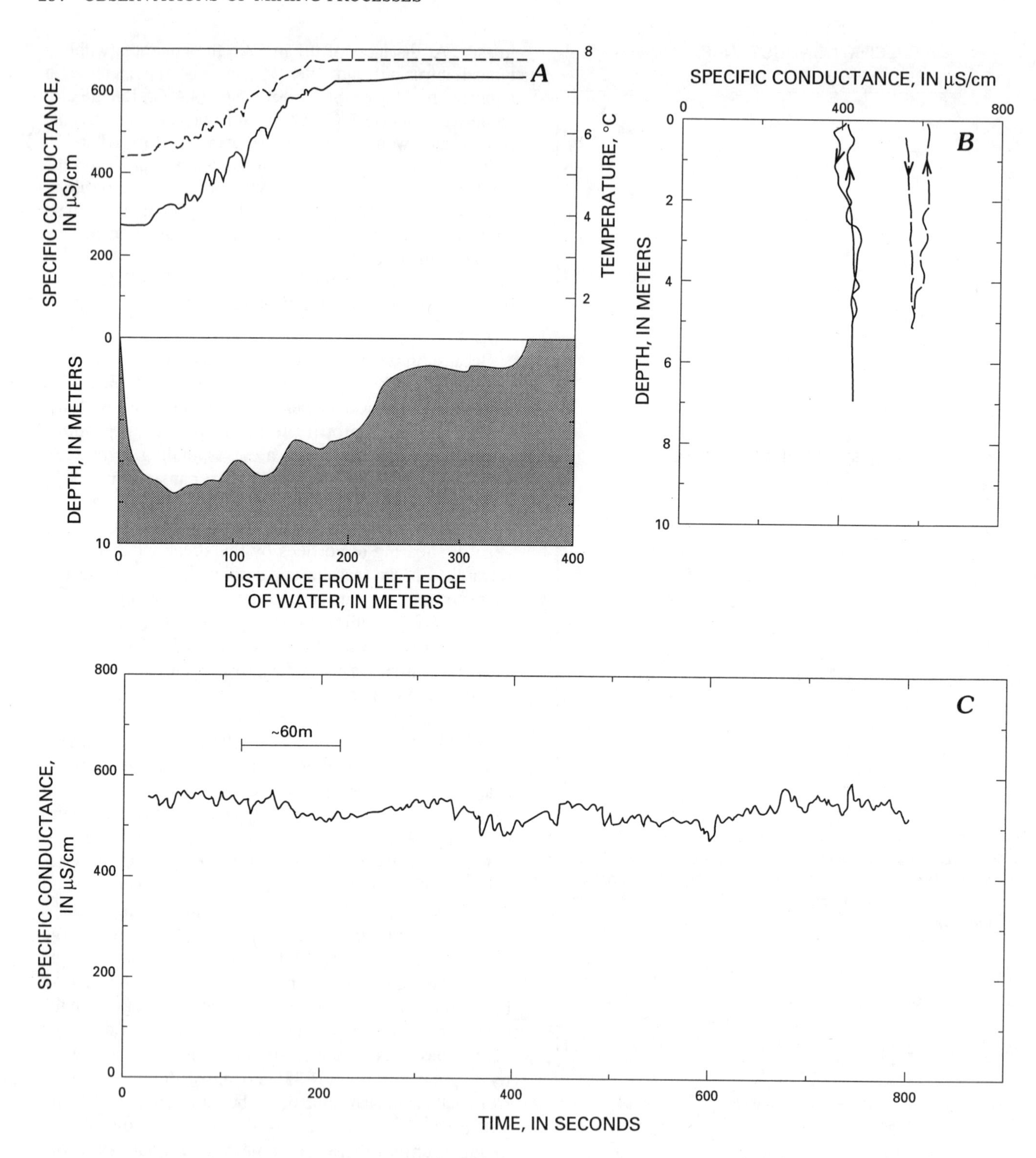

Fig. 8. Horizontal and vertical profiles and time series measured 3400 meters downstream from the confluence. A. Temperature (dashed line) and specific conductance (solid line) plotted above the shaded cross-channel bottom profile. B. Vertical profiles of specific conductance within the mixing zone (arrows indicate the direction of travel of the CTD profiler). The profiles on the left and right were located at 78 m and 176 m from the left edge of water. C. Time series at l-m depth in the mixing zone. The length scale is based on an estimated velocity of 0.60 m/s.

so that changes in temperature and specific conductance indicative of double diffusion may not be recorded. The CTD profiler measures an average of many smaller intrusions; this average measurement only indicates that the water is more uniformly mixed as shown by the TS-diagram at 900 m (Figure 7C). It is hypothesized, however, that double diffusion is still going on inside the smaller intrusions but at a slower rate because the temperature and specific-conductance gradients have been reduced, and a situation closer to uniform mixing has developed.

Farther downstream, the smaller intrusions become uniformly mixed and even smaller in size until they approach the microscopic size of a water "particle." By the time intrusions reach 3400 m downstream, the water is vertically homogenous; there are smooth monotonic changes in temperature and specific conductance across the channel and very small deviations of temperature and specific conductance as a function of time. At this distance downstream (about 8 river widths), the water has left the near-field zone and entered the far-field zone where other mixing processes such as 2-dimensional horizontal eddies, bank eddies, and wind-driven circulation dominate over lateral shear and double diffusion.

6. SUMMARY

Mixing in the near-field zone downstream from the confluence of the Mississippi and St. Croix Rivers consisted of stirring and diffusion processes. The vertical interface between the two different water masses began as an initial step-like gradient with differences of temperature (3.4 °C), specific conductance (500 μS/cm), and velocity (0.18 m/s). Lateral shear across the interface provided the stirring mechanism resulting in shear instabilities which appeared to have 3-dimensional structures similar to the streamwise vortices observed in laboratory experiments. The streamwise vortices may have appeared between 0.05 and 2 river widths downstream from the confluence point as interleaving intrusions of water with vertical scales less than 0.5 m. Intrusions decreased in longitudinal size in the downstream direction. The approximate longitudinal size at 0.05 river widths downstream was 10-80 m; at 2 river widths downstream it was 5-40 m, and at 8 river widths it had decreased to about 4 m. The intrusions were large enough near the confluence so that the CTD profiler was able to record data inside the intrusions that suggests mixing at the edges of intrusions may be double diffusive. Double diffusion was identified by numerous temperature and specific-conductance values within intrusions that deviated from the uniform mixing line in a TS-diagram and that were consistent with predictions based on non-advective double diffusion. An estimate of the gradient Richardson number suggests that small-scale density-driven flow could be present and may be a mechanism for increasing the surface area of intrusions such that diffusion was a more effective mixing process during the time that the intrusions are advected downstream. The rms temperature deviation from the uniform mixing line was used as a measure of the magnitude of double diffusion. In the well-mixed, pure Mississippi or St. Croix water, the rms value was 0.01°C, but 0.05 river widths downstream from the confluence the rms value was about 0.48°C. The rms value decreased to 0.18°C at 2 river widths and finally to 0.03°C at 8 river widths downstream from the confluence, where intrusions were too small to be resolved by the CTD profiler. At this distance downstream from the confluence, the cross-channel variations of temperature and specific conductance were approximately smooth monotonic functions of cross-channel distance, indicating a transition to the far-field mixing zone.

Acknowledgments. Collecting water and instrument depths, distances, temperatures, specific conductances, and velocities simultaneously from a small boat using digital and analog methods required two people and John Garbarino provided not only another pair of hands on a cold day, but also some valuable suggestions. To help us get a clear perspective of the mixing zone Bob Meade provided numerous photographs from various heights above the water. Alan Brandt, John Flager, Mary Kidd, Deborah Martin, Bob Meade, Jon Nelson, Jim Smith, and two other reviewers all made excellent suggestions that improved each iteration of this paper.

REFERENCES

Alavian, V., and V. H. Chu, Turbulent exchange flow in a shallow compound channel, *Proc., 21st Congress of Int. Assoc. of Hydraulic Res.*, 3, 446-451, 1985.

Babarutsi, S., and V. H. Chu, Dye-concentration distribution in shallow recirculating flows, *J. Hydr. Eng.*, 117, 5, 643-659, 1991.

Bernal, L. P., and A. Roshko, Streamwise vortex structure in plane mixing layers, *J. Fluid Mech., 170*, 499-525, 1986.

Best, J. L., and A. G. Roy, Mixing-layer distortion at the confluence of channels of different depth, *Nature*, 350, 411-413, 1991.

Browand, F. K. and T. R. Troutt, The turbulent mixing layer: geometry of large vortices, *J. Fluid Mech., 158*, 489-509, 1985.

Bruno, M. S., M. Muntisov, and H. B. Fischer, Effect of buoyancy on transverse mixing in streams, *J. Hydr. Eng., 116*, 1484-1494, 1990.

Chen, C. T., and F. J. Millero, Precise thermodynamic properties for natural waters covering only the limnological range, *Limnology and Oceanography, 31*, 657-662, 1986.

Chu, V. H., J. H. Wu, and R. E. Khayat, Stability of turbulent shear flows in shallow channel, *Proc. 20th Congress of IAHR*, Moscow, USSR, *3*, 128-133, 1983.

Chu, V. H., and S. Babarutsi, Confinement and bed-friction effects in shallow turbulent mixing layers, *J. Hydr. Eng., 114*, 1257-1273, 1988.

Clesceri, L. S., A. E. Greenberg, and R. R. Trussell, Eds., Standard methods for the examination of water and wastewater: American Public Health Association, Washington, D. C., Chap. 2, p. 2-57 to

2-61, 1989.

Eckart, C., An analysis of the stirring and mixing processes in incompressible fluids, *J. Mar. Res., VII*, 265-275, 1948.

Fischer, H. B., The effect of bends on dispersion in streams, *Water Resour. Res.*, 5, 496-506, 1969.

Jimenez, J., M. Cogollos, and L. P. Bernal, A perspective view of the plane mixing layer, *J. Fluid Mech., 152*, 125-143, 1985.

Krishnappan, B. G., and Y. L. Lau, Transverse mixing in meandering channels with varying bottom topography, *J. Hydr. Res., 15*, 351-371, 1977.

Lasheras, J. C., J. S. Cho, and T. Maxworthy, On the orgin and evolution of streamwise vortical structures in a plane, free shear layer, *J. Fluid Mech., 172*, 231-258, 1986.

Mackay, J. R., Lateral mixing of the Liard and Mackenzie rivers downstream from their confluence, *Can. J. Ear. Sci., 7*, 111-124, 1970.

Matsui, E., F. Salati, I. Friedman, and W. L. F. Brinkman, Isotopic hydrology of the Amazonia, 2, Relative discharges of the Negro and Solimões rivers through ^{18}O concentrations, *Water Resour. Res., 12*, 781-785, 1976.

Ruddick, B., Intrusive mixing in a Mediterranean salt lens--intrusion slopes and dynamical mechanisms, *J. Phys. Oceanogr., 22*, 1274-1285, 1992.

Sayre, W. W., and F. M. Chang, A laboratory investigation of open-channel dispersion processes for dissolved, suspended, and floating dispersants, *U. S. Geological Surv. Prof. Pap., 433-E*, 1968.

Stallard, R. F., Cross-channel mixing and its effect of sedimentation in the Orinoco River, Water Resour. Res., 23, 1977-1986, 1987.

Thorpe, S. A., P. K. Hutt, and R. Soulsby, The effect of horizontal gradients on thermohaline convection, *J. Fluid Mech., 38*, 375-400, 1969.

Turner, J. S., *Buoyancy effects in fluids*, 368 pp., Cambridge University Press, New York, 1973.

Webel, G., and M. Schatzmann, Transverse mixing in open channel flow, *J. Hydr. Eng.*, 110, 423-435, 1984.

Weibezahn, F. H., A. Heyvaert, and M. A. Lasi, Lateral mixing of the waters of the Orinoco, Atabapo, and Guaviare Rivers, after their confluence, in southern Venezuela, *Ecologia, 40*, 263-270, 1989.

Williams, A. J., III, The role of double diffusion in a Gulf Stream frontal intrusion, *J. Geo. Res., 86*, 1917-1928, 1981.

Winant, C. D., and F. K. Browand, Vortex pairing: the mechanism of turbulent mixing-layer growth at moderate Reynolds number, *J. Fluid Mech., 63*, 237-255, 1974.

Yotsukura, N., H. B. Fischer, and W. W. Sayre, Measurement of mixing characteristics of the Missouri River between Sioux City, Iowa, and Plattsmouth, Nebraska, *U. S. Geol. Surv. Water-Sup. Pap., 1899-G*, 1970.

John A. Moody, U. S. Geological Survey, Mail Stop 413, Denver Federal Center, Lakewood, CO 80225.

A Model of the Ocean Thermocline Stepwise Stratification Caused by Double Diffusion

Ye. Yu. Kluikov, and L. N. Karlin

Department of Oceanology, Russian State Hydrometeorological Institute, St.Petersburg, Russia

Conditions of convection formation in the regimes of salt fingers and diffusion are considered, resulting in characteristics discontinuities and convective mixed layers on the temperature and salinity profiles. A mathematical model for simulation of the water temperature and salinity evolution in the mixed layer structure formation above and below the interface boundaries with the intrusion process, heat and salt fluxes in the vertical direction as well as effective vertical salt and heat exchange coefficient, is proposed.

INTRODUCTION

In the upper layer of the World ocean there are widely spread stratification conditions facilitating convective instability caused by double diffusion. The processed climatic data on the vertical temperature and salinity distribution all over the World ocean in the layer between 200 m to 250 m deep have shown that about two thirds of the water mass have stratification favoring development of salt fingers, 1/6 of the water mass convection in the diffusive regime, and only 1/6 is stratified in a convectively stable way (Figures 1,2) [*Karlin et al,* 1988]. The convectively unstable state of the stratification is often accompanied by the stepwise stratification of the temperature T and salinity S profiles.

We can propose the following physical explanation of this stratification. Suppose in a layer with the stratification favoring development of a convective instability caused by some disturbance there starts development of a convection either in the regime of salt fingers or in a diffusive one (Figure 3). Both types of this convection are known to be accompanied by a flow of negative buoyancy. The magnitude of this flow is comparable to that which usually occurs on the ocean surface at night. This flow conditions intensive mixing leading to the emergence of a mixed layer. As the convection develops, the mixed layer expands. Because of entrainment of undisturbed liquid there occur steep water temperature and salinity gradients at the mobile boundary of the convective layer. Conditions for forming either salt finger or diffusive - regime convection also develop there. As a result, a convective mixed layer develops from discontinuity layer in the way desribed above. Thus this process can take place many times. Eventually there emerges a stepwise stratification of the temperature and salinity profiles which were formerly smooth.

We propose a mathematical model for the phenommenon described. The heat and salt transfer equations integrated between the limits of the mixed layer have the form:

$$(dT_0/dt)h = q_0^T - q_h^T \tag{1}$$

$$(dS_0/dt)h = q_0^S - q_h^S \tag{2}$$

where T_0, S_0 are water temperature and salinity in a mixed layer of a convective origin which are taken as vertically constant; h is thickness of a mixed

Double-Diffusive Convection
Geophysical Monograph 94

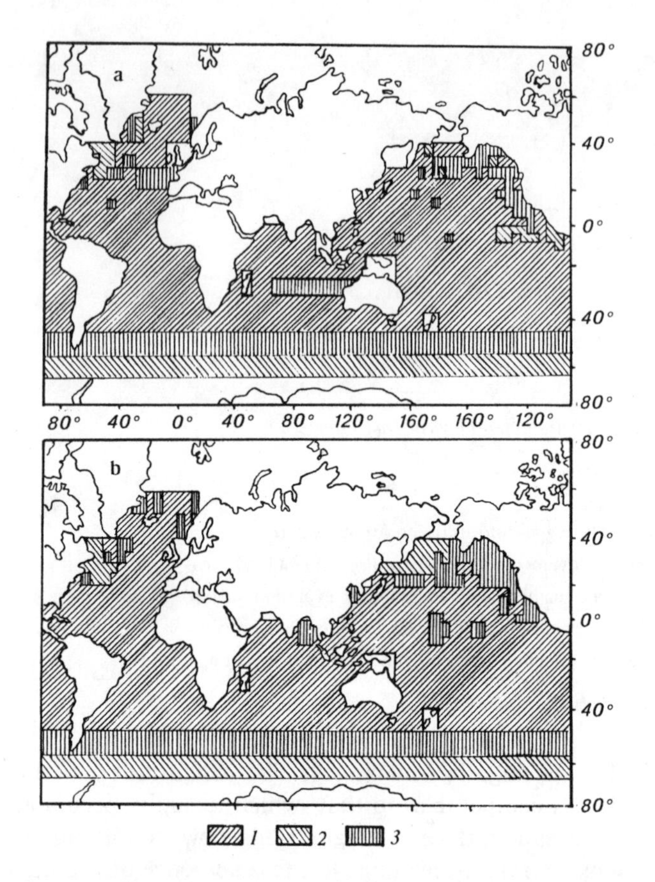

Fig.1. Conditions of convective stability in the layer between 200 m to 250 m deep: *1* - conditions with instability in the form of salt fingers; *2* - conditions with diffusive instability; *3* - convectively stable conditions. *a*. February; *b*. July.

layer; q_0^T, q_0^S, q_h^T, q_h^S are heat and salt fluxes at the upper and lower boundaries of a mixed layer. The heat and salt fluxes are normalized on $c_p r_0$ and r_0 respectively (c_p, r_0 are sea water specific heat and density, respectively).

The heat and salt fluxes across the boundary of a mixed layer are presented as a sum of two fluxes, one caused by entrainment and one caused by convection due to double diffusion, that is:

$$q_h^T = q_{h1}^T + q_{h2}^T \tag{3}$$

$$q_h^S = q_{h1}^S + q_{h2}^S \tag{4}$$

where q_{h1}^T, q_{h1}^S are heat and salt fluxes caused by entrainment; q_{h2}^T, q_{h2}^S are heat and salt fluxes due to convection with double diffusion.

The heat and salt fluxes caused by the entrainment from below are estimated employing traditional relationships:

$$q_{h1}^T = -(dh/dt)\,\Delta T \tag{5}$$

$$q_{h1}^S = -(dh/dt)\,\Delta S \tag{6}$$

where ΔT, ΔS are steep temperature and salinity gradients at the mobile boundary of a mixed layer.

The fluxes caused by double diffusive convection have been calculated according to the known empirical relationships. With the salt fingers [*Schmitt*, 1979]:

$$q_{h2}^S = c(g\mu^T \beta)^{1/3} (\Delta S)^{4/3} \tag{7}$$

where g is free fall acceleration; μ^T is coefficient of molecular heat exchange; β is coefficient of salt compression; c is a coefficient, it is a function of the the density relationship, R.

The type of the dependence of the coefficient c on the density relationship is set so that when $R \to 1$ the value of $c>0.1$ ($\sim$0.12), when $R\sim 2$, c equals 0.1, and with increasing R c decreases; when $R>>3.5$ c remains almost constant and equals $\sim$0.05.

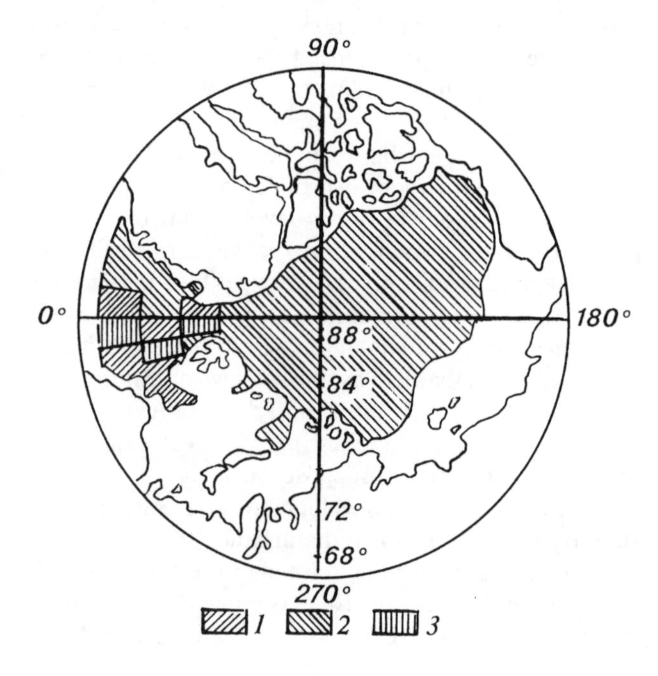

Fig.2. Conditions of convective stability in the layer between 200 m to 250 m deep: *1* - conditions with instability in the form of salt fingers; *2* - conditions with diffusive instability; *3* - convectively stable conditions.

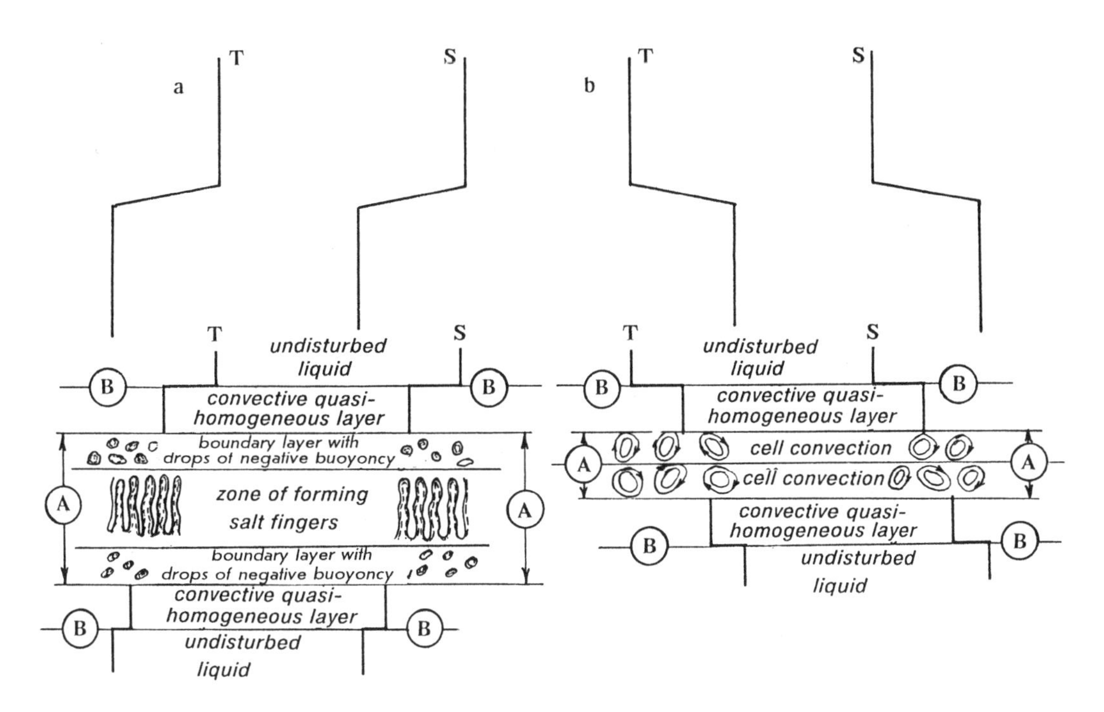

Fig.3. Schematic formation of thermocline stepwise structure: *a* - with salt fingers; *b* - with diffusive convection. Letter *B* indicates places where the same processes as in zone *A*, take place. Above, initial profiles.

With diffusive convection [*Marmorine and Caldwell*, 1976]:

$$q^{T}_{h2} = 0.085 f(R)\, \mu^{T} (g\alpha/\mu\mu^{T})^{1/3} (\Delta T)^{4/3} \quad (8)$$

where α is the coefficient of sea water thermal expansion; μ is the coefficient of molecular viscosity; R is the density relationship; $f(R)=0.101 \exp\{4.6 \times [\exp[-0.54(R-1)]\}$ [*McDougall*, 1983].

To complete the system we employ the turbulence energy balance equation written for the level $z = h$:

$$(db/dt) - Diff + Diss + g(-\alpha q^{T}_{h} + \beta q^{S}_{h}) = 0 \quad (9)$$

where b is energy of turbulent pulsations in the mixed layer.

In Equation 9 the first term is the change of intensity of turbulent pulsations in time. The second term is diffusion of turbulence energy. The third term is dissipation of turbulence energy. The last term in (9) allows for the influence of the buoyancy force on the speed of the change in turbulent pulsation intensity.

Diffusion of the turbulence energy from one mixed layer into another is presented in the form:

$$Diff = c_1\{[(b^{3/2}/h) - (b_2^{3/2}/h_2)]\} \quad (10)$$

where h_2 and b_2 are thickness of a second overlying mixed layer and turbulence energy in it; c_1 is an empirical constant.

We allow for dissipation by the traditional relationship:

$$Diss = c_5\,[(b^{3/2})/h] \quad (11)$$

where c_5 is an empirical constant.

To estimate the magnitude of the energy of turbulent pulsations we assume that the turbulence source in a mixed layer is a flow of negative buoyancy causing convection. Then we have for b:

$$b = c_4\,[gh(-\alpha q^{T}_{0} + \beta q^{S}_{0})]^{2/3} \quad (12)$$

$$b_2 = c_4\,[gh(-\alpha q^{T}_{h2} + \beta q^{S}_{h2})]^{2/3} \quad (13)$$

After making some transformations we get the formula for the thickness of a mixed layer:

$$(dh/dt)^{2/3} = (2/3)\,[(-c_4^{3/2})\,g(-\alpha q^{T}_{0} + \beta q^{S}_{0})(c_1 - c_5) + (-\alpha q^{T}_{h2} + \beta q^{S}_{h2})(c_1 c_4^{3/2} - 1)] \,/\, [(2/3) c_4 g^{2/3} (-\alpha q^{T}_{0} + \beta q^{S}_{0})^{2/3} + g h^{1/3} (\alpha \Delta T - \beta \Delta S)] \quad (14)$$

Proceeding from (14) (when ΔT and ΔS are small) we can obtain for the initial stage:

$$h(t)=(c_1-c_5)^{3/2}g^{1/2}(\left|-\alpha q_0^T+cq_0^S\right|)^{1/2}t^{3/2}c_4^{3/4} \quad (15)$$

The described system makes it possible to reproduce the water temperature and salinity evolution in a mixed layer and its thickness. It should be pointed out that existence of the first convective layer creates conditions for development of the second one, and existence of the second layer - for the third one and so on. There may be several layers simultaneously. They can expand with various speeds, overtake each other, which results in absorption of a slowly expanding layer by a faster one. Because of this, the given formulas should be written for each forming layer, employing the respective steep temperature and salinity gradients at the boundaries.

Several initial states have been selected for numeric experiment. For instance, structurization from the interface of a warm salt intrusion. With the initially linear temperature and salinity profiles of water above it in the process of the evolution the latter break into a great number of layers divided by high-gradient strata. As a rule, the layer thickness decreases when moving away from the intrusion. In some time the system attains a steady state (Figure 4). The intensity of structurization here is defined by the magnitude of the steep temperature gradient at the upper intrusion boundary and by the background stratification of the water temperature and salinity fields. The steeper the initial temperature gradient, the thicker the total water layer where a fine structure is forming (h^*). This relationship is close to a linear one (Figure 5a). The more pronounced is the background stratification, the thinner is the layer where structurization takes place, ΔT_1 being equal. The average thickness of the mixed layers ($\bar{h}$) is also linearly dependent on the initial ΔT_1, and the weaker the background stratification of waters, the larger the average thickness (Figure 5b). The thicker the layer of the fine structure, the longer the time of attaining a steady state. This relationship is also linear (Figure 5c). Also, the slope is not the same with various background stratification. Well pronounced is the tendency to increase the time necessary for formation of a stepwise structure of equal thickness, background stratification of waters increasing.

Similar relationships have been obtained for water temperature and salinity profiles when T and S decrease with depth (Figures 6,7).

With structurization due to the mechanisms considered above the heat and salt transfers are carried out in the vertical direction. Proceeding from the numeric experiments there have been calculated changes in heat and salt content in liquid when a fine

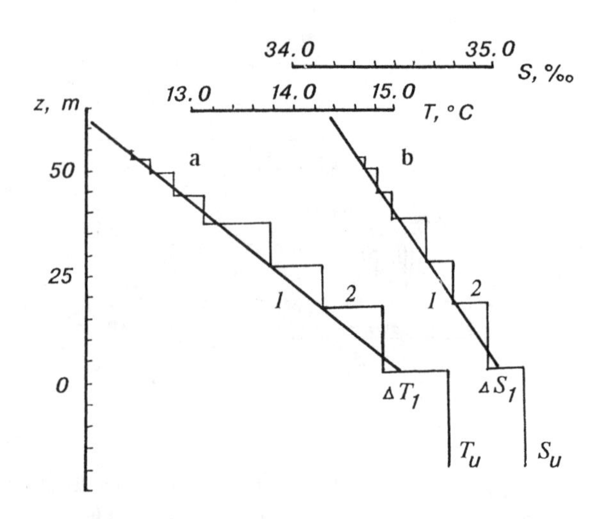

Fig.4. Temperature (a) and salinity (b) of water (R calculated by the background G^T and G^S, tends to 1, ΔT_1 =0.48 °C): 1 - initial profiles; 2 - steady state (attained after 184 hrs since initial moment). Regime of diffusive convection.

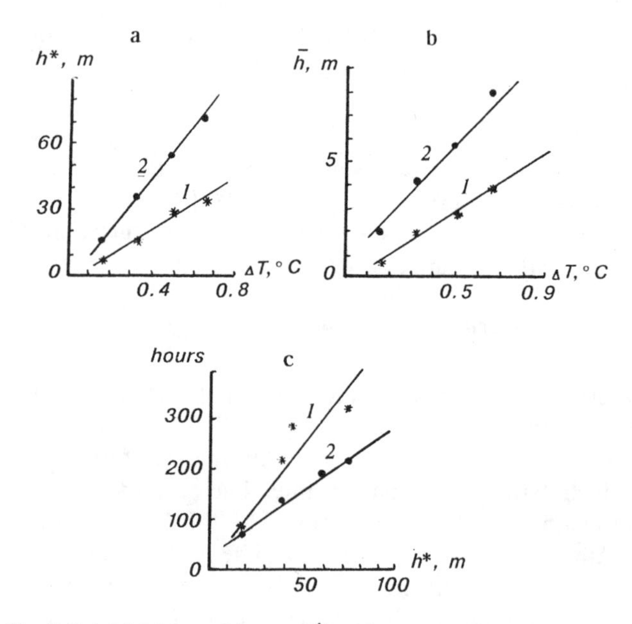

Fig.5. Total thickness of layer (h^*) with structurization (a), average thickness ($\bar{h}$) of mixed layer of convective origin (b) and time of attaining steady stepwise state depending on h^* (c): 1 - R=2; 2 - $R \rightarrow 1$. * - are values calculated with R=2; • - are values calculated with $R \rightarrow 1$. Regime of diffusive convection.

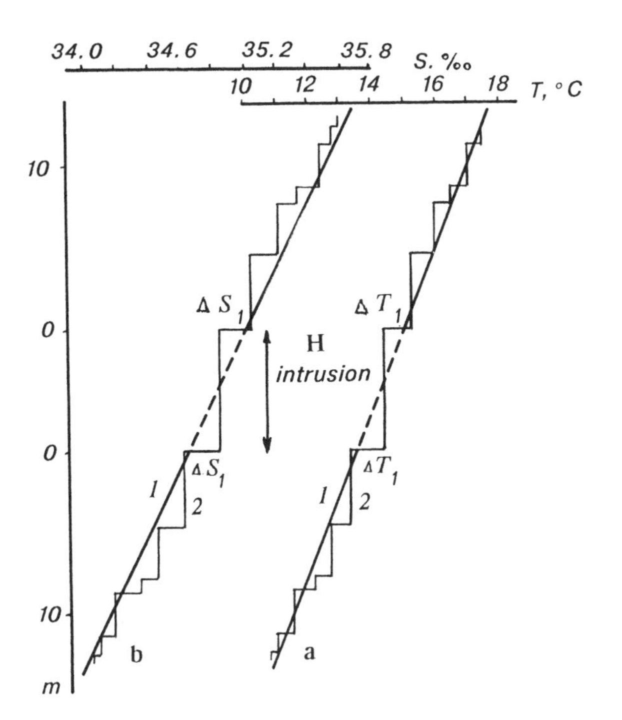

Fig.6. Temperature (*a*) and salinity (*b*) of water (*R* calculated by the background G^T and G^S, tends to 1, ΔS_1 =0.16‰): *1* - initial profiles; *2* - steady state (attained after 184 hrs since initial moment). Regime of salt fingers.

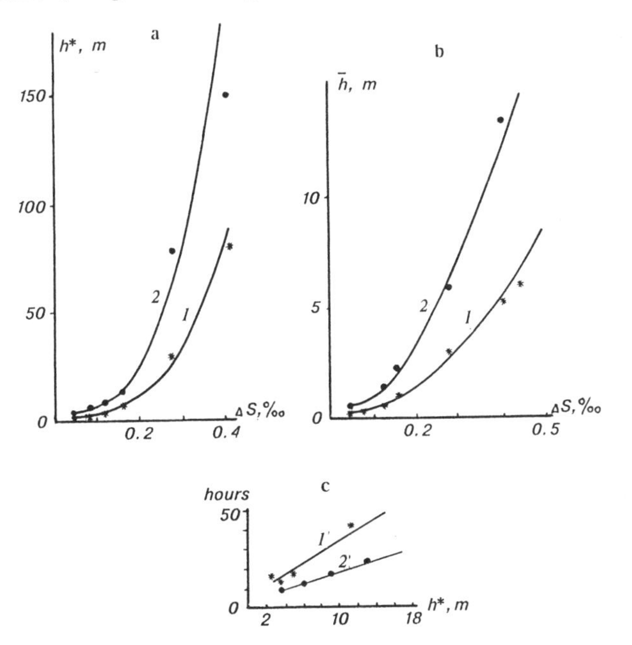

Fig.7. Total thickness of layer (h^*) with structurization (*a*), average thickness ($\bar{h}$) of mixed layer of convective origin (*b*) and time of attaining steady stepwise state depending on h^* (*c*): *1* - *R*=2; *2* - $R \rightarrow 1$. * - are values calculated with *R*=2; • - are values calculated with $R \rightarrow 1$. Regime of salt fingers.

structure was forming. According to these changes and taking into account the time necessary for forming a fine structure, heat and salt fluxes have been estimated in the vertical direction. It enabled us to calculate the coefficients of effective vertical heat and salt exchange (k^T and k^S, respectively), taking into account the initial preset values of the temperature and salinity gradients, G^T and G^S. With structurization due to diffusive convection the values of the exchange coefficients depend on the magnitude of the initial temperature gradient and background ground value *R* (plane ratio). The larger T_1, the larger k^T and k^S (Figure 8). When the initial value of *R* increases, k^T and k^S decrease. When *R*--1, the values of k^T and k^S are similar, other conditions being equal (Figure 9). Also, k^S is somewhat less than k^T.

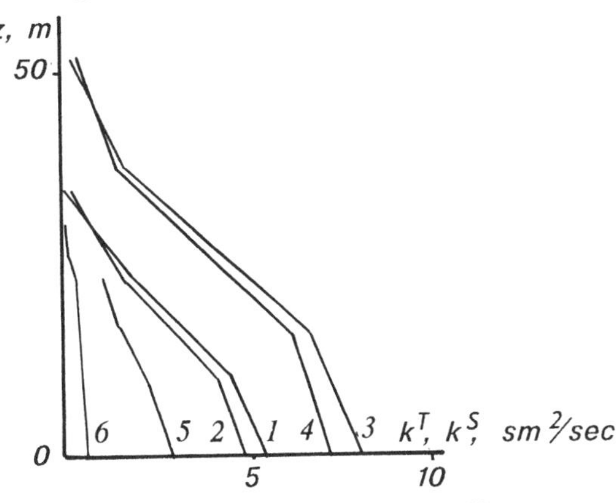

Fig.8. Coefficients of effective vertical heat (k^T) and salt (k^S) exchange with structurization due to convection in diffusive regime: *1* - k^T, *2* - k^S ($R \rightarrow 1, \Delta T_1 = 0.32\,°C$); *3* - k^T, *4* - k^S ($R \rightarrow 1, \Delta T_1 = 0.48\,°C$); *5* - k^T, *6* - k^S ($R=2, \Delta T_1 = 0.32\,°C$).

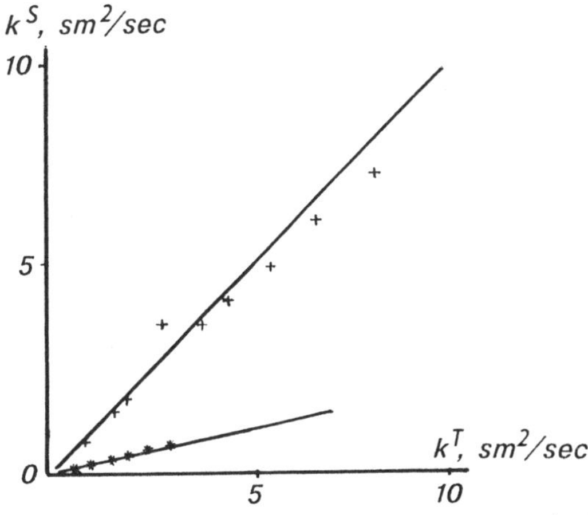

Fig.9. Ratio of coefficients of effective vertical heat and salt exchange with structurization due to convection in diffusive regime: + - with *R*--1; * - with *R*=2.

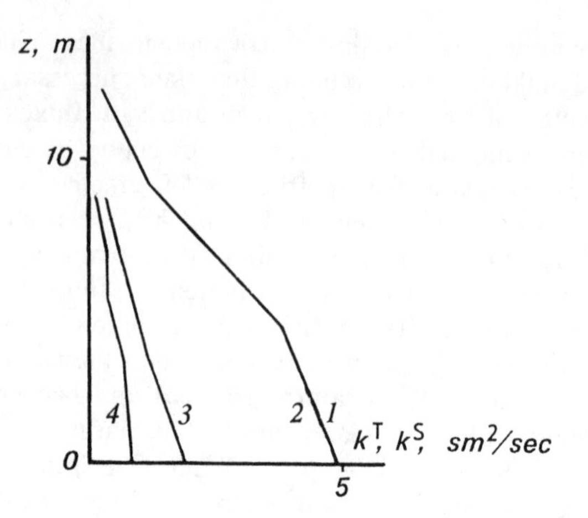

Fig.10. Coefficients of effective vertical heat (k^T) and salt (k^S) exchange with structurization due to salt fingers: *1* - k^T, *2* - k^S ($R\to 1, \Delta S_1=0.16‰$); *3* - k^T, *4* - k^S ($R=2, \Delta S_1=0.16‰$).

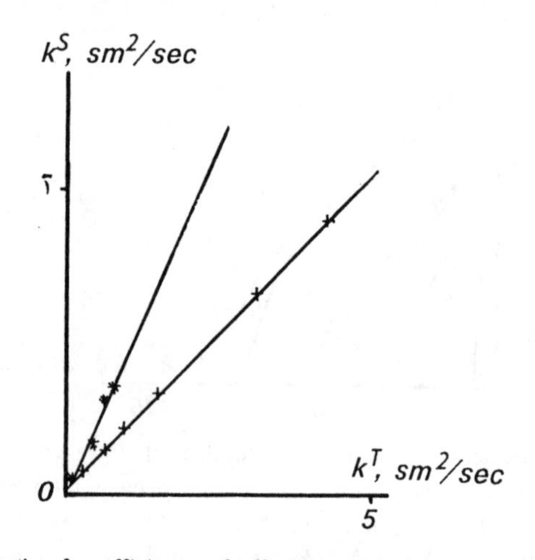

Fig.11. Ratio of coefficients of effective vertical heat and salt exchange with structurization due to salt fingers: + - with $R\to 1$; * - with R=*2*.

When R=2, k^T is about five times larger than k^S. Similar results were obtained in laboratory experiments of J.Turner. True, when R=2, his ratio k^T/k^S is somewhat smaller than ours.

With structurization due to salt fingers the values of k^T and k^S depend on the background value of R and initial salinity gradient. The larger R, the less the intensity of effective vertical exchange (Figure 10). When the salinity gradient increases, k^T and k^S increase. The exchange intensity decreases when moving away from the steep gradient (Figure 10). When R--1 k^T and k^S are practically equal (Figure 11). When R increases, the salt transfer is more intensive than that of heat (Figure 11). The coefficients of effective vertical salt exchange are about two times larger than those of heat exchange with instability in the form of salt fingers.

REFERENCES

Karlin L. N., Kluikov Ye. Yu., Kut'ko V. P., Small-scale structure of hydrophysical fields of the ocean upper layer, 162 pp., Gydrometeoizdat Publishers, Moscow, 1988.

Marmorine G. J., Caldwell D. R. Heat and salt transport through a diffusive thermohaline interface, *Deep-Sea Res, 23,* 59-67, 1976.

McDougall T. J. Double-diffusive convection caused by coupled molecular diffusion, *J.Fluid Mech, 126,* 379-397, 1983.

Schmitt R. W. Flux measurements on salt fingers at an interface, *J.Mar.Res, 37,* 419-436, 1979.

Ye. Yu. Kluikov, and L. N. Karlin, Department of Oceanology, Russian State Hydrometeorological Institute, 98, Malookhtynsky av., St.Petersburg, 195196, Russia

Wind-Driven, Double-Diffusive Convection in Ocean Surface Layers

George H. Knightly and D. Sather

Department of Mathematics, University of Massachusetts, Amherst, Massachusetts
Department of Mathematics, University of Colorado at Boulder, Boulder, Colorado

A qualitative study is initiated for wind-driven, double-diffusive convection in an ocean surface layer when the Stokes drift has a crosswind component. It is shown that, even in the simplest case of rolls where the Rayleigh numbers for heat and solute are small and one might expect monotonic convection, the basic state corresponding to constant temperature and concentration gradients always undergoes Hopf bifurcation losing stability to an asymptotically stable periodic traveling wave. This leads, e.g., to periodic traveling wave corrections to the density in the layer. Estimates are given for the critical value of the Langmuir number and the drift velocity of the rolls.

1. INTRODUCTION

It is well-known that certain wind-induced instabilities in surface layers of the ocean are analogous to double-diffusive instabilities. Such wind-induced instabilities include, e.g., Langmuir circulations which arise in the study of the density-stratified form of the C–L equations [*Craik et al.*, 1977; *Leibovich*, 1977b, p. 566; *Leibovich*, 1983, p. 399]. The C–L equations constitute a rational model that describes the development of mean currents and Langmuir circulations as a single system driven by a prescribed wind stress and surface wave field. The use of the C–L model in the study of roll-like solutions independent of the wind direction leads to a set of equations [*Leibovich*, 1983, p. 409] that is formally identical with those of double-diffusive convection [e.g., *Joseph*, 1976, Chap. IX], provided that one considers the disturbance velocity component in the wind direction as temperature and the disturbance temperature as concentration. By making use of this analogy, one finds that a number of features of double-diffusive convection have their counterparts in Langmuir circulation problems and vice versa. In this way a number of striking analytical and numerical results relevant to both areas have been obtained such as those described in [*Cox et al.*, 1992a,b; *Knobloch et al.*, 1986; *Leibovich*, 1985; *Leibovich et al.*, 1989; *Moore et al.*, 1983].

In the present paper we develop instead a theory of wind-driven, double-diffusive convection in ocean surface layers in which the velocity components, temperature and concentration play their traditional roles. As in the C–L theory of Langmuir circulations, a rational model is obtained that describes the interaction of surface waves, wind, ocean currents and the diffusivity of temperature and concentration as a single system driven ultimately by the wind. The derived equation, e.g., for the velocity, v, is similar to the velocity equation appearing in standard three-dimensional double-diffusive convection problems except for the presence of a term involving a "vortex force" [*Leibovich*, 1983, p. 400] of the form $u_s \times [\nabla \times v]$. Here u_s is the Stokes drift which is assumed to be generated by a prescribed irrotational surface wave field, $u_w = \nabla \varphi_w$. If the Stokes drift is aligned with the wind as in the references listed above, then even in this special case the dynamics of the problem are quite complex with the possibility of steady states, traveling waves, or standing waves depending on whether there is monotonic or overstable convection. It is expected, of course, that even more complex dynamics are likely when the Stokes drift has a cross-wind component because in this case there is an additional symmetry breaking; if γ is a measure of the cross-wind component of the Stokes drift, one easily sees that the model equations (5) obtained in §3 have $O(2)$ invariance when $\gamma = 0$ but only $SO(2)$ invariance when $\gamma \neq 0$.

In the present paper we begin the qualitative study of

Double-Diffusive Convection
Geophysical Monograph 94

wind-driven, double-diffusive convection when the Stokes drift has a cross-wind component by showing that, even in the simplest case of rolls where the Rayleigh numbers for heat and solute are small so that one might expect monotonic convection, the basic state corresponding to constant temperature and concentration gradients *always* undergoes Hopf bifurcation losing stability to an asymptotically stable, periodic traveling wave. The presence of such traveling waves, in turn, leads to the possibility of some new phenomena in double-diffusive convection including periodic traveling wave corrections to the density in the layer. Thus, various physical models for ocean surface layers based upon constant layered density properties may have to be re-evaluated in situations in which there is wind-driven, double-diffusive convection with a Stokes drift having a substantial cross-wind component. Such a situation could occur, e.g., when there is a mean current present that is not aligned with the wind.

The outline of the paper is as follows. The general equations for wind-driven, double-diffusive convection in a stratified layer are described in Section 2 and the simplified equations for linear Stokes drift and roll-like solutions are given in Section 3. Linearized stability and critical Langmuir numbers (i.e., inverse Reynolds numbers) are then discussed in Section 4. In Section 5 the problem of finding disturbance solutions in an infinite-dimensional space is reduced to a single Landau-type amplitude equation by the use of symmetry methods and a center manifold. The resultant amplitude equation is then solved in Section 6 to obtain periodic traveling wave disturbances for the velocity, temperature and concentration. Section 7 contains a discussion of some physical interpretations of these results including estimates of the critical Langmuir numbers and the cross-wind drift velocity of the system of rolls.

The emphasis throughout the paper is on the description of qualitative results for periodic traveling waves in wind-driven, double-diffusive convection. Complete mathematical proofs of these qualitative results as well as some more quantitative numerical results will appear in forthcoming papers.

2. WIND-DRIVEN, DOUBLE-DIFFUSIVE CONVECTION

The equations for wind-driven, double-diffusive convection in a stratified layer are described in this section. The derivation of these equations follows closely the multiple-scale approach used in [*Leibovich*, 1977a,b; *Knightly et al.*, 1995] so that the physical motivation and details given in those papers are not repeated here.

The velocity equations derived in [*Leibovich*, 1977a,b] for Langmuir circulations also apply in the present situation, provided that the Boussinesq approximation for density (see (3) below) is suitably modified to include solute concentration. In physical situations in which either three-dimensional motions are considered or the Stokes drift, u_s, has a cross-wind component, however, the temperature equation for θ_0 given in [*Leibovich*, 1977b, p. 566] is not correct. In the derivation of the equation for heat balance there is an error (the term $-u_w \cdot \nabla\theta_0$ should have been included on the right side of the equation for $\frac{\partial\theta_1}{\partial t}$, whose correction ultimately leads to the inclusion of an important additional term $u_s \cdot \nabla\theta_0$. Similarly, it is clear that the diffusivity of the solute concentration in the layer also is affected by the presence of the Stokes drift but one does not necessarily know to what order. However, if one introduces an "eddy solute diffusivity" that is analogous to the "eddy temperature diffusivity" used in [*Leibovich*, 1977b, 1983] to obtain a "second order theory" for Langmuir circulations, then a similar analysis leads to the following system of equations for wind-driven, double-diffusive convection in a stratified layer, $|z| \leq 1/2$,

$$\begin{aligned} 0 &= \frac{\partial V}{\partial t} - u_s \times (\nabla \times V) + V \cdot \nabla V + (U_1 + z)\frac{\partial V}{\partial x} \\ &\quad + \nabla p + (V \cdot \hat{k})\hat{\imath} - \nu_0 \Delta V - g\beta_T \Theta \hat{k} + g\beta_C \chi \hat{k}, \\ 0 &= \frac{\partial \Theta}{\partial t} + (u_s + V) \cdot \nabla\Theta + (U_1 + z)\frac{\partial \Theta}{\partial x} \\ &\quad + (V \cdot \hat{k})T' - \alpha_T \Delta\Theta, \\ 0 &= \frac{\partial \chi}{\partial t} + (u_s + V) \cdot \nabla\chi + (U_1 + z)\frac{\partial \chi}{\partial x} \\ &\quad + (V \cdot \hat{k})C' - \alpha_C \Delta\chi, \\ 0 &= \nabla \cdot V, \\ 0 &= \frac{\partial V_1}{\partial z} = \frac{\partial V_2}{\partial z} = V_3 = \Theta = \chi \text{ on } z = \pm 1/2, \end{aligned} \tag{1}$$

where $\nabla = (\frac{\partial}{\partial x}, \frac{\partial}{\partial y}, \frac{\partial}{\partial z}, 0, 0)^\top$, $^\top$ denotes transpose and $\Delta = \frac{\partial^2}{\partial x^2} + \frac{\partial^2}{\partial y^2} + \frac{\partial^2}{\partial z^2}$. Here V, p, Θ and χ are the velocity, pressure, temperature and concentration disturbances, respectively, representing perturbations of a basic state

$$((U_1 + z)\hat{\imath}, p_0, T_0 + T'z, C_0 + C'z)^\top, \tag{2}$$

where U_1 is an arbitrary constant, T' and C' are *positive* constants (i.e., the layer is heated and salted from above) and $\hat{\imath}, \hat{\jmath}$ and $\hat{k}$ are unit vectors in the x, y and z directions, respectively; the basic solution here represents a constant stress, unidirectional current aligned with the wind. In (1) g is the gravitational constant, u_s denotes the Stokes drift, the constants ν_0, α_T and α_C are proportional to the (eddy) kinematic viscosity, thermal diffusivity and solute diffusivity, respectively, and β_T and β_C are constants proportional to those used in the standard Boussinesq approximation [e.g., *Joseph*, 1976, §54]

$$\rho = -\rho_0[\beta_T^0(T'z + \Theta) - \beta_C^0(C'z + \chi)]. \qquad (3)$$

The x-axis is chosen in the direction of the wind, y is the horizontal cross-wind variable and z measures depth in the layer with $z = \frac{1}{2}$ corresponding to the upper mean free surface. If $C' = 0$ and $\chi = 0$ in (1), then the equations reduce to those for wind-driven convection in a stratified layer [*Knightly et al.*, 1995]; in this case if, in addition, the Stokes drift has no cross-wind component and if one seeks roll-like solutions aligned with the wind direction, then the equations in (1) further reduce to those given in [*Leibovich*, 1983, p. 399]. However, in all other cases the terms $u_s \cdot \nabla\Theta$ and $u_s \cdot \nabla\chi$ represent important contributions to the disturbances and must be included in the model equations.

The boundary conditions in (1) are derived under the assumptions that the temperature and concentration are known and constant on $z = \pm\frac{1}{2}$, there is no flow across $z = -\frac{1}{2}$ and the mean stress on $z = -\frac{1}{2}$ is that of the basic flow. As observed in [*Leibovich et al.*, 1989, p. 476], these boundary conditions model a stratified layer adjacent to the surface and overlying a much deeper and heavier body of water with the density difference so large that vertical motions of the interface may be ignored. Additional physical assumptions related to stress-free boundary conditions on $z = -\frac{1}{2}$ and the neglect of internal waves in the problem also are discussed in [*Leibovich et al.*, 1989, p. 476]. More general boundary conditions that do not assume stress-free boundary conditions for the velocity disturbances are described in [*Cox et al.*, 1993]. Since in typical situations these more general boundary conditions may be considered as perturbations of stress-free boundary conditions, it is expected that results similar to those obtained here would hold also for this more general situation.

3. LINEAR STOKES DRIFT AND ROLL-LIKE SOLUTIONS

We assume throughout the remainder of the paper that the layer depth is small compared with the length of the surface waves and that the Stokes drift is linear in z and of the form

$$\begin{aligned} u_s &= U_s(z)(\hat\imath + \gamma\hat\jmath), \\ U_s(z) &= \bar U_s + U_s' z, \; \bar U_s > 0, \; U_s' > 0, \end{aligned} \qquad (4)$$

where γ is a measure of the cross-wind component of the Stokes drift. Note that $\bar U_s(\hat\imath + \gamma\hat\jmath)$ is the average of u_s over $-\frac{1}{2} \le z \le \frac{1}{2}$.

We seek roll-like solutions of (1) that are independent of the wind-direction, i.e., the x-direction. Then $u_s \cdot \nabla\Theta = \gamma U_s \frac{\partial\Theta}{\partial y}$, $u_s \cdot \nabla\chi = \gamma U_s \frac{\partial\chi}{\partial y}$ and

$$u_s \times (\nabla \times V) = \nabla[u_s \cdot V] - (V_1 + \gamma V_2)U_s'\hat k - \gamma U_s \frac{\partial V}{\partial y}.$$

Substituting these equations into (1), making the change of variables

$$\begin{aligned} t &= t^*/\nu_0, \; V_1 = \tilde c_0 w_1, \; V_2 = \nu_0 w_2, \; V_3 = \nu_0 w_3, \\ p &= u_s \cdot V + \nu_0^2 q, \\ \Theta &= \tilde d_1 w_4, \; \chi = \tilde d_2 w_5, \; \tilde c_0 = -\nu_0 (U_s')^{-1/2}, \\ \tilde d_1 &= -\nu_0 \left(\frac{T'}{g\beta_T}\right)^{1/2}, \; \tilde d_2 = -\nu_0 (C'/g\beta_C)^{1/2}, \end{aligned}$$

setting $w = (w_1, w_2, w_3, w_4, w_5)^\top$ and dropping the * on t^*, one obtains the following system for roll-like disturbance flows:

$$\begin{aligned} 0 &= \frac{\partial w}{\partial t} + \mathcal{L}_0 w - \lambda\mathcal{L}_1 w - \mathcal{L}_2 w \\ &\quad - \lambda\gamma(\mathcal{M}_1 + \mathcal{M}_2 w + w \cdot \nabla w + \nabla q, \\ 0 &= \nabla \cdot w, \\ 0 &= \frac{\partial w_1}{\partial z} = \frac{\partial w_2}{\partial z} = w_3 = w_4 = w_5 \text{ on } z = \pm\frac{1}{2}. \end{aligned} \qquad (5)$$

Here ∇ now denotes the operator $(0, \frac{\partial}{\partial y}, \frac{\partial}{\partial z}, 0, 0)^\top$, $\Delta = \nabla^2$,

$$\begin{aligned} \mathcal{L}_0 &= -\Delta \begin{pmatrix} 1 & 0 & 0 & 0 & 0 \\ 0 & 1 & 0 & 0 & 0 \\ 0 & 0 & 1 & 0 & 0 \\ 0 & 0 & 0 & P_T^{-1} & 0 \\ 0 & 0 & 0 & 0 & P_C^{-1} \end{pmatrix}, \\ \mathcal{L}_1 &= \begin{pmatrix} 0 & 0 & 1 & 0 & 0 \\ 0 & 0 & 0 & 0 & 0 \\ 1 & 0 & 0 & 0 & 0 \\ 0 & 0 & 0 & 0 & 0 \\ 0 & 0 & 0 & 0 & 0 \end{pmatrix}, \\ \mathcal{L}_2 &= \begin{pmatrix} 0 & 0 & 0 & 0 & 0 \\ 0 & 0 & 0 & 0 & 0 \\ 0 & 0 & 0 & -\mu_1 & \mu_2 \\ 0 & 0 & \mu_1 & 0 & 0 \\ 0 & 0 & \mu_2 & 0 & 0 \end{pmatrix}, \\ \mathcal{M}_1 &= -(U_s')^{-1/2} \bar U_s \frac{\partial}{\partial y}, \\ \mathcal{M}_2 &= -(U_s')^{1/2} \left[\begin{pmatrix} 0 & 0 & 0 & 0 & 0 \\ 0 & 0 & 0 & 0 & 0 \\ 0 & 1 & 0 & 0 & 0 \\ 0 & 0 & 0 & 0 & 0 \\ 0 & 0 & 0 & 0 & 0 \end{pmatrix} + z\frac{\partial}{\partial y} \right], \end{aligned} \qquad (6)$$

$$\lambda = (U_s'/\nu_0^2)^{1/2}, \ \mu_1 = (g\beta_T T'/\nu_0^2)^{1/2}, \qquad \mu_2 = (g\beta_C C'/\nu_0^2)^{1/2}, \tag{7}$$

where $P_T = \nu_0/\alpha_T$ and $P_C = \nu_0/\alpha_C$ are the (turbulent) Prandtl and Schmidt numbers. Note that μ_1 and μ_2 are Rayleigh-type numbers for heat and solute, λ is a Reynolds number and λ^{-1} is a Langmuir number [e.g., *Leibovich*, 1977a, p. 724]. Since the model equations in (5) have $O(2)$ invariance when $\gamma = 0$ but only $SO(2)$ invariance when $\gamma \neq 0$ (the operator $\mathcal{M}_2$ is not invariant under $y \to -y$), one would hope under certain conditions to find solutions of (5) when $\gamma \neq 0$ in the form of periodic traveling waves [e.g., *Knightly et al.*, 1994, 1995].

As in [*Leibovich et al.*, 1989, p. 478], we now suppose that some mechanism exists (as part of the imposed surface wave field or the physical limitation of the system of parallel rolls) which serves to quantize the wavelengths that can be realized by the system; in fact, we assume that the admissible wave numbers are limited to multiples of a *positive* fundamental wave number $a_0 = 2\pi/\ell$, where ℓ is the lateral extent of the system of rolls. Thus, we consider only $w = w(y, z, t)$ in (5) that are periodic in y of period $2\pi/a_0$; the corresponding period rectangle in y, z space is denoted by

$$R = \left\{(y,z) : 0 < y < 2\pi/a_0, \ -\frac{1}{2} < z < \frac{1}{2}\right\}.$$

It is convenient throughout the remainder of the paper in describing the results to make use of some geometrical ideas in inner product spaces.

Let $\mathbb{L}^2(R)$ denote the Hilbert space of square-integrable 5-vectors defined on R and let $(\cdot,\cdot)$ denote the inner product on $\mathbb{L}^2(R)$ given by

$$(f,g) = \int_R f \cdot \bar{g} \equiv \sum_{j=1}^{5} \int_R f_j \bar{g}_j dydz, \quad f, g \in \mathbb{L}^2(R).$$

The norm, $\|\cdot\|$, on $\mathbb{L}^2(R)$ is given by $\|w\|^2 = (w,w)$. Let $\mathcal{H} \subset \mathbb{L}^2(R)$ be the Hilbert space defined as the closure in the $\mathbb{L}^2(R)$-norm of smooth 5-vectors w such that $\nabla \cdot w = 0$, $w_3 = 0$ on $z = \pm\frac{1}{2}$, and $\int_R w_1 = \int_R w_2 = 0$. If Π denotes the orthogonal projection of $\mathbb{L}^2(R)$ onto $\mathcal{H}$, then $I - \Pi$ is the projection onto the subspace $\mathcal{G}$ of $\mathbb{L}^2(R)$ consisting of gradients. Any gradient in $\mathbb{L}^2(R)$ can then be written as $\nabla p = (I - \Pi)u$ for some $u \in \mathbb{L}^2(R)$; it follows that $\Pi\nabla p = 0$ because $\Pi^2 = \Pi$. Thus, using that Π is the projection onto divergence free vectors in $\mathbb{L}^2(R)$, one sees that the system (5) can be replaced by a single operator equation *in* $\mathcal{H}$ (without the divergence condition), namely

$$0 = \frac{dw}{dt} + L_0 w - \lambda L_1 w - L_2 w - \lambda\gamma(M_1 + M_2)w + B(w), \tag{$*$}$$

where $L_j = \Pi\mathcal{L}_j$ $(j = 0,1,2)$, $M_j = \Pi\mathcal{M}_j$ $(j = 1,2)$ and $B(w) = \Pi w \cdot \nabla w$. To give a rigorous formulation of $(*)$, it is necessary to define another subspace, $\mathcal{D}$, of $\mathcal{H}$ that incorporates all of the boundary conditions in (5). The subspace $\mathcal{D}$ is then used as the domain for the various operators in $(*)$ and one seeks solutions of $(*)$ such that $w(t) \in \mathcal{D}$. The appropriate subspaces and the mathematical details of such a formulation of $(*)$ are similar to those given in [*Knightly et al.*, 1995].

4. LINEARIZED STABILITY AND CRITICAL LANGMUIR NUMBERS

In this section we consider the linearized stability problems for the operator equation $(*)$ when $\gamma = 0$ and $\gamma \neq 0$, and determine the critical Langmuir number when $\mu = (\mu_1, \mu_2)$ and γ are sufficiently small.

4.1. Linearized Stability when $\gamma = 0$

The linearized time-independent system associated with $(*)$ when $\gamma = 0$ is given by

$$0 = L_0 u - \lambda L_1 u - L_2 u, \qquad u \in \mathcal{H}. \tag{8}$$

We regard $\mu = (\mu_1, \mu_2)$ here as a 2-vector parameter and determine the critical values of λ for which (8) has nontrivial solutions.

Note first of all that solving problem (8) *in* $\mathcal{H}$ is equivalent to solving the classical problem for smooth u, q with $\int_R u_1 = \int_R u_2 = 0$ obtained by setting $\gamma = 0$ and omitting the nonlinear terms in (5), namely

$$\begin{aligned} 0 &= \mathcal{L}_0 u - \lambda\mathcal{L}_1 u - \mathcal{L}_2 u + \nabla q, \\ 0 &= \nabla \cdot u, \\ 0 &= \frac{\partial u_1}{\partial z} = \frac{\partial u_2}{\partial z} = u_3 = u_4 = u_5 \\ &\quad \text{on } z = \pm\frac{1}{2}, \\ 0 &= \int_R u_1 = \int_R u_2. \end{aligned} \tag{9}$$

The mean value conditions in (9) are part of the definition of $\mathcal{H}$ and imply that the fluid layer cannot be displaced parallel to the x, y plane. We consider only $u = u(y, z)$ in (9) that are periodic in y with period $2\pi/a_0$.

The eigenvalue problem (9) is essentially a linear Bénard-type problem [*Chandrasekhar*, 1961] which for $k \neq 0$ has solutions in the form

$$\begin{aligned} u_j &= e^{iky}\varphi_j(z), \quad j = 1,3,4,5 \\ u_2 &= ia^{-2}ke^{iky}\varphi_3'(z), \quad a = |k| \\ q &= a^{-2}e^{iky}D^2\varphi_3'(z). \end{aligned} \tag{10}$$

Here $D^2 = \frac{d^2}{dz^2} - a^2$, a prime denotes $\frac{d}{dz}$ and $\varphi = (\varphi_1, \varphi_3, \varphi_4, \varphi_5)^\top$ satisfies

$$\begin{aligned} \mathcal{A}(\lambda, \mu)\varphi = 0, \quad -\frac{1}{2} < z < \frac{1}{2}, \\ \varphi_1' = \varphi_3 = D^2\varphi_3 = \varphi_4 = \varphi_5 = 0, \\ \text{at } z = \pm\frac{1}{2}, \end{aligned} \tag{11}$$

where $\mu = (\mu_1, \mu_2)$ and

$$\mathcal{A}(\lambda, \mu_1, \mu_2) = \begin{pmatrix} D^2 & \lambda & 0 & 0 \\ \lambda & -a^{-2}D^4 & -\mu_1 & \mu_2 \\ 0 & \mu_1 & P_T^{-1}D^2 & 0 \\ 0 & \mu_2 & 0 & P_C^{-1}D^2 \end{pmatrix}.$$

The adjoint problem for (11) is

$$\begin{aligned} \mathcal{A}^*(\lambda, \mu_1, \mu_2)\varphi = 0, \ -\frac{1}{2} < z < \frac{1}{2}, \\ \varphi_1' = \varphi_3 = D^2\varphi_3 = \varphi_4 = \varphi_5 = 0, \\ \text{at } z = \pm\frac{1}{2}, \end{aligned} \tag{12}$$

where $\mathcal{A}^*(\lambda, \mu_1, \mu_2) = \mathcal{A}(\lambda, -\mu_1, \mu_2)$. Thus, for given (λ, μ), $\varphi = (\varphi_1, \varphi_3, \varphi_4, \varphi_5)^\top$ satisfies (11) if and only if $\varphi^* = (\varphi_1, \varphi_3, -\varphi_4, \varphi_5)$ satisfies (12). Note also that since e^{iky} must have period $2\pi/a_0$, $a \equiv a_m = ma_0$, where m is a *positive* integer; the case $m = 0$ is excluded here because (9) has no nontrivial constant solutions when $a = 0$.

We now regard (11) as a perturbation of the eigenvalue problem for $\mu = 0$. When $\mu = 0$, φ_4 and φ_5 are uncoupled from φ_1, φ_3 in (11) and are easily seen to satisfy $\varphi_4(z) \equiv \varphi_5(z) \equiv 0$. The remaining equations for $\lambda, \varphi_1, \varphi_3$ define an eigenvalue problem that can be treated as in [*Iudovich*, 1969] using the theory of oscillation kernels. One finds that for each positive integer m the eigenvalues consist of a countable sequence of positive simple eigenvalues satisfying $0 < \lambda_1(ma_0) < \lambda_2(ma_0) < \cdots$. The calculations in [*Leibovich et al.*, 1989, §2.2] for $\mu_2 = 0$ suggest that the minimum critical eigenvalue λ_{00}, occurs when $m = 1$; we henceforth assume this to be the case. Thus $\lambda_{00} = \lambda_1(a_0) > 0$ with a corresponding real eigenfunction $\varphi^{00} = (\varphi_1^{00}, \varphi_3^{00})^\top$ that is even in z. One can now show that for each $|\mu|$ sufficiently small there exists a positive simple minimum critical eigenvalue, $\lambda_0(\mu)$, of problem (11) such that when $\gamma = 0$ the trivial solution $w = 0$ of (5) is stable for $\lambda < \lambda_0(\mu)$ and unstable for $\lambda > \lambda_0(\mu)$. (For larger values of $|\mu|$ one expects overstability, a supposition that is verified for the non-solute case as part of the extensive numerical calculations in [*Leibovich et al.*, 1989, §2.2].) In fact, since λ_{00} is simple, by making use of the implicit function theorem as in [*Knightly et al.*, 1995], one can determine a solution $(\lambda_0(\mu), \varphi^0(\mu))$ of (11) of the form

$$\begin{aligned} \lambda_0(\mu) &= \lambda_{00} + |\mu|^2\sigma(\mu), \\ \varphi^0(\mu) &= \varphi^{00} + |\mu|\Phi^0(\mu) \equiv (\varphi_1^{00}, \varphi_3^{00}, 0, 0)^\top \\ &\quad + |\mu|\varphi^1(\mu), \end{aligned} \tag{13}$$

depending continuously on μ and such that $\lambda_0(0) = \lambda_{00}$ and $\varphi^0(0) = \varphi^{00}$; the solvability condition here is just

$$\int_{-\frac{1}{2}}^{\frac{1}{2}} \varphi_1^{00}(z)\varphi_3^{00}(z)dz \neq 0. \tag{14}$$

Moreover, since $\lambda_0(\mu)$ is simple, $\varphi^0(\mu)(-z) = \pm\varphi^0(\mu)(z)$ so that the parity of the eigenfunction is preserved under perturbation with $(\varphi^0(\mu))_j$ being even $(j = 1, 3, 4, 5)$.

It now easily follows from (10) that, for each *fixed* μ, $0 \leq |\mu| < \mu_0$ and μ_0 sufficiently small, the eigenvalue problem (9) has a positive minimum critical eigenvalue, $\lambda_0 \equiv \lambda_0(\mu)$, of multiplicity two with corresponding eigenfunctions $\bar{\psi}^0 \equiv \overline{\psi^0(\mu)}$ and

$$\psi^0 \equiv \psi^0(\mu) = e^{ia_0y}\Phi^0(\mu)(z), \tag{15}$$

where $\Phi^0(\mu) = (\varphi_1^0(\mu), ia_0^{-1}\frac{d}{dz}\varphi_3^0(\mu), \varphi_3^0(\mu), \varphi_4^0(\mu), \varphi_5^0(\mu))$ and $q^0 \equiv q^0(\mu)$ is determined by (10) with $a = a_0$. Note that $(\psi^0)_j$ is even in z $(j = 1, 3, 4, 5)$ and $(\psi^0)_2$ is odd. Also note that $(\Phi^0)_4, (\Phi^0)_5$ and, hence, $(\psi^0)_4, (\psi^0)_5$ are of order $O(|\mu|)$ as $\mu \to 0$ because $\varphi_4^{00} = \varphi_5^{00} = 0$. The adjoint problem for (9) is obtained by replacing μ by $-\mu$ in (9). It then follows from the remark after (12) that the adjoint problem for (9) has eigenfunctions $(\bar{\psi}^0)^* \equiv (\overline{\psi^0(\mu)})^*$ and $(\psi^0)^* \equiv (\psi^0(\mu))^* = e^{ia_0y}(\Phi^0(\mu))^*(z)$, where $(\psi^0)_j^* = (\psi^0)_j$, $(j = 1, 2, 3, 5)$ and $(\psi^0)_4^* = -(\psi^0)_4$.

To complete the discussion of the critical eigenvalue for (9), one needs to show that λ_0 is semi-simple (i.e., no generalized eigenfunctions), provided that $|\mu|$ is sufficiently small. In doing so it is useful to introduce some subspaces of $\mathcal{H}$, where $\mathcal{H}$ is the Hilbert space defined at the end of Section 3.

Note first of all that ψ^0 in (15) belongs to $\mathcal{H}$ and that (λ_0, ψ^0) satisfies $L\psi^0 = 0$, where

$$L \equiv L_0 - \lambda_0 L_1 - L_2, \qquad \lambda_0 = \lambda_0(\mu). \tag{16}$$

The null space, $\mathcal{N}$, of L is defined to be the subspace of $\mathcal{H}$ consisting of linear combinations of $\psi^0, \bar{\psi}^0$ with complex coefficients. The set, $\mathcal{N}^\perp$, of vectors in $\mathcal{H}$ that are orthogonal to $\mathcal{N}$ is given by

$$\mathcal{N}^\perp = \{w \in \mathcal{H} : (w, \psi^0) = (w, \bar{\psi}^0) = 0\}.$$

Similarly, if $\mathcal{N}_*$ denotes the null space of L^* with $L^* = L_0 - \lambda_0 L_1 + L_2$, then $\mathcal{N}_*^\perp$ is the orthogonal complement of $\mathcal{N}_*$. As is well-known [*Kato*, 1966] $\mathcal{N}_*^\perp = \mathcal{R}$, where $\mathcal{R}$ is the range of L in $\mathcal{H}$. Thus, if $(\psi^0, (\psi^0)^*) \neq 0$, then $\mathcal{R} \cap \mathcal{N} = \{0\}$ and any vector $w \in \mathcal{H}$ can be written uniquely as $w = u + v$, where $u \in \mathcal{N}$ and $v \in \mathcal{R}$, i.e., $\mathcal{H} = \mathcal{N} \oplus \mathcal{R}$. The condition that λ_0 be semi-simple is just the condition $(\psi^0, (\psi^0)^*) \neq 0$. Since $\psi_j^0 = O(|\mu|)$, $(j = 4, 5)$ as $|\mu| \to 0$, this last condition follows from (10) and (13) provided that $|\mu|$ is sufficiently small, i.e.,

$$
\begin{aligned}
(\psi^0, (\psi^0)^*) &= \int_R \left[\sum_{j=1}^{3} |\psi_j^0|^2 - |\psi_4^0|^2 + |\psi_5^0|^2 \right] dydz \\
&= 2\pi a_0^{-1} \left[\int_{-\frac{1}{2}}^{\frac{1}{2}} \left[(\varphi_1^{00})^2 \right.\right. \\
&\quad \left.\left. + a_0^{-2} \left(\frac{d}{dz} \varphi_3^{00} \right)^2 \sum_0^1 + (\varphi_3^{00})^2 \right] dz \right] \\
&\quad + O(|\mu|^2) > 0.
\end{aligned} \tag{17}
$$

The spectral properties of L in (16) required in the next section are summarized in the following lemma; the proof given in [*Knightly et al.*, 1995, Lemma 3.3] for the non-solute case is easily extended to the present case. The subspace $\mathcal{D}$ of $\mathcal{H}$ in the following lemma is that introduced at the end of Section 3 and $\mathcal{R}e$ and $\mathcal{I}m$ denote the real and imaginary parts of a complex number, respectively.

Lemma 4.1. There exists $\mu_0 > 0$ such that if $|\mu| < \mu_0$, then all of the following properties of L hold.

(i) The operator L has a semi-simple eigenvalue $\rho_0 = 0$ having a two-dimensional null space, $\mathcal{N}$, spanned by $\psi^0, \bar{\psi}^0$ in (15).

(ii) There exists $c_0 > 0$, depending only on μ_0, such that

$$
\mathcal{R}e\,(Lv, v) \geq c_0 \|v\|^2, \; v \in \mathcal{D} \cap \mathcal{R}
$$
$$
|\mathcal{I}m\,(Lv, v)| \leq |\mu| \|v\|^2, \; v \in \mathcal{D}.
$$

(iii) The spectrum of L consists of only isolated eigenvalues, $\{\rho_n\}_{n \geq 0}$, of finite multiplicity with $\mathcal{R}e\,\rho_n \geq c_0 > 0$ $(n \geq 1)$.

4.2. Linearized Stability when $\gamma \neq 0$

If $\gamma = 0$ and μ is sufficiently small, the discussion in §4.1 shows that the eigenvalue problem (9) has a positive critical, semi-simple eigenvalue, $\lambda_0 = \lambda_0(\mu)$, of multiplicity two. Considering λ_0 at a *fixed* value of μ as given, we now determine the critical Langmuir number $\lambda_c^{-1}(\mu, \gamma)$, for the system (5) when γ is sufficiently small.

The problem for linearized stability associated with (*) (and, hence (5)) when $\gamma \neq 0$ is given by

$$
\begin{aligned}
\mathcal{L}(\lambda, \mu, \gamma)u &\equiv L_0 u - \lambda L_1 u - L_2 u - \lambda\gamma(M_1 + M_2)u \\
&= \zeta u, \; \zeta = \xi + i\eta.
\end{aligned} \tag{18}
$$

One seeks to determine solutions $(u(\lambda, \mu, \gamma), \zeta(\lambda, \mu, \gamma))$ of (18) for (λ, μ, γ) near $(\lambda_0, \mu, 0)$. The perturbation scheme in [*Knightly et al.*, 1993, Lemma 3.1] can be easily modified (see the discussion in the next paragraph) to show that for μ, γ sufficiently small there exists a unique smallest positive value of λ, $\lambda_c \equiv \lambda_c(\mu, \gamma)$, at which the spectrum of $\mathcal{L}(\lambda_c, \mu, \gamma)$ consists of eigenvalues with finite-multiplicity having positive real parts except for a unique pair of complex conjugate eigenvalues $\zeta(\lambda_c, \mu, \gamma), \bar{\zeta}(\lambda_c, \mu, \gamma)$ with real part $\xi(\lambda_c, \mu, \gamma) = 0$, i.e., $\zeta(\lambda_c, \mu, \gamma) = i\gamma\eta(\mu, \gamma)$ with eigenvector $\Psi(\mu, \gamma)$ such that

$$
\mathcal{L}(\lambda_c, \mu, \gamma)\Psi(\mu, \gamma) = i\gamma\eta(\mu, \gamma)\Psi(\mu, \gamma).
$$

Moreover, $\lambda_c(\mu, \gamma)$ and $\eta(\mu, \gamma)$ satisfy

$$
\begin{aligned}
\lambda_c(\mu, \gamma) &= \lambda_0(\mu) + b(\mu)\gamma^2 + \gamma^3 \Lambda(\mu, \gamma), \\
\eta(\mu, 0) &\equiv \eta_0(\mu) = a_0 (U_s')^{-1/2} \bar{U}_s \lambda_0(\mu),
\end{aligned} \tag{19}
$$

where $b(\mu)$ is a calculable constant for each μ. (To calculate $\eta_0(\mu)$ in a formal way, one notes that $\mathcal{M}_1$ in (6) and ψ^0 in (15) satisfy $\mathcal{M}_1 \psi^0 = -ia_0 (U_s')^{-1/2} \bar{U}_s \psi^0$.) The solvability condition used to establish the above results is $(L_1 \psi^0, (\psi^0)^*) \neq 0$. To see that this condition holds, one makes use of (13) and (15) to obtain

$$
\begin{aligned}
d_1 &= (L_1 \psi^0, (\psi^0)^*) = \lambda_0^{-1} [(L_0 \psi^0, (\psi^0)^*) \\
&\quad - (L_2 \psi^0, (\psi^0)^*)] \\
&= \lambda_0^{-1} \int_R \left[- \sum_{j=1}^{3} (\psi^0)_j^* \Delta \psi_j^0 - P_T^{-1} (\psi^0)_4^* \Delta \psi_4^0 \right. \\
&\quad \left. - P_C^{-1} (\psi^0)_5^* \Delta \psi_5^0 \right] + O(|\mu|^2) \\
&= \lambda_0^{-1} \int_R \sum_{j=1}^{3} |\nabla \psi_j^0|^2 + O(|\mu|^2)
\end{aligned} \tag{20}
$$

which is positive provided that $|\mu| < \mu_0$ and μ_0 is sufficiently small. The condition $d_1 > 0$ corresponds to $\mathcal{R}e\,\zeta(\lambda, \mu, \gamma)$ decreasing through zero as λ increases through $\lambda_c(\mu, \gamma)$; this is the expected situation for $w = 0$ to lose stability at $\lambda = \lambda_c(\mu, \gamma)$.

One can show also that $\Psi(\mu, \gamma)$ has the form $\Psi(\mu, \gamma) = \psi^0(\mu) + \gamma \widetilde{\Psi}(\mu, \gamma)$ and that, for fixed μ, $\lambda_c(\mu, \gamma)$, $\eta(\mu, \gamma)$,

$\Lambda(\mu,\gamma)$ and $\widetilde{\Psi}(\mu,\gamma)$ are analytic in γ. In addition, the adjoint problem for $\mathcal{L}$ at $\lambda = \lambda_c$,

$$\mathcal{L}^*(\lambda_c,\mu,\gamma)\Psi^*(\mu,\gamma) = -i\gamma\eta(\mu,\gamma)\Psi^*(\mu,\gamma),$$

has a solution where Ψ^* is again analytic in γ for each μ and of the form $\Psi^*(\mu,\gamma) = (\psi^0)^* + \gamma\widetilde{\Psi}^*(\mu,\gamma)$. Here the adjoint operator is

$$\mathcal{L}^*(\lambda,\mu,\gamma) = L_0 - \lambda L_1 - L_2^* - \lambda\gamma(-M_1 + M_2^*),$$

where $L_2^* = \Pi\mathcal{L}_2^*$ and $M_2^* = \Pi\mathcal{M}_2^*$ with

$$\mathcal{L}_2^* = \begin{pmatrix} 0 & 0 & 0 & 0 & 0 \\ 0 & 0 & 0 & 0 & 0 \\ 0 & 0 & 0 & \mu_1 & \mu_2 \\ 0 & 0 & -\mu_1 & 0 & 0 \\ 0 & 0 & \mu_2 & 0 & 0 \end{pmatrix},$$

$$\mathcal{M}_2^* = -(U_s')^{1/2}\left[\begin{pmatrix} 0 & 0 & 0 & 0 & 0 \\ 0 & 0 & 1 & 0 & 0 \\ 0 & 0 & 0 & 0 & 0 \\ 0 & 0 & 0 & 0 & 0 \\ 0 & 0 & 0 & 0 & 0 \end{pmatrix} - z\frac{\partial}{\partial y}\right].$$

One then easily sees from (17) that Ψ, Ψ^* satisfy $(\Psi,\Psi^*) \neq 0$, provided that μ,γ are sufficiently small. Thus, $\lambda_c(\mu,\gamma)$ is semi-simple and, in addition, we may choose Ψ, Ψ^* so that $(\Psi,\Psi^*) = 1$.

The modifications required in establishing the above results consist essentially of replacing M_1, L_0, ψ_0 and various inner products $(\cdot,\psi_0)$ in [*Knightly et al.*, 1993, Lemma 3.1] by M_1 in (6), L in (16), ψ^0 in (15) and $(\cdot,(\psi^0)^*)$, respectively.

As a final remark on linear stability, we recall that $\lambda_c(\mu,\gamma)$ gives the smallest value of λ for which $\mathcal{L}(\lambda,\mu,\gamma)$ has a unique pair of complex conjugate eigenvalues $\zeta, \bar{\zeta}$ with $\mathcal{R}e\,\zeta = 0$. Thus $\lambda_c(\mu,\gamma)$ is the critical value of λ in problem (18) for μ,γ sufficiently small, and the corresponding Langmuir number La $= \lambda^{-1}$ in (7) is the critical Langmuir number for the possible onset of overstable motions.

5. REDUCTION TO A LANDAU-TYPE AMPLITUDE EQUATION

In Section 4 many of the properties and estimates were established for small μ. Henceforth, we regard μ as *fixed*, $0 < |\mu| < \mu_0$, with μ_0 chosen sufficiently small so that all such previous results are valid. Although certain "constants" in our analysis actually depend upon μ, such dependence is suppressed in this and succeeding sections. Thus, for *fixed* μ, $0 < |\mu| < \mu_0$, the center manifold approach developed in [*Vanderbauwhede et al.*, 1992] can be used with only slight modifications to reduce the problem of finding solutions of $(*)$ to solving a single amplitude equation. Hypothesis (Σ), in particular, is satisfied with $\mathcal{K}$ as the intermediate space and $\alpha \in (0,1)$ (see [*Iooss*, 1972, p. 305; 1984, p. 164]); the required properties of the spectrum of L are given in Lemma 4.1 and the decomposition used is $\mathcal{H} = \mathcal{N} \oplus \mathcal{R}$ introduced at the end of Section 4.1.

Let P denote the projection of $\mathcal{H}$ onto $\mathcal{N}$ and let Q denote the projection of $\mathcal{H}$ onto $\mathcal{R}$. Let $X = Pw$ and $Y = Qw$, $w \in \mathcal{H}$, and define $\lambda - \lambda_0 = \gamma\delta$, where $\lambda_0 = \lambda_0(\mu)$ is given in (13). (This choice of scaling for $\lambda-\lambda_0$ and others to follow emphasize the special nature of the parameter γ and lead to a second order theory in γ.) Then for all $k \geq 1$ there exists [*Vanderbauwhede et al.*, 1992, p. 136] a neighborhood $\mathcal{O} \times \mathcal{I}^2$ of $(0,0,0)$ in $\mathcal{D} \times \mathbb{R}^2$ on which is defined a $\mathbb{C}_b^k$ map $V : P\mathcal{O} \times \mathcal{I}^2 \to Q\mathcal{D}$ such that $V(0,\delta,\gamma) = 0$, $(\delta,\gamma) \in \mathcal{I}^2$, $D_X V(0,0,0) = 0$, and if

$$\mathcal{M} \equiv \mathcal{M}(\delta,\gamma) = \{X + V(X,\delta,\gamma) : X \in P\mathcal{O}, (\delta,\gamma) \in \mathcal{I}^2\},$$

then $\{(Z,\delta,\gamma) : Z \in \mathcal{M}, (\delta,\gamma) \in \mathcal{I}^2\}$ is a local center manifold for the system consisting of $(*)$ and the scalar equations $\dfrac{d\delta}{dt} = \dfrac{d\gamma}{dt} = 0$. Moreover, for each $(\delta,\gamma) \in \mathcal{I}^2$, $\mathcal{M}$ is a local center manifold for $(*)$, $\dim \mathcal{M} = \dim \mathcal{N} = 2$ and the flow on $\mathcal{M}$ is governed by an equation in $\mathcal{N}$, namely

$$\begin{aligned} \frac{dX}{dt} &= P[-(L - \gamma\delta L_1 - \gamma(\lambda_0 + \gamma\delta)(M_1 + M_2))(X + \\ &\qquad V(X,\delta,\gamma)) - B(X + V(X,\delta,\gamma))] \\ &\equiv F(X,\delta,\gamma), \end{aligned} \tag{21}$$

where $F \in \mathbb{C}_b^k(P\mathcal{O} \times \mathcal{I}^2; \mathcal{N})$. Since the system (5) with μ fixed and $\lambda = \lambda_0 + \delta\gamma$ is invariant under translations in y, i.e., $\tau_\alpha w(y,z) = w(y+\alpha,z)$, the function F in (21) also satisfies (e.g., see [*Chossat et al.*, 1985, §II])

$$\tau_\alpha F(X,\delta,\gamma) = F(\tau_\alpha X,\delta,\gamma), \quad X \in \mathcal{N}. \tag{22}$$

To obtain real solutions of (21) we set

$$\begin{aligned} X &= \beta\psi^0 + \bar{\beta}\bar{\psi}^0, \\ F(X,\delta,\gamma) &= f(\beta,\bar{\beta},\delta,\gamma)\psi^0 + \overline{f(\beta,\bar{\beta},\delta,\gamma)}\bar{\psi}^0, \end{aligned}$$

where $\beta, f(\beta,\bar{\beta},\delta,\gamma)$ are complex and ψ^0 is defined in (15). One now easily sees that

$$\begin{aligned} S\psi^0 &= (S\varphi)e^{-ia_0 y} = \bar{\psi}^0, \\ \tau_\alpha\psi^0 &= e^{ia_0\alpha}\psi^0. \end{aligned}$$

Here we have used also that ψ^0 in (15) is of the form $\varphi = (\varphi_1, (i/a_0)\varphi_3', \varphi_3, \varphi_4, \varphi_5)^\top$, where $\varphi_1, \varphi_3, \varphi_4$ and φ_5 are real. Thus, solving (21) is equivalent to solving the scalar equation

$$\frac{d\beta}{dt} = f(\beta, \bar{\beta}, \delta, \gamma), \quad \beta \in \mathbb{C}, \ (\delta, \gamma) \in \mathbb{R}^2, \tag{23}$$

where, by (22), f satisfies (e.g., see [*Chossat et al.*, 1985, §III])

$$f(\beta e^{i\alpha}, \bar{\beta} e^{-i\alpha}, \delta, \gamma) = e^{i\alpha} f(\beta, \bar{\beta}, \delta, \gamma). \tag{24}$$

Since α is arbitrary, setting $\beta = \varepsilon e^{i\psi}$, $(\varepsilon, \psi) \in \mathbb{R}^2$, and choosing first $\alpha = -\psi$ and then $\alpha = \pi$ in (24), one sees that (23) is necessarily of the form

$$\frac{d\beta}{dt} = \beta g(|\beta|^2, \delta, \gamma), \qquad \beta \in \mathbb{C}, \ (\delta, \gamma) \in \mathbb{R}^2, \tag{25}$$

where g is smooth.

To obtain the lowest order terms in the Landau amplitude equation (25) and to emphasize the special nature of the parameter γ, we set

$$\beta = \gamma\xi, \quad \delta = \gamma\theta \quad \text{and} \quad s = \gamma t. \tag{26}$$

Making use of some special mapping properties of M_1, M_2 and B [*Knightly et al.*, 1995, §4], one can then show that (25) has a solution (β, δ, γ) whenever (ξ, θ, γ) is a solution of

$$\begin{aligned} \frac{d\xi}{ds} &= -id_0\xi + \gamma[d_1\theta\xi + \lambda_0^2(b_1 + ib_2)\xi - c\xi|\xi|^2 \\ &\quad + \gamma\xi h(|\xi|^2, \theta, \gamma)], \end{aligned} \tag{27}$$

where $h(|\xi|^2, \theta, \gamma) = O(1)$ uniformly for bounded ξ and θ, provided that $|\gamma| < \gamma_0$ with γ_0 sufficiently small. The constants in (27) satisfy [*Knightly et al.*, 1995, §4] $c > 0$, $d_1 = (L_1\psi^0, (\psi^0)^*)$,

$$\begin{aligned} b_1 + ib_2 &= (M_2 A^{-1} M_2 \psi^0, (\psi^0)^*), \ b_1, b_2 \in \mathbb{R}, \\ d_0 &= \lambda_0 a_0 (U_s')^{-1/2} \bar{U}_s > 0, \end{aligned}$$

where A^{-1} denotes the bounded inverse of the restriction of L to $\mathcal{R}$; the crucial calculation here is to show $c > 0$. Note that (20) implies $d_1 > 0$ provided that $|\mu| < \mu_0$ and μ_0 is sufficiently small. We shall see in Section 6 that $d_1 > 0$ is the usual solvability condition for Hopf bifurcation and that $c > 0$ yields the stability of the bifurcating traveling wave.

Remark 5.1 It is clear from the construction of the center manifold and the form of the substitutions in (26) that smooth solutions $\{\xi, \theta\}$ of (27) yield smooth solutions $\{w = \gamma(u + \gamma v), \ \lambda = \lambda_0 + \gamma^2\theta\}$ of $(*)$, where $u \in \mathcal{N}$ and $v \in \mathcal{R}$. Thus, we see that the use of the structure parameter γ yields a *second order* theory in γ for solutions $\{w, \lambda\}$ of $(*)$ with γ playing the role of an amplitude parameter.

6. PERIODIC TRAVELING WAVE SOLUTIONS

In this section it is shown that there exists a family of bifurcating supercritical, periodic orbits of $(*)$ depending continuously on γ up to $\gamma = 0$. Using known arguments to determine the form of the bifurcating orbits (e.g., see [*Chossat et al.*, 1985]), one then sees that these asymptotically stable states of $(*)$ are actually periodic traveling waves.

To obtain a standard Hopf bifurcation problem from (27), we set

$$\xi = \varepsilon e^{i(\omega_1 s + \gamma\psi)}, \qquad (\varepsilon, \psi) \in \mathbb{R}^2,$$

and choose ω_1 to be

$$\omega_1 = -d_0 \equiv -\lambda_0 a_0 (U_s')^{-\frac{1}{2}} \bar{U}_s. \tag{28}$$

Then (27) reduces to the system

$$\begin{aligned} \frac{d\varepsilon}{ds} &= \gamma\varepsilon[d_1\theta + \lambda_0^2 b_1 - c\varepsilon^2 + \gamma \mathcal{R}e\, h(\varepsilon^2, \theta, \gamma)], \\ \frac{d\psi}{ds} &= \lambda_0^2 b_2 + \gamma \mathcal{I}m\, h(\varepsilon^2, \theta, \gamma), \end{aligned} \tag{29}$$

where $\mathcal{R}e\, h$ and $\mathcal{I}m\, h$ denote the real and imaginary parts of h. One now determines solutions of the autonomous system (29) by first solving the equation

$$0 = d_1\theta + \lambda_0^2 b_1 - c\varepsilon^2 + \gamma \mathcal{R}e\, h(\varepsilon^2, \theta, \gamma), \tag{30}$$

and then integrating to find ψ. Since $d_1 > 0$ (see (20)), one can solve (30) by the implicit function theorem for $\theta = \theta(\varepsilon^2, \gamma)$ and then (29) for $\psi = \omega s$ with

$$\omega = \lambda_0^2 b_2 + \gamma \mathcal{I}m\, h(\varepsilon^2, \theta(\varepsilon^2, \gamma), \gamma).$$

In fact, we have the following result in which the μ dependence is suppressed.

Theorem 6.1. Given $\Theta_0 > 0$ and $\varepsilon_0 > 0$, there exists $\gamma_0 > 0$ such that, for $|\theta| < \Theta_0$, $0 < \varepsilon < \varepsilon_0$ and $0 < |\gamma| < \gamma_0$, Equation (27) has a solution $\{\xi = \varepsilon e^{i\gamma(\omega_1 + \gamma\omega)s}, \theta\}$ of the form

$$\begin{aligned} \theta &= d_1^{-1}(-\lambda_0^2 b_1 + \theta_0(\gamma) + \varepsilon^2[c + \tilde{\theta}(\varepsilon^2, \gamma)]), \\ \omega &= \lambda_0^2 b_2 + \omega_0(\gamma) + \varepsilon^2 \tilde{\omega}(\varepsilon^2, \gamma), \end{aligned} \tag{31}$$

that is uniformly bounded, continuous in γ up to $\gamma = 0$, and *unique (up to translations in s) in* $|\theta + d_1^{-1}(\lambda_0^2 b_1 - c\varepsilon^2)| < K_1\gamma_0$, $|\omega - \lambda_0^2 b_2| < K_2\gamma_0$. *Here* K_1, K_2 *depend only on* Θ_0 *and* ε_0, *and* $\theta_0(\gamma)$, $\omega_0(\gamma)$, $\tilde{\theta}(\cdot,\gamma)$ *and* $\tilde{\omega}(\cdot,\gamma)$ *are* $O(\gamma)$, *uniformly for* $0 < \varepsilon < \varepsilon_0$.

Remark 6.2. Note that $c > 0$ and the form of the solution θ in (31) ensure that for each *fixed* γ, $0 < |\gamma| < \gamma_0$, we have supercritical bifurcation with the unique turning point of the "parabola" at $\theta^* = -d_1^{-1}\lambda_0^2 b_1 + \theta_0(\gamma)$. Thus, in the following theorem the asymptotic stability of the bifurcating periodic orbit for each *fixed* γ, $0 < |\gamma| < \gamma_0$, is a standard result for supercritical bifurcation (e.g., see [*Iooss*, 1972, Theorem 4]).

The solution $\{\xi, \theta\}$ of (27) described in the last theorem leads directly to the existence of periodic orbits and, hence, periodic traveling wave solutions of $(*)$.

Theorem 6.3. Given $\varepsilon_0 > 0$, *there exists* $\gamma_0 > 0$ *such that, for* $0 < \varepsilon < \varepsilon_0$ *and* $0 < |\gamma| < \gamma_0$, *Equation* $(*)$ *has a unique branch of periodic orbits bifurcating from* $w = 0$ *at* $\lambda = \lambda_c(\mu,\gamma) = \lambda_0(\mu) + \gamma^2\theta^*(\mu,\gamma)$. *The periodic orbit* $w(\varepsilon,\gamma)(t)$ *at* $\lambda = \lambda(\varepsilon^2,\gamma)$ *has frequency* $\gamma(\omega_1 + \gamma\omega)$ *in* t *and depends continuously on* γ *up to* $\gamma = 0$. *Moreover,* $w(\varepsilon,\gamma) \in \mathbb{C}_b^1(\mathcal{D})$ *and*

$$\begin{aligned} w(\varepsilon,\gamma)(t) &= \gamma\varepsilon[w_0(\gamma(\omega_1+\gamma\omega)t) \\ &\quad + \gamma W(\varepsilon,\gamma)(\gamma(\omega_1+\gamma\omega)t)], \\ \lambda(\varepsilon^2,\gamma) &= \lambda_0 + \gamma^2\theta(\varepsilon^2,\gamma), \end{aligned}$$

where $w_0(s) = e^{is}\psi^0 + e^{-is}\bar{\psi}^0$, $\omega_1 = -\lambda_0 a_0 (U_s')^{-\frac{1}{2}}\bar{U}_s$, $\theta = \theta(\varepsilon^2,\gamma)$ *and* $\omega = \omega(\varepsilon^2,\gamma)$ *are given by (31), and* $W(\varepsilon,\gamma) = 0(1)$, *uniformly in* ε *and* γ. *For each* fixed γ, $0 < |\gamma| < \gamma_0$, *the branch of periodic orbits* $w(\varepsilon,\gamma)$ *bifurcates supercritically from* $\lambda_c(\mu,\gamma)$, *is asymptotically stable in* $\mathcal{H}$, *and determines a wave* $w(\varepsilon,\gamma) = w(a_0 y + \gamma(\omega_1 + \gamma\omega)t, z)$, *periodic in* $a_0 y + \gamma(\omega_1+\gamma\omega)t$ *of period* 2π.

The proof of Theorem 6.3 follows from Remark 5.1 and Theorem 6.1; the stability of the bifurcating supercritical orbits for each *fixed* γ, $0 < |\gamma| < \gamma_0$, follows from Remark 6.2. Finally, since X is of the form (up to an arbitrary phase)

$$\begin{aligned} X &= \gamma\varepsilon[\Phi^0(z)e^{i(a_0 y + \gamma(\omega_1+\gamma\omega)t)} \\ &\quad + \overline{\Phi^0(z)e^{-i(a_0 y+\gamma(\omega_1+\gamma\omega)t)}}], \qquad (32) \\ \Phi^0(z) &\equiv \Phi^0(\mu)(z), \end{aligned}$$

and, by the construction of the center manifold in Section 5, $w = X + V(X,\gamma\theta,\gamma)$, it follows that w depends only on the two variables z and $a_0 y + \gamma(\omega_1 + \gamma\omega)t$. Thus, the bifurcating periodic orbits are actually periodic traveling waves of the indicated form.

Remark 6.4. Note that, since ω_1 is negative, for each *fixed* (μ,γ) sufficiently small, the periodic wave determined in Theorem 6.3 travels in the negative y-direction when $\gamma < 0$ and in the positive y-direction when $\gamma > 0$. Moreover, as we shall see in Section 7, the velocity of the cross-drift of the rolls is completely determined by the magnitude and direction of the cross-wind component of the Stokes drift, a fact that was not known even for Langmuir circulations until recently (see [*Knightly et.*, 1995]).

7. DISCUSSION AND CONCLUDING REMARKS

In this section we discuss some of the physical interpretations of the results obtained above for the model equations (5). In particular, we determine estimates of the critical Langmuir numbers and the cross-wind drift velocity of the system of rolls.

We suppose as in [*Leibovich*, 1977a,b; *Knightly et al.*, 1995] that the prescribed irrotational surface wave field, $\tilde{u}_w$, has frequency σ, wave slope ε and velocity $\varepsilon\sigma d$, where d is the depth of the layer. If "$\tilde{\ }$" denotes dimensional quantities, then the velocity vector of the mean current, $\tilde{q}$, temperature, $\widetilde{T}$, concentration, $\widetilde{C}$, and Stokes drift, $\tilde{u}_s = \widetilde{U}_s(\tilde{z})(\hat{\imath}+\gamma\hat{\jmath})$, are related to the corresponding dimensionless quantities in (4) and (5) by (see [*Knightly et al.*, 1995])

$$\tilde{t} = (d^2/\nu_T)t, (\tilde{x},\tilde{y},\tilde{z}) = d(x,y,z-\frac{1}{2}), \quad (33)$$

$$\widetilde{U}_s(\tilde{z}) = (u_0^2 d/\nu_T)\, U_s(z), \quad (34)$$

$$\begin{aligned} \tilde{q} &= \varepsilon\sigma d u_w + (u_0^2 d/\nu_T)(U_1 + z)\hat{\imath} \\ &\quad + (\nu_T/d)(-(U_s')^{-1/2}w_1, w_2, w_3)^\top, \\ \widetilde{T} &= T_r(1 + T'z - \nu_T(T'/g\beta_T^0 d^3)^{1/2} w_4), \\ \widetilde{C} &= C_r(1 + C'z - \nu_T(C'/g\beta_C^0 d^3)^{1/2} w_5). \end{aligned}$$

Here ν_T is the constant kinematic (eddy) viscosity and u_0 is the water friction velocity induced by the wind [*Leibovich*, 1977b, p. 564]. One then easily sees from (34) that

$$\widetilde{U}_s'((0) = (u_0^2/\nu_T)U_s' \quad (35)$$

$$\begin{aligned} (u_0^2 d/\nu_T)\bar{U}_s &= \int_{-1/2}^{1/2} (u_0^2 d/\nu_T)(\bar{U}_s + U_s' z)dz \\ &= d^{-1}\int_{-d}^{0} \widetilde{U}_s(\tilde{z})d\tilde{z} \equiv (\widetilde{U}_s)_{\text{ave}}. \quad (36) \end{aligned}$$

If one now defines the Reynolds number, $\widetilde{R}$, and the Rayleigh numbers, S_1 and S_2, to be

$$\widetilde{R} = \frac{u_0^4 d^4 U_s'}{\nu_T^4} = \frac{u_0^2 d^4 \widetilde{U}_s'(0)}{\nu_T^3} \quad (37)$$

$$S_1 = \frac{\beta_T^0 g T' d^3}{\nu_T^2}, \; S_2 = \frac{\beta_C^0 g C' d^3}{\nu_T^2}, \qquad (38)$$

then, in (5), $\lambda^2 = \widetilde{R}$, $\mu_1^2 = S_1$ and $\mu_2^2 = S_2$. Thus, since $\widetilde{R}$ and S_1 are the same parameters "$R = \widetilde{R}$ and $S = S_1$" in [*Leibovich et al.*, 1989 , p. 477] and since the linear problem discussed there is equivalent to the linear problem in (5) with $\mu_2 = \gamma = 0$ and $P_T = P_C = 6.7$, one sees that the critical Reynolds number $\lambda_0(\mu_1, 0)$ in §4.1 is completely determined by $\widetilde{R}_c$ for *fixed* q and a wide range of μ_1^2 values; here $q = 2/\ell$ $(= a_0/\pi)$ denotes the "aspect ratio" of the roll. One sees, e.g., that for values of $q < 0.4$ and $\mu_1^2 < 500$ the onset of instability is to monotonic convection as required in §4.1 [*Leibovich et al.*, 1989, Figure 1]. For larger values of q, overstable convection occurs for smaller values of $\mu_1^2 = S^D(q)$, where $S^D(q)$ decreases as q increases [*Leibovich et al.*, 1989, p. 478]; if, e.g., $q = 1$ then monotonic convection still occurs but only for $\mu_1^2 < S^D(1)$, where $S^D(1) \approx 78$ and $\lambda_0(\sqrt{78}, 0) \approx 28$. The variation of $S^D(q)$ with q may be inferred from Figure 1 in [*Leibovich et al.*, 1989] in which the branch with onset to monotonic convection is labeled M.

Once the possibility of overstable convection has been eliminated for $\mu_2 = \gamma = 0$ and $\mu_1^2 < S^D(q)$, it suffices for purposes of this paper to make use of (19) and the form of $\mathcal{L}_2$ in (6) to obtain essentially the same numerical estimates for the critical values of the Reynolds number $\lambda_0(\mu) = \lambda_0(\mu_1, \mu_2)$ in §4.1 and the Langmuir number $\lambda_c^{-1}(\mu, \gamma)$ in §4.2, provided that μ_2 and γ are sufficiently small. Numerical solutions of the general linear problem as well as the nonlinear problem in (5) will be considered in subsequent papers.

To determine the cross-wind drift velocity of the system of rolls, we make use of the formula for the frequency ω_1 in (28) to determine the speed of the traveling wave in (32). Thus, to first order in γ, the wave speed in dimensionless variables is given by

$$v_d = \gamma \lambda_0(\mu)(U_s')^{-1/2}\bar{U}_s. \qquad (39)$$

It then follows easily from (33)–(37) that the wave speed in (32) in dimensional variables is

$$\tilde{v}_d = (\nu_T/d)v_d = \gamma(u_0^2 d/\nu_T)\bar{U}_s = \gamma(\widetilde{U}_s)_{\text{ave}}. \qquad (40)$$

Thus, the wave speed of $\tilde{v}_d$ is completely determined by the average of the cross-wind component of the Stokes drift. Note, in particular, that (at least to first order in γ) $\tilde{v}_d$ is independent of the basic wave number a_0. This is of some importance since the value of a_0 here and in [*Leibovich et al.*, 1989, §2.2] has been assigned in a somewhat arbitrary way. The same result for $\tilde{v}_d$ holds, of course, for a system of k rolls with $a_0 = k\pi/\ell$. Thus, one obtains the expected result that the cross-wind drift velocity is independent of both the size and the number of rolls in the system. However, one obtains also the perhaps surprising result that $\tilde{v}_d$ is, to first order in γ, completely determined by the average of the cross-wind component of the Stokes drift which, in turn, is determined *a priori* by the surface wave field. Since the C–L theory is based in part upon an assumption that the rotational currents and the Stokes wave drift are of comparable order but otherwise act as independent parts of the current system (e.g., see [*Craik et al.*,1977, p. 411; *Leibovich*, 1983, pp. 399–400]), the result in (40) for $\tilde{v}_d$ offers strong evidence that such an assumption is consistent with the order to which the theory holds.

The existence of traveling waves obtained in the present paper shows clearly the difference in wind-driven, double diffusive convection between physical situations in which the Stokes drift has a cross-wind component and in which it does not. If, e.g., $\gamma = 0$ then, as shown in §4.1 for μ_1 and μ_2 sufficiently small, one does not expect overstability to occur. However, if $\gamma \neq 0$, then even for μ_1 and μ_2 small the basic state *always* undergoes Hopf bifurcation losing stability to asymptotically stable periodic traveling waves. Thus, because of the form of the Boussinesq approximation in (3), one must also *always* expect traveling wave corrections to the density in the layer when the Stokes drift has a cross-wind component.

Acknowledgements. The research of G. Knightly was supported in part by ONR Grant N00014–90–J–1031 and that of D. Sather by ONR Grant N00014–94–1–0194.

REFERENCES

Chandrasekhar, S., *Hydrodynamic and Hydromagnetic Stability*, Clarendon Press, Oxford, 1961.

Chossat, P., and G. Iooss, Primary and secondary bifurcations in the Couette–Taylor problem, *Japan J. Applied Math.*, 2, 37–68, 1985.

Cox, S.M., and S. Leibovich, Langmuir circulations in a surface layer bounded by a strong thermoline, *J. Phys. Oceaneog.*, 23, 1330–1345, 1993.

Cox, S.M., S. Leibovich, I.M. Moroz and A. Tandon, Nonlinear dynamics in Langmuir circulations with $O(2)$ symmetry, *J. Fluid Mech.*, 241, 669–704, 1992a.

Cox, S.M., S. Leibovich, I.M. Moroz, and A. Tandon, Hopf bifurcations in Langmuir circulations, *Physica D*, 59, 226–254, 1992b.

Craik, A.D.D., and S. Leibovich, A rational model for Langmuir circulations, *J. Fluid Mech.*, 73, 401–426, 1977.

Iooss, G., Existence et stabilité de la solution périodique secondaire intervenant dans les problèmes d'évolution du type Navier–Stokes, *Arch. Rational Mech. Anal.*, 47, 301–329, 1972.

Iooss, G., Bifurcation and transition to turbulence in hydrodynamics, *Bifurcation Theory and Applications*, (L. Salvadori, ed.), Lecture Notes in Mathematics, Vol. 1057, Springer–Verlag, New York, 152–201, 1984.

Iudovich, V.I., On the origin of convection, *J. Appl. Math. Mech.*, 30, 1193–1199, 1969.

Joseph, D.D., Stability of Fluid Motions, *Springer Tracts in Natural Philosophy*, 28, Springer–Verlag, New York, 1976.

Kato, T., *Perturbation Theory for Linear Operators*, Springer Verlag, New York, 1966.

Knightly, G.H., and D. Sather, Periodic waves in rotating plane Couette flow, *Z. angew. Math. Phys.*, 44, 1–16, 1993.

Knightly, G.H., and D. Sather, Continua of periodic waves in rotating plane Couette flow, *Euro. J. Mech. B/Fluids*, 13, 511–526, 1994.

Knightly, G.H., and D. Sather, Langmuir circulations when the Stokes drift has a cross-wind component, *Euro. J. Mech. B/Fluids*, in press, 1995.

Knobloch, E., D.R. Moore, J. Toomre, and N.O. Weiss, Transitions to chaos in two-dimensional double-diffusive convection, J. *J. Fluid Mech.*, 166, 409–448, 1986.

Leibovich, S., On the evolution of the system of wind drift currents and Langmuir circulations in the ocean. Part I. Theory and averaged current, *J. Fluid Mech.*, 79, 715–743, 1977a.

Leibovich, S., Convective instability of stably stratified water in the ocean, *J. Fluid Mech.*, 82, 561–581, 1977b.

Leibovich, S., The form and dynamics of Langmuir circulations, *Ann. Rev. Fluid Mech.*, 15, 391–427, 1983.

Leibovich, S., Dynamics of Langmuir circulations in a stratified ocean, *The Ocean Surface*, (Y. Toba and H. Mitsuyasu, eds.), Reidel, 457–464, 1985.

Leibovich, S., S.K. Lele, and I.M. Moroz, Nonlinear dynamics in Langmuir circulations and in thermosolutal convection, *J. Fluid Mech.*, 198, 471–511, 1989.

Moore, D.R., J.Toomre, E. Knobloch, and N.O. Weiss, Period doubling and chaos in partial differential equations for thermosolutal convection, *Nature*, 303, 663–667, 1983.

Vanderbauwhede, A., and G. Iooss, Center manifold theory in infinite dimensions, *Dynamics Reported*, 1, New Series, 125–163, 1992.

George H. Knightly, Department of Mathematics, University of Massachusetts, Amherst, MA 01003–4515.

D. Sather, Department of Mathematics, University of Colorado, Boulder, CO 80309–0395.

The Salt Finger Wavenumber Spectrum

Colin Y. Shen

Naval Research Laboratory, Washington, D.C. 20375

Raymond W. Schmitt

Woods Hole Oceanographic Institution, Woods Hole, Massachusetts 02543

A model for the salt finger wavenumber spectrum is presented. This model describes the +2 power dependence of the horizontal temperature gradient spectrum on the horizontal wavenumber which has been found for salt fingers in the ocean, in the laboratory experiments, and in numerical simulations. The ability of the model to predict the finger spectral amplitude is verified with direct numerical simulations. The model also shows that the Cox number for salt fingers varies inversely with the mean vertical temperature gradient similar to that observed in the ocean.

1. INTRODUCTION

Temperature measurements in the ocean have consistently shown that a +2 power dependence of the horizontal temperature gradient spectrum on the horizontal wavenumber clearly characterizes those regions with salt fingering [*Gargett and Schmitt*, 1982; *Marmorino*, 1987; *Mack*, 1989]. This power dependence is noticeably weaker than that predicted by the theoretical salt finger spectrum obtained by *Schmitt* [1979] for a model of temporally growing fingers, although the peak of the gradient spectrum has been associated with the fastest growing finger mode. Despite the difference between the measurement and the model, the observed +2 power dependence in the fingering favorable region can be considered a clear evidence for the presence of salt fingering, since direct laboratory measurement of salt fingers [*Taylor and Bucens*, 1989] has obtained essentially the same power dependence for the gradient spectrum, and the same has also been obtained form direct numerical computation of salt fingers [*Shen*, 1993].

The inability of the previous model to account for the +2 power dependence is not totally unexpected. As the spectrum of temporally growing fingers is dominated by the fastest growing finger mode, the spectral distribution of temperature variance across wavenumbers can not be constant, like that required of a gradient spectrum with +2 power dependence, but must peak at the wavenumber of the fastest growing mode. Consequently, the model for temporally growing fingers tends to over predict the slope of the gradient spectrum in the low wavenumber range where +2 power dependence has been observed, though this is a function of the amount of time allowed for growth. In actual convection, the temporally growing fingers are certain to play a role. However, ultimately, their growth has to be limited either by the vertical extent of the fingering region, such as in the case of salt fingering between two well-mixed fluid layers, or by internal finger instability such as that identified in the studies by *Stern* [1969], *Holyer* [1984] and *Howard and Veronis* [1992]. Such amplitude limiting will have the effect of lessening the disparity between the spectral amplitudes of the fastest growing mode and those of the slower growing ones. The consequence can be a spectral slope less than that predicted based on continually growing fingers.

In the following, the amplitudes of the temporally growing fingers are given, and then we constrain the amplitude growth to the maximum vertical extent of the fingering zone to obtain amplitudes limits. Such a constraint has been discussed previously in *Shen* [1989] for the fastest growing mode. Here, it is extended to all finger modes. Finger instability is known to play an important role in salt fingering. Its effect as a constraint on finger amplitude can only be approximately indicated, given that so little is

Double-Diffusive Convection
Geophysical Monograph 94

known except for a recent direct numerical simulation of the finite amplitude finger instability. The temperature gradient wavenumber spectrum subjected to the above mentioned amplitude constraint will be shown to have a +2 spectral slope consistent with the observed spectrum. The adequacy of this model spectrum is further verified with the salt finger wavenumber spectrum computed from direct numerical simulations. The spectrum will be shown, in its most basic form, to be invariant at a given constant heat-to-salt density ratio, R_ρ, so that it should be useful for synthesizing measured spectra and their interpretation. The spectrum also clarifies the relationship between the spectral peak of the gradient spectrum and the fastest growing mode. Additionally, the model spectrum can be integrated to give an estimate of the Cox number. The model shows that the Cox number varies with the mean vertical temperature gradient to the -1.5 power, in support of the observed inverse relationship between these two quantities, a relationship not obtained in previous finger models.

2. SPECTRAL AMPLITUDES

We begin with the amplitudes of temporally growing fingers, whose vertical structures are uniform and horizontal structures are those of periodic Fourier modes. Such uniform periodic fingers in an unbounded constant T and S gradient region are known to satisfy the nonlinear governing equation exactly, and the amplitudes of each of these finger modes' vertical velocity (w), temperature (T′) and salt (S′) fluctuations grow exponentially in time. Specific details can be found in *Stern* [1975] and *Schmitt* [1979]. For this discussion, the finger amplitudes w, T′ and S′, in the dimensionless form, are

$$w=w_o e^{\lambda t}\sin(\mathbf{k}\cdot\mathbf{x}+\theta) \tag{1a}$$

$$T'=\frac{-w}{(\lambda+k^2)} \tag{1b}$$

$$S'=\frac{-w}{(\lambda+\tau k^2)} \tag{1c}$$

where T′ and S′ are nondimensionalized with $(\partial\overline{T}_*/\partial z_*)d$ and $(\partial\overline{S}_*/\partial z_*)d$, respectively. The dimensional mean vertical gradients $\partial\overline{T}_*/\partial z_*$ and $\partial\overline{S}_*/\partial z_*$ are constants, and the length scale d is defined as

$$d=\left[\frac{\nu\kappa_T}{g\alpha(\partial\overline{T}_*/\partial z_*)}\right]^{1/4}$$

where ν is the kinematic viscosity, κ_T is the heat diffusivity, g is the gravitational constant, and α is the thermal expansion coefficient. w is nondimensionalized with κ_T/d, having an initial ampltude w_o. Without loss of generality, the sine function has been chosen in (1) as the periodic structure, and θ is an arbitrary phase between 0 and 2π. The horizontal wavenumber vector **k** has been scaled by d^{-1}. It has components k_x and k_y, and $k^2=k_x^2+k_y^2$. The finger amplitude growth rate λ has been scaled by κ_T/d^2. The growth rate and the wavenumber magnitude are related by the dispersion relation

$$\lambda^3+(\sigma+1+\tau)k^2\lambda^2+[(\tau+\tau\sigma+\sigma)k^4+\sigma(1-R_\rho^{-1})]\lambda$$
$$+\sigma k^2[\tau k^4+\tau-R_\rho^{-1}]=0 \tag{2}$$

which has a positive real root for λ in the range $R_\rho<\tau^{-1}$ and for $k<[(1/R_\rho\tau)-1]^{1/4}$. The other dimensionless parameters are the Prandtl number $\sigma=\nu/\kappa_T$, the salt-to-heat diffusivity ratio $\tau=\kappa_S/\kappa_T$, and the density ratio $R_\rho=(\alpha\partial\overline{T}_*/\partial z_*)/(\beta\partial\overline{S}_*/\partial z_*)$, where κ_S is the salt diffusivity, and β is the density contraction coefficient due to salt.

For the moment, we consider (1a-c) to be applicable to a finite height fingering zone (or interface); we return to address the appropriateness of this application later in the discussion. Next, we proceed to identify finger w, T′ and S′ with specific fluid particles and consider the changes of these properties with the particles' vertical displacements across the fingering zone. Let the particles be labeled by the coordinates $\mathbf{x}_o$ and z_o, whose physical locations are given by

$$\mathbf{x}=\mathbf{x}_o$$

$$z=\eta(t)\sin(\mathbf{k}\cdot\mathbf{x}+\theta)+z_o$$

Each particle has a speed dz/dt=w. It follows that $\eta(t)=w_o e^{\lambda t}/\lambda$, or

$$w=\lambda(z-z_o) \tag{3}$$

Thus, with respect to each moving particle, w is proportional to the displacement, $z-z_o$. For a fluid particle that traverses across the fingering zone, its vertical displacement increases as $z-z_o$, and so does its velocity as given by (3). The maximum vertical distance that the particle travels in the fingering zone is the fingering zone height H itself, i.e., $z-z_o=H$. This maximum vertical distance means that w can have at most the speed λH across the fingering zone.

Therefore, we see that the finger's w amplitude in a finite height fingering zone is necessarily limited, on account of this consideration of the fluid particle's displacement, rather than growing continuously as implied by (1a). Because the T' and S' amplitudes are directly proportional to w in (1a,b), we see that the T' and S' amplitudes are also necessarily limited. It follows that in a finite height fingering zone, each of the finger amplitudes has a finite maximum value, being limited by the vertical physical extent of the fingering zone.

The physical limit applied to the amplitudes by the fingering zone height H has meaning here clearly only if H is either fixed or evolves at a rate slower than the rate λ at which the displacement $z-z_o$ increases. Direct numerical solution of the nonlinear governing equations showed that the height H is related to the height of the stabilizing T stratification [*Shen*, 1989, 1993]. The heat flux transported by the particle initially increases the height of the T stratification at the same rate as the displacement, $z-z_o$. However, the heat flux is exceeded by the density-destabilizing salt flux along the edges of the T stratification zone, because of the higher diffusive heat loss by the particle. The result is that the density along the edges becomes gravitationally unstable, and finger structure is prevented from forming. From that point on, the edges of the T-stratification zone or the fingering zone do not advance with the particle displacement $z-z_o$, and correspondingly, H increases at a rate slower than λ. This simulation result is in agreement with the use of the H constraint; moreover, the simulation result shows that the geometric appearance of the constraint has its underlying cause in the gravitational instability occurring along the edge of the T stratifiction zone.

The amplitudes (1a,b,c) to which the height constraint is applied have been obtained for the unbounded fingering system. *Howard and Veronis* [1987] pointed out that the salt diffusion in a finite H fingering zone usually acts too slowly to alter S' of the fluid that passes through it, with the result that S' in the fingering zone is essentially constant, as opposed to that given by (1c). The S' amplitude (1c) thus cannot be used for finite H. Nevertheless, we will continue to use (1a,b) for the w and T' amplitudes, an approximation which the analysis by *Shen* [1993] has shown to be still reasonable for finite H. Moreover, we will continue to use the dispersion relation (2), although it is applicable only if salt diffusion is significant. This limitation turns out to be not critical to the heat-salt system as shown in the same study. We will return to this point later in the concluding section

We now obtain the T' amplitude subject to the height constraint. In (1b), T' has been scaled by $(\partial\overline{T}_*/\partial z_*)d$ which is appropriate for an unbounded system. For the fingering zone of finite H, it is common to scale the temperature fluctuation with ΔT, the total temperature change across the fingering zone. Let T'_* denote the dimensional temperature fluctuation, and let $\hat{T}'=T'_*/\Delta T$. By definition, $T'=T'_*/(\partial\overline{T}_*/\partial z_*)d$. This means that $\hat{T}'=[T'_*/(\partial\overline{T}_*/\partial z_*)d][(\partial\overline{T}_*/\partial z_*)d/\Delta T]=T'/(H_*/d)=T'/H$, where $H_*=\Delta T/(\partial\overline{T}_*/\partial z_*)$ is the definition for the dimensional fingering zone height H_*. Since, for $z-z_o=H$, the w amplitude limit is $w=\lambda H$ from (3), substituting this w into (1b) and multiplying (1b) by 1/H yield

$$\hat{T}'=\frac{-\lambda}{(\lambda+k^2)} \tag{4}$$

a result which is independent of the finger zone height. The dimensional w, on the other hand, depends on $H_*^{1/2}$, specifically,

$$w_*=(\kappa_T g\alpha\Delta T/\nu)^{1/2}\lambda H_*^{1/2}$$

This is obtained by multiplying the two sides of (3) by κ_T/d and setting $z-z_o=H$.

At this point, it appears that the finite H is the sufficient limiting condition for the finger amplitudes. However, a second constraint needs to be imposed, because as just shown, w varies proportional to $H_*^{1/2}$. This means that the vertical kinetic energy can increase without bound with increasingly larger H_*, which is clearly physically unrealistic. For large H_*, the salt finger instability can provide the second constraint. Previous work on salt finger instability by *Stern* [1969], *Holyer* [1984], and *Howard and Veronis* [1992] have shown that long fingers are intrinsically unstable. Direct numerical calculation of the finite amplitude evolution of this instability [*Shen*, 1994] shows that the instability increases viscous dissipation and eventually limits the w amplitudes. For sufficiently large H_*, we can expect the finger instability to have similarly limiting effect on the w amplitude. How to carry out this next step, to impose this constraint on the fingers, is problematic. Numerical simulations such as that given in *Shen* [1993] showed that the finger instability is not limited to large H_*; even at relatively small H_* fingers show sign of instability, in the form of vertical undulation of finger cell structure. The effect of this instability on w is most probably small at small H_* but can be expected to increase with increasing H_*, as the increasing vertical distance allows the instability to evolve more fully. A knowledge of how the instability constraint varies with H is thus required to implement the constraint. This is not known except in the large H_* case studied with numerical simulations. Hence, for now, we will simply adopt this second constraint as suggested by the simulations by defining a critical dimensional height, H_{*c}, above which finger instability limits the w amplitude such that w is a constant independent of H_*, being fully aware that instability could also affect w at heights less than H_{*c}. The value of H_{*c} can be inferred from the equilibrium w obtained previously in numerical simulations of the instability by setting the equilibrium value of w to λH_c and solving for H_c. This yielded $H_c = H_{*c}/d \sim O(10^2)$. Note that from the definition of d, the Rayleigh number at this critical height is $Ra = (H_{*c}/d)^4 \sim O(10^8)$.

Since $\hat{T}' = T'/H$ and $w = \lambda H_c$, the existence of a critical height H_c also means that (4) must be multiplied by H_c/H or H_{*c}/H_* for $H_* > H_{*c}$. Also, note that in the fingering zone, $\hat{T}'$ ranges between 0 and the maximum (4) which is achieved when $z - z_o = H$. The more appropriate $\hat{T}'$ value to prescribe to the fingering zone should be the average or 1/2 of (4). With these two modifications, we write the final $\hat{T}'$ amplitude as

$$\hat{T}'_{\mathbf{k}} = \frac{1}{2}\frac{\lambda}{(\lambda + k^2)}\varepsilon \tag{5}$$

for all wavenumbers k up to $[(1/R_\rho \tau) - 1]^{1/4}$, at which $\lambda = 0$, and the ε factor is

$$\varepsilon = \begin{cases} 1 & \text{for } H_* \le H_{*c} \\ \dfrac{H_{*c}}{H_*} & \text{for } H_* > H_{*c} \end{cases}$$

In the above the subscript **k** has been added to T' to formally identify (5) with the spectral amplitude of the finger mode at wavenumber **k**.

This completes the specification of the finger spectral amplitude. Our initial choice of the Fourier sine function in (1) has been a matter of convenience. The amplitude (5) obtained in this section applies equally well to complex Fourier basis function, $\exp(i\mathbf{k}\cdot\mathbf{x})$. Since the spectrum is conventionally calculated in terms of the complex Fourier basis, hereafter we will associate the amplitude (5) with the complex basis function.

3. THE WAVENUMBER SPECTRUM

The spectral density for the temperature variance at each wavenumber is given by the product of the spectral amplitude and its complex conjugate, which in the present case is just (5) squared,

$$B(\mathbf{k}) = \hat{T}'^2_{\mathbf{k}} = \frac{1}{4}\left[\frac{\lambda}{(\lambda + k^2)}\right]^2 \tag{6}$$

where the ε factor in (5) has been set to unity. This wavenumber spectrum $B(\mathbf{k})$ is isotropic in the horizontal directions, as it depends on only the wavenumber magnitude k. To get some idea of how B varies with k, the function $\lambda(k)$ in (2) may be simplified as in *Stern* [1975] by noting that the following inequalities may be applied to the heat-salt system,

$$\tau \approx 0.01 \ll 1 \ll \sigma \approx 7$$

This lets one to neglect the τ terms as well as those not multiplied by σ in the dispersion relation (2). These simplifications reduces (2) to a quadratic equation, from which the growth rate can be obtained in a closed form approximately as,

$$\lambda \approx \frac{k^2}{R_\rho(k^4 + 1) - 1}$$

Substituting this into (6) and rearranging terms yield the wavenumber spectrum $B(\mathbf{k})$ explicitly in terms of k,

$$B(\mathbf{k}) = \frac{1}{4}\left[\frac{1}{R_\rho(k^4 + 1)}\right]^2 \tag{7}$$

This shows that the spectral level is nearly constant at low wavenumbers with $B \to (2R_\rho)^{-2}$ as $k \to 0$, and the spectrum decays to zero at large k.

The ocean salt finger spectrum is typically obtained from one-dimensional, horizontally towed measurements. The appropriate form of the model spectrum for comparison with the towed measurements is not (7) but (6) integrated along the horizontal direction normal to the towed direction. Since (6) is horizontally isotropic, the direction of integration is not particularly relevant. Here, we simply integrate (6) along k_y assuming the tow is in the k_x direction. This integration

$$\hat{B}(k_x) = \frac{1}{2\pi}\int_{-\infty}^{+\infty} \hat{T}'^2_{\mathbf{k}}\, dk_y$$

has been carried out numerically. Figure 1 shows the

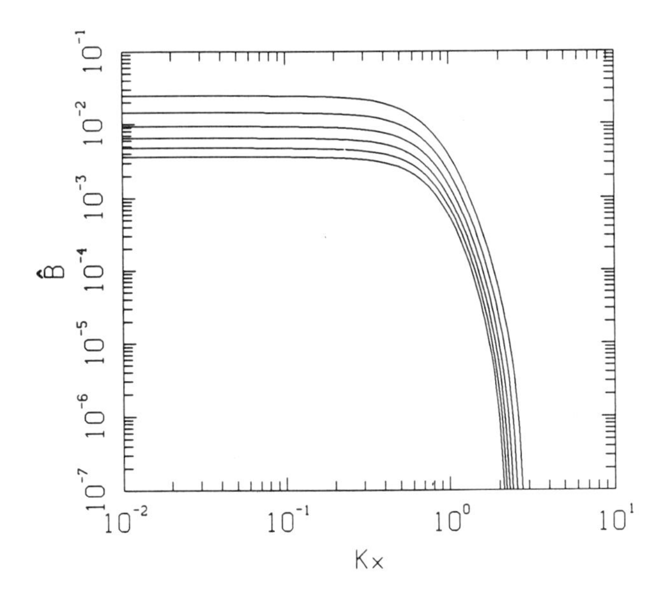

Fig. 1. The model variance wavenumber spectrum for salt finger temperature B(k). The spectra are shown from top to bottom for R_ρ=1.5, 2.0, 2.5, 3.0, 3.5, 4.0, respectively.

resulting one-dimensional spectrum for several different R_ρ. The spectral shape is not different from that of the two-dimensional spectrum (7) described above, i.e., the prominant feature being the constant spectral density at low wavenumbers and the decay at high wavenumbers. Given that spectral density is constant at low wavenumbers, it is clear that the horizontal T gradient spectrum must have +2 slope, since the gradient spectrum $\hat{B}x(k_x)=k_x^2\,\hat{B}(k)$. Figure 2 shows the T gradient spectra for the corresponding five cases in Figure 1. All have a +2 slope similar to the observations shown by *Marmorino* [1987], *Mack* [1989], and *Mack and Schoeberlein* [1993]. Notice that the value of the dimensionless λ in (2) is only a function of R_ρ for a given system with fixed σ and τ, such as the heat-salt system considered here. The temperature variance spectrum (6) with the temperature variance being scaled by ΔT^2 thus depends only on R_ρ. This suggests that the measured salt finger wavenumber spectra may be synthesized into one common form by scaling the variance with ΔT^2 appropriate to each measured spectrum.

With the present model wavenumber spectrum, we can further address the question of whether or not the spectral peak in the gradient spectrum is related to the fastest growing finger mode. It has been a common practice to identify the spectral peak with the fastest growing finger mode in the interpretation of ocean salt fingers. Table 1 lists the wavenumbers of the spectral peak and those of the fastest growing mode at different R_ρ values It is seen that in all cases spectral peak wavenumbers are less by about 10-20% than the fastest growing finger wavenumbers. This difference might be too small to be detected confidently in the measured wavenumber spectrum. Nevertheless, *Marmorino* [1987] has noted in his data that the wavenumber of the measured spectral peak tends to be, systematically, slightly smaller than that predicted of the fastest growing finger mode, a tendency consistent with that shown here. An indication of this difference is also present in *Gargett and Schmitt's* data [1982].

A more accurate test of the foregoing model spectrum is obtained by comparing it against the wavenumber spectrum derived from the direct numerical simulation of salt fingers, in which the finger structure is known precisely. The numerical simulations reported in the literature so far have been carried out strictly in a two-dimensional vertical plane, neglecting one of the horizontal directions. The equivalent one-dimensional form of the model spectrum is (6) with one of the wavenumber components, say k_y, set to zero. To

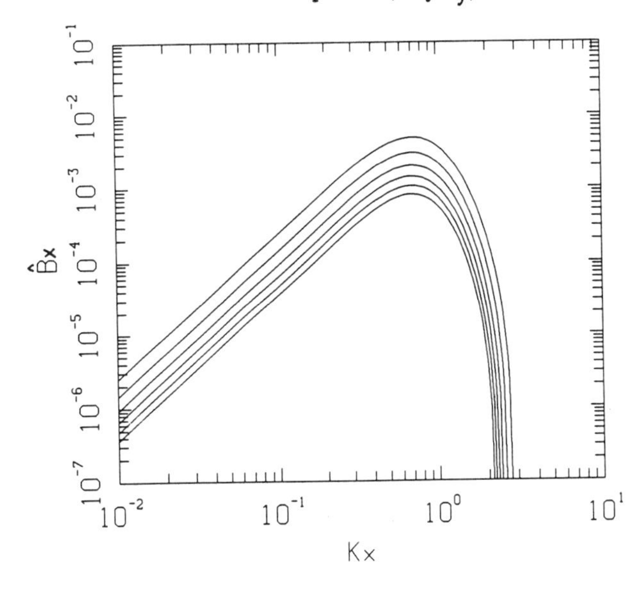

Fig. 2. The model variance wavenumber spectrum for horizontal temperature gradient Bx(k). The spectra are shown in the same order as in Figure. 1.

TABLE 1. k_f, k_p, and c_1 as a function of R_ρ. k_f = wavenumber of the fastest growing mode; k_p= wavenumber of the gradient spectral peak; c_1 = coefficient for the Cox number.

R_ρ	k_f	k_p	c_1
1.5	0.74	0.68	1.2×10^{-2}
2.0	0.84	0.69	6.8×10^{-3}
2.5	0.85	0.70	4.4×10^{-3}
3.0	0.91	0.72	3.1×10^{-3}
3.5	0.92	0.69	2.2×10^{-3}
4.0	0.89	0.71	1.9×10^{-3}

compare with the simulation spectrum obtained from discrete representation, we need to convert the one-dimensional model spectrum to the following discrete form,

$$\tilde{B}(k_x)=\frac{1}{4}\left[\frac{\lambda}{(\lambda+k_x^2)}\right]^2 \Delta k_x^2$$

where $k_x=n\Delta k_x$, $\Delta k_x=2\pi d/L$, and n is an integer; the discrete wavenumber k_x has been nondimensionalized with 1/d, and L is the dimension of the simulation domain which is also the horizontal extent of the fingering zone. Figures 3 and 4 show the simulation spectra (solid curves) sampled at different times and the model spectra (dashed curves) for $R_\rho=2$ and $R_\rho=4$, respectively. The salt finger simulations used to obtain the spectra are those described in *Shen* [1993]. The model spectra can be seen to agree well with the simulation results.

Finally, we present results on the finger Cox number computed using the model spectrum. The Cox number is defined as

$$Co=\left\langle(\nabla T'_*)^2\right\rangle/\left(\partial\overline{T}_*/\partial z_*\right)^2$$

where <> denotes spatial averaging. In terms of the present dimensionless T′, this number is

$$Co=\left\langle\left(\frac{\partial T'}{\partial x}\right)^2+\left(\frac{\partial T'}{\partial y}\right)^2\right\rangle$$

$$=\frac{H^2}{(2\pi)^2}\int_{-\infty}^{+\infty}\int_{-\infty}^{+\infty}\frac{k^2}{4}\left[\frac{\lambda}{(\lambda+k^2)}\right]^2 dk_x dk_y=c_1H^2$$

where H enters because T′ is scaled by $(\partial\overline{T}_*/\partial z_*)d$, and c_1 denotes the value of the integral divided by $(2\pi)^2$. The c_1 values from direct numerical integration of the integral are listed in Table 1 as a function of R_ρ.

The above expression of the Cox number shows the dependence of the number on the fingering zone height, H. The Co values measured in the salt fingering regions have been presented in terms of the vertical mean temperture gradient [*Marmorino*, 1989; *Fluery and Lueck*, 1991]. The above expression can be rewritten in this form by noting that H is also the Rayleigh number to the 1/4 power, i.e.,

$$H=\frac{H_*}{d}=\left[\frac{g\alpha(\partial\overline{T}_*/\partial z_*)H_*^4}{\nu\kappa_T}\right]^{1/4}.$$

Moreover, the dimensional height has the definition, $H_*=\Delta T_*(\partial\overline{T}_*/\partial z_*)^{-1}$. It follows that

$$Co=c_1H^2=c_1\Delta T_*^2\left(\frac{g\alpha}{\nu\kappa_T}\right)^{1/2}\left(\frac{\partial\overline{T}_*}{\partial z_*}\right)^{-3/2} \quad (8)$$

where ΔT_* is the constant temperature difference across the fingering zone. This result shows that Co decreases

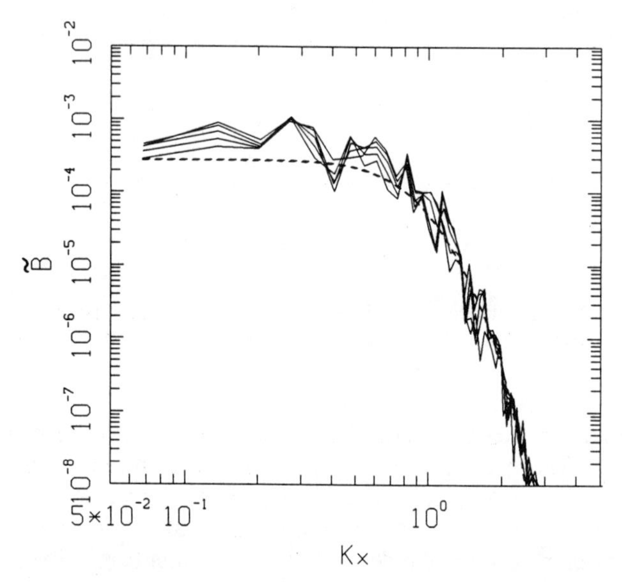

Fig. 3. Temperature variance wavenumber spectrum from the numerical simulation (solid curves) and from the model (dashed curve) at $R_\rho=2.0$.

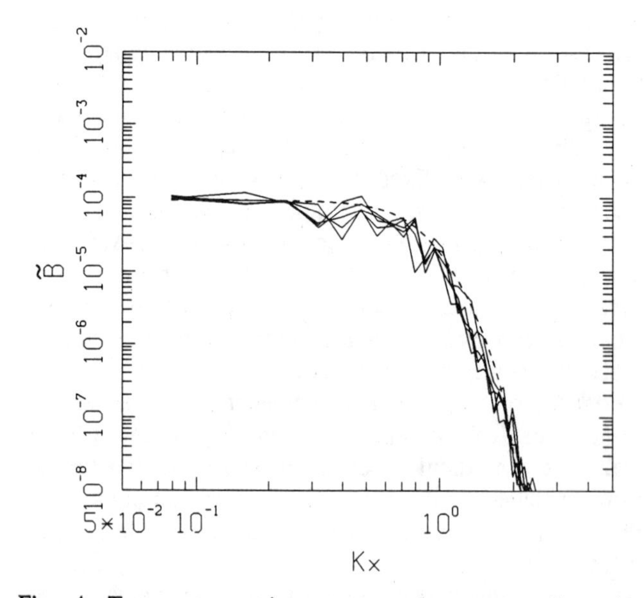

Fig. 4. Temperature variance wavenumber spectrum from the numerical simulation (solid curves) and from the model (dashed curve) at $R_\rho=4.0$.

with increasing mean vertical gradient. The measurements referenced above indicate an approximately -1 power dependence on the mean gradient. The result here gives a larger -1.5 dependence. It is difficult to assess the significance of this difference given the large scatter of the measured Co values.

The actual value of Co can be estimated from (8). Instead of using the right-most expression of (8) which requires measurement parameters unavailable to us, we estimate Co directly from H with some knowledge about the dominant finger wavelength in the measurement area. The estimate proceeds with the approximation that the dominant finger mode, be it the fastest growing mode or the spectral peak mode, has a dimensionless wavenumber $k \approx 1$. The dimensional wavelength is thus $b=2\pi d$, and

$$H=\frac{H_*}{d}=\frac{2\pi H_*}{b}$$

For $b \sim 10$cm and $H_*=1$m, the Co value at $R_\rho=1.5$ is approximately 40, using the c_1 value from Table 1. The estimated Co is within the range of measured values reported.

It is of interest to note the critical height H_c imposed in (5) corresponds to approximately $H_{*c}/b \sim 16$ or $H_{*c} \sim 2$m, a height at which the finger amplitude could be completely limited by finger instability. For heights greater than this, Co would be independent of the mean T gradient because Co would have to be multiplied by the factor ε^2 which is inversely proportional to H^2.

4. CONCLUDING REMARKS

The model spectrum (6) is able to describe the +2 power dependence of the horizontal temperature gradient spectrum on the horizontal wavenumber which has been found for salt fingers in the ocean, in laboratory experiments, and in numerical simulations. The model has been also able to predict the spectral amplitudes to some extent based on tests with direct numerical simulations, and it is able to predict the ocean Cox number to some extent as well. The major unknown presently concerning this spectral model is the effect of the finger instability on the spectral amplitudes in a finite height fingering zone. As pointed out in section 2, the finger instability can be present in a fingering zone of almost any height H, and its effect on fingers could vary with H. Unable to quantify the relation of the instability effect with H, the present model has simply ignored the instability except at very large H where its effect is known from the recent numerical study. This ignorance of the finger instability will probably result in over estimating the finger spectral amplitudes for $H<H_c$, the critical height inferred in section 2. The Cox number will similarly be overestimated. Understanding of the finger instability in a finite height fingering zone is presently very much needed.

The basis for the present formulation of the wavenumber spectrum has been the salt finger amplitudes for vertically uniform fingers in an unbounded fluid. As pointed out in section 2, in a finite height fingering zone, the effect of salt diffusion is negligible, which renders the salt amplitude of the uniform fingers inapplicable to finite H. The negligible salt diffusion affects also the dispersion relation (2) for λ. The analysis by *Shen* [1993] of salt fingering subject to varying degree of salt diffusion, from zero to complete diffusion, has shown that the growth rate λ can vary from a value given approximately by (2) to a value of zero, i.e., a state of steady convection. However, the important result from that study is that in this limit of steady convection, the T' amplitude approaches the magnitude,

$$T'=\frac{H}{2}\frac{1}{R_\rho(k^4+1)}$$

which is Equation (12c) in his paper, where S_o is equivalent to H/2 in this paper. Converting the above amplitude to $\hat{T}'$ unit and assuming that the spectrum can be given again by squaring $\hat{T}'$, one finds that (7) is reproduced exactly, even though (7) is derived via a different route with $\lambda \neq 0$. The wavenumber spectrum formulated here thus does not appear to be sensitive to the finger amplitude model used.

Acknowledgments. This research was supported by the Office of Naval Research, Physical Oceanography Program and the Naval Research Laboratory. The numerical computation was supported by a Naval Research Laboratory computer grant. RWS specifically acknowledges Grant N00014-92-J-1323 from ONR.

REFERENCES

Fluery, M., and R. G. Lueck, Fluxes across a thermohaline staircase, *Deep-Sea Res.*, *38*, 745-769, 1991.

Gargett, A. E., and R. W. Schmitt, Observations of salt fingers in the central waters of the eastern North Pacific, *J. Geophys. Res.*, *87*, 8017-8030, 1982.

Holyer, J. Y., The stability of long steady, two-dimensional salt fingers, *J. Fluid Mech.*, *147*, 169-185, 1984

Howard, L. N., and G. Veronis, The salt-finger zone, *J. Fluid Mech.*, *183*, 1-23, 1987.

Howard, L. N., and G. Veronis, Stability of salt fingers with negligible salt diffusivity, *J. Fluid Mech.*, *239*, 511-522, 1992.

Mack, S. A., Towed chain measurement of ocean microstructure, *J. Phys. Oceanogr.*, *19*, 1108-1129, 1989.

Mack, S. A., and H. C. Schoeberlein, Discriminating salt fingering from turbulence-induced microstructure: Analysis of towed temperature-conductivity chain data, *J. Phys. Oceanogr., 23*, 2073-2106, 1993.

Marmorino, G. O., Substructure of oceanic salt finger interfaces, *J. Geophys. Res., 94*, 4891-4904, 1989.

Marmorino, G. O., Observations of small-scale mixing processes in the seasonal thermocline. Part I: Salt-fingering, *J. Phys. Oceanogr., 17*, 1339-1347, 1987.

Schmitt, R. W., The growth rate of supercritical salt fingers, *Deep Sea Res., 26A*, 23-44, 1979.

Shen, C. Y., The evolution of the double-diffusive instability: salt fingers, *Phys. Fluids A, 1*, 829-844, 1989.

Shen, C. Y., Heat-salt finger fluxes across a density interface, *Phys. Fluids A, 5*, 2633-2643, 1993.

Shen, C. Y., Equilibrium salt-fingering convection, in press, 1994.

Stern, M. E., Ocean circulation physics, 246 pp., Academic, New York, 1975.

Stern, M. E., Collective instability of salt fingers, *J. Fluid Mech., 35*, 209-218, 1969.

Taylor, J. R., and P. Bucens, Laboratory experiments on the structure of salt fingers, *Deep-Sea Res., 36*, 1675-1704, 1989.

C. Y. Shen, Code 7250, Naval Research Laboratory, Washington, D. C. 20375

R. W. Schmitt, Woods Hole Oceanographic Institution, Woods Hole, Massachusetts 02543

Quantifying Salt-Fingering Fluxes in the Ocean

Eric Kunze

School of Oceanography, University of Washington, Seattle, Washington

We review attempts to quantify salt-fingering fluxes in the ocean. Laboratory experiments find flux ratios $R_F = \alpha\langle w\delta T\rangle/\beta\langle w\delta S\rangle$ consistent with the theoretical values for fastest-growing fingers, eliminating steady fingers and fingers of maximum buoyancy-flux from the parameter range of interest. Application of the laboratory $\Delta S^{4/3}$ flux law to the ocean is confounded by the lack of well-defined salinity steps ΔS in most fingering-favorable regions. Even in a well-defined thermohaline staircase such as the one east of Barbados, microstructure measurements find fluxes much smaller than the $\Delta S^{4/3}$ law prediction because the high-gradient interfaces between the layers are too thick. A Stern number constraint $\langle wb\rangle/\nu N^2 \leq 1.0$, or equivalently a finger Froude number constraint $|\nabla w|/N \leq 1.0$, produces fluxes of the right magnitude. However, these criteria imply Cox numbers independent of the background temperature gradient $\bar{T}_z$ while observations find Cox numbers depending inversely on temperature gradient. A hybrid wave/finger Froude number stability criterion $U_z|\nabla w|/N^2 < 1.0$, which involves background (internal wave) shear U_z and finger shear $|\nabla w|$, reproduces the observed temperature gradient dependence but remains untested.

1. INTRODUCTION

Of principal interest to oceanographers studying salt fingers are the divergences of the fingers' fluxes of heat $F_T = \langle w\delta T\rangle$, salt $F_S = \langle w\delta S\rangle$, and buoyancy $F_b = g\langle w(\alpha\delta T - \beta\delta S)\rangle$, since these are responsible for water-mass modification in the ocean. The role fingers play in mixing waters of different temperature T and salinity S may have important consequences for maintaining observed T and S properties. For example, *Gargett and Holloway* [1992] found they could reproduce the intermediate salinity minimum in a numerical model by using differing diffusivities for heat and salt as would result from double-diffusive mixing.

Both absolute finger fluxes and their magnitude relative to turbulent mixing are of interest. Thermohaline staircases in regions of low density ratio $R_\rho = \alpha\bar{T}_z/\beta\bar{S}_z \leq 1.7$ — such as below the Mediterranean salt tongue, the Tyrrhenian Sea, and east of Barbados — are thought to be sites where finger fluxes dominate since double-diffusive buoyancy-fluxes are destabilizing, that is, in the right sense to produce steppy finestructure [*Schmitt*, 1994]. But, though most of the world's Central Waters are fingering-favorable [*Schmitt*, 1981], persistent staircases are found at only a few locales. This may imply that turbulent mixing predominates and smooths most of the ocean pycnocline. It is now recognized that for typical (GM) internal-wave shear levels in the pycnocline, eddy diffusivities are 0.05 - 0.15×10^{-4} m^2 s^{-1} [*Gregg*, 1989; *Ledwell et al.*, 1993; *Polzin et al.*, 1995]. Therefore, where staircases form, either turbulent mixing, and thus internal wave shear, must be weaker (as under the ice in the Arctic [*Padman and Dillon*, 1987]), or fingering fluxes must be stronger.

This note reviews recent efforts to quantify salt-fingering fluxes in the ocean. It is deliberately sketchy and the reader is referred to the cited literature for details. Past efforts have all been indirect as ocean finger fluxes are too small to be measured directly. Guidance comes from theory, laboratory experiments, and ocean microstructure observations

Double-Diffusive Convection
Geophysical Monograph 94

of measurable salt-finger properties. Theory has provided a framework for studying fingers, providing relations between finger growth rate σ, wavenumber k, finger-induced temperature δT, salinity δS, and vertical velocity w anomalies as functions of density ratio R_ρ, and background salinity gradient $\bar{S}_z$ [*Stern*, 1960; 1975; *Schmitt*, 1979a; *Kunze*, 1987]. Laboratory and numerical estimates of the flux ratio $R_F = \langle w\alpha\delta T\rangle / \langle w\beta\delta S\rangle \sim 0.6$ [*Turner*, 1967; *Schmitt*, 1979b; *Taylor and Bucens*, 1989; *Shen*, 1993] are consistent with fastest-growing fingers ($\sigma = \sigma_{max}$) dominating the fluxes. This allows us to focus on fastest-growing fingers and ignore steady fingers ($\sigma = 0$) and fingers of maximum buoyancy-flux [*Stern*, 1976; *Howard and Veronis*, 1992)]. Layer density ratios $\alpha\Delta T/\beta\Delta S = 0.85 \pm 0.02$ in the thermohaline staircase east of Barbados [*Schmitt et al.*, 1987] were argued by *Schmitt* [1988] to be identical to the flux ratio through the interfaces. However, *McDougall* [1991] demonstrates that nonlinearity of the equation of state coupled with very weak interface migration (with $R_F = R_\rho = 1.6$) would boost the layer density ratio slightly, accounting for the observed flux ratio being higher than the fastest-growing finger flux ratio found theoretically and in the laboratory. Alternatively, *Marmorino* [1990] argued that turbulent mixing (with $R_F = R_\rho = 1.6$) is sufficiently strong to raise the layer density ratio.

2. THE $\Delta S^{4/3}$ FLUX LAWS

Two-layer rundown experiments in the laboratory [*Turner*, 1967; *Linden*, 1973; *Schmitt*, 1979b; *McDougall and Taylor*, 1984; *Taylor and Bucens*, 1989] and numerical simulations [*Shen*, 1993] find the fingering heat- and salt-fluxes proportional to $\Delta S^{4/3}$, where ΔS is the salinity step across the interface. *Stern and Turner* [1969] reported that initially smooth fingering-favorable gradients broke down into a sequence of layers and interfaces with fluxes that satisfied the $\Delta S^{4/3}$ law across each interface.

Application of the laboratory $\Delta S^{4/3}$ flux law is frustrated in most fingering-favorable parts of the ocean by the absence of well-defined interfaces and salinity steps ΔS. In the three well-defined thermohaline staircases mentioned in the introduction, *Lambert and Sturges* [1977] and *Schmitt* [1981] used the $\Delta S^{4/3}$ law to estimate finger diffusivities of 10 - 40 $\times$ 10^{-4} m^2 s^{-1}, much larger than typical turbulent eddy diffusivities. This would argue that salt-finger fluxes completely dominate mixing in the ocean pycnocline. However, microstructure measurements in the staircase east of Barbados found fluxes thirty times *smaller* than the $\Delta S^{4/3}$ law prediction [*Gregg and Sanford*, 1987; *Lueck*, 1987; *Fleury and Lueck*, 1991]. *Kunze* [1987] argued that the weak fluxes arose because the interface gradients were too weak to support $\Delta S^{4/3}$ law fluxes, that is, the interfaces were too thick [see also *Linden*, 1978]. Thus, the $\Delta S^{4/3}$ flux laws do not appear to be applicable in the ocean.

3. EQUATIONS OF MOTION

Assuming that the time rate of change can be expressed as an exponential growth rate $\partial/\partial t = \sigma$, the derivative in terms of a horizontal wavenumber $\nabla = ik$, and that the fingers are tall and narrow so the vertical derivative can be ignored ($\nabla^2 \simeq \partial^2/\partial x^2 + \partial^2/\partial y^2$), the equations of motion for growing salt fingers in a quiescent fluid can be expressed as

$$(\sigma + \nu k^2)w = b = g(\alpha\cdot\delta T - \beta\cdot\delta S)$$

$$(\sigma + \kappa_T k^2)\cdot\delta T + wT_z = 0$$

$$(\sigma + \kappa_S k^2)\cdot\delta S + wS_z = 0 \qquad (1)$$

where ν is the molecular viscosity, κ_T the molecular diffusivity of heat, and κ_S the molecular diffusivity of salt. In the following, it is assumed that $\nu >> \kappa_T >> \kappa_S$, as is the case for a heat-salt fluid, so that many of the expressions below are oceanic approximations not valid in general. The reader is referred to *Schmitt* [1979a, 1983] for more general expressions. For finite-length fingers (Figure 1), the finger-induced temperature and salinity anomalies δT and δS can be expressed in terms of the finger height h, the background gradients $\bar{T}_z$ and $\bar{S}_z$ and nondimensional anomalies δ_T and δ_S,

$$\delta T = -\frac{\bar{T}_z h \delta_T}{2} \text{ and } \delta S = -\frac{\bar{S}_z h \delta_T}{2}, \qquad (2)$$

as can the vertical velocity

$$w = w_o + \frac{\sigma h}{2} \qquad (3)$$

from continuity where w_o is the initial finger velocity. The vertical gradients inside the fingers are weakened by advection (Figure 1)

$$T_z = \bar{T}_z(1 - \delta_T)\,,\ S_z = \bar{S}_z(1 - \delta_S). \qquad (4)$$

For fastest-growing fingers, it can be shown that the growth rate

$$\sigma = \frac{1}{2}\sqrt{\frac{(\kappa_T - R_\rho\kappa_S)g\beta\bar{S}_z}{\nu}} \times (\sqrt{R_\rho} - \sqrt{R_\rho - 1}) \quad << \nu k^2 \qquad (5)$$

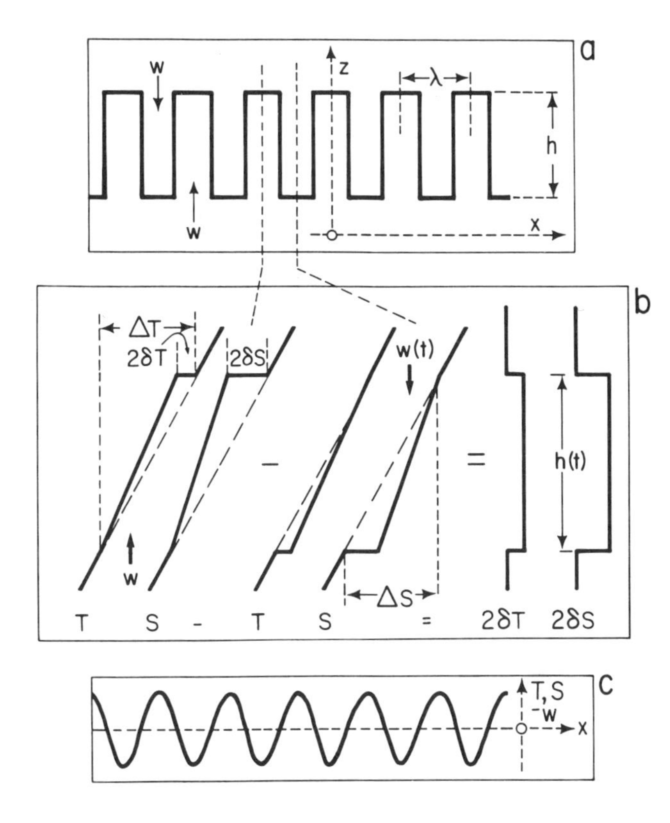

Figure 1: A schematic describing the major features of *Kunze's* [1987] model of growing, finite-length fingers. In the vertical cross-section through a fingering zone (a), thick solid lines demark the strong gradients between adjacent fingers and at their intruding tips. Profiles through upgoing (left) and downgoing (center) fingers (b) illustrate that temperature and salinity are continuous at the inlet to a finger, vary smoothly over its length, and have discontinuities at the intruding tip. Upgoing fingers are lighter than the surrounding fluid, downgoing fingers heavier, and vertical gradients inside the fingers (solid lines) weaker than in the unperturbed fluid (dashed lines). The fingers lengthen and their anomalies δT and δS increase in time (right). Sinusoidal horizontal structure is assumed for temperature, salinity, and vertical velocity (c).

so that vertical acceleration can be neglected in the vertical momentum equation (1). The fastest-growing wavenumber squared is

$$k^2 \simeq \sqrt{\frac{g\beta \bar{S}_z(R_\rho - 1)}{\nu\kappa_T}} = \frac{\bar{N}}{\sqrt{\nu\kappa_T}} , \qquad (6)$$

which corresponds to wavelengths of a few centimeters for most oceanic conditions. The nondimensional finger-induced anomalies are

$$\delta_T \simeq \frac{\sqrt{R_\rho} - \sqrt{R_\rho - 1}}{2\sqrt{R_\rho}} \quad \text{and} \quad \delta_S \simeq \frac{1}{2} \qquad (7)$$

so that the flux ratio

$$R_F = \frac{\alpha <w\delta T>}{\beta <w\delta S>} \simeq \sqrt{R_\rho}(\sqrt{R_\rho} - \sqrt{R_\rho - 1}) \qquad (8)$$

[e.g., *Stern*, 1975; *Schmitt*, 1979a; *Kunze*, 1987]. Quantifying the fluxes requires a constraint on the amplitude of fingers, expressed above in terms of the finger height h. Ideally, such a constraint should arise from a physical understanding of the mechanisms that disrupt growing fingers. However, the most effective approaches to date [*Stern*, 1969; *Kunze*, 1987; 1994] have been heuristic in nature.

4. THE STERN OR FINGER FROUDE NUMBER CONSTRAINT

For fingers growing in a uniform background gradient $\bar{S}_z$, *Stern* [1969; 1975] guessed that their amplitude would be limited by the nondimensional constraint $<wb>/\nu N^2 < O(1)$, limiting the maximum finger fluxes. This ratio has come to be known as the Stern number. *Kunze* (1987) showed that this constraint was equivalent to a finger Froude number criterion

$$Fr_f = \frac{|\nabla w|}{N} \leq O(1) \qquad (9)$$

for high Prandtl number ν/κ_T fluids, where $|\nabla w|$ is the finger shear and N the buoyancy frequency in the fingers. It can also be shown to be equivalent to a finger Reynolds number constraint $w/(\nu k) \leq O(1)$, where w is the vertical finger velocity and k the wavenumber.

Application of constraint (9) to typical interfaces found in the staircase east of Barbados (salinity step ΔS = 0.1‰; interface thickness l_i = 2 m [*Boyd and Perkins*, 1987]) — assuming that fingers grow from small perturbations until $|\nabla w|/N$ = 2, then are disrupted by instability, and disperse until weak enough to allow new fingers to grow from small perturbations (Figure 2a) — produces average buoyancy-fluxes of 0.13×10^{-9} W kg^{-1} for R_ρ = 1.6 [*Kunze*, 1987], consistent with the average microstructure estimates of *Gregg and Sanford* [1987] and *Lueck* [1987] east of Barbados (top panel in Figure 3). Assuming blob-shedding from the fingers' tips (Figure 2b) would produce fluxes larger by σt_{max} ~ 3-6 where σ is the finger growth rate. Laboratory studies suggest that fingers maintain marginal stability by shedding blobs. Colin Shen's simulations (this volume) show a field of densely-packed high-aspect-ratio blobs (worms) wiggling upward

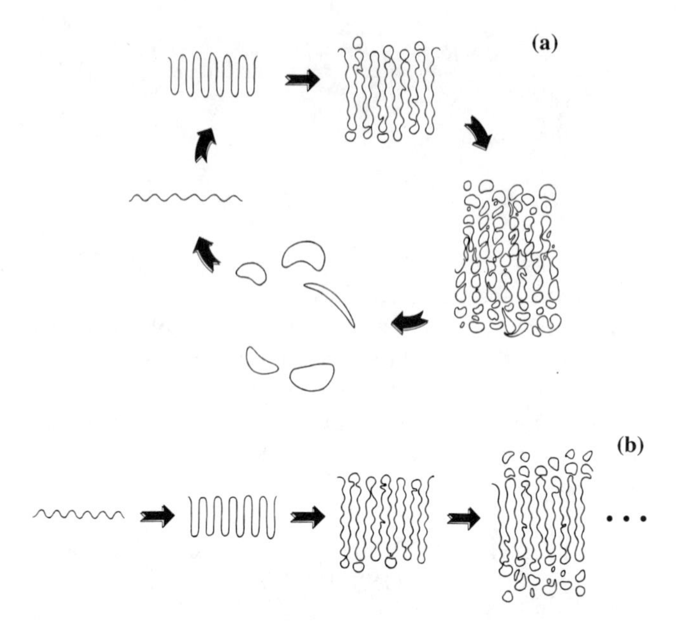

Figure 2: Two ways that salt fingers could grow and become unstable. In (a), the fingers start as small perturbations, grow until violating some stability criterion, then are disrupted and dispersed until their fluctuations are weak enough for new fingers to start. In (b), initially small perturbations grow until their amplitudes exceed a stability criterion, then shed blobs off the finger tips to maintain marginal stability.

and downward around each other. However, other processes, such as intermittent turbulence (see text discussing Figure 5), may favor periodic disruption and re-initialization of fingers (Figure 2a) in the ocean.

The laboratory $\Delta S^{4/3}$ fluxes can be reproduced with the Stern or finger Froude number constraint by setting the interface thickness l_i equal to the maximum finger height h_{max} (bottom panel of Figure 3). The associated maximum finger heights $h_{max} \sim 0.3$ m are an order of magnitude thinner than observed ocean interface thicknesses. This suggests that interfaces in the laboratory have thicknesses constrained by the maximum finger height while ocean interfaces are thicker, and that the reason why ocean fluxes are so much smaller than the $\Delta S^{4/3}$ law prediction is because ocean interfaces are too thick. It does not explain why ocean interfaces are so much thicker than those formed in the lab. Possibly, competition with other finescale ocean processes, such as turbulent mixing or internal wave shear and strain, prevents ocean staircase interfaces from thinning to $\Delta S^{4/3}$ law thicknesses. *Kelley* [1984] was able to quantify the layer thicknesses for diffusively unstable staircases. Substituting the inferred C-SALT salt diffusivity for the molecular heat diffusivity in his expression (2), his parameterization underestimates the layer thicknesses by an order of magnitude. Interestingly, the observed layer thicknesses are reproduced for $\Delta S^{4/3}$ law diffusivities. This may imply that the staircase is a relict of past stronger fluxes when its interfaces were thinner.

5. A HYBRID WAVE/FINGER FROUDE NUMBER

The Stern or finger Froude number constraint successfully reproduces both the laboratory $\Delta S^{4/3}$ fluxes and the *average* buoyancy-flux in the staircase east of Barbados. However, this stability criterion implies temperature Cox numbers $(\nabla \delta T)^2/\bar{T}_z^2 \simeq 10$, independent of the background temperature gradient $\bar{T}_z$ [*Kunze*, 1987)] while towed microstructure measurements in the same staircase [*Marmorino*, 1989; *Fleury and Lueck*, 1991] find Cox numbers depending inversely on temperature gradient $\bar{T}_z$ (linearly on interface thickness l_i for a constant temperature step ΔT).

There are several important differences between the ocean environment and the laboratory, principally, the presence of additional processes such as finescale internal waves and turbulence. When turbulence is present, it will completely disrupt salt-finger fluxes [*Linden*, 1971]. At any given time, turbulent patches are typically present in 10% of the ocean pycnocline, generated by wave-induced shear instability roughly every ten buoyancy periods [*Kunze et al.*, 1990]. Turbulence was weaker and less frequent (1%) in the staircase east of Barbados [*Gregg and Sanford*, 1987; *Marmorino*, 1990; *Fleury and Lueck*, 1991].

Internal-wave shear and strain will deform fingers and modify finger-stability conditions. In unsheared conditions, square planform fingers with horizontal structure proportional to $\sin(k_x x)\sin(k_y y)$ dominate [*Shirtcliffe and Turner*, 1970; *Chen and Sandford*, 1976]. But in even weak vertical shear $U_z \geq 0.1N$, fingers form sheets $\sin(k_y y)$ aligned with the shear [*Linden*, 1974; *Kunze*, 1990]. As a result, the finger shear component $w_x = 0$. *Kunze* [1994] examines the stability of sheared fingers heuristically, looking for a ratio of timescales that reproduces the observed Cox number's linear dependence on interface thickness. He identified three stabilizing timescales: the buoyancy frequency N, viscous damping νk^2, and the finger growth rate σ, all of which have the same dependence on interface thickness. He also identified three destabilizing timescales: the background shear U_z, the finger shear w_y, and the buoyancy-gradient anomaly $\sqrt{\nabla b}$. Noting that the near-inertial velocity steps across the interfaces ΔU were independent of interface thickness east of Barbados [*Gregg and Sanford*, 1987] so that the background near-inertial shear $U_z = \Delta U/l_i$ depends inversely

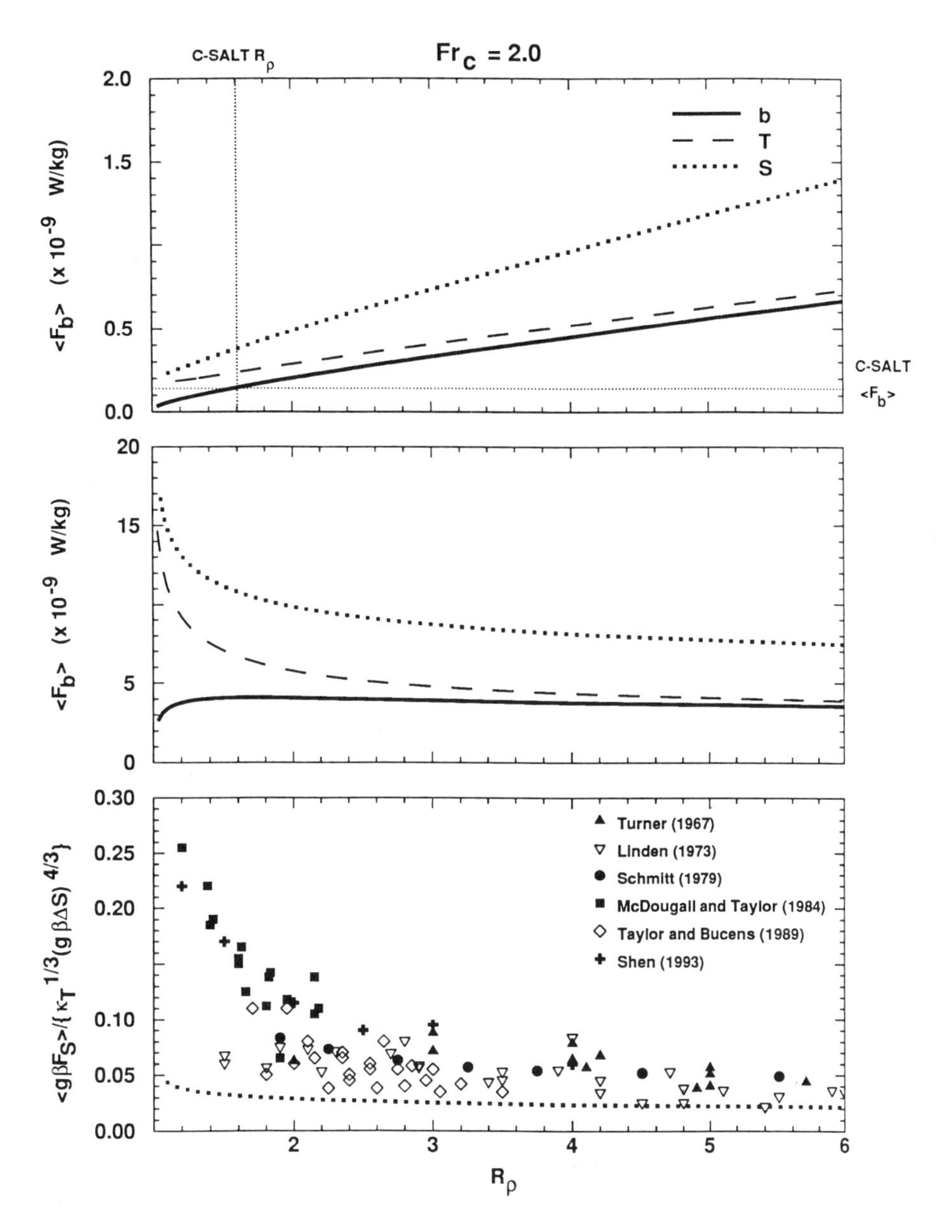

Figure 3: Heat, salt, and total buoyancy-fluxes as a function of density ratio R_ρ for conditions found in the thermohaline staircase east of Barbados and a critical finger Froude number $|\nabla w|/N = 2$. In the upper panel, the background salinity gradient $\bar{S}_z = 0.05‰\ \mathrm{m}^{-1}$ independent of R_ρ. The model buoyancy-flux at $R_\rho = 1.6$ is consistent with the C-SALT measurements of *Gregg and Sanford* [1987] and *Lueck* [1987]. In the central panel, the salinity step is $\Delta S = 0.1$ ‰ and the interface thickness fixed at the maximum finger height ~ 0.3 m. In the lower panel, the salt-flux from the central panel is normalized by the $\Delta S^{4/3}$ flux law and compared with laboratory and numerical estimates. Theory and data are in good agreement for $R_\rho > 1.8$ but *McDougall and Taylor* [1984] and *Shen* [1993] find higher salt-fluxes at lower density ratios.

on interface thickness l_i, he finds that a hybrid wave/finger Froude number constraint

$$Fr_{w/f} = Fr_w Fr_f = \frac{U_z w_y}{N^2} \leq 1.0 \qquad (10)$$

reproduces the observed Cox number C_T dependence on interface thickness (Figure 4) observed by *Marmorino* [1989] and *Fleury and Lueck* [1991]. Recalling that background shear suppresses finger w_x, this constraint

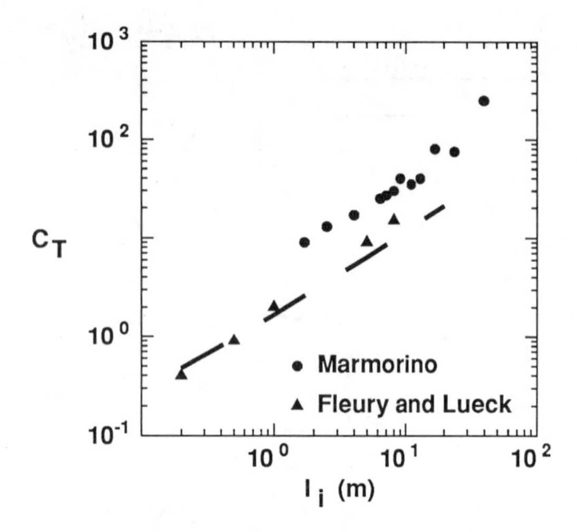

Figure 4: Comparison of the observed temperature Cox number's C_T dependence on interface thickness l_i (▲, ●) with the model prediction for a critical $U_z w_y / N^2 = 1.0$ constraint (dashed line). Also shown is the predicted salinity Cox number C_S (dotted line). The predicted Cox number C_T has the same linear dependence on interface thickness l_i as the observations. The predicted level is consistent with the *Fleury and Lueck* [1991] microscale thermistor data. The *Marmorino* [1989] microconductivity data are a factor of three higher.

might be generalizable as $|\chi \cdot \eta|/N^2 \leq 1.0$, where $\chi = w_y - v_z$ and $\eta = u_z - w_x$ are the two horizontal vorticity components, by the argument that the horizontal vorticity must be strong enough to overcome the stratification to induce instability. For unsheared fingers, this reduces to $Fr_f^2 = w_x w_y / N^2 \leq 1.0$, reproducing, to within a factor of two, the finger Froude number constraint of *Kunze* [1987].

Taking $\bar{S}_z$ to be the largescale gradient smoothed over the staircase following *Schmitt* [1981] and applying constraint (10) to sheared salt sheets, the finger-induced diffusivity for salt can be shown to be

$$K_{S_f} = \frac{<w\delta S>}{\bar{S}_z} = \frac{\nu}{Fr_w^2} \frac{\sqrt{R_\rho - 1}}{\sqrt{R_\rho} - \sqrt{R_\rho - 1}} \left(\frac{l_o + l_i}{l_i}\right) \quad (11)$$

as detailed in *Kunze* [1994], where ν is the molecular viscosity, l_o and l_i are the layer and interface thicknesses ($l_o = 0$ in a continuously-stratified fluid), and $Fr_w = U_z / N$ is the wave Froude number.

Figure 5 from *Kunze* [1994] displays the finger diffusivity (11) for a variety of oceanic conditions as a function of density ratio R_ρ. For typical continuously-stratified ($l_o = 0$) pycnocline wave Froude numbers $Fr_w = 1.0$ [*Eriksen*, 1978; *Kunze et al.*, 1990], the finger diffusivity *increases* from ~ 0.01 × 10^{-4} m^2 s^{-1} at low R_ρ to ~ 0.1 × 10^{-4} m^2 s^{-1} at $R_\rho = 5$ (solid curve), so only becomes as large as typical turbulent diffusivities (stippling) at higher density ratios. For a lower wave Froude number, $Fr_w = 0.4$, the finger diffusivity is a factor of six larger (thin dashed), still at most comparable to typical turbulent mixing rates for $R_\rho < 2$. For a layered pycnocline with the homogeneous layer thickness l_o ten times the interface thickness l_i, as is typical for thermohaline staircases, the finger diffusivity $K_{S_f} \sim 10^{-4}$ m^2 s^{-1} is an order of magnitude larger than typical turbulent diffusivities. Persistent staircases are only found for $R_\rho < 1.7$ [*Schmitt*, 1994].

Finally, the dotted curve in Figure 5 assumes that growing fingers are disrupted every ten buoyancy periods by turbulence. The finger diffusivity is then

$$K_{S_f} = \frac{\pi\sqrt{\nu\kappa_T}}{160} \times \left[\exp\left[20\pi\sqrt{\frac{\kappa_T - R_\rho\kappa_S}{\nu}\,\frac{\sqrt{R_\rho} - \sqrt{R_\rho - 1}}{\sqrt{R_\rho - 1}}}\,\right] - 1.0\right] \quad (12)$$

assuming that the fingers initially have aspect ratio one,

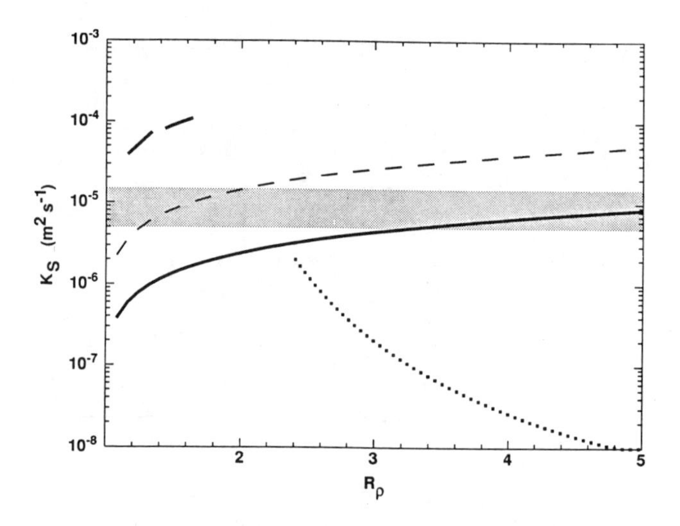

Figure 5: Model finger salt diffusivity K_S as a function of density ratio R_ρ (11) under continuously-stratified conditions for wave Froude numbers $Fr_w = 1.0$ (thick solid) and $Fr_w = 0.4$ (thin dashed). The band of stippling shows the typical range of turbulent diffusivities in the pycnocline [*Gregg*, 1989; *Ledwell et al.*, 1993; *Polzin et al.*, 1995]. Higher finger diffusivities occur at higher density ratios. If finger growth is disrupted by turbulence every ten buoyancy periods (12), the finger diffusivity is depressed for density ratios $R_\rho > 2$ (dotted curve). In thermohaline staircases (found only at $R_\rho < 1.7$) such as the one east of Barbados where the layer thickness l_o is ten times the interface thickness l_i, the diffusivity is elevated (thick dash).

$h_o = \lambda/4 = \pi/(2k)$. For $R_\rho > 2.4$, turbulent disruption prevents fingers from ever achieving $U_z w_y / N^2 = 1.0$, depressing their fluxes to well below turbulence levels.

Turning of the inertial wave shear on a timescale f^{-1} will introduce shear normal to the salt sheets which will tilt them and also reduce their fluxes [*Kunze*, 1990], but the impact of this is difficult to quantify.

6. CONCLUSIONS

Microstructure measurements collected in the thermohaline staircase east of Barbados during the 1985 C-SALT experiment [*Schmitt et al.*, 1987] dramatically advanced our understanding of the role of salt fingering in ocean mixing, revising estimates of salt-fingering fluxes [*Gregg and Sanford*, 1987; *Lueck*, 1987; *Fleury and Lueck*, 1991] downward to levels comparable to those of shear-driven turbulence [*Gregg*, 1989; *Ledwell et al.*, 1993; *Polzin et al.*, 1995]. Further advances in determining the role salt fingers play in modifying ocean water-masses will likely require understanding how interactions with turbulence and finescale internal-wave shear and strain impact finger stability.

Even in thermohaline staircases with well-defined salinity steps, the $\Delta S^{4/3}$ flux law derived from laboratory studies overestimates the fluxes by more than an order of magnitude because the high-gradient interfaces are too thick. The Stern or finger Froude number constraint produces fluxes of the right magnitude but predicts temperature Cox numbers independent of interface thickness while measurements find a linear dependence [*Marmorino*, 1989; *Fleury and Lueck*, 1991]. A hybrid wave/finger Froude number constraint (10) in which internal-wave shear modulates finger stability, reproduces the observed Cox number's dependence on interface thickness (Figure 4). This criterion remains untested by laboratory or numerical experiments. The effects of disruptive turbulence and of internal-wave strain and shear on salt-fingering fluxes in the ocean remain relatively unexplored despite their potential importance.

The ocean seems best-described (Figure 5) by weakly sheared staircases (11) for density ratios $R_\rho < 1.7$ (thick dashed) and finger fluxes disrupted by turbulence (12) rather than self-instability for $R_\rho > 2$ (thick dotted). The shape of the resulting dependence of finger diffusivity on density ratio, with dramatically elevated fluxes for $R_\rho < 1.7$, is remarkably similar to that inferred by *Schmitt* [1981, his Figure 5], albeit two orders of magnitude smaller. It is consistent with thermohaline staircases appearing only for density ratios $R_\rho \leq 1.7$.

Acknowledgments. This work was supported under NSF grant OCE-92-404-93. Helpful comments were provided by Dan Kelley and Ray Schmitt.

REFERENCES

Boyd, J.D., and H. Perkins, Characteristics of thermohaline steps off the northwest coast of South America, *Deep-Sea Res.*, *34*, 337-364, 1987.

Chen, C.F., and R.D. Sandford, Sizes and shapes of salt fingers near the marginal state, *J. Fluid Mech.*, *78*, 601-607, 1976.

Eriksen, C.C., Measurements and models of finestructure, internal gravity waves and wave breaking in the deep ocean, *J. Geophys. Res.*, *83*, 2989-3009, 1978.

Evans, D.L., Velocity shear in a thermohaline staircase, *Deep-Sea Res.*, *28A*, 1409-1415, 1981.

Fleury, M., and R.G. Lueck, Fluxes across a thermohaline interface, *Deep-Sea Res.*, *38*, 745-769, 1991.

Gargett, A.E., and G. Holloway, Sensitivity of the GFDL ocean model to different diffusivities for heat and salt, *J. Phys. Oceanogr.*, *22*, 1158-1177, 1992.

Gregg, M.C., Scaling turbulent dissipation in the thermocline, *J. Geophys. Res.*, *94*, 9686-9698, 1989.

Gregg, M.C., and T.B. Sanford, Shear and turbulence in thermohaline staircases, *Deep-Sea Res.*, *34*, 1689-1696, 1987.

Howard, L.N., and G. Veronis, Stability of salt fingers with negligible diffusivity, *J. Fluid Mech.*, *239*, 511-522, 1992.

Kelley, D.E., Effective diffusivities within oceanic thermohaline staircases, *J. Geophys. Res.*, *89*, 10,484-10,488, 1984.

Kunze, E., Limits on growing, finite-length fingers: A Richardson number constraint, *J. Mar. Res.*, *45*, 533-556, 1987.

Kunze, E., The evolution of salt fingers in inertial wave shear, *J. Mar. Res.*, *48*, 471-504, 1990.

Kunze, E., A proposed constraint for salt fingers in shear, *J. Mar. Res.*, *52*, 999-1016, 1994.

Kunze, E., A.J. Williams III, and M.G. Briscoe, Observations of shear and vertical stability from a neutrally-buoyant float, *J. Geophys. Res.*, *95*, 18,127-18,142, 1990.

Lambert, R.B., and W. Sturges, A thermohaline staircase and vertical mixing in the thermocline, *Deep-Sea Res.*, *24*, 211-222, 1977.

Ledwell, J.R., A.J. Watson, and C.S. Law, Evidence of slow mixing across the pycnocline from an open-ocean tracer-release experiment, *Nature*, *364*, 701-703, 1993.

Linden, P.F., Salt fingers in the presence of grid-generated turbulence, *J. Fluid Mech.*, *49*, 611-624, 1971.

Linden, P.F., On the structure of salt fingers, *Deep-Sea Res.*, *20*, 325-340, 1973.

Linden, P.F., Salt fingers in a steady shear flow, *Geophys.*

Fluid Dyn., *6*, 1-27, 1974.

Linden, P.F., The formation of banded salt-finger structure, *J. Geophys. Res.*, *83*, 2902-2912, 1978.

Lueck, R.G., Microstructure measurements in a thermohaline staircase, *Deep-Sea Res.*, *34*, 1677-1688, 1987.

Marmorino, G.O., Substructure of oceanic salt-finger interfaces, *J. Geophys. Res.*, *94*, 4891-4904, 1989.

Marmorino, G.O., "Turbulent mixing" in a salt-finger interface, *J. Geophys. Res.*, *95*, 12,983-12,994, 1990.

McDougall, T.J., Interfacial advection in the thermohaline staircase east of Barbados, *Deep-Sea Res.*, *38*, 357-370, 1991.

McDougall, T.J., and J.R. Taylor, Flux measurements across a finger interface at low values of the stability ratio, *J. Mar. Res.*, *42*, 1-14, 1984.

Padman, L., and T.M. Dillon, Vertical heat-fluxes through the Beaufort Sea thermohaline staircase, *J. Geophys. Res.*, *92*, 10,799-10,806, 1987.

Polzin, K., J.M. Toole, and R.W. Schmitt, Finescale parameterizations of turbulent dissipation, *J. Phys. Oceanogr.*, in press, 1995.

Schmitt, R.W., The growth rate of supercritical salt fingers, *Deep-Sea Res.*, *26A*, 23-44, 1979a.

Schmitt, R.W., Flux measurements at an interface, *J. Mar. Res.*, *37*, 419-436, 1979b.

Schmitt, R.W., Form of the temperature-salinity relationship in the Central Water: Evidence for double-diffusive mixing, *J. Phys. Oceanogr.*, *11*, 1015-1026, 1981.

Schmitt, R.W., The characteristics of salt fingers in a variety of fluid systems, including stellar interiors, liquid metals, oceans and magmas, *Phys. Fluids*, 26, 2373-2377, 1983.

Schmitt, R.W., Mixing in a thermohaline staircase, in *Smallscale Turbulence and Mixing in the Ocean*, edited by J.C.J. Nihoul and B.M. Jamart, pp. 435–452, Elsevier, The Netherlands, 1988.

Schmitt, R.W., Double diffusion in oceanography, *Ann. Rev. Fluid Mech.*, *26*, 255-285, 1994.

Schmitt, R.W., H. Perkins, J.D. Boyd, and M.C. Stalcup, C-SALT: An investigation of the thermohaline staircase in the western tropical North Atlantic, *Deep-Sea Res.*, *34*, 1655-1666, 1987.

Shen, C.Y., Heat-salt finger fluxes across a density interface, *Phys. Fluids A*, *5*, 2633-2643, 1993.

Shirtcliffe, T.G.L., and J.S. Turner, Observations of the cell structure of salt fingers, *J. Fluid Mech.*, *41*, 707-719, 1970.

Stern, M.E., The "salt-fountain" and thermohaline convection, *Tellus*, *12*, 172-175, 1960.

Stern, M.E., Collective instability of salt fingers, *J. Fluid Mech.*, *35*, 209-218, 1969.

Stern, M.E., *Ocean Circulation Physics*, pp. 191-203, Academic Press, New York, 1975.

Stern, M.E., Maximum buoyancy-flux across a salt-finger interface, *J. Mar. Res.*, *34*, 95-110, 1976.

Stern, M.E., and J.S. Turner, Salt-fingers and convecting layers, *Deep-Sea Res.*, *16*, 497-511, 1969.

Taylor, J., and P. Bucens, Laboratory experiments on the structure of salt fingers, *Deep-Sea Res.*, *36*, 1675-1704, 1989.

Turner, J.S., Salt fingers across a density interface, *Deep-Sea Res.*, *14*, 599-611, 1967.

E. Kunze, School of Oceanography, WB-10, University of Washington, Seattle, WA 98195

An Investigation of Kunze's Salt Finger Flux Laws – Are They Stable?

David Walsh and Barry Ruddick

Department of Oceanography, Dalhousie University, Halifax, Nova Scotia B3H 4J1

A simple model is used to investigate the salt finger diffusivity formulations proposed by *Kunze* [1987, 1994]. To provide a framework for the discussion, we also discuss the flux laws of *Schmitt* [1981]. The different density ratio dependences of these formulations causes them to behave very differently. We find that T-S anomalies governed by Schmitt's *ad hoc* flux laws decay rapidly, but that Kunze's flux laws are unstable, with predicted growthrates that grow without bound as the vertical scale of the disturbance approaches zero (an "ultraviolet catastrophe").

1. INTRODUCTION

Conditions are favorable for the growth of salt fingers over much of the ocean, yet surprisingly little is known about the role they play in the general circulation. It is possible that the fluxes of salt, heat, and density carried by fingers have an important influence on larger scale fields. *Schmitt* [1981] has suggested that fingers play an important part in determining the vertical distributions of heat and salt in the ocean. Schmitt proposes that finger diffusivities depend upon local T-S gradients, and that the form of this dependence has important implications for the form of the T-S curve. For a diffusivity that decreases with density ratio R_ρ, where

$$R_\rho = \frac{\alpha T_z}{\beta S_z}\,, \qquad (1)$$

Schmitt found that T-S anomalies decayed at an accelerated rate, and he suggested that this could provide a mechanism for maintaining the nearly uniform values of R_ρ found in the North Atlantic Central Waters.

The fluxes of heat and salt produced by salt fingering are traditionally assumed to be proportional to $\Delta S^{4/3}$, as discussed by *Turner* [1967]. Laboratory flux formulations based on the "4/3" flux laws [*Schmitt*, 1979; *McDougall and Taylor*, 1984] have been used to estimate the fluxes through T-S interfaces in the ocean. *Schmitt* [1981] has used data from thermohaline staircases in conjunction with laboratory flux laws to estimate the effective diffusivities produced by salt fingering. His analysis indicated that the salt diffusivity K_S depends strongly on R_ρ, with the diffusivity varying from 5.7 x $10^{-4}\mathrm{m^2s^{-1}}$ at a density ratio $R_\rho = 1.6$ to just $10^{-5}\mathrm{m^2s^{-1}}$ when $R_\rho = 1.94$. However, the oceanic relevance of the 4/3 flux laws used by Schmitt and others has been called into question. Data from the C-SALT experiment [*Schmitt et al.*, 1987] seem to show that effective diffusivities due to fingering are much smaller than had previously been thought, as thick oceanic interfaces are apparently incapable of supporting the large fluxes predicted by lab flux laws. As a result of the uncertainty about the magnitude of the finger fluxes, it is unclear at this point how important fingering is as a mechanism of water mass conversion in the ocean.

Various models have been advanced to predict the fluxes of heat and salt caused by salt fingering at a T-S interface. The first such model is due to *Stern* [1969,1975], who found that growing salt fingers eventually undergo a secondary instability; this "collective" instability places an upper bound on the fluxes that can be carried by fingers. *Kunze* [1987] has formulated a related model, which apparently explains why the C-SALT interfaces were unable to support the large fluxes predicted by the laboratory flux laws. Kunze's model is based on the premise that finger growth is regulated by

Double-Diffusive Convection
Geophysical Monograph 94

a finger Froude number, defined by

$$Fr_f = \frac{|\nabla_H w|}{N}, \tag{2}$$

which is formally equivalent to Stern's collective instability constraint [*Kunze*, 1987]. Kunze assumes that finger fluxes equilibrate when the finger Froude number reaches a critical value. He then relates the finger fluxes to basic state quantities (*e.g.*, R_ρ), allowing an effective diffusivity to be computed. His results suggest that the fluxes depend upon the length of the fingers relative to the thickness of the T-S interface. If the fingers extend through the entire T-S interface Kunze's model reproduces the 4/3 flux laws, but significantly smaller fluxes are predicted when the interface is much thicker than a finger length; these reduced fluxes through "thick" interfaces are perhaps more representative of typical oceanic conditions than the large fluxes predicted by the laboratory flux laws. *Kunze* [1987] demonstrated that linking finger growth to the Froude number (2) leads to a temperature Cox number

$$C_T = \frac{(\vec{\nabla} T)^2}{\bar{T}_z^2}$$

which is independent of the interfacial temperature gradient, while recent observations [*Marmorino*, 1989; *Fleury and Lueck*, 1991] indicate that the Cox number is a decreasing function of the interfacial temperature gradient. Motivated by these observations, *Kunze* [1994] proposed that finger growth is instead regulated by a hybrid wave-finger Froude number

$$Fr_{w/f} = \frac{U_z \nabla_H w}{N^2},$$

which allows his model to reproduce the observed relation between the Cox number and interfacial temperature gradients.

In contrast with Schmitt's diffusivities, both Kunze's models predict a diffusivity which grows with R_ρ. It will be shown that this seemingly minor difference between the flux formulations has important consequences for the behavior of the fluxes: rather than predicting an accelerated decay rate for T-S anomalies, Kunze's laws predict that small disturbances grow exponentially in time. It is possible that this mechanism plays a role in oceanic layer formation, producing step-like T-S finestructure. In the next section we formulate a simple model to examine the implications of the flux laws due to Schmitt and Kunze for the small-scale fluxes of heat and salt produced by salt fingers.

2. MODEL FORMULATION

We shall assume that the temperature and salinity fields evolve solely as a consequence of vertical flux divergences produced by salt fingering. Thus, the equations governing the evolution of T and S are:

$$\begin{aligned} S_t &= -F_z^{(S)} \\ T_t &= -F_z^{(T)}, \end{aligned} \tag{3}$$

where $F^{(S)}$, $F^{(T)}$ represent vertical fluxes of salt and heat, respectively. The salt flux is specified according to a simple gradient flux law. Following *Stern* [1967], the heat flux is linked to the salt flux using a salt finger flux ratio γ

$$\gamma = \frac{\alpha F^{(T)}}{\beta F^{(S)}},$$

which is assumed to be constant. In contrast with Stern's model, the diffusivity is taken to be a function of the density ratio, so the fluxes can be written

$$\begin{aligned} F^{(S)} &= -K_S(R_\rho) S_z \\ F^{(T)} &= -\gamma K_S(R_\rho)\frac{\beta}{\alpha} S_z . \end{aligned} \tag{4}$$

These flux laws are similar to those of *Schmitt* [1981], who considered the influence of non-constant (R_ρ-dependent) diffusivity as a possible explanation for the near uniformity of R_ρ over the North Atlantic Central Waters. *Zhurbas et al.* [1987] have considered somewhat more sophisticated flux laws, with additional parametrizations for diffusive sense and unstable stratifications, and find that such flux formulations can produce step-like T-S profiles. Using (3) and (4) it can be verified that the linearized equations governing the evolution of the temperature and salinity fields are

$$\begin{aligned} \beta\tilde{S}_t &= (K_S - \underline{\bar{R}_\rho K_S'})\beta\tilde{S}_{zz} + \underline{K_S'\alpha\tilde{T}_{zz}} \\ \alpha\tilde{T}_t &= \gamma[(K_S - \underline{\bar{R}_\rho K_S'})\beta\tilde{S}_{zz} + \underline{K_S'\alpha\tilde{T}_{zz}}], \end{aligned} \tag{5}$$

where $\tilde{S}$ and $\tilde{T}$ are small perturbations, and the primes denote differentiation with respect to R_ρ (the diffusivities and their derivatives are evaluated at $R_\rho = \bar{R}_\rho$). The basic state temperature and salinity fields $\bar{T}(z)$ and $\bar{S}(z)$ are assumed to be linear in z. The underlined terms in (5) result from vertical diffusivity variations caused by small density ratio anomalies. It is clear that non-constant diffusivity introduces an important modification to the system. By taking the appropriate linear combinations of (5a) and (5b) one obtains

$$
\begin{aligned}
(\alpha\tilde{T} - \gamma\beta\tilde{S})_t &= 0 \\
(\beta\tilde{S} + \frac{K'_S}{K_S^{(s)}}\alpha\tilde{T})_t &= K_S^{eff}(\beta\tilde{S} + \frac{K'_S}{K_S^{(s)}}\alpha\tilde{T})_{zz} \,,
\end{aligned} \tag{6}
$$

where the "effective diffusivity" K_S^{eff} is defined by

$$K_S^{eff} \equiv K_S + (\gamma - \bar{R}_\rho)K'_S \,. \tag{7}$$

We have also used the notation

$$K_S^{(s)} \equiv K_S - \bar{R}_\rho K'_S \,.$$

This quantity represents the effective diffusivity for salt if cross-diffusion effects are neglected. From (6a) it follows immediately that

$$\frac{\partial\alpha\tilde{T}/\partial t}{\partial\beta\tilde{S}/\partial t} = \gamma \,, \tag{8}$$

so that any perturbation evolves along a line of slope γ in the $(\beta S, \alpha T)$ plane. It is important to note that the result (8) does not rely upon the small amplitude assumptions that were made in deriving (5). It is a fully nonlinear result. We now derive some exact solutions to the set (5). Looking for solutions of the form:

$$[\tilde{S}, \tilde{T}] = \mathrm{Re}\left([\hat{S}(t), \hat{T}(t)]e^{imz}\right) \,,$$

it can be shown that an initial perturbation $[\hat{S}_0, \hat{T}_0] \equiv [\hat{S}(0), \hat{T}(0)]$ will evolve according to

$$
\begin{aligned}
\begin{bmatrix} \hat{S}/\hat{S}_0 \\ \hat{T}/\hat{T}_0 \end{bmatrix} &= \frac{K_S^{(s)} + K'_S\alpha\hat{T}_0/\beta\hat{S}_0}{K_S^{eff}} \begin{bmatrix} 1 \\ \gamma\frac{\beta\hat{S}_0}{\alpha\hat{T}_0} \end{bmatrix} e^{-m^2K_S^{eff}t} \\
&+ \frac{\gamma - \alpha\hat{T}_0/\beta\hat{S}_0}{K_S^{eff}} \begin{bmatrix} K'_S \\ -\frac{\beta\hat{S}_0}{\alpha\hat{T}_0}K_S^{(s)} \end{bmatrix} .
\end{aligned} \tag{9}
$$

The associated density perturbations are of the form

$$
\begin{aligned}
\hat{\rho} &\equiv \rho_0(\beta\hat{S} - \alpha\hat{T}) \\
&= -\rho_0\beta\hat{S}_0\Phi(1-\gamma)\left[\frac{\alpha\hat{T}_0/\beta\hat{S}_0 - \gamma}{1-\gamma}\right. \\
&\quad \left. - \left(\frac{K_S^{(s)} + K'_S\alpha\hat{T}_0/\beta\hat{S}_0}{K_S + (1-\bar{R}_\rho)K'_S}\right)e^{-m^2K_S^{eff}t}\right] ,
\end{aligned} \tag{10}
$$

where

$$\Phi = \frac{K_S + (1-\bar{R}_\rho)K'_S}{K_S^{eff}} \,.$$

Equation (10) shows that T-S anomalies don't decay to zero under the action of salt finger fluxes – there is in general a residual anomaly left as $t \to \infty$.

3. SCHMITT'S DIFFUSIVITY FORMULATION

Based upon his analysis of data from thermohaline staircases, *Schmitt* [1981] proposed the following *ad hoc* model for the salt diffusivity K_S:

$$K_S = \frac{10^{-3}m^2s^{-1}}{\left[1 + \left(\frac{R_\rho}{1.7}\right)^{32}\right]} + 5 \times 10^{-6}m^2s^{-1} \,, \tag{11}$$

which is plotted in Figure 1. Notice that the diffusivity decreases rapidly with R_ρ – this led *Schmitt* [1981] to propose that flux divergences due to diffusivity gradients may dwarf those that would be expected from estimates which assume constant diffusivities. Schmitt's diffusivity formulation predicts enhanced decay rates for T-S anomalies, as both self-diffusion terms in (5) ($K_S^{(s)}$ and $\gamma K'_S$) are positive. The formulation (11) is consistent with the idea that finger fluxes should mimic linear growthrates, which increase as R_ρ decreases toward one. Because $K_S^{eff} > 0$, the system approaches an equilibrium as $t \to \infty$. This equilibrium is described by

$$
\begin{aligned}
\hat{S}^\infty/\hat{S}_0 &= -(\alpha\hat{T}_0/\beta\hat{S}_0 - \gamma)\frac{K'_S}{K_S^{eff}} \\
\hat{T}^\infty/\hat{T}_0 &= +(\alpha\hat{T}_0/\beta\hat{S}_0 - \gamma)\frac{\beta\hat{S}_0}{\alpha\hat{T}_0}\frac{K_S^{(s)}}{K_S^{eff}} \\
\hat{\rho}^\infty/\hat{\rho}_0 &= \frac{K_S + (1-\bar{R}_\rho)K'_S}{K_S^{eff}}\left(\frac{\gamma - \alpha\hat{T}_0/\beta\hat{S}_0}{1 - \alpha\hat{T}_0/\beta\hat{S}_0}\right) \\
&= \Phi\frac{\gamma - \alpha\hat{T}_0/\beta\hat{S}_0}{1 - \alpha\hat{T}_0/\beta\hat{S}_0} \,, \qquad \hat{\rho}_0 \neq 0 \,.
\end{aligned} \tag{12}
$$

In this equilibrium state the flux-divergence due to the local curvature of the salinity and temperature fields exactly counterbalances that due to the depth-variation of the diffusivity:

$$K_S S_{zz} + K_{S,z}S_z = (K_S S_z)_z = 0 \,.$$

Notice that for any perturbation for which $1 > \alpha\hat{T}_0/\beta\hat{S}_0 > \gamma$, $\hat{\rho}^\infty/\hat{\rho}_0$ is negative, so that the finger fluxes may cause the density anomaly to change sign.

The graphical solution to the set (6) is sketched in Figure 2, which shows the evolution of a pair of anomalies in the $(\beta S, \alpha T)$ plane, using *Schmitt's* [1981] flux formulation. The heavy line in the figure represents the basic state, which has uniform R_ρ. Each vector represents the imposed perturbation at a given depth in the water column. The lines labeled γ and η are the eigenvectors of the system (5), and have slopes γ and $-K_S^{(s)}/K'_S$,

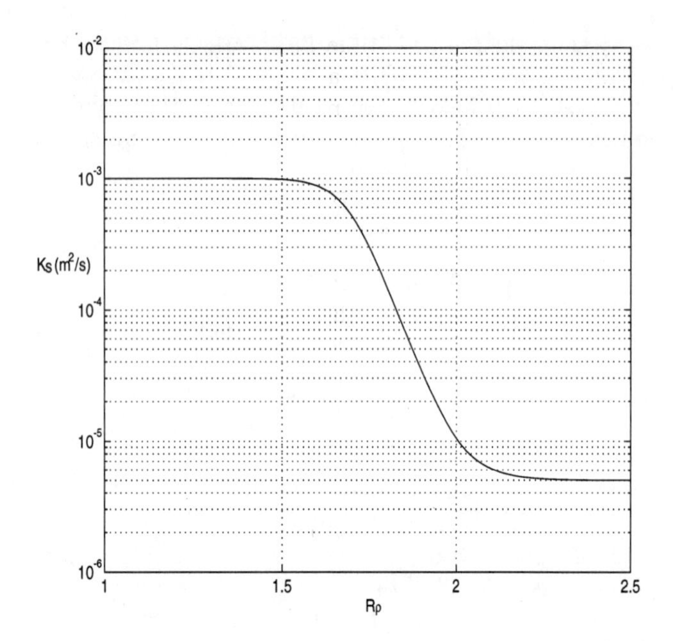

Fig. 1. The salt diffusivity formulation proposed by *Schmitt* [1981], as given by equation (11). The diffusivity is a monotonically decreasing function of the density ratio.

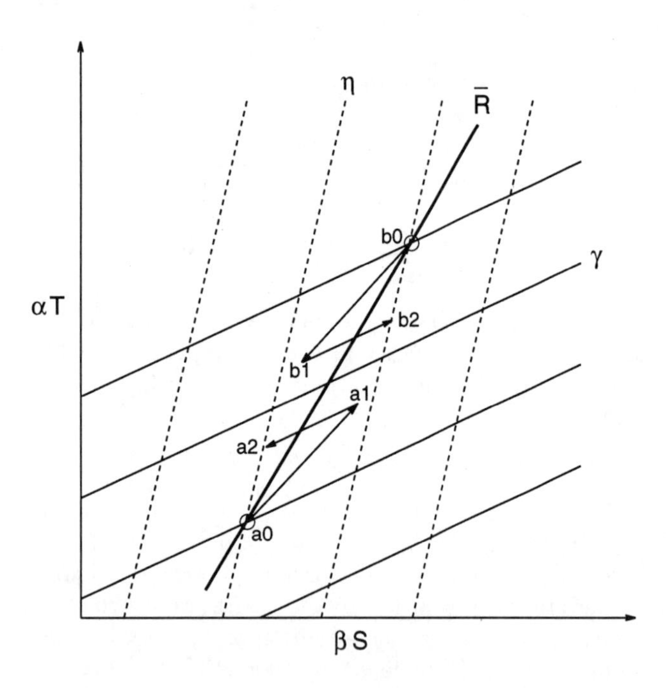

Fig. 2. Evolution of equal and opposite perturbations in a fluid for which R_ρ is initially uniform. When fluid initially at point 'a0' is perturbed to 'a1', finger fluxes cause it to decay to the point 'a2' as $t \to \infty$. Similarly, the perturbation 'b1' approaches 'b2' as $t \to \infty$. The density ratio R_ρ is nonuniform in the final equilibrium state. The lines labelled γ (flux ratio) and η are the eigenvectors of the system (5).

respectively; the quantities $(\beta\tilde{S} + K_S'\alpha\tilde{T}/K_S^{(s)})$ and $(\alpha\tilde{T} - \gamma\beta\tilde{S})$ in (6) are proportional to distance measured along the eigenvector axes. The set (6) shows that the component of a perturbation parallel to the η axis is preserved, while the component parallel to the γ axis vanishes as $t \to \infty$. Thus, when a perturbation is imposed such that a fluid parcel initially at "a0" is moved to "a1", salt finger fluxes cause the system to approach "a2" as $t \to \infty$. Similarly, when fluid initially at "b0" is perturbed to "b1", double-diffusive fluxes cause it to decay toward the point "b2". As $t \to \infty$, all perturbations asymptotically approach the equilibrium state $K_S^{(s)}\beta\tilde{S} = -K_S'\alpha\tilde{T}$. Because the perturbations do not disappear entirely, the final equilibrium state will not in general have uniform R_ρ unless the dashed lines in Figure 2 have slope $\bar{R}_\rho$, or the initial perturbation has slope γ. From (12a) and (12b) it follows that the equilibrium density ratio is given by

$$R_\rho^\infty = \bar{R}_\rho + \frac{\tilde{S}_{0,z}}{\bar{S}_z}\left(\frac{\alpha\hat{T}_0}{\beta\hat{S}_0} - \gamma\right)\frac{K_S}{K_S^{eff}} . \tag{13}$$

This is in contrast with the results of *Schmitt* [1981], who predicted that all density ratio anomalies should diffuse away as $t \to \infty$. The apparent disagreement can be reconciled by noting that Schmitt considered the limit $|K_S'| \gg K_S$, in which case $-K_S^{(s)}/K_S' \to \bar{R}_\rho$. In this limit the η lines in Figure 2 have slope $\bar{R}_\rho$, and it follows that finger fluxes will cause any density ratio anomalies (but not T-S anomalies) to disappear completely as $t \to \infty$.

Equation (6b) shows that the evolution of the system is governed by a diffusion law with a diffusivity given by K_S^{eff}; this 'effective diffusivity' is larger than K_S if $K_S' < 0$ (in agreement with *Schmitt* [1981]). The diffusivity enhancement results from the fact that, for fixed T_z, K_S is an increasing function of S_z. The effect can be understood physically by considering the evolution of a sinusoidal perturbation $(\beta\tilde{S}, \alpha\tilde{T})$. If the perturbation is decomposed into components parallel to the eigenvector axes in Figure 2, then the η component produces a density ratio perturbation which is *in phase* with $\tilde{S}_z$, since

$$\tilde{R}_\rho = \frac{\tilde{S}_z}{\bar{S}_z}\left(\frac{\alpha\hat{T}}{\beta\hat{S}} - \bar{R}_\rho\right) , \tag{14}$$

and the quantity in parentheses is positive if $\alpha\hat{T}/\beta\hat{S} = -K_S^{(s)}/K_S'$ (*i.e.*, if the perturbation is parallel to an η-line). However, this component of the perturbation is in equilibrium, so we need only consider the γ component to get the time evolution of the system. The reason for the diffusivity enhancement can now be illustrated

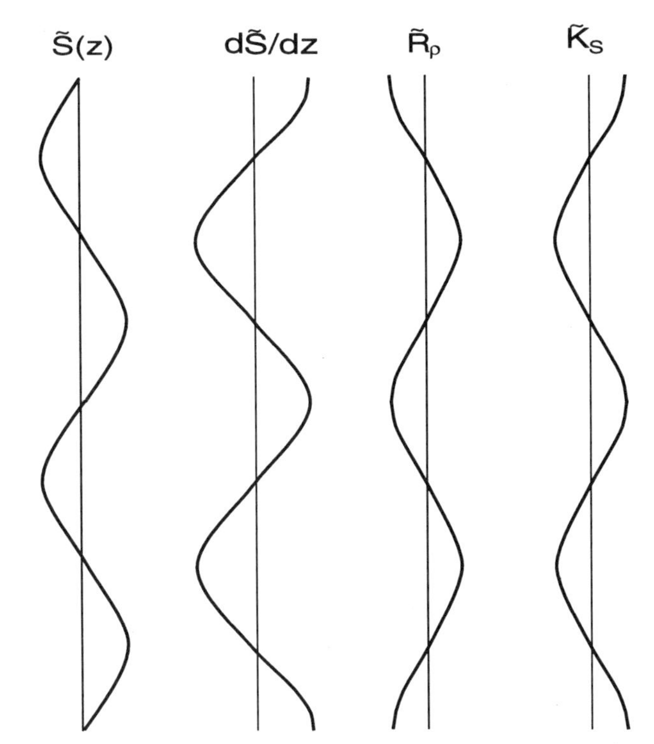

Fig. 3. Illustration of the flux enhancement mechanism for Schmitt's diffusivity formulation. The enhancement results from the dependence of the diffusivity on R_ρ. If the salinity field is perturbed so that $S = \bar{S} + \tilde{S}$, there is a corresponding diffusivity perturbation $\tilde{K}_S$. The flux of salt is given by $F_S = -[\bar{K}_S \bar{S}_z + \bar{K}_S \tilde{S}_z + \tilde{K}_S \bar{S}_z + h.o.t.]$, so that the flux is increased by an amount $-\tilde{K}_S \bar{S}_z$ over the constant diffusivity case.

quite simply by sketching the fluxes resulting from an imposed anomaly. If $K_S = (\bar{K}_S + \tilde{K}_S)$ and $S = (\bar{S} + \tilde{S})$, then the salt flux is given by

$$F_S = -[\bar{K}_S \bar{S}_z + \bar{K}_S \tilde{S}_z + \tilde{K}_S \bar{S}_z + \text{ higher order terms}] .$$

The quantities which contribute to the flux divergence are sketched in Figure 3 for an anomaly with slope $\alpha\tilde{T}/\beta\tilde{S} = \gamma$, so that R_ρ perturbations are out of phase with $\tilde{S}_z$. Because K_S is assumed to be a monotonically decreasing function of R_ρ, $\tilde{K}_S(z)$ is in phase with $\tilde{S}_z$, and it follows that the second and third terms on the right hand side reinforce one another. The first term on the right hand side is divergence-free, so the flux-divergence $F_{S,z}$ is larger by an amount $-\tilde{K}_{S,z}\bar{S}_z$ than the constant diffusivity value of $-\bar{K}_S \tilde{S}_{zz}$.

4. KUNZE'S MODELS

We now examine the formulations proposed by *Kunze* [1987], who used a finger Froude number constraint to limit the growth of salt fingers, allowing him to predict the fluxes and diffusivities associated with fully developed salt fingers. Kunze's model predicts a diffusivity of the form

$$K_S^{(87)} \approx 2\nu \left(\sqrt{R_\rho} + \sqrt{R_\rho - 1}\right)^2 \qquad (15)$$

for fully developed salt fingers, assuming that the fingers are double sinusoids, and that the fingers stop growing after attaining a critical Froude number of 2. This is plotted in Figure 4 (solid line).

More recently, *Kunze* [1994] has modified his earlier theory to account for observations that show an inverse relationship between the Cox number and the interfacial temperature gradient T_z [*Marmorino*, 1989; *Fleury and Lueck*, 1991] . Using a hybrid wave-finger Froude number to constrain finger growth rather than the finger Froude number used in his earlier study, Kunze predicts a diffusivity of the form

$$K_S^{(94)} \approx \nu \frac{\sqrt{R_\rho - 1}}{\sqrt{R_\rho} - \sqrt{R_\rho - 1}} , \qquad (16)$$

where ν is the molecular diffusivity of water, and it is assumed that the internal wave Froude number $| U_z | /N \approx 1$. The diffusivity (16) is plotted in Figure 4 (dashed line). Both of Kunze's diffusivity formulations

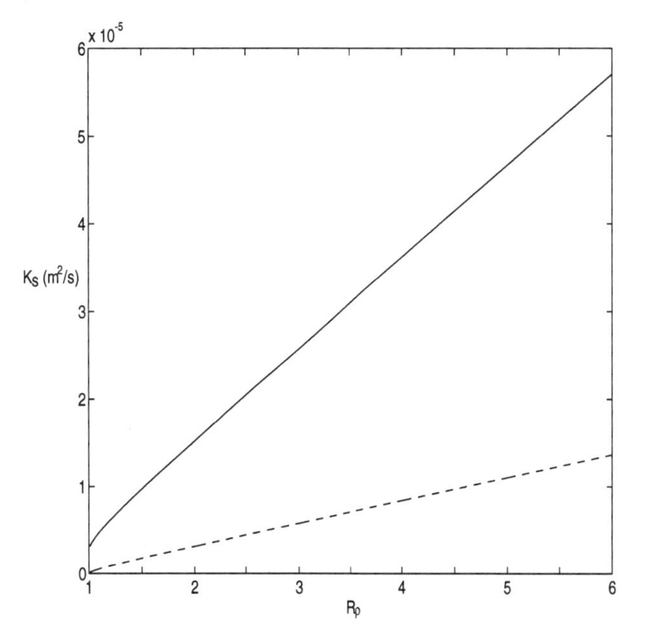

Fig. 4. The two diffusivity formulations proposed by *Kunze* [1987, 1994]. The solid line shows the diffusivity predicted by Kunze's 1987 model; the dashed line shows the diffusivity for the revised (1994) model. In both cases the diffusivities increase monotonically with R_ρ.

predict that the diffusivity increases with R_ρ. This is in contrast with *Schmitt* [1981], whose analysis of data from thermohaline staircases indicates that salt finger diffusivities decrease rapidly with R_ρ. According to *Kunze* [1987], the different behavior of his model diffusivities is due to the fact that for fixed S_z, a larger R_ρ implies greater stability, so the fingers have larger fluxes when they equilibrate.

Figure 5 shows the growthrates (scaled by m^2) of the system (5) using Kunze's diffusivity formulation. Figure 5a shows the growthrates for the 1987 flux laws; Figure 5b shows the growthrates for Kunze's revised (1994) flux laws. In each case there is one marginally stable mode (dashed line), and a second which may be unstable (solid line). The growthrates are the roots of the characteristic polynomial for the system (5), which is obtained by setting $(\tilde{S}, \tilde{T}) = \mathrm{Re}\left((\hat{S}_0, \hat{T}_0)e^{\lambda t + imz}\right)$:

$$\lambda(\lambda + m^2 K_S^{eff}) = 0 \,. \qquad (17)$$

From (17), we see that the solid curves in Figure 5 give the negative of the "effective diffusivity" K_S^{eff} (as defined by (7)). The effective diffusivity is always negative for Kunze's revised [1994] flux laws, and is negative when $R_\rho < 1.8$ for his original [1987] flux laws. It is relatively easy to determine the cause of the instability – it results from the form of the diffusivity, which ensures that (for fixed temperature gradient) the magnitude of the salt flux *decreases* as the salinity gradient increases. In terms of the earlier discussion in connection with Figure 3, Kunze's flux laws predict a diffusivity perturbation $\tilde{K}_S$ which is out of phase with $\tilde{S}_z$. Thus, the flux "correction" resulting from diffusivity gradients is opposite in sign (and larger in magnitude) than the component $\bar{K}_S \tilde{S}_z$ associated with the mean diffusivity. Using either (15) or (16), it is readily verified that K_S increases sufficiently rapidly with R_ρ that the coefficient $[K_S - \bar{R}_\rho K_S']$ in (5a) (an effective salt diffusivity) is negative for all R_ρ. This will tend to

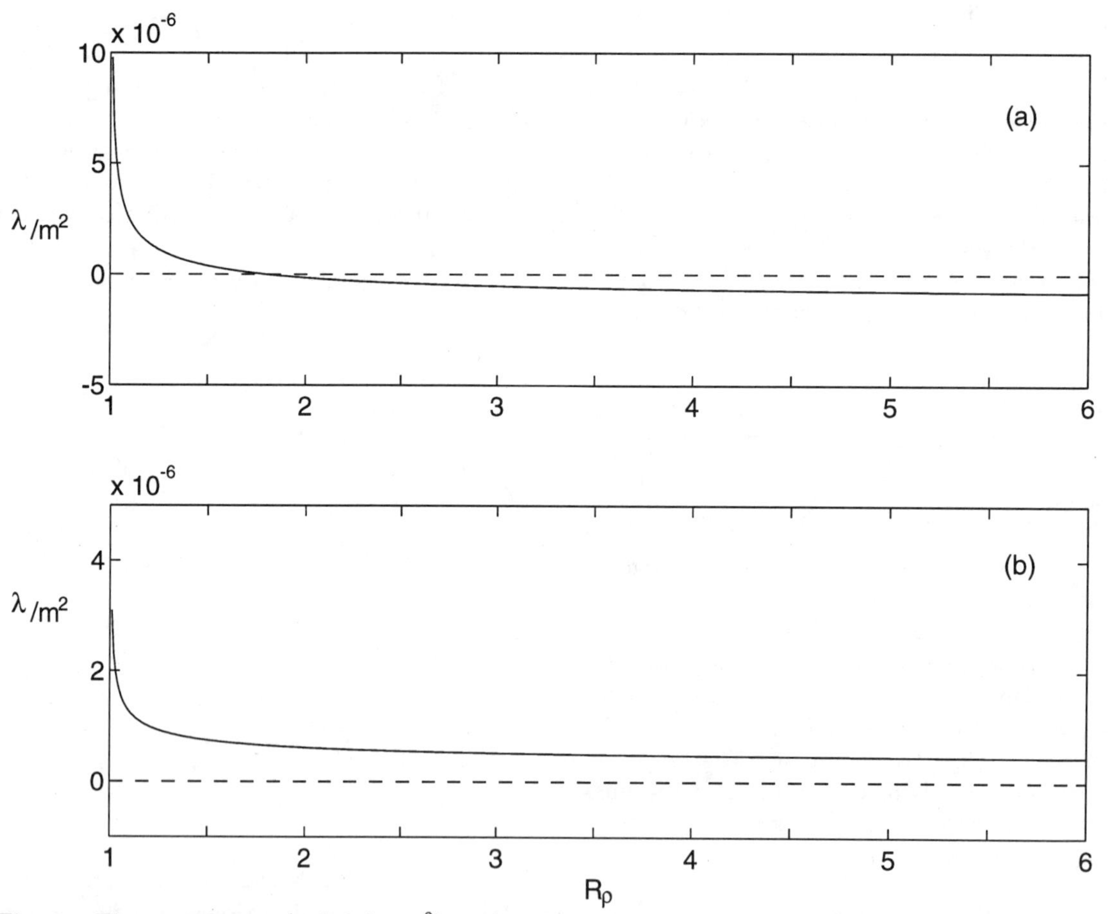

Fig. 5. The growthrates (scaled by m^2) predicted for Kunze's flux formulations. Fig. 5a shows the growthrates predicted by the model for the flux laws proposed by *Kunze* [1987]; Fig. 5b gives the growthrates for Kunze's revised (1994) flux laws. In each case there is one marginally stable mode and a second which may be unstable.

make salinity anomalies grow with time, but the system can be stabilized by cross-diffusion effects (as happens with the 1987 flux laws when $R_\rho > 1.8$). The instability mechanism is shown schematically in Figure 6, where a small anomaly is imposed on a linear basic-state salinity gradient. According to Kunze's flux laws, the diffusivity K_S will increase above the anomaly (since R_ρ increases there) and will decrease below the anomaly. The fluxes are more strongly influenced by the diffusivity changes than by the altered salinity gradient, so the adjusted fluxes reinforce the initial perturbation.

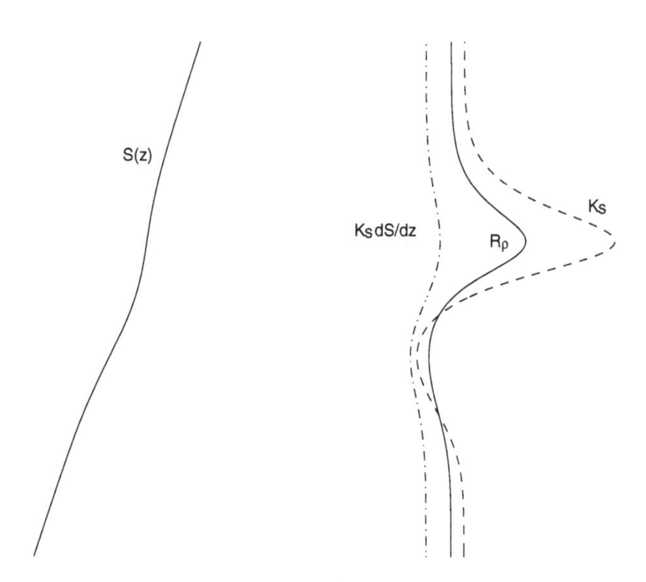

Fig. 6 Illustration of the instability mechanism for Kunze's flux laws. Introducing a small salinity anomaly leads to an enhanced downward flux above and a decreased downward flux below the anomaly, causing the initial perturbation to grow.

5. DISCUSSION

We have verified that the *ad hoc* flux laws of *Schmitt* [1981] are stable, and that they predict an accelerated decay of temperature, salinity, and density ratio anomalies as a result of the enhanced vertical flux divergences produced by vertical diffusivity gradients. Our model predicts that small temperature and salinity anomalies should persist even as $t \to \infty$. In contrast with *Schmitt* [1981], we find that while anomalies in R_ρ decay rapidly, they do not disappear altogether. A residual density ratio anomaly remains as $t \to \infty$. The predicted persistence of T,S, and R_ρ anomalies is an interesting result, given the nearly ubiquitous finescale structure found in the thermocline. The character of the equilibrium predicted by our model depends upon the initial perturbation and on the functional dependence of the diffusivity on R_ρ.

In contrast with Schmitt's diffusivity formulation, Kunze's models predict diffusivities which increase with R_ρ. We have seen that this introduces the possibility of a *negative* effective diffusivity for small amplitude T-S anomalies, in which case all wavenumbers are unstable, and the highest wavenumbers grow most quickly (an "ultraviolet catastrophe"). It was shown that the flux laws of *Kunze* [1994] are unstable for all R_ρ, while the flux formulation proposed by *Kunze* [1987] is unstable if R_ρ is sufficiently small. Thus, any fully developed field of salt fingers obeying Kunze's constraints should be unstable for some values of R_ρ.

Kunze [1994] has also considered the influence of intermittent turbulent mixing on salt finger fluxes. He finds that diffusivities are essentially unaffected by turbulence for $R_\rho < 2$ (when finger growthrates are large), but decrease rapidly with R_ρ when $R_\rho > 2$. In this case we would expect exponential growth of small perturbations for $R_\rho < 2$, and stable behavior similar to that found for Schmitt's formulation when $R_\rho > 2$. It is possible that the unstable behavior predicted by Kunze's flux laws is related to the process of layer formation, as "steppy" T-S profiles are generally observed at low values of R_ρ [*e.g. Schmitt*, 1981, Figure 4].

It is worthwhile to explore the possible consequences of our analysis for the interpretation of oceanographic data. We do this by investigating the predicted correlations between T-S anomalies. Making use of (12a,b) it can be shown that the equilibrium ratio of $\tilde{T}_z$ to $\tilde{S}_z$ is given by:

$$\frac{\alpha \tilde{T}_z^\infty}{\beta \tilde{S}_z^\infty} = \bar{R}_\rho - \frac{K_S}{K_S'} \,. \tag{18}$$

This relation suggests that the functional dependence of the diffusivity on R_ρ might be inferred from oceanographic data by examining correlations between temperature and salinity anomalies. However, it is possible that many observed T-S perturbations are due to thermohaline intrusions, for which lateral advective effects are important, so we must also discuss the signal expected for such intrusions. We divide the observations into T-S perturbations large enough to create inversions in T and S, and those which are not. If small perturbations without inversions are lateral intrusions, experience with a numerical model of thermohaline intrusions shows that the ratio of $\tilde{T}_z$ to $\tilde{S}_z$ perturbations stays very close to the value predicted by linearized perturbation theory right up to the point when inversions first appear. That value is [*Walsh and Ruddick*, 1994]:

$$\frac{\alpha \tilde{T}_z^{\infty}}{\beta \tilde{S}_z^{\infty}} = 1 + \frac{1-\gamma}{2\Phi - 1 + \Phi^{-1/2}\sqrt{\frac{\bar{R}_\rho - \gamma}{\bar{R}_\rho - 1}}} , \quad (19)$$

where $\Phi = (K_S + (1 - \bar{R}_\rho)K_S')/K_S^{eff}$. Large perturbations are necessarily bounded by salt fingering on one side and diffusive convection on the other. If these are intrusions, they were created by lateral advection in a lateral T-S gradient, and advection will be important to the steady state achieved. If they are not intrusions, then lateral advection is not important to their evolution and the combined action of the diffusive and finger fluxes will decrease the T-S anomaly until the inversion disappears [*Ruddick*, 1984]. After this time, the anomaly will evolve in the manner described by the one dimensional model discussed above, leading to T and S gradients as given by equation (18).

In summary, if an anomaly is large enough to have temperature and salinity inversions, then it is either an intrusion or is a transient phenomenon evolving to a one-dimensional balance. If the anomaly is too small to have inversions, then it will have a perturbation gradient ratio given by (19) if it is an intrusion, or by (18) if it is not. As an example, if we assume that $|K_S'| \gg K_S$ (as in *Schmitt* [1981]), and take $\bar{R}_\rho = 2$, we find that (18) predicts $\tilde{T}_z/\tilde{S}_z \approx 2$, while (19) predicts that $\tilde{T}_z/\tilde{S}_z \approx 1.2$.

Although we have not considered the effect of turbulent mixing on the double-diffusive phenomena investigated here, it seems likely that many of our conclusions would continue to hold if turbulence were included. Unfortunately, given the intermittent nature of turbulent mixing events and the unknown nature of the interactions between double-diffusive convection and turbulence, it is not obvious how turbulence should be incorporated. The most straightforward and simplistic way to model the mixing due to turbulence is to incorporate a constant eddy diffusivity K_{turb} into the flux parameterizations (4). This modification introduces a turbulence-modified flux ratio which increases with R_ρ:

$$\gamma_{EFF} = \frac{\gamma K_S + K_{turb} R_\rho}{K_S + K_{turb}} . \quad (20)$$

While incorporating this non-constant γ_{EFF} into the flux laws will modify the flux convergences and hence alter the results of our calculation somewhat, the fundamental conclusions about the instability of Kunze's formulation will not be changed, because the instability mechanism shown in Figure 6 will still be operative.

Acknowledgements. We wish to thank Eric Kunze for making helpful comments and suggestions during the preparation of this manuscript.

REFERENCES

Fleury, M., and R.G. Lueck, Fluxes across a thermohaline interface. *Deep Sea Res., 38*, 745–769, 1991.

Kunze, E., Limits on growing, finite-length salt fingers: a Richardson number constraint. *J. Mar. Res., 45*, 533–556, 1987.

Kunze, E., The evolution of salt fingers in inertial wave shear. *J. Mar. Res., 48*, 471–504, 1990.

Kunze, E., A proposed flux constraint for salt fingers in shear. *J. Mar. Res., 52*, 999–1016, 1994.

Marmorino, G.O., Substructure of oceanic salt finger interfaces. *J. Geophys. Res., 94*, 4891–4904, 1989.

McDougall, J.T., and J. Taylor, Flux measurements across a finger interface at low values of the stability ratio. *J. Mar. Res., 42*, 1–14, 1984.

Ruddick, B., The life of a thermohaline intrusion, *J. Mar. Res., 42*, 831–852, 1984.

Schmitt, R. W., Form of the temperature-salinity relationship in the central water: evidence for double-diffusive mixing. *J. Phys. Oceanogr., 11*, 1015–1026, 1981.

Schmitt, R.W., H. Perkins, J.D. Boyd, and M.C. Stalcup, C-SALT: An investigation of the thermohaline staircase in the western tropical North Atlantic. *Deep Sea Res., 34*, 1655–1666, 1987.

Stern, M.E., Lateral mixing of water masses. *Deep Sea Res., 14*, 747–753, 1967.

Stern, M.E., Collective instability of salt fingers. *J. Fluid Mech., 35*, 209–218, 1969.

Stern, M.E., Ocean Circulation Physics, Academic Press, NY, 191–203, 1975.

Turner, J.S., Salt fingers across a density interface. *Deep Sea Res., 14*, 599–611, 1967.

Walsh, D., and B. Ruddick, Double-diffusive interleaving: the influence of non-constant diffusivities, *J. Phys. Oceanogr., 25* (3), 348–358, 1995.

Zhurbas, V.M., N.P. Kuz'mina, and YE. Kul'sha, Step-Like Stratification of the ocean thermocline resulting from transformations associated with thermohaline salt finger intrusions (numerical experiment). *Oceanology, 27*, 277–281 (English Translation), 1987.

D. Walsh and B. Ruddick, Department of Oceanography, Dalhousie University, Halifax, Nova Scotia, B3H 4J1, Canada

Observations of the Density Perturbations which drive Thermohaline Intrusions

Barry Ruddick and David Walsh

Department of Oceanography, Dalhousie University, Halifax, Canada

We present correlated salinity and density traces from the intrusive frontal zone of Meddy Sharon. The warm and salty intrusive laminae tend to have positive density anomalies in the upper part of the Meddy, and negative anomalies in the lower part, and the change in behaviour occurs at the depth where the intrusions were observed by *Ruddick* [1992] to change slope. This is argued to be consistent with the hypothesis that the density perturbations were created, and the intrusions driven, by vertical double-diffusive buoyancy fluxes.

1. INTRODUCTION

When *Stommel and Fedorov* [1967] first discovered salinity-compensated temperature inversions in CTD casts, they deduced that these features were indicative of lateral mixing, and that these intrusive layers blend into their surroundings via vertical mixing. It was discovered both theoretically [*Stern*, 1967] and in laboratory experiments [*Turner*, 1978] that vertical mixing by double-diffusion can create density perturbations that drive lateral intrusive motions. Many investigators have discovered intrusions at oceanic fronts, and these were presumed to cause significant cross-frontal transports of salt and heat.

Armi et al. [1989] described the structure and evolution of a Mediterranean salt lens (Meddy) over a two year period. The region outside the core of the Meddy had thermohaline intrusions that eroded 30 km into the Meddy core during the first year of observation. Both the salinity front and the vorticity front moved inward with the intrusions, and *Armi et al.* [1989] concluded the intrusions were responsible for mixing the Meddy into the surrounding ocean.

Ruddick [1992] examined temperature and salinity data from a series of closely-spaced "tow-yo" stations within the Meddy intrusions, and was able to track the layers over a horizontal distance of several kilometres. He found that in the upper part of the Meddy, above the warm salty core, warm and salty intrusions sloped downwards as they moved outward from the lens center. Below the core, intrusions sloped in the opposite sense. This pattern was consistent with that expected from *McIntyre*'s [1970] instability for Prandtl number less than one. This version of the instability uses the preferential diffusion of mass over momentum (possible for double-diffusion, but unlikely for turbulence), to release the kinetic energy of the shear. However, *Ruddick* [1992] found that the slopes were too small in magnitude for the energy release mechanism to function. The pattern and magnitude of intrusion slopes *was* consistent with double-diffusive driving, as described by *Turner* [1978], and indicated that salt fingers dominated the vertical fluxes in the lower part of the Meddy, while diffusive convection dominated in the upper part. These results showed that the intrusions that mixed the Meddy were driven solely by double-diffusion involving both salt finger and diffusive fluxes, and that the azimuthal velocity or baroclinic density field did not provide energy to the intrusions.

Turner's [1978] laboratory model and its interpretation, and theoretical models of intrusions [e.g., *Toole and Georgi*, 1981] indicate thermohaline intrusions are driven by density perturbations that are produced by double-diffusive buoyancy flux divergences. Can these density perturbations be observed? In this note we

Double-Diffusive Convection
Geophysical Monograph 94

present observations of intrusive density perturbations which are consistent with the dynamical picture of intrusions described in the next section.

2. WHAT DENSITY SIGNAL IS EXPECTED?

Figure 1 shows the structure of finite-amplitude thermohaline intrusions in the case where diffusive (upper) and finger (lower) buoyancy fluxes dominate. In both cases, warm and saline water (i.e., the Meddy Core) is on the left. First we consider the advective-diffusive balance for the finger-dominant case (lower panel). As warm, salty water is advected over cool fresh water, finger fluxes cause it to become cooler, fresher, and less dense. It therefore rises with respect to isopycnal surfaces as it moves from left to right. Conversely, the cool fresh water becomes more dense, and sinks as it moves from right to left. *Turner* [1978] and *Ruddick and Turner* [1979] explained the observed slope of laboratory intrusions in this manner. Theoretical models of intrusions which suppose that salt fingers dominate and drive a countergradient buoyancy flux [cf *Toole and Georgi*, 1981] also predict intrusions sloping in this manner.

When salt finger buoyancy flux dominates over the diffusive buoyancy flux, the density anomaly that drives the lateral motion is the net result of the tendencies for lateral advection to increase and salt finger fluxes to decrease the density of the warm salty layers. The feedback loop is the following:

- quasi-horizontal advection produces T-S anomalies.
- vertical T-S gradients drive salt finger fluxes.
- the finger fluxes cause the warm salty layers to become less dense, and the cool fresh layers more dense.
- the buoyancy perturbations, in combination with the layer slopes, produce pressure perturbations which drive along-layer advection. Since the warm, salty layers move upslope to the right, they must be anomalously light. The cool, fresh layers that move downslope to the left must be anomalously dense. [See also fig. 4 of *Ruddick and Hebert*, 1988]. These arguments result in the predictions for temperature, salinity, and density structure shown at the right of the diagram, which suggest that salinity and density, and their gradients, should be anticorrelated.

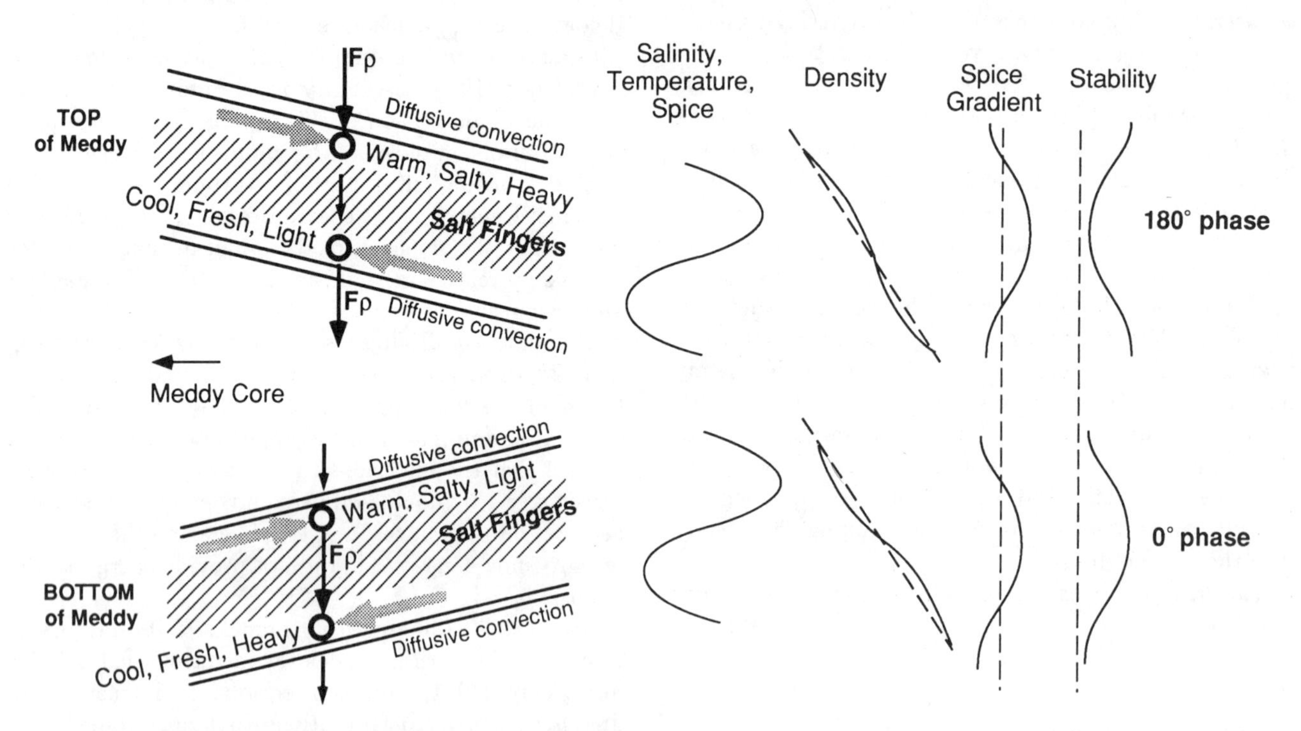

Fig. 1. Illustration of the vertical buoyancy fluxes and expected density perturbations which drive large amplitude thermohaline intrusions.

In a situation where diffusive buoyancy fluxes dominate over finger fluxes (upper panel of figure 1), the arguments are reversed. Warm salty water parcels become cooler, fresher, and more dense due to the diffusive fluxes from above, and sink as they advect from left to right. Cool fresh parcels become lighter and rise as they flow in the opposite direction. Hence the intrusion slope must be in the opposite sense, as shown. In order to drive the motion as shown, warm, salty parcels must be anomalously dense, and cool fresh ones light.

3. CTD DATA PROCESSING

Observations of temperature and salinity from the Meddy frontal zone were processed so as to identify warm/salty and cool/fresh intrusions, and to calculate the density anomaly associated with those intrusions. First, the potential temperature (θ_1) and potential density (σ_1) were computed using a reference pressure of 1000 dbar. Then, σ_1 and salinity were low-pass filtered using a fourth order Butterworth filter with a 67 m 3 db cutoff. (The cutoff was chosen to reject the 10-30 m wavelength intrusions [*Ruddick*, 1992, figure 14], but capture the larger-scale T/S profile of the Meddy, which is O(1000 m) thick.) This produced a smooth $\overline{S}$ vs $\overline{\sigma}_1$ profile which preserved the basic structure of the T-S curve, but rejected fluctuations due to the 10-25 m wavelength intrusions [*Ruddick*, 1992]. This procedure is illustrated on the $\theta_1 - S$ curve of figure 2, in which the smoothed T/S curve is shown as a dashed curve. Next, the "spiciness deviation" from this curve is found by computing $2\beta(S(\sigma_1) - \overline{S}(\sigma_1))$. *Jackett and McDougall* [1985] show that this is approximately the distance from the smoothed $\theta_1 - S$ curve measured in a direction parallel to density contours, as shown in the inset to figure 2, and is an excellent indicator of the water property anomaly. A positive spiciness deviation indicates a warm and salty intrusion.

The density anomaly corresponding to each spiciness deviation was computed as $\sigma_1 - \overline{\sigma}_1$, the difference between the original and smoothed potential density. Positive density anomaly means relatively dense water.

The advantage of computing the spiciness deviation in this manner is that it is insensitive to the effects of internal wave strain on the vertical density structure. Vertical straining motions will alter the spacing between isopycnals but will not alter either the $\theta_1 - S$ relation or the corresponding $\sigma_1 - S$ relation. However, internal wave strain will affect the density profiles, and so the density anomaly is expected to contain a component due to internal waves.

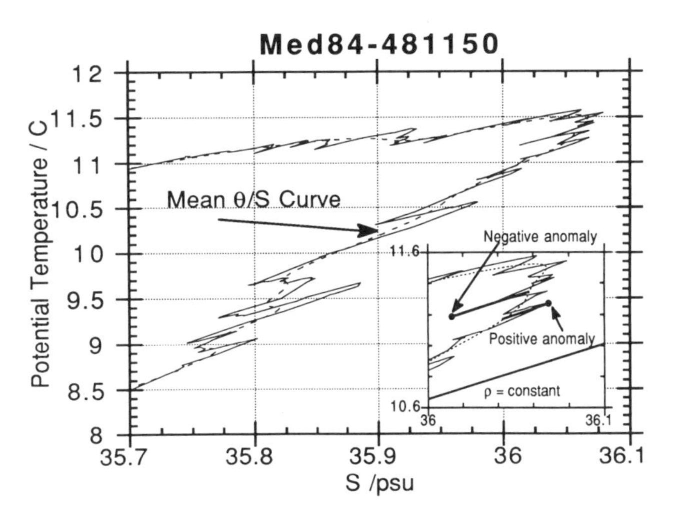

Fig. 2. Potential Temperature-Salinity diagram illustrating the actual (solid) and smoothed (dashed) $\theta - S$ profiles, a contour of constant potential density, and the computed $\theta - S$ deviation (as shown by the + and − segments in the inset), from which a $\theta - S$ anomaly profile was calculated. A positive anomaly indicates a warm, salty intrusion, and vice-versa. Internal wave strain acting on the vertical T-S gradient will not produce such anomalies.

4. RESULTS

Figure 3 shows the results for the 750-1050 m intrusive depth range of one of the Meddy tow-yo casts discussed by *Ruddick* [1992]. The original and smoothed salinity and density profiles are shown in the first panel. The density and salinity deviations are visibly correlated in the upper part, and anti-correlated in the lower part. The middle panel shows the buoyancy gradient (thin solid curve), the spiciness deviation (dashed) and the density anomaly (solid). The density anomaly and spiciness deviation are correlated in the upper part, and anti-correlated in the lower part. This indicates that warm salty intrusions are anomalously heavy in the upper part and light in the lower, as expected of thermohaline intrusions. The relationship changes sign at about 900 m, where the salinity is a maximum, and where the slope of these intrusions with respect to isopycnals was found by *Ruddick* [1992] to change sign.

The correlation between density anomaly and spiciness deviation is imperfect. Several factors may be causing this:

1. An isotropic field of internal waves creates density perturbations which are not related to spiciness deviation.

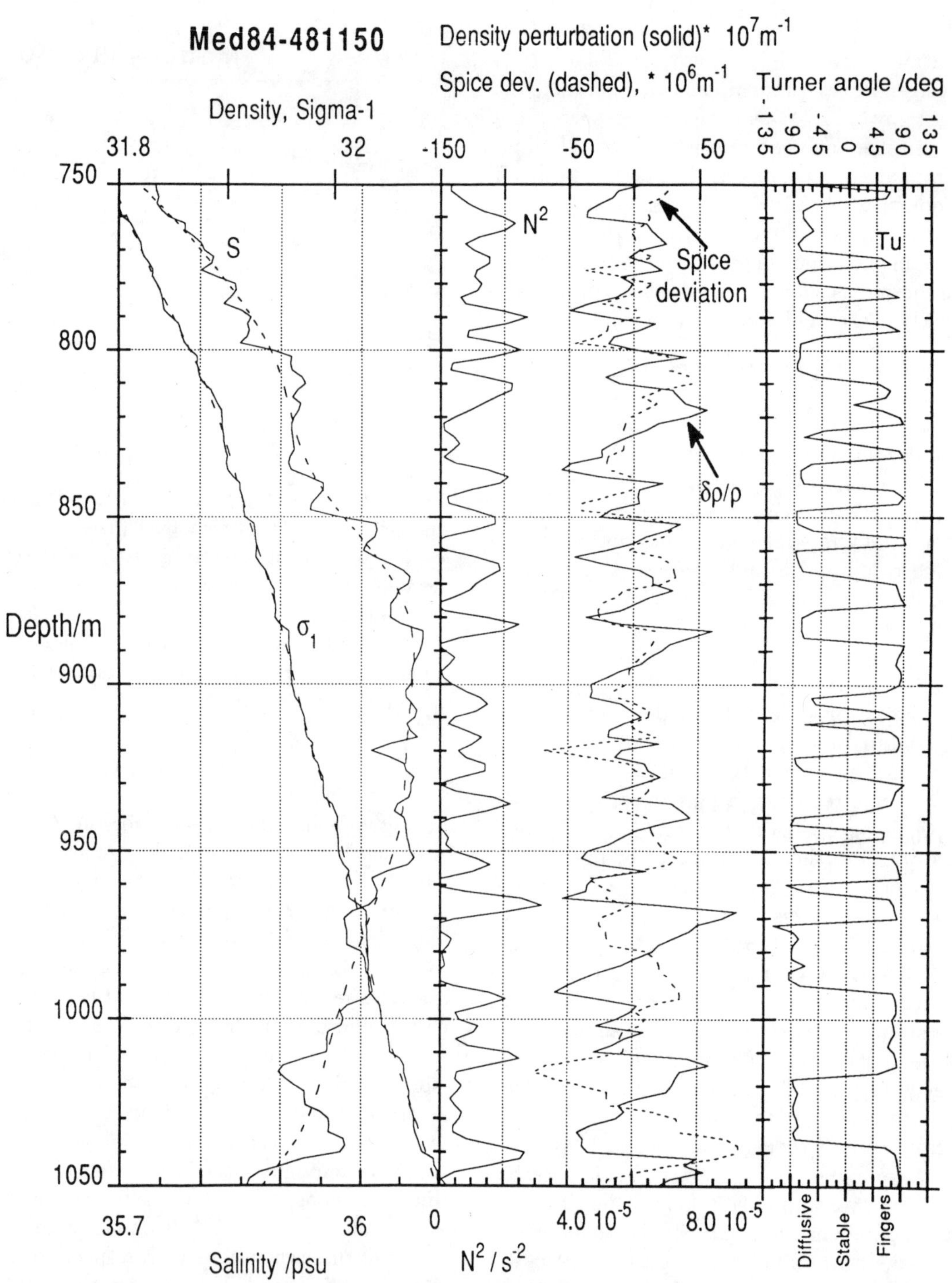

Fig. 3. Finescale profiles of thermohaline quantities versus depth, taken from cast 481150 (radius 37 km) of the tow-yo shown in figure 1 of *Ruddick* [1992].
First frame: density and salinity profiles (showing smoothed and original traces),
Second frame: N^2 (solid light curve), density perturbation (solid) and T-S deviation (dashed),
Third frame: Turner angle (Ruddick, 1983), indicating finger ($45° < Tu < 90°$) and diffusive ($-90° < Tu < -45°$) regions.
800–900 m: The density perturbation is correlated with the T-S deviation, indicating that warm, salty intrusions are relatively dense.
900–1050 m: The density perturbation is anti-correlated with the T-S deviation, indicating that warm, salty intrusions are relatively light.

2. Relatively large negative density gradients occur in both the finger and diffusive regions. The sign of the density anomaly is governed by which of these regions dominates.

3. The step-like nature of the density profile causes the density perturbations to have a sawtooth profile, while the spiciness deviations do not exhibit the same tendency.

Scatter diagrams

We compare the density gradient variations with the intrusive layer temperature-salinity structure by plotting the density gradient ($N^2 = -\frac{g}{\rho}\frac{\partial\rho}{\partial z}$) against Turner angle

$$\tan(Tu) = \frac{\beta\frac{\partial S}{\partial z} + \alpha\frac{\partial T}{\partial z}}{\beta\frac{\partial S}{\partial z} - \alpha\frac{\partial T}{\partial z}}$$

(Note that z is positive upwards for these definitions), which indicates the local (estimated on 12 m scales) tendency toward double-diffusion [*Ruddick*, 1983]. Salt fingering is possible for $45° < Tu < 90°$, diffusive convection occurs for $-45° > Tu > -90°$, gravitational instability is possible for $|Tu| > 90°$, and the water column is stable for $|Tu| < 45°$. Figure 4 shows the upper part of the cast (750-900m) in the upper panel, and the lower part (900-1050 m) in the lower panel. The distributions of N^2 vs Tu are different in the two panels, with the regions of small gravitional stability tending to occur in the range $75° < Tu < 90°$ in the upper panel, and in the range $-70° > Tu > -90°$ in the lower panel. This is in accord with the predictions of figure 1, which indicate that the regions of highest stability should occur in the finger (diffusive) regions in the lower (upper) part of the Meddy.

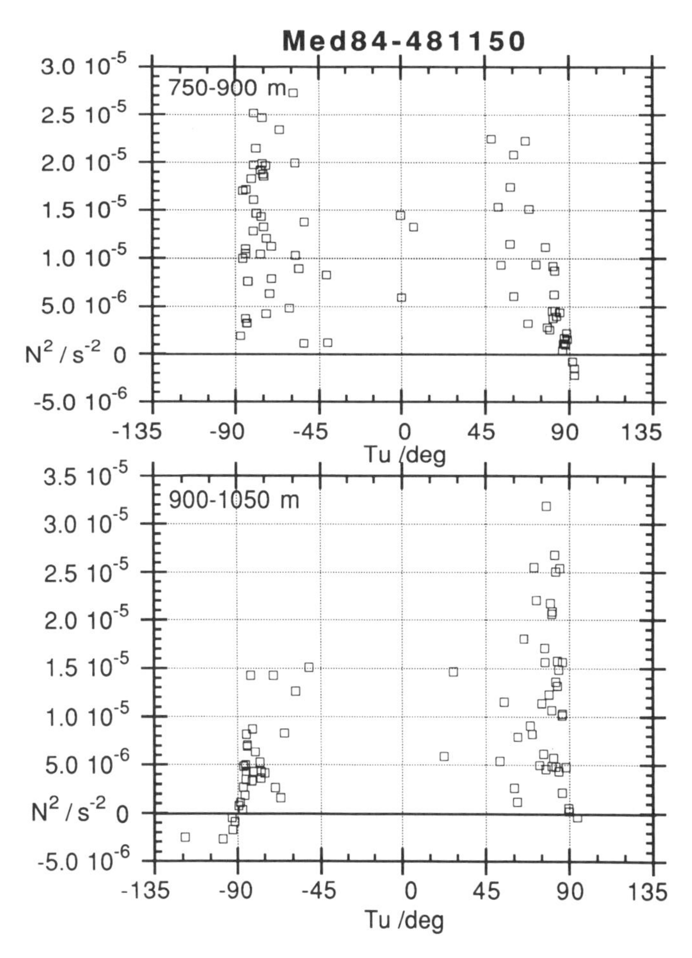

Fig. 4. N^2 versus Turner angle. In the 750–900 m part of the Meddy (upper panel), finger stratified regions ($45° < Tu < 90°$) are less stable than diffusively stratified regions ($-45° > Tu > -90°$). In the 900–1050 m part of the Meddy (lower panel), the opposite is true.

5. DISCUSSION

We have presented correlated salinity and density traces from the intrusive frontal zone of Meddy Sharon. The warm and salty intrusive laminae tend to have positive density anomalies in the upper part of the Meddy, and negative anomalies in the lower part, and the change in behaviour occurs at the depth where the intrusions were observed by *Ruddick* [1992] to change slope. This is consistent with the hypothesis that the density perturbations were created, and the intrusions driven, by vertical double-diffusive buoyancy fluxes.

Ruddick [1992] found that the observed intrusion slopes were consistent in sign with McIntyre's instability for Prandtl number (momentum diffusivity/mass diffusivity) less than 1, but were too small in magnitude for the mechanism to operate. It is worth noting that these motions are inertial in nature, so the density perturbations are created by advection. Thus, upward-moving McIntyre intrusions should be anomalously dense, and downward-moving intrusions should be light. The observed density perturbations in figure 3 are therefore inconsistent with McIntyre intrusions.

The observation of density structure coherent with intrusive layer structure not only supports the hypothesis that double-diffusion can cause systematic lateral transports of salt and heat via frontal intrusions, but gives valuable clues as to the mechanisms which serve to limit the structure and strength of finite-amplitude thermohaline intrusions. The systematic appearance of regions of very low gravitational stability suggests that the dominant type of double-diffusion (diffusive convection in the upper part, fingers in the lower part) can drive part of the water column nearly to overturning. Perhaps this process helps to limit the salinity anomaly and quasi-horizontal velocity attained by intrusions.

The data in figure 2 are from one station selected for its strong and clearly correlated signals. Analysis of other stations is presently being undertaken in hopes of putting these findings on firmer statistical ground.

Acknowledgements. The authors wish to thank Eric Kunze and two referees for helpful comments which led to significant improvements in the manuscript. The beginnings of this work were funded and encouraged by the U.S. Office of Naval Research, through Alan Brandt. Our work is presently funded through the Canadian Natural Sciences and Engineering Research Council.

REFERENCES

Armi, L., D. Hebert, N. Oakey, J. Price, P. Richardson, T. Rossby, and B. Ruddick. Two years in the life of a Mediterranean Salt Lens. *J. Phys. Oceanogr. 19*, 354-370, 1989.

Jackett, D.R., and T.J. McDougall. An oceanographic variable for the characterization of intrusions and water masses. *Deep-Sea Res.* 32, (10) 1195-1207, 1985.

McIntyre, M.E. Diffusive destabilization of the baroclinic circular vortex. *Geophys. Fluid Dyn.*, *3*, 321-345, 1970.

Ruddick, B.R. A practical indicator of the stability of the water column to double-diffusive activity. *Deep-Sea Res. 30* (10a) 1105-1107, 1983.

Ruddick, B.R. Intrusive mixing in a Mediterranean salt lens -- intrusion slopes and dynamical mechanisms. *J. Phys. Oceanogr. 22*(11) 1274-1285, 1992.

Ruddick, B., D. Hebert. The mixing of Meddy "Sharon". In: Small-scale Mixing in the Ocean, Elsevier Oceanography Series, *46*, 249-262, J.C.J. Nihoul and B. M. Jamart, Eds., 1988.

Ruddick, B.R., and J.S. Turner. The vertical length scale of double-diffusive intrusions. *Deep-Sea Res.*, *26A,* 903-913, 1979.

Stommel, H, and K.N. Fedorov. Small scale structure in temperature and salinity near Timor and Mindanao. *Tellus XIX*, (2), 306-325, 1967.

Toole, J.M., and D.T. Georgi. On the dynamics and effects of double-diffusively driven intrusions. *Prog. Oceanogr.*, *10*, 121-145, 1981.

Turner, J.S. Double-diffusive intrusions into a density gradient. *J. Geophys. Res.*, *83*, 2887-2901, 1978.

Barry Ruddick and Dave Walsh, Department of Oceanography, Dalhousie University, Halifax N.S., Canada B3H 4J1